Lecture Notes in Computer Science 1327
Edited by G. Goos, J. Hartmanis and J. van Leeuwen

Springer-Verlag Berlin Heidelberg GmbH

Wulfram Gerstner Alain Germond
Martin Hasler Jean-Daniel Nicoud (Eds.)

Artificial Neural Networks – ICANN '97

7th International Conference
Lausanne, Switzerland, October 8-10, 1997
Proceeedings

Springer

Series Editors

Gerhard Goos, Karlsruhe University, Germany

Juris Hartmanis, Cornell University, NY, USA

Jan van Leeuwen, Utrecht University, The Netherlands

Volume Editors

Wulfram Gerstner
Alain Germond
Martin Hasler
Jean-Daniel Nicoud
Swiss Federal Institute of Technology
École Polytechnique Fédérale de Lausanne
CH-1015 Lausanne, Switzerland
E-mail: wulfram.gerstner@di.epfl.ch
germond@de.epfl.ch
hasler@de.epfl.ch
nicoud@di.epfl.ch

Cataloging-in-Publication data applied for

Die Deutsche Bibliothek - CIP-Einheitsaufnahme

Artificial neural networks : 7th international conference ; proceedings / ICANN '97, Lausanne, Switzerland, October 8 - 10, 1997. Wulfram Gerstner ... (ed.). - Berlin ; Heidelberg ; New York ; Barcelona ; Budapest ; Hong Kong ; London ; Milan ; Paris ; Santa Clara ; Singapore ; Tokyo : Springer, 1997
(Lecture notes in computer science ; Vol. 1327)

CR Subject Classification (1991): F.1.1, C.2.1, C.1.3, I.2, G.1.6, I.5.1, B.7.1, J.1, J.2

ISSN 0302-9743

ISBN 978-3-540-63631-1 **ISBN 978-3-540-69620-9 (eBook)**
DOI 10.1007/978-3-540-69620-9

Originally published by Springer-Verlag Berlin Heidelberg New York in 1997.

Typesetting: Camera-ready by author
SPIN 10545612 06/3142 – 5 4 3 2 1 0 Printed on acid-free paper

Preface

This book is based on the papers presented at the *International Conference on Artificial Neural Networks,* ICANN'97, which was hosted by the Swiss Federal Institute of Technology in Lausanne, Switzerland. The ICANN conferences were initiated in 1991 and have since become the major European meetings in the field of neural networks.

From 365 submitted papers 190 were accepted for publication. In addition there were several invited papers. The process of paper selection for ICANN'97 relied heavily upon the work of our referees whom we would like to thank for their great effort. We also owe a warm 'thank you' to the members of the technical program committee, F. Blayo, M. Cottrell, F. de Viron, C. Jutten, E. Mayoraz, and K. Pawelzik who spent two days in Lausanne for the final selection of the papers and the preparation of the conference program.

The conference and the proceedings would not have been possible without the enormous work of M. Dubois and A. Moinat who handled all the organizational aspects of paper submission and registration. M.-J. Pellaud generously helped out whenever it was necessary. Financial support was provided by the Foundation 'Latsis' and the Swiss National Science Foundation.

Lausanne, July 1997

Wulfram Gerstner
Alain Germond
Martin Hasler
Jean-Daniel Nicoud

Referees and Advisory Board

Part I: Coding and Learning in Biology

Part II: Cortical Maps and Receptive Fields

Part III: Learning: Theory and Algorithms

Part IV: Signal Processing: Blind Source Separation, Vector Quantization, and Self-Organization

Part V: Robotics, Adaptive Autonomous Agents, and Control

Part VI: Speech, Vision, and Pattern Recognition

Part VII: Prediction, Forecasting, and Monitoring

Part VIII: Implementations

Part I:
Coding and Learning in Biology

Reward Responses of Dopamine Neurons: A Biological Reinforcement Signal

Wolfram Schultz

Institute of Physiology, University of Fribourg, CH-1700 Fribourg, Switzerland

Abstract. A class of reinforcement models termed Temporal Difference (TD) models has been developed from theoretical grounds as effective algorithms for various learning situations. Based on the observation that learning depends on the unpredictability of primary motivating events, these models use errors in the prediction of reinforcing events as teaching signals. Independent of the theoretical work, neurophysiological experiments have revealed that neurons in the mammalian midbrain using the neurotransmitter dopamine process information about rewards and reward-predicting stimuli in a very similar manner as the teaching signal of TD models.

1 Dopamine Responses to Reward-Related Stimuli

1.1 Description of Data. The optimal stimulus for activating 75-80% of dopamine neurons consists of unpredicted food or liquid rewards. This occurs when monkeys touch a small morsel of hidden food during spontaneous movements or when they receive a drop of liquid at the mouth outside of any behavioral task or while learning a task [16, 18, 25, 29]. The responses are specific for the rewards and do not occur when similarly shaped non-food objects are touched or when a fluid valve is operated without actually delivering liquid. By contrast, dopamine neurons are depressed in their activity when a predicted reward fails to occur [15, 29].

Reward responses occur only when reward arrives unexpectedly and disappear when a conditioned phasic stimulus precedes the reward. Most dopamine neurons then respond to the conditioned stimulus. This response transfer occurs in single dopamine neurons tested both with unpredicted rewards and with reward-predicting conditioned visual or auditory stimuli in a reaction time task. Likewise, the response to the unconditioned reward is transferred to the reward-predicting conditioned stimulus while a behavioral task is being learned. After extensive overtraining, the response to conditioned stimuli can be greatly reduced. Dopamine neurons discriminate between reward predicting and non-predicting stimuli, unless the stimuli are very similar and the animal does an orienting response to both [30]. The majority of responses is specifically appetitive, and only about 15% of neurons are also activated by a conditioned aversive light or sound stimulus in an air puff avoidance task, and in response to other arousing stimuli [19]. Dopamine neurons only show minor or no activations prior to and during the execution of movements and are not activated during the delays of typical frontal cognitive tasks [29, 33].

Unexpected novel stimuli are also effective in activating dopamine neurons as long as they elicit behavioral orienting reactions (e.g. ocular saccades). Neuronal responses subside together with orienting reactions after several stimulus repetitions [16].

Similar responses of dopamine neurons to unconditioned high intensity or novel stimuli in parallel with orienting reactions have been described in cats [35].

The responses of dopamine neurons to these different stimuli are remarkably similar. They consist of phasic increases of activity and occur with latencies of 50-120 ms, last less than 200 ms, and are composed of occasionally a single impulse or, more often, a short burst of <10 impulses. They are polysensory and occur independent of the side of visual, auditory or somatosensory stimulus presentation relative to the body axis. Effective stimuli in most situations activate the majority of the population of dopamine neurons, and different dopamine neurons in groups A8, A9 and A10 respond to the same stimuli in the same manner, such that clearly separate response types cannot be associated with different populations of neurons. This considerable homogeneity of neuronal responses suggests that dopamine neurons respond in parallel as a population rather than displaying response profiles that are well differentiated among each other.

1.2 Interpretation of Data. The optimal stimuli for dopamine neurons are unconditioned rewards and conditioned reward-predicting stimuli (commonly called reward-related events). Exceptions to this are the responses to novel or arousing stimuli and to conditioned aversive stimuli. Novel stimuli are potential rewards or reward predictors and might be included in the class of reward-related events. The relatively small group of aversive neuronal responses might in fact constitute a subpopulation of less specific dopamine neurons responding to a larger spectrum of behaviorally important stimuli.

The responses of dopamine neurons depend entirely on the unpredictability of stimuli. Dopamine neurons respond to primary rewards only when the reward occurs unpredictably, either outside of a task or during learning. By contrast, a fully predicted reward does not elicit a dopamine response. In situations requiring repeated learning, dopamine neurons are activated by primary rewards during each learning phase but stop responding to the reward when the learning curve reaches its asymptote and reward becomes fully predicted [29]. Inversely, when a fully predicted reward suddenly fails to occur, dopamine neurons are depressed in their activity exactly at the time at which the reward would have occurred. It thus appears that dopamine neurons code the deviation or 'error' between the prediction and the actual occurrence of reward [27, 32]. This holds also for conditioned, reward-predicting stimuli which are only effective in activating dopamine neurons when they occur unpredictably [29]. Thus, the principle of coding of a prediction error by dopamine applies to both primary and conditioned appetitive stimuli.

The observed neuronal response transfer from the unconditioned reward to the conditioned stimulus predicting reward resembles the stimulus substitution process of Pavlovian learning in which the intrinsically neutral stimulus acquires behavioral values of the unconditioned event. This suggests that the dopamine response itself is subject to learning.

2 Impact of Dopamine Message on Striatal Processing

Clues to the function of dopamine neurons may be derived from two particular properties, their activities in comparison to those of the target structures of dopamine neurotransmission, such as the striatum (caudate nucleus, putamen and ventral striatum including nucleus accumbens), and the properties of the neuronal architecture in which they are embedded.

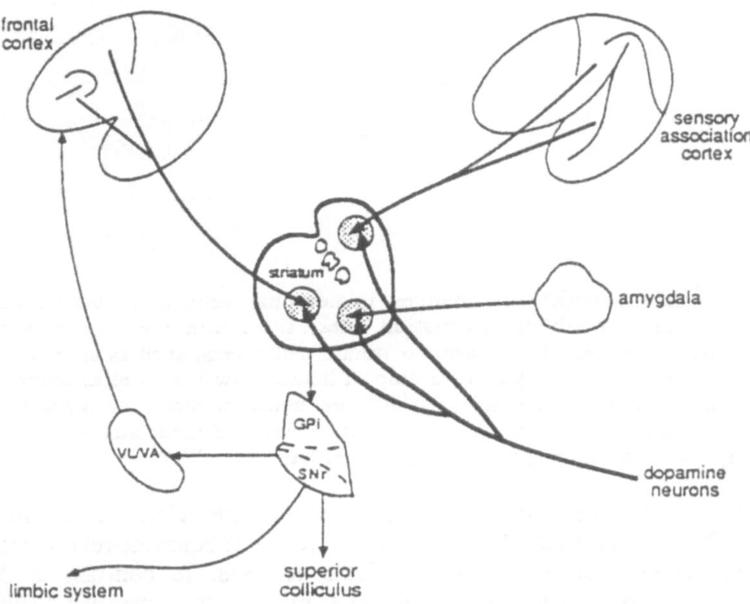

Fig. 1. Highly simplified schema showing how dopamine neurons act on the striatum which is linked through major circuits to other brain centers controlling behavioral output. The striatum denotes the dorsal caudate, the putamen and the ventral striatum including nucleus accumbens. For reasons of simplicity, the dopamine innervation to the frontal lobe (upper left: prefrontal cortex, premotor cortex, motor cortex) has been omitted.

2.1 Large Circuits. Dopamine neurons project principally to the striatum and the frontal cortex. These structures globally subserve the organization of behavioral output. The striatum is a part of loops involving the frontal cortex and in addition receives input from postcentral sensory and association cortex and from limbic cortical and subcortical structures (Fig. 1). Besides the loops with the cortex, basal ganglia output is directed to limbic structures and to the superior colliculus which are involved in affective behavioral components, memory and simple motor output.

Thus, dopamine neurons are in a position to influence the control of behavioral output by their projections to structures which are involved in highly differentiated information processing. The limited spectrum of their homogeneous activity suggests that dopamine neurons contribute a particular component of neuronal processing that is commonly important for their target structures.

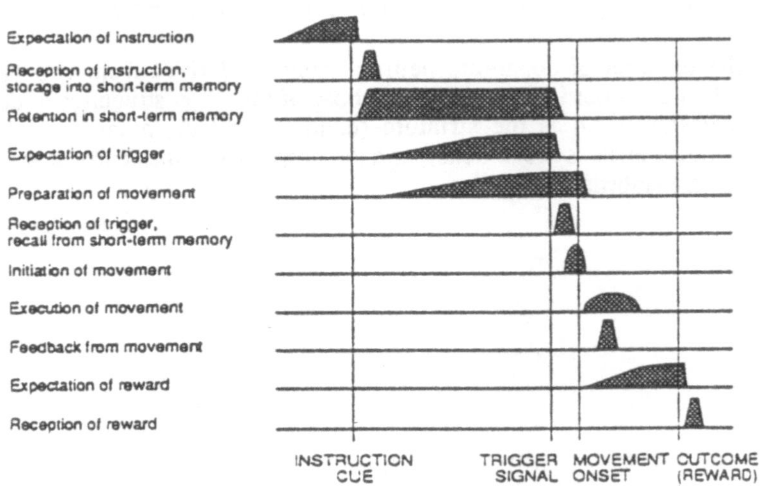

Fig.2. Schematic overview of changes in neuronal activity in the striatum during sensorimotor tasks involving instruction cues, short-term memory, movements and rewards. There are neuronal responses to stimuli and events, such as an initial instruction cue, a movement-triggering signal or a drop of liquid reward, as well as activity preceding these events, including movements, which are being expected or prepared following extended experience in the task. A given striatal neuron, if modulated in the task, usually shows only one or occasionally two of these changes.

2.2 Striatal Neurons Process Detailed Information about Behavioral Actions. There could hardly be a greater difference of behavior-related activity than that seen between dopamine and striatal neurons. In contrast to the rather homogeneous responses to reward-related stimuli, striatal neurons display a high variety of very distinct and specific behavioral relationships [e.g. 1, 5, 13, 14, 24, 31]. Three main categories of activity are found (Fig. 2). (i) Phasic responses to stimuli of behavioral significance, such as mnemonic instruction cues, trigger stimuli, conditioned reward predictors, and primary rewards. Since the striatum does not receive primary sensory input, these responses may be due to inputs from frontal cortex (instruction and trigger stimuli), amygdala (reward) and other cortical, thalamic and limbic structures. (ii) Sustained activations occur during the short-term retention of specific past events and during the expectation and preparation of forthcoming events known to the animal through its experience. They occur in relation to instruction and target cues, trigger stimuli, moving external stimuli, movements of the limbs and eyes, and reward. These activities may be due to inputs from several cortical areas in which sustained delay activity is frequently found, but thalamic and even amygdala input may also contribute. Alternatively, they may be generated through reverberating activity in closed loops involving the basal ganglia. The origin of sustained activity is unknown, in particular for the slow onset expectation-related activity, but clues to this may lie in the notorious instability of membrane potentials of striatal neurons through which sustained activity may be generated by small deviations from subthreshold variations and subsequent build up in loops. (iii) Striatal neurons are activated immediately prior to and particularly during the execution of limb and eye movements where they may encode specific aspects of direction, speed, velocity and target position. These activities may be predominantly due to motor, premotor and somatosensory cortical input. Neurons in the ventral striatum are mainly activated in relation to rewards [28].

The other main projection structure of dopamine neurons is the frontal cortex. More than 20 years of research in behavioral neurophysiology on primate prefrontal and premotor cortex have revealed similarly differentiated neuronal relationships to various behavioral acts as seen in the striatum. Neurons in orbitofrontal cortex show a preponderance of different kinds of reward-related activity (Tremblay and Schultz, in preparation).

2.3 Local Circuitry in Target Structures. The basic arrangement of synaptic influences of dopamine neurons on striatal and frontal cortex neurons consists of a triad, comprising the dendritic spine of the postsynaptic neuron, excitatory cortical terminal at the tip of the dendritic spine and the dopamine presynaptic varicosity contacting the same dendritic spine (Fig. 3) [10, 12]. Every medium size striatal neuron receives about 1,000 dopaminergic synapses at its dendritic spines and about 10,000 cortical synapses [7]. It is unlikely that all cortical inputs to a single striatal neuron come from the same cortical neuron, and it would be interesting to know which cortical areas give rise to converging inputs for striatal neurons. It has been shown that homotopical somatosensory and motor cortical areas project into common regions of the striatum [9]. This arrangement of convergence would allow dopamine neurons to influence the synaptic effects of cortical inputs to striatal neurons.

The immediate effect of dopamine is probably a reduction in postsynaptic excitability. This results in a general reduction of corticostriatal processing, thus focusing striatal activity onto the strongest inputs whereas weaker activity is lost [8, 39]. However, the dopamine effects do no seem to operate in the same millisecond time range as corticostriatal excitatory potentials, because striatal and cortical dopamine receptors are invariably associated with G-proteins. This suggests onsets and durations of postsynaptic membrane effects in the time range of seconds. In addition, corticostriatal neurotransmission is subject to posttetanic potentiation and depression, the latter of which is dependent upon intact dopamine transmission [2, 3]. Recent experiments show that phasically applied dopamine leads to potentiation of corticostriatal transmission [40].

Taken together, this suggests that corticostriatal neurons are informed by dopamine neurons about the occurrence of an unconditioned or conditioned reward-related event, without receiving specific informations about its modality, body side or the exact nature of the reward. At the same time, very detailed information about the same event arrives in the striatum and cortex, with different aspects of the event being encoded by different neurons, according to the striatal data described above. The released dopamine may act on the striatal and cortical neurons in several possible ways. (i) The focusing effect would only let the strongest of the convergent dendritic inputs pass through to the impulse generating mechanism at the cell body, whereas weaker activity would be lost, thus giving priority processing to all information pertaining to the curren reward event. (ii) The slower time course of dopamine membrane action may leave a short term trace of the reward event and influence all subsequent activity for a short while. (iii) Dopamine release modifies information travelling in cortico-basal ganglia and other loops and thus may ultimately influence neurons in the cortex and in limbic structures involved in structuring behavioral output. (iv) The presence of striatal dopamine-dependent plasticity and the manner of reward-induced activation of dopamine neurons may suggest that dopamine responses are involved in plasticity changes in postsynaptic neuronal circuitry underlying reward-directed learning. The reward signal could be the decisive signal upon whose reception the striatal and cortical synapses concurrently activated by the events leading to reward would be strengthened. This last possibility will be elaborated in the following chapter.

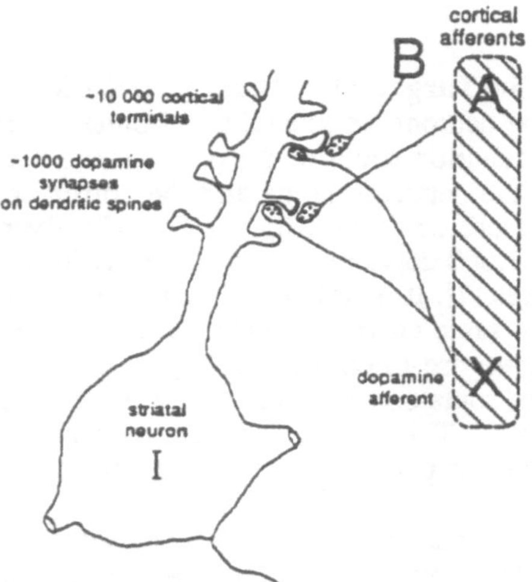

Fig. 3. Basic design of hypothetical dopamine-dependent heterosynaptic plasticity induced by a reward-related event. The graph shows the synaptic arrangement of inputs from dopamine neurons and cortex to medium size spiny striatal neuron. In the synaptic triad, dendritic spines are contacted by dopamine varicosities from a single dopamine axon and by a cortical axon. Cortical neurons A and B converge at the tip of dendritic spines on a single striatal neuron I. These connections are modifiable by increased use, eg. in the form of posttetanic potentiation. However, the modification occurs only when dopamine input X coming indiscriminately to the stems of the same dendritic spines is active at the same time. In the present example, cortical input A, but not B, is active at the same time as dopamine neuron X (shaded area), e.g following a reward-related event. This leads to a modification of the A->I transmission, but leaves the B->I transmission unaltered. Anatomical drawing modified from Smith and Bolam [34].

3 Theoretical Considerations

According to current learning theories, the unpredictability of primary reward is a necessary condition for the acquisition of appetitive behavioral reactions to novel stimuli [17, 22, 23]. Associative learning of behavioral reactions to external stimuli follows the equation

$$\Delta V = \alpha \beta (\lambda - V),$$

with ΔV as the change in associative strength of conditioned stimulus (=learning), α and β as the salience of the two stimuli, respectively, λ as the amount of processing of the unpredicted unconditioned stimulus, V as the associative strength of the conditioned stimulus (=amount learned) [see 6]. Thus, the rate of learning is proportional to the degree of unpredictability of reward (maximum amount of processing of reward minus current predictive power of conditioned stimulus: $\lambda - V$), and the asymptote is reached when the unconditioned stimulus becomes entirely predictable by the conditioned stimulus ($V = \lambda$). This form in general corresponds to the delta rule of neuronal network models in which learning is proportional to the error or unpredictability of the outcome.

Dopamine neurons respond to reward only under the condition that it occurs unpredictably, and the responses are absent when the reward becomes predictable by a preceding conditioned stimulus. The response is negative when a predicted reward fails to occur. Thus, the dopamine prediction error signal very much resembles the error term of the Rescorla-Wagner rule, indicating that the dopamine signal could be suitable as a teaching signal.

The error signal in neuronal network models indicates the difference between the intended output and the actual output. It is used in different versions of learning networks for adjusting synaptic weights in order to reduce the error. It might be interesting to consider how the responses of dopamine neurons could be involved in modifying the synaptic weights of cortical inputs to striatal neurons. In analogy to the synaptic triad described by anatomy, let A and B be two inputs, each of which contacts a dendritic spine on a single striatal neuron I (Fig. 3). The synaptic weights between A and I and between B and I are Hebbian modifiable. The same spines are indiscriminately contacted by the global reward prediction error signal from dopamine input X. When a reward-related signal is encountered, both neuron X and neuron A are activated, whereas neuron B does not modify its activity. Neuron X transmits the message that some rewarding event has occurred without giving specific details, whereas neuron A sends a message about one of several detailed aspects of the reward-related event, such as color, texture, position, surrounding etc. of the stimulus and may code a movement leading to obtaining the reward. The weigths of striatal synapses could be modified according to the learning rule

$$\Delta \omega = \varepsilon \, r \, i \, o,$$

with ω as synaptic weight, ε as learning constant, r as dopamine prediction error signal, i as input activation, o as activation of striatal neuron. Thus, through the simultaneity or near simultaneity of activity in A and X, the activity of neuron X may induce a change in weight at the active synapse between A and I, but would leave the synaptic weights between B and I unchanged since this synapse is inactive. If neuron A is subsequently activated again by a stimulus that shares some features with the stimulus activating both neurons A and X without necessarily activating also neuron X, the response in neuron I would be modified, whereas input from neuron B leads to an unchanged postsynaptic response. This model would constitute a dopamine-dependent heterosynaptic plasticity (A->I) induced by a reward-related event (coming from neuron X). Or, the synaptic plasticity of A->I and B->I neurotransmission is conditional on X being conjointly active with A or B. This mechanism could work at the local level of the synaptic triad and without involving larger networks. Thus, the key functions of dopamine neuron X would be to signal the event (reward) that may be particularly important for behavior and according to which striatal neurotransmission needs to be tuned. The reward signal from dopamine neurons would influence as a kind of learning gate the highly structured activity circulating in cortico-striatal and limbic-striatal connections engaged in organizing and controlling behavioral output.

The dopamine signal closely resembles the teaching signal of TD reinforcement models [37, 38]. Although the described biological aspects are still speculative, biologically inspired TD models using dopamine-like teaching signals have been used to efficiently train neuronal networks to perform ocular reactions on the basis of learned predictions, the reinforcement module itself being adaptive like the dopamine response [11, 20]. A teaching signal modelled in explicit analogy to dopamine responses mediates decision-making in close correspondence with human action choice behavior [21]. In close analogy to animal experiments, TD models simulate the learning of approach behavior, stimulus discrimination and spatial delayed responding of primates [4, 36].

Acknowledgements

Preparation of this article was supported by grants from the Swiss National Science Foundation (3100-43331.95; 4038-43997), the European Community via the Swiss Office of Education and Science (CHRX-CT94-0463 via 93.0121; BMH4-CT95-0608 via 95.0313-1) and the James S. McDonnell Foundation (94-39).

References

1. Alexander, G.E. and Crutcher, M.D.: Neural representations of the target (goal) of visually guided arm movements in three motor areas of the monkey. J. Neurophysiol. 64: 164-178, 1990
2. Calabresi, P., Maj, R., Mercuri, N.B. and Bernardi, G.: Coactivation of D1 and D2 dopamine receptors is required for long-term synaptic depression in the striatum. Neurosci. Lett. 142: 95-99, 1992
3. Calabresi, P., Pisani, A., Mercuri, N.B. and Bernardi, G.: Long-term potentiation in the striatum is unmasked by removing the voltage-dependent magnesium block of NMDA receptor channels. Europ. J. Neurosci. 4: 929-935, 1992
4. Contreras-Vidal, J.L. and Schultz, W.: A neural network model of reward-related learning, motivation and orienting behavior. Soc. Neurosci. Abstr. 22: 2029, 1996
5. Crutcher, M.D. and DeLong, M.R.: Single cell studies of the primate putamen. II. Relations to direction of movement and pattern of muscular activity. Exp. Brain Res. 53: 244-258, 1984
6. Dickinson, A.: Contemporary animal learning theory. Cambridge University Press, Cambridge 1980
7. Doucet, G., Descarries, L. and Garcia, S.: Quantification of the dopamine innervation in adult rat neostriatum. Neuroscience 19: 427-445, 1986

8. Filion, M., Tremblay, L. and Bédard, P.J.: Abnormal influences of passive limb movement on the activity of globus pallidus neurons in parkinsonian monkey. Brain Res. 444: 165-176, 1988
9. Flaherty, A.W. and Graybiel, A.: Two input systems for body representations in the primate striatal matrix: experimental evidence in the squirrel monkey. J. Neurosci. 13: 1120-1137, 1993
10. Freund, T.T., Powell, J.F. and Smith, A.D.: Tyrosine hydroxylase-immunoreactive boutons in synaptic contact with identified striatonigral neurons, with particular reference to dendritic spines. Neuroscience 13: 1189-1215, 1984
11. Friston, K.J., Tononi, G., Reeke, G.N.Jr., Sporns, O. and Edelman, G.M.: Value-dependent selection in the brain: simulation in a synthetic neural model. Neuroscience 59: 229-243, 1994
12. Goldman-Rakic, P.S., Leranth, C., Williams, M.S., Mons, N. and Geffard, M.: Dopamine synaptic complex with pyramidal neurons in primate cerebral cortex. Proc. Natl.Acad. Sci. USA 86: 9015-9019, 1989
13. Hikosaka, O., Sakamoto, M. and Usui, S.: Functional properties of monkey caudate neurons. III. Activities related to expectation of target and reward. J. Neurophysiol. 61: 814-832, 1989
14. Kimura, M.: Behaviorally contingent property of movement-related activity of the primate putamen. J. Neurophysiol. 63: 1277-1296, 1990
15. Ljungberg, T., Apicella, P. and Schultz, W.: Responses of monkey midbrain dopamine neurons during delayed alternation performance. Brain Res. 586: 337-341, 1991
16. Ljungberg, T., Apicella, P. and Schultz, W.: Responses of monkey dopamine neurons during learning of behavioral reactions. J. Neurophysiol. 67: 145-163, 1992
17. Mackintosh, N.J.: A theory of attention: Variations in the associability of stimulus with reinforcement. Psychol. Rev. 82: 276-298, 1975
18. Mirenowicz, J. and Schultz, W.: Importance of unpredictability for reward responses in primate dopamine neurons. J. Neurophysiol. 72: 1024-1027, 1994
19. Mirenowicz, J. and Schultz, W.: Preferential activation of midbrain dopamine neurons by appetitive rather than aversive stimuli. Nature 379: 449-451, 1996
20. Montague, P.R., Dayan, P., Nowlan, S.J., Pouget, A. and Sejnowski, T.J.: Using aperiodic reinforcement for directed self-organization during development. In: Neural Information Processing Systems 5 (Eds. S.J. Hanson, J.D. Cowan and C.L. Giles). pp. 969-976. Morgan Kaufmann, San Mateo, 1993
21. Montague, P.R., Dayan, P. and Sejnowski, T.J.: A framework for mesencephalic dopamine systems based on predictive Hebbian learning. J. Neurosci. 16: 1936-1947, 1996
22. Pearce, J.M. and Hall, G.: A model for Pavlovian conditioning: variations in the effectiveness of conditioned but not of unconditioned stimuli. Psychol. Rev. 87: 532-552, 1980
23. Rescorla, R.A. and Wagner, A.R.: A theory of Pavlovian conditioning: Variations in the effectiveness of reinforcement and nonreinforcement. In: Classical Conditioning II: Current Research and Theory (Eds. Black, A.H. and Prokasy, W.F.) New York: Appleton Century Crofts, pp. 64-99, 1972

24. Rolis, E.T., Thorpe, S.J. and Maddison, S.P.: Responses of striatal neurons in the behaving monkey. 1. Head of the caudate nucleus. Behav. Brain Res. 7: 179-210, 1983
25. Romo, R. and Schultz, W.: Dopamine neurons of the monkey midbrain: Contingencies of responses to active touch during self-initiated arm movements. J. Neurophysiol. 63: 592-606, 1990
26. Schultz, W.: Activity of dopamine neurons in the behaving primate. Sem. Neurosci. 4: 129-138, 1992
27. Schultz, W., Dayan, P. and Montague, R.R.: A neural substrate of prediction and reward. Science 275: 1593-1599, 1997
28. Schultz, W., Apicella, P., Scarnati, E. and Ljungberg, T.: Neuronal activity in monkey ventral striatum related to the expectation of reward. J. Neurosci. 12: 4595-4610, 1992
29. Schultz, W., Apicella, P. and Ljungberg, T.: Responses of monkey dopamine neurons during performance of a delayed response task. J. Neurosci. 13: 900-913, 1993
30. Schultz, W. and Romo, R.: Dopamine neurons of the monkey midbrain: Contingencies of responses to stimuli eliciting immediate behavioral reactions. J. Neurophysiol. 63: 607-624, 1990
31. Schultz, W. and Romo, R.: Role of primate basal ganglia and frontal cortex in the internal generation of movements: comparison with instruction-induced preparatory activity in striatal neurons. Exp. Brain Res. 91: 363-384, 1992
32. Schultz, W., Romo, R., Ljungberg, T., Mirenowicz, J., Hollerman, J.R. and Dickinson, A.: Reward-related signals carried by dopamine neurons. In: Models of Information Processing in the Basal Ganglia (Eds. J.C.Houk, J.L.Davis and D.G.Beiser) MIT Press, Cambridge, MA, pp. 233-248, 1995
33. Schultz, W., Ruffieux, A. and Aebischer, P.: The activity of pars compacta neurons of the monkey substantia nigra in relation to motor activation. Exp. Brain Res. 51: 377-387, 1983
34. Smith, A.D. and Bolam, J.P.: The neural network of the basal ganglia as revealed by the study of synaptic connections of identified neurones. Trends Neurosci. 13: 259-265, 1990
35. Steinfels, G.F., Heym, J., Strecker, R.E and Jacobs, B.L.: Behavioral correlates of dopaminergic unit activity in freely moving cats. Brain Res. 258: 217-228, 1983
36. Suri, R. and Schultz, W.: A neural learning model based on the activity of primate dopamine neurons. Soc. Neurosci. Abstr. 22: 1389, 1996
37. Sutton, R.S. and Barto, A.G.: Toward a modern theory of adaptive networks: expectation and prediction. Psychol. Rev. 88: 135-170, 1981
38. Sutton, R.S. and Barto, A.G.: Time-derivative Models of Pavlovian Reinforcement. In: Learning and Computational Neuroscience: Foundations of Adaptive Networks (Eds. M. Gabriel and J. Moore). MIT Press, Cambridge, pp. 497-537, 1990
39. Toan, D.L. and Schultz, W.: Responses of rat pallidum cells to cortex stimulation and effects of altered dopaminergic activity. Neuroscience 15: 683-694, 1985
40. Wickens, J. and Kötter, R.: Cellular models of reinforcement. In: Models of Information Processing in the Basal Ganglia (Eds. J.C.Houk, J.L.Davis and D.G.Beiser) MIT Press, Cambridge, MA, pp. 187-214, 1995

The Information Content of Action Potential Trains A Synaptic Basis

Henry Markram and Misha Tsodyks

Department of Neurobiology, Weizmann Institute of Science, Rehovot 76100, Israel.

Abstract. Electrical recordings from three neurons revealed that the same spike train emitted by one neuron had markedly different effects on two target neurons. A spike train from a single neocortical pyramidal neuron produced synaptic responses in two target pyramidal neurons that differed in response strength and rates of activity-dependent depression of synaptic transmission. When a pyramidal neuron targeted another pyramidal neuron as well as an interneuron, then responses were also qualitatively different. The responses onto the pyramidal neuron displayed marked activity-dependent depression while those onto the interneuron displayed marked activity-dependent facilitation. The results suggest that each target could have a unique response to the same presynaptic signal. The information contained within the spike train therefore appears to be fragmented and re-integrated into the network at specific locations. The degree to which the specific fragment extracted by each synapse, will influence the spiking activity of the neuron, depends the ongoing integration of input from other presynaptic neurons. It is therefore proposed that differential synaptic transmission enables the neocortex to encode and decode the information contained within spike trains in an associative manner.

1 A multitude of targets - a computational dilemma

The axonal tree of most neurons forms several thousand synapses which can contact hundreds of neurons of different types, in different locations and that could have different functions. At the most simplistic level, if 1000 synapses each produce a binary response, the spatial structure (or spatio-functional equivalent) of the response generated by a single action potential emitted from one neuron could be one of 2^{1000} possibilities. This could either indicate an astronomical computational potential or noise. Noisy or unreliable transmission is a popular view and many potential mechanisms could reduce noise, for example, by innervating several neurons that have redundant physiological functions or innervating the same neuron with many synapses to obtain an averaged effect or by averaging across many spikes (i.e. rate coding) [1, 2, 3]. The less traveled view is that of an astronomical computational potential. One approach towards this view is to consider transmission at discrete moments in time and as a function of time. In other words, the nature of synaptic transmission is such that any one of the 21000 possible spatial structures that is evoked by one AP will influence subsequent choices.

2 Dynamic synapses

A binary choice to release or not may be probabilistic, but the consequences of a failure to release or a success in release may be deterministic and transient. Release probability (Pr) can change from one spike to the next [4] and in some synapses, typically where Pr is low to start with, Pr is transiently increased by each arriving spike and in others this mechanism of facilitation is poorly developed. In both cases, Pr decreases after successful release. A facilitating synapse therefore signals that the choice was yes and no, while the depressing synapse only signals that the choice was yes. The signaling of the choice made by the synapse is mediated by the influence on subsequent responses. Thus, the average synaptic response is a direct product of either activity-dependent depression of transmission (see Fig. 1), facilitation or a combination of both [5]. An examination of the properties of activity-dependence therefore carries with it possible computation performed as stochastic binary devices. However, while formulation of activity-dependent synaptic transmission implicitly incorporates stochastic synaptic transmission, the explicit role of stochasticity remains to be formulated and is not discussed here.

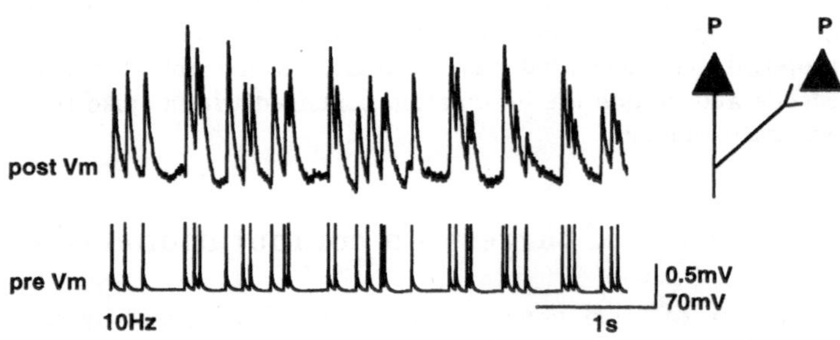

Fig. 1. Dynamic synaptic transmission: The postsynaptic response in a pyramidal neuron to an irregular train of presynaptic action potentials generated in a neighboring pyramidal neuron.

2.1 Modeling dynamic synapses

Activity-dependent depression was previously modeled with three parameters [6]. Briefly, the absolute synaptic efficacy of the connection (A_{SE}), is an activity-independent parameter that describes the maximum potential synaptic response that could be produced by an AP in the soma. Biophysically,

$$A_{SE} = qnE_f, \tag{1}$$

where q is the quantal size, n is the number of release sites and E_f is the correction for electrotonic filtering. An utilization of synaptic efficacy parameter, U_{SE}, represents the fraction of A_{SE} that an AP U_{SE}s on average to produce a synaptic response. Biophysically, U_{SE} is equivalent to Pr at low frequencies. The time constant τ_{rec} underlies the rate of recovery from depression. Note that this model holds because of a rigid rule at these synapses that a successful release event is invariably followed by depression. A modification allowing U_{SE} for each AP to be selected from a distribution enables simulation of stochastic responses indistinguishable from experimental traces (not shown).

To include facilitation, we propose that U_{SE} is changed instantaneously by each spike. The amplitude of the change is U_{SE}^1 which is equivalent to U_{SE} at very low frequencies. Between spikes its value decays with a time constant of τ_{facil}. This process might be due to the relaxation of intra-terminal [Ca2+] which has several time constants [7], but for generalization of the main properties of synapses, a single time constant is sufficient and does not run the risk of over interpreting the data. The formulation of the model is then,

$$\frac{dU_{SE}}{dt} = -\frac{U_{SE}}{\tau_{facil}} + U_{SE}^1(1 - U_{SE})\delta(t - t_{AP}), \tag{2}$$

where t_{AP} is the time of arrival of the presynaptic AP and U_{SE}^1 is the parameter describing the facilitation of U_{SE} after each spike. This model serves to describe strikingly different synaptic behaviors with the same set of 4 parameters and can therefore be used to examine, mathematically, the computational significance of dynamic synapses.

3 Differential signaling onto a homogeneous class of pyramidal neuron

The potential for pyramidal neurons in the neocortex to differentially signal their pyramidal targets was examined in neocortical slices by recording simultaneously from 3 pyramidal neurons [5]. The target pyramidal neurons selected were of the same morphological and electrophysiological class as the projecting neuron. This enabled also a comparison of the anatomical innervation. Two target neurons were innervated with a different number of synaptic contacts and synapses were located on different dendritic branches (branch orders and types) indicating that no two pyramidal neurons are innervated in the same manner. A_{SE} varied on average 5.72 fold, U_{SE} varied 1.27 fold and τ_{rec} varied by a factor of 1.63. The heterogeneity in A_{SE}, U_{SE} and τ_{rec} indicate that each synaptic connection transmits unique features of the same spike train (Fig. 2A)

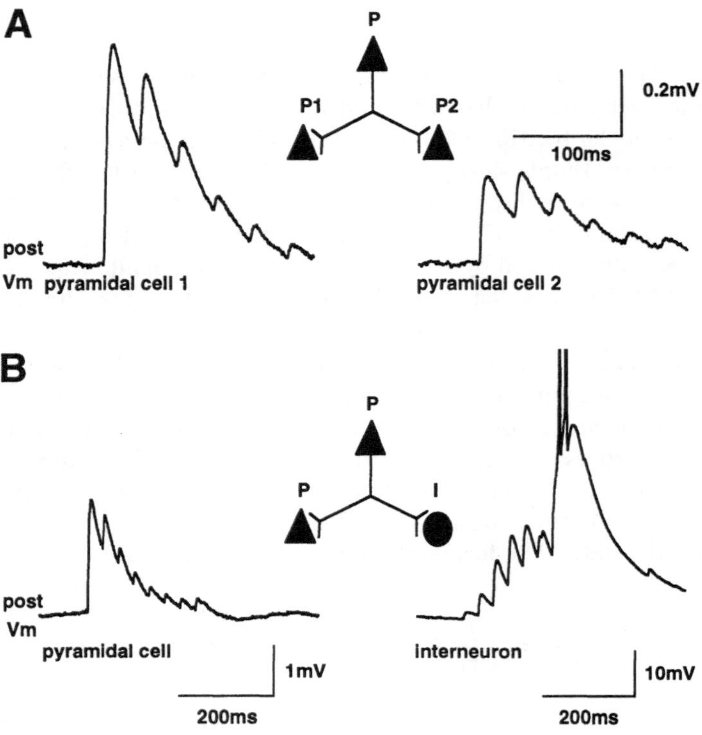

Fig. 2. Differential synaptic transmission via the same axon: (A) The responses to a high frequency train of presynaptic APs in a tufted layer 5 pyramidal neuron recorded simultaneously in two target pyramidal neurons. (B) The responses recorded simultaneously in a pyramidal neuron and an interneuron.

3.1 Differential signaling of pyramidal neurons and interneurons

A train of spikes in a single pyramidal neuron generated qualitatively different responses in a pyramidal neuron and an interneuron ([5]; Fig. 2B). The response onto interneurons displayed powerful activity-dependent facilitation (see also [8]) while depression was observed in the pyramidal neuron. In some cases these facilitating synapses were so powerful that the pyramidal neuron could discharge the interneuron. As the presynaptic frequency is increased, the interneuron discharged at progressively earlier times, suggesting that the precise time of interneuron discharge could encode the presynaptic frequency. It is then not surprising to find that the inhibitory synapse back onto the pyramidal neurons are rapidly depressing synapses, since this enables them to convey more reliably the time of discharge (see also [9]).

4 Comparison of the computational properties of facilitating and depressing synapses

In a previous report [6, 10] we showed that as the presynaptic frequency was increased beyond a certain frequency (defined as the limiting frequency, λ) which is given by,

$$\frac{1}{U_{SE}\tau_{rec}} \tag{3}$$

the stationary EPSPs ($EPSP_{STAT}$) decreased inversely proportional to the frequency (Fig. 2C) (see also [11]). For facilitating synapses $EPSP_{STAT}$ exhibits a tuning curve as the frequency is increased where the peak of the tuning curve (defined as the peak frequency, θ; Fig. 2D) is approximately,

$$\frac{1}{\sqrt{U_{SE}\tau_{facil}\tau_{rec}}}. \tag{4}$$

For these synapses both U_{SE} and τ_{rec} are small and hence λ is large.

Signaling via the depressing and facilitating synapses is different for certain frequency ranges and similar for others. At low frequencies (below λ) depressing synapses transmit the absolute discharge rate of the presynaptic neuron (linear regime) and above λ, transitions of rates (derivatives) are transmitted (sublinear regime). Facilitating synapses exhibit also a supra-linear regime below θ where rate transmission is amplified by a facilitating factor which is equivalent to the integral of rates (r) weighted with a decaying kernel of time constant τ_{facil}, which is given by,

$$\int_{-\infty}^{t} r(t')e^{-(t-t')/\tau_{facil}}\,dt' \tag{5}$$

An approximation of the facilitating factor is the number of spikes emitted during the time window of τ_{facil}. Like depressing synapses, absolute discharge rates are transmitted between θ and λ and the derivative of rates is transmitted above λ.

5 Transient excitation and delayed inhibition (TE-DI)

At presynaptic frequencies below θ, facilitating synapses transmit in the supra-linear regime, thus enable a delayed but amplified synaptic response in the interneurons. Direct evidence for delayed inhibition was found in a prevalent circuit in the neocortex (Fig. 3a). In this circuit, a pyramidal neuron transiently excites (TE) a neighboring pyramidal neuron with a direct monosynaptic connection which displays the characteristic depression and the same axon also contacts a neighboring interneuron with a facilitating synaptic connection. At high pyramidal frequencies, the interneuron discharges and inhibits the pyramidal neuron (DI). Differential synaptic transmission via the same axon enables the delayed excitation of the interneuron and subsequent inhibition of the pyramidal neuron.

These circuits are referred to as TE-DI circuits for "transient excitation-delayed inhibition". Evidence was also found for a TE-DE circuits in which delayed excitation (DE) either via excitatory interneurons or another class of pyramidal neurons was observed (not shown).

5.1 Functions of TE-DI circuits

Delayed inhibition could be fundamental to neocortical network dynamics. It provides a frequency-dependent time window ($\sim \frac{1}{r^2}$) for excitation to spread within the population of pyramidal neurons before inhibition is recruited. This could subserve many cortical functions: (1) TE-DI circuits may prevent explosive excitation [12]. (2) They could enforce synchrony since only highly synchronous bursts of activity could propagate without decrement through the pyramidal population without "waking up" these interneurons. On the other hand, poorly synchronous network activity would be strongly influenced by inhibition. DI loops therefore operate synergistically with TE connections between pyramidal neurons to favor synchrony [13, 14]; (3) Many adaptive mechanisms of the neocortex [15, 16] and the influence of background activity on responseproperties could be due to the synergistic action of growing inhibition and decreasing excitation; (4) In experiments when 4 pyramidal neurons were recorded simultaneously, we found that high frequency excitation of any one pyramidal neuron could induce DI in the other three neurons (Fig. 3b). TE-DI within recurrent networks could therefore also subserve a "winner take all" mechanism for these pyramidal neurons, since the first pyramidal cell to reach the critical frequency to discharge the interneurons can inhibit the pyramidal neurons located only 10's of microns away. This could be analogous to surround inhibition, but operates within the network of neurons which could be receiving common input.

5.2 TE-DI circuits encode spatio-temporal structure of network activity

We also considered the dynamics that could be generated by TE-DI circuits in a more general context where the current generated in a single target pyramidal neuron was examined as a function of the spatio-temporal patterns in a presynaptic population of pyramidal neurons connected to the pyramidal neuron through the TE-DI circuit. To this end, we simulated an input from 10 presynaptic pyramidal neurons that targeted a single pyramidal neuron and a population of 10 interneurons, which all projected onto the target pyramidal neuron (Fig. 4). The presynaptic population was undergoing a transition from 0Hz to 40Hz random Poisson firing either instantaneously or gradually over the course of up to 80 msec. As in Fig. 3, rapid input evoked biphasic response in the target, consisting of initial excitatory response followed by the delayed inhibition (Fig. 4A, solid curve). The same presynaptic frequency activated gradually over 80ms time window resulted in the diminished excitatory response followed by stronger inhibition. This result shows that TE-DI circuits may differentially respond to various patterns of activation depending on their temporal course.

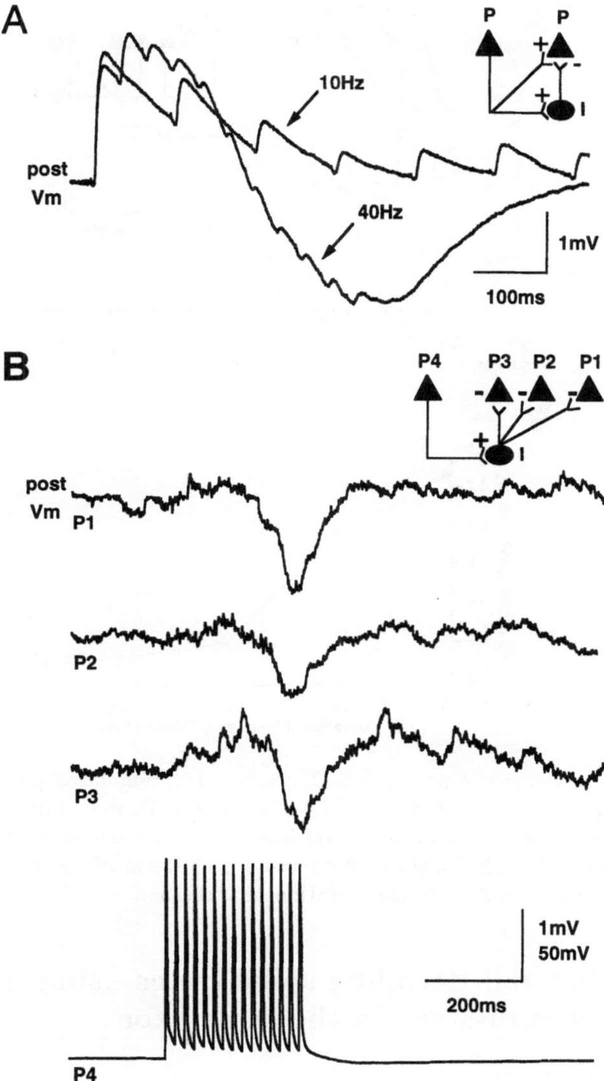

Fig. 3. Transient Excitation-Delayed Inhibition Circuits in the neocortex: (A) The response of a pyramidal neuron to a 10 Hz and 40 Hz presynaptic train of action potentials. The polysynaptic loop via the interneuron is not activated at 10 Hz, but is at 40 Hz. Average of 15 sweeps (B) Simultaneous recording from 4 pyramidal neurons reveals extensive divergent innervation from an interneuron activated by a single pyramidal neuron. Single sweeps are shown to demonstrate the simultaneity of the inhibition, suggesting that 1 interneuron was recruited.

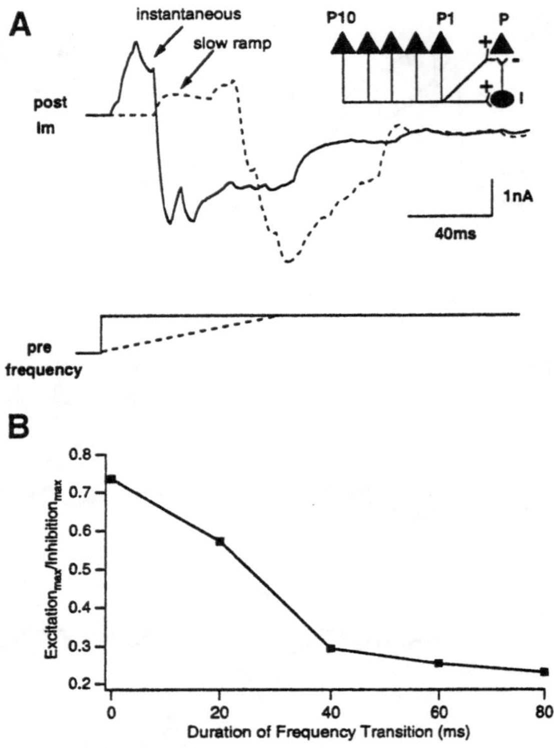

Fig. 4. Computation performed by TE-DI circuits: (A) Simulated postsynaptic current generated in a pyramidal target due to 40Hz random Poisson firing of a presynaptic population of 10 pyramidal neurons. Solid line - instantaneous frequency onset; dashed line - ramped onset with duration of 80 msec. (B) Ratio of excitatory to inhibitory peak responses as a function of the duration of transition.

6 Encoding and decoding spike trains using differential synaptic transmission via the same axon

The computational dilemma of extensive divergent synaptic innervation can now be re-addressed from the view point of synaptic dynamics. Each synapse can be characterized in terms of its absolute synaptic efficacy and its dynamics of transmission. The computational significance of each synapse's dynamics can be derived from the relationship between frequency and $EPSP_{STAT}$ which can reveal potentially three signaling regimes; supra-linear, linear and sub-linear. For an irregular train of APs, each of the several thousand synapses could transmit unique mixtures of features from all regimes. A single presynaptic signal is therefore fragmented by differential synaptic transmission. In a network that developed a specific non-random architecture either through developmental or learning processes, unique fragments of the signal would then be transferred to specific points (neurons) in the network. Each target neuron could integrate all the fragments received and biase the integration according to the specific biophysical properties of the neuron. The relevance of each fragment to the specific neuron in the context of all fragments (the network context) is therefore computed. Network-irrelevant fragments are eliminated in the integration and network-relevant fragments contribute towards discharging the neuron. Network-relevant fragments therefore co-operate to inject another spike into the network. A sequence of "fragment selection and elimination" therefore could occur with each time step after external input to the network (an iteration). Differential synaptic transmission may therefore enable the neocortex to both encode and decode APs in a massively associative manner.

7 Redistribution of synaptic efficacy

In a network where neurons are connected with dynamic synapses it is necessary to consider how these dynamics may change in addition to how the strength of activity-independent transmission changes. In order to examine the dynamics of transmission it is necessary to stimulate synapses with trains of high frequency spikes. Using the TM model [6], it is possible to derive the 4 key parameters governing the strength and dynamics and thereby to predict responses to arbitrary spike trains. These four parameters can also be monitored during experiments to determine which property of the synapse changed. Hebbian paring, for example, results in a change in U_{SE} and not in A_{SE} or τ_{rec} [6] which is observed phenomenologically as a redistribution of synaptic efficacy (RSE) [17], termed such to reflect a change in dynamic transmission since each spike in a given train is able to utilize the existing synaptic efficacy in a different manner. Essentially any changes in the kinetic parameters changes produce RSE.

8 RSE - the effect of modifying dynamics of dynamic synapses

As the computational significance of synaptic dynamics can be quantified so can those of RSE. Changing any of the kinetic parameters (U_{SE} , τ_{rec}, τ_{facil}) can change the peak frequency, θ and or the limiting frequency, λ. This changes the frequency ranges for the three signaling regimes and at the level of a given Poisson spike train, a change in the mixture of information from the these regimes. Functionally, this can lead to complex effects on the network. First, if U_{SE} is increased for depressing synapses then the rate of depression increases and the time window for optimal summation (coincidence window) is reduced [6]. Second, if U_{SE} is increased for facilitating synapses then θ is less and the interneuron discharges sooner, thus reducing the frequency-dependent time window for pyramidal neurons to propagate excitation among themselves. Indeed, after pairing of the pre and postsynaptic pyramidal neurons (which also excited the interneuron), TE-DI circuit dynamics were amplified such the TE and DI responses were larger (Fig. 5). Third, in the context of TE-DI and TE-DE circuits, modifications of any of the kinetic parameters leads to changes in the decoding of the temporal structure of pyramidal spiking activity (Fig.4). Finally, at a more general level, changing the dynamics of synaptic transmission not only changes the nature of the fragments of information extracted by each synapse but also the network-relevance of the fragments, since the dynamics themselves determine the network relevance. RSE could thus also change the information transmitted by the same action potential train.

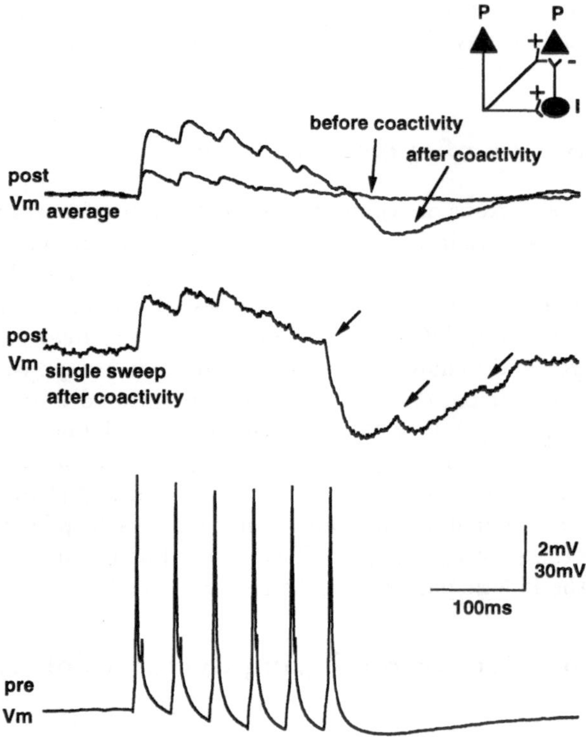

Fig. 5. Synaptic learning in TE-DI circuits: Before coactivity, a presynaptic train of APs (23 Hz) did not activate the polysynaptic inhibitory loop. After coactivity of the two pyramidal neurons which also excited the interneuron, the transient excitation was increased and polysynaptic inhibitory responses were observed. The upper traces show the average of 30 sweeps and the middle traces show a single sweep response. The arrows mark the onset of IPSPs.

9 Conclusion

Differential synaptic signaling onto hundreds of targets within a network is proposed as a mechanism used by neuronal networks to encode and decode the information contained within the spike train in a massively associative manner.

References

1. Allen, C. & Stevens, C. F. Proc Nat Acad Sci USA 91, 10380-10383 (1994).
2. Softky, W. R. Curr Opin Neurobiol 5, 239-47 (1995).
3. Shadlen, M. N. & Newsome, W. T. Curr Opin Neurobiol 4, 569-79 (1994).
4. Betz, W. J. Physiol (Lond.) 206, 629-644 (1970).
5. Markram, H., Tsodyks, M. & Wang, Y. Differential signalling via the same axon from neocortical layer 5 pyramidal neurons, submitted (1997).
6. Tsodyks, M. & Markram, H. Proc Nat Acad Sci (USA) 94, 719-723 (1997).
7. Bertram, R., Sherman, A. & Stanely, E. F. Journ Neurophysiol 75, 1919-1931 (1996).
8. Thomson, A. M., Deuchars, J. & West, D. C. Neuroscience 54, 347-359 (1993).
9. Thomson, A. M. & Deuchars, J. Trends in Neuroscience 17, 119-126 (1994).
10. Tsodyks, M. & Markram, H. Lect Notes Comput Sci 1112, 445-450 (1996).
11. Abbott, L. F., Varela, J. A., Sen, K. & Nelson, S. B. Science 275, 220-224 (1997).
12. Grinvald, A., Frostig, R. D. & Lieke, E. Proc Nat AcadSciences (USA) 68, 1285-1366 (1988).
13. Gray, C. M., Konig, P., Engel, A. K. & Singer, W. Nature 338, 334-7 (1989).
14. Abeles, M., Prut, Y., Bergman, H. & Vaadia, E. Prog. Brain. Res.102, 395-404 (1994).
15. Albrecht, D. G., Farrar, S. B. & Hamilton, D. B. J. Physiol. (Lond.) 347, 713-39 (1984).
16. Maffei, L., Fiorentini, A. & Bisti, S. Science 182, 1036-8 (1973).
17. Markram, H. & Tsodyks, M. Nature 382, 807-810 (1996).

Cortical Cell Assemblies, Laminar Interaction, and Thalamocortical Interplay

Robert Miller

Department of Anatomy and Structural Biology, University of Otago, New Zealand.

Abstract. Hebb's concept of cell assemblies was formulated in a highly schematized and simplified version of the cerebral cortex. This paper present two parallel hypotheses of how cell assemblies might actually be realised in the cortex. The first hypothesis concerns interaction between laminae II/III and lamina V of the cortex. The second concerns interactions between cortex and thalamus. In both hypothesis, laminae II and III are envisaged to form a relatively inactive "library" in which information is stored as strengthened connections. The other components are responsible for "priming" this store to allow registration and retrieval of memory. Specially, lamina V performs this role for local cortical interactions while the thalamus does so on a larger scale, which includes temporal coordination of neuronal firing across the cortex.

1. Introduction

Hebb's concept of the cell assembly [1] remains the best we have for bridging between the structure/function of nervous tissues at the level of single cells, and the integrated performance of the brain at the psychological level. Hebb envisaged that cell assemblies were formed within the cerebral cortex, and that the members of an assembly become connected together by strengthening of excitatory connections between cortical neurones. This is plausible, because the principal neurones of the cortex are indeed connected together by excitatory connections (which make up about 80% of all cortical synapses)(2], and these connections are subject to strengthening, as Hebb had postulated (3).

A significant elaboration of Hebb's concept occurred as a result of the demonstration by (4,5) and others that there may be precise temporal structure in the timing of action potentials distributed amongst several neurones recorded simultaneously. This has led Abeles to put forward the concept of "synfire chains". Like Hebbian cell assemblies these are envisaged to be the vehicle for cortical representation, but unlike Hebbian assemblies they may represent temporally changing patterns of information rather than just static ones.

Information available about the cerebral cortex has grown enormously since Hebb's day. However, the full wealth of information now available about cortical anatomy and neuronal dynamics has not impinged much on conceptual developments of cell assembly theory, or on the synfire chain hypothesis. The aim of this paper is to introduce a number of facts coming from recent empirical research which allow the cell assembly and synfire chain concepts to be formulated as they might apply to the real cortex, rather than to a simplified and idealized cortex. The paper consists of two parts: The first explores the local interaction between different laminae in the cortex, with particular reference to the Hebbian assembly concept. The second develops the argument further, bringing in the manner of interaction on a larger scale between cortex and thalamus, and makes proposals applicable both to Hebbian cell assemblies and synfire chains. These two hypotheses have been expounded in greater detail in two recent papers (6,7).

2. Local Laminar Interactions and Hebbian Cell Assemblies

Cortico-cortical connections, clearly of central importance in cell assemblies, originate or terminate in most laminae of the cortex. However, the pyramidal cells in laminae II and III are distinguished in that they have almost exclusively cortico-cortical connections (8). The excitatory transmitter of such connections appears to be glutamate. The NMDA glutamate receptor is one of those mediating synaptic actions at such connections, and is thought to be of special importance for Hebb-type modification of the efficacy of excitatory synapses (9). This receptor is found in greatest density in laminae II and III (10). One is thus led to conclude that laminae II and III are of special importance in the formation of Hebbian cell assemblies, forming a store of information within cell assemblies, figuratively a neuronal "library". These laminae would have a large information storage capacity, related to the vast number of cortico-cortical synapses.

Pyramidal cells in these laminae have a very low level of spontaneous activity (11,12,13), and may be essentially silent for seconds at a time. Presumably such neurones usually maintain a resting membrane potential well below firing threshold. Thus, convergence of many active afferents from within laminae II and III and elsewhere would be required before action potentials are triggered. Adequate convergence would be quite rare. This creates a problem for the formation of cell assemblies, since Hebbian strengthening requires convergence of a number of active afferents.

However, the neuronal library of laminae II and III does not function in isolation. Neurones in lamina V receive connections from laminae II and III and send projections back to these laminae (14). In this lamina, frequencies of spontaneously occurring action potentials in the waking state are very much higher than in laminae II/III (11,12,13). Such pyramidal cells appear to be poised just below the threshold for firing, such that a single active afferent, or a very few, may trigger an action potential. Given this, the population of lamina V pyramidal cells can be regarded as having a "priming" or "enabling" role for the more superficial laminae, permitting it to function in formation and maintenance of cell assemblies in a way not possible without lamina V. The argument supporting this hypothesis is as follows:

Consider two neurones within lamina II/III with an anatomical connection from one to the other. They may be referred to as the "transmitting" and the "target" neurone. It is envisaged that the connection from one to the other has not yet been strengthened. In order for such strengthening to occur, impulses along the connection must consistently and repeatedly coincide with impulses in other connections to the target neurone. These additional afferent activations are unlikely to occur in afferents from within laminae II and III because of the very low levels of spontaneous activity therein. However, it may be envisaged that the transmitting neurone has axonal collaterals providing a variety of synaptic contacts to pyramidal cells in the deeper layer V. Whenever an action potential is triggered in the transmitting neurone in lamina II/III, it is likely to trigger action potentials in a number of lamina V pyramidal cells to which the transmitting neurone projects. Each of these cells will in turn distribute action potentials to all their postsynaptic targets, including a significant population in laminae II and III. While the direct connection from the transmitting to the target neurone in laminae II/II specifies just a single target neurone, the indirect pathway, via lamina V will distribute activity to a much larger number of cells in lamina II/III. Because of this multiplicative spread of synaptic activation, it is likely that the target in lamina II/III will be amongst the population indirectly activated via lamina V. As a result, the probability of the target neurone receiving a number of simultaneously converging afferent activations sufficient to produce an action potential or to activate the Hebbian synaptic modification process will be enhanced by the participation of lamina V. If this happens, the multiple convergence will occur repeatedly and consistently, rather than in a haphazard chance fashion, and thus will lead to the progressive strengthening of the connection between the transmitting and the target neurone. One or more neurones in lamina V will have played an essential "catalytic" role in this process. These relationships are shown in the illustration below. (The “target” neurone is depicted as part of a cell assembly, shown as filled circles.)

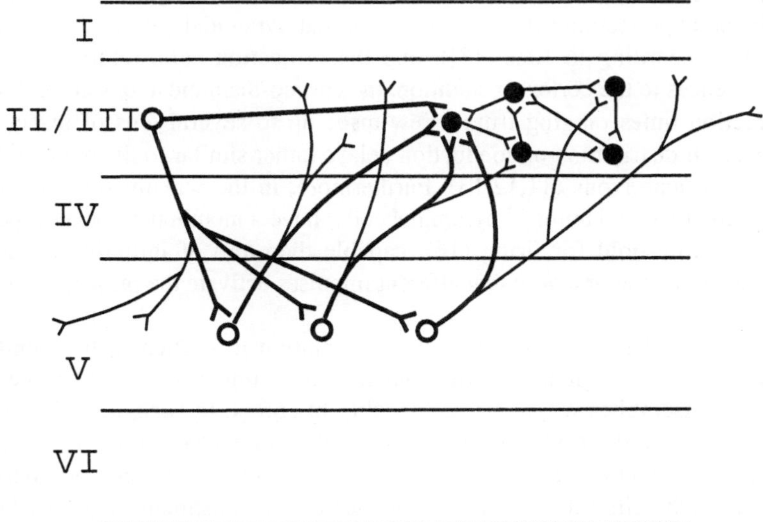

3. Larger-Scale Integrations and Thalamocortical Interaction

The argument presented above made the implicit assumption that any local interactions between laminae involve no significant conduction delays. Thus, impulses transmitted to the target neurone via the direct and the indirect route would inevitably arrive within the same time interval for neuronal integration, and no loss of amplification would arise due to asynchrony of converging afferent impulses. This assumption is realistic because conduction distances in the small block of tissue considered would be 2 mm at most, and the slowest possible conduction velocity is about 0.2 m/sec (13). Thus, maximum conduction time via the indirect route would not exceed 10 msec. This is shorter than the EPSP duration in a cortical pyramidal cell (15), the effective integration time within which converging afferent influences can summate.

For larger scale cell assembly interactions, the assumption of synchronous convergence can no longer be made: Cortico-cortical connections between distant cortical regions have a range of conduction times from a few msec up to 40 msec in small brained animals (11,12,13), and probably much longer in large brained animals such as primates. Given that axonal conduction delays are distributed throughout this range, a signal originating in one locus is subject to temporal dispersion along the various axons collaterals by which it spreads its influence. Synchrony of afferent activity originating from cortical one locus cannot then be guaranteed. This means that the conditions for synaptic strengthening of a connection between an active afferent neurone and members of a cell assembly located in laminae II/III in distant cortical regions are less likely to be met that when using local connections. A more powerful "priming" mechanism is therefore required than for local interactions.

The thalamus is in reciprocal relation with the cortex: The majority of excitatory inputs to thalamic projection cells come from cortical pyramidal cells (mainly located in lamina VI, according to Jones [8]), and the projection cells return excitatory synaptic influences to the cortex. In addition, the cortico-thalamic axons are known to have conduction times ranging from a few msec up to several tens of msec, thus having an overall distribution of conduction delays rather similar to the population of long cortico-cortical axons (11,12,13). Furthermore, in the waking state, thalamic projection cells (like the lamina V pyramidal cells) have a membrane potential poised just below the threshold for firing (16), capable therefore of initiating an action potential when only one or a very few afferent impulses activate the projection cell.

Given these facts, the role of the thalamus in relation to cortical cell assemblies distributed over distant regions of cortex may be rather similar to that of lamina V in relation to cell assemblies within a small locality of cortex. In each case, a group of cells initially outside the assembly, but reciprocally connected with it is poised to discharge action potentials in response to a quite low degree of excitatory synaptic convergence. Such cells can "prime" or "enable" the establishment of strengthened transcortical connections, which would not otherwise form. Such neurones (in lamina V or in the thalamus) would then become an integral part of the assembly. The greater amplification provided for large-scale integration of cell assemblies by the trans-thalamic mechanisms, compared to the indirect pathway via lamina V for local integration, may be accounted for by the larger number of synapses provided in the thalamus by each corticothalamic neurone, compared with those distributed in lamina V by lamina II/III neurones. This larger amplification makes it possible for the strengthening of connections which are selective not only with respect to the target, but also with respect to conduction delays: A necessary condition for Hebbian strengthening to occur is the synchronization (within a single neuronal integration interval) of synaptic activation via the direct and the transthalamic route. Given this, the thalamus can serve the function of large scale integration of cell assemblies, which enhances not only the chances of "spatial integration" of cell assembly function, but also the security of transmission in the temporal sequences described by Abeles (4,5).

The interlaminar and cortico-thalamic interactions described above can be stated easily in informal and qualitative terms. Quantitative specification of the conditions required are more difficult to define. In each case an important principle is illustrated: It is feasible to use a high amplification indirect pathway to catalyze the establishment of connections between two loci in the cortical mantle, for which direct connections have a very low amplification factor. This means that the neuronal library, an excitatory network in which information is represented as combinations of strengthened connections, can operate in storage and retrieval while itself having a low level of activity. Therefore, despite its being a network of mutually excitatory neurones, it need not be prone to "explosive" instabilities of neural activity.

4. Strategies of Theoretical Neuroscience

Science can be idealized as a fruitful interplay between ideas and experiments, between theories and experiments. When this fruitful interaction occurs it represents the most significant scientific advances. However, such an idealized process occurs rather rarely. In the field of neural network modelling, this author is struck by two disturbing facts: (i) Those who model neural networks use sophisticated mathematics, but rarely base their modelling on the best available biological data. (ii) Those who study brains empirically use sophisticated experimental techniques, but rarely use them to produce the quantitative data which the network modellers really need. As a result of these two facts, real progress in understanding the brain, which could come about by the fruitful interaction between experimentalists and theoreticians, occurs painfully slowly. The author believes that the synergy between experiment and theory could occur in a far more effective fashion.

References

1. D.O. Hebb. *The organization of behavior: a neuropsychological theory.* Wiley, New York, 1949.
2. V. Braitenberg and A. Schuz. Anatomy of the cortex: statistics and geometry. *Studies on Brain Function No. 18.*, Springer, Heidelberg, 1991.
3. T. Tsumoto. Long-term potentiation and long-term depression in the neocortex. *J.Comp. Neurol.*, 300:47-60, 1992.
4. M. Abeles. Role of the cortical neuron: integrator or coincidence detector? *Isr J Med Sci* , 18:83-92, 1982a.
5. M. Abeles. Local cortical circuits. *Studies in Brain Function No 6..* Springer, Heidelberg, 1982b.
6. R. Miller. Neural assemblies and laminar interactions in the cerebral cortex. *Biol. Cybern.,* 75:253-261, 1996a.
7. R. Miller. Cortico-thalamic interplay and the secutiry of operation of neural assemblies and temporal chains in the cerebral cortex. *Biol. Cybern.*, 75: 263-275, 1996b.
8. E.G. Jones. Laminar distribution of cortical efferent cells. In:*Cerebral Cortex,* Jones EG, Peters A (eds), Volume V. Plenum, New York pp. 521-553, 1984.
9. K. Fox and N.W. Daw. Do NMDA receptors have a critical function in visual cortical plasticity. *Trends Neurosci.,* 16:116-122, 1993.
10. D.T. Monaghan and C.W. Cotman. Distribution of N-methyl-D-aspartate-sensitive L-[3H]Glutamate binding sites in rat brain. *J.Neuroscience.*, 5:2909-2919, 1985.
11. H.A. Swadlow. Efferent neurons and suspected interneurons in binocular visual cortex of awake rabbits: receptive fields and binocular properties. *J Neurophsyiol.,* 59:1162-1187, 1988.
12. H.A. Swadlow. Efferent neurons and suspected interneurons in S-1 vibrissal cortex: receptive fields and binocular properties. *J Neurophysiol.*, 62: 288-308, 1989.
13. H.A. Swadlow. (1994) Efferent neurons and suspected interneurons in motor cortex of the awake rabbit: axonal properties, sensory receptive fields and subthreshold synaptic inputs. *J Neurophysiol* ., 71:437-453, 1994.
14. E.L. White. *Cortical circuits: synaptic organization of the cerebral cortex. Structure, function and theory.* Birkhauser, Boston, 1989.
15. A. Mason, A. Nicoll and K. Stratford. Synaptic transmission between individual pyramidal neurons of the rat visual cortex in vitro. *J.Neuroscience.*, 11:72-84, 1991.
16. J.C. Hirsch, A. Fourment and M.E. Marc. Sleep-related variations of membrane potenital in the lateral geniculate body relay neurons of the cat. *Brain Res.,* 259:308-312, 1983.

Cross-Correlations in Sparsely Connected Recurrent Networks of Spiking Neurons

Nicolas Brunel

LPS**, Ecole Normale Supérieure, 24 rue Lhomond, 75251 Paris Cedex 05, France

Abstract. We study the dynamics of sparsely connected recurrent networks composed of excitatory and inhibitory integrate-and-fire (IF) neurons firing at low rates, and in particular cross-correlations (CC) between spike times of pairs of neurons using both numerical simulations and a recent theory. CCs exhibit damped oscillations with a frequency which depends on synaptic time constants. Individual CCs are shown to depend weakly on synaptic connectivity. They depend more strongly on the firing rates of individual neurons.

1 The cortical network model

We consider the model of cortical module introduced by Amit and Brunel [4]. It is a recurrent network composed of a large number N of neurons, divided in two populations. 80% are excitatory (E) neurons, representing pyramidal cells. The remaining 20% are inhibitory (I) neurons which represent interneurons. The recurrent collaterals are set randomly with connection probability ϵ. The network receives input from outside the network: a background input representing spontaneous activity from other cortical areas, modelled as a Poisson spike train with a specified frequency to all external synapses.

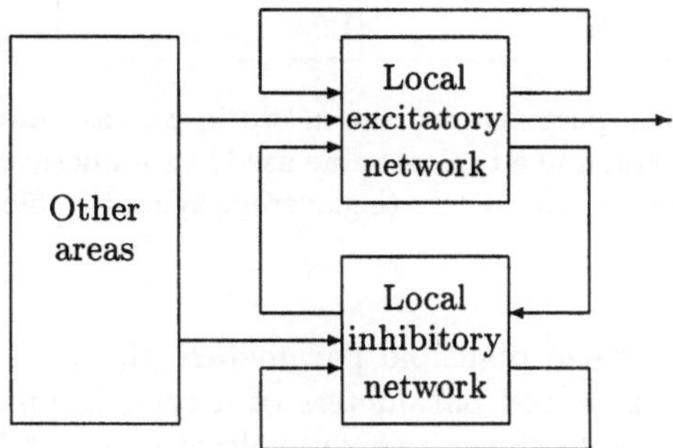

Fig. 1. Architecture of the model.

** Associated with CNRS, Paris 6 and Paris 7 universities.

Neurons are integrate-and-fire (IF) units[10]. It allows large-scale simulations and permits analytical estimates[4]. Neuron i ($i = 1, \ldots, N$) is characterized by its membrane depolarization $V_i(t)$, its afferent synaptic currents $I_i(t)$, an integration time constant τ, a threshold for spike emission θ, a post-spike reset potential V_r, and an absolute refractory period τ_0. Below threshold, the membrane depolarization integrates the synaptic currents according to $\tau \dot{V}_i = -V_i + I_i$. If $V_i(t) \geq \theta$, neuron i emits a spike. The synaptic current I_i is the sum of the recurrent synaptic contribution I_i^{rec} and of the external contribution I_i^{ext}. The recurrent synaptic contribution is

$$I_i^{rec}(t) = \sum_{j \in E,k} \rho_j(t) J_{ij} \tau \delta(t - \tau_{ij} - t_j^k) - \sum_{j \in I,k} \rho_j(t) J_{ij} \tau \delta(t - \tau_{ij} - t_j^k)$$

where the first (second) sum on the r.h.s. is the sum of E (I) spike contributions. $\rho_j(t)$=1,0 is the 'transmission probability' of a spike at synapse j. J_{ij} is the synaptic efficacy of synapse $j \to i$, or in other words the PSP amplitude provoked by a spike arriving at that synapse. τ_{ij} is the synaptic delay. t_j^ks are the emission times of spikes of neuron j.

Parameter	Cortex	Typical simulation
number of excitatory neurons	80,000–800,000	6,000
number of inhibitory neurons	20,000–200,000	1,500
connection probability	0.01–0.1	0.2
fraction of local synapses	~ 0.5	0.5
spike emission thresholds (mV)	10.–20.	20.
reset potentials (mV)	?	10.
membrane decay time constants, E (ms)	10.–30.	10
membrane decay time constants, I (ms)	5.–20.	5.
average synaptic delay (ms)	1.-5.	1.
average synaptic efficacies E→E (mV)	0.05–0.5	0.2
average synaptic efficacies I→E (mV)	?	0.6
average synaptic efficacies E→I (mV)	?	0.35
average synaptic efficacies I→I (mV)	?	1.05

Table 1. Table of parameters. For each parameter of the network, we have indicated plausible values in a cortical environment, and a typical value used in a numerical simulation. Note the large uncertainty in many parameters (e.g.average synaptic efficacies).

Table 1 shows a comparison between plausible parameters [6, 1, 11] for a cortical module of about 1mm^2 section and parameters of a typical numerical simulation on a workstation. This table shows two difficulties inherent in the simulation of neural systems: first, the impossibility to perform extensive simulations of networks even approaching the numbers of neurons present in such modules. Second, since there is a large uncertainty in many parameters, one should explore a large parameter space before any meaningful conclusion can be reached. These two facts point out the importance of having a simple analytical theory, which allows, on one hand to predict the behaviour of a large system, and on the other hand makes possible a full exploration of parameter space. The roles of the simulation are to investigate quantities of the system for which no theory is (yet) available (much as in an experiment, see e.g. the discussion in Amit [3]), and to test the theory, which in general makes simplifying assumptions on the system, with parameters for which a simulation is possible.

In both a simulation of duration T and theory we are interested in the following quantities, which are obtained from the spike trains $S_i(t) = 0,1$ of all neurons, discretized in time bins dt ~1ms: The global activity of the E(I) population, defined as

$$n_{E,I}(t) = \sum_{i \in E,I} S_i(t);$$

frequencies of individual neurons and mean frequencies in each population, the autocorrelation of the global activity

$$C_{\alpha\beta}(\tau) = \frac{1}{N_\alpha N_\beta (T-\tau)} \sum_{t=1}^{T-\tau} n_\alpha(t) n_\beta(t+\tau)$$

and finally the CC between spike times of a pair of neurons

$$C_{ij}(\tau) = \frac{1}{T-\tau} \sum_{t=1}^{T-\tau} S_i(t) S_j(t+\tau).$$

2 Distribution of frequencies in stationary states: simulations and theory

The theory of [4] describes the mean frequencies in stable states of activity of the network, given its parameters. It was then extended such as to obtain the full distribution of frequencies, and shown to reproduce very well the results of the simulations[5]. Recently it has been understood how to extend it to dynamical quantities (see Section 4 and [7]).

The main results of the theory have been to clarify the conditions necessary to obtain a low stable spontaneous activity: recurrent inhibition has to dominate locally recurrent excitation, otherwise a low activity state is unstable and the network may go in a high activity ('epileptic') state. Spontaneous activity is found to be crucial for the network to have a fast reaction to incoming stimuli: since in this state neurons are depolarized most of the time near threshold, they can react much more rapidly to stimuli of weak contrast than in a silent network. The theory has also clarified the conditions necessary to obtain stable 'memory states' (i.e. self-sustained firing above spontaneous activity levels of a subset of neurons selective for the memorized object), as recorded in delayed response tasks in several areas of primate cortex[8], after unsupervised Hebbian learning of stimuli takes place.

3 Network dynamics and CCs: simulations

CC measurements using multiple unit recordings has become an increasingly popular technique in recent years[2, 9]. However, to date there is no convincing theory describing CCs in a network of spiking neurons, and therefore it remains unclear how to interpret the experimental results.

Some questions we can ask about these CCs are: what is the origin of the observed structure in CCs observed in cortex? What information gives these CCs on the observed network? To begin with our investigation of CCs, first note that the average of CCs over all pairs of neurons is equal to the autocorrelation (AC) of global activity. Thus a first step in the understanding of CCs is to study the behaviour of the global activity.

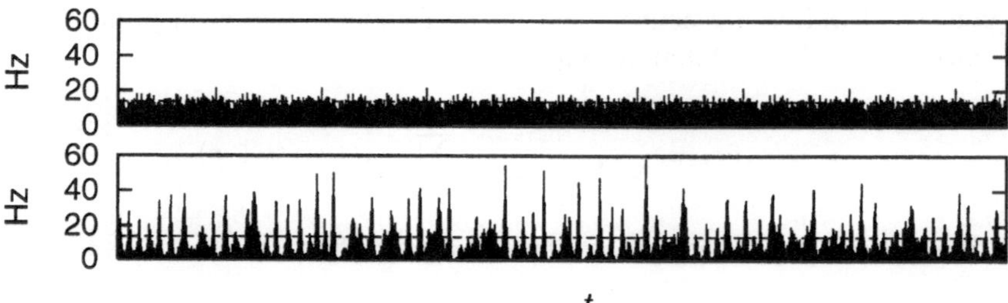

Fig. 2. Average activity in the entire excitatory population during 500ms. Simulation of a network with parameters given in Table 1: lower strip. For comparison is shown a simulation of uncorrelated Poisson processes with the same rate as in the simulation (upper strip)

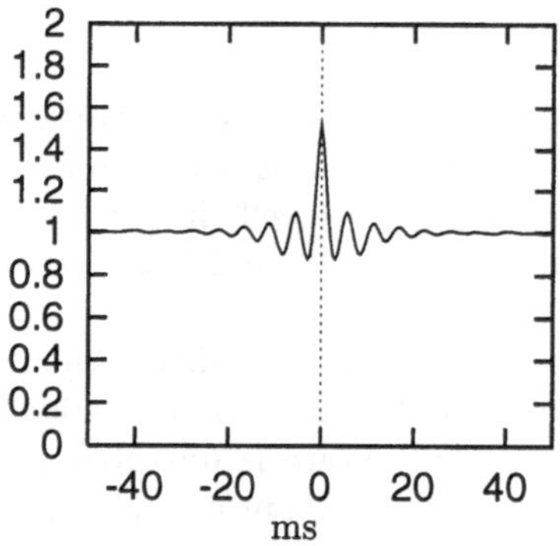

Fig. 3. Autocorrelation of the E global activity, or average CCs between E-E, neurons.

The evolution of the global activity in a simulated network is shown in Fig. 2. Note the very sharp and brief fluctuations, to be compared with the relatively smooth behavior of a network of uncorrelated Poisson processes with rates equal to the average rates (horizontal line) in the simulated network (upper strips, same scale). The inhibitory network exhibits the same kind of very irregular activity, in phase with the E one. These fluctuations are reflected in the autocorrelation of the E global activity, shown in Fig. 3. Note the central peak with an amplitude of about 1.5 relative to the background, and the damped oscillations with a time constant of about 5ms.

Simulations show that the size of the central peak relative to background, which can be taken as a measure of the degree of synchonization of the network, increases linearly on the connection probability ϵ, at low ϵ; and increases linearly with the fraction of local synapses. It thus mainly reflects the global 'architecture' of the system. On the other hand, the temporal structure depends crucially on the 'microscopic' synaptic time constants: it increases with synaptic delays.

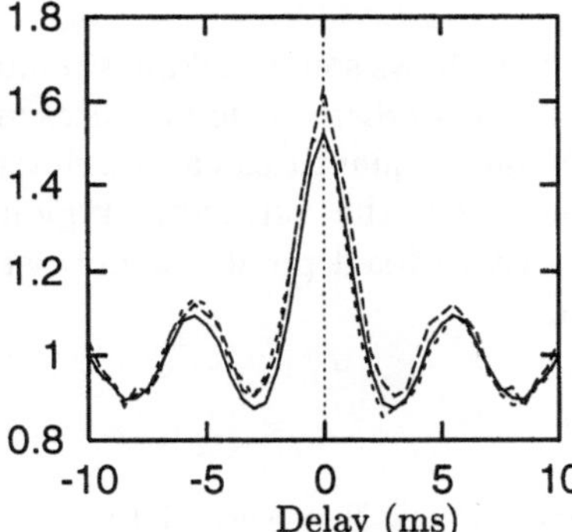

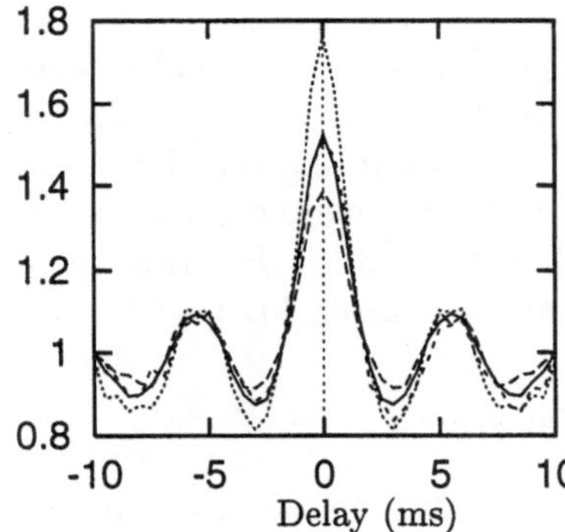

Fig. 4. Dependence of average CCs on synaptic connectivity (left) and on rates (right). Left: Full line: average CCs. Long-dashed line: CCs of reciprocally connected neurons. Short-dashed line: CCs of one-way connected neurons. CCs of unconnected cells indistinguishable from average CCs (see text). Right: Full line: average CCs; long-dashed line: CCs of pairs having both high rates ($>30s^{-1}$); short dashed line: CCs of pairs one with low rate ($<10s^{-1}$), the other high rate ($>30s^{-1}$), asymmetric; dotted line: CCs of pairs with both low rates ($<10s^{-1}$).

Individual CCs have a structure which is very similar to the AC of the global activity, indicating that the major part of the correlations between spike times stems from the temporal structure of the global activity. However they do depend on monosynaptic contacts and on firing rates of both units. These effects are shown in Fig. 4. The weak dependence on the presence of monosynaptic contacts between the recorded neurons would make it very difficult to extract information about monosynaptic contacts from CCs. On the other hand CCs depend more on the frequency of recorded neurons. Neurons with low frequencies are typically more correlated than high frequency neurons. This is an experimental prediction that could be easily checked given enough CC data.

4 Network dynamics and CCs: theory

We have recently developed a theory that captures the temporal evolution of the global activity in such a network, and enables to determine analytically its AC[7]. The theory is in quantitative agreement with the simulations. Increasing further the inhibition a Hopf bifurcation is observed: the network settles in an oscillatory limit cycle. In this regime CCs do not decrease anymore when N increases at parity of average number of connections per neuron ϵN, but rather increase with N. The period of the oscillations is shown to depend exclusively on the distribution of synaptic delays; for a wide delay distribution, this period is of order five times the average delay. Note that in both regimes neuronal firing is essentially Poisson, apart from the absolute refractory period.

5 Conclusions

The model has a very rich phenomenology, which shows some striking similarities with available experimental data, both at the level of distribution of rates and of CCs. It is rewarding that both static and dynamic quantities can be described by a relatively simple theory[4, 7], which identifies the parameter regions for which a given state (low rate spontaneous activity fixed point, memory states, low rate oscillatory limit cycle) can be found.

References

1. Abeles, M.,: Corticonics, Cambridge University Press, New York (1991)
2. Abeles, M., Bergman, H., Gat, I., Meilijson, I., Seidemann, E., Tishby, N., Vaadia, E.: Cortical activity flips among quasi-stationary states, Proc. Natl. Acad. Sci. USA **92** (1995) 8616
3. Amit, D.J.: Simulation in neurobiology – theory or experiment?, submitted. See http://www.fiz.huji.ac.il/staff/acc/faculty/damita/papers.html.
4. Amit, D.J., Brunel, N.: Model of global spontaneous activity and local structured activity during delay periods in the cerebral cortex, Cerebral Cortex **7** (1997) 237-252
5. Amit, D.J., Brunel, N.: Dynamics of recurrent networks of spiking neurons before and following learning, to be published in Network. See http://www.lps.ens.fr/~brunel/publi.html.
6. Braitenberg, V., Schüz, A.: Anatomy of the cortex, Springer-Verlag, Berlin (1991)
7. Brunel, N., Hakim, V., in preparation.
8. Fuster, J.: Memory in the cerebral cortex, MIT press (1995)
9. Kreiter, A.K., Singer, W., Stimulus dependent synchronization of neuronal responses in the visual cortex of the awake macaque monkey, J. Neurosci., **16** (1996) 2381
10. Knight, B.W.: Dynamics of encoding in a population of neurons, J. Gen. Physiol., **59** (1972) 734
11. Mason, A., Nicoll, A., Stratford, K.: Synaptic transmission between individual pyramidal neurons of the rat visual cortex *in vitro*, J. Neurosci., **11** (1991) 72

A Comparative Study of Pattern Detection Algorithm and Dynamical System Approach Using Simulated Spike Trains

Igor V. Tetko[1,2] and Alessandro E. P. Villa[1]

(1) Laboratoire de Neuro-heuristique, Institut de Physiologie, UNIL, Rue du Bugnon 7, Lausanne, CH-1005, Switzerland
(2) Institute of Bioorganic and Petroleum Chemistry, Murmanskaya, 1, Kiev-660, 253660, Ukraine, tetko@bioorganic.kiev.ua

Abstract. We apply two different approaches--pattern detection algorithm and dynamical system analysis--to study sets of simulated spike trains produced by chaotic attractors and Poisson processes. We show that both algorithms are able to detect a deterministic activity in the chaotic spike trains and they are tolerant to the presence of noise in input data. A method for noise filtering in input data series is proposed and its application is demonstrated for the simulated data sets.

1 Introduction

Recent studies introduced an application of dynamical system approach (DSA) for analysis of sequences of action potentials recorded in neurophysiological experiments [1]. It was shown that a considerable number of spike trains (up to 25%) recorded in the subsantia nigra and thalamus of rats can be characterized by deterministic dynamics. The finding of deterministic chaos confirm an existence in the brain of some delicate mechanisms able to support repetitive firing of neurons for a long time.

An attractor behavior of spiking neural networks has been theoretically predicted and analyzed in detail by several researches [2, 3]. Such behavior of the networks is considered as an important mechanism for representation of learned stimuli and can be simulated in large scale neural networks with simple but reasonable assumptions of interactions between neurons.

Another conception that proposes a possible mechanism able to account for sustained repeated activity of neurons is the synfire chain theory [4]. The simple synfire hypothesis suggests that whenever the same process repeats, the same spatio-temporal firing patterns should appear. Synfire chains may exhibit structures in which a group of neurons excite themselves and maintain elevated firing rates for a long period. The same neuron can participate in many different synfire chains simultaneously. The synfire theory predicts the existence of repeatable spatio-temporal patterns of spikes. This prediction can be tested by the pattern detection algorithm [5, 6] (PDA). An application of this algorithm revealed highly repeatable temporal patterns of spikes in many brain structures [7, 8]. These data provide firm experimental confirmation of the theory.

Let us note, that the aforementioned theories are based on completely opposite assumptions about mechanisms of information coding in the brain. The synfire chain theory suggests an importance of precise timing of spikes (precise temporal coding), while theories of attractor neural networks do not require it (noisy rate coding).

Since both approaches--PDA and DSA--characterize a repetitive dynamic of activity of neurons in the brain, the question can be raised if there are some relations between these approaches? Another question of importance is how the performance of these algorithms can be affected by a presence of noise. We address these queries using simulated spike trains and show that both algorithms are able to detect a deterministic behavior within chaotic time series even in presence of significant noise in input data. We also propose to use PDA to filter noise that can be present in chaotic time series.

2 Discrete Time Series

Two types of data sets were used in the current study.

Abeles's set. The first data sets consisted of simulated spike train *files sim.002-sim.007* produced according to ref. 5. The spike trains represented stationary and nonstationary Poisson processes. These data were originally used to test ability of PDA to correctly estimate expected number of patterns.

Chaotic time series. Five files *a-e* were generated according to the mapping equations of the standard map: and the Hénon map:

$$\begin{cases} y_{n+1} = y_n + a \cdot \sin(x_n) \\ x_{n+1} = x_n + y_{n+1} \end{cases} \quad (1) \qquad \begin{cases} x_{n+1} = -a \cdot x_n^2 + y_n + 1 \\ y_{n+1} = b \cdot x_n \end{cases} \quad (2)$$

Here $\{x_{n+1};y_{n+1}\}$ refer to coordinate of the point on step $n+1$, if its coordinates on step n are $\{x_n;y_n\}$. The sets are vigorously studied in literature [9] and it has been shown that these maps are characterized by chaotic behavior within certain range of input parameters. In the current study the values of parameters were arbitrary fixed as shown in Table 1. The corresponding simulated spike train series were derived by taking the absolute values of $|x_{n+1} - x_n|$. In order to have data comparable with those recorded in usual neurophysiological experiments, the time series were scaled to have on average a base frequency of spikes about 3 Hz (1 Hz = 1 spike/sec). One thousand spikes in each series were generated approximately corresponding to 330 sec of recording time.

Table 1. Initial and Dynamical System Parameters of the Chaotic Time Series

file	no of points	initial parameters x	y	a	b	theoretical ν	d	estimated ν	d
			2-dimensional standard map (eq. 1)						
a	1,000	0	0.62355	0.1	-	1	2	1.00	2
b	1,000	0	0.576	0.01	-	1	2	0.99	2
c	1,000	0	0.62355	0.8	-	2	2	1.81	2
*c**	3,000	0	0.62355	0.8	-	2	2	1.98	2
			2-dimensional Hénon map (eq. 2)						
d	1,000	0.5	0.2	1.6	0.1	1.1	2	1.06	2
e	1,000	0.67	0.18	1.6	0.1	1.1	2	1.09	2

3 Description of the Algorithms

3.1 Pattern Detection Algorithm

A critical point of the application of the PDA is the careful estimation of the number of patterns that are expected to be found by chance. Our previous study [6] introduced two computationally efficient approaches, the combinatorial and probabilistic-combinatorial methods, able to provide fast and accurate estimations of the expected number of patterns in multiple spike trains. Both approaches are significantly faster and they require significantly less memory than the ad-hoc algorithm originally developed in ref. 5. The differences between ad-hoc and other algorithms become important only for analysis of multiple spike trains, while the same results are calculated by them for single spike train--the analysis applied in the current study. We used the combinatorial algorithm, because the operational speed of this algorithm is faster than that of the other methods. Let us remember, that two parameters--window duration, *W*, and jitter--are required for work of the PDA. The first parameter determines a maximal duration of spike patterns to be considered in the analysis, while the second parameter determines the timing resolution. These parameters were selected for each particular analysis as indicated further in the article.

3.2 Dynamical System Approach

The algorithm of Grassbergr and Procaccia (GP) [10] was used to calculate two quantities of the chaotic spike data train: the dimension d of the embedding space and the correlation dimension v. Both quantities were estimated using the correlation integral [1, 10]. The minimal allowed number points for analysis was N_{min}=700.

4 Calculated Results and Discussion

The DSA is usually applied for analysis of single time series (i.e., single spike train), while PDA can be applied for analysis of multiple spike trains. In the current study in order to have a quantitative comparison of performance of the algorithms, the both approaches were used to analyze single spike trains only, i.e. each neuron was analyzed independently from other neurons.

Abeles's set. An application of PDA for Abeles's data sets did not detect any significant excess of found patterns over the number of patterns estimated by chance. This result is in complete agreement with our previous calculation, where all spikes were analyzed simultaneously.
An application of GP algorithm to this data, however, detected a low dimensional attractor with embedding space d=6 and correlation dimension v=3.1 for spike time series no. 2 from simulated data set *sim.005a*. This data file was generated according to nonstationary Poisson processes with the time fluctuation constant of 50 ms. Careful inspection of the data shows, that this simulated neuron had a number of spikes (743) near the limit N_{min}=700, used by us a lowest value considered for the algorithm. After increasing of the limit to N_{min}=800 (limit used by us in ref. 1), the problem was corrected and no deterministic behavior was found for analyzed data sets. This result indicates that usage of GP with low number of points in the time series can produce erroneous estimations.

Chaotic Time Series. An application of GP algorithm for analysis of the chaotic time series provided in 4 out of 5 cases (except data set *c*) an estimation of correlation dimensions and embedding spaces that are in good agreement with theoretical data (Table 1). The estimation of the correlation dimension for the data set *c* was lower than the theoretical value. However, the correlation dimension was correctly estimated for this series when the number of points was increased up to 3,000 (see file *c**).

The PDA algorithm applied for the same time series analysis determined a significant number of patterns within each series. The number of detected patterns increased for large window and jitter. Patterns of complexity 16 were found for W=3000 ms (data not shown).

Table 2. Influence of addition or/and deletion of some percentage of spikes in chaotic time series on analyzed algorithms

	added, %			deleted , %			added (%) and deleted (%)		
	10	20	30	10	20	30	10	20	30
				Dynamical System Analysis					
a	0.99 (2)	1.05 (2)	*1.14 (2)*	*0.82 (2)*	*0.85 (2)*	*0.85 (1)*	0.96 (2)	1.04 (2)	-
b	1.08 (2)	*1.11 (2)*	*1.07 (3)*	*0.86 (2)*	*0.85 (3)*	-	0.93 (2)	*1.17 (2)*	*1.26 (2)*
c	1.80 (2)	*2.01 (3)*	*2.14 (3)*	1.87 (2)	*1.84 (3)*	*1.74 (3)*	1.95 (2)	*2.57 (5)*	-
d	1.15 (2)	1.24 (2)	*1.35 (2)*	1.09 (2)	*1.16 (3)*	-	1.19 (2)	*1.99 (5)*	-
e	1.20 (2)	*1.35 (2)*	*1.31 (3)*	1.12 (2)	0.97 (2)	*1.18 (2)*	1.13 (2)	-	-
				Pattern Detection Algorithm					
a	19	19	19	12	12	5	13	10	7
b	93	85	78	72	49	42	60	42	35
c	2	1	1	1	0	1	1	0	0
d	10	10	10	8	6	4	8	7	5
e	10	10	9	6	4	3	6	5	4

The detected correlation dimension and dimension of the embedding space (in parentheses) were calculated by GP approach [1, 10]. The cases with incorrectly estimated embedding space or cases with > 10% error in estimation of the correlation dimension are shown in italic. The performance of PDA was estimated as a number of records with significant excess (at 99% confidence limit) of found patterns over estimation. The window duration for PDA was W=500 ms and no jitter was used.

Effect of Noise.

Real experimental data are characterized by different sources of noise. Usually, in our recording session we use an analog template matching sorter [11]. This technique isolate up to four distinct single units from the same electrode. However, it is able to lose some spikes generated by the same unit, if the forms of spikes is changed. This effect can take place especially during burst activity of spike units. Our observation demonstrated (unpublished data) that in some experiments up to 20-30% of spikes can be lost due to this effect. The opposite effect can consist in detection of the spikes from different cells as spikes from the same cell, or both effects can take place simultaneously.

Three series of files were generated to study the effect of noise on simulated data. The noise was simulated by inclusion or deletion of some spikes in the time series. The spikes to be deleted were selected by chance amid all available spikes in the original time series. The added spikes were generated according to uniform distribution on the actual interval and were inserted in the time series.

The calculated results in Table 2 show a wide tolerance of the algorithms for the presence of noise in the input data set. Both algorithm were able to detect regularities in input data files even after as much as 30% of spikes were added or/and deleted in the simulated files. The PDA was less sensitive to the amount of added or deleted spikes and was able to determine significant patterns of spikes even in some cases, where GP algorithm did not detect any deterministic behavior.

Noise Filtering Algorithm.

The performance of PDA was more strongly affected by deletion of spikes compared to cases where additional spikes were included into analysis (Table 2). We observed, that the spikes representing the added noise usually did not participate in the formation of significant patterns. This property was used by us to filter out added noise in the spike trains. The idea of such filtering is based on estimation of a significance of each spike in the time series according to the numbers when the spike belong to the highly significant patterns. This estimated significance *e(i)* of spike *i* was calculated as

$$e(i) = \min_j \left({}^{n_j}\!\!\Big/\!{}_{N_j} \right)^k$$

where n_j is expected and N_j is a found number of spike patterns of given complexity and repetition, while *k* counts how many time the spike *i* participated to such patterns. We expected that the noise spikes will have on average lower values *e(i)* compared to the spikes from the chaotic time series. A bootstrap method was used to estimate a value of *e(i)* to be used as a limit for noise filtering as follows. An additional 100 noise spikes were added to analyzed time series and corresponding *e(i)* were calculated for all spikes following application of PDA. The *e(i)* values found for noise spikes were sorted in increasing order from minimal to maximal value and formed an array *e_lim(i)*, *i*=1,..,100. Let us consider, for example, *i*=10. If we remove all spikes with *e(j)>e_lim(10)*, we can expect that 90% of original noise spikes will be filtered out by such procedure.

Table 3 Number of Original and Noise Spikes (in parentheses) Restored by PDA as a Function of Window Duration

file*W* (ms)	50	100	200	500	1,000	3,000
a	0 (0)	0 (0)	31 (1)	991 (42)	998 (61)	-
b	0 (0)	7 (0)	54 (2)	998 (45)	998 (30)	-
c	0 (0)	0 (0)	16 (1)	56 (12)	193 (29)	349 (61)
*c**	15 (1)	121 (13)	230 (32)	1108 (163)	1623 (224)	1957 (181)
d	6 (0)	6 (0)	26 (1)	145 (50)	665 (52)	755 (56)
e	0 (0)	11 (0)	39 (0)	149 (46)	742 (49)	814 (59)

See the numbers of spikes generated by chaotic time series in Table 1. The added noise spikes constituted 30% of the original spikes. The allowed amount of noise in restored time series was fixed to be less than 20% of the original noise.

The number of spikes from chaotic time series removed by the proposed algorithm for the same percentage of filtered noise is changed when the parameters of PDA are varied. An application of this algorithm should consist in finding parameters of PDA such that the number of removed spikes for the same level of filtered noise is minimal. An analysis of simulated data sets showed that the quality of the filtering procedure usually increases with window duration (Table 3). The application of the algorithm to files *a* and *b* restored 99% of the spikes from original time series while the amount of noise in the time series was decreased more than 5 times. The lowest performance was calculated for data file *c*, where only 36% of spikes (359 out of 1000) from the chaotic time series were restored by the algorithm. The percentage of the restored spikes increased to 65% when the bigger number of spikes was used in the time series (*c**).

5 Conclusion

This results show that for time series generated according to chaotic attractors, there is a strict relation between PDA and DSA, in the sense that the both algorithms are able to determine some regularities in the analyzed data files. On the contrary, no deterministic behavior and no significant patterns were observed when analyzing simulated data generated according Poisson distribution. A more detailed study is required to provide a theoretical basis for such relations.

The time series generated by chaotic attractors are able to produce patterns of spikes, that are detected by PDA as significant even for some cases in which corresponding DSA algorithms are unable to determine a chaotic behavior of analyzed time series. This result suggests why a big number of significant patterns can be found in brain structures, despite the number of files with detected deterministic behavior being relatively low. Let us remember that application of DSA has an additional restriction on minimal number of spikes for analysis, and use of too small number of them can produce erroneous result, as it is demonstrated in study with simulated data. This finding raises the question whether significant patterns of spikes detected in experimental data are the product of attractor behavior of neural networks (the possibility demonstrated in this study) or if they are produced solely by synfire chains.

We also propose a filtering algorithm be used to significantly decrease amount of noise spikes in time series. The filtering can improve the quality of data for subsequent study, for example by DSA. This method can be of considerable interest for specialists working with practical application of time series analysis.

Acknowledgments

We thank Moshe Abeles and Alessandra Celletti for providing us a source code of their programs and simulated data. We thank Brian Hyland for his helpful suggestions. This study was supported by Swiss National Science Foundation grant FNRS 31-37723.93, HFSPO fellowship ST 421/95 and INTAS-Ukraine grant 95-0060.

References

1) Celletti A, Villa AEP (1996) Low dimensional chaotic attractors in the rat brain. Biol Cybern 74: 387-393

2) Amit DJ, Brunel N (1997) Model of global spontaneous activity and local structured activity during delay periods in the cerebral cortex. Cerebral Cortex. 7(3):237-252

3) Herrmann M, Ruppin E, Usher M (1993) A neural model of the dynamic activation of memory. Biol Cybern 68: 455-63

4) Abeles M (1991) Corticotronics: neural circuits of the cerebral cortex. Cambridge University Press, Cambridge

5) Abeles M, Gerstein GL (1988) Detecting spatiotemporal firing patterns among simultaneously recorded single neurons. J Neurophysiol 60: 909-924

6) Tetko IV, Villa AEP (1997) Fast combinatorial methods to estimate the probability of complex temporal patterns of spikes Biol Cybern (in press).

7) Villa AEP, Abeles M (1990) Evidence for spatiotemporal firing patterns within the auditory thalamus of the cat. Brain Res. 509: 325-327

8) Villa AEP, Fuster JM (1992) Temporal correlates of information processing during short-term memory. NeuroReport 3: 113-116

9) Bai-Lin H (1989) Chaos. World Scientific, Singapore

10) Grassberger P, Procaccia I (1983) Estimation of the Kolmogorov entropy from a chaotic signal. Phys Rev A 28: 2591-2593

11) Villa AEP (1990) Functional differentiation within the auditory part of the reticular nucleus of the cat, Brain Res. Rev. 15: 25-40

Spatio-Temporal Pattern Recognition with Neural Networks: Application to Speech

Jean ROUAT[1,2]

[1] ERMETIS, Sciences Appliquées, Université du Québec à Chicoutimi***
[2] Neuro-Heuristique, I.P., Faculté de Médecine, Université de Lausanne

Abstract. The processing or the recognition of non stationary process with neural networks is a challenging and yet unsolved issue. The paper discuss the general pattern recognition framework using neural networks in relation with the understanding of the peripheral auditory system. We propose a short-time structure representation of speech for speech analysis and recognition. We give examples of neural networks architecture and applications that are designed to take into account the time structure of the process to be analysed.

1 Introduction

Most of the contemporary pattern processing or recognition techniques assume that the pattern or the time series to be recognised are stationary. Furthermore, in real life applications, the information is most of the time corrupted, partial or noisy (image, speech, etc.). Therefore, the pattern recognisers have also to be robust.

2 Speech and the Auditory System

Speech is a good example of a complex signal that is very difficult to analyse or recognise by computers when produced in real life situations. The best contemporary technology is not able to propose speech recognition systems with acceptable performance when the speech is corrupted or noisy.

2.1 Structures of Speech

One of the reason is that speech is a fuzzy structured signal. It is structured in reference to the production and hearing systems that impose the constraints and structures on the speech (formants, voiced/unvoiced, spectral and temporal distributions, etc.). For speech scientists those constraints and structures seem to be somewhat fuzzy when analysing the signal without any a priori knowledge about the production and hearing systems.

*** Jean_Rouat@uqac.uquebec.ca

2.2 Where is the Recognition Performed?

Another reason is that the perceptive system does not process speech as pattern recognition systems usually do. To a certain extent, it is true that the cochlear nucleus, the superior olivary complex and the colliculus, for example, are apparently specialised and they might perform 'signal processing' tasks. But there is evidence that a partial 'recognition' is already made at the level of the intermediate auditory system. In fact, the time constants and the best modulation transfer functions of the cortical auditory neurons reveal that the time locking to periodicities with frequencies greater than 100Hz is very rarely observed in the auditory cortex. This is lower to pitch and formant frequencies. Therefore, pitch and periodic stimulus are probably processed at the peripheral or intermediate level of the auditory system. Periodic stimulus with different frequencies can elicit similar evoked potential responses in the auditory cortex. On the opposite, the response of the cortical neurones is very strong when the stimulus are transients. The afferent pathways to the primary auditory cortex have preserved the neural timing of transient stimulus events. In the auditory cortex, neural activity patterns in response to tones show a low-frequency periodic response with strong ON and OFF responses. Those responses are the result of a collaborative work between hundred of cells. The global activity has a complex dynamic that encode the information.

The auditory system seems to process speech by taking into account the specific speech structure and the recognition task is probably performed at all stages of the auditory processing (continuum) in contrast with standard pattern recognisers where the parameter extraction is first made and then the recognition is performed.

2.3 Short-time Structures of Speech

We define the short-time structures of speech as the characteristics of speech observed through a peripheral auditory system on very short-time scales (a few ms). We are interested in the short-time structure observed at the output of a cochlear filter bank in terms of amplitude modulation (AM). This structure is typical of speech. Generally speaking, the structures depend very much on the way speech is produced. The production and hearing systems are closely related and adapted to each others. The hearing system provides time structured representations of speech that are representative and characteristic of the speech production modes. These structures can not be observed via conventional speech analysis techniques and are important to speech analysis and recognition.

3 Modulation in the Auditory System: an Example of Structure

The modulation information is one of the main cues extracted by the auditory system. Various work is made regarding the physiology of AM processing

[2] [4] [7] [8]. For example, Schreiner and Langner [4] [8] show that the inferior colliculus of cats contains a highly systematic topographic representation of AM parameters and maps showing 'best modulation frequency' have been determined. Shannon [9] reports experiments on listeners that can understand artificial speech. This artificial speech is obtained by preserving the envelope and deleting the fine structure of the original speech in three adjacent spectral bands.

For the nerve fibres whose characteristic frequency (CF) is close to a formant frequency, a phase-coupling to the formant frequency or to an adjacent harmonic is observed with little or no envelope modulation as the discharge pattern of the fibre is dominated by a single large harmonic component. Other fibres may show modulations corresponding to harmonic interactions.

Today, most of the speech recognisers extract acoustical parameters from the signal based on spectral representations of speech. By doing so, one has to assume that the signal is stationary under the analysis window. This yields a representation of speech that is an estimate of the time-averaged parameter values. Therefore, the short-time structure of speech is partially hidden by the analysis and the AM fine structure can not be seen.

3.1 The Short-term AM Structure

An example of a short-term AM structure is given below. A bank of 24 filters centred on 330Hz to 4700Hz is used. The output of each filter is a bandpass signal with a narrow-band spectrum centred around f_i where f_i is the central frequency (CF) of channel i. The output signal $s_i(t)$ from channel i can be considered to be modulated in amplitude and phase with a carrier frequency of f_i.

$$s_i(t) = A_i(t) cos[\omega_i t + \phi_i(t)] \tag{1}$$

$A_i(t)$ is the modulating amplitude (envelope), $\phi_i(t)$ is the modulating phase and $\omega_i = 2\pi f_i$.

An image representation of the structure can be obtained by plotting $A_i(t)$ versus time and central frequency. The x axis is the time and the y axis is expressed in Hertz according to the ERB (equivalent rectangular bandwidth) scale [5]. The image colour is the $A_i(t)$ variable.

Fig.1. shows the envelope structure for a signal segment comprising 23 ms of speech. A female pronounced the letter /i/ along with a male speaker saying /a/. In that example, a segregation of the speakers is possible when looking to the structure along the time dimension (x axis) and across the channels (y axis). The low frequency channels (centre frequency for channels 1-13: 330–1500 Hz) are dominated by the signal from the male speaker (/a/) while the high frequency channels are dominated by the female voice (/i/). The most interesting modulations occur during the glottal explosion or just after. The figure includes approximately 3 glottal explosions from the male voice and 7 glottal explosions for the woman. Modulations due to harmonics interactions occur during a pitch period just after the glottal explosion.

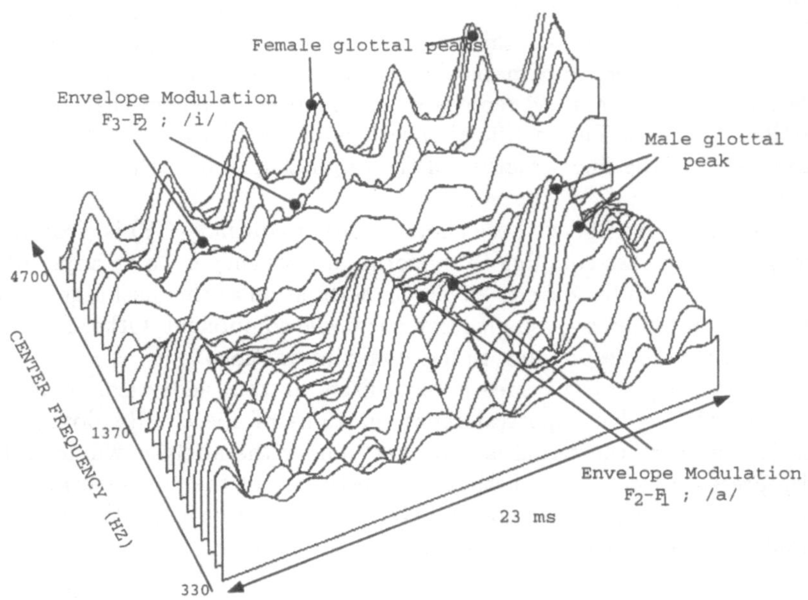

Fig. 1. Envelopes for a two speakers speech segment: /a/ from a male speaker, /i/ from a female speaker.

3.2 The Recognition

Previous sections introduced a short-term AM structure of speech that can not be observed by using conventional analysis techniques. To perform the recognition, neural networks have to take into account the internal time structure of the representation. We propose two examples on speech. The first uses an associative neural network memory for a speech recognition task and the second is based on an oscillatory neural network for a speaker verification task.

The Associative Neural Network: Dystal.[1] Dystal (DYnamically STable Associative Learning) is proposed by Alkon et *al.* and is inspired from a marine snail and from the hippocampus of a rabbit. We assume that clusters are characterised by an explicit encoding of the reference patterns in dendritic patches of neurons [1]. The network associatively learns correlations and anticorrelations between time events occurring in pre synaptic neurons. Those neurons synapse on the same element of a common post synaptic neurone. A learning rule modifies the cellular excitability at dendritic patches. These synaptic patches are postulated to be formed on branches of the dendritic tree of vertebrate neurons. Weights are associated to patches rather than to incoming connection. After learning, each patch characterises a pattern of activity on the input neurons.

In comparison with most commonly used networks, the weights are not used to store the patterns and the comparison between patterns is based on normalised correlations instead of projections between the network input vectors and the neurone weights. Based on the Dystal network, a prototype vowel recognition system has been designed and preliminary results show that the short-time AM structure carries information that can be used for recognition of voiced speech [6].

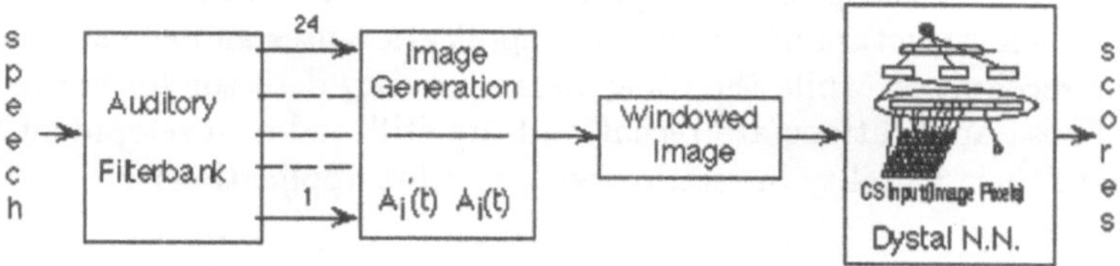

Fig. 2. Architecture of a speech recogniser prototype.

An image representation is obtained by plotting the product $A_i(t) \cdot A_i(t)'$ versus time and central frequency. That representation enhances transitions and amplitude modulations of the envelope. A sliding and synchronised to glottal peak window is moved on the image representation of speech and is used as input to the associative memory. A preliminary experiment has been sucessfully conducted on four vowel clusters: /a/, /i/, /y/ and /ϵ/. Detailed results are presented in [6].

The Oscillatory Neural Network: a Novelty Detector. The topology of the network is inspired from cortical layer IV. The neurone model is based on the integrate and fire model. Each neurone receives excitatory or inhibitory inputs from their neighbourhood. A global inhibitor neurone is connected to each neurone in the network. It regulates and stabilises the network activity. Synaptic rules are used to adapt the weights of the local and global connections.

There is no difference between 'training' and recognition. 'Training' is always performed except when the network is stable. As soon as signals are presented to the network, it learns by modifying the synaptic weights. During training, the network oscillates. When changes on weights are too small to modify the dynamic of the network, we assume that 'training' and recognition have been completed. The network is then declared to be in a stable state. The time necessary to reach that state is the relaxation time. It is used as the novelty detection criteria.

The clusters are not explicitly encoded. The relaxation time of the network is used to characterise the input pattern. A short relaxation time implies that the pattern has been already 'seen' by the network. This paradigm allows the creation of novelty detection systems with a high degree of robustness against noise. Such network is used for the recognition of noisy numbers [3] and experiments on speech are currently made with larger networks. For example, a speaker verification system is being designed. The network uses the short-term AM structure to verify if the speaker belongs to the 'authorised' cluster.

4 Conclusion

The recognition of non stationary spatio-temporal process is very difficult when using formal neural networks. A better understanding of the processing of information in the brain should ease the introduction of new paradigms in pattern recognition with Neural Networks.

We gave two examples with applications to speech that are based on the short-term AM structure of speech. One application uses an associative memory for vowel recognition while the other uses a novelty detector for speaker verification. Those applications are recent and are still under development in order to evaluate their viability in relation with real life applications.

Acknowledgements

This work has been supported by the NSERC of Canada and by the fondation from Université du Québec à Chicoutimi. Many thanks to Yves de Ribaupierre and Alessandro Villa for stimulating discussions.

References

1. Alkon, D.L., Blackwell, K.T., Barbourg, G.S., Rigler, A.K. and Vogl, T.P.: Pattern-recognition by an artificial network derived from biologic neuronal systems. *Biological Cybernetics, vol.* **62** (1990) 363–376
2. Frisina, R. D. et al.: Differential encoding of rapid changes in sound amplitude by second-order auditory neurons. *Exp. Brain Res., vol.* **60** (1985) 417–422
3. Ho, T. V. and Rouat, J.: A Novelty Detector using a Network of Integrate and Fire Neurons.*ICANN97.*
4. Langner, G. and Schreiner, C.E.: Periodicity coding in the inferior colliculus of the cat. Neuronal mechanisms. *Journal of Neurophysiology, vol.***60**, 6, (1988) 1799–1822
5. Patterson, R.D.: Auditory filter shapes derived with noise stimuli. *Journal of the Acoustical Society of America, vol.* **59**, 3 (1976) 640–654
6. Rouat, J. and Garcia, M.: A prototype speech recogniser based on associative learning and nonlinear speech analysis. In Proc. of the Workshop on *Computational Auditory Scene Analysis*, International Joint Conference (IEEE-ACM) on Artificial Intelligence, (1995) 7–12. To be published in, *Readings In Computational Auditory Scene Analysis*, Edited by H. Okuno and D. Rosenthal, Erlbaum.
7. Schreiner, C. E. and Urbas, J. V.: Representation of amplitude modulation in the auditory cortex of the cat. I. The anterior auditory field (AAF). *Hearing Research, vol.* **21** (1986) 227–241
8. Schreiner, C.E. and Langner, G.: Periodicity coding in the inferior colliculus of the cat. Topographical organization. *Journal of Neurophysiology, vol.* **60**, 6 (1988) 1823–1840
9. Shannon, R. V et al.: Speech Recognition with Primarily Temporal Cues. *Science, vol.* **270** (1995) 303–304

Noise in Integrate-and-Fire Models of Neuronal Dynamics

Petr Lánský[1] and Vera Lánská[2]

[1] Institute of Physiology, Academy of Sciences of the Czech Republic, Videnska 1083, 142 20 Prague 4, CZECH REPUBLIC

[2] Institute for Clinical and Experimental Medicine, Videnska 800, 140 00 Prague 4, CZECH REPUBLIC

Abstract. The sequence of action potentials produced by a neuron is best characterized in terms of a stochastic point process. In this contribution we will be primarily concerned with different variants of stochastic leaky-integrator models for the membrane potential. The point process representation is then achieved by the first passage time transformation of the underlying membrane potential model. Different sources of the noise in the diffusion neuronal models resulting from the stochastic leaky-integrator model will be discussed.

1 Introduction

The experimental data recorded from very different neuronal structures and under different experimental conditions suggest the presence of stochastic variables in neuronal activity. We may assume that there is a random component, generally considered as a noise, contained in the input and/or output signal. The term noise usually denotes something negative and blurring the signal processing, however, in some cases it could be a message by itself or a highly desirable part of the message important for its processing. From a mathematical point of view the introduction of stochasticity into the description of neuron represents an increase of complexity. On the other hand, from a point of view of biophysical reality it simplifies the task substantially as all the features considered at the current stage as marginal can be attributed as a system noise.

From a physical point of view, the models of single neuron reflect the electrical properties of its membrane. The corresponding circuit models can be written in terms of differential equations for the membrane voltage. After reducing them, we can obtain models of the integrate-and-fire type [1]. These models, despite they are often considered as oversimplified, play an indisputable role in computational neuroscience because they serve as reference points and test cases to models which exist only in computer implementation. The simplification implies that spike duration and shape are neglected and all neuronal activity is represented by uniform events appearing in time, which defines a point process [2]. Most of the models aiming at describing the dynamics of interspike intervals are based on one-dimensional representation of the time evolution of the neuronal membrane potential. The spike trigger zone serves as the reference point and all the other properties of the neuron have to be integrated into it. This one-point representation, of course, induces several strong limitations [3], which cannot be removed unless the spatial properties of neuron are taken into account.

The simplest realistic model which has been used for describing the intensity of stimulus is the deterministic leaky-integrator model,

$$\frac{dx(t)}{dt} = -\frac{x(t)}{\tau} + \mu(t), x(0) = x_0, \tag{1}$$

where $x(t)$ represents the cell membrane voltage, $\mu(t)$ being an input signal, and $\tau > 0$ is a time constant governing the decay of the voltage back to a resting level, which for notational simplicity is set to zero [4]. Model (1) has been often used for modeling sensory neurons under external periodic stimulation by applying a periodic signal $\mu(t)$. The term $\mu(t)$ appearing in (1) is a representation of external signal (light, sound, odorant, or a sequence of incoming action potentials) transformed into an internal generator potential, a quantity having dimension of voltage by itself. Due to the simplicity of (1), the action potential generation is not an inherent part of the model like in more complex models and the firing threshold S, such that $S > x_0$, has to be imposed here. The model neuron fires whenever the threshold is reached; then the voltage $x(t)$ is reset to its initial value. For a constant $\mu = \mu(t)$, such that $\mu > S/\tau$, the relation between the intensity of stimulation and frequency is

$$f_S(\mu) = \frac{1}{t_{ref} - \tau \ln\left(\frac{\mu\tau - S}{\mu\tau - x_0}\right)} \tag{2}$$

assuming the saturation frequency $f_{sat} = t_{ref}^{-1}$ equal to the reciprocal value of an absolute refractory period for $\mu \to \infty$. An important feature of the relationship between firing frequency and input intensity (2) is a discontinuity of the derivative of $f_S(\mu)$ at point S/τ. For recent results on leaky integrate-and-fire model from a computational point of view, see [5, 6].

An input to a neuron is either a sequence of pulses coming from other neurons or a stimulation arriving from the external environment. From the modeling point of view the source of the input is not a relevant information. The simplest stochastic neuronal model may assume that any incoming pulse, or any incoming quanta of external stimulation is reflected by the generation of an output spike. If the incoming stimulation has *many independent* sources, then the output of such a pooling device is described by a Poisson process. However, there are other mechanisms leading to the Poisson model of a neuron. In any case, the Poisson model represents pure randomness without any memory as the probability of firing is constant and independent of the past at any time instant. The consequences of the assumptions are the linearity of the input-output curve and too high firing frequency for low levels of excitation, which can be eliminated only phenomenologically by imposing the absolute refractory period. This handicaps the Poisson process as an acceptable neuronal model. On the contrary, we may easily consider the input to a neuron as Poissonian. It appears to be an appropriate imitation mainly for spontaneous activity or for evoked activity due to a constant stimulus of long duration. Even for dynamically stimulated system this assumption is well established, only the constant intensity has to be replaced by a function of time properly mimicking the time evolution of the stimulation.

2 Stochastic Neuronal Models

A phenomenological way how to introduce stochasticity into the deterministic leaky-integrator model is simply by assuming an additional noise term in (1),

$$\frac{dX(t)}{dt} = -\frac{X(t)}{\tau} + \mu(t) + F(t)\,, X(0) = x_0\,, \tag{3}$$

where $F(t)$ represents a Gaussian and δ correlated noise with zero mean and strength 2σ; we will mention later why the white noise $F(t)$ or a stream of Poissonian pulses are the only ones suitable for this purpose. The membrane potential X makes random excursions to the firing threshold S, commonly taken to be a deterministic function of time, but most often a constant. As soon as the threshold is reached, a firing event occurs and the membrane potential is reset deterministically to its starting point $X(0)$. The interspike intervals are identified with the first passage time of X across S,

$$T_S = inf\{t \geq 0, X(t) > S \mid X(0) = x_0 < S\}. \tag{4}$$

The properties of the random variable T_S are studied and finally they can be compared with properties of interspike intervals. In general we investigate the distribution of T_S represented for example by the probability density function. When the distribution is too difficult to obtain, the analysis is usually restricted to its moments, $M_n(S \mid x_0)$, primarily the mean $M_1(S \mid x_0) = E(T_S)$ and the variance $Var(T_S) = M_2(S \mid x_0) - M_1^2(S \mid x_0)$. The reciprocal relationship between the instantaneous frequency on one hand and the interspike interval on the other leads to plotting of reciprocal value of $M_1(S \mid x_0)$ versus the intensity of stimulation as a stochastic counterpart of relation (2).

Irrespectively of the selected model X in (4), if the input is constant, the neuronal output is a renewal process (intervals between threshold crossings are independent and identically distributed). This is caused by unidimensionality of the description in conjecture with input constancy. There can be only two kinds of information lasting after the spike generation. At first, the information is accumulated on the neuron, but under this scenario it is deleted by the reset after spike generation, as only a single-variable function X is available. At second, the information which is contained in the incoming signal, but it is against our assumption of the constant (time unstructured) input, in other words, the input noise can be only the white noise or Poissonian. Obtaining non-renewal output from a single-point model can be achieved only by a more or less apparent introduction of the variable (time structured) input or spatial properties of the neuron [3].

The simplest, biologically acceptable and most common way how to derive model (3) is to start from Stein's model of the membrane potential fluctuation [7]. Stein's model is characterized as a one-dimensional stochastic process, which can be expressed in the form

$$dX(t) = -\frac{X(t)}{\tau} + a dP^{+}(t) + i dP^{-}(t)\,;\; X(0) = x_0 \tag{5}$$

where $\tau > 0$ plays the same role as in (1). $i < 0 < a$ are constants; $P^+(t)$, $P^-(t)$ are two independent homogeneous Poisson processes with intensities λ and β, respectively. Following the model (5) the values a and i represent the amplitudes of excitatory and inhibitory postsynaptic potentials, as they contribute to the membrane potential at the trigger zone. Properties of model (5) are as follows: synaptic activation of a neuron leads to a postsynaptic potential which is characterized by a short rise time. Therefore, the corresponding membrane potential change is modeled by a step discontinuity. We use the standard terminology of Stein's model but reinterpretation into the terms of first-order sensory neurons is straightforward [8]. The first and second infinitesimal moments of X defined by (5) are

$$M_1(x) = \lim_{\Delta \to 0} \frac{E[\Delta X(t) | X(t) = x]}{\Delta} = -\frac{x}{\tau} + \lambda a + \beta i \,, \tag{6}$$

$$M_2(x) = \lim_{\Delta \to 0} \frac{E[(\Delta X(t))^2 \,|\, X(t) = x]}{\Delta} = \lambda a^2 + \beta i^2 \,, \tag{7}$$

where $\Delta X(t) = X(t + \Delta) - X(t)$.

In stochastic diffusion models, which are the models of integrate-and-fire type characterized by continuous trajectories, the membrane potential is described by a scalar diffusion process $X(t)$ given by the Itô-type stochastic differential equation

$$dX(t) = \nu(X(t))dt + \sigma(X(t))dW(t)\,;\; X(0) = x_0 \tag{8}$$

where ν and σ are real-valued functions (called drift and infinitesimal variance) of their arguments satisfying certain regularity conditions and $W(t)$ is a standard Wiener process (Brownian motion). The first two infinitesimal moments of the process (8) are $M_1(x) = \nu(x)$ and $M_2(x) = \sigma^2(x)$. Let us remind how (3) can be obtained from (8). A sequence of models X_n given by (5) characterized by a quadruplet $\{\lambda_n, \beta_n, a_n, i_n\}$ is needed such, that with $\lambda_n \to +\infty$, $\beta_n \to +\infty$, $a_n \to 0_+$, $i_n \to 0_-$ the quantities (6) and (7) converge to the drift and infinitesimal variance of the Ornstein-Uhlenbeck process (3), whereas the higher infinitesimal moments tend to zero. For recent results on diffusion approximation see [9].

The simplest diffusion neuronal model is the Wiener process defined by (8) with the constant infinitesimal moments

$$\nu(x) = \mu > 0\,,\; \sigma(x) = \sigma > 0 \tag{9}$$

where positivity of the drift is a substantial condition not needed in (3). Creating input-output curve for model (9), we see that it is a linear function of μ, independently of the noise and other characteristics of the model, so analogous to Poisson description of the neuron. The most common diffusion model proposed for nerve membrane behavior is the Ornstein-Uhlenbeck process (3). This model can be defined by (8) with infinitesimal moments

$$\nu(x) = -\frac{x}{\tau} + \mu\,,\; \sigma(x) = \sigma > 0 \tag{10}$$

The input-output curve of model (10) is analogous to (2) and removes undesirable features of it; namely the second order discontinuity at S/τ and a narrow coding range. An exponential trend of the first-passage-time density for the Ornstein-Uhlenbeck process appears when the relative distance between the asymptotic depolarization $\mu\tau$ and the threshold increases with respect to the infinitesimal variance σ. Mathematically the same effect is induced by decreasing the variability σ while keeping the difference $S - \mu\tau$ constant, which is biologically more relevant. Experimentally it has been observed many times and in different types of neurons, that the interspike interval distribution becomes closer to the exponential distribution as the firing rate decreases.

It is a well know fact that the change of the membrane depolarization by a synaptic input depends on its actual value. The depolarization of the membrane caused by an excitatory postsynaptic potential decreases with decreasing distance of the membrane potential from the excitatory reversal potential, V_E. In the same manner, the hyperpolarization caused by inhibitory postsynaptic potential is smaller if the membrane potential is closer to the inhibitory reversal potential, V_I. Due to the transformation of the resting level to zero, we have $V_I < 0 < S < V_E$, As for the basic model, also for its modification with the reversal potentials, the analysis is complicated and thus the diffusion variants have been examined [10, 11]. While Stein's model has been always substituted with the Ornstein-Uhlenbeck process, there is a whole class of diffusion processes which can be substituted for its variants with reversal potentials. Two of these substitutes have been studied in detail. The first one considers both reversal potentials and is given by (8) with the infinitesimal moments

$$\nu(x) = -\frac{x}{\tau} + \mu_1(V_E - X) + \mu_2(X - V_I), \ \sigma(x) = \sigma\sqrt{(V_E - X)(X - V_I)} \quad (11)$$

while the second one stresses the importance of the inhibitory reversal potential

$$\nu(x) = -\frac{x}{\tau} + \mu_1(V_E - X) + \mu_2(X - V_I), \ \sigma(x) = \sigma\sqrt{X - V_I} \quad (12)$$

where $\mu_1 > 0$, $\mu_2 < 0$ are constants. The results for model (11) were established in [12] and comparison between the Ornstein-Uhlenbeck model and (12) was studied in [13]. The effect of the inclusion of reversal potentials into the diffusion models is apparent when comparing (11) or (12) with (10). From a qualitative point of view it means that the infinitesimal variance becomes non-constant while the drift preserves its linearity. However, the parameters in the drift term are entirely different. There is constant "leakage term" τ^{-1} in (10) while for the models with reversal potentials the leakage is input dependent $(\tau^{-1} + \mu_1 - \mu_2)$. Further, the absolute term of the drift is multiplied by the reversal potentials in models (11) and (12). We should stress that the diffusion approximations of the model which takes into account the existence of the reversal potentials lead always to the models with multiplicative noise. As noted in [14], an additive noise is generated by events outside the transmitted message whereas the multiplicative noise accompanies the passage of the message either from point to point in the network or inside the processing unit, i.e. inside the system.

From a modeling point of view, the variety of forms for the infinitesimal variance and the linear form of the infinitesimal mean are not unexpected. These models are trying to reflect by an "equivalent" noisy ordinary differential equation, the properties at a single location, of a spatially distributed neuron with noisy inputs, i. e., a stochastic partial differential equation. The linear mean term describes the passive electrical circuit properties of the membrane at the trigger zone and the mean effect of the noisy input. The infinitesimal variance, on the other hand, must not only take into account the diversity of spatial configurations for different neurons, but the location and type of synaptic input on that neuron as well. Hence, a variety of forms for this term in the diffusion equation are appropriate.

The integrate-and-fire models operate in two relatively distinct regimes. In the first one the signal (μ term) is large enough so the firing events occur even in the absence of noise. The noise activated regime corresponds to the situation when the drift term alone is not sufficient to cause a firing and it is the noise which activates the firing. The "positive" role of noise in information transfer and processing within the nervous system, and especially in sensory neurons, has been noted for decades (see [15] for an extensive review).

ACKNOWLEDGMENTS: This work was supported by Grant 309/95/0627 from grant agency of the Czech Republic and by Academy of Sciences Grant No. A7011712/1997

References

1. L.F. Abbot and T.B. Kepler. Model neurons: From Hodgkin-Huxley to Hopfield. In: L. Danto (ed.) *Statistical Mechanics of Neural Networks*. Springer, Berlin, 1990.
2. D.H. Johnson. Point process models of single-neuron discharges. *J. Comput. Neurosci.* 3:275-300, 1996.
3. P. Lánský and J.-P. Rospars. Ornstein-Uhlenbeck neuronal model revisited. *Biol. Cybernet.* 72:397-406, 1995.
4. H.C. Tuckwell. *Introduction to Theoretical Neurobiology*. Cambridge Univ. Press, Cambridge, 1988.
5. G. Bugmann. Summation and multiplication: two distinct operation domains of leaky integrate-and-fire neurons. *Network* 2:489-509, 1991.
6. D. Tal and E.L. Schwartz. Computing with the leaky integrate-and-fire neuron: Logarithmic computation adn multiplication. *Neural Computation* 9:305-318, 1997.
7. R.B. Stein. A theoretical analysis of neuronal variability. *Biophys. J.* 5:173-195, 1965.
8. P. Lánský and J.-P. Rospars. Coding of odor intensity. *BioSystems* 31:15-38, 1993.
9. P. Lánský. Sources of periodical force in noisy integrate-and-fire models of neuronal dynamics. *Phys. Rev. E* 55:2040-2043, 1997.
10. G. Kallianpur and R.L. Wolpert. Weak convergence of stochastic neuronal models, In: M. Kimura, G. Kallianpur, T. Hida (eds.) *Stochastic Methods in Biology*. Springer, Berlin, 1987.
11. P. Lánský and V. Lánská. Diffusion approximations of the neuronal model with synaptic reversal potentials, *Biol. Cybernet.* 56:19-26, 1987.
12. V. Lánská, P. Lánský and C.E. Smith. Synaptic transmission in a diffusion model for neural activity. *J. theor. Biol.* 166:393-406, 1994.
13. P. Lánský, L. Sacerdote and F. Tomassetti. On the comparison of Feller and Ornstein-Uhlenbeck models for neural activity. *Biol. Cybernet.* 75:457-465, 1995.
14. W.J. McGill and M.C. Teich. Alerting signals and detection in a sensory network. *J. Math. Psychol.* 39:146-163, 1995.
15. J.P. Segundo, J.-F. Vibert, K. Pakdaman, M. Stiber and O. Diez Martinez. Noise and the neurosciences: A long history, a recent revival and some theory. In: K.H. Pribram (ed.) *Origins: Brain & Self Organization*. Lawrence Erlbaum, Hillsdale, 1994.

Coarse Coding Accounts for Improvement of Spatial Discrimination after Plastic Reorganization in Rats and Humans

Christian W. Eurich[1]*, Hubert R. Dinse[2], Ulrike Dicke[1], Ben Godde[2] and Helmut Schwegler[1]

[1] Institut für Theoretische Physik, Universität Bremen
D-28334 Bremen, Germany
[2] Institut für Neuroinformatik, Ruhr-Universität Bochum
D-44780 Bochum, Germany

Abstract. We reported that a protocol of associative (Hebbian) pairing of tactile stimulation (APTS) evokes cortical plastic changes. In rats, we found a selective enlargement of the areas of cortical neurons representing the stimulated skin fields and of the corresponding receptive fields (RFs). Using an analogous APTS protocol in humans revealed an increase of spatial discrimination performance indicating that fast plastic processes based on coactivation patterns act on a cortical and a perceptual level [4]. Here we use the coarse coding scheme as a model for parallel information processing in neural populations to calculate the resolution changes of sensory neurons before and after plastic reorganization. We demonstrate that an increase of RF size that is paralleled by an increase of RF overlap and an enlargement of cortical representational area result in a substantial improvement of sensory resolution which is in the same range as the psychophysically obtained improvement of discrimination in humans.

1 Introduction

As a rule, perceptual thresholds are lower than corresponding tuning properties of single neurons. This has been exemplified for the 3-dimensional visual object localization in salamanders [10] and the auditory direction localization in barn owls [7]. These observations can only be explained by neural network properties.

Several mechanisms have been proposed for the neural coding of the position of a stimulus in a sensory space, X [6,12,11]. In the *coarse coding scheme* [6], neurons have receptive fields which are large compared to the sensory resolution, resulting in a receptive field overlap everywhere on X. The position $\boldsymbol{x_0} \in X$ of the stimulus is encoded by the ensemble of neurons contributing to the overlap at $\boldsymbol{x_0}$. The neuron in the coarse coding scheme is binary in nature: it has the output 1 if a stimulus is within its receptive field, and otherwise the output 0. Eurich [2] and Eurich and Schwegler [3] calculated the resolution achieved by a population

* e-mail: eurich@physik.uni-bremen.de

of receptive fields with an arbitrary spatial distribution and an arbitrary size distribution by mapping the density of receptive field centers onto the density of receptive field boundaries. The calculations revealed that large receptive fields yield a higher resolution than small receptive fields if the dimension of the sensory space exceeds one (cf. Snippe and Koenderink [12], where a similar result is obtained only for $\dim X > 2$ in the case of firing rate coding. In such systems, the resolution is essentially determined by the neurons' noise properties).

In the present paper, the coarse coding model is applied to the somatosensory system of adult rats. We estimate the spatial resolution of a population of tactile neurons in primary somatosensory cortex (SI) before and after the induction of plastic changes following a coactivation paradigm as explained later. The calculations utilizes the electrophysiologically measured size and spatial distributions of the neurons' receptive fields on the skin surface. The simulation results are compared to the changes in perceptual thresholds in human subjects after the application of a similar stimulation protocol.

2 Experimental Results

According to the Hebbian postulate [5, p. 62], the temporal coincidence of neural events and thus the characteristics of the input statistics are hypothesized to be crucial parameters for the induction of changes of synaptic excitability. This protocol has been extensively used in cellular studies of synaptic plasticity [8]. A similar degree of plasticity has been postulated to account for the alterability of cortical receptive fields and cortical maps, not only during the critical period of developmental, but also in adult nervous systems, to which we refer as ***postontogentic plasticity***. We therefore studied the capacity of rapid in vivo cortical reorganizations induced by associative (Hebbian) pairing of natural, i.e. tactile stimulation through temporally coherent coactivation patterns. To address the question in how far plastic reorganization induced by pure variation of the input statistics without providing any reinforcement bears relevance for a perceptual level, we initiated a parallel study to test in humans psychophysically the impact of an analogous APTS protocol by measuring spatial discrimination performance [4,1].

To induce plastic changes, the centers of two non-overlapping receptive fields on two selected digits or on one digit and one pad were simultaneously stimulated (APTS) for 6 or 15 hours in 23 rats according to the following protocol: A train of 8 different interstimulus intervals between 100 ms and 3000 ms were used randomly followed by a pause of 15 s. Before and after PPTS, spatial extent and topography of the hindpaw representation in SI were mapped with extracellular recordings. The sensory neurons' receptive field sizes on the skin surface were also determined quantitatively by means of response planes. For details of the experimental procedure, see Recanzone et al. [9] and Godde et al. [4].

As a result of the APTS, the cortical area representing the skin fields on the stimulated digits increased from 0.06–0.1 mm^2 to 0.2–0.61 mm^2. The receptive field size increased from $45(\pm 4.6)\,\mathrm{mm}^2$ to $79.9\,(\pm 8.22)\,\mathrm{mm}^2$ of skin surface, and at the same time, receptive field overlap increased from 19.9 (±3.11) % to 34.0 (±4.18) % [4,1].

In order to explore the perceptual consequences of APTS-induced short-term plastic processes, we tested 35 right-handed human subjects before and after a 2 or 6 hours APTS protocol in a two-alternative forced-choice tactile spatial discrimination task. After 2 or 6 h of APTS, discrimination thresholds decreased from 1.37 (±0.21) mm to 1.12 (±0.22) mm. Thresholds returned to normal 12 hours after terminating APTS (1.37 (±0.25) mm), indicating a reversibility. For details, see Godde et al. [4] and Dinse et al. [1].

Under the assumption that in rats and humans APTS leads in principal to identical cellular and perceptual changes, the experiments indicate that an increase in RF size together with an increase of RF overlap and an increase in the cortical representational sensory map result in a better sensory resolution. This finding cannot be easily explained on the basis of single neuron properties. We therefore utilize the coarse coding scheme to demonstrate that parallel information processing within the entire cortical network can indeed explain the experimental data.

3 The Resolution on the Rat Hindpaw

3.1 The Calculation of Local Resolution

In a sensory space, X, the *resolution* A is defined as the minimal distance between two points which have different representations in the encoding neural population. In the coarse coding scheme, a stimulus at $\boldsymbol{x}_0 \in X$ is represented by the activity of those neurons whose receptive fields overlap at $\boldsymbol{x}_0$. Two points can be resolved if there is a receptive field boundary running between them. This leads to the expression

$$A(\boldsymbol{x}_0) = 1/L(\boldsymbol{x}_0)\,, \tag{1}$$

where $L(\boldsymbol{x}_0)$ the *density of receptive field boundaries* at $\boldsymbol{x}_0$. It follows from (1) that

$$A(\boldsymbol{x}_0) \propto 1/N \tag{2}$$

(N: number of receptive fields), because $L(\boldsymbol{x}_0) \propto N$.

In order to obtain $L(\boldsymbol{x}_0)$ for an arbitrary $\boldsymbol{x}_0$ from a finite population of neurons with a given distribution of receptive field centers, we proceed in two steps:

1. From the distribution of receptive field centers, a continuous density of receptive fields, $\varrho(\boldsymbol{x})$, is calculated. $\varrho(\boldsymbol{x})$ has the following properties:

$$\varrho(\boldsymbol{x}) \geq 0\,, \qquad \int_X \varrho(\boldsymbol{x})\, d\boldsymbol{x} = N\,. \tag{3}$$

   All receptive fields are assumed to have the mean receptive field diameter.
2. The density of receptive field boundaries at $\boldsymbol{x}_0$, $L(\boldsymbol{x}_0)$, is determined by integrating $\varrho(\boldsymbol{x})$ over a curve κ, consisting of the centers of all receptive fields which contribute with a boundary at $\boldsymbol{x}_0$. The concrete form of the integral depends on the topology of X.

3.2 Application to the Rat Hindpaw

Figure 1 shows a schematic view of the rat hindpaw. For the calculations, the pad is idealized to a rectangle with sides $a = 21.0\,\mathrm{mm}$ and $b = 12.5\,\mathrm{mm}$, and the digits are approximated by cylinders (radii $R = 1.25\,\mathrm{mm}$, lengths $l_1 = l_5 = 2.5\,\mathrm{mm}$, $l_2 = l_3 = l_4 = 7.5\,\mathrm{mm}$). In order to reduce errors resulting from boundary effects at the line between the digits and the pad, digits 3 and 4 are shortened by 1.5 mm in the critical region (dashed line). Digits 1 and 5 are too short to ignore boundary effects, so only digits 2, 3 and 4 will be considered in the calculations.

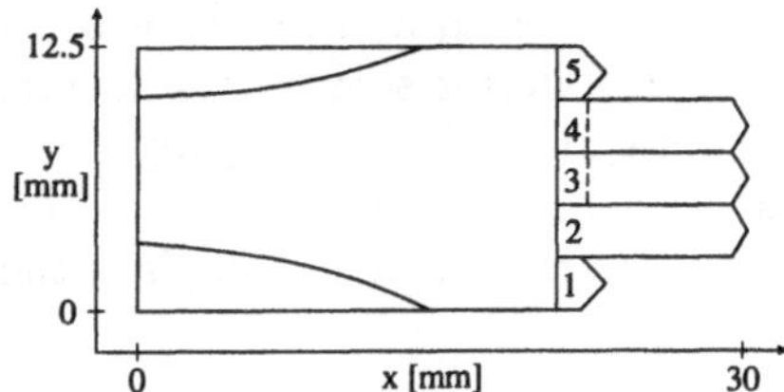

Fig. 1. Schematic view of the rat hindpaw.

The Pad. Consider N circular receptive fields of radius r.
Step1. For the construction of $\varrho(\boldsymbol{x})$, we follow a recipe given in Eurich ([2]). Each receptive field center $\boldsymbol{x}_i$ is replaced by a Gaussian

$$w(\boldsymbol{x}, \boldsymbol{x}_i) = \frac{1}{\sqrt{2\pi\sigma^2}} \exp\{-\frac{(\boldsymbol{x} - \boldsymbol{x}_i)^2}{2\sigma^2}\} \tag{4}$$

with variance $\sigma^2 = (a^2 + b^2)/4$, and $\varrho(\boldsymbol{x})$ is defined as the normalized sum over all Gaussians,

$$\varrho(\boldsymbol{x}) = N \frac{\sum_{i=1}^{N} w(\boldsymbol{x}, \boldsymbol{x}_i)}{\int_X \sum_{i=1}^{N} w(\boldsymbol{x}', \boldsymbol{x}_i) d\boldsymbol{x}'} . \tag{5}$$

$\varrho(\boldsymbol{x})$ has the properties (3).
Step 2. The local density of receptive field boundaries is given by

$$L(\boldsymbol{x}_0) = \int_0^{2\pi} \varrho(\boldsymbol{x}(\beta))\, r \,|\cos\beta| d\beta\,, \tag{6}$$

where β parametrizes the curve κ, a circle of radius r around $\boldsymbol{x}_0$ which consists of all centers of receptive fields whose boundaries contain the point $\boldsymbol{x}_0$. $|\cos\beta|$ is a factor which is derived from the local geometry of the integration path. For further details, see Eurich [2] and Eurich and Schwegler [3].

The Digits. Each digit is represented by a cylinder of radius R and length l; a coordinate system is chosen which lies parallel to the z-axis. The center of the cylinder has the coordinates $(0, R, l/2)$. For simplicity of calculation, the receptive fields are assumed to be squares of length r on the surface, two sides of which are parallel to the cylinder axis.

Step1. The density of receptive fields is again given by (4) and (5), but the Gaussians have the variance $\sigma^2 = (R^2\pi^2 + l^2/4)$.
Step 2. We consider the resolution parallel to the cylinder axis only. In this case, the curve κ consists of two line segments of length r, perpendicular to the cylinder axis. Introducing an angle φ s.t., for points on the cylinder, $x(\varphi) = R\sin(\varphi)$ and $y(\varphi) = R - R\cos(\varphi)$, the expression for the local density of receptive field boundaries is

$$L(\boldsymbol{x_0}) = \int_{-r/(2R)}^{r/(2R)} R\{\varrho(x(\varphi_0+\varphi), y(\varphi_0+\varphi), z_0-\frac{r}{2})+\varrho(x(\varphi_0+\varphi), y(\varphi_0+\varphi), z_0+\frac{r}{2})\}d\varphi, \tag{7}$$

where φ_0 is the angle corresponding to $\boldsymbol{x_0}$.

4 Results and Discussion

Equations (5), (6) and (7) are numerically evaluated for the empirically determined receptive field centers $\boldsymbol{x_i}$ before and after APTS; the receptive fields are assumed to have the mean receptive field sizes given in Sect. 2. The resolution is subsequently calculated with (1).

Figures 2a and 2b show the ratio of the resolutions after APTS and before APTS, $D = A_{\text{post}}(\boldsymbol{x_0})/A_{\text{prae}}(\boldsymbol{x_0})$, for $\boldsymbol{x_0} = (x, y)^T$ on the pad and on digit 2, respectively. In both cases, perceptual thresholds are lower after APTS, corresponding to an increase in resolution due to the enlargement of the receptive fields only. Threshold decrease to 70% on the pad (Fig. 2a), to 50% on digit 2 (Fig. 2b), and to 50% and 40% on digits 3 and 4, respectively (data not shown).

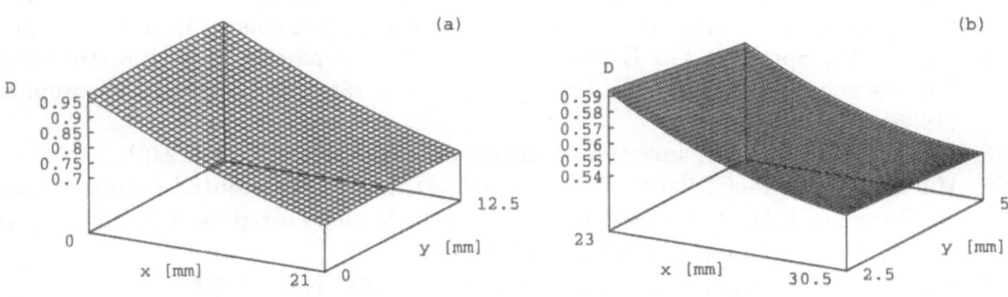

Fig. 2. Ratio of the resolutions after PPTS and before PPTS, D, as a function of somatic position on the pad (a) and on digit 2 (b).

In addition to the enlargement of receptive fields, the cortical area representing the stimulated skin fields increases (Sect. 2). Under the assumption that the number of encoding sensory neurons is proportional to the cortical surface they occupy, (2) implies that this increase of representational area leads to another increase in resolution, approximately by a factor of 4. Thus our model predicts that the combination of both effects – enlargement of receptive fields and increase in cortical area – result in a decrease of sensory thresholds to 15% – 20% on the pad and to 10% – 15% on the digits.

So far, we used the electrophysiological data to estimate a possible gain of spatial resolution due to changes in cortical RFs and maps. These calculations of the resolution provide estimations about parallel perceptual changes. Under the assumption that the protocol used for plastic reorganization in rats and humans leads to equivalent physiological changes, our simulations can be compared with the results of the APTS experiments in humans [4]. These experiments revealed an improvement of spatial discrimination performance which is in the range of improvement that is predicted by the coarse coding scheme based on changes of cortical organization. We therefore conclude that the coarse coding scheme provides an appropriate approach for the explanation of sensory properties based on the underlying anatomy and physiology. The theoretical understanding of the APTS-induced physiological and psychophysical changes may lead to more insight into the mechanisms which are responsible for the flexibility of the nervous system and its ability to adapt to changing environments.

References

1. Dinse, H. R., Godde, B., Zepka, R. F., Shuishi-Haupt, S., Spengler, F., Hilger, T.: Short-term functional reorganization of cortical and thalamic representations and its implication for information processing. Adv. Neurol. **173** (1997) 159–178
2. Eurich, C. W.: Objektlokalisation mit neuronalen Netzen. Harri Deutsch, Frankfurt/Main (1995)
3. Eurich, C. W., Schwegler, H.: Coarse coding: calculation of the resolution achieved by a population of large receptive field neurons. Biol. Cybern., in press (1997)
4. Godde, B., Spengler, G., Dinse, H. R.: Associative pairing of tactile stimulation induces somatosensory cortical reorganization in rats and humans. Neuroreport **8** (1996) 281–285
5. Hebb, D. O.: The Organization of Behavior. Wiley, New York (1949)
6. Hinton, G. E., McClelland, J. L., Rumelhart, D. E.: Distributed representations. In: Rumelhart, D. E., McClelland, J. L. (eds): Parallel Distributed Processing, Vol. 1. MIT Press, Cambridge MA (1986) 77–109
7. Konishi, M.: Listening with two ears. Sci. Am. **268**, No. 4 (1993) 34–41
8. Nicoll, R. A., Kauer, J. A., Malenka, R. C.: The current excitement in long-term potentiation. Neuron **1** (1988) 97–103
9. Recanzone, G. H., Merzenich, M. M., Dinse, H. R.: Expansion of the cortical representation of a specific skin field in primary somatosensory cortex by intracortical microstimulation. Cerebral Cortex **2** (1992) 181–196
10. Roth, G.: Visual Behavior in Salamander. Springer, Berlin (1987)
11. Seung, H. S., Sompolinsky, H.: Simple models for reading neuronal population codes. Proc. Natl. Acad. Sci. USA **90** (1993) 10749–10753
12. Snippe H. P., Koenderink, J. J.: Discrimination thresholds for channel-coded systems. Biol. Cybern. **66** (1992) 543–551

Analogue Resolution in a Model of the Schaffer Collaterals

Simon Schultz[1], Stefano Panzeri[1], Alessandro Treves[2] and Edmund T. Rolls[1]

[1] Department of Experimental Psychology, University of Oxford, South Parks Rd., Oxford OX1 3UD, U.K.
[2] Programme in Neuroscience, SISSA, via Beirut 2-4, 34013 Trieste, Italy

Abstract. We have analytically and numerically solved the mutual information expression for a quantitative model of the Schaffer collateral projections from the CA3 to the CA1 pyramidal cells within the hippocampus. Here we discuss in particular results from the model on the effect of analogue coding levels in the Schaffer collaterals, and the fact that this depends upon the sparseness of firing rate distributions in the hippocampus.

1 Introduction

Recent advances in techniques for the formal analysis of neural networks [1, 2, 3, 4] have introduced the possibility of detailed quantitative analyses of real brain circuitry. This approach is particularly appropriate for regions such as the hippocampus, which show distinct structure and for which the microanatomy is relatively simple and well known [5].

The Schaffer collateral model describes, in a simplified form, the connections from the N CA3 pyramidal cells to the M CA1 pyramidal cells. Most Schaffer collateral axons project into the stratum radiatum of CA1, although CA3 neurons proximal to CA1 tend to project into the stratum oriens [6]; in the model these are assumed to have the same effect on the recipient pyramidal cells. Inhibitory interneurons are considered to act only as regulators of pyramidal cell activity. The perforant path synapses to CA1 cells are here ignored, as are the few CA1 recurrent collaterals. The system is considered for the purpose of analysis to operate in two distinct modes: storage and retrieval. During storage the Schaffer collateral synaptic efficacies are modified using a Hebbian rule reflecting the conjunction of pre- and post-synaptic activity. This modification has a slower time-constant than that governing neuronal activity, and thus does not affect the current CA1 output. During retrieval the Schaffer collaterals relay a pattern of neural firing with synaptic efficacies which reflect all previous storage events.

For reasons of space the full description of the model and calculation cannot be given here; see [7] for more details. Each CA3 pyramidal cell is taken to code for independent information (a condition which will be relaxed elsewhere) with a firing rate distribution $P(\eta)$. Pyramidal cells are modelled using a threshold

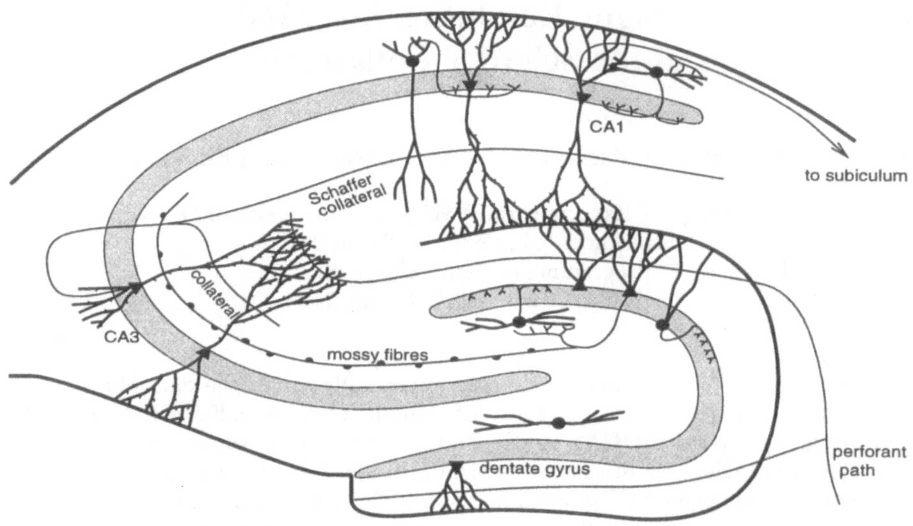

Fig. 1. A schematic diagram of the hippocampal formation. Information enters the hippocampus from layer 2 entorhinal cells by the perforant path, which projects into dentate gyrus, CA3 and CA1 areas. In addition to its perforant path inputs, CA3 receives a lesser number of mossy fibre synapses from the dentate granule cells. The axons of the CA3 pyramidal cells project commissurally, recurrently within CA3, and also forward to area CA1 by the Schaffer collateral pathway. Information leaves the hippocampus via backprojections to the entorhinal cortex from CA1 and the subiculum, and also via the fornix to the mammillary bodies and anterior nucleus of the thalamus.

linear transfer function with a gaussian fast noise distribution. The synaptic matrix is very sparse, as each CA1 cell receives inputs from C (of the order of 10^4) CA3 cells. A plasticity model for the Schaffer collaterals is used which corresponds to gradual decay of memory traces.

The aim of the analysis is to calculate how much, on average, of the information present in the original CA3 pattern $\{\eta_i\}$ is still present in the effective output of the system, the CA1 pattern $\{U_j\}$, i.e. to average the mutual information

$$i(\{\eta_i\},\{U_j\}) = \int \prod_i d\eta_i \int \prod_j dU_j P(\{\eta_i\},\{U_j\}) \ln \frac{P(\{\eta_i\},\{U_j\})}{P(\{\eta_i\})P(\{U_j\})} \quad (1)$$

over the interaction variables of the system. This is achieved by making use of the replica trick and saddle-point approximation in the thermodynamic (infinite system size) limit.

2 Results

Specification of the probability density $P(\eta)$ allows different distributions of firing rates in CA3 to be considered in the analysis. Clearly the distribution of firing rates that should be considered in the analysis is that of the firing of CA3 pyramidal cells, computed over the time-constant of storage (which we can assume to be the time-constant of LTP), during those periods in which biophysical conditions are appropriate for learning to occur. It is reasonable to assume that the distribution of firing rates during storage is graded, sparse, and exponentially tailed [8, 9]. This accords with the observations of neurophysiologists. The easiest way to introduce this to the current investigation is by means of a discrete approximation to the exponential distribution, with extra weight given to low firing rates. This allows quantitative investigation of the effects of analogue resolution on the information transmission capabilities of the Schaffer collateral model.

The required CA3 firing rate distributions were formed by the mixture of the unitary distribution and the discretized exponential, using as mixture parameters the offset between their origins, and relative weightings. The distributions were constrained to have first and second moments $\langle\eta\rangle$, $\left\langle\eta^2\right\rangle$, and thus sparseness $\langle\eta\rangle^2 / \left\langle\eta^2\right\rangle$, equal to a. In the cases considered here a was allowed values of 0.05, 0.10 and 0.20 only. The width of the distribution examined was set to 3.0, and the number of discretized firing levels contained in ths width parameterized as l. The binary distribution was completely specified by this; for distributions with a large number of levels, there was some degree of freedom, but its numerical effect on the resulting distributions was essentially negligible. Those distributions with a small number of levels ≥ 2 were non-unique, and were chosen fairly arbitrarily for the following results, as those that had entropies interpolating between the binary and large l situations. Some examples of the distributions used are shown in Fig. 2a.

The total entropy per cell of the CA3 firing pattern, given a probability distribution characterised by L levels, is

$$i(\eta) = -\sum_{l=1}^{L} P_{\eta_l}(\eta_l) \ln P_{\eta_l}(\eta_l). \tag{2}$$

The results are shown in Fig. 2b–d. The entropy present in the CA3 firing rate distributions is marked by asterisks. The mutual information conveyed by the retrieved pattern of CA1 firing rates, which must be strictly less than the CA3 entropy, is represented by circles. It is apparent that maximum information efficiency occurs in the binary limit. More remarkably, even in absolute terms the information conveyed is maximal for low resolution codes, at least for quite sparse codes. The results are qualitatively consistent over sparsenesses a ranging from 0.05 to 0.2; obviously with higher a (more distributed codes), entropies are greater. For more distributed codes (i.e. with signalling more evenly distributed over neuronal firing rates), it appears that there may be some small absolute increase in information with the use of analogue signalling levels.

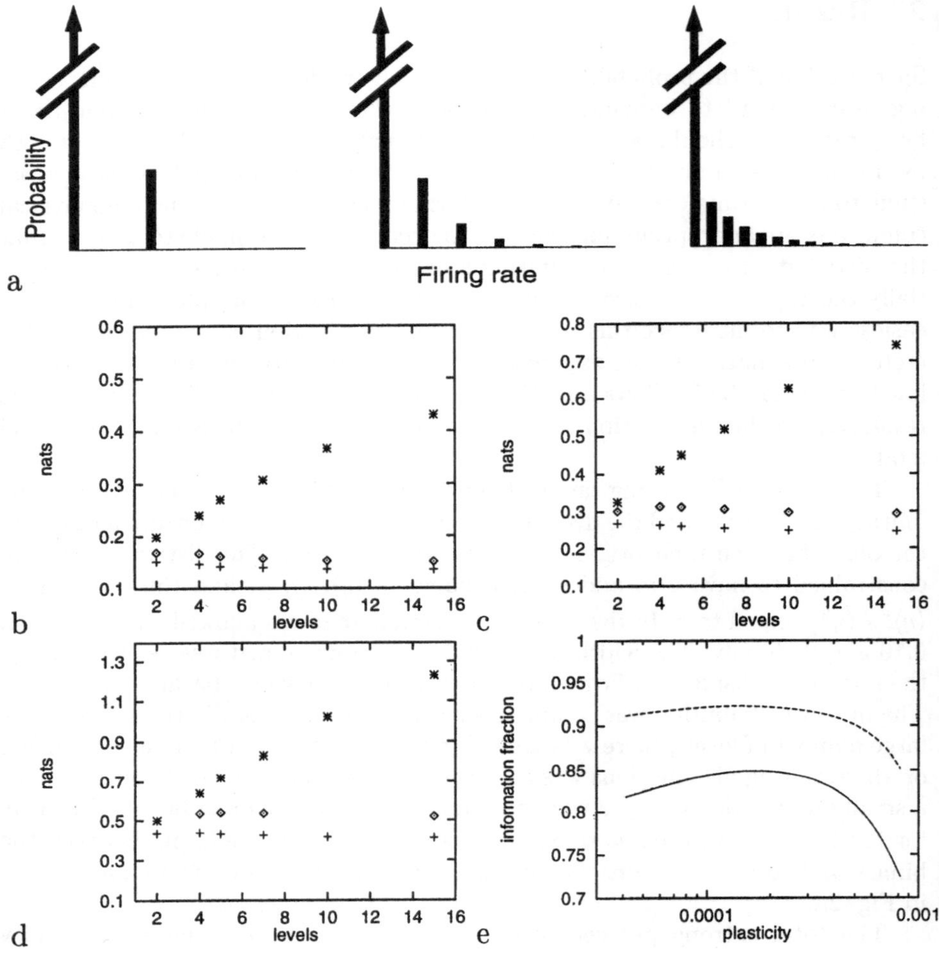

Fig. 2. **a** Some of the CA3 firing rate distributions used in the analysis. These are, in general, formed by the mixture of a unitary distibution and a discretized exponential. **b** – **d** The mutual information between patterns of firing in CA1 and patterns of firing in CA3, expressed in natural units (nats). Asterisks represent the entropy of the CA3 pattern distribution, diamonds the CA1 retrieved mutual information, and crosses the CA1 information during the storage phase. The horizontal axis parameterizes the number of discrete levels in the input distribution: for codes with fine analogue resolution, this is greater. **b** is for $a = 0.05$ (sparse), **c** for $a = 0.10$, and **d** for $a = 0.20$ (slightly more distributed). **e** The dependence of information transmission on the degree of plasticity in the Schaffer collaterals, for $a = 0.05$ (solid) and $a = 0.10$ (dashed). A binary pattern distribution was used in this case.

For comparison, the crosses in the figures show the information stored in CA1. This was computed using a simpler version of the calculation, in which the mutual information $i(\{\eta_i\}, \{\zeta_j\})$ was calculated. Obviously, in this simpler calculation, the CA3 and CA1 retrieval noises are not present; on the other hand, neither is the Schaffer collateral memory term. Since the retrieved CA1 information is in every case higher than that stored, we can conclude that for the parameters considered, the additional Schaffer memory effect outweighs the deleterious effects of the retrieval noise distributions.

It follows from the forgetting model used that information transmission is maximal when the plasticity (mean square contribution of the modification induced by one pattern) is matched in the CA3 recurrent collaterals and the Schaffer collaterals [10]. It can be seen in Fig. 2e that this effect is robust to the use of more distributed patterns.

3 Discussion and summary

This chapter has presented quantitative results, for a model of the Schaffer collaterals, of the effect of analogue resolution on the total amount of information that can be transmitted using relatively sparse codes. What can these results tell us about the actual code used to signal information in the mammalian hippocampus? In themselves, of course, they can make no definite statement. It could be that there is a very clear maximum for information transmission in using binary codes for the Schaffer collaterals, and yet external constraints, such as CA1 efferent processing, might make it more optimal overall to use analogue signalling. So results from a single component study must be viewed with due caution. However, these results can provide a clear picture of the operating regime of the Schaffer collaterals, and that is after all a major aim of any analytical study.

The results from this paper reiterate some previously known points, and bring out others. For instance, it is very clear from Fig. 2 that, while nearly all of the information in the CA3 distribution can be transmitted using a binary code, this information fraction drops off rapidly with analogue level. The total amount of information transmitted is similar regardless of the amount of analogue level to be signalled – but this is a well known and relatively general fact, and accords with common sense intuition. However, the total amount of information that can be transmitted is only *roughly* constant. It appears, from this analysis, that while the total transmitted information drops off slightly with analogue level for very sparse codes, the maximum moves in the direction of more analogue levels for more evenly distributed codes. This provides some impetus for making more precise measurements of sparseness of coding in the hippocampus.

Clearly it is essential to further constrain the model by fitting the parameters as sufficient neurophysiological data becomes available. As more parameters assume biologically measured values, the sensible ranges of values that as-yet unmeasured parameters can take will become clearer. It will then be possible to address further issues such as the quantitative importance of the constraint upon dendritic length (i.e. the number of synapses per neuron) upon information processing.

In summary, we have used techniques for the analysis of neural networks to quantitatively investigate the effect of analogue resolution of signalling on information transmission by the Schaffer collaterals. We envisage that these techniques, developed further and applied in a wider context to networks in the medial temporal lobe, will yield considerable insight into the organisation of the mammalian hippocampal formation.

References

1. D.J. Amit, H. Gutfreund, and H. Sompolinsky. Statistical mechanics of neural networks near saturation. *Ann. Phys. (N.Y.)*, 173:30–67, 1987.
2. E. Gardner. The space of interactions in neural network models. *J. Phys. A: Math. Gen.*, 21:257–270, 1988.
3. A. Treves. Threshold-linear formal neurons in auto-associative nets. *J. Phys. A: Math. Gen.*, 23:2631–2650, 1990.
4. J-P. Nadal and N. Parga. Information processing by a perceptron in an unsupervised learning task. *Network*, 4:295–312, 1993.
5. D. G. Amaral, N. Ishizuka, and B. Claiborne. Neurons, numbers and the hippocampal network. In J. Storm-Mathisen, J. Zimmer, and O. P. Ottersen, editors, *Understanding the brain through the hippocampus*, volume 83 of *Progress in Brain Research*, chapter 17. Elsevier Science, 1990.
6. N. Ishizuka, J. Weber, and D. G. Amaral. Organization of intrahippocampal projections originating from CA3 pyramidal cells in the rat. *J. Comp. Neurol.*, 295:580–623, 1990.
7. S. Schultz, S. Panzeri, E. T. Rolls, and A. Treves. Quantitative analysis of a Schaffer collateral model. In R. Baddeley, P. Foldiäk, and P. Hancock, editors, *Information Theory and the Brain.* Cambridge University Press, Cambridge, U.K., 1997.
8. S. Panzeri, E. T. Rolls, A. Treves, R. G. Robertson, and P. Georges-Francois. Efficient encoding by the firing of hippocampal spatial view cells. Society for Neuroscience Abstracts Volume 23, 1997.
9. C. A. Barnes, B. L. McNaughton, S. J. Mizumori, B. W. Leonard, and L. H. Lin. Comparison of spatial and temporal characteristics of neuronal activity in sequential stages of hippocampal processing. *Prog. Brain Res.*, 83:287–300, 1990.
10. A. Treves. Quantitative estimate of the information relayed by the Schaffer collaterals. *J. Comput. Neurosci.*, 2:259–272, 1995.

Modeling Networks with Linear (VLSI) Integrate-and-Fire Neurons

Maurizio Mattia and Stefano Fusi

INFN, Sezione di Roma 1, Dipartimento di Fisica
Università di Roma, La Sapienza, P.le Aldo Moro 2, Rome Italy

Abstract. We analyse in detail the statistical properties of a "canonical" integrate-and-fire neuron with a linear integrator as often used in VLSI implementations [1]. We show that a network of such elements can maintain both stable spontaneous activity and selective (stimulus specific) activity, contrary to current opinion. The spike statistics appears to be qualitatively the same as in networks of conventional (exponential) integrate-and-fire neurons that in turn, exhibit a wide variety of characteristics observed in cortical recordings[2].

Introduction

The integrate-and-fire (IF) neuron has become popular as a simplified neural element in modeling the dynamics of large scale networks of spiking neurons. A simple version of IF neuron integrates the input current as an RC circuit (with a leakage current proportional to the depolarization) and emits a spike when the depolarization crosses a threshold. We will refer to it as to '*RC neuron*'. Networks of neurons schematized in this way exhibit a wide variety of characteristics observed in single and multiple neuron recordings in cortex in vivo. These networks can maintain stable spontaneous activity and when subjected to Hebbian learning, show a coexistence of selective (stimulus specific) activity distributions with the underlying spontaneous activity [2].

In VLSI the natural IF neuron has been canonized by Mead[1] and, since it integrates linearly the input current, we will refer to it as to '*LIF neuron*' (Linear IF neuron). This VLSI implementation of the neuron has many desirable features: it operates with current generators and hence minimal currents and very low power consumption, an essential feature for integrating a large number of neurons on a single chip. It is also a natural candidate for working with transistors in a sub-threshold regime which brings another significant reduction in consumption[1]. One can also implement a VLSI dynamic synapse with similar attractive electronic characteristics[3].

Here we will concentrate on the statistical properties of the spikes generated by a *LIF neuron*, as a function of the statistics of the input current, and on the effect on network dynamics of composing networks of *LIF neurons*, keeping distributions of synaptic efficacies fixed. We ask the following question: given that the depolarization dynamics of the *LIF neuron* is significantly different from that of the *RC neuron*, can the collective dynamics found in a network of *RC neurons* be reproduced in networks of neurons of the VLSI type? The collective behaviour we examine includes the coexistence of low rate, highly variable, spontaneous activity and selective activity stabilized by learning [2].

RC Neuron vs *LIF Neuron*

The difference between the two types of neurons can be summarized as follows: the *RC neuron* below threshold, is an RC circuit integrating the input current with a decay proportional to the depolarization of the neuron $V(t)$: $\dot{V}(t) = -V(t)/\tau + I(t)$, where $I(t)$ is the net charging current, produced by afferent spikes. A spike is emitted when $V(t_0) = \theta$, followed by hyperpolarization.

The depolarization below threshold of the *LIF neuron* is:

$$\dot{V}(t) = -\beta + I(t), \tag{1}$$

with the constraint that if $V(t)$ is driven below 0, it remains 0. β is a constant decay that, in absence of afferent currents, drives $V(t)$ to 0. The spiking condition remains unmodified. An absolute refractory period τ_{arp} after a spike is emitted, the depolarization is reset to 0.

The question about the collective behaviour of such 'linear' neurons is underlined by the following consideration: the linear integrator dynamics in the 'positive drift' regime (when on average the RHS of Eq.(1) is positive and the variance of the current is much smaller than the mean) leads to current to rate dynamics which is threshold-linear for a wide range of values of the input current [4, 6] and the coexistence of spontaneous activity with structured selective activity is not possible. Each of the two types of behaviour is implementable in a network with 'linear' neurons [6], but not both.

In contrast, if the statistics of the input current is such that the *LIF neuron* can operate also is in its 'negative drift' regime the transduction of the neuron is non-linear and mean-field theory exhibits the coexistence of the two collective states.

Current-to-Rate Transduction Function

We derive the current-to-rate transduction function for a *LIF neuron* with a noisy source $I(t)$. At any time t, the current is drawn randomly from a Gaussian distribution with mean $\mu_I(t)$ and variance $\sigma_I^2(t)$ per unit time, so that Eq. (1) can be rewritten as: $dV = \mu(t)dt + \sigma(t)z(t)\sqrt{dt}$, where $\mu(t) = -\beta + \mu_I(t)$, $\sigma(t) = \sigma_I(t)$ and $z(t)$ is a stochastic variable normally distributed with zero mean and unit variance.

If $p(v,t)$ is the probability density that at time t the neuron has a depolarization v, it obeys the Fokker-Plank equation (see e.g. [5]). This equation must be completed by restricting the process to the interval $[0,\theta]$, which is done by imposing appropriate boundary conditions: at $v = 0$ a reflecting barrier, since no process can pass below 0 and at $v = \theta$ an absorbing barrier that resets processes crossing the threshold to 0. Formally, this is equivalent to the conditions that $p(v,t) = 0$ at $v = \theta$ (see e.g. [5]), and that no process is lost when absorbed at θ or reflected at $v = 0$. This implies that the rate at which processes are crossing the threshold must be the same as the rate at which they re-enter from 0, i.e.:

$$\sigma^2 \partial_v p\big|_{v=\theta} = -2\nu(t) \qquad \left[\sigma^2 \partial_v p - 2\mu p\right]_{v=0} = -2\nu(t) \tag{2}$$

where $\nu(t)$ is the probability per unit time of crossing the threshold θ.

If $\tau_{arp} > 0$, then the realizations that drive the neuron above the threshold θ must be delayed before coming back to the reset value. This is obtained by imposing that the neuron, after crossing the threshold, has to walk on a fictitious interval (which can be arbitrarily set to $[\theta, \theta + 1]$) at constant drift, in such a way that the process of getting from θ to $\theta + 1$ takes a time equal to τ_{arp}. At the end of the interval the depolarization is reset to 0. In terms of a diffusion process we have that, for $v > \theta$, $-\mu_{arp}\partial_v p = \partial_t p$ where $\mu_{arp} = 1/\tau_{arp}$.

The complete set of boundary conditions is now given by:

$$\sigma^2 \partial_v p\big|_{v=\theta} = -2\nu(t) \qquad \left[\sigma^2 \partial_v p - 2\mu p\right]_{v=0} = -2\nu'(t) \qquad \mu_{arp}\, p\big|_{v=\theta+1} = \nu'(t)$$

In the case of a steady statistics of the input current (i.e. $\mu(t) = \mu, \sigma(t) = \sigma$ for any t) the asymptotic solution ($p(v) = \lim_{t\to\infty} p(v,t)$) of Fokker-Planck equation is:

$$p(v) = \frac{\nu}{\mu}\left[1 - \exp\left(-\frac{2\mu}{\sigma^2}(\theta - v)\right)\right] \qquad \text{for } v \in [0, \theta]$$

and $p(v) = \nu\tau_{arp}$ for $v > \theta$. ν is determined by imposing that the integral of $p(v)$ over $[0, \theta + 1]$ is 1:

$$\nu = \Phi(\mu, \sigma) = \left[\tau_{arp} + \frac{\sigma^2}{2\mu^2}\left(\frac{2\mu\theta}{\sigma^2} - 1 + e^{\frac{-2\mu\theta}{\sigma^2}}\right)\right]^{-1}$$

which gives the mean rate of the *LIF neuron* as a function of the mean and the variance of the input current.

Inter-Spike Interval Distribution. The probability density of the inter-spike intervals is computed following [5]: the first passage time T is a random variable with a p.d.f. $g(v_0, T)$ that depends on the initial value v_0 (i.e. the hyperpolarization) and satisfies a backward Kolmogorov diffusion equation (see e.g. [5]). The Laplace transform $\gamma(v_0, s)$ of the $g(v_0, T)$, calculated in $v_0 = 0$ is

$$\gamma(0, s) = \frac{z e^{\theta C}}{z \cosh(\theta z) + C \sinh(\theta z)}$$

where $C \equiv \mu/\sigma^2$ and $z \equiv \sqrt{\mu^2 + 2s\sigma^2}/\sigma^2$.

Positive and Negative Drift Regimes

If the total mean drift μ is positive and σ is small, Φ is essentially linear over a wide range of μ (see Fig.3): $\nu = \Phi(\mu, \sigma) \simeq \mu\left[\theta + \tau_{arp}\mu\right]^{-1}$. The non-linearity shows only at high frequencies, comparable to $1/\tau_{arp}$. In this case the random walk is dominated by the drift which is the deterministic part of the current, and the neuron fires quite regularly (see Figs.1,2). The probability density $p(v)$ is almost uniform (see Fig.2) because the neuron tends to go from 0 to θ at constant speed.

In contrast, in the 'negative drift' regime, the neuron spends most of the time fluctuating near the reflecting barrier, and emits only when, by chance, a large fluctuation in the input current drives the depolarization above the threshold

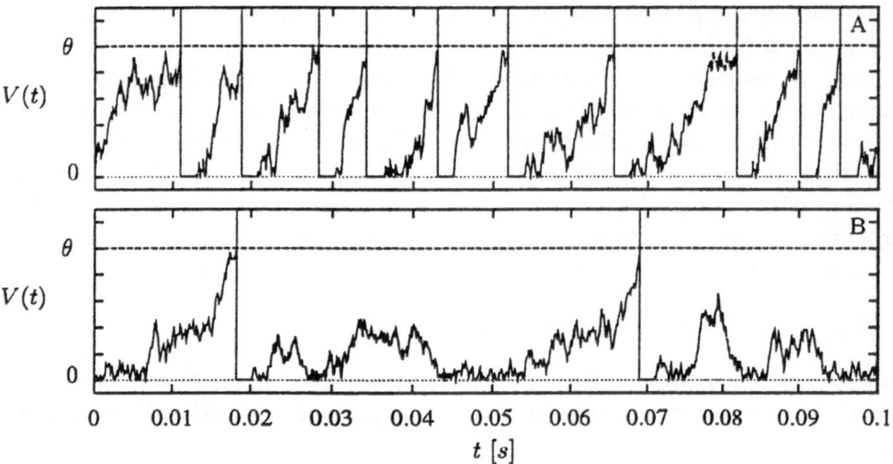

Fig. 1. Simulation of depolarization dynamics at positive (A) and negative drift (B). Parameters: (A) $\mu = 102\theta/s$, $\sigma = 5.3\theta/s^{1/2}$, producing a mean firing rate $\nu = 94Hz$; (B) $\mu = -10.1\theta/s$, $\sigma = 3.8\theta/s^{1/2}$, mean rate $\nu = 8.1Hz$. τ_{arp}=2ms in both cases. At positive drift the process is dominated by the deterministic part of the input current. The noisy linear ramp is clearly visible. At negative drift the depolarization fluctuates under threshold, waiting for the large fluctuation of the input current to drive $V(t)$ above threshold.

(see Fig.1). Since the fluctuations are random and uncorrelated, the neuron fires irregularly and the ISI distribution is wide (see Fig.2). In this regime the process is essentially dominated by the variance of the afferent current.

In principle it is possible to have such a regime also for $\mu > 0$, provided σ is large enough. This would require a fine tuning of the parameters.

Network Dynamics: Double Fixed Point

The extended mean-field theory approach [2] allows to study the dynamics of a network of identical neurons randomly interconnected, provided that one knows the current-to-rate transduction function. In the most general case the afferent current to any neuron is composed of two parts: one from spikes emitted by other neurons in the same network, and the other from outside. If the mean number of afferent connections is large and the mean charging current produced by the arrival of a single spike (the mean synaptic efficacy) is small, then the current $I(t)$ is Gaussian and: $\mu(t) = a_\mu \nu(t) + b_\mu(t)$, $\sigma^2(t) = a_\sigma \nu(t) + b_\sigma(t)$, where the as and bs are constants depending on the statistics of the connectivity, the synaptic efficacy, the decay β and the external afferents (see e.g. [2]). In order to have a fixed point, the self-consistency equation must be satisfied: $\nu = \Phi(\mu(\nu), \sigma(\nu))$. If the Φ function is linear in ν, as in the case of a neuron operating in a 'positive drift' regime, then only one stable fixed point is possible [6].

The non-linearity in the 'negative drift' regime is sufficient to allow for a double fixed point (see Fig.4).

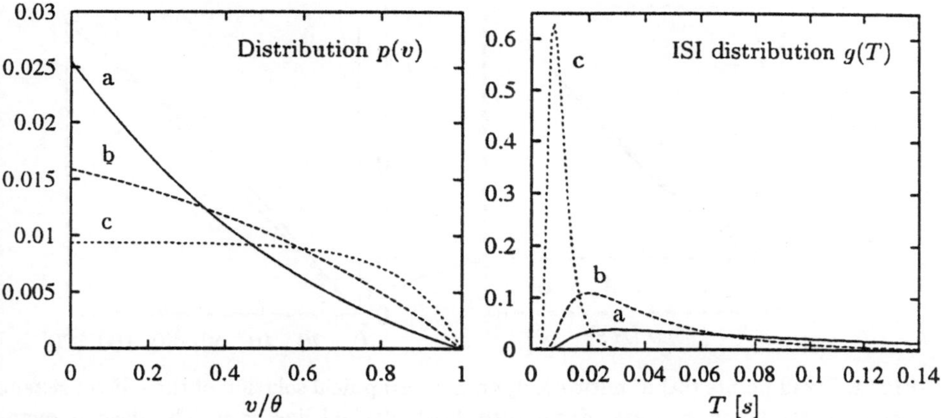

Fig. 2. Probability density function $p(v)$ (left) and ISI distribution $g(T)$ (right) at (a) negative, (b) intermediate and (c) positive drift. Parameters: (a) $\mu = -10.1\theta/s$, $\sigma = 3.8\theta/s^{1/2}$; (b) $\mu = 10.0, \theta/s$, $\sigma = 4.0\theta/s^{1/2}$ and (c) $\mu = 102\theta/s$, $\sigma = 5.3\theta/s^{1/2}$. At negative drift $p(v)$ is concentrated well below the threshold, near to the reset potential. As μ increases the curve changes concavity and tends to a uniform distribution. The ISI distribution is widespread for negative drift and tends to a peaked distribution as μ goes to positive values. The variability is clearly much higher in the negative drift regime.

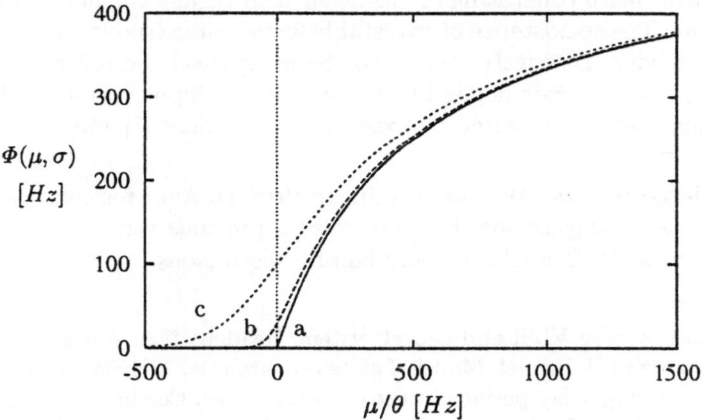

Fig. 3. Current-to-rate transduction function $\Phi(\mu)$ for different variances of afferent current: (a) $\sigma = 0$, (b) $\sigma = 5.6\theta/s^{1/2}$ and (c) $\sigma = 11\theta/s^{1/2}$. The firing rate in the region around $\mu = 0$, in the so called *negative drift regime*, is sensible to changes in the variance: the $\phi(\mu)$ passes from a threshold-linear function at $\sigma = 0$ to a non-linear function when $\sigma > 0$. If $\mu \gg \sigma$, the transduction function is almost independent on σ. The non-linearity that appears for large μ is due to τ_{arp}: Φ tends to the asymptotic frequency $1/\tau arp$.

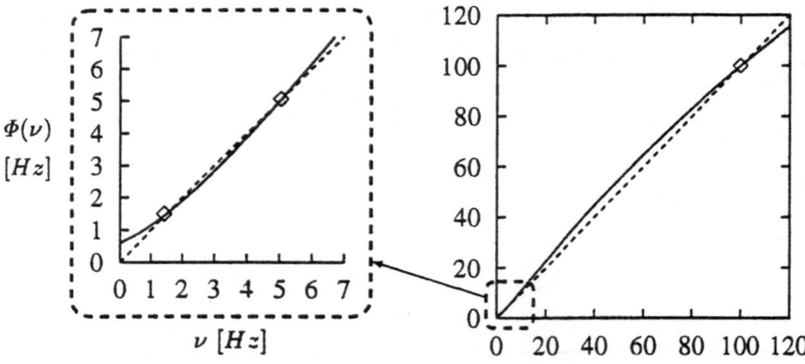

Fig. 4. Fixed points (◇) of network dynamics: graphical solution of the self-consistency equation: solid line = mean firing rate $\Phi(\nu)$; dashed line = ν. The dashed rectangle on the left is an enlargement of the low frequency region. Drift and variance: $\mu(\nu) = (-2.55 + 1.5\nu)\theta/s$ and $\sigma(\nu)^2 = (1.88 + 0.03\nu)\theta/s$. There are 3 intersections between $\Phi(\nu)$ and ν: two correspond to stable fixed points ($\nu = 1.5Hz$ at negative drift and $99Hz$ at positive drift) and one to an unstable fixed point ($5.0Hz$).

Conclusions

Considering fully the statistical properties, the collective dynamics of a network of *RC neurons* can be reproduced in a network of *VLSI (linear) neurons*. The existence of a reflecting barrier in 0 turned out to be fundamental in order to have the *LIF neuron* operating in 'negative' drift regime and not only in 'positive' drift regime. The coexistence of two stable fixed points (spontaneous activity and selective activity) is strictly related to the presence of the reflecting barrier and to the existence of a not negligible variance in the input current. All the results of this paper have been tested by extensive simulations [7] and will be published in a future work.

Acknowledgements. We are grateful to Prof. D. Amit for encouragement and for many useful suggestions that improved a previous version of the paper. We are indebted to N. Brunel for many helpful discussions.

References

1. C. Mead: Analog VLSI and neural system. Reading, MA: Addison Wesley (1989)
2. D.J. Amit and N. Brunel: Model of global spontaneous activity and local structured activity during delay periods in the cerebral cortex. Cerebral Cortex, **7**, 3 (1997)
3. M. Annunziato: Hardware implementation of an attractor neural network with IF neurons and stochastic learning, thesis (in italian) (1995)
4. G.L. Gerstein and B. Mandelbrot: Random walk models of the spike activity of a single neuron. Biophysical Journal, **4** (1964) 41–68
5. Cox Miller: D.R. Cox and H.D. Miller, The theory of stochastic processes. London, Methuen and Co LTD, (1965)
6. C.A. van Vreeswijk and M.E. Hasselmo, Self-sustained memory states in a simple model with excitatory and inhibitory neurons. unpublished
7. M. Mattia, Dynamics of an IF linear neurons network, thesis (in italian) (1997)

An Information-Theoretic Analysis of Temporal Coding Strategies by Spiking Central Neurons

Gustavo Deco and Bernd Schürmann
Siemens AG, ZT IK 4, Otto-Hahn-Ring 6, 81739 Munich, Germany

Abstract: The brain encodes information in the intervals between the spikes which characterize neural firing events. Therefore it is relevant to study in a timing code how many spikes are necessary for reliably encoding input signals. We analyze the transmission of information, the reliability of signal detection and the coding strategy for the case of central neurons which contrary to peripheral sensory neurons handle input signals assumed to be given by a combination of Poisson spike trains. We consider an integrate-and-fire model of a central neuron which combines diffusion and jump processes. In order to obtain analytical results, we introduce in addition a new Rényi-Information based measure for the discrimination ability of single neurons, which is investigated in the framework of a simple spike response model.

1 Introduction

In neurobiology it is generally agreed upon that the brain encodes information by the action potentials or "spikes" which characterize neural firing events [1-2]. Historically, the first proposal of a code mechanism was made by Adrian [3], who introduced the concept of *rate coding*. The idea is that the mean firing rate alone encodes the signal, while the variability about the mean is noise. Softky and Koch [4] have shown that the spike trains of cortical cells in the visual areas V1 and MT display a high degree of variability. This variability is characterized by the coefficient of variation $CV = \sigma_{ISI}/\mu_{ISI}$, where σ_{ISI} and μ_{ISI} are the standard deviation and mean value of the interspike intervals (ISI), respectively. Cortical cells have CV s in the range 0.5-1 as reported by Softky and Koch [4]. Thus, these authors suggested that high ISI variability may be more consistent with the idea of Abeles [5] of neurons acting as *coincidence detectors* rather than rate encoders. An alternative philosophy is to consider that it is variability itself that encodes the information contained in the input signal [6]. This concept corresponds to the notion that the precise sequences of time in which the spikes are emitted encodes the signal, yielding a *timing coding*. The experiments of Bialek [2] on the H1 movement detector neuron of the fly offer solid support for this concept. Recently, also Mainen and Sejnowsky [7] have shown the reliability of single spike coding using recordings from neurons in rat neocortical slices. Therefore, the relevant question is to study in a timing code how many spikes are necessary for reliably encoding input signals. The appropriate framework to study this problem from a theoretical point of view is given by information theory [8-10]. The aim of this paper is to analyze by use of information-theoretic concepts the transmission of information, the reliability of signal detection and the coding strategy for the case of central neurons, which contrary to peripheral sensory neurons handle input signals assumed to be given by a combination of Poisson spike trains.

2. Information and Coding

We consider first an integrate-and-fire model of a central neuron which combines diffusion and jump processes. The model combining diffusion and discontinuous development of the membrane potential is formulated in Musial and Lánsky [11]. The motivation is the revelation of experimental and theoretical studies that the effect of synaptic input on neural excitability decreases with the distances between the synapse and the cell body where spike generation is initiated. Somatic synapses cause changes in the membrane potential which are a large fraction of the threshold depolarization, while signals impinging on the periphery of an extensive dendritic tree evoke small potential changes at the soma. The changes induced by the inputs in the dendritic tree structure are well described by the continuous approach of Stein which is given by the Ornstein-Uhlenbeck diffusion process used in neural modeling. The model combining diffusion (dendritic tree) and jump process (soma-synapses-spikes) can be expressed by the Itô-type stochastic differential equation

$$dV(t) = \left(-\frac{V(t)}{\tau} + \mu\right)dt + \sigma dW(t) + w dS(t) . \tag{1}$$

In eq. (1), $dW(t)$ is a standard Wiener process. The constant τ describes the decay of the membrane potential in the absence of input signals. Here, $S(t)$ is a homogeneous Poisson process, i.e. $s(t) = \partial_t(S(t)) = \sum_i \delta\left(t - t_i\right)$ where the t_i are Poisson distributed random instants with mean value $1/\lambda$. The soma-synaptic-strength is denoted by w and the mean value by μ which is taken equal to zero. A spike is generated when the membrane potential $V(t)$ reaches a prefixed threshold θ. After the generation of the spike, the model is reset to a given initial potential $V(0)$ (in this paper taken to be equal to zero). The output spike train is therefore described by the spike generation times $t'_1, \ldots, t'_k, \ldots$, and is given by $o(t) = \sum_k \delta\left(t - t'_k\right)$. The interspike intervals (ISIs) of the output train are independent because the model is reset ("leaky integration") and the input signal is uncorrelated in time. The entropy per spike of the input spike train corresponds to the entropy of a Poisson Process given by $(1 - \ln(\lambda\varepsilon))$ where ε is the time precision assumed. In this paper we choose a time precision of 0.1 ms. Considering that the rate of the input spike train is λ, the entropy per unit time is $H_{in} = \lambda(1 - \ln(\lambda\varepsilon))$. Due to the independence of the output ISIs, the mutual information between the input and output spike train per unit time is given by

$$\begin{aligned} I_{io} &= I(s(t);o(t)) = I\left(\{t_1, \ldots, t_i, \ldots\};\{t'_1, \ldots, t'_k, \ldots\}\right) \\ &= I\left(\{t_1, \ldots, t_i, \ldots\};\{t'_1 - t'_2, \ldots, t'_k - t'_{k-1}, \ldots\}\right) \\ &= R \cdot \mathrm{I}\left(\{t_1, \ldots, t_i, \ldots\};T'\right). \end{aligned} \tag{2}$$

where $R = \langle T'\rangle^{-1}$ is the rate of the output spikes and $\mathrm{I}\left(\{t_1, \ldots, t_i, \ldots\};T'\right)$ is the mutual information per output spike, T' being the ISI of the output train. The mutual information per spike between the input signal and a single output ISI is $\mathrm{I}\left(\{t_1, \ldots, t_i, \ldots\};T'\right) = H(T') - H\left(T' | \{t_1, \ldots, t_i, \ldots\}\right)$. An upper bound of the output entropy is given by assuming a Poisson distribution of the output ISIs with the same rate R, i.e $p(T') = Re^{-RT'}$. This upper bound for the entropy per unit of time is therefore $H_{max}(T') = R(1 - \ln(R\varepsilon))$. A measure of the loss of information during the transmission is given by comparing the transmitted information and the input entropy. We define the loss of information $L = (H_{in} - I_{io})/H_{in}$. When L is equal to one, the whole information is lost and when L is minimal, maximal transmission is achieved.

3 .Simulations

We integrate the diffusion and jumping processes given by eq. (1) numerically by discretizing it in the following fashion:

$$V(t+\Delta t) = V(t) + \left(-\frac{V(t)}{\tau} + \mu\right)\Delta t + \sigma\sqrt{\Delta t}\ \upsilon + w\Delta S(t) \qquad (3)$$

where υ is standard Gaussian noise and $\Delta S(t) = \int_t^{(t+\Delta t)} \left[\sum_i \delta\left(t - t_i\right)\right] dt$ is the number of input spikes t_i between t and $t+\Delta t$. This Monte Carlo simulation involves the statistics of the Poisson input train and of the noise. We compute the probability distribution $p(T')$ for a given resolution ε via the construction of a histogram for T' by generating new realizations of both the input train and of the noise. $p(T'|t_1, \ldots, t_i, \ldots)$ is computed for a fixed realization of the input and different realizations of the noise, i.e. of the integration process. Afterwards we change the fixed realization $\{t_1, \ldots, t_i, \ldots\}$ of the input spike train in order to compute the mean value $\langle\ \rangle_{\{t_1, \ldots, t_i, \ldots\}}$. In the calculation of the discriminability, the probabilities $p(T'|s_i)$ have been calculated via the construction of a histogram for T' by generating new realizations of both the input spike train for a fixed process s_i, and of the noise. We choose $\tau = 20$ ms and $\theta = 20$ and an absolute refractory time of 3 ms.

Figure 1 displays the transmitted information from input spike trains to output spike trains as a function of the intensity of the input signal given by its mean value. For large values of λ^{-1} the transmitted information reaches the upper bound given by the Poisson assumption. In this Poisson regime, the maximum value achievable is reached but the efficiency of the transmission is very bad. This fact can be studied by observing the plots of the loss of information L. The minimum of information loss and therefore the maximum of efficiency is achieved *before* the Poisson regime is reached. It is interesting to observe that in the case when the synapse strength is such that the transmission is more efficient (see the deep minimum in the case of $w = 9$), the CV of the output ISIs exceeds the value 0.5 according to the experimental results of Softky and Koch [4]. Neurons which transmit efficiently information operate in a range of parameters such that they are not in the Poisson regime but in a regime of CV of the output ISIs between 0.5 and 1. A large value of CV does not necessarily mean that a rate code is correct but merely that a timing code is using the large irregularity at the output spike train for coding input signals as efficiently as possible.

4 . Discriminability

Let us now concentrate on the capacity of the neuron to detect two different signals. We suppose that there are two different input spike trains s_1 and s_2, corresponding to two different Poisson processes with rate λ_1 and λ_2 respectively. Let us further

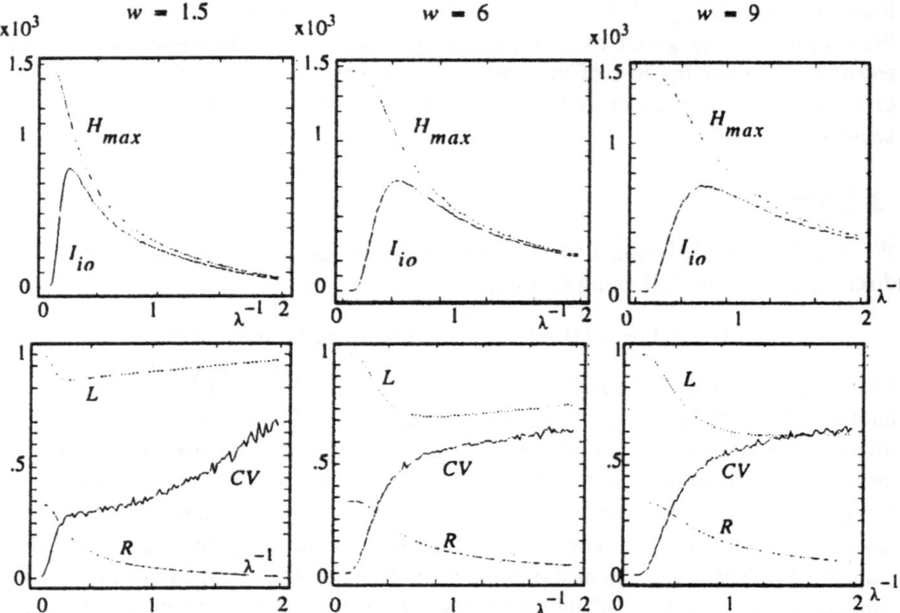

Figure 1: $H_{max}, I_{io}, L, CV, R$ for a single neuron as function of the input mean value (intensity of the input) for three different synapsis strength.

assume that the two signals are presented with uniform probability $p = 0.5$. We denote by s the random variable which corresponds to the class of the signal, i.e. the outcomes of s are s_1 and s_2 with equal probability $p = 0.5$. We can measure the discriminability by calculating the mutual information between the random variable s and the output spike train, i.e. by

$$I_{dis}(n) = I\left(s;\{t'_1,\dots,t'_n\}\right) = H(t'_1,\dots,t'_n) - \langle H(t'_1,\dots,t'_n|s)\rangle_s. \tag{4}$$

The problem of using the mutual information for calculations is the presence of the term $\ln(p(x))$ in the entropies which causes $I_{dis}(n)$ for large n to be numerically intractable. In the context of nonlinear dynamics, Pompe [12] introduced a generalized measure of statistical dependences that he phrased generalized mutual information, based in the second order Rényi entropies. The generalized mutual information of order 2 is defined by the expression

$$\begin{aligned} I^{(2)}{}_{dis}(n) &= I^{(2)}\left(s;\{t'_0,\dots,t'_n\}\right) = I^{(2)}\left(s;\{t'_1-t'_0,\dots,t'_n-t'_{n-1}\}\right) \\ &= H^{(2)}(T'_1,\dots,T'_n) + H^{(2)}(s) - H^{(2)}(T'_1,\dots,T'_n,s). \end{aligned} \tag{5}$$

In eq. (5), the Rényi entropy of order 2 is given by $H^{(2)}(x) = -\ln\int dxp^2(x)$. Pompe [12] has demonstrated that when one of the variables with N outcomes is uniformly distributed, i.e. each outcome has a probability $1/N$, as in our case the variable s, the generalized mutual information $I^{(2)}{}_{dis}(n)$ fulfills $0 \le I^{(2)}{}_{dis}(n) \le \ln N$. Even more, the lower bound is attained if and only if the variables $\{T'_1,\dots,T'_n\}$ and s are independent, and the maximum if and only if there is a function f such that $s = f(T'_1,\dots,T'_n)$, i.e. meaning an absolutely reliable classification of the input signals based in the output ISIs. This means that for our task of classification we can use the second order generalized mutual information which allows analytical calculation, and therefore the study of discriminability between two signals even in cases where the number of output ISIs required is large. The involved entropies can be calculated by considering the fact that for a given input signal s_i the output ISIs are independent, i.e.

$$p(t'_1, \dots, t'_n | s_i) = \prod_{k=1}^{n} p(t'_k | s_i) \quad , \quad p(t'_1, \dots, t'_n) = \frac{1}{2}\prod_{k=1}^{n} p(t'_k | s_1) + \frac{1}{2}\prod_{k=1}^{n} p(t'_k | s_2) \tag{6}$$

The maximum value of I_{dis} is given again by the entropy of the random variable s, i.e. $H^{(2)}(s) = \ln 2$. In order to obtain analytical results we use for our calculation a simple neural model which captures the principipal effects of real neurons and which is simple enough to permit analytical calculations. The model that we use consists of different versions of the spike response model introduced by Gerstner and van Hemmen [13]. In contrast to integrate-and-fire models which are given essentially by a differential equation, the spike response model is based on response kernels which describe the integrated effect of spike reception or emission on the membrane potential. In this model, spikes are generated by a threshold process, i.e the firing time t' is given by the condition that the membrane potential $h(t')$ reaches the firing threshold θ, i.e. $h(t') = \theta$. The membrane potential is given by

$$h(t') = J\sum_i \Theta(t' - t_i)\Theta(t_i - t'_{last})\psi(t' - t_i) \tag{7}$$

where $\Theta(\ )$ is the Heavyside function. We ignored the refractory time and consider a neuron with only one input spike train, which is Poisson. The calculation of $p(T'|s_i)$ is then reduced to the solution of the first passage time for the spike response model with an input given by a Poisson spike train. After some cumbersome algebra we obtain:

$$p(T'|s_i) = \frac{\lambda_i^{m+1} T'^m e^{-\lambda T'}}{m!} \tag{8}$$

where $m = \lfloor \theta/J \rfloor$ is the greater integer contained in θ/J. We calculate the Rényi entropies involved in the measure of discriminability given by the generalized mutual information, obtaining:

$$H^{(2)}(T'_1, \dots, T'_n) = -\ln\left(\frac{(2m!)^n}{4(m!)^{2n}}\left(\frac{\lambda_1^n}{2^{(m+1)n}} + \frac{\lambda_2^n}{2^{(m+1)n}} + \frac{2(\lambda_1\lambda_2)^{(m+1)n}}{(\lambda_1+\lambda_2)^{(2m+1)n}}\right)\right), \tag{9}$$

$$H^{(2)}(T'_1, \dots, T'_n, s) = -\ln\left(\frac{(2m!)^n}{4(m!)^{2n}}\left(\frac{\lambda_1^n + \lambda_2^n}{2^{(2m+1)n}}\right)\right). \tag{10}$$

Figure 2 shows the dependences of the generalized mutual information between the random variable s and the output spikes as a function of the number of spikes for different classification cases. The maximum value of the generalized mutual information means that the input signal can be classified with certainty, i.e. the output spikes contains the required information for a perfect and reliable distinction of the signals. The results can be interpreted in the following fashion: In cases where the two signals to be separated are very similar, i.e. small $\Delta\lambda = \lambda_2 - \lambda_1$, the convergence of the generalized mutual information to the maximal value $\ln(2)$ is very slow meaning that a large number of output spikes is required for a reliable classification of the signals. On the contrary, in cases where the signals are very different, i.e. $\Delta\lambda$ is big, the task of classification is more simple and can be achieved with a small number of spikes.

In conclusion, the information-theoretic analysis of single central neurons which process input spike trains teaches us that: 1) The maximum efficiency in the transmission of information is not reached in the Poisson regime but just before it, and in regions of high output CV. 2) An information-theoretic first principle (namely Infomax) is useful for defining a learning algorithm for fixing the optimal w for the task of discrimination of input signals or just for the efficient transmission of information. 3) The timing code can be rigorously studied in the framework of the Rényi Information concept. In fact, a small amount of output spikes suffices for efficient discrimination of input signals if the separation is easy; a large amount of output spikes is required in the hard cases of separation of very similar input signals.

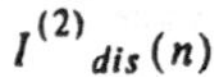

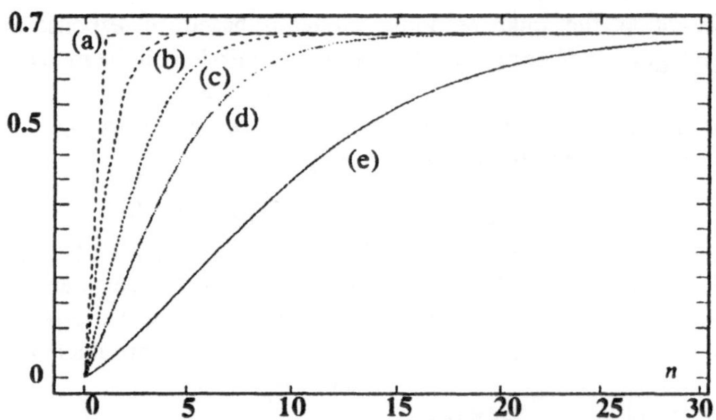

Figure 2: Discriminability vs. number of output spikes considered for the spike response model. The classification task shown correspond to input signals with $\lambda_1^{-1} = 4$ and λ_2^{-1}: (a) 40; (b) 10; (c) 7; (d) 6; (e) 5.

References

[1] H. Tuckwell, *Introduction to Theoretical Neurobiology* (Cambridge Press, 1988).

[2] F. Rieke, D. Warland, R. de Ruyter van Steveninck, and W. Bialek, *Spikes: Exploring the Neural Code* (The MIT Press, Cambridge, 1997).

[3] E. Adrian, *The Physical Background of Perception;* (Oxford Univ. Press, 1947).

[4] W. Softky and C. Koch, *J. Neuroscience* **13**, 334 (1993).

[5] M. Abeles, *Israel J. Med. Sci.* **18**, 83 (1982).

[6] W. Softky, *Current Opinion in Neurobiology* **5**, 239 (1995).

[7] Z. Mainen and T. Sejnowski, *Science* **268**, 1503 (1995).

[8] G. Deco and B. Schürmann, *Physical Review* **E 51**, 1780 (1995a).

[9] G. Deco and B. Schürmann, *Physical Review* **E 52**, 6580 (1995b).

[10] G. Deco and D. Obradovic, *An Information Theoretic Approach to Neural Computing* (Springer, New York, 1996).

[11] M. Musila and P. Lánsky, *Int. J. Biomed. Comput.* **31**, 233 (1992).

[12] B. Pompe, *Chaos, Solitons and Fractals*, 4, 83 (1994).

[13] W. Gerstner and J. van Hemmen, *Phys. Rev. Lett.*, **71**, 312 (1993).

Correlation Coding in Stochastic Neural Networks

Raphael Ritz and Terrence J. Sejnowski

Computational Neurobiology Laboratory
The Salk Institute for Biological Studies
10010 North Torrey Pines Road, La Jolla, CA 92037, USA

Abstract. Stimulus–dependent changes have been observed in the correlations between the spike trains of simultaneously–recorded pairs of neurons from the auditory cortex of marmosets even when there was no change in the average firing rates. A simple neural model can reproduce most of the characteristics of these experimental observations based on model neurons having leaky integration and fire–and–reset spikes and with Poisson–distributed, balanced input. The source of the synchrony in the model was common sensory input. The outputs of neurons in the model appear noisy (almost Poisson) owing to the stochastic nature of the input signal, but there is nevertheless a strong central peak in the correlation of the output spike trains. The experimental data and this simple model clearly demonstrate how even a noisy–looking spike train can convey basic information about a sensory stimulus in the relative spike timing between neurons.

1 Introduction

It is commonly believed that the neural code used by nerve cells to transmit information in the cerebral cortex is the mean firing rate of action potentials. Whereas there is solid evidence for this coding scheme at the neuromuscular junction, where this concept originated, the temporal averaging involved in the decoding process causes problems at the cortical level, where neurons usually fire at rates too low to allow for a sufficiently long decoding time. As a possible solution to this problem it has been proposed that cells could also perform a spatial average instead of, or in addition to, temporal averaging. But this form of population code also assumes that information is coded in a firing *rate*—whether spatial or temporal—and that a neuron simply reflects changes in its input firing rates by modulating its output firing rate. This is the underlying assumption allowing the common reduction to a transfer function used by most artificial neural network models to describe single neuron processing.

Recently, deCharms and Merzenich [3] presented evidence for a different form of coding in the primary auditory cortex of marmosets. They showed that rapidly adapting cells responded to elongated tone stimuli with a fast transient onset response returning quickly to spontaneous firing rates. Thus, these cells cannot convey information about a steady–state stimulus by their firing rate. However, these cells do show an increase in their tendency to fire *simultaneously* as revealed by correlation analysis if they are tuned to the presented stimulus frequency. Nevertheless, each spike train looked almost like it was randomly generated and there was no stimulus–locked component as shown by a flat shift predictor.

Most characteristics of these experimental findings can be reproduced in a simple neuronal model using leaky integrate–and–fire units with Poisson–distributed, balanced input, as shown below.

2 The Random Walk Model

Assume that the generation of action potentials relies on the membrane potential $u_i(t)$ of cell i $(1 \le i \le N)$ at time t crossing a firing threshold θ and that deviations from the resting potential (set to 0 here) are due to an input current $C_i(t)$ and given these deviations decay exponentially with the membrane time constant τ_{m}. The following equation governs the temporal evolution of the membrane potential:

$$\frac{\mathrm{d}}{\mathrm{d}t}u_i(t) = -\frac{1}{\tau_{\mathrm{m}}}u_i(t) + C_i(t) \ . \tag{1}$$

A spike occurs when $u_i(t) = \theta$, and u_i is reset to its resting level. To avoid unrealistically large hyperpolarizations, we also introduce a negative saturation limit θ^{inh}, i. e., we assure $u_i(t) \ge \theta^{\mathrm{inh}}$ for all t. To specify the input current, $C_i(t)$, assume that this input can be subdivided into a background and a stimulus component, $C_i^{\mathrm{bg}}(t)$ and $C_i^{\mathrm{stim}}(t)$ respectively

$$C_i(t) = C_i^{\mathrm{bg}}(t) + C_i^{\mathrm{stim}}(t) \ , \tag{2}$$

and that each of these components consists of excitatory as well as inhibitory parts

$$C_i^{\mathrm{bg,stim}}(t) = E_i^{\mathrm{bg,stim}}(t) - b^{\mathrm{bg,stim}} I_i^{\mathrm{bg,stim}}(t - \Delta^{\mathrm{inh}}) \ , \tag{3}$$

where b denotes a balancing factor indicating the relative strength of the inhibition with respect to the excitation and Δ^{inh} represents a delay.

To introduce noise in the model, assume that all excitatory and inhibitory signal components are realizations of an ideal Poisson process, i. e.,

$$E_i^{\mathrm{bg}}(t) = k \ \text{ with probability } \ p(k) = \frac{\lambda^k}{k!}e^{-\lambda} \tag{4}$$

where k is drawn at each time step for every component independently. The parameter λ denotes the mean and the variance of the distribution. Here, it can be interpreted as $\lambda = n_{\mathrm{aff}} \cdot p_{\mathrm{f}}$ the product of the number of afferents times the probability of firing in a single time step, thus fixing the input firing rate. For $\lambda = 10$ and a basic time step of 1 ms, this might correspond to 100 afferents each firing at a rate of 100 Hz.

Due to the randomness in the input the membrane potential undergoes a sort of a random walk with renewal [4].

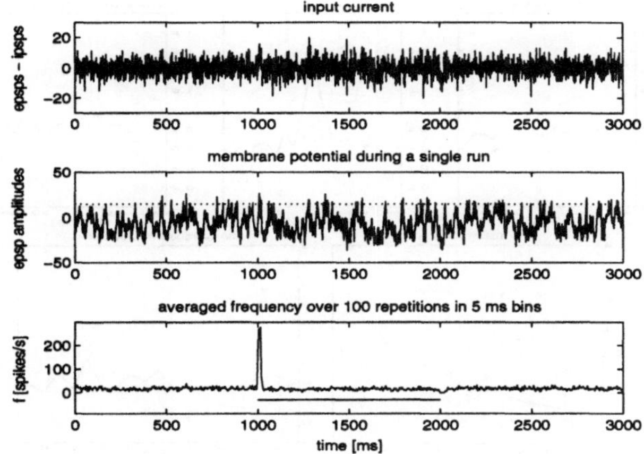

Fig. 1. Single neuron model. (Top) The total input to the neuron is a sum of four Poisson processes, each with intensity $\lambda = 10$, consisting of excitatory and inhibitory background activity during the whole run and excitatory and delayed inhibitory signals ($\Delta^{\text{inh}} = 20\,\text{ms}$) from $t = 1000$ to $2000\,\text{ms}$. (Middle) Membrane potential at the receiving neuron (time constant $\tau_{\text{m}} = 10\,\text{ms}$); The dotted line indicates the firing threshold ($\theta = 15$). (Bottom) A spike histogram computed using 100 trials and 5 ms bins scaled to represent the mean firing rate in spikes per second. Note the pronounced onset response together with a rapid decay to the spontaneous rate even during stimulation. The stimulus duration is indicated by the horizontal bar.

3 Simulation Results

For simulations, we used the following set of parameters. The thresholds were set to $\theta = 15$ and $\theta^{\text{inh}} = -30$, both in units of single EPSP amplitudes. The time scale was fixed by $\tau_{\text{m}} = 10\,\text{ms}$ and $\Delta^{\text{inh}} = 20\,\text{ms}$. We solved (1) using a simple forward Euler method with a time step size of 1 ms. The input was specified by $\lambda = 10$ for all four components and the balancing factors are $b^{\text{bg}} = 1$ and $b^{\text{stim}} = 1.1$. A typical simulation run lasted for three seconds where a stimulus was switched on after the first second and turned off after the second.

3.1 Single Cell Properties

Consider first the firing of a single cell. As seen in the top row of Fig. 1, the total input current fluctuates vigorously. The resulting membrane potential (Fig. 1, middle) is smoother due to the temporal integration. Threshold crossings of the membrane potential resulting in spike emission were only driven by fluctuations except for the stimulus onset period, where there was an excess of excitatory input due to the delayed arrival of the balancing inhibitory input. This can clearly be seen in the spike histogram (lower part of Fig. 1) obtained by averaging over

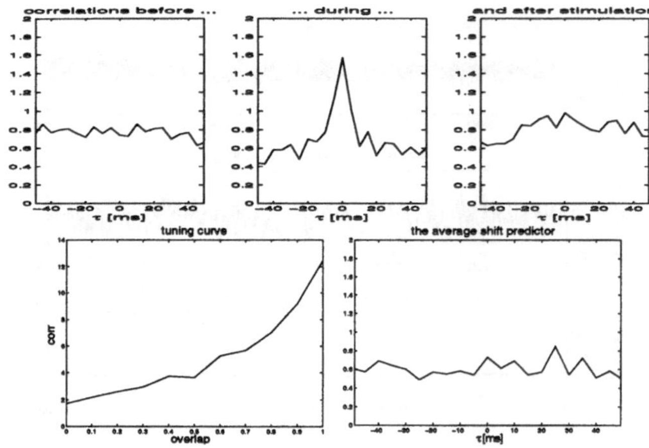

Fig. 2. A pair of cells receiving a common input. (Top row) Average correlations from three different time periods: (Left) From $t = 250$ to $750\,\text{ms}$ – before stimulation; (Middle) From $t = 1250$ to $1750\,\text{ms}$ – during stimulation but after the onset response and (Right) from $t = 2250$ to $2750\,\text{ms}$ – after stimulation. There is a clear peak at $\tau = 0$ for the stimulation period indicating that these two neurons have a tendency to fire in synchrony during presence of the stimulus. Correlations were calculated for every trial using 5 ms bins and averaged afterwards. (Bottom, left) Tuning curve: Height of the central peak in the correlation during stimulation as a function of the fraction of identical input. The peak height increases with the overlap. (Bottom, right) Due to the overall noisy structure of the observed response, the shift predictor, correlating responses from different trials, is flat. Thus, there was no stimulus–locked activity during the tonic phase of the response.

100 repetitions of the same experiment (but using a different seed for initializing the random number generator each time). The mean firing rate stayed constant throughout the whole run except for a pronounced burst at stimulus onset and a reduction of firing after stimulus offset where the experimental recordings show a transient increase of the firing rate. A large trial–to–trial variability in firing was observed as well (data not shown).

3.2 Multiple Cell Properties

In the experiments of [3] simultaneous recordings of spike trains from pairs of cells were analyzed. We simulated two cells getting independent background signals but sharing identical stimulus components in their input ($C_1^{\text{stim}}(t) = C_2^{\text{stim}}(t)$ for all t) assuming that cells with similar tuning properties are to some extend enervated by the same afferent neurons. The top row of Fig. 2 shows correlations between the firing times of the two cells calculated for every single trial for three different periods of time (before, during, and after stimulus presentation) and

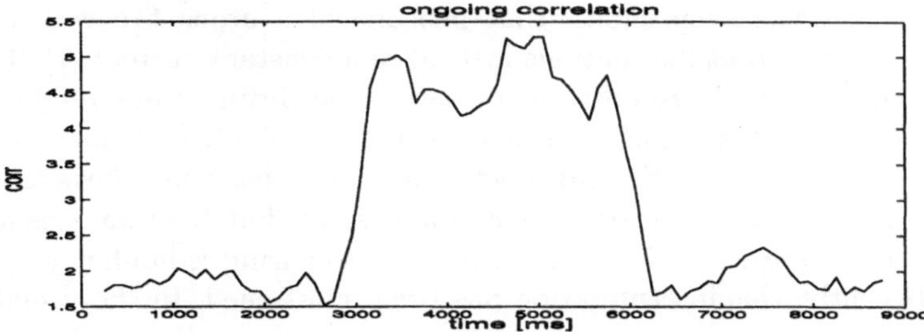

Fig. 3. Time course of the average correlation calculated using a 500 ms time window sliding over a 9 s simulation run in 100 ms time steps. A stimulus was presented from $t = 3000$ to 6000 ms.

averaged afterwards. There was a strong peak at zero time shift only during common stimulation indicating an increased tendency of the two cells to fire simultaneously.

This is not a surprising result, because one might expect the common input to drive both cells to firing threshold simultaneously, but it is worth noticing since only a fraction of the emitted spikes are affected. These synchronous spikes happen to occur at random times and are not stimulus–locked, as indicated by the flat shift predictor in the lower–right part of Fig. 2. The height of the central peak in the correlation depends mainly on the amount of common input relative to the total input to both cells as shown in Fig. 2 (bottom left). The overlap here is defined as the ratio of common versus total input, ranging from zero (no common input) to one (absolutely identical input).

Finally, the time course of the correlation peak is shown in Fig. 3. Correlations were calculated from a 500 ms time window sliding over the entire run in 100 ms time steps leading to five times oversampling following [3]. During the entire stimulation period, there was a pronounced increase in the correlations, which disappeared when the stimulus was turned off.

4 Discussion

In contrast to the common belief that neurons code information only in their mean firing rate, deCharms and Merzenich have shown that there is another possibility of coding, based on the relative timing of spikes from different neurons [3]. We have replicated their results in a neural model. Conceptually, this idea is not new, and the underlying firing pattern may be even more complicated than just synchronous firing, as in synfire chains [1] or arbitrary firing patterns [5] or with respect to an internal neuronal clock [6].

What is new here is the observation that relative spike timing might be used in a noisy mode of operation. For this regime, it has commonly been assumed that the only way to get at reliable information transmission should be based on a rate code [9]. But there is increasing evidence for the possibility of temporal codes. First, it has been shown that neocortical neurons fire very reliably if driven mainly by input fluctuations instead of a constant current [7]. Therefore, the well known high variability in cortical spike firing times might reflect a high variability in the input to a neuron instead of intrinsic noise due to the spike generation process. Second, correlations in firing times between neurons tuned to similar stimulus features are omnipresent, but they have usually been interpreted as an artifact of common stimulation causing redundancy and having no use. Recently, this interpretation has been questioned. In the visual system, correlations seem to improve stimulus representation on the level of the retina [8] as well as the LGN [2]. In the auditory system, [3] provided evidence for the crucial role of correlations in stimulus representations. Their study was the starting point for the model presented here. We do not claim to have reproduced every single detail of their data. For this, a biophysically more realistic model should be appropriate. But we have shown here how such a code might work naturally and reliably even in a noisy environment.

The final question, however, whether this type of coding is really used in the brain (i.e., read out at the next level) remains to be experimentally examined. Correlations are easily read out by neurons and they play a central role in learning, so there is every reason to continue along this line of investigation.

Acknowledgments Supported by DFG (grant Ri 821/1-1) and The Howard Hughes Medical Institute.

References

1. M. Abeles, H. Bergman, E. Margalit, and E. Vaadia. Spatiotemporal firing patterns in the frontal cortex of behaving monkeys. *J. of Neurophysiology*, 70:1629–1638, 1993.
2. Y. Dan, J. J. Atick, and R. C. Reid. Efficent coding of natural scenes in the lateral geniculate nucleus: Experimental test of a computational theory. *J. Neurosci.*, 16(10):3351–3362, 1996.
3. R. C. deCharms and M. M. Merzenich. Primary cortical representation of sounds by the coordination of action–potential timing. *Nature*, 381:610–613, 1996.
4. G. L. Gerstein and B. Mandelbrot. Random walk models for the spike activity of a single neuron. *Biophysic. J.*, 4:41–68, 1964.
5. W. Gerstner, R. Ritz, and J. L. van Hemmen. Why spikes? Hebbian learning and retrieval of time–resolved excitation patterns. *Biol. Cybern.*, 69:503–515, 1993.
6. J. J. Hopfield. Pattern recognition computation using action potential timing for stimulus representation. *Nature*, 376:33–36, 1995.
7. Z. F. Mainen and T. J. Sejnowski. Reliability of spike timing in neocortical neurons. *Science*, 268:1503–1506, 1995.
8. M. Meister, L. Lagnado, and D. A. Baylor. Concerted signaling by retinal ganglion cells. *Science*, 270:1207–1210, 1995.
9. M. N. Shadlen and W. T. Newsome. Noise, neural codes and cortical organization. *Curr. Opin. Neurobiol.*, 4:569–579, 1994.

Two-Dimensional Hodgkin-Huxley Equations for Investigating a Basis of Pulse-Processing Neural Networks

Akira Hirose

Research Center for Advanced Science and Technology (RCAST), University of Tokyo
4-6-1 Komaba, Meguro-ku, Tokyo 153, Japan
e-mail: ahirose@ee.t.u-tokyo.ac.jp

Abstract. Pulse-processing neural networks are expected to perform more delicate and essential signal processing than conventional firing-rate networks. The architecture is based on biological nerve systems much more and, hence, it is very important as a basis to investigate the dynamics of pulse signals in biological neural networks such as interactions of action potential (signal pulses). In this paper, two-dimensional Hodgkin-Huxley equations are proposed for a mesoscopic analysis and synthesis of membrane potential activities such as propagation and interaction. The proposed method will enable us to describe signal pulse dynamics by a universal method.

1 Introduction

Pulse-processing neural networks are based on a more precise model of human neural systems than conventional firing-rate networks. Therefore, they are expected to be applied wider and more usefully to complicated adaptive signal processing systems such as human nerval interfaces [1]. On the other hand, because of their delicate behavior, it becomes significantly important to investigate their basis: i.e., the pulse dynamics in biological neural networks.

The recent development of measurement technology of biological neural signals enables us to observe various new temporal and spatial dynamics of the membrane potential propagation and interaction; e.g., synchronization or anti-synchronization of neural pulses and backward propagation of pulses on dendrites. Such mesoscopic dynamics [2] cannot be analyzed by conventional neuron models.

For elucidating such delicate but important phenomena in neural networks, a new analysis method is required. In time domain, a set of nonlinear equations proposed by Hodgkin and Huxley [3] fundamentally describes precise potential dynamics at a point on the membrane in terms of ion channel activities. After their work, the equations were extended into the cable theory [4] and the compartmental model [5], as shown in Table 1, where the pulse propagation can be analyzed in space one dimensionally [6].

In these years, such one-dimensional models are also applied to investigations of dendritic propagation of the EPSP or IPSP (excitatory / inhibitory postsynaptic potential) [7],[8]. However, the potential activity is caused by spatially-

Table 1. Spatial evolution of microscopic potential dynamics equations based on Hodgkin-Huxley equations (H-H eqns).

0-dimensional H-H eqns :	Original dynamics equations (Hodgkin, Huxley, 1952) [3]
1-dimensional H-H eqns :	Cable theory (Rall, 1959) [4]
	Compartmental model (Rall, 1964) [5]
2-dimensional H-H eqns :	General spatial and temporal dynamics (This work)

extending two-dimensional active membrane on which the potential propagates and interacts.

In this paper, two-dimensional analysis method of the membrane potential is proposed from a view point that the potential activity is generated essentially by two-dimensional membrane. Using this method, we can investigate the mesoscopic dynamics of biological neural networks, both in space and time, such as potential propagations and interactions on the dendritic and somatic membrane and the pulse coincidental effects between nerval circuits.

A general dynamics will be described universally by using the geometry of the neural cell and a set of parameters such as channel density distribution on the membrane. Then, in the future, we will be able to explain the reason why a neural cell under measurement has to be shaped as it is. Moreover, such numerical analysis and synthesis may predict an existence of new functions unknown so far. In this paper, a result of a preliminary calculation is also reported.

2 Theory

From a microscopic view point, somas, dendrites, and axons are formed essentially by two-dimensional membrane and, hence, the potential propagates and interacts everywhere on the cell obeying the properties of the two-dimensional membrane. Although parameters such as channel density may be variable depending on membrane parts and positions, the principle of a potential diffusion effect should be universal.

Consequently, we extend the original Hodgkin-Huxley equations into a two-dimensional version by introducing a two-dimensional Laplacian on the membrane ∇^2 ($\equiv \frac{\partial^2}{\partial x^2} + \frac{\partial^2}{\partial y^2}$ for an orthogonal curvilinear coordinate) as

$$\tau \frac{\partial V}{\partial t} = \lambda^2 \nabla^2 V + \frac{\tau}{C_{\mathrm{m}}} \{ i - g_{\mathrm{Na}} m^3 h (V - V_{\mathrm{Na}}) - g_{\mathrm{K}} n^4 (V - V_{\mathrm{K}}) - g_{\mathrm{L}} (V - V_{\mathrm{L}}) \} \quad (1)$$

where

$$\frac{dn}{dt} = \alpha_{\mathrm{n}}(1-n) - \beta_{\mathrm{n}} n \quad (2)$$

$$\frac{dm}{dt} = \alpha_{\mathrm{m}}(1-m) - \beta_{\mathrm{m}} m \quad (3)$$

$$\frac{dh}{dt} = \alpha_{\mathrm{h}}(1-h) - \beta_{\mathrm{h}} h \quad (4)$$

In these equations, V, C_{m}, and i denote membrane potential, membrane capacity per unit area, and injection current density equivalent to a synaptic excitation, respectively. $\lambda^2 \equiv DR_{\mathrm{m}}/R_{\mathrm{i}}$ and $\tau \equiv R_{\mathrm{m}} C_{\mathrm{m}}$ are a characteristic length and a time constant determined by the effective conductive-layer thickness inside the membrane D, the intracellular and membrane resistivity R_{i} and R_{m}, and the membrane capacity C_{m}, respectively. g_{Na}, g_{K}, and $g_{\mathrm{L}} \equiv 1/R_{\mathrm{m}}$ denote, respectively, maximum conductances of sodium and potassium ion channels and a leak conductance.

In (2)-(4), n, m, and h express open probabilities of ion channels which are fitted in Ref.[3] as

$$\alpha_{\mathrm{n}} = 1.00 \times 10 \ \times (0.010 - v)/(\exp[(0.010 - v)/0.016] - 1.0) \quad (5)$$

$$\beta_{\mathrm{n}} = 1.25 \times 10^2 \times \exp[-v/0.080] \quad (6)$$

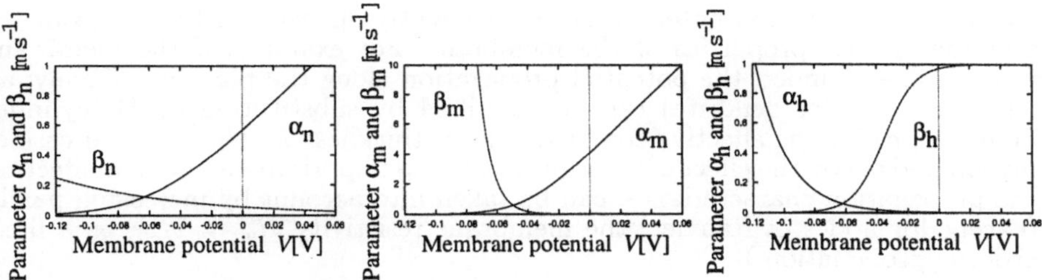

Fig. 1. Parameters for ion channels dependent on membrane potential.

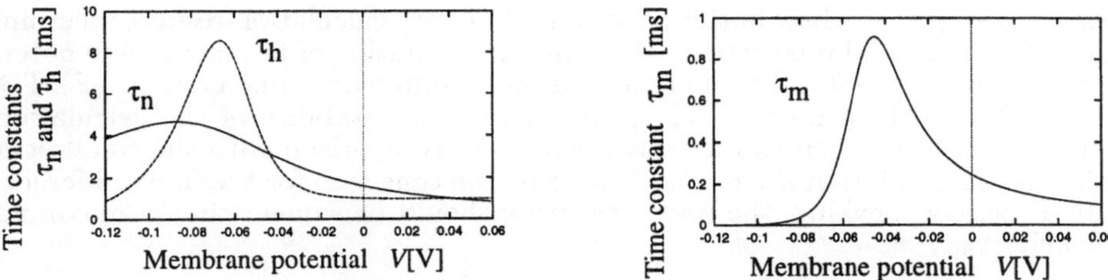

Fig. 2. Time constants of ion channels calculated with the parameters in Fig.1 as functions of membrane potential.

$$\alpha_{\rm m} = 1.00 \times 10^2 \times (0.025 - v)/(\exp[(0.025 - v)/0.010] - 1.0) \tag{7}$$

$$\beta_{\rm m} = 4.00 \times 10^3 \times \exp[-v/0.008] \tag{8}$$

$$\alpha_{\rm h} = 7.00 \times 10 \times \exp[-v/0.020] \tag{9}$$

$$\beta_{\rm h} = 1.00 \times 10^3 \;/(\exp[(0.030 - v)/0.010] + 1.0) \tag{10}$$

where $v \equiv V - V_{\rm L}$[V] and the dimension of alphas and betas is [s^{-1}]. These parameters are shown as functions of the membrane potential in Fig.1. Channel time constants $\tau_{\rm n}$, $\tau_{\rm m}$, and $\tau_{\rm h}$ are also calculated as functions of potential V, respectively, as

$$\tau_{\rm n} = \frac{1}{\alpha_{\rm n} + \beta_{\rm n}}, \qquad \tau_{\rm m} = \frac{1}{\alpha_{\rm m} + \beta_{\rm m}}, \qquad \tau_{\rm h} = \frac{1}{\alpha_{\rm h} + \beta_{\rm h}} \tag{11}$$

Figure 2 shows the time constants calculated numerically by using (5)-(10).

Taking into consideration the relations $\lambda^2 \equiv DR_{\rm m}/R_{\rm i}$ and $\tau \equiv R_{\rm m}C_{\rm m}$, the dynamics (1) can be rewritten as

$$C_{\rm m}\frac{\partial V}{\partial t} = \frac{D}{R_{\rm i}}\nabla^2 V + \{i - g_{\rm Na}m^3h(V - V_{\rm Na}) - g_{\rm K}n^4(V - V_{\rm K}) - g_{\rm L}(V - V_{\rm L})\} \tag{12}$$

where the both side of (12) has a dimension of [A/m^2]. This equation (12) expresses the potential diffusion on the membrane and the nonlinear conductivity associated with (2)-(10).

According to (12), a potential at a certain membrane point diffuses around on the membrane nonlinearly. The propagation and interaction characteristics of

EPSP, IPSP, or action potential are consequently determined by the geometry and the electric properties of the membrane. For example, if the membrane forms a fine cylinder, the potential propagation along the pipe (or, namely, an axon or a slender dendrite) can be described by substituting $(d/4)$ (cylinder diameter: d) for the effective conductive-layer thickness D. (In a special case of myelinated nerve, the effect of a myelin sheath, an important component affecting the propagation characteristics, can be taken into account by increasing partly (except for nodes of Ranvier) the membrane resistivity $R_m = 1/g_L$ as a first-order approximation.)

3 Preliminary Calculation Experiment

As an example, we show in this paper a preliminary calculation result concerning an action potential interaction. The temporal evolution of the membrane potential can be calculated by applying the finite-difference time-domain (FDTD) method to (12). In linear FDTD, for ensuring the stability of the calculation, the wave velocity v should be small enough in comparison with the smallest finite spatial resolution Δx or Δy divided by the time step for the finite difference Δt. Precisely speaking, the wave velocity v should be smaller than a maximum velocity $v_{\max}$ as:

$$v < v_{\max} \equiv \frac{1}{\Delta t \sqrt{(\frac{1}{\Delta x})^2 + (\frac{1}{\Delta y})^2}} \tag{13}$$

As an example, we analyze action potential waves propagating and interacting on the thick-dendritic and somatic membrane of a single neuron shown in Fig.3. It is assumed that there are 4 thick dendritic roots on the left-hand side of the soma, and one axon root on the right-hand side.

It is also assumed that the injection current on the dendrites equivalent to a synaptic excitation is large (20[A/m^2], 0.1[ms] duration) so that action potentials are generated at the injection points in stead of EPSP. Generally speaking, such a situation is not supposed to be observed so often. In most cases, contrarily, a summed EPSP causes a depolarization large enough to generate a single action potential at a point near to the axon. However, by choosing the multi action-potentials situation mentioned above, we can observe clearer what is happening on the membrane. In these years, moreover, such multiple action-potential phenomena have come to be observed in recent physiological experiments.

The membrane parameters are assumed to be constant for all the parts of the membrane in this preliminary experiment. They are listed in Table 2. The diameter of the cell body is chosen at 50[μm]. A modified (nonlinear) spherical coordinate is adopted three-dimensionally on the membrane with a 32$\times$16 mesh dimension whereas the time step is chosen as Δt=0.01[μs]. In this case, the smallest spatial resolution is about 0.25[μm] at around the root of the axon. Then, from (13), the upper limit of the wave velocity $v_{\max}$ treatable in the FDTD is estimated at $v_{\max} \approx 25$[m/s]. The maximum velocity value $v_{\max}$ is almost as large as the reported conduction velocity of action potential on axons with myelin laminae (the fastest in the nerval system) of a few tens [m/s], although the nonlinearity of the present calculation may cause some modification in a detailed estimation.

Figures 4-10 show the potential evolution on the membrane. The center shape in Fig.4, for example, corresponds to the schematic geometry of the neuron cell under investigation (Fig.3) with an axon root at right-hand side and dendrite roots left-hand side. On its right side is an axon-side view, and left side a dendrite-side view.

Table 2. Parameters used in the experiment expressed in SI units.

Parameter	Value
Membrane resistivity $R_m(\equiv 1/g_L)$:	$0.1\Omega m^2$
Membrane capacity C_m :	$1\mu F/m^2$
Intracellular resistivity R_i :	$1.0\Omega m$
Effectively conductive thickness inside soma & dendrites D :	$0.02\mu m$
Equilibrium potential of Na^+ V_{Na} :	55mV
Equilibrium potential of K^+ V_K :	-72mV
Stationary potential V_L :	-65mV
Maximum Na^+-channel conductance g_{Na} :	$1200S/m^2$
Maximum K^+-channel conductance g_K :	$360S/m^2$
Soma radius (without dendrites) r_0 :	$25\mu m$
Time step in FDTD calculation Δt :	$0.01\mu s$
Spatial dimension of sampling points (modified spherical coordinate) :	32×16
Treatable maximum wave velocity v_{max} :	~25m/s

At a measurement time of $t = 0.1 \sim 0.2$[ms], a stimulatory current equivalent to a synaptic excitation is injected at a dendrite point **a**, and also at $t = 1.1 \sim 1.2$[ms] at dendrite points **b** and **c** with a same current density mentioned above. Brighter parts on the membrane in Figs.5-10 indicate a higher (depolarized) potential (up to about +40[mV]), whereas darker parts around a stationary potential (about -65[mV]). (Black speckles on the membrane are caused merely by graphical defectiveness. The membrane covers the cell interior smoothly and completely in the calculation.)

It is oberved that an action potential is generated at point **a** (Fig.5). Its risetime after injection is ≈ 0.65[ms]. The pulse spreads first around (Fig.6), and then get confluent with those around points **b** and **c** (Fig.7). Afterward, the united action potential extends quickly out over the soma including the axon and other dendtites (Figs.8–10).

In another experiment generating a single action-potential pulse, the pulse-peak traveling time required for the present soma after the pulse rise is measured about 2.1[ms]. Figure 11 shows a traveling time T required for the cell mentioned above in terms of intracellular resistivity R_i calculated numerically with other parameters unchanged. It is found that the propagation time T is related to the resistivity R_i approximately as $T \propto R_i^{1/3}$ in this membrane parameter-region, whereas they are intrinsically nonlinear. The details will be reported elsewhere.

4 Conclusion

Two-dimensional Hodgkin-Huxley equations have been proposed for the mesoscopic analysis of membrane potential. A preliminary experimental result has been reported. The proposed method enables us to investigate all the temporally and spatially mesoscopic dynamics of biological neural networks quantitatively in various aspects. This method can be applied not only to single-neuron dynamics but also to multiple-neuron circuit dynamics, by which the newly oberved neural pulse phenomena such as the coincidence / anti-coincidence will be elucidated. A detailed dynamics, which are the basis of pulse-processing neural networks, will be described systematically and universally using the presented method.

References

1. H.Napp-Zinn, M.Jansen, R.Eckmiller, Biol. Cybern. 74 (1996) 449-453
2. A.Hirose, Joint Conference of Information Sciences (JCIS) '97, Proc. 2 (1997, Research Triangle Park, U.S.A.) 42-45
3. A.L.Hodgkin, A.F.Huxley, J. Physiol., 117 (1952) 500-544
4. W.Rall, Experimental Neurol. 1 (1959) 491-527
5. W.Rall, *in Neural Theory and Modeling,* ed. R.F.Reiss, Palo Alto: Stanford University Press (1964)
6. W.Rall, *"The Theoretical Foundation of Dendritic Function,"* I.Segev, J.Rinzel, G.M.Shephard, eds., MIT Press (1995)
7. G.Stuart, M.Haeusser, Neuron, 13 (1994) 703-712
8. N.Spurston, D.B.Jaffe, D.Johnston, TINS, 17 (1994) 161-166

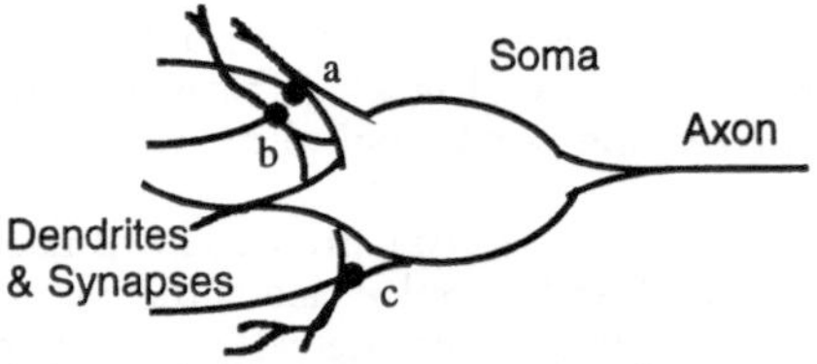

Fig. 3. Schematic geometry of a neuron.

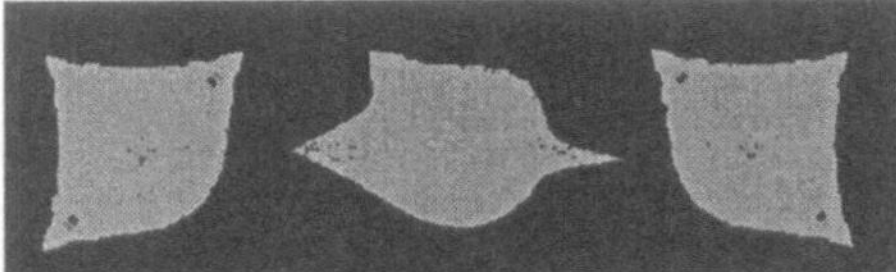

Fig. 4. t=0 [ms] (Black speckles are merely because of graphical defects. Center: corresponding to Fig.3, Left: dendrite-side view, Right: axon-side view.)

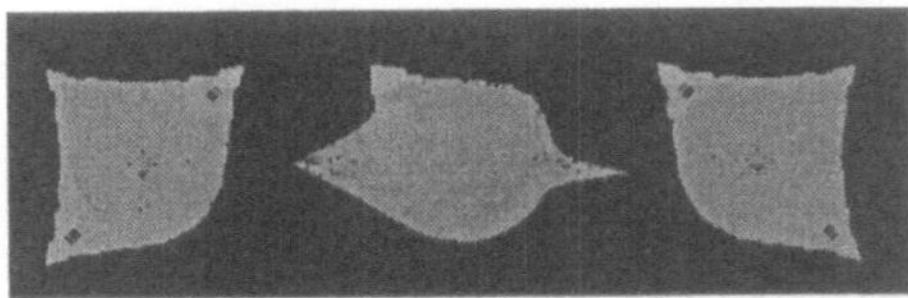

Fig. 5. t=0.5 [ms]

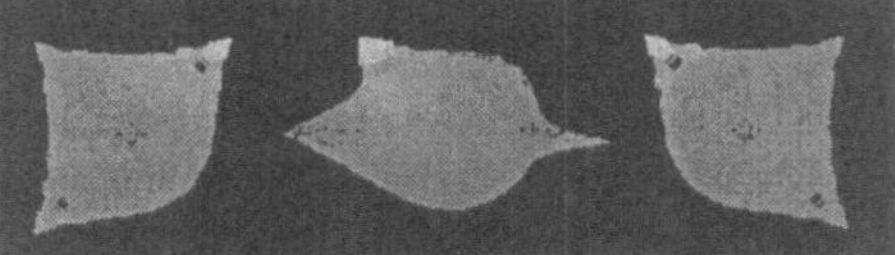

Fig. 6. t=1.0 [ms]

Fig. 7. t=1.5 [ms]

Fig. 8. t=2.0 [ms]

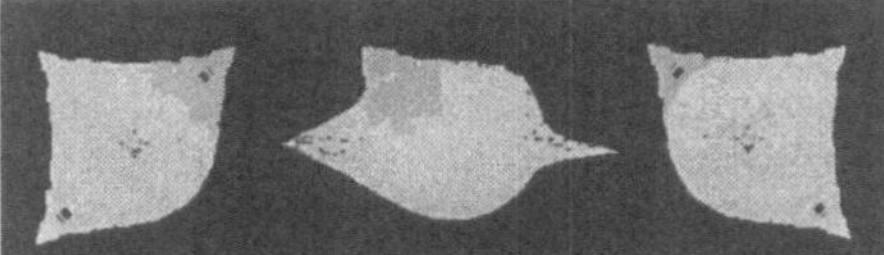

Fig. 9. t=2.5 [ms]

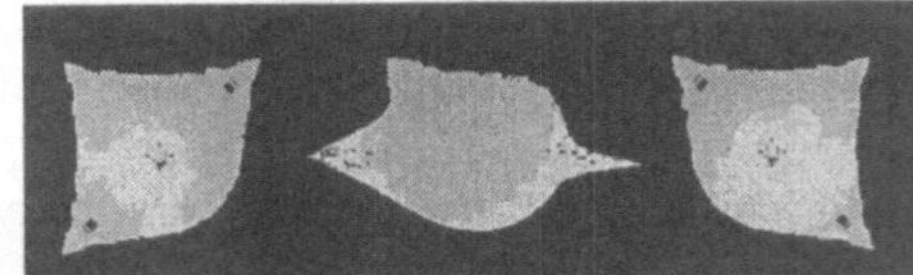

Fig. 10. t=3.0 [ms]

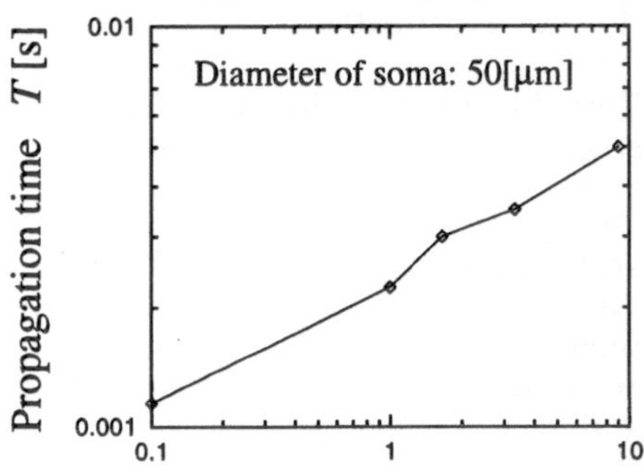

Fig. 11. Propagation time T over the cell versus intracellular resistivity R_i.

Concurrent Parallel-Sequential Processing in Gamma Controlled Cortical-Type Networks of Spiking Neurones

Edgar Koerner and Ursula Koerner

HONDA R&D Europe (Germany) Ltd., Future Technology Division, Offenbach/Main, Germany.
email: Edgar.Koerner@hre-ftr.f.rd.honda.co.jp

Abstract. In sensory recognition, rapid activation of an initial hypothesis based on few but reliably detected features, and the discrimination of the object by refined analysis of the sensory input are two conflicting requirements. We show that these two aspects of sensory recognition can be optimally served by a neocortical firmware model, the stereotypically structured columnar architecture, and that the resulting ensemble temporal encoding performed by such a modular unit enables both rapid and robust recognition by utilisation of the dimension of time to encode the reliability of a decision. The model suggests a novel functional interpretation of gamma oscillations in cortical processing.

1. How to trigger and refine a hypothesis in neural systems?

The marvelous performance of vision must be based on an active search for interpretation guided by top-down feedback that integrates the details of sensory analysis into a globally consistent hypothesis. However, bi-directional processing by itself does not resolve the problem of combinatorial explosion. For making a proper prediction, a hypothesis on the sensory input is both a prerequisite and result of processing. The activation of an initial hypothesis in forward processing for a subsequent bi-directional refinement must be made as rapid as possible, and, with a high probability, it has to be correct. First, activation of an initial hypothesis should be based on evaluation of afferent stream by ensemble coincidence detection since the required time interval for observation of the input for either average spike rates or spike sequences would by far exceed the experimentally observed ones: Face selective cells at inferotemporal cortex (IT) respond with a latency of about 100 ms, which is barely several milliseconds more than the time required for pure signal transmission from retina to IT without any additional delay [7]. The principal decision on the input is seemingly done with arrival of the very first spikes at any hierarchic level by an ensemble coincidence detection [8], as it is expected from the short integration interval of cortical neurones. Second, for a both rapid and correct initial hypothesis, the barrage of afferent signals must be reduced to a sparse set of the most reliably signaled features only. Thorpe suggested that cortical neurons emit a spike in response to an afferent input with a latency that is the shorter the better the input matches the tuning characteristics of the neuron. Then, rapidly transmitted best matches trigger the recall, while partial matches are delayed or even suppressed in a multilevel hierarchy of forward coincidence [1].

But there are two serious flaws in this scheme: 1) Encoding the degree of matching into relative latency of spiking can be evaluated at the next higher level only then if all of the local decision processes (not the spikes!) at the lower level are in synchrony to each other. 2) The coding capacity of that scheme is rather low, and the analog information provided by the partial matches cannot be used effectively.

A model for encoding of analog information of partial matches into relative time advance of spiking that provides the necessary coding capacity has been proposed by Hopfield [3], but it bears the cost of fixed latency for decision making at any level of processing. Furthermore, this model lacks a mechanism to enforce the unitary time relation between local decision processes over the entire system which is actually the necessary precondition for its proper function.

In this paper, we propose a solution to the problem in combining a code that supports fast processing but has low capacity, with a code that has an extremely high encoding capacity but requires comparatively long time for decoding at any processing level. As we will show, neocortical columns could provide the firmware control architecture to embody the proposed concurrent processing utilising the two qualitatively different codes.

2. Encoding patterns and pattern sequences at any columnar module -- Computational hypothesis

2.1. A cortical-type neuron model

The model neurons evaluate the ensemble of spikes at their input in a short integration interval (coincidence detector characteristic). Depending on basic columnar neuron types, the share of afferent, efferent and lateral contribution to the input is different. The threshold is set high to enable the release of a spike only if a sufficient number of inputs provide a spike during the integration interval that constitute the feature the neuron is tuned to. In case of best match, the spike is released with minimum latency. This latency increases with lower but still sufficient degree of matching. Non-specific depolarizing modulation of the membrane potential (MP) of all neurons of the global system enables the signaling of partial matches.

2.2. Columnar modules as elementary processing nodes in neocortical-type neural networks

The neocortex seems to be composed of complex nodes, columnar units, the basic internal organisation of which is rather similar regardless of the information represented there. We propose that this unitary architecture of columnar units represents not the structure of the knowledge stored there, but the control that forces the system to make that representations [4]. Columns are clustered to functional units, the macrocolumns. In the ventral visual pathway, a macrocolumn at the lower, still topographic representation levels, assembles all features the system has acquired to describe the respective local spot at the retinal input (like the set of oriented edges in V1), and at the higher non-topographic levels of processing a macrocolumn seems to represent a class of features and its variations [10]. We formalise any columnar unit into 3 functional different submodules (Fig. 1): Submodules A define the prototypes encoded in a macrocolumn and perform a parallel categorisation, submodules B decode a sequence of activated categories, and submodules C represent patterns of selected activation trajectories at all submodules B of a macrocolumn.

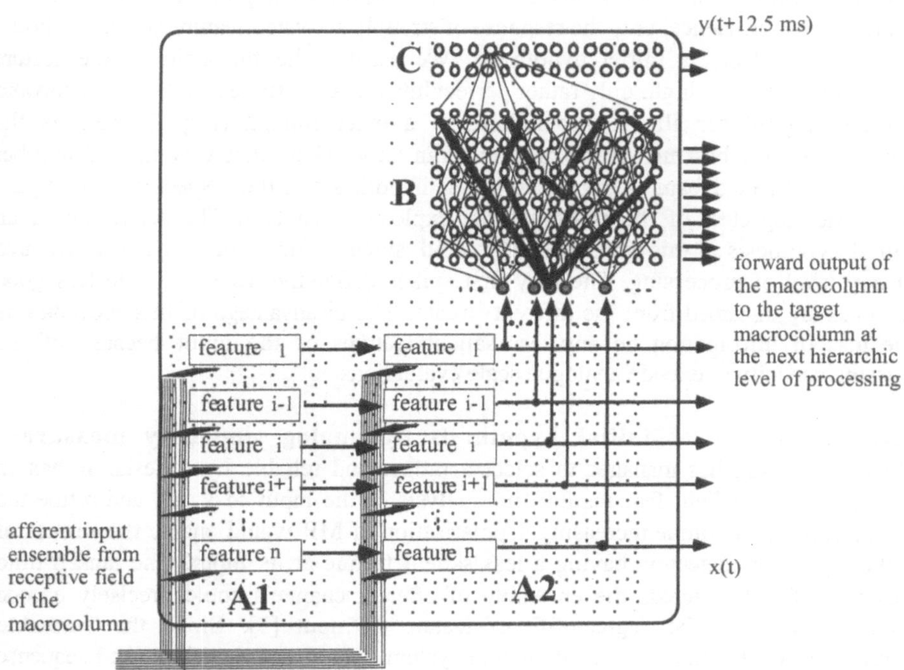

Fig. 1: Model macrocolumn composed of n minicolumnar units.
Submodule A1 generates the initial hypothesis that is transmitted by A2, and during MP modulation, A2 generates a sequence of partial matches. Submodule B consist of a population of variable tapped delay lines that re-code any spike sequence into a localised object representation at submodules C. A1 is related to the spiny stellate of layer 4 (SS4). SS4 make multisynaptic contacts to selected pyramids of layer 3 (Py3), which we interpret as a strong driving input that elicits a spike at a Py3 if the respective SS4 fires. SS4 make contacts within the local processing module only. Py3 forward its output both to the next hierarchic level of processing and (via its ascending axon collateral's) to the supragranular pyramids of the same macrocolumn. Afferent inputs contact not only SS4 (A1), but to a weaker degree also lower Py3 (A2). We relate B to the upper neocortical layer 3 because the system of ascending axon collateral's of supragranular pyramids that make en-passant-type connections to other supragranular pyramids within the macrocolumn could serve as such tapped delay lines. Neurones in C (which we relate to layer 2) represent those traces as sparse and very specific parallel object descriptions which are signalled with a fixed delay after the start of MP modulation.

2.3. Rapid activation of initial hypothesis

Neuronal feature detectors give a graded response to many input configurations. For the next level decision making unit, only a perfect matching of the input to the feature the respective neuron is tuned to delivers by itself a definite decision on the sensory input. The lower the analog activation of a neuronal feature detector, the more the brain needs an already established context to properly interpret the decision of that feature detector. Hence, only the response of strongly activated feature detectors should be used to select the initial hypothesis. We assume the threshold of the feature detectors in A is set high: only rather perfect hits release a spike at all. A fast forward inhibition pool for all A1 submodules of a macrocolumn (Fig. 1) permits the signaling of one best match per macrocolumn only. Then, that very limited number of reliably detected features trigger an initial hypothesis at the respective next higher level, steering clear of the combinatorial explosion problem. The activation of an initial hypothesis in this system is very rapid, since no time for decoding is required at any level of processing. The very first spikes define the decision while less good matches are discarded from the afferent stream. The disadvantage of this procedure is the loss of information on any less salient features of the input because of the missing capacity to encode analogue scale evaluations.

2.4. Refinement of initial hypothesis by analog similarity measures

Once the system has managed to select a proper and reliable hypothesis, it has to include the less salient features into its analysis of the input to verify and refine the hypothesis. A monotone increasing depolarisation of MP would enable the release of spikes by feature detectors having a less salient feature at its input. The ranked time sequence of spikes across the ensemble of afferent channels could precisely encode their difference of the degree-of-fit of feature to input [3]. Since the ensemble coincidence detection in our A submodule system cannot encode this ranked sequence of partial matches, an additional module has to be provided at any local processing node. We propose that the sequence of partial matches generated at any macrocolumn in submodules A is decoded by submodules B and C (Fig. 1).

2.5. Clocking of decision intervals in cortical processing

To make such a dual encoding scheme work in a large-scale neural system, three necessary conditions have to be guaranteed: 1) To rectify local decisions according to a globally consistent hypothesis and to follow rapid fluctuations of the input, the decision process has to be repeated several times with a sufficient high frequency. 2) To enable the evaluation of the sequence of partial matches, a time interval must be prepared in-between any recurrence of the decision process since decoding of spike sequences bears the cost of a fixed delay at any stage of processing. 3) To encode analogue values into relative latency in a large-scale neural system, the modulation must be synchronous over all the system. Moreover, this definition of a globally synchronous time interval for refined analysis must be in a definite relation to the initial hypothesis which has to be refined.

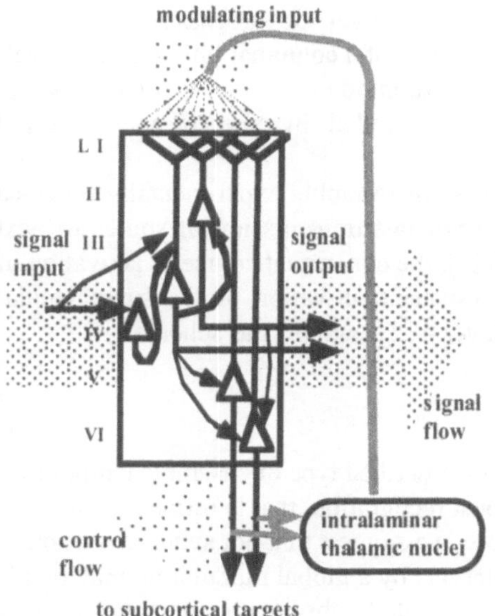

Fig. 2: Global synchronisation of decision intervals by gamma modulation

We propose that the initial wave of most reliably detected features in the A-submodule system triggers gamma activation of the thalamic intralaminar nuclei (ILN) and that the ILN modulate the MP of the cortical pyramids to enforce a synchronous decision interval for analogue encoding onto the cortex (Fig. 2). The ILN receive inputs from infragranular pyramids from all over the cortex, and return a non-specific modulating feedback to apical dendrites of pyramids in layer 1 [5]. All principal neurones of cortical columns extent apical dendrites to layer 1 (and may undergo the here proposed modulation by ILN), except those that are the major recipients of afferent input (like spiny stellate of layer 4, SS4).

3. Ensemble temporal encoding by columnar modules and a novel interpretation of gamma oscillations

We hypothesise that feature detector neurones of layer 4 in a macrocolumn perform a competitive recall of a best matching feature, the winners of which activate a lower Py3 that rapidly transmits the "express" feature to the next hierarchic level of processing. Fast forward activation of inhibitory neurones in layer 4 prevents any further activation of SS4, and is assumed to cause there a gamma period hyperpolarization [11]. The highly coherent wave of spikes signalling "express" features is assumed to trigger additionally via ILN a gamma period depolarisation of the MP of principal neurones of neocortical columns except SS4. Without modulation, lower Py3 can fire only by driving input from SS4, since the share of afferent inputs is too low to trigger a spike by itself. Because of the weak inputs by en-passant contacts of ascending axon collaterals of Py3, submodules B and C cannot be activated without MP modulation. Only the coherent wave of "express" features propagates upwards the hierarchy of the submodule A system. After this initial hypothesis at any local processing node of each level of processing, the A submodules are blocked for the remainder of the gamma period, while with a certain delay caused by the conduction time from cortex to ILN and back, the ranked sequence of partially matching features are released by the lower Py3 (A2) within the 12.5 ms of raising depolarisation of MP. The global modulation of MP enables the analysis and re-coding of this ranked sequence in a refinement process at all local processing units as proposed above. Since supragranular pyramids send their axons to the next hierarchic level and to more distant targets of the same level of processing, and since infragranular pyramids feedback the processing state of a columnar unit to lower levels of processing, the refinement process in B and C submodule system is a recurrence of an indeed parallel optimisation guided and controlled by the preceding initial hypothesis in the A submodule system.

Resulting activity patterns in the proposed system resemble experimentally observed ensemble temporal coding [9]. The time course of fast transmission of spikes to next hierarchic level by lower pyramids of layer 3 [1], the concurrent course of activation of neighboured hierarchic levels [6], and the subsequent activation of the supragranular pyramids in an interval of 12...15 ms after layer 4 [2], support our scheme.

4. Conclusion

We claim that the proposed architecture and the described type of ensemble temporal encoding enables a novel type of fast but robust recognition, that is refined in an analysis-by-synthesis manner [4]. Furthermore, we suggest that not synchronisation of spikes, but synchronisation of decision intervals by a global modulation may be a biologically plausible explanation for the experimentally observed gamma oscillations.

References

1. S. Celebrini, S. Thorpe, Y. Trotter & M. Imbert. Dynamics of orientation coding in area V1 of the awake primate. Visual Neuroscience 10 (1993): 811-825
2. R.J. Douglas & K.A.C. Martin. A functional microcircuit for cat visual cortex. Journal of Physiology 440 (1991): 735-769
3. J.J. Hopfield. Pattern recognition computation using action potential timing for stimulus representation. Nature 376 (1995): 33-36
4. E. Koerner, H. Tsujino & T. Masutani. A cortical-type modular neural network for hypothetical reasoning. Neural Networks (in press)
5. R. Llinas & U. Ribary. Coherent 40-Hz oscillation characterizes dream state in humans. Proc.Natl.Acad.Sci.USA, 90 (March 1993): 2078-2081
6. L.G. Nowak, M.H.J. Munk, P. Girard & J. Bullier. Visual latencies in areas V1 and V2 of the macaque monkey. Visual Neuroscience 12 (1995): 371-384
7. M.W. Oram & D.I. Perret. Time course of neural responses discriminating different views of the face and the head. Journal of Neurophysiology 68 (1992): 70-84
8. S. Thorpe, D. Fize & C. Marlot. Speed of processing in the human visual system. Nature 381 (1996): 520-522
9. E. Vaadia, I. Haalman, M. Abeles, H. Bergman,Y. Prut, H. Slovin& A.Aertsen. Dynamics of neuronal interactions in monkey cortex in relation to behavioral events. Nature 373 (1995): 515-518
10. G. Wang, K. Tanaka & M. Tanifuji. Optical imaging of functional organization in the monkey temporal cortex. Science 272 (1996): 1665-1668
11. M.A. Whittington, R. D. Traub & J.G.R. Jeffreys. Synchronized oscillations in interneuron networks. Nature 373 (1995): 612-615

A Noise-Robust Auditory Modelling Front End for Voiced Speech

Leslie S. Smith

Department of Computing Science and Mathematics, University of Stirling, Stirling FK9 4LA, Scotland

Abstract. A method for detecting and displaying voiced elements of speech using amplitude modulated pulses due to unresolved harmonics of the excitation frequency (fundamental) is presented. It uses an auditory model consisting of a gammatone filterbank (modelling the basilar membrane), simple rectification (modelling the organ of Corti inner hair cells), envelope bandpass filters (modelling some spiral ganglion neuron effects) and amplitude modulation detectors (modelling certain cell populations in the cochlear nucleus). We demonstrate that it can display a pattern of activity across the spectrum and across time that describes the energy distribution in voiced speech, and that this pattern degrades slowly in the presence of non-speech noise.

1 Aims

This work aims to contribute to the task of streaming a foreground speech signal from a background non-speech signal. This is one aspect of the streaming problem [2], and one which humans currently perform far better than machines. The particular task addressed is producing a representation of voiced speech which is robust in the face of non-speech noise by taking advantage of the amplitude modulation caused by unresolved harmonics. Such a representation can show where in the upper part of the frequency spectrum the voiced speech energy is concentrated, while maintaining the fine time-structure of the signal.

2 Background

We use techniques based loosely on early mammalian auditory processing. Quite apart from the fact that the task is one which humans perform easily, we believe that streaming problems are not well served by techniques which interpret the coded audio signal by immediately integrating features across the spectrum. We believe (following Allen [1]) that human speech recognition uses an *across-time* processing scheme before using an *across-frequency* one. In this sense, auditory processing is unlike visual processing where lateral interaction of sensor outputs occurs on the retina.

But when should across-frequency integration take place? In the mammalian auditory system, detection of sound is performed at the inner hair cells of the Organ of Corti, and these signals are kept separate until the cochlear nucleus.

Even there, the tonotopic organisation of the auditory nerve is maintained, so that it appears that the feature detection which takes place in this nucleus uses information only from some range of auditory nerve fibres. The feature detecting cells of this nucleus (such as the onset cells, or the chopper cells [4]) process some spectral range of auditory signals across time. By using local temporal and local frequency clustering features from the same source can be grouped. Only once this level of feature detection has taken place are features integrated across the whole spectrum. In this way, accidental near co-occurrences in different parts of the spectrum are less likely to become grouped into spurious features. We have already applied this to onset features [6]. The techniques we have used are most similar to those used to produce the stabilised auditory image [8]: however, we have maintained neurobiological plausibility by relating the processing to cells in the cochlear nucleus.

In this paper, we use local temporal and local frequency clustering. The particular features clustered are amplitude modulation (AM) pulses, caused by unresolved harmonics of the fundamental frequency (F_0). These are detected in channels of sufficiently high centre frequency (f_c) to have a bandwidth containing more than one harmonic of f_c using across-time processing. They are grouped using across-frequency processing to retain those believed to originate from the voiced speech source.

3 Technique Used

The processing stages are outlined in figure 1. The Gammatone cochlear filterbank [8] provides the initial bandpass filtering. This filterbank has f_c distributed approximately logarithmically. It has a fixed Q for all the filters. Channel Q should be about 9.265 for single quiet tones [3], but falls for wideband sounds (such as speech) at medium sound pressure levels (SPL). We have generally used a lower value for Q as this causes unresolved harmonics to start at a lower f_c.

The overall aim of the AM detection and AM pulse clean-up stages is to take advantage of the nature of AM in voiced speech. Most animal noises share the same characteristics, being produced by selective filtering of a low frequency excitation which is rich in harmonics. The AM pulse detection stage performs across-time processing on each channel independently. This stage produces a spike output for each ampitude modulated pulse. This is achieved by rectifying the signal (corresponding to the rectification performed by the inner hair cells of the Organ of Corti). This is followed by an envelope bandpass filter which applies a difference of Gaussians transform. The output of this stage is bipolar, being positive at the start (onset) of a pulse, and negative at the end (offset) of a pulse. These onset and offset elements are separated out and individually compressed, resulting in two non-overlapping positive-going outputs, one of which rises at the start of a pulse, and the other at the end of a pulse. The pulse sequence detector emits a spike when it detects a rise in the compressed onset signal followed after a suitable interval by a rise in the compressed offset signal: it selects pulses within the correct range of durations. More details are in [5].

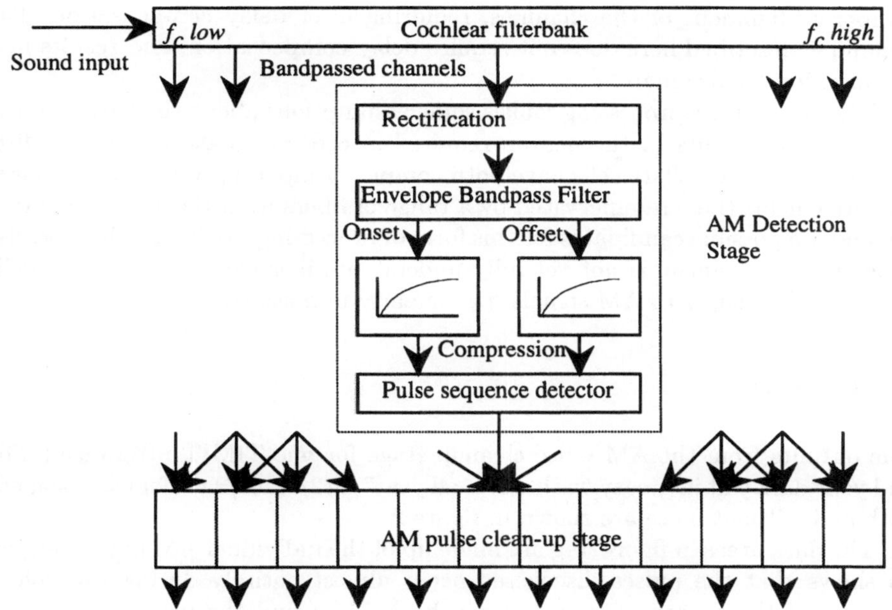

Fig. 1. Stages in the processing of the sound signal. For clarity, only one AM detection stage is shown. See text for details.

The envelope bandpass and compression sections could be implemented neurally by applying the rectified signal to two different synapses on a dendrite. If one synapse was fast-acting and excitatory, and the other slower acting and inhibitory this would implement an approximation to the difference of Gaussians on the dendrite resulting in the onset signal. The compression occurs because of the nonlinear nature of the membrane depolarisation and neuron firing. Transposing the speed and nature of the synapses would give the compressed offset signal.

The AM pulse clean-up stage determines whether to discard pulses by performing both across-time and across-frequency processing. The across-time processing checks whether the previous AM pulse in this channel occurred within a time-frame corresponding to a relevant AM frequency. This is followed by the across frequency processing, discarding the pulse if there were insufficient pulses in adjacent channels within some (short) time period. Though currently implemented algorithmically, this could be implemented neurally using lateral inhibition. In each channel, neither the total excitatory input from adjacent channels nor the input to the neuron from its own channel would be sufficient to make the neuron fire. However, their sum would be sufficient to cause the neuron to fire. The technique used is an improvement on the simple summation technique used in [5] in which the across-time processing consisted of summing the compressed onset and offset signals across all channels. This needs precise temporal alignment of the channels, requiring filter delay compensation. The technique described here does not require delay compensation, and results in a 2-dimensional pulse map.

The across-time processing makes some assumptions about the nature of the timing of the inputs to the neuron required in order to make the neuron fire. Cochlear nucleus stellate cells have both complex temporal properties, and large receptive fields (i.e. are innervated by a range of fibers from the auditory nerve) [9], and are possible candidates for this form of processing. Although their precise processing mechanism is not yet fully understood, it is clear is that these cells do respond strongly to AM signals, particularly in noise [7].

4 Results

The outputs from the AM pulse cleanup stage for one TIMIT utterance ("She had your dark suit in greasy wash water all year", dr2/feac0/sa1 (female speaker)) with and without noise are shown in figure 2.

The dark areas in figure2b-g are made up of the individual AM pulses. Figure 2b shows that the pulses discovered occur almost entirely during the voiced portions of the speech. The pulses can be used to find the voiced sections of the speech, and to estimate F_0, using only the upper part of the spectrum of the speech. This is described in [5], using the different across-time processing described earlier. We have done some experiments using a simple bucket-based summation technique on the AM pulse cleanup output. Space does not permit detailed discussion of the results. In the absence of noise, the two techniques have similar performance: in white noise, the earlier technique misses more of the voiced sections, while the new technique tends to misinterpret more of the unvoiced sections. Work on an improving the decision rule for discovering voicing from the AM pulse cleanup output is continuing.

The dark areas in figure2b-g display a characteristic shape, corresponding to the intensity of the harmonics in that region of the spectrum at that time. Though some of this shape can be seen, (e.g. the /iy/ pulses occur at the high end of the spectrum, the /eh/ pulses are broadband, the /aa/ pulses occur both at the high and low end of the spectrum) the low Q used in figure2b-f hides the detail of the shape. Using a higher Q (figure 2g), only the highest part of the spectrum is available, but the shape becomes more visible, and some internal structure can be seen (lighter and darker horizontal bands).

The effect of non-speech noise is to hide some of the spectrum. In figure 2c (5dB SNR, white noise), some of the shape is masked, but it remains visible. In figure 2d (0dB SNR, white noise), the shape is harder to see, though the voiced elements remain visible. When the added signal is a simple tone, the degradation is less pronounced. With a 400Hz tone (figure 2d), little degradation occurs, and degradation is concentrated at the 400Hz end of the spectrum. With a 900Hz tone, the low Q (figure 2f) results in the area of the spectrum near 900Hz being degraded (but not lost entirely): using higher Q, the structure at the high end of the spectrum is undamaged.

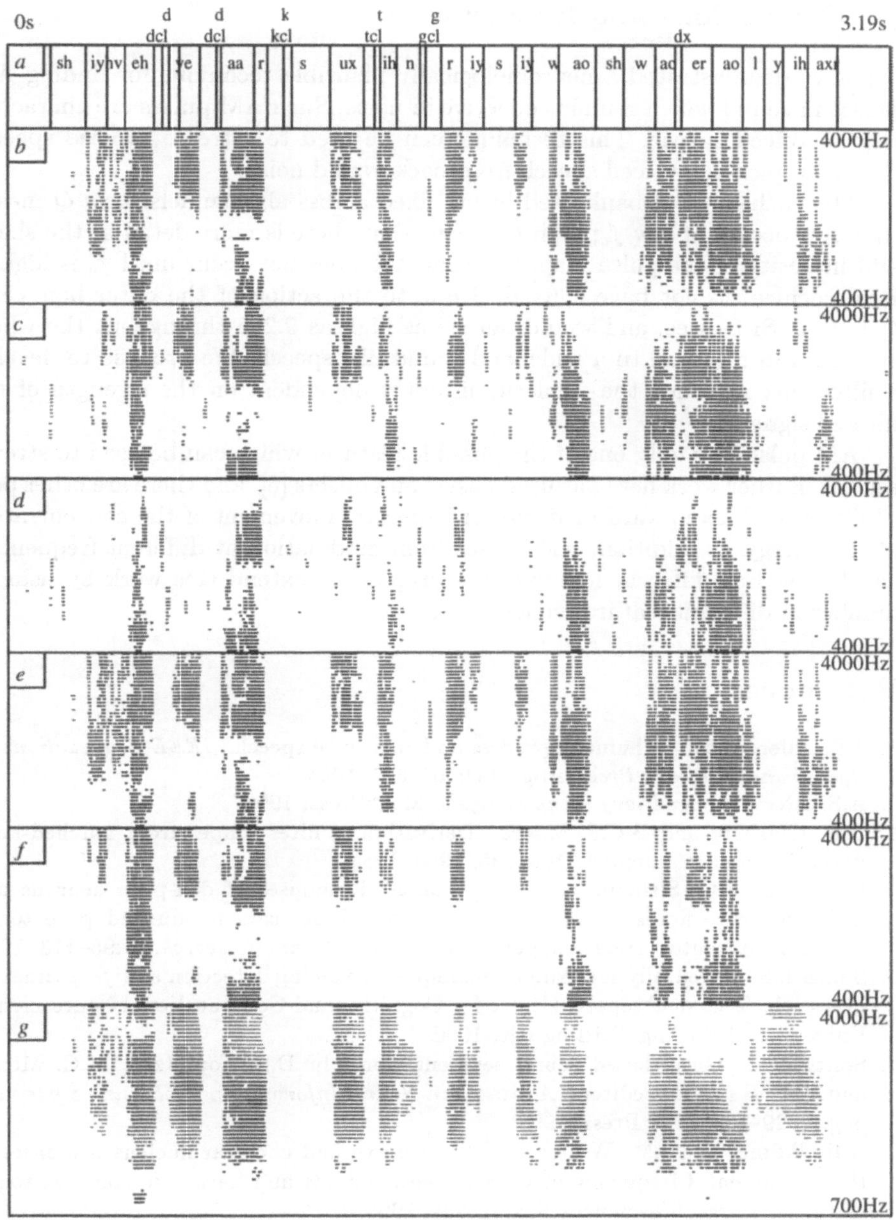

Fig. 2. *a*: Phonetic segments for the utterance. *b*: Cleaned-up AM pulse map. Cochlear filterbank has 50 channels, 400-4000Hz, $Q = 2.3$, AM frequency range 150-250Hz. *c*: as *a*, but original signal has white noise added to give 5dB mean SNR. *d*: as *c*, but 0dB mean SNR. *e*: as *a*, but original signal has 400Hz tone added to give 10dB peak signal: peak noise ratio, approximately 1dB mean SNR. *f*: as *e* but 900Hz tone added. *g*: as *f*, but with $Q = 9.265$, and 700-4000Hz range, 56 channels.

5 Conclusions and further work

We have demonstrated a neurobiologically plausible technique for finding AM pulses in sound which remains effective in noise. Such AM pulses are characteristic of voiced speech. This technique can be used to "stream" voiced speech, that is to pick out voiced speech from background noise.

The cochlear filterbank used has Q fixed across all channels. Low Q means that AM occurs at low f_c; high Q means that there is more detail in the shape and intensity of the pulse map, but that AM does not occur until f_c is higher. Real cochlea do not have a fixed Q due to the action of the outer hair cells: Q falls as SPL rises, and would not be as high as 9.265 throughout the whole spectrum in response to a wideband sound like speech. We propose to develop a filterbank in which the Q of the filters is dependent on the strength of the filtered signal.

AM pulses are only one of the possible features which can be used to stream sounds. Earlier work here has used onsets and offsets [6], and there are other possibilities, such as upward or downward spectral movement of the concentration of the energy. In addition, AM pulses from modulation at different frequencies can be used as different features. We propose to extend this work by using a number of different features concurrently.

References

1. J.B. Allen. How do humans process and recognize speech. *IEEE Transactions on Speech and Auditory Processing*, 2(4):567–577, 1994.
2. A.S. Bregman. *Auditory scene analysis.* MIT Press, 1990.
3. B.R. Glasberg and B.C.J. Moore. Derivation of filter shapes from notched-noise data. *Hearing Research*, 47:103–138, 1990.
4. D.O. Kim, J.G. Sirianni, and S.O. Chang. Responses of dcn-pvcn neurons and auditory nerve fibres in unanesthetized decerebrate cats to am and pure tones: analysis with autocorrelation/power-spectrum. *Hearing Research*, 45:95–113, 1990.
5. Smith L.S. A neurally motivated technique for voicing detection and f_0 estimation in speech. Technical report, Centre for Cognitive and Computational Neuroscience, University of Stirling, Stirling UK, 1996.
6. Smith L.S. Onset-based sound segmentation. In D.S. Touretzky, M.C. Mozer, and M.E. Hasselmo, editors, *Advances in Neural Information Processing Systems 8*, pages 729–735. MIT Press, 1996.
7. A.R. Palmer and I.M. Winter. Cochlear nerve and cochlear nucleus responses to the fundamental frequency of voiced speech sounds and harmonic complex tones. *Advances in the Biosciences*, 83:231–239, 1992.
8. R.D. Patterson, M.H. Allerhand, and C. Giguere. Time-domain modelling of peripheral auditory processing: A modular architecture and a software platform. *Journal of the Acoustical Society of America*, 98:1890–1894, 1995.
9. I.M. Winter and A.R. Palmer. Level dependence of cochlear nucleus onset unit responses and facilitation by second tones or broadband noise. *Journal of Neuroscience*, 73(1):141–159, 1995.

A Novelty Detector Using a Network of Integrate and Fire Neurons

Tuong Vinh HO[1,2] and Jean ROUAT[2]

[1] Ecole Polytechnique de Montréal, Canada, Dept. de Génie Informatique
[2] Université du Québec à Chicoutimi, Canada, Dept. Des Sciences Appliquées
email: vho@uqac.uquebec.ca Jean_Rouat@uqac.uquebec.ca

Abstract. Information in the nervous system has often been considered as being represented by simultaneous discharges of a large set of neurons. We propose a learning mechanism for neural information processing in a simulated cortex model. Also, a new paradigm for pattern recognition by oscillatory neural networks is proposed. The relaxation time of the oscillatory networks is used as a criterion for novelty detection.

1 Introduction

Representation of information in the nervous system has often been considered as being contained in simultaneous discharges of a large set of neurons. How does a neural system use that kind of information representation while performing learning and pattern recognition ? Recent studies on nonlinear cooperative complex dynamics in neural systems provide various kinds of models that describe the cooperative behavior such as synchronization and chaos. In [6], Thiran and Hasler present a valuable overview on this approach. Hayashi [2] presents an interesting characteristic of an oscillatory network: a limit cycle near a memory pattern (memory retrieval with ambiguous fluctuation) for an input closed to it, and a chaotic orbit wandering among memory patterns (autonomous search) for an input far from them. It is not easy to identify dynamical behavior. Stassinopoulos and Bak [5] propose a self-organizing model with a capability to interact with the surrounding environment. Self-organizing behavior arises by interaction between non-fixed threshold neurons with feedback from environment. Although the model displays a rich dynamical behavior, it is still not clear how to associate patterns to the network's states. Dayhoff [1] proposes a learning mechanism that allows a Hopfield network having a rich dynamic behavior including fixed point, limit cycle and chaotic attractors. She also shows that the network can have many attractors and it overcomes the limitation of the original Hopfield model. However, we still do not know how to associate patterns to these behaviors, in other words, how to apply this network model to recognition problems. In fact, we need a method to manipulate the chaotic behaviors.

Hill and Villa [3] developed neural models to study the spatiotemporal pattern generation properties in a simulated "cortical neural network". The model uses integrate-and-fire neurons as elementary units. Furthermore, the topology is inspired from that of layer IV in the cortex. Although this model helps to observe the evolution of spatiotemporally organized activity in a simulated cortex, the learning rule is not yet proposed. In this paper, we propose a learning mechanism for neural information processing in the simulated cortex model. Along with the learning mechanism, we propose a new paradigm for pattern recognition by oscillatory neural networks. The relaxation time of oscillatory behavior is used as a criterion for novelty detection.

2 Neuronal Model

Our neuron model is inspired from the integrate-and-fire neuronal model proposed by Hill and Villa [3] with refractory period and post-synaptic potential decay. The state of the neuron at time t, is deterministically modeled by a control potential, U as:

$$S_i(t) = \begin{cases} 0 & \text{if } (t - t_{spike}) < \rho, \\ \mathcal{H}[U_i(t) - \theta] & \text{otherwise,} \end{cases} \tag{1}$$

where $\mathcal{H}$ is the Heaviside function defined as $\mathcal{H}[x] = 1$ for $x > 0$ otherwise $\mathcal{H}[x] = 0$. The value t_{spike} represents the last firing time for unit i. The value ρ denotes the absolute refractory period. Refractoriness corresponds to the period following the production of a spike or action potential, during which the cellular biochemical mechanisms cannot generate another signal, regardless of the strength of the stimulation. The control potential is defined as the integration of all afferent postsynaptic potential at time t:

$$U_i(t+1) = \sum_i C_{ij} S_j(t) + U_i(t) + s_i \tag{2}$$

where the indices i and j indicate the units, C_{ij} is the connection strength, and s_i is the input signal.

In order to introduce the influence of the firing frequency into the neuron's behavior, we add a variable firing frequency factor f to the neuron model. Simulating experiments showed that this factor has a strong influence in the neuron's behavior. Thus, equation (2) becomes

$$U_i(t+1) = \sum_i C_{ij} S_j(t) + U_i(t) + s_i + f_i. \tag{3}$$

3 Network Architecture

The network architecture is inspired from an oversimplified model of cortical layer IV [3]. This model defines a single two-dimensional sheet of excitatory and inhibitory neurons with recurrent connections. The layer consists of two populations of neurons interspersed within the plane. These neurons are positioned according to a space-filling pseudo-random Sobol distribution. Each neuron has a set of interconnections chosen according to a square neighborhood, centered at the neuron itself and with a radius depending on whether the neuron is excitatory or inhibitory. From this topology, we can say that this model uses an interactivity at local level to create a self-organizing evolution. Here, we want to modify the model by introducing an interactivity at global level with a global inhibitor (Fig. 1). By this approach, we can create an interactivity between all neurons in the network. The global inhibitor is actually a trigger whose state is either active, i.e. firing, or inactive depending on a control mechanism. The control mechanism is based on a threshold for the total number of firing neurons. Whenever the number of firing neurons at time t is above the threshold, the global inhibitor fires and it generates a negative feedback signal to every neuron in the network. Otherwise, if the number of firing neurons is below the threshold, the global inhibitor generates a positive feedback signal. Thus, the global inhibitor plays a role of regulating neuron activity at global level. With a feedback signal h, equation (3) becomes:

$$U_i(t+1) = \sum_i C_{ij} S_j(t) + U_i(t) + s_i + f_i + h. \tag{4}$$

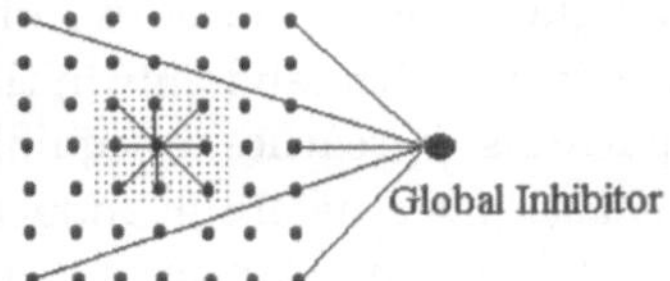

Fig. 1. The architecture of the neural network

4 Learning Rule

The learning rule, that modifies the coupling strengths, is widely used in many kinds of neural network models. In our model, there are two kinds of coupling: one is between neighboring neurons and the other is between the global inhibitor and all neurons. The Hebbian rule is used as updating rule for the coupling weights. The following equation is used for the coupling weights between a pair of neurons:

$$C_{ij}(t+1) = C_{ij}(t) + \alpha C_{ij}(1 - C_{ij}) S_i(t) S_j(t) \tag{5}$$

where α is the learning rate. The coupling weights between each neuron and the global inhibitor are updated according to:

$$G_i(t+1) = G_i(t) + \beta G_i(t)(1 - G_i(t)) S_i(t) h(t) \tag{6}$$

where G_i is the coupling weight of the neuron i, β is the update rate and h is the feedback signal from the global inhibitor.

Learning phase starts when the network is stimulated by an input signal. The network begins to oscillate. At each instant, the coupling weights of a neuron are updated if this neuron fires. It receives also a feedback signal. The later is either negative or positive depending on the number of firing neurons above or below the threshold of firing neurons. The network is considered to reach a stable state when its local coupling weights do not change anymore or when they change in a very small given range. When the network reaches a stable state, the learning phase terminates.

5 Novelty Detection by this Model

The proposed network model is a non linear dynamical system. How can the evolution of non linear dynamical systems be associated with the execution of cognitive tasks ? We need to find a paradigm that can be used to characterize dynamical evolution inside the system so that it can be applied to pattern recognition. We observed from simulating experiments that the dynamical network can reach a stable state very fast if the input signal has already been seen. From this observation, we propose a new paradigm for novelty detection by this network model. The paradigm is comprised of two phases:

+ *Learning phase*: the network with randomly initialized connection strengths is trained with training patterns. It reaches an equilibrium state after learning.

+ *Novelty detection phase*: patterns are introduced to the trained network. The network reaches an equilibrium state after a relatively small number of iterations if these patterns have been learned before. Otherwise, it takes a long time for the network to reach an equilibrium state. Based on the relaxation time, novelty detection can be done by our neural network model.

In the following, we present an example of this paradigm to novelty detection by our neural network model. A set of 0-9 digits is used as a pattern set in this paper. As seen in Fig. 2, each digit pattern is coded by using a 7x5 binary pixel matrix. In order to test the robustness of the network, a set of noisy patterns obtained from the original patterns is also used. The noisy patterns are created by adding a certain amount of noise to the original pattern images [Fig 2., right]. In other words, given an amount of noise (by percentage), a number of pixels in the patterns are changed. The pixels are randomly chosen with a uniform probability distribution. As in this experiment, with a 20% of noise, 7 pixels in each 7x5 pixel image have their value changed. Though the pattern images used herein are binary images, our network can manipulate analog images (i.e. real numbers can be manipulated by the network). For the experiments in this paper, we used a 7x7 dimension network with 70% population being excitatory neurons and the remaining 30% being inhibitory neurons. The neighborhood radius is 2 for excitatory neurons and 1 for inhibitory neurons. The pattern image is positioned at the center of the network plane.

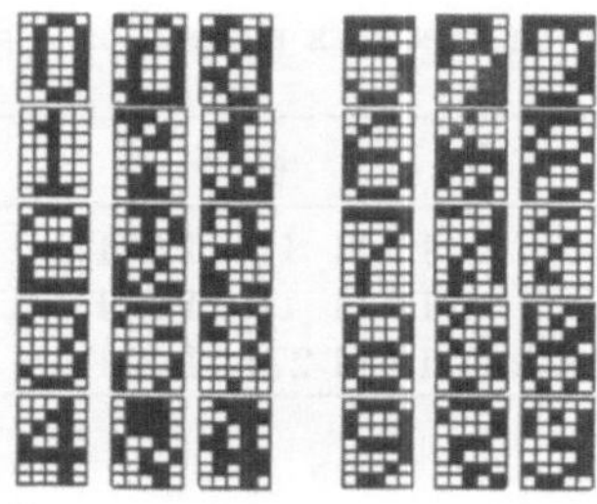

Fig. 2. Patterns of 0-9 digits with 20% noise

The digit patterns from 0 to 4 are used to train the network. Each pattern is presented sequentially to the network. The network oscillates and reaches an equilibrium state. When the network reaches a stable state, a new pattern is fed into it. In this experiment, the set of 5 patterns (0-4 digits) is presented to the network only one time. Oscillation times of patterns 0-4 are 351, 290, 321, 11 and 307 iterations respectively. Note that oscillation time of the network is often dependent on the sequence of training data. After learning phase, we use either the noisy versions of learned patterns or a set of "never seen" patterns (5-9 digits) to test the ability of novelty detection of the network. According to the proposed paradigm, we use the relaxation time (in term of number of iterations) of the network during testing phase as a criterion to decide whether a pattern is "seen" or "never seen" by the network. A short relaxation time means that the pattern has been seen before. Otherwise, the pattern has never been seen before. Table 1 shows that the network has a short relaxation time (11 iterations) when the testing patterns are either the learned patterns or the noisy patterns of the learned pattern. In contrast, the network has a significant long relaxation time (271 or 162 iterations) when the testing patterns have never been seen before. The network made recognition mistakes on 3 patterns (noisy version 2 of patterns 3, 7 and noisy version 1 of pattern 9), i.e. with an error rate of 10% (3/30).

In order to examine the network's performance, another test based on the initial training on digits [5-9] and testing on digits [0-9] was also done. During training, oscillation times of patterns 5-9 are 183, 132, 148, 11 and 11 iterations respectively. The testing result is given in Table 2. The network made more recognition mistakes than previous test with an error rate of 23% (7/30). On-going works are focused to improve recognition performance of the network. In addition, theoretical analysis is left as a future work.

6 Conclusion

A new paradigm for pattern recognition by non linear systems is proposed in this study. This paradigm is based on a criterion that is the time of oscillation of the network when a pattern is injected into it. In other words, the relaxation time is

Table 1. Relaxation time of the network trained on digits [0-4] and tested on [0-9]

Patterns	0	1	2	3	4	5	6	7	8	9
Original version	11	11	11	11	11	271	271	162	271	270
Noisy version 1	11	11	11	11	11	271	271	162	271	11
Noisy version 2	11	11	11	271	11	271	271	11	271	270

Table 2. Relaxation time of the network trained on digits [5-9] and tested on [0-9]

Patterns	0	1	2	3	4	5	6	7	8	9
Original version	11	152	153	11	143	11	11	11	11	11
Noisy version 1	153	152	153	11	152	153	11	11	11	153
Noisy version 2	11	153	153	11	143	11	11	11	11	11

used to decide whether a pattern has ever been seen before. A short relaxation time implies that the pattern has been already seen. Otherwise, a long relaxation time implies that the pattern has never been seen. This paradigm allows us to develop novelty detection systems based on the proposed network model with capability against noise as well as spatiotemporal transformation.

Acknowledgments

This work has been supported by the NSERC of Canada, by the "Fondation" from Université du Québec à Chicoutimi. We would like to thank Alessandro Villa and Sean Hill (Université de Lausanne, Swiss) for their cooperative discussions concerning this work.

References

1. Dayhoff J.E. et al.: "Developing Multiple Attractors in a Recurrent Neural Networks", *Proc. of WCCN'94*, San Diego, Jun. 1994, Vol. 4, pp. 710-715.
2. Yukio Hayashi: "Oscillatory Neural Networks and Learning of Continuously Transformed Patterns", *Neural Networks*, 1994, Vol. 7, No 2, pp. 219-231.
3. Hill S., Villa A.: "Global Spatiotemporal Activity Influenced by Local Kinetics in a Simulated "Cortical" Neural Network, *Workshop on Supercomputing in Brain Research: from topography to neural networks*, 1995, World Scientific, pp. 371-375.
4. Matsuno .Tet al.: "Periodic Signal Learning and Recognition in Coupled Oscillators", *Journal of Physical Society of Japan*, Vol. 63, No. 3, March, 1994, pp. 1194-1204.
5. Stassinopoulos D., Bak P.: "Self-organization in a Simple Brain Model", *Proc. of WCNN'94*, San Diego, Jun, 1994, Vol. 1, pp. 4-26.
6. Thiran P., Hasler M: "Information storage using stable and unstable oscillations: an overview", *Int. Journal of Circuit Theory and Applications*, Vol. 24, 57-67, 1996.

Derivation of Pool Dynamics from Microscopic Neuronal Models

J. Eggert and J.L. van Hemmen

Physik-Department, Technische Universität München,
D-85747 Garching bei München, Germany

Abstract. Starting from single, spiking neurons, we derive a system of differential equations for the description of the dynamics of pools of extensively many neurons. The derivation is exact and axonal delays and memory effects such as refractory behavior are taken into account. Simulations show a good quantitative agreement with microscopically modeled pools both in a quasistationary and in a non-stationary dynamical regime including fast transients and oscillations.

1 Pool Model

What kind of mathematical models should be chosen to study large, biologically realistic neural networks? A growing number of models describe neurons at the single cell level. Other models describe the joint activity of a group of functionally equivalent neurons. Between these two modeling levels, only in few cases a connection has been made, e.g. see [1,4]. To bridge this gap between the microscopic and the "assembly" or "pool" level, we start from a single cell model and derive a set of differential equations that describe the activity of a large group of equivalent neurons.

Contrary to previous work on pool dynamics, the derivation is exact for an extensive number of pool neurons (no time averaging is necessary). One important consequence of this fact is that the model is suited to describe both quasi-stationary and *fast, transient dynamics* such as oscillatory behavior.

First we introduce the pool model that will be used for the derivation of the collective dynamics. Let $A(t)$ be the total spike density of a pool of equivalent neurons, that is, $A(t)\Delta t$ is the total number of spikes released by all pool neurons in the time interval $[t - \Delta t, t]$. After the release of a spike at time t^*, each pool neuron undergoes a *refractory phase*, described by an *activation probability* $p_A(t - t^*)$ smaller than 1. All pool neurons feel a common input field $h(t)$ due to synaptic input from other neurons. Neurons that feel an input field $h(t)$ during an interval $[t, t+\Delta t]$ *and are activated* release a spike with a probability $p_F[h(t); \Delta t]$. The total spiking probability $p_S[h(t), t-t^*; \Delta t]$ during the interval is taken to be the spiking probability under the condition that it fired the last time at $t^* < t$,

$$p_S[h(t), t - t^*; \Delta t] = p_F[h(t); \Delta t]\, p_A(t - t^*) \ . \tag{1}$$

Additionally, we introduce a field-dependent time constant $\tau^{-1}(h)$ and define $p_F(h; \Delta t) := \tau^{-1}(h)\Delta t$ for small time intervals of length Δt.

In a pool of extensively many equivalent neurons, we can calculate the number of neurons $N_I(t)$ that are presently in an inactivated state. In a subgroup of pool neurons that all spiked the last time at t^* (i.e. that are all in the same refractory state), a fraction $1-p_A(t-t^*)$ is inactivated. The number of neurons that are in the same refractory state at time t is given by the number of neurons that fired at time t^* *and did not spike again* until t. In our notation this quantity reads $D_h(t,t^*)A(t^*)\Delta t^*$, with the "survival function" (e.g. [2])

$$D_h(t,t^*) = \exp\left\{-\int_{t^*}^{t} \mathrm{d}t'\, \tau^{-1}[h(t')]\, p_A(t'-t^*)\right\} . \tag{2}$$

Integrating the number of inactivated neurons that spiked at t^*, $1-p_A(t-t^*)\, D_h(t,t^*)A(t^*)$, over all possible refractory states of the pool neurons gives the total number of inactivated neurons $N_I(t)$,

$$N_I(t) = \int_0^\infty \mathrm{d}s\, \{1-p_A(s)\}\, D_h(t,t-s)A(t-s) \qquad \text{with } s = t-t^* . \tag{3}$$

The total number of pool neurons is N. The total number of activated neurons that can contribute to the activity is given by $N - N_I(t)$. For the next time step $t+\Delta t$, the total pool activity is then given by the firing probability of the activated neurons multiplied by the number of activated neurons, $A(t+\Delta t)\Delta t = p_F[h(t);\Delta t]\,\{N - N_I(t)\}$. Using $p_F(h;\Delta t) = \tau^{-1}(h)\Delta t$, the main equation for the pool spike density reads in the limit $\Delta t \to 0$:

$$A(t) = \tau^{-1}[h(t)] \left\{N - \int_0^\infty \mathrm{d}s\, [1-p_A(s)] D_h(t,t-s)A(t-s)\right\} . \tag{4}$$

Although this integral equation already provides an accurate description of the pool dynamics, the question arises whether we can gain a differential equation system that still approximates the dynamics reasonably well and is better suited to numerical simulations. In the next section we will derive such an equation system from the integral equation (4) for a special set of functions $p_A(s)$. The real behavior of neurons can then be approximated using these functions $p_A(s)$.

2 Macroscopic Pool Dynamics

The refractory function $p_A(s)$ with $s = t-t^* > 0$ describes the time course of the activation probability of a neuron. We divide the refractory function into two parts. After releasing a spike, the neuron is first in an absolute refractory state, with an activation probability zero. After a fixed time interval of length γ^{abs}, the neuron enters a relative refractory period, with an activation probability greater than zero. In the relative refractory state, i.e., for $s \geq \gamma^{abs}$, we assume that the activation probability $p_A(s)$ rises monotonically and continuously from some value $0 \leq c \leq 1$ to 1 according to a function $P_A(s)$. Between the two refractory periods, we allow a discontinuity in the function $p_A(s)$ at $s = \gamma^{abs}$,

$$p_A(s) = \begin{cases} 0 & \text{if } 0 \leq s < \gamma^{abs} \\ P_A(s) & \text{if } s \geq \gamma^{abs} \end{cases} . \tag{5}$$

In the derivation of the differential equations we restrict ourselves to a fixed choice of functions $P_A(s)$. We will treat the exponential case, the case of a sigmoidal relative refractory function, and the case of an inverse decay,

$$P_A^{exp}(s) = 1 - p_0 e^{-\frac{(s-\gamma^{abs})}{\tau_{ref}}}, \quad P_A^{sig}(s) = 1 - p_0 \frac{1}{1 + e^{\frac{(s-s_0)}{\tau_{ref}}}}, \quad P_A^{inv}(s) = 1 - \frac{\tau_{ref}}{s - s_0} . \tag{6}$$

The constants p_0, τ_{ref} and s_0 are free parameters of the refractory function that can be used for the approximation of a neuron's realistic refractory behaviour.

All neurons of a pool i are equivalent in the sense that they all have the same refractory function $p_A(s)$, the same $\tau^{-1}(h)$ and they all feel the same input field $h_i(t)$. We assume a linear summation of all the inputs that arrive at a neuron of the pool i from other pools j,

$$h_i(t) = \sum_{\text{pools } j} J_{ij} \int_0^\infty \mathrm{d}s\, \alpha_{ij}(s)\, A_j(t-s) \tag{7}$$

with a summation kernel $\alpha_{ij}(s)$. We choose an "alpha"-function-like time course for the kernel $\alpha_{ij}(s)$,

$$\alpha^k(s) = \Theta(s - \Delta^{ax}) \frac{s^k}{c_k} \exp\{-(s - \Delta^{ax})/\tau_\alpha\} \tag{8}$$

with $s = (t - t^*)$, rise time τ_α, normalization factor c_k, axonal delay Δ^{ax} between two pools and Heavyside Step function $\Theta(x)$.

We proceed with the main equation (4) and study the spike density dynamics. Since the main equation can be written in terms of the input field $h(t)$ and the number of inactivated neurons $N_I(t)$, it suffices to find the dynamics for these two variables.

First we consider the field dynamics. The input field of pool i is given by (7) and (8) with an alpha kernel $\alpha_{ij}(s) := \alpha_{ij}^k(s)$, $k \in \mathbf{N}_0^+$. We define additionally

$$h_i^l(t) := \sum_{\text{pools } j} J_{ij} \int_0^\infty \mathrm{d}s\, \alpha_{ij}^l(s)\, A_j(t-s) \qquad \text{for } 0 \le l \le k \tag{9}$$

so that $h_i(t) = h_i^k(t)$. Differentiating (9) yields the equation system

$$\frac{\mathrm{d}}{\mathrm{d}t} h_i^l(t) = l \frac{c_{l-1}}{c_l} h_i^{l-1}(t) - \frac{1}{\tau_\alpha} h_i^l(t) \qquad 1 \le l \le k \tag{10}$$

$$\frac{\mathrm{d}}{\mathrm{d}t} h_i^0(t) = \sum_{\text{pools } j} J_{ij} A_j(t - \Delta_{ij}^{ax}) - \frac{1}{\tau_\alpha} h_i^0(t) \tag{11}$$

which completes the dynamics of the spike densities $A_j(t)$ together with (4).

The dynamics of the number of inactivated neurons can be derived as follows. We start by defining quantities $N^0 := N$ (the number of pool neurons), $N^1(t) := N_I(t)$ (the number of inactivated neurons of the pool) and $N^\infty(t) = M(t) :=$

$\int_0^{\gamma^{\text{abs}}} \mathrm{d}s\, A(t-s)$ (the number of inactivated neurons for absolute refractory period only). We extend the definitions of N^0, $N^1(t)$ and $N^\infty(t)$ by the definition of additional time-dependent inactivation variables so that

$$N^m(t) := \int_0^\infty \mathrm{d}s \,\{1 - p_{\text{A}}(s)\}^m \, D_h(t, t-s) A(t-s) \qquad \text{for } m \in \mathbf{N}_0^+ \ . \tag{12}$$

The $N^m(t)$ obey the relationship

$$N^0 \geq N^1(t) \geq N^2(t) \geq \ldots \geq N^m(t) \geq \ldots \geq N^\infty(t), \qquad \forall t. \tag{13}$$

With these definitions, for each pool i we have a set of "recovery variables" $N_i^0, N_i^1(t), \ldots, N_i^\infty(t)$ that describe its actual state. Using (3), the properties of the recovery variables (12), the refractory functions (6), and the survival function (2), we get the following recurrent set of differential equations

$$\begin{aligned} \frac{\mathrm{d}}{\mathrm{d}t} N_i^m(t) &= A_i(t) - \{1 - [1 - p_{\text{A}}(\gamma^{\text{abs}})]^m\} A_i(t - \gamma^{\text{abs}}) \\ &\quad -\tau^{-1}[h(t)]\{N_i^m(t) - N_i^{m+1}(t)\} \\ &\quad - \begin{cases} \frac{m}{\tau_{\text{ref}}}[N_i^m(t) - M_i(t)] & \text{for exponential } p_{\text{A}}(s) \\ \frac{m}{\tau_{\text{ref}}}[N_i^m(t) - M_i(t) - \frac{N_i^{m+1}(t) - M_i(t)}{p_0}] & \text{for sigmoidal } p_{\text{A}}(s) \\ \frac{m}{\tau_{\text{ref}}}[N_i^{m+1}(t) - M_i(t)] & \text{for inverse } p_{\text{A}}(s) \end{cases} \end{aligned} \tag{14}$$

The last recovery variable $N_i^\infty = M_i(t)$ increases with the number of spiking neurons and decreases with the number of neurons that are released from their absolute refractory phase,

$$\frac{\mathrm{d}}{\mathrm{d}t} M_i(t) = A_i(t) - A_i(t - \gamma^{\text{abs}}) \ . \tag{15}$$

This completes our derivation of the dynamics. The pool dynamics is defined by the field dynamics given by equations (10) and (11) together with the dynamics of the recovery variables given by equations (14) and (15). The spike density acts only as an auxiliary variable that is calculated using the main equation

$$A_i(t) = \tau^{-1}[h_i^k(t)]\{N_i - N_i^1(t)\} \ . \tag{16}$$

Other pools j influence the dynamics equation system of pool i through equation (11). Delays appear in (11) and in the dynamics of the recovery variables variables due to the discontinuity of $p_{\text{A}}(s)$, but the functional differential equation system (10), (11), (14), and (15) can be approximated systematically by a system without delays using other alpha kernels and refractory functions.

Because of property (13) of the recovery variables, the approximation of the present procedure consists in breaking the infinite series of differential equations at the smallest desired recovery variable $N_i^{n+1}(t)$ and in introducing an appropriate dynamics for this quantity. There are two sensible ways of approximating $N_i^{n+1}(t)$. First, for large n, the influence of the relative refractory field on the $(n+1)$-th recovery variable is negligible, and we can set $N_i^{n+1}(t) = M_i(t)$. For fast, transient pool dynamics $M_i(t)$ is then calculated according to its dynamics (15). Second, for quasistationary dynamics, we can approximate $N_i^{n+1}(t)$ by its stationary value for the momentary input field $h_i(t)$.

3 Connection with Graded-Response Models

The exact stationary spike density for a constant input field can be calculated from the differential equation system or directly from equation (4). With the ansatz $\tau^{-1}[h] = \tau_0^{-1}\exp\{2\beta(h-\theta)\}$ (spike-rate at threshold τ_0^{-1}, noise parameter β and threshold θ, see e.g. [2]) we get

$$A(h) = \frac{1}{\gamma^{\text{abs}}}\,\frac{1}{1+\exp\{-2\beta(h-\theta')\}+\kappa_h/\gamma^{\text{abs}}} \tag{17}$$

with the modified threshold $\theta' = \theta + 1/(2\beta)\ln(\tau_0/\gamma^{\text{abs}})$ and the relative refractory period time constant $\kappa_h = \int_{\gamma^{\text{abs}}}^{\infty} \mathrm{d}s\,\{1-p_A(s)\}\,D_h(t,t-s)$. In the case we have only an absolute refractory period, κ_h vanishes and we get an equation of the same form as the standard logistic gain function $\mathrm{g}[h] = 1/\{1+\exp[-2\beta(h-\theta)]\} = 1/2\{1+\tanh[\beta(h-\theta)]\}$. The pool spike rate saturates at $1/\gamma^{\text{abs}}$ as it is bounded by the inverse length of the absolute refractory period. We see that the present model lets us understand the gain function quantitatively in terms of the ***microscopic neuronal parameters*** γ^{abs}, κ_h, τ_0, β and θ.

Graded-response models can now be motivated as follows. We look at the normalized form ($A \to A/N$) of equation (4). In a quasistationary regime we define a dynamics by an exponential relaxation towards the stationary solution (17),

$$\tau\frac{\mathrm{d}}{\mathrm{d}t}A(t) = -A(t) + \tau^{-1}[h(t)]\{1-(\gamma^{\text{abs}}+\kappa_h)A(t)\}\ . \tag{18}$$

This equation, and its simpler variant $\tau\mathrm{d}/\mathrm{d}t\,A(t) = -A(t) + \tau^{-1}[h(t)]$, is of the form of an assembly averaged "graded-response" model ([1,4]). Although we introduced the relaxation dynamics by definition, graded-response models as in (18) may present a valid approach if the assembly dynamics is always close to the stationary state calculated from the microscopical parameters. For fast transient dynamics (such as the development of oscillatory activity), the full differential equation system is to be used instead.

4 Connection with Models of Spiking Neurons

In the Spike Response Model (SRM, see e.g. [2]), the response of a neuron is determined by a total field that has two contributions: one from the synaptic inputs from other neurons and a second one that accounts for the neuron's refractory behaviour due to the release of action potentials.

$$h^{\text{total}}(t) = h^{\text{syn}}(t) + h^{\text{ref}}(t) \tag{19}$$

The neuron fires if the total field reaches a fixed threshold θ from below. The fields $h^{\text{syn}}(t)$ and $h^{\text{ref}}(t)$ are calculated with an alpha–kernel $\alpha(s)$ and a refractory field kernel $\eta(s)$. For neurons l and firing times t_l^f we have

$$h_k^{\text{syn}}(t) = \sum_l J_{kl}\sum_f \alpha(t-t_l^f) \quad\text{and}\quad h_k^{\text{ref}}(t) = \sum_f \eta(t-t_k^f)\ . \tag{20}$$

The $\alpha(s)$ and $\eta(s)$ kernels of the Spike Response Model can be used to model a broad range of types of neuronal models, e.g., of the Integrate & Fire type.

Noise is introduced into the SRM by defining a spiking probability

$$p_{\mathrm{S}}[h^{\mathrm{total}}(t); \Delta t] = \Delta t\, \tau_0^{-1} \exp\{2\beta[h^{\mathrm{total}}(t) - \theta]\} \ . \tag{21}$$

The present differential equation system for pool dynamics is exact in the limit of pools composed of infinitely many equivalent spike-response neurons with alpha kernels $\alpha(s) = \alpha^k(s)$ (8) and refractory field kernels of the form

$$\eta(s) = \frac{1}{2\beta} \ln[p_{\mathrm{A}}(s)] \ , \tag{22}$$

with $p_{\mathrm{A}}(s)$ being one of the previously presented refractory functions (6). Since Integrate & Fire type models are a subgroup of the SRM, pools of the former category can be simulated as well with the model presented in this paper.

5 Results and Summary

Starting from a single neuron threshold model, we have derived a system of differential equations with and without delays that describe the spike density dynamics of a pool or assembly of equivalent neurons. Contrary to previous derivations of pool dynamics from microscopic models, the derivation is exact for any dynamical range. This means that the model can operate equally well in a near stationary condition or, when fast, transient dynamics is required. For numerical simulations, an approximation is introduced by breaking the chain of differential equations at the desired depth. Simulations show a good quantitative agreement of the pool dynamics with the dynamics of pools of neurons described by microscopic models of the spike-response or the Integrate & Fire type. Pools modeled by the presented system of equations show the capability of developing oscillatory behavior in the parameter regimes predicted by the "locking–theorem" [3] for spike-response neurons. The present model also reveals the connection of graded-response models with the microscopical single neuron models. In summary, it closes the gap between the microscopic and the macroscopic modelling levels.

References

1. J. L. Feldman and J. D. Cowan. Large-scale activity in neural nets I: Theory with application to motoneuron pool responses. *Biol. Cybern.*, 17:29–38, 1975.
2. W. Gerstner and J. L. van Hemmen. Coding and information processing in neural networks. In E. Domany, J. L. van Hemmen, and K. Schulten, editors, *Models of Neural Networks II*, pages 1–93, New York, 1994. Springer-Verlag.
3. W. Gerstner, J. L. van Hemmen, and J. D. Cowan. What matters in neuronal locking? *Neural Computation*, 8(8):1653–1676, 1996.
4. H. R. Wilson and J. D. Cowan. Excitatory and inhibitory interactions in localized populations of model neurons. *Biophysical J*, 12:1–24, 1972.

How a Single Purkinje Cell Could Learn the Adaptive Timing of the Classically Conditioned Eye-Blink Response

Volker Steuber and David J. Willshaw

Centre for Neural Systems,
Centre for Cognitive Science, Edinburgh University,
Edinburgh EH8 9LW,
Scotland, U.K.
email: {V.Steuber, D.Willshaw}@cns.ed.ac.uk

Abstract. Experimental evidence supports the view that the cerebellum is involved in the adaptive timing of the classically conditioned eye-blink response. Previous modelling studies have demonstrated that a group of cerebellar Purkinje cells can learn the adaptive timing of the eye-blink response if the cells in the group have predefined response latencies which cover the range of conditionable interstimulus intervals (ISIs). Here we show how the timing can be learnt by a *single Purkinje cell.* Phosphorylation of metabotropic glutamate recptors (mGluRs) in our model causes the time delay between parallel fibre input and voltage response to be adaptive and makes it unnecessary to specify a conditionable ISI for each cell in advance. The model is able to learn conditioned responses (CRs) for delay conditioned ISIs between 200 and 1000 msec. Modification of parts of the intracellular signalling network might represent a general mechanism for neurons to learn the timing between input and output.

1 Introduction

Several experimental results indicate an involvement of the cerebellum in the adaptive timing of the classically conditioned eye-blink response [1, 2, 3]. Most of the evidence supports the following scenario: conditioned stimulus (CS) and unconditioned stimulus (US) reach cerebellar Purkinje cells through parallel and climbing fibres, respectively. Appropriately timed parallel and climbing fibre inputs result in Long-Term Depression (LTD) of AMPA receptors and/or phosphorylation of calcium dependent potassium (KCa) channels, leading to a decreased Purkinje cell response to CS presentation. The decreased Purkinje cell activation results in disinhibition of interpositus neurons, opening a timed gate that allows CS presentation after training to cause an eye-blink conditioned response (CR). The CR is adaptively timed and peaks at about the time of the US onset during training. Repeated CS presentation will result in AMPA receptor potentiation or KCa channel dephosphorylation and extinction of the CR.

2 The Spectral Timing Model

A possible explanation for the adaptively timed CR is a variable time delay between CS/parallel fibre input and voltage reponse in the Purkinje cell. Fiala et al.'s Spectral Timing Model [4] demonstrates that the metabotropic glutamate receptor (mGluR) second messenger network in the Purkinje cell can implement such a variable time delay. In the Spectral Timing Model, the CS results in release of glutamate (Glu) from parallel fibres which binds to inactive mGluRs to produce an active receptor B. The active receptor B can be inactivated by dissociation of glutamate or phosphorylation by PKC (C) which produces a phosphorylated receptor A. With a total concentration of available receptors B_a, the change of activated receptors is given by:

$$\frac{dB}{dt} = k_1(B_a - B - A)[Glu] - k_{-1}B - k_2BC \tag{1}$$

Activated mGluRs trigger the activation of G-proteins which in turn stimulate the production of inositol trisphosphate (IP3) and diacylglycerol (DAG) by Phospholipase C (PLC). IP3 binding to receptors in the ER membrane leads to release of calcium into the cytosol and to an increase in the intracellular calcium concentration. The $[Ca^{2+}]$ increase has three major effects:

1. Ca^{2+} and DAG activate Protein Kinase C (PKC).
2. Ca^{2+} efflux through the $3Na^+/Ca^{2+}$ exchanger leads to Na^+ influx and depolarization of the Purkinje cell membrane.
3. Activation of KCa channels causes hyperpolarization of the membrane.

These processes are represented by ordinary differential equations (ODEs) similar to equation 1.

The net effect of parallel fibre input and mGluR stimulation on the membrane potential depends on the peak conductance of the KCa channels g_{KCa}: a large g_{KCa} will lead to hyperpolarization of the Purkinje cell and a decrease in the simple spike firing rate. In the model, $[Ca^{2+}]$ increase, PKC activation and hyperpolarization or depolarization happen virtually at the same time. Their latency with respect to CS and parallel fibre activation is dependent on the concentration of available mGluRs B_a which is assumed to vary in different Purkinje cells.

Learning only takes place in Purkinje cells with a $[Ca^{2+}]$/PKC/voltage response latency equal to the ISI between parallel fibre input (CS) and climbing fibre input (US). Climbing fibre stimulation is assumed to cause (via Purkinje cell depolarization, depolarization of neighbouring basket cells and NO production in the basket cells) a rise of cGMP in the Purkinje cell and activation of Protein Kinase G (PKG) almost immediately after US presentation. PKG phosphorylates G-substrate and inhibits Protein Phosphatase 1 (PP-1) which counteracts PKC. Coinciding PKC and PKG activation in Purkinje cells with latencies equal to the CS-US ISI will result in persistent PKC phosphorylation of KCa channels and an increase in g_{KCa}. Thus, Purkinje cells with a B_a that corresponds to a latency equal to the ISI will respond with hyperpolarization to CS presentations after training.

Fiala et al. assume (a) there is a group of Purkinje cells with appropriately specified B_a values covering the conditionable range of $[Ca^{2+}]$/PKC/voltage response latencies, (b) all Purkinje cells in the group receive parallel fibre input at the same time and, after the ISI, climbing fibre input at the same time and (c) the population response of the Purkinje cell group determines if and when the interpositus neurons are disinhibited and a CR will occur. Thus, combined CS and US presentations during conditioning will lead to a hyperpolarizing response of the Purkinje cell in the group that is tuned to the ISI, to a hyperpolarization peak of the population response at the US time and to an adaptively timed CR.

3 The Adaptive Timing Model

It is not easy to see how assumption (b) in the Spectral Timing Model could be implemented in the cerebellar cortex, given that a single parallel fibre activates Purkinje cells in sequence and not simultaneously. However, neither of the assumptions is necessary if we postulate an adaptive process that adjusts the value of B_a to the conditioned ISI. We suggest this adaptive process could be the phosphorylation of mGluRs by PKC and PKG (figure 1).

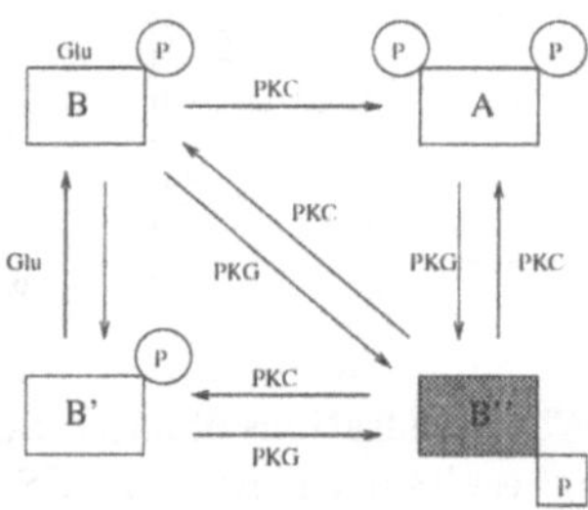

Fig. 1. Four state model of mGluRs that forms the basis of the adaptive timing. PKG phosphorylation of mGluRs leads to production of an inactive form B". The inactivation is reversed by PKC phosphorylation.

In our model, phosphorylation of mGluRs by PKG gives rise to inactive receptors B'' which are unavailable for activation by glutamate. The inactivation is counterbalanced by PKC phosphorylation of the inactive receptors which increases the concentration of available receptors B_a. Persistent phosphorylation by PKC requires concurrent inactivation of PP-1 through phosphorylation of G-substrate by PKG. We represent active PKG by the concentration of its activator cGMP and replace B'' by $B_{max} - B_a$. Thus, the change of available mGluRs B_a is given by:

$$\frac{dB_a}{dt} = k_f(B_{max} - B_a)[cGMP]C - k_b B_a[cGMP] \tag{2}$$

The activation of mGluRs by glutamate and all other processes were modelled as in the Spectral Timing Model. We implemented the model in $C++$, using a fifth order Runge-Kutta algorithm with an adaptive step size. Parameters were taken from Fiala et al., with the exception of k_b and k_f which were set to 0.08 and 1.0, respectively. Simulations were performed according to a delay conditioning paradigm with a glutamate pulse as CS.

4 Simulation Results

Learning in the model can be divided into two overlapping phases: an *adaptive timing phase* and a *hyperpolarization phase*. During the adaptive timing phase, US presentation leads to PKG activation and a decrease in the concentration of available mGluRs B_a. The decrease in B_a results in a slower activation of mGluRs by glutamate and an increase of the Ca^{2+}/PKC/voltage response latency (figure 2).

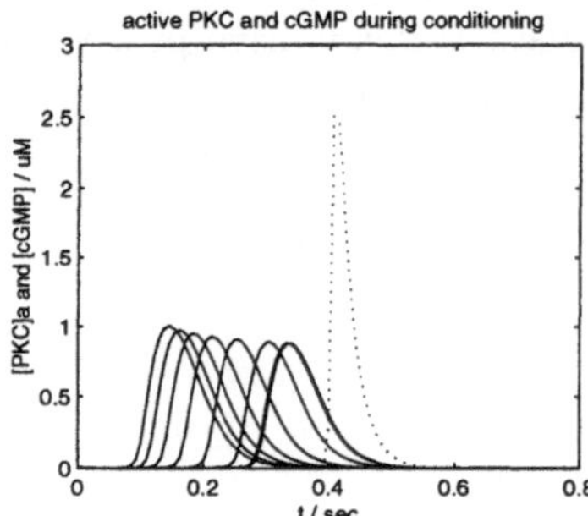

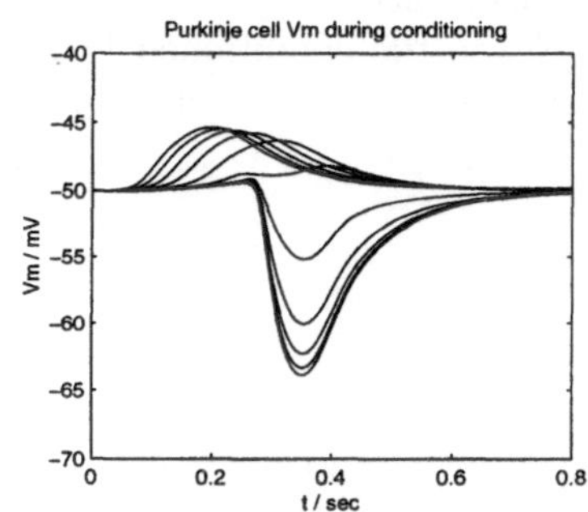

Fig. 2. Simulation results for 75 presentations of a CS starting at t=0, followed by a US peak at t=0.4sec. Conditioning results in a shift of the CS induced PKC peak (solid line) towards the US induced cGMP peak (dotted line) and in a shift of the voltage response from a depolarization at t=0.15sec to a hyperpolarization at t=0.36sec. Results are shown for every 5th trial.

Once the latency reaches the ISI the PKG and PKC peaks overlap, leading to an equilibrium between PKG induced B_a decrease and PKC induced B_a increase, a constant B_a and a stable latency. Coinciding PKC and PKG activation during the hyperpolarization phase causes persistent KCa channel phosphorylation and hyperpolarization of the Purkinje cell membrane at the US time (figure 2 right). Thus, CS presentation after training leads to disinhibition of interpositus neurons at the US time and an adaptively timed CR. The model is able to learn adaptively timed CRs for ISIs between 200 and 1000 msec (figure 3a), compared to experimentally observed delay conditionable ISIs between 100 and 4000 msec [5]. The CR strength (measured as the extent of the hyperpolarization in the model) decreases with increasing ISIs (figure 3b, c), in accordance with experimental results for $\geq 200msec$ ISIs [6]. A prediction of the model is faster extinction of CRs for longer ISIs (figure 3d).

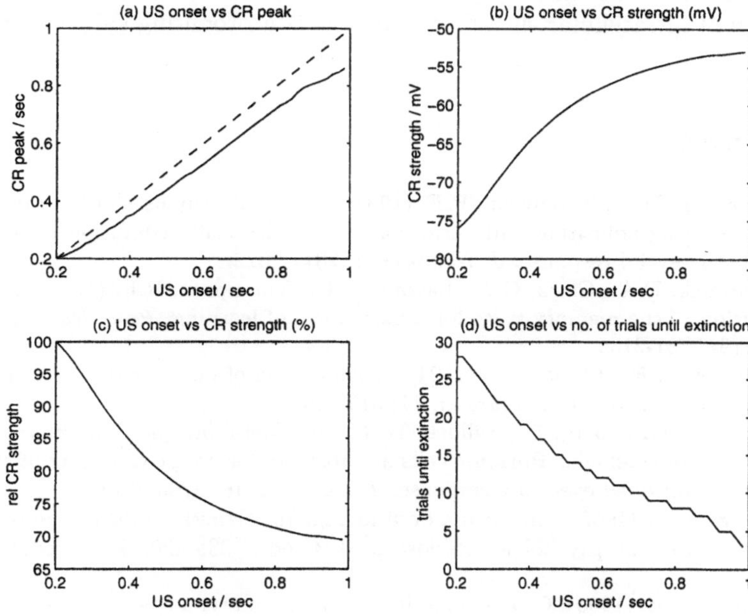

Fig. 3. Simulation results for delay conditioning to a CS starting at t=0 and USs after varying times. (a) The CR (solid line) occurs shortly before the US onset (broken line: y=x). (b) CR strength in mV hyperpolarization, (c) relative CR strength compared to the maximum for 200msec ISIs, (d) number of CS presentations necessary for extinction of the CR.

5 Conclusions

We have shown that a single Purkinje cell can learn the adaptive timing of the classically conditioned eye-blink response. Substrate for the adaptive timing is a four-state model of metabotropic glutamate receptors (mGluRs), based on the assumption that PKC phosphorylation increases and PKG phosphorylation decreases the number of mGluRs which are available for glutamate activation. PKC phosphorylation of mGluRs has been described [7]. No evidence for PKG phosphorylation of mGluRs is known to date. However, PKG phosphorylation is difficult to investigate experimentally, and several PKG substrates in the brain have yet to be identified [8, 9]. Alternatively, PKG could exert its effect by phosphorylating PP-1 or any other protein that interferes with PKC phosphorylation of mGluRs. Further characterization of PKG substrates in the brain should clarify the plausibility of our model.

Finally, it should be noted that the model might have a relevance beyond the study of classical conditioning or cerebellar Purkinje cells. Modification of parts of the intracellular signalling network may represent a general mechanism for neurons to learn a time delay between input and output. The computational capability of intracellular signalling networks is a promising subject for future research.

References

1. McCormick, D.A., Thompson, R.F. (1984). Neuronal responses of the rabbit cerebellum during aquisition and performance of a classically conditioned nictitating membrane-eyelid response. *J. Neurosci.* 4: 2811-2822.
2. McCormick, D.A., Clark, G.A., Lavond, D.G., Thompson, R.F. (1982). Initial localization of the memory trace for a basic form of learning. *Proc. Natl. Acad. Sci. USA* 79: 2731-2735.
3. Thompson, R.F., Krupa, D.J. (1994). Organization of memory traces in the mammalian brain. *Ann. Rev. Neurosci.* 17: 519-549.
4. Fiala, J.C., Grossberg, S., Bullock, D. (1996). Metabotropic glutamate receptor activation in cerebellar Purkinje cells as substrate for adaptive timing of the classically conditioned eye-blink response. *J. Neurosci.* 16: 3760-3774.
5. Gormezano, I. (1966). Classical Conditioning. In: Experimental methods and instrumentation in psychology (Sidowski, J.B. ed.), 385-420. McGraw-Hill, New York.
6. Steinmetz, J.E. (1990). Classical nictitating membrane conditioning in rabbits with varying interstimulus intervals and direct activation of cerebellar mossy fibres as the CS. *Behav. Brain Res.* 38: 97-108.
7. Kawabata, S., Tsutsumi, R., Kohara, A., Yamaguchi, T., Nakanishi, S., Okada, M. (1996). Control of calcium oscillations by phosphorylation of metabotropic glutamate receptors. *Nature* 383: 89-92.
8. Wang, X., Robinson, P.J. (1995). Cyclic GMP dependent protein kinase substrates in rat brain. *J. Neurochem.* 65: 595-603.
9. Tegge, W., Frank, R., Hofmann, F., Dostmann, W.R.G. (1995). Determination of cyclic nucleotide-dependent protein kinase substrate specificity by the use of peptide libraries on cellulose paper. *Biochemistry* 34: 10569-10577.

An Algorithm for Synaptic Modification Based on Exact Timing of Pre- and Post-synaptic Action Potentials

Walter Senn[1,2], Misha Tsodyks[2] and Henry Markram[2]

[1] Physiologisches Inst., Universität Bern, Bühlplatz 5, 3012 Bern, Switzerland
[2] Dept. of Neurobiology, The Weizmann Institute, Rehovot, 76100, Israel

Abstract. The timing between individual pre- and post-synaptic action potentials is known to play a crucial role in the modification of the synaptic efficacy during activity. Based on stimulation protocols of two synaptically connected neurons, we infer an algorithm which explains the data by modifying the probability of neurotransmitter discharge as a function of the pre- and postsynaptic spike delays. The characteristics of this algorithm is its asymmetry with respect to the delays: if the postsynaptic spike arrives *after* the presynaptic spike, the probability of discharge is up-regulated while it is down-regulated if the postsynaptic spike arrives *before* the presynaptic spike. The algorithm allows to predict stimulation protocols which induce maximal up- and down-regulation of the discharge probability.

1 Introduction

Since the work of D. Hebb [1], several attempts were made to formulate precise 'learning rules', which determine the change in synaptic efficacies from the known activities of neurons [2, 3, 4]. In these rules neuronal activities are represented by an analog variable reflecting the average firing rates of the neurons. Such formulations were used in training neural networks to perform various computational tasks.

Recent experimental developments indicate that a novel approach to modeling synaptic plasticity is needed. In [5], it was found that synaptic modification is very sensitive to the relative timing between the spike trains. If postsynaptic spikes occur 10ms after the presynaptic one, synaptic responses were increased. The same pattern of stimulation with the opposite time delay resulted in a decrease of responses. In [6] it was shown that the synaptic modification is a complex *redistribution* of synaptic efficacy between the spikes in the train without uniform strengthening of the connections. This redistribution can result from the increase in the probability of neurotransmitter release [7].

Experiments of [5] for the first time provide the experimental basis for formulating the learning rules based on individual spikes rather then firing rates. In the current contribution, we present what we believe is the minimal phenomenological model which reproduces the experimental results and which allows the computation of the synaptic modification for arbitrary patterns of spikes. This model can now be tested against other experimental paradigms and provides a useful foundation for computational models which utilize exact spike timing for information processing [8].

2 The algorithm

The learning rule enables adaptation of the probability of neurotransmitter release resulting from simultaneous activity of pre- and post-synaptic neurons. Specifically, we adapt the probability that a presynaptic spike discharges a vesicle which is ready at the site of release. We refer to this probability as the probability of discharge, P_{dis}. The biophysical processes involved in modifying P_{dis} are triggered either by pre- or post-synaptic spikes, by a spontaneous presynaptic release or by elevation of the postsynaptic membrane potential. We assume that the up-regulation of P_{dis} is only induced by a postsynaptic spike following a presynaptic release. Down-regulation of P_{dis}, on the other hand, is only induced by a presynaptic release following a postsynaptic spike or voltage increase.

Practically, long-lasting synaptic modification does not instantaneously follow the pairing but develops slowly to peak within $\approx$ 20min. Accordingly, P_{dis} is kept fixed during the simulated pairing and effects of each spike are summed up to determine the overall change in the *limit* probability P_{dis}^{∞} (Fig. 1). The convergence of P_{dis} to the limit probability P_{dis}^{∞} evolves with time constant τ_{mod}^{P} = 20min. This work does not include a short lasting up-regulation of P_{dis} analogous to post tetanic potentiation because this phenomenon is not clearly evident at these depressing synapses. Neither considered is the decay of any change in P_{dis} (i.e. natural decay of P_{dis}^{∞}) which could occur on a time scale of hours.

In detail, the synaptic modification works as follows. The primary events for up- and down-regulation are mediated by the NMDA-receptors located at the postsynaptic membrane. These receptors may be in 3 different states: the recovered state, N_{rec}, the state saturated with glutamate, N_u, and the state altered by intracellular calcium, N_d (Fig. 1). The secondary messenger for *up*- and *d*own-regulation may be in an active state, S_u and S_d, or in an inactive state, $\bar{S}_u$ and $\bar{S}_d$, respectively. If a vesicle of neuro-transmitter discharges, either spontaneously or due to a presynaptic spike, glutamate is released and bound by postsynaptic NMDA-receptors ($N_{rec} \rightarrow N_u$). Being in a state saturated by glutamate, the NMDA-receptors will open when a back-propagating postsynaptic spike arrives or, more generally, when the the postsynaptic membrane potential increases allowing calcium to flow through NMDA-channels into the postsynaptic cell. This calcium activates a secondary messenger ($\bar{S}_u \rightarrow S_u$) which diffuses to the presynaptic site and up-regulates the probability of discharge ($\bar{P}_{dis}^{\infty} \rightarrow P_{dis}^{\infty}$). If, on the other hand, there is first an increase of the postsynaptic membrane potential e.g. due to a postsynaptic spike, voltage activated calcium-channels open, calcium flows in through these channels and binds to the NMDA-receptors [9], altering or redirecting their function ($N_{rec} \rightarrow N_d$). Subsequently released glutamate now activates an altered NMDA receptor which leads to the activation of down-regulating secondary messenger ($\bar{S}_d \rightarrow S_d$). This messenger diffuses to the presynaptic location and the probability of discharge is down-regulated ($P_{dis}^{\infty} \rightarrow \bar{P}_{dis}^{\infty}$).

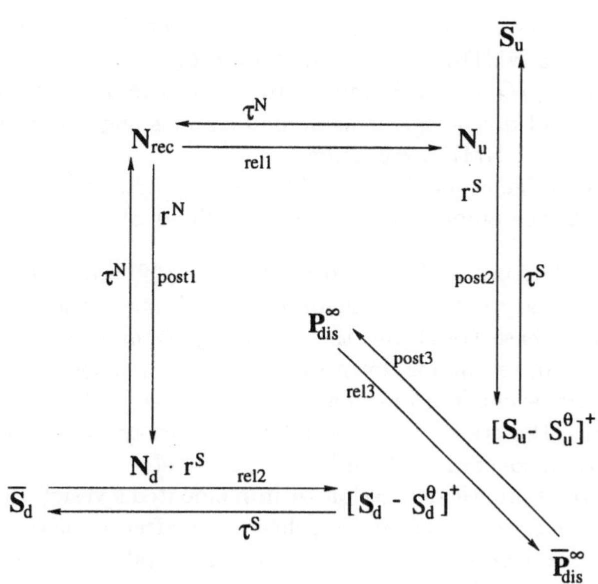

Fig. 1. *The kinetic scheme for modification of the limit probability of discharge, P_{dis}^{∞}. Up-regulation of P_{dis}^{∞} is mediated through the states N_u and S_u while* down-regulation *is mediated through the states N_d and S_d. These states decay naturally with time constants τ^N and τ^S, respectively. Transitions labeled with* reli *and* posti *(i=1,2,3) occur instantaneously at either a* presynaptic *release or at a* post*synaptic spike. These instantaneous transitions are weighted by the factors written onto their arrows. For instance, at a postsynaptic spike, the state N_d is increased by $N_{rec} \cdot r^N$* (post1) *and the state S_u is increased by $\bar{S}_u \cdot N_u \cdot r^S$* (post2)*. At a presynaptic release, for instance, P_{dis}^{∞} is decreased by $P_{dis}^{\infty} \cdot [S_d - S_d^{\theta}]^+$* (rel3)*.*

The above scenario can be summarized in the following 'learning algorithm'. Whenever a postsynaptic action potential arrives at the synaptic site, 3 different processes are induced (indicated by *post1* - *post3* in the scheme):

post1 The fraction r^N of N_{rec} is moved to N_d. This describes the altering of NMDA-receptors due to calcium flowing into the postsynaptic site through voltage activated channels.

post2 The fraction $r^S N_u$ of $\bar{S}_u$ is moved to S_u ($\bar{S}_u = 1 - S_u$). This describes the activation of up-regulating secondary messenger proportional to the amount of NMDA-receptors saturated with glutamate.

post3 The limit probability P_{dis}^{∞} is increased by $\bar{P}_{dis}^{\infty} [S_u - S_u^{\theta}]^+$, where $\bar{P}_{dis}^{\infty} = 1 - P_{dis}^{\infty}$ and S_u^{θ} denotes the threshold to trigger up-regulation ($[x]^+ =$

max$(x, 0)$). Thus, P_{dis}^{∞} is pushed towards 1 proportional to the amount of secondary messenger above threshold.

A release of a vesicle at the presynaptic site induces, completely symmetrically, the following 3 processes (indicated by *rel1* - *rel3* in the scheme):

rel1 The available amount of N_{rec} is moved to N_u . This describes the saturation of the recovered NMDA-receptors with glutamate.

rel2 The fraction $r^S N_d$ of $\bar{S}_d$ is moved to S_d $(\bar{S}_d = 1 - S_d)$. This describes the activation of down-regulating secondary messenger proportional to the amount of altered NMDA-receptors.

rel3 The limit discharge probability is reduced by $P_{dis}^{\infty}\,[S_d - S_d^{\theta}]^+$, i.e. proportional to P_{dis}^{∞} and to the amount of S_d above threshold S_d^{θ} .

In a temporal order, up- and down-regulation of P_{dis}^{∞} are each mediated by a primary and secondary event (cf. scheme): The primary event for up-regulation is a presynaptic release (*rel1*) and the secondary event is a *post*synaptic spike (*post2*). The primary event for down-regulation is a postsynaptic spike (*post1*) and the secondary event is a presynaptic release (*rel2*). Beside these instantaneous transitions, the states N_u, N_d and S_u, S_d decay exponentially with the corresponding time constants τ^N and τ^S, respectively.

To describe the synaptic depression we implemented a vesicle depletion model according to which a vesicle recovers stochastically after a release with time constant $\tau_{rec}^P = 800$ms. Due to this recovery process, a spike following shortly after a previous discharge has less chance to encounter an occupied site of release and thus less chance to discharge a vesicle. Averaging over an ensemble of synapses, the model turns into the deterministic model of a depressing synapse [7] and P_{dis} represents the *use of synaptic efficacy* (U_{SE}).

3 Application of specific stimulation protocols

The algorithm is based on dual whole-cell voltage recordings of neocortical pyramidal cells [5]. It was tested against the following 3 experiments:

Experiment1: Pre- and postsynaptic spike trains of 10Hz and 5 spikes were paired with a time difference between post- and presynaptic spikes of +10 and −10ms, respectively. The pairing was repeated 10 times every 4s and the change in P_{dis} was recorded for the next 60min. The increase in P_{dis}^{∞} for the 10ms retarded postsynaptic spikes and the decrease for the 10ms advanced postsynaptic spikes was faithfully reproduced by the algorithm (simulation results Fig. 2a, experimental data [5, Fig. 3C]).

Experiment2: Paired pre- and postsynaptic spikes trains of 5 spikes were triggered with a postsynaptic delay of 2ms. The simulation was performed for different frequencies ranging from 2 to 40Hz and the final change in the probability of discharge, P_{dis}^{∞}, is evaluated. The main characteristics of the learning curve, the steep upstroke at 10Hz and the saturation at higher frequencies, are well reproduced (simulation results Fig. 3a, experimental data [5, Fig. 2C]).

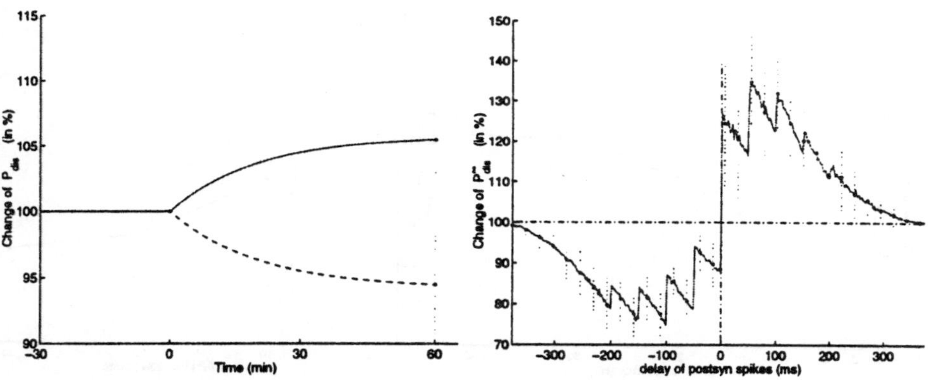

Fig. 2. *Modification of P_{dis} as a function of the postsynaptic spike delays.* **a** *Exp1: The evolution of P_{dis} towards P^{∞}_{dis} induced by repeated pairings at 10Hz with delay of 10ms (upper trace) and −10ms (lower trace). The parameters in all our simulations are $r^N = .5$, $\tau^N = 300ms$, $r^S = .7$, $\tau^S = 600ms$. The thresholds for the secondary messengers were set to $S^{\theta}_u = r^S$ and $S^{\theta}_d = r^N r^S$. As starting values for the discharge probabilities we chose $P_{dis} = P^{\infty}_{dis} = .5$.* **b** *A prediction: Change of P_{dis} after pairing of 20Hz trains of 5 spikes with delays ranging from −350 to 350ms. Pointed lines represent standard deviations.*

Experiment3: Pre- and post-synaptic spike trains of 20Hz were paired with a postsynaptic spike delay of 2ms. The number of spikes in the paired trains were varied from 2 up to 20 and for each number P^{∞}_{dis} was determined. The astonishing fact in this experiment was that the change in P^{∞}_{dis} did not accumulate but was rather neutralized by the following spikes (Markram, unpublished). Simulations of the model are compatible with these results (Fig. 3b).

Apart from reproducing existing experiments, the algorithm produces hypotheses about the outcome of new experiments. One of the remaining questions is the change in P_{dis} for pre- and postsynaptic spike trains of a given frequency and varying delays. Figure 2b shows the relative change in P^{∞}_{dis} for spike trains of 5 spikes at 20Hz with different delays of the postsynaptic train. Notice that down-regulation is maximal at a presynaptic delay of 100ms while up-regulation is maximal at a postsynaptic delay of 50ms. The delay of −100ms is explained by the fact that at a postsynaptic spike only the fraction r^N of N_{rec} is moved to N_d and the delay of 50ms is explained by the fact that not each presynaptic spike induces a neurotransmitter release.

4 Conclusion

The presented algorithm predicts the change in a specific synaptic property as a result of coherent activity of neurons and reproduces experimental observations of [5]. It confirmed three characteristics of the synaptic modification: 1. the asymmetry with respect to delays, 2. the monotonic increase with respect to the

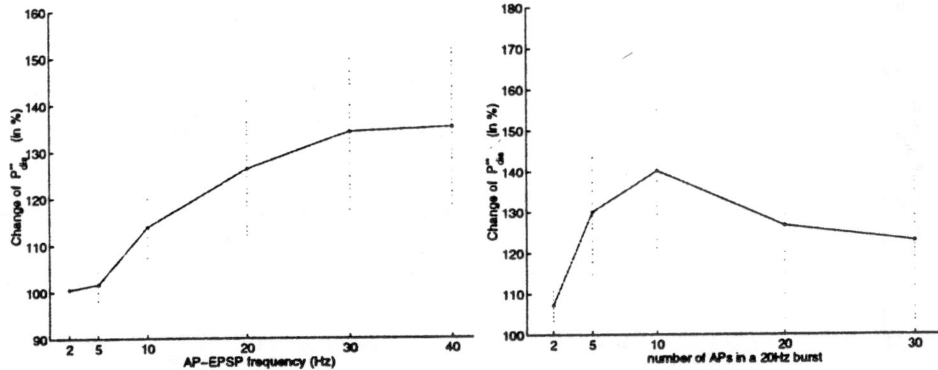

Fig. 3. *The change in* P_{dis} *shown* 60 *minutes after pairing.* **a** *Exp2: Dependence on the frequency of the paired spike trains, each composed of* 5 *spikes.* **b** *Exp3: Dependence on the number of spikes within the paired* 20*Hz spike trains.*

common pre- and postsynaptic frequency above 5Hz and 3. the non-monotonic dependence of the number of spikes in the paired trains. In agreement with [6], the algorithm does not predict the uniform increase of the synaptic strength.

One could speculate that due to additional computation such as the integration of a third coincident signal provided by growth factors or neuromodulators (e.g. synaptic growth or unmasking of postsynaptic receptors [10]) is required for the induction of real synaptic strengthening. Specifying these conditions remains an important challenge for a future research.

Acknowledgment We thank Josef Kleinle for critically reading the manuscript. This study was supported by an ONR grant and a SNF grant (5002-037939) for WS.

References

1. Hebb, D.O. *The organization of behavior*, J. Wiley and Sons, New York (1949)
2. Bienenstock, E.L. & Cooper, L.N. & Munro, P.W. *J. Neuroscience* **2**(1), 32-48 (1982)
3. Sejnowski, T. *J. Math. Biol.* **4**, 303-321 (1977)
4. Artola, A. & Singer, W. *TINS* 16(11), 480-487 (1993)
5. Markram, H. & Lübke, J & Forster, M & Sakmann, B. *Science* **275**, 213-215 (1997)
6. Markram, H. & Tsodyks, M. *Nature* **382**, 807-810 (1996)
7. Tsodyks, M. & Markram, H. *Proc. Natl. Acad. Sci. USA* **94**, 719-723 (1997)
8. Hopfield, J.J. *Nature* **376**, 33-36 (1995)
9. Mayer, M.L. & McDermott, A.B. & Westbrook, G.L. & Smith, S.J. & Barker, J.L. *J. Neuroscience* **7**, 3230-3244 (1987)
10. Liao, D. Z. & Hessler, N. A. & Malinow, R. *Nature* **375**, 400-404 (1995).

Modelling Plasticity in Rat Barrel Cortex Induced by One Spared Whisker

Lubica Benušková

Slovak University of Technology, Department of Computer Science and Engineering, 81219 Bratislava, Slovakia, e-mail: benus@decef.elf.stuba.sk

Abstract. We have extended the previous model [1,2] of experience-dependent plasticity in the rat cortical representation of the whiskers (the barrel cortex) to include excitatory and inhibitory interactions within a barrel-column. Recent experimental observations [3] indicate that after trimming all but one whisker on one side of the rat snout, the spared and deprived whisker dominance distribution was changed significantly for neurons located in layers *II/III* of barrel-columns surrounding the spared whisker column. The present simulations suggest that the experimental observations may be explained by means of the Bienenstock, Cooper and Munro (BCM) theory of synaptic plasticity and by the masking effect of the intralayer feedback inhibition.

Keywords: Barrel Cortex, Plasticity, Experience, BCM Modification Threshold, Inhibition.

1 Introduction

We model experience-dependent plasticity in adolescent rat barrel cortex observed in electrophysiological experiments [3] in which whisker *D1* was left intact while all other whiskers were cut unilaterally (Fig. 1).

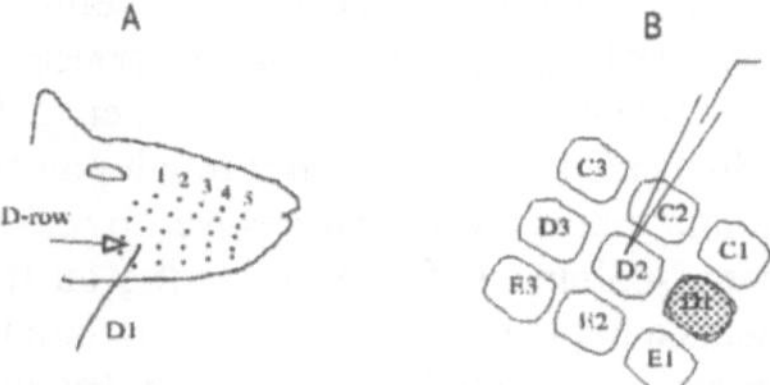

Fig. 1. (A) In rats, the facial whiskers are aligned in 5 rows (row A is dorsal and row E is ventral) and the whiskers within a row are numbered from caudal to rostral. Each facial whisker projects, via the trigeminal nuclei and VPM (ventral posterior medial) barreloids, to a separate cortical barrel, a cluster of neurons in layer *IV* [4,5]. (B) In the barrel cortex, these barrels (as well as barrel-columns) form a topographical representation of the rat snout. From barrel-columns surrounding the *D1* barrel-column, the recordings were made.

The response characteristics of the layer *II/III* cells in barrel-columns surrounding the *D1* barrel-column were measured after 7, 20 and 60 days of whisker trimming (see Fig. 5); and significant changes in the sensory-evoked activity were found. Our present model is based on the BCM theory [6,7], and it is the same model as the one used recently for explanation of the course of experience-evoked plasticity in different layers of the rat barrel cortex [8]. In this model we have used a biologically realistic inhibitory circuits (see Fig. 2) in which feedback inhibition, as described below, acts differently upon different temporal components of the sensory-evoked response.

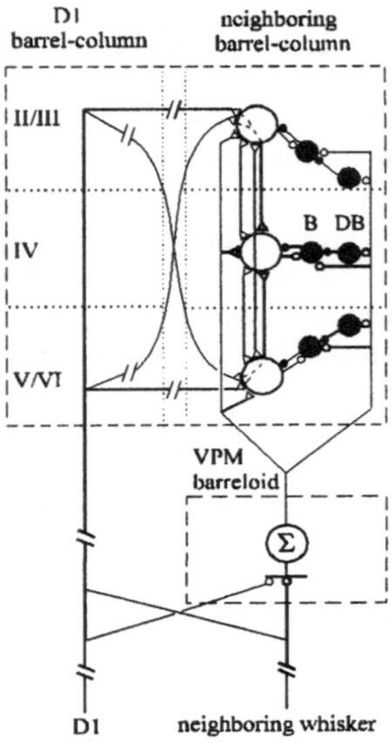

Fig. 2. Summary of the inputs converging on BCM neurons in the present model barrel-column. For the model circuit, whisker *D1*, and its neighboring whisker, converge through polysynaptic pathways (broken lines) upon a cell in VPM barreloid. The summation sign in the VPM barreloid denotes the multiwhisker nature of the VPM cell's input. The axon arising in the VPM barreloid synapses directly on the cortical cells of interest in barrel-column mediating a major short latency (<10 ms poststimulus) response to the neighboring principal whisker and a minor short latency response to whisker *D1*. Long latency (>10 ms poststimulus) responses to whisker *D1* are mediated through polysynaptic connections (broken lines, going through perigranular zones) which are relayed by neurons in the upper and deep layers of the neighboring barrel-column before terminating on the barrel cell. Long latency responses to the principal whisker are mediated mainly by circuits within the same barrel-column [9]. Modifiable synapses are denoted by small open triangles. Inside a barrel-column, in layers *II/III, IV* and *V/VI*, there are feedback inhibitory circuitries. DB denotes the double bouquet cells and B denotes the basket cells. At present, we do not consider potential modifications of subcortical synapses or excitatory synapses on inhibitory neurons (small open circles) or inhibitory synapses (small filled circles), or other polysynaptic pathways (broken lines), although their modifiability should not be excluded. The thickness of a synapse reflects the relative weight of the input with respect to other depicted inputs.

In the simulated circuit, the VPM axon synapses on excitatory cortical neurons at three levels, the layer *IV*, the upper layers (*II/III*), and the deep layers (*V/VI*). The excitatory neurons in the upper and deep layers correspond to pyramidal neurons, whereas the layer *IV* barrel neuron most likely would correspond to a spiny stellate neuron. Also within layers *II/III*, *IV* and *V/VI* are included two types of inhibitory neurons, one phasically active double bouquet (DB) cell, and the other a tonically active basket (B) cell [10–12]. The DB cell and the B cell both receive inputs from VPM and the B cell receives input from the spiny stellate cell (or from the pyramidal cells if we are in the layers *II/III* or *V/VI*, resp). The DB cell in turn projects onto B cells, which have feedback projections to excitatory cells. Prior to the stimulus (i.e. when all neurons are firing spontaneously) the B cell tonically inhibits the excitatory cell. When a VPM afferent sensory volley arrives in the cortical layer, the DB cell is quickly activated and inhibits the B cell. Inhibited by the DB cell, the B cell interrupts its firing. As a result, we suppose that during the first 10 ms poststimulus, the excitatory neurons in the upper, middle and deep layers do not receive the feedback inhibition. Then, during the long latency period (>10 ms), excitatory cells in the upper, middle and deep layers receive the intracortical activation due to the intra- and inter-columnar spread of activity, and activate the B inhibitory neurons. After reaching the threshold of excitation, the feedback inhibitory neurons inhibit the corresponding excitatory neurons in proportion to the magnitude of their long latency activity (Fig. 4). As shown below, this functional circuit may help explain some crucial experimental observations.

2 BCM Model of the Barrel-Column and Simulation Results

Synaptic plasticity is modeled according to the BCM synaptic modification rule [6,7]. If we consider the case of a linear cell, the modification of the ith synapse with the weight m_i at time t is proportional to the product of input activity at the ith synapse, $d_i(t)$, and a function ϕ, in such a way that

$$\frac{dm_i(t)}{dt} = \eta\, \phi\left[c(t), \theta_M(t)\right] d_i(t) \tag{1}$$

The "modification rate", η, is equal to the magnitude of the synaptic modification for the ith input in one time step, when $\phi = 1$ and $d_i(t)=1$. According to [7], ϕ is a parabolic function of the cell's current firing rate $c(t)$ and modification threshold $\theta_M(t)$, i.e. $\phi[c(t), \theta_M(t)] = c(t)\,[c(t) - \theta_M(t)]$. Firing rate is in turn linearly proportional to the weighted sum of incoming activity, so that $c(t)=\Sigma m_i(t) d_i(t)$. The dynamic modification threshold $\theta_M(t)$ is proportional to the postsynaptic response averaged over some recent past time, such that $\theta_M(t) = \langle c^2(t)\rangle_\tau / c_o$ where c_o is positive constant.

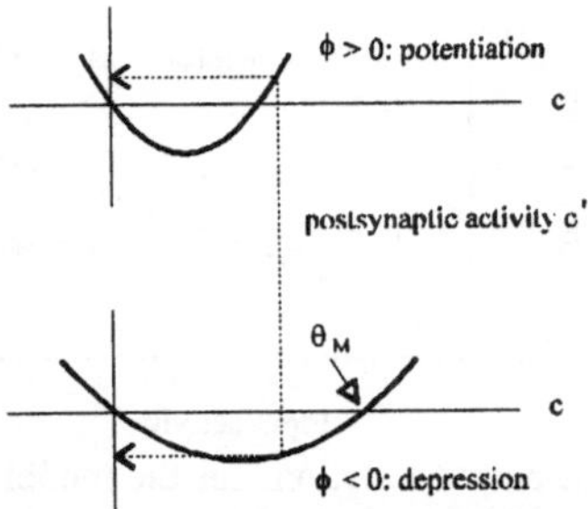

Fig. 3. Schematic illustration of how the same level of postsynaptic activity c' can result in synapse potentiation or depression depending on the current value of the synaptic modification threshold θ_M..

The averaged cell activity over some recent past $\langle c^2(t)\rangle_\tau$ is determined by the following integral:

$$\langle c^2(t)\rangle_\tau = \frac{1}{\tau}\int_{-\infty}^{t} c^2(t')\, e^{-\left[\frac{(t-t')}{\tau}\right]} dt' \tag{2}$$

where τ is the averaging period. From these relations it follows that when postsynaptic activity $c(t)$ is greater than zero but less than the modification threshold $\theta_M(t)$, all active synapses (i.e. $d_i(t)>0$) weaken. On the other hand, when postsynaptic activity $c(t)$ is greater than $\theta_M(t)$, all active synapses potentiate. Since $c(t)=\Sigma m_i(t)d_i(t)$, the correlation (synchronicity, pairing) of excitatory inputs plays a crucial role in driving the postsynaptic cell activity above the modification threshold $\theta_M(t)$ (Fig. 3).

The activity $c(t)$ of a model layer *II/III* cell consists of a sum of the short latency c_{SL}, (VPM-to-layer) and long latency c_{LL} (intracortical) components, i.e.

$$c(t) = c_{SL}(t) + c_{LL}(t) =$$
$$= m^{vpm}(t)\sum_{i=1}^{V}\left[I_i^{vpm} d(t) + n_i^{vpm}(t)\right] + \sum_{i=1}^{C} g\left(m_i^{cor}(t)\left[I_i^{cor} d(t) + n_i^{cor}(t)\right]\right) \tag{3}$$

In this equation, $d(t)$ is equal to either 1 or 0, depending on whether or not the ith whisker is deflected. $0< I_i^{vpm} <1$ and $0< I_i^{cor} <1$ are the input strength constants of the ith whisker input conveyed through the VPM barreloid and intracortical synapses, respectively. The noise, either $n_i^{cor}(t)$ or $n_i^{vpm}(t)$, is defined as a random variable uniformly distributed in the interval [$-A$(noise),$+A$(noise)], where $A<1$ is the noise amplitude. V is the number of inputs converging onto the VPM cell, and C is the number of intracortical inputs converging on the layer *II/III* cell. The 2nd term (>10 ms poststimulus) is inhibited, which is expressed by the function g (Fig. 4):

$$g(x) = \begin{cases} x - \mu\log(1+x) & if \quad x > x_0 \\ x & if \quad x \le x_0 \end{cases} \tag{4}$$

Parameter μ measures the strength of a feedback inhibition inside the corresponding layer when the long latency output of excitatory cells reaches the inhibitory threshold x_0.

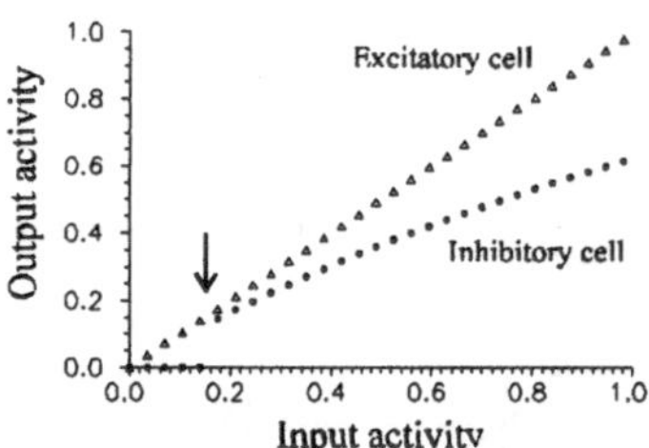

Fig. 4. Illustration of the input-output functions for the model excitatory and inhibitory cells. The arrow points to the threshold when inhibition is being switched on. Then it follows a logarithmic relation. Other relations like linear, or a power function with the exponent around 0.8 would also work.

Analogous equations can be written for the deep and barrel cells, except that in the latter case in the 2nd (cortical *cor*) term, we replace $[\,I_i^{cor} d(t) + n_i^{cor}(t)\,]$ by

$g\left(m_i^{cor}(t)[I_i^{cor}d(t)+n_i^{cor}(t)]\right)$, that is by already inhibited intracortical input. Whisker trimming was simulated by setting input activities of the cells in the model neighboring barrel-column to the noise level, i.e. $d_i^{vpm}(t)=n_i^{vpm}(t)$ and $d_i^{cor}(t)=n_i^{cor}(t)$, except for the only input that represented untrimmed whisker *D1*, for which $d_i^{vpm}(t)=I_i^{vpm}d(t)+n_i^{vpm}(t)$ and $d_i^{cor}(t)=I_i^{cor}d(t)+n_i^{cor}(t)$, as shown in equation (3).

After computer simulation of the above relations we got the following results (see Fig. 5), when we computed the output as $out = \sigma(c_{SL}) + \sigma(c_{LL},)$, where σ is the sigmoid function, $\sigma(x) \in (-1, 1)$.

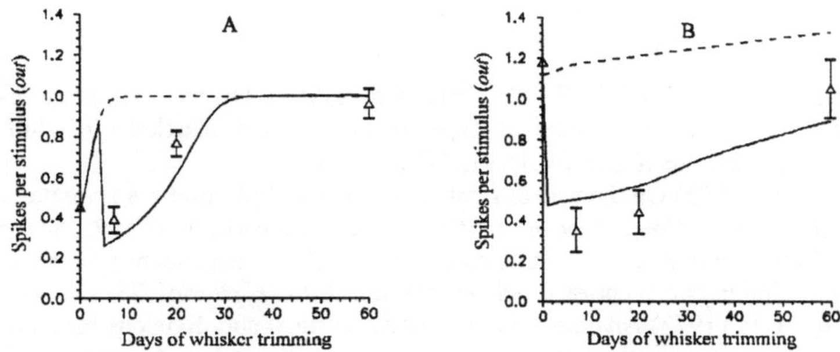

Fig. 5. (A) Effect of deprivation on response of layer *II/III* cells in neighboring barrel-column to spared whisker stimulation, and (B) to principal (cut) whisker stimulation. The solid lines denote the results of computer simulations of the BCM model of the neighboring barrel-column with the feedback inhibition in the layers *II/III*, *IV*, and *V/VI* (see Fig. 2). The dashed curves were obtained in the simulation of the model with no inhibitory interactions. Open triangles are values obtained in *in vivo* recordings after 7, 20 or 60 days of whisker trimming [3]. All the values of model parameters for the modeled cells were the same as in [8], except for the inhibitory parameters that were: μ=0.7, 0.8, 0.9, for the *V/VI*, *IV* and *II/III* layer, respectively, and x_0=0.171.

3 Conclusion

The model accurately simulates the response evolution of the cells located in layers *II/III* of barrel-columns immediately surrounding the spared whisker *D1* column. In the model circuit (Fig. 2), thalamocortical connections ending upon excitatory cells, intracolumnar connections relaying the long latency response to the principal (cut) whisker, and intercolumnar connections relaying the long latency response to the spared whisker *D1* – they all potentiate during the course of sensory deprivation, due to the low value of the modification threshold θ_M, which is in turn the result of a low average activity of the cells. However, as suggested by the recent simulations, the feedback inhibition may act as a kind of mask, which modulates the response evolution of the layer *II/III* excitatory cells. That is, increase in the strengths of the indicated excitatory synapses is not enough to overcome the inhibitory influences. We consider to be important that the same model with the same values of parameters (except for the values of inhibitory parameters), was successful in modelling the different experiment (when 2 whiskers were left intact) of a different group of experimenters [8,13].

Acknowledgments

This publication is based on work sponsored by the GAV grant 95/5195/605, and by the U.S.–Slovak Science and Technology Joint Fund in cooperation with the Department of Health and Human Services in the U.S.A. and the Ministry of Education in Slovakia, under Project No. 015-95.

References

1. Benušková L., Diamond M.E. & Ebner F.F. (1994) Dynamic synaptic modification threshold: Computational model of experience-dependent plasticity in adult rat barrel cortex. *Proc. Natl. Acad. Sci. USA*, 91, 4791–4795.
2. Benušková L. (1995) On the role of inhibition in cortical plasticity: a computational study. *Proc. International Conference on Artificial Neural Networks ICANN'95*, 521–526.
3. Glazewski S. & Fox K. (1996) Time course of experience-dependent synaptic potentiation and depression in barrel cortex of adolescent rats. *J. Neurophysiol.* 75, 1714-1729.
4. Killackey H. P. (1973) Anatomical evidence for cortical sub-divisions based on vertically discrete thalamic projections from the ventral posterior nucleus to cortical barrels in the rat. *Brain Res.*, 51, 326–331.
5. Jensen K. F. & Killackey H. P. (1987) Terminal arbors of axons projecting to the somatosensory cortex of adult rats. I. The normal morphology of specific thalamocortical afferents. *J. Neurosci.*, 7, 3529–3543.
6. Bienenstock E. L., Cooper L. N & Munro P. W. (1982) Theory for the development of neuron selectivity: orientation specificity and binocular interaction in visual cortex. *J. Neurosci.*, 2, 32–48.
7. Intrator N. & Cooper L. N (1992) Objective function fomulation of the BCM theory of visual cortical plasticity: Statistical onnections, stability conditions. *Neural Networks*, 5, 3–17.
8. Benušková L., Ebner F.F., Diamond M.E. & Armstrong-James M. (1997) Computational study of plasticity in different layers of rat barrel cortex. (Submitted to J. Computational Neurosci.)
9. Armstrong-James M., Callahan C. A. & Friedman M.A. (1991) Thalamo-cortical processing of vibrissal information in the rat. I. intracortical origins of surround but not centre-receptive fields of layer IV neurones in the rat S1 barrel field cortex. *J. Comp. Neurol.*, 303, 193–210.
10. Kawaguchi Y. (1993) Groupings of nonpyramidal and pyramidal cells with specific physiological and morphological characteristics in rat frontal cortex. *J. Neurophysiol.*, 69, 416–431.
11. Kawaguchi Y. & Kubota Y. (1993) Correlation of physiological subgroupings of nonpyramidal cells with parvalbumin-immunoreactive neurons in layer V of rat frontal cortex. *J. Neurophysiol.*, 70, 387–395.
12. Salin P.A. & Prince D.A. (1996) Electrophysiological mapping of $GABA_A$ receptor-mediated inhibition in adult rat somatosensory cortex. J. Neurophysiol., 75, 1589–1600.
13. Diamond M.E., Huang W. & Ebner F.F. (1994) Laminar comparison of somatosensory cortical plasticity. *Science*, 265,1885–1888.

Mathematical Analysis of Competition Between Sensory Ganglion Cells for Nerve Growth Factor in the Skin

Raymond Kohli[1] and Peter G.H. Clarke[2]

[1] Ecole Polytechnique Fédérale de Lausanne, Département de mathématiques, CH-1015 Lausanne, Switzerland

[2] Université de Lausanne, Institut de biologie cellulaire et de morphologie, Rue du Bugnon 9, CH-1005 Lausanne, Switzerland

Abstract. We model the competition between sensory axons for nerve growth factor (NGF) produced in the periphery. Previous models predicted the loss of all but one of the axons innervating a given region, owing to the unlimited growth of the "fittest" axon. We have imposed an upper limit to axon growth, thereby introducing new equilibria, and we show by LaSalle's theorem that several axons can then survive, depending on the rate of NGF production.

1 Introduction

Throughout the nervous system, competition between developing axons is a major determinant of how they share their target territory between them, and strongly influences which of their parent cell bodies will survive the acute phase of massive cell death that eliminates 30 - 70 % of the neurons in most neuronal populations [3]. In sensory ganglia, neuronal death is regulated by competition between neurons for a neurotrophic factor ("nerve growth factor" - NGF) produced in their axonal target territory and according to a classic hypothesis it serves to match the number of neurons to the "size" (or trophic factor producing capacity) of the target [2]. We formalize this hypothesis in an explicit model and we test whether competition for NGF is indeed a feasible means of matching neuron numbers to their target. Our model is concerned only with axonal competition in the periphery, which is believed to be the essential event determining ganglion cell survival. When an axon has diminished, through competition, to the point of receiving no NGF, we assume its parent cell body will likewise die.

2 Formulation of the model

2.1 The role of NGF receptors

NGF is not merely a survival factor, but has immediate positive effects on the growth and maintenance of axon terminals. These effects are due to the binding of NGF to specific receptors. We shall assume that the growth promoting effect of NGF is a function of the proportion of bound receptors.

2.2 Equations for binding of NGF

We assume the concentration S(t) of NGF in the extracellular space to be sufficiently uniform for us to ignore any spatial variation. Then, the proportions of bound receptors can conventionally be expressed by the Michaelis-Menten (or Langmuir) equation as $P(S) = S/(K_d + S)$, K_d being the equilibrium dissociation constants of receptors. Despite the plausibility of the Michaelis-Menten equation, we shall not need to assume its exact validity, but shall make the weaker assumption that $P(S)$ is an increasing function of S. Also, it follows from the definitions of $P(S)$ that $P(0) = 0$ and $P(\infty) = 1$.

2.3 Equation for axonal growth or regression

Of the many axonal parameters that grow or decay, we choose to deal with the number of receptors, since it is these that mediate axonal growth and NGF-removal. Let $x_i(t)$ be the number of receptors on axon i. We assume that the rate of growth will depend on the level of some second messenger in the receptor-bearing (terminal) region of the axon, and that this will be an increasing function of the proportion $P(S)$ of bound receptors. Since $P(S)$ is likewise an increasing function of S, the level of second messenger $F = F(S)$ is one also. But since growth is negative in the presence of a low or zero level of NGF, we assume that the rate of growth will be proportional to $F(S) - F(S_i^T)$, S_i^T being the concentration of NGF required for zero growth of axon i. We assume that S_i^T is a constant.

Clearly $\frac{dx_i}{dt}$ will also depend on x_i. We therefore write

$$\frac{dx_i}{dt} = H_i(x_i)[F(S) - F(S_i^T)]$$

We assume also that the growth rate will be zero when $x_i = 0$ or when x_i becomes so large that the capacity of the cell body and axon to maintain it becomes saturated. To express these conditions, we write $H_i(x_i) = x_i(X_i - x_i)G_i(x_i)$, X_i being a constant and $G_i(x_i)$ any continuously differentiable, strictly positive function of x_i. Hence,

$$\frac{dx_i}{dt} = x_i(X_i - x_i)G_i(x_i)[F(S) - F(S_i^T)]$$

2.4 Equation for the changing concentration of NGF

Since the cutaneous production of NGF is not modulated by the innervating axons, and since the cutaneous cells express few if any NGF receptors, we shall assume that the rate of NGF production Q is constant, and that the only means of NGF elimination are by the innervating axons and by passive processes such as diffusion and degradation.

NGF removal by axons involves receptor-mediated endocytosis. We assume that the endocytosis is proportional to the number of bound receptors. Hence, the axonal removal rate is $R_a(S) = \sum_{i=1}^{n} x_i[KP(S)]$. Let $W(S) = KP(S)$ and $W_i^T = KP(S_i^T)$. We express the passive removal rate as $R_p(S) = AS$. Hence,

$$\frac{dS}{dt} = Q - R_p(S) - R_a(S) = Q - AS - W(S)\sum_{i=1}^{n} x_i$$

3 Mathematical analysis

Let us rewrite the system of equations we wish to analyse:

$$\begin{cases} \frac{dS(t)}{dt} = Q - AS(t) - W(S(t))\sum_{i=1}^{n} x_i(t), \\ \frac{dx_i(t)}{dt} = x_i(t)(X_i - x_i(t))G_i(x_i(t))(F(S(t)) - F(S_i^T)),\ i = 1, \ldots, n \end{cases} \tag{1}$$

Let us determine first all the equilibria inside $D = \{(S, x_1, \ldots, x_n) \in \mathbb{R}^{n+1} \mid S \geq 0, 0 \leq x_i \leq X_i, i = 1, \ldots, n\}$, the biological domain. As it can be seen directly from (1), the set E of equilibria inside D is the union of the two following subsets $E_1 = \{(S^0, x_1^0, \ldots, x_n^0) \in D | x_i^0 = 0 \text{ or } x_i^0 = X_i,\ i = 1, \ldots, n \text{ and } S^0 \text{ such that } Q - AS^0 = W(S^0)\sum_{i=1}^{n} x_i^0\}$ and $E_2 = \cup_{r=1}^{n}\{(S^0, x_1^0, \ldots, x_n^0) \in D | S^0 = S_r^T, x_i^0 = 0 \text{ or } x_i^0 = X_i,\ i = 1, \ldots, n, i \neq r \text{ and } x_r^0 = \frac{Q - AS^0}{W(S^0)} - \sum_{i=1, i\neq r}^{n} x_i^0\}$.

We remark that at equilibrium some axons have no receptor ($x_i^0 = 0$) and others have the maximal number of receptors ($x_i^0 = X_i$); there can however be one axon with a number of receptor between 0 and X_i. Axons with no receptor are "dead"; hence the number of surviving axons is given by the number of axons with $x_i^0 \neq 0$. The next theorem shows that we can determine explicitly the number of surviving axons at the end of development as a function of the trophic factor production (Fig. 1.).

Theorem 1 *If $0 < S_1^T < S_2^T < \ldots < S_n^T < \infty$, then the solutions of system (1) for initial conditions inside the biological domain D have the following properties:*

(i) If $0 \leq Q \leq AS_1^T$, then

$$\lim_{t\to\infty} S(t) = S_1^* = \frac{Q}{A},$$
$$\lim_{t\to\infty} x_i(t) = 0 \ for\ 1 \leq i \leq n;$$

(ii) If $AS_1^T < Q < AS_1^T + W_1^T X_1$, then

$$\lim_{t\to\infty} S(t) = S_1^T,$$
$$\lim_{t\to\infty} x_1(t) = x_1^* = \frac{Q - AS_1^T}{W_1^T},$$
$$\lim_{t\to\infty} x_i(t) = 0 \ for\ 2 \leq i \leq n;$$

(iii) If $AS_{r-1}^T + W_{r-1}^T \sum_{i=1}^{r-1} X_i \leq Q \leq AS_r^T + W_r^T \sum_{i=1}^{r-1} X_i$ *for* $2 \leq r \leq n$*, then*

$$\lim_{t\to\infty} S(t) = S_r^* \textit{ with } W(S_r^*) = \frac{Q - AS_r^*}{\sum_{i=1}^{r-1} X_i} ,$$
$$\lim_{t\to\infty} x_i(t) = X_i \textit{ for } 1 \leq i \leq r-1 ,$$
$$\lim_{t\to\infty} x_i(t) = 0 \textit{ for } r \leq i \leq n ;$$

(iv) If $AS_r^T + W_r^T \sum_{i=1}^{r-1} X_i < Q < AS_r^T + W_r^T \sum_{i=1}^{r} X_i$ *for* $2 \leq r \leq n$*, then*

$$\lim_{t\to\infty} S(t) = S_r^T ,$$
$$\lim_{t\to\infty} x_i(t) = X_i \textit{ for } 1 \leq i \leq r-1 ,$$
$$\lim_{t\to\infty} x_r(t) = x_r^* = \frac{Q - AS_r^T}{W_r^T} - \sum_{i=1}^{r-1} X_i ,$$
$$\lim_{t\to\infty} x_i(t) = 0 \textit{ for } r+1 \leq i \leq n ;$$

(v) If $AS_n^T + W_n^T \sum_{i=1}^{n} X_i \leq Q < \infty$*, then*

$$\lim_{t\to\infty} S(t) = S_{n+1}^* \textit{ with } W(S_{n+1}^*) = \frac{Q - AS_{n+1}^*}{\sum_{i=1}^{n} X_i}$$
$$\lim_{t\to\infty} x_i(t) = X_i \textit{ for } 1 \leq i \leq n .$$

In words, the meaning of this theorem is as follows. The system is globally stable, with a unique equilibrium that depends on the parameters but not on the initial conditions. If the production, Q, is varied but the other parameters are held constant, the equilibrium values vary as follows. At low Q, no axons survive but the final concentration of NGF, S^*, is proportional to Q until a critical value above which one axon can survive. Further increases in Q leave S^* unchanged but increase linearly the final number of receptors on this axon until its upper limit is reached. Still further increases in Q raise either S^* or the final number of receptors on new axons, but not both at the same Q value, until all axons survive with their maximal number of receptors.

To prove this theorem, we use LaSalle's theorem [5]. Space does not permit us to give a complete proof, but the essence was to find a Liapunov function. A different one is required for each part of our theorem. For parts (i), (ii), (iii), (iv) and (v), the corresponding Liapunov functions are:

(i) $V(S, x_1, \ldots, x_n) = \int_{S_1^*}^{S} \frac{F(s)-F(S_1^*)}{W(s)} ds + \sum_{i=1}^{n} \int_0^{x_i} \frac{dx}{(X_i-x)G_i(x)}$,

(ii) $V(S, x_1, \ldots, x_n) = \int_{S_1^T}^{S} \frac{F(s)-F_1^T}{W(s)} ds + \int_{x_1^*}^{x_1} \frac{(x-x_1^*)dx}{x(X_1-x)G_1(x)} + \sum_{i=2}^{n} \int_0^{x_i} \frac{dx}{(X_i-x)G_i(x)}$,

(iii) $V(S, x_1, \ldots, x_n) = \int_{S_r^*}^{S} \frac{F(s)-F(S_r^*)}{W(s)} ds - \sum_{i=1}^{r-1} \int_{X_i}^{x_i} \frac{dx}{xG_i(x)} + \sum_{i=r}^{n} \int_0^{x_i} \frac{dx}{(X_i-x)G_i(x)}$,

(iv) $V(S, x_1, \ldots, x_n) = \int_{S_r^T}^{S} \frac{F(s)-F_r^T}{W(s)} ds - \sum_{i=1}^{r-1} \int_{X_i}^{x_i} \frac{dx}{xG_i(x)} + \int_{x_r^*}^{x_r} \frac{(x-x_r^*)dx}{x(X_r-x)G_r(x)} + \sum_{i=r+1}^{n} \int_0^{x_i} \frac{dx}{(X_i-x)G_i(x)}$,

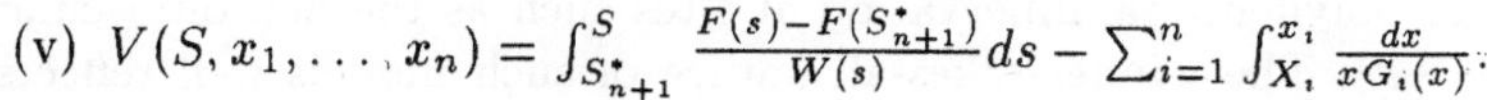

(v) $V(S, x_1, \ldots, x_n) = \int_{S_{n+1}^*}^{S} \frac{F(s)-F(S_{n+1}^*)}{W(s)} ds - \sum_{i=1}^{n} \int_{X_i}^{x_i} \frac{dx}{xG_i(x)}$.

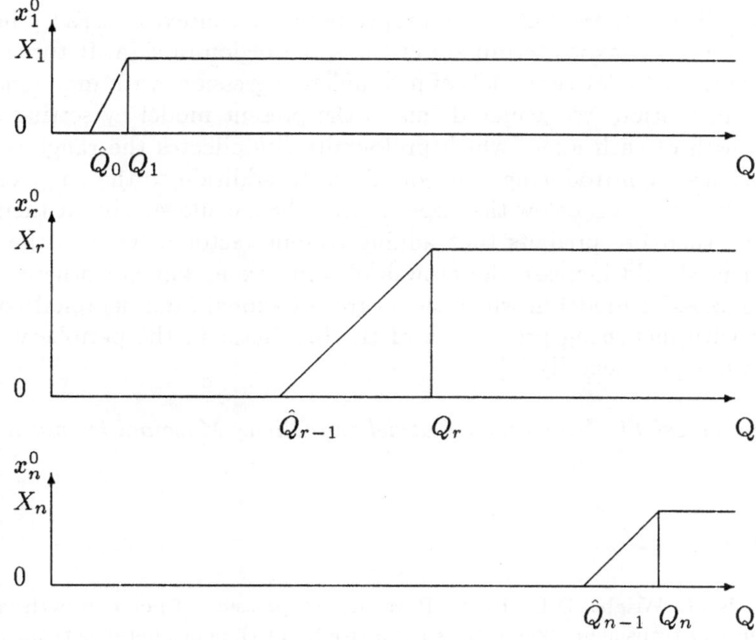

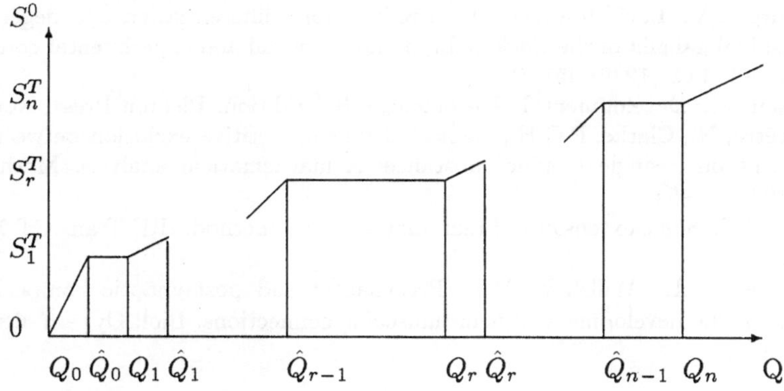

Fig. 1. The graphs of final values of the x_i and S as a function of the parameter Q with $Q_0 = 0$, $\hat{Q}_0 = AS_1^T$, $Q_1 = W_1^T X_1 + AS_1^T$, $\hat{Q}_1 = W_2^T X_1 + AS_2^T$, $\hat{Q}_{r-1} = W_r^T \sum_{i=1}^{r-1} X_i + AS_r^T$, $Q_r = W_r^T \sum_{i=1}^{r} X_i + AS_r^T$, $\hat{Q}_r = W_{r+1}^T \sum_{i=1}^{r} X_i + AS_{r+1}^T$, $\hat{Q}_{n-1} = W_n^T \sum_{i=1}^{n-1} X_i + AS_n^T$, $Q_n = W_n^T \sum_{i=1}^{n} X_i + AS_n^T$.

4 Discussion

All previous attempts at modelling interaxonal competition were intended to explain the loss of polyneuronal innervation at sites such as the neuromuscular junction ([6] and [4]). The winner-takes-all nature of such models well reflects certain biological situations, where the fittest axon becomes so strong that it eliminates all the others, but is inappropriate in the context of sensory innervation, where several axons terminate in the same region of skin. It turns out to be quite difficult to devise models of incomplete regression with maintenance of multiple innervation. We achieved this in the present model by setting a limit to the growth of each axon, which profoundly complicates the range of possible behaviours by introducing new equilibria. In addition to the one axon that could survive at a level below the upper limit, others could survive at their limit. Our model therefore predicts that adding trophic factor in vivo at the end of development should increase the growth of some axons but not others. This is the first successful model in which the degree of axonal (and neuronal) survival increases with increasing production of trophic factor in the periphery, as has been shown experimentally [1].

We are grateful to Jean-Pierre Gabriel and Olivier Maggioni for much helpful advice.

References

1. Albers, K.M., Wright, D.E., Davis, B.M.: Overexpression of nerve growth factor in epidermis of transgenic mice causes hypertrophy of the peripheral nervous system. J.Neurosci. **14** (1994) 1422–1432.
2. Hamburger, V., Levi-Montalcini R.: Proliferation, differentiation and degeneration in the spinal ganglia of the chick embryo under normal and experimental conditions. J.Exp.Zool. **111** (1949) 457–501.
3. Jacobson, M.: Developmental Neurobiology, 3rd Edition. Plenum Press, New York.
4. Jeanprêtre, N., Clarke, P.G.H., Gabriel, J.P.: Competitive exclusion between axons dependent on a single trophic substance: A mathematical analysis. Math.Biosci. **135** (1996) 23–54.
5. LaSalle, J.P.: Some extensions of Liapunov's second method. IRE.Trans.CT **7** (1960) 520–527.
6. Rasmussen, C.E., Willshaw, D.J.: Presynaptic and postsynaptic competition in models for the development of neuromuscular connections. Biol. Cybern. **68** (1993) 409–419.

Competition Amongst Neurons for Neurotrophins

Arjen van Ooyen and David J. Willshaw

Centre for Neural Systems,
Centre for Cognitive Science, University of Edinburgh,
2 Buccleuch Place, Edinburgh EH8 9LW, United Kingdom.
e-mail: A.vanOoyen@cns.ed.ac.uk

Abstract. In the development of nerve connections, neurons are believed to compete for target-derived neurotrophic factors which support their survival and maintenance of their synapses. We introduce a mathematical framework for neurotrophin release and its uptake by the innervating neurons. We explore the idea that central to the action of neurotrophins is their capacity to upregulate their own receptors.
Using nerve growth factor (NGF) as the paradigm case, we show theoretically how the form of the upregulation determines the nature and outcome of the competitive process. Under some conditions, the target structure becomes singly innervated; under others, multiple innervation results, the amount of multiple innervation depending on the supply of neurotrophins. The finding that electrical activity increases the numbers of receptors means that competition for neurotrophin amongst synapses leads to the survival of the more active ones. Reduction in receptor upregulation or in the supply of neurotrophin (which may occur in ageing and disease-related neurodegeneration), can lead to a complete loss of innervation.
Our model encompasses previous models of neuronal competition during development and couples the field of neurobiology to that of population biology, where the notion of competition is better developed.

1 Introduction

During the development of the nervous system, neurons become assembled into networks by making synaptic connections with other neurons. In many cases there is not a simple, inflexible pairing together of particular pre- and postsynaptic elements but there is competition among neurons. The crucial question concerns the nature of this competition.

In order to survive, neurons and synapses need to obtain sufficient amounts of so-called neurotrophic factors. These survival-promoting substances are released in limited amounts by the neurons' target elements. They are taken up by the innervating neurons, which are thought to compete for them. Neurons or synapses that fail to obtain sufficient amounts will die. The notion of competition has been used to explain axonal elimination during, for instance, the innervation of mammalian skeletal muscle [1], cerebellar Purkinje cells [2], autonomic ganglion cells [3] and the formation of ocular dominance columns [4].

An important class of neurotrophic factors are the neurotrophins, with NGF (nerve growth factor) being its best characterised member [5]. Neurotrophins bind to their specific receptors at the synaps, resulting in activation of the receptor. The ligand-receptor complex is then internalised, and the neurotrophin and activated receptor are retrogradely transported to the cell body. The activated receptor, by triggering a signalling cascade, has direct local effects in the synapse on growth and maintenance, and on somatic processes such as neuronal survival.

A key observation is that neurotrophins can upregulate their own receptors: exposure to NGF causes a slow increase in the number of its receptors [5, 6, 7, 8, 9]. Correspondingly, the amount of receptors increases markedly as initial target contact is made [10]. Thus neurons that get a good supply of neurotrophin increase their number of receptors and become even more competent to bind neurotrophin, and so might outcompete neurons that do not. The number of neurotrophin receptors is also increased by membrane depolarization in the innervating neurons [11].

Although the notion of competition is commonly used, it is usually applied in a rather unspecified manner to cover any situation where neurons appear to hinder each other. The underlying mechanisms and the nature of the competitive process are largely unknown. Only a few formal models exist, for a few systems only, such as the neuromuscular junction [12]. In contrast, in the field of theoretical population biology the concept of competition (in the sense of different consumers competing for the same resources) is well developed, and has been studied by means of many mathematical models [13]. Benefiting from the insights obtained in population biology, we have developed a novel model to investigate the nature of neuronal competition.

2 Model

A single target is considered, at which there are n synapses, each from a different neuron. The maintenance of the synapses depends on the neurotrophic factor released by the target. Although all neurotrophins have many properties in common, the model is specifically based on NGF (nerve growth factor). We model the neurotrophin concentration (L) in the extracellular medium of the target cell, assumed to be uniform, and for each synapse i the concentration of unoccupied (R) and occupied receptors (C). The dynamics, which apply to any type of neurotrophin, are described by the following set of equations:

$$
\begin{aligned}
\frac{dC_i}{dt} &= k_i^a L R_i - k_i^d C_i - \rho_i C_i \\
\frac{dR_i}{dt} &= \phi_i - \gamma_i R_i - k_i^a L R_i + k_i^d C_i \\
\frac{dL}{dt} &= \sigma - \delta L - \textstyle\sum_{j=1}^{n} (k_j^a L R_j - k_j^d C_j),
\end{aligned}
\tag{1}
$$

There are parameters for the rate of neurotrophin release (σ), degradation and diffusion away of neurotrophin (δ), production of receptors (ϕ), turnover of unoccupied receptors (γ), association and dissociation of neurotrophin to the receptor (k^a and k^d), and degradation of the occupied receptor (ρ). As the functional response of neurotrophins is mediated by receptor activation (see Introduction), the state of a synapse is determined by the amount of occupied (i.e., activated) receptor, C; if $C = 0$ the synapse cannot be maintained.

As exposure to neurotrophins has been found to upregulate their own receptors, the production rate of receptors, ϕ, is not a fixed parameter but a dynamical variable that changes in response to neurotrophin. We model it with

$$\tau \frac{d\phi_i}{dt} = f_i(C_i) - \phi_i, \tag{2}$$

where $f_i(C_i)$ is the dose-response function determining the steady-state value of ϕ_i. The time constant τ is of the order of many hours or even days, reflecting the fact that the number of receptors changes slowly in response to neurotrophin, since it involves slow processes such as the expression of genes. The dynamics of C, R, and L are orders of magnitude faster, and on the time scale of the dynamics of ϕ they are essentially in equilibrium. In the analysis of the model we make a quasi steady-state approximation for these variables.

3 Results

If there were no upregulation, i.e., $f_i(C_i)$ is a constant or a decreasing function, all synapses would survive. Although f_i has been investigated experimentally [14], its exact functional form is not known. We therefore studied three basic and well-known types of dose-response curves consistent with upregulation. The values of the parameters of the curves could differ between innervating neurons, and may depend on, among other things, the level of their electrical activity.

Type I: monotonically increasing function. We use $f_i(C_i) = \alpha_i C_i$. We find that, generically, at most one C_i can be non-zero, i.e., at most one synapse can be maintained in equilibrium (see Appendix and Fig. 1a). This is irrespective of initial conditions or supply of neurotrophins (value of σ). Our result is analogous to the well-known "competitive exclusion" principle from population biology saying that k different resources can maximally sustain k different species ($k = 1$ in our case). The synapse with the highest value of $\beta_i \equiv (\alpha_i - \rho_i)k_i^a/\gamma_i\delta(k_i^d + \rho_i)$ will outcompete all others. It can, however, never be maintained in equilibrium if $\beta_i \leq 1/\sigma$.

Competitive exclusion could explain that during development cells lose all but one of their innervating axons, as is the case for skeletal muscle, cerebellar Purkinje cells, and autonomic ganglion cells lacking many dendrites.

Type II: Michaelis-Menten function. This dose-response curve, described by $f_i(C_i) = \alpha_i C_i/(K_i + C_i)$, is also monotonically increasing but gradually saturates towards an upper bound. Depending on the parameter settings, either one synapse will survive (competitive exclusion), or, which is now also possible, several different synapses (coexistence, see Fig. 1b). The more synapses can coexist the larger σ and the smaller K_i. In the limiting case, all synapses will survive. Again, there is no dependence on initial conditions. For large K (with values of α increased correspondingly), the model becomes equivalent to the model with the type I curve. When only one synapse survives, it will the one with the highest value of $\beta_i' \equiv (\alpha_i/K_i - \rho_i)k_i^a/\gamma_i\delta(k_i^d + \rho_i)$. A synapse can, however, never be maintained in equilibrium if $\beta_i' \leq 1/\sigma$. The order in which synapses survive if σ is increased is determined by the values of β_i'.

A type II curve could explain that, in addition to single innervation, many kinds of neurons retain several input axons.

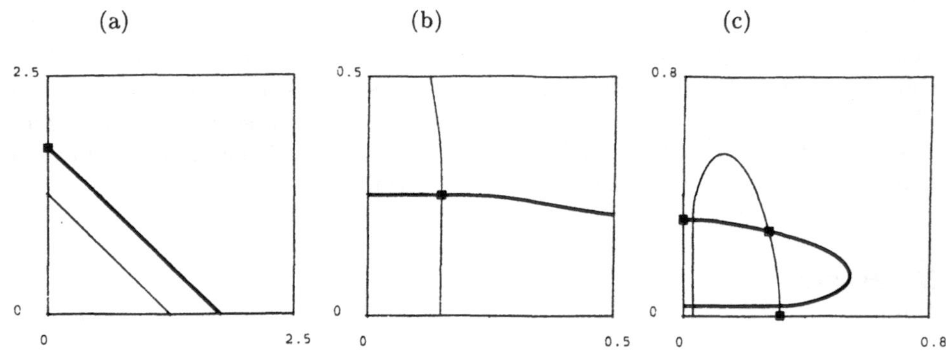

Fig. 1. For $n = 2$ the nullclines $\frac{d\phi_1}{dt} = 0$ (thin line) and $\frac{d\phi_2}{dt} = 0$, with quasi-steady state conditions of the other variables; x-axis: ϕ_1, y-axis: ϕ_2; ■ indicate stable equilibrium point. **a.** Type I: competitive exclusion. **b.** Type II: coexistence. **c.** Type III: coexistence and exclusion. Notice that $\phi_i = 0$ means $C_i = 0$.

Type III: sigmoidal function. We use the Hill equation, $f_i(C_i) = \alpha_i C_i^m/(K_i + C_i^m)$ with a Hill coefficient, $m = 2$, which also saturates but has a particularly low response at low C. Independent of the parameter settings, for each synapse there is a stable equilibrium point where only this synapse is present and all others are extinct. In contrast to the case with the type II function, coexistence equilibrium points can be present at the same time (see Fig. 1c), depending on the parameter settings. Which equilibrium point is approached, a coexistence equilibrium point (if present) or one where only one of the synapses survives, depends on the initial values of ϕ_i. The nullcline configurations (Fig. 1c) are similar to those found in the dual constraint model of the neuromuscular junction [12], which describes competition among motor neurons in the innervation of muscle fibres. Changing the values of β_i' changes the basin of attractions of the equilibrium points.

Electrical activity. For a neuron i the actual values of the parameters of the dose-response function, as well as those of the other parameters, will depend on, among many other things, the level of the neuron's electrical activity. Membrane depolarization increases the number of receptors [9], which implies a higher ϕ_i (and thus a higher α_i or lower K_i), or a lower γ_i. A high level of activity thus means a high β_i (β_i'), giving active innervating neurons a competitive advantage.

With both Type I and Type II curves, a synapse can displace existing synapses if its value of β_i. is higher. As a consequence, the average value of β_i and the level of electrical activity for the surviving synapses will increase during development. This selection process is reminiscent of a competitive mechanism for affinity selection of T-cell clones for antigen during the course of an immune response. With a Type III curve a synapse with a higher β_i' can also invade provided its initial value of ϕ_i is high enough.

4 Discussion

The model shows how the dynamics of neurotrophin signalling leads to neuronal competition. The model integrates other models of competition (e.g. [12] and [15]), and opens up an important area of neurobiology to the concepts from population biology. Notice that our model involves a continuous supply of neurotrophin (resource) that is used up and constantly required by the synapses. Models that are based on a fixed amount of neurotrophin that becomes divided up, however, also falls into our general framework. We obtain this by choosing $\sigma = \delta = 0$, and $\rho_i = 0$ for all i, and thus having the conservation law $L = L_0 + \sum_i^n C_i$, where L_0 is the total amount of neurotrophin. With this choice, we obtain very similar results for all types of dose-response curves.

The nature of the competitive interactions is determined by the way neurotrophins regulate their own uptake. Experimental studies should focus on measuring this, also for other neurotrophins than NGF. Differences in dose-response functions for different cell types or neurotrophins could explain different innervation patterns.

To illustrate the main points, we studied simplified neurons without explicitly modelling neuritic extensions. Target neurons with an extensive dendritic tree, resulting in a non-uniform distribution of neurotrophin through local release, local uptake and diffusion, may mitigate competition. In this way, the morphology of the dendritic tree could also determine the pattern of innervation.

5 Appendix

Making a quasi steady-state approximation for L, R, and C (i.e., $\frac{dL}{dt} = \frac{dR}{dt} = \frac{dC}{dt} = 0$) yields $\phi_i = C_i[\rho_i + b_i/(\sigma - a)]$, where $a = \sum_j^n \rho_j C_j$ with n the total number of synapses, and $b_i = \gamma_i \delta(k_i^d + \rho_i)/k_i^a$. We consider the dose-response function $f_i(C_i) = \alpha_i C_i$. In equilibrium all synapses i have to satisfy $\frac{d\phi_i}{dt} = 0$, and thus for each i (using the quasi steady-state expression for ϕ_i), $C_i = 0$ or

$\alpha_i - \rho_i - b_i/(\sigma - a) = 0$. When considered for all i, the latter equation is a linear system of n equations in one unknown, a. Generically, at most one of these can be satisfied. Hence, at most one C_i can be non-zero, i.e., at most one synapse can be maintained in equilibrium.

References

1. Jansen, J. K. S., Fladby, T. (1990) The perinatal reorganization of the innervation of skeletal muscle in mammals. *Prog. Neurobiol.* 34: 39-90.
2. Crepel, F. (1982) Regression of functional synapses in the immature mammalian cerebellum. *Trends Neurosc.* 5: 266-269.
3. Purves, D. (1988) *Body and Brain: A Trophic Theory of Neural Connections*, Harvard Univ. Press, Cambridge, MA.
4. Wiesel, T. N. (1982) Postnatal development of the visual cortex and the influence of the environment. *Nature* 299: 583-591.
5. Bothwell, M. (1995) Functional interactions of neurotrophins and neurotrophin receptors. *Ann. Rev. Neurosc.* 18: 223-253.
6. Bernd, P., Greene, L. A. (1984) Association of I^{125}-nerve growth factor with PC12 pheochromocytoma cells. Evidence for internalization via high affinity receptors only and for long-term regulation by nerve growth factor of both high- and low-affinity receptors. *J. Biological Chemistry* 259 (24): 15509-15516.
7. Holtzman, D. M., Li, Y., Parada, L. F., Kinsman, S., Chen, C.-K., Valletta, J. S., Zhou, J., Long, J. B., Mobley, W. C. (1992) p140^{trk} mRNA marks NGF-responsive forebrain neurons: evidence that *trk* gene expression is induced by NGF. *Neuron* 9: 465-478.
8. Verge, V. M. K., Merlio, J.-P., Grondin, J., Ernfors, P., Persson, H., Riopelle, R. J., Hokfelt, T., Richardson, P. M. (1992) Colocalization of NGF binding sites, trk mRNA, and low-affinity NGF receptor mRNA in primary sensory neurons: responses to injury and infusion of NGF. *J. Neurosc.* 12 (10): 4011-4022.
9. Zhou, J., Valletta, J. S., Grimes, M. L., MObley, W. C. (1995) Multiple levels for regulation of TrkA in PC12 cells by nerve growth factor. *J. Neurochemistry* 65: 1146-1156.
10. Wyatt, S., Shooter, E. M., Davies, A. M. (1990) Expression of the NGF receptor gene in sensory neurons and their cutaneous targets prior to and during innervation. *Neuron* 2: 421-427.
11. Black, I. B. (1993) Environmental regulation of brain trophic interactions. *Int. J. Dev. Neurosc.* 11: 403-410.
12. Rasmussen, C. E., Willshaw, D. J. (1993) Presynaptic and postsynaptic competition in models for the development of neuromuscular connections. *Biol. Cybern.* 68: 409-419.
13. Yodzis, P. (1989) Introduction to Theoretical Ecology, Harper and Row, New York.
14. Doherty, P., Seaton, P., Flanigan, T. P., Walsh, F. S. (1984) Factors controlling the expression of the NGF receptor in PC12 cells. *Neuroscience Letters* 92: 222-227.
15. Jeanpretre, N., Clarke, P. G. H., Gabriel, J.-P. (1996) Competitive exclusion between axons dependent on a single trophic substance: a mathematical analysis. *Mathematical Biosciences* 135: 233-54.

Implementing Hebbian Learning in a Rank-Based Neural Network

Manuel Samuelides[1,2], Simon Thorpe[3] and Emmanuel Veneau[1,3]

[1] Ecole Nationale Supérieure de l'Aéronautique et de l'Espace, BP 4032, 31055 ,Toulouse Cedex, France.
[2] ONERA-CERT, Computer Science Department, BP 4025, 31055 Toulouse Cedex France.
[3] Centre de Recherche Cerveau et Cognition, CNRS, Faculté de Médecine, 133 route de Narbonne - 31062, Toulouse, France.

Abstract. Recent works have shown that biologically motivated networks of spiking neurons can potentially process information very quickly by encoding information in the latency at which different neurons fire, rather than by using frequency of firing as the code.In this paper, the relevant information is the rank vector of latency order of competing neurons. We propose here a Hebbian reinforcement learning scheme to adjust the weights of a terminal layer of decision neurons in order to process this information. Then this learning rule is shown to be efficient in a simple pattern recognition task. We discuss in conclusion further extensions of that learning strategy for artificial vision.

1 Introduction

In the vast majority of artificial neural-network architectures, the activation state of the individual units is either a binary variable (as in the original McCulloch-Pitts formulation), or a continuous function, typically taking values between 0 and 1. Continuous activation functions are generally believed to correspond to the firing rates of biological neurones. However, as [2] pointed out, there are situations in which the speed of neural computation is too fast to be able to make use of firing rate codes, simply because individual neurones will only have enough time to generate one spike. One way of overcoming such temporal constraints is to take advantage of the fact that even the most simple integrate-and-fire neurones can be effectively thought of as analog-to-delay convertors in that the time needed for such a neuron to reach threshold and generate a spike will depend on the strength of its input Such ideas have received increasing interest in the last few years [1, 3, 6]. Recently it has been proposed in [4] that, instead of using the relative latency values directly, one could make use of the order in which the neurons fire.
In this paper, after providing a mathematical description of this model, we propose a Hebb-like reinforcement learning mechanism with constant learning rate and test it in a one-dimensional pattern classification problem. Then

we conclude by discussing the extension of this learning algorithm to multi-layered networks.

2 Model of a rank-based neural network

We use a simple integrate and fire model to describe spike generation by the input neurons to mimic the spike emission. The evolution equation of the activation potential of a neuron is

$$v(t+1) = \begin{cases} (1-\alpha)v(t) + f[u(t)] & \text{when } v(t) < v_f \\ v_0 & \text{when } v(t) \geq v_f \end{cases} \quad (1)$$

where v_f is the neuron firing threshold, v_0 is the neuron ground potential, α is the leakage rate per time unit, and f is a transfer integrating function of the neuron input $u(t)$ at time t.

The network architecture under consideration is a feed-forward architecture with two layers. The first layer is a preprocessing layer which is activated by a particular intensity profile, considered to be constant during processing. In the model we present here, the neurons of the first layer are in a one-to-one correspondance with the pixels of the presented pattern. So the transfer integrating function is an increasing sigmoidal function of the pixel intensity u. Since the architecture we study is dedicated to a recognition task, the neuron of the second layer are decision neurons. They receive input signals from all the neurons of the first layer. These signals are ponderated by synaptic weights and summed to form the input of the transfer integration function of the decision neuron. We shall propose in section 3 a Hebbian learning rule to adjust these weights

We define the *latency* λ of a neuron as the time of the first spike. From the equation (1) it is easy to compute the expression of the latency (in the low α approximation) $\lambda = \frac{v_f - v_0}{f(u)}$

However, although measuring delays between the arrival times of spikes originating from different sources is used in a great number of sensory systems (including echolocation in bats, electric fish, and auditory localisation), it tends to require very large amounts of neuronal machinery. In contrast, determining the order in which the inputs to a neurone arrive is in principle much easier. It was proposed in [4] to use a mechanism similar to the sorts of activity-dependent synaptic depression reported recently [5] to progressively desensitise the target neuron as a function of the number of inputs that have already been activated. According to this hypothesis, the response of a post-synaptic neurone to one of its inputs would depend not only on the effective weight of the synapse, but also on a modulatory factor that controls the neurons sensitivity. In our model, we implement the following mechanism: the synaptic efficiency w_i^j of the synapse from neuron j to neuron i is decreased by a given modulation rate β each time a spike is received by neuron i. After the image is processed, it is reset at its original value. Then the modulated input received by neuron i at time t from neuron j is $u_{ij}(t) = \beta^{r_j} x_j(t)$ where r_j is the rank of the activity of neuron j classified in a decreasing order among other afferent neurons. So if the leakage rate of the target neuron is neglected, the total activity of the target neuron i at time t is

$$v_i(t) = \sum_j w_{ij} \beta^{r_j} x_j(t) \tag{2}$$

At the end of the process,the activity of the target neuron depends only on the order of the activities of the contributing afferent neurons.
This rank-based coding is quite robust against perturbations or distortions of the stimulus. Indeed most perturbations will not affect the ranks if the neuronal activities are clearly distinct or will exchange ranks of two successive neuronal spikes. Some values for r_j in equation (2) may be shifted of plus one or minus one but this will have little effect on the final activation level of the target neuron. Moreover, the response of the output neurons will be effectively invariant to global shifts in input intensity or to reductions in contrast.
However, the mechanism of response is quite selective. To illustrate this, let the connection weights be equally spaced in [0,1] and let $\beta = 0.8$. Suppose a random input feeds the first layer, then the law of $v = \sum_j w_j \beta^{r_j}$ follows approximately a normal law with mean 2.5 and standard deviation 0.5 provided the number of weights is large enough. An explicit computation shows that the optimal distribution of ranks (for the given weights) generates an activity more than 3 standard deviations away from the mean.

3 Reinforcement learning of prototypes

Here we show that a Hebbian reinforcement learning rule is capable of adjusting the synaptic weights in order to perform a pattern classification task.The Hebbian learning rule states that the synaptic weights of a firing neuron are increased whenever they convey signals from afferent neuron that contributed to the firing. In the context of the rank coding model, this learning rule is translated into the following formula (the origin of time is the instant at which the stimulus have been presented):

$$w_{ij}(t+1) = w_{ij}(t) + \gamma x_i(t) \left(\sum_{k=0}^{k=t} \alpha^k u_{ij}(t-k) - w_{ij}(t) \right) \tag{3}$$

The constant γ is the learning rate and takes a value such that $0 < \gamma < 1$. This rule may be implemented in a pure unsupervised mode or in a reinforcement learning context. We shall study here reinforcement learning. In that case, during the learning process, a reinforcement signal is sent to a specific output neuron to force it to fire a fixed time interval after each presentation of a pattern selected in the training class of that neuron.

The class of patterns to be learnt by a decision neuron is represented by a probability distribution over the space of patterns. In the reinforcement learning scheme we present, there is no modification of the synaptic weights if other patterns are presented to the system. If a relevant pattern for the target neuron i is presented to the system at learning step τ ,then, if we neglect the leakage rate of target neuron i, the afferent weights are changed according the stochastic recurrence

$$w_{ij}(\tau) = w_{ij}(\tau - 1) + \gamma[\beta^{r_j(\tau)} - w_{ij}(\tau - 1)] \tag{4}$$

where r_j is the rank of the activity of neuron j when the pattern is presented at the learning step τ. Since the relevant patterns form an independant sample of the probability distribution defining the class to be learnt, the random variables $(\beta^{r_j(\tau)})$ are independant identically distributed variables with mean m_{ij} and standard deviation σ_{ij}. Then for given (i, j), $(w_{ij}(\tau))$ is an ergodic Markov chain. Its stationary law has a mean equal to m_{ij} and a variance equal to $\frac{\sigma_{ij}^2 \gamma}{2-\gamma}$ It follows that for a small learning rate, the weights will fluctuate weakly around the mean value of their input synaptic signal over the class of patterns of interest . Moreover, for a small learning rate, the approximate form of the stationnary law for the weights is gaussian, so it is possible to control precisely the size of the fluctuation by the learning rate.

4 Simulation results

To check the practical interest of the hebbian learning rule proposed here, we tested it on a simple problem of discrimination between six intensity profiles. They are coded on 30 pixels. (see figure 1 below) We are interested by

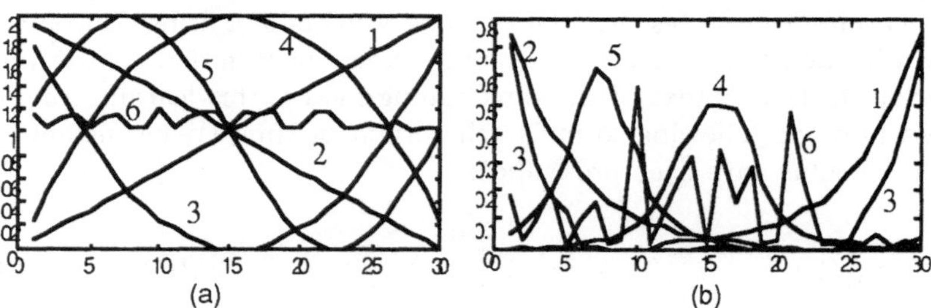

Fig. 1. Diagrams of the prototypes (a) and of the associate synaptic weights (b)

the general shapes, the rough localization of maxima and minima. So the six classes to be learnt are defined around these prototypes by multiplicative and additive perturbation by white noise. The learning process is rather fast: with a learning rate of 0.05, 50 learning steps are enough to reach a close neighbourhood of the stationary state. We present here results for 100 learning steps. Recall that the convergence speed of learning process is controlled and does not depend on the size of the network.

Let us examine first how the weights are coding the shapes. Note that the coded shapes are a rough reproduction of the maxima of the original profiles (a kind of adaptive thresholding). We verified that the coding is contrast invariant. The thresholding effect depends on the modulation rate β of the synaptic transmission.In that simulation, we choose $\beta = 0.8$. This rate is adapted to the size of the retina and to our discrimination task. Higher values would give more faithful codes of the original profiles but would not necessarily improve the discrimination ability of the whole system. A more difficult discrimination task would require a more faithful encoding of the patterns. A richer set of attributes may be more efficient than a higher β

The results of simulation are presented in the figure 2. They represent the

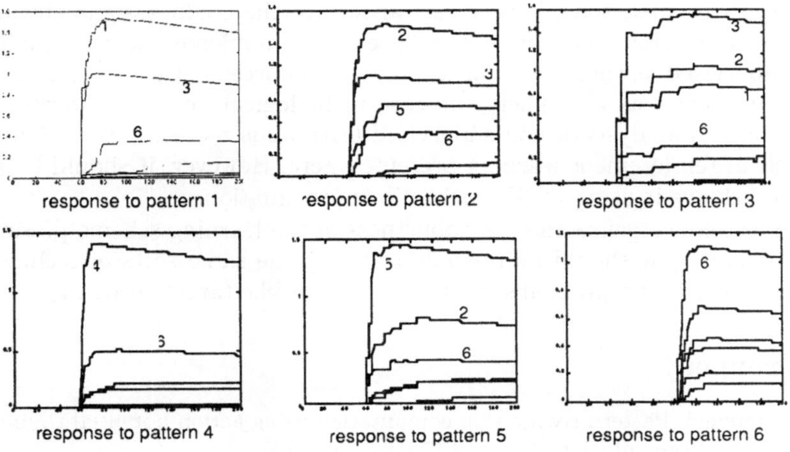

Fig. 2. Generalization response of trained target neurons. The curves represent the variations of neuronal activation potentials versus time. A neuron is labelled by the associate pattern during the learning process

response of the specific decision neurons when we presented corrupted versions of the prototypes (additive and multiplicative noise). The system is generalizing quite well. As soon as the maxima of the profile are correlated to the "blobs" of the associated code (see figure 1b), the pattern is correctly detected. One may notice that the response to the convex parabola (pattern 3) and to the monontonous lines (patterns 1 and 2) are strongly correlated . The response of decision neuron 6 (trained to detect a white noise realization) is not clear since the system detects noise in other patterns.

5 Discussion and Conclusion

In this paper, a Hebbian learning algorithm has been proposed to endow latency based feedforward neural networks with the ability of learn. The originality of this algorithm is that the weight increment depends only on the rank of spikes arriving from a particular input and not directly on the intensity of the stimulus. The type of learning considered here is reinforcement learning. This algorithm has been proved to be efficient and robust on small recognition tasks.
Moreover, it has been shown versatile enough to be used in more complex architectures where several attributes are contributing to the decision. To test more complex and plausible architecture, we fed the decision layers with several different attribute maps. For instance, in the previous example, by adding a second layer to encode low intensity locations, we improved the discrimination. The next challenge is to implement learning in multilayer architectures where the information is processed through several layers. In such systems, the reinforcement learning scheme cannot be applied to the lower level layers because there is no direct answer to the credit assignment problem. So the Hebbian learning scheme need to be implemented by combining non- supervised learning at lower levels and reinforcement learning at higher processing levels as it is likely the case in biological neural networks. The mathematical analysis of non supervised learning is more complex than the analysis of reinforement learning presented here. However, it should be possible to achieve it using ODE stochastic approximation.
So the fast convergence and the robustness of the learning scheme presented here mean that it should be possible to apply the same sorts of techniques to more challenging problems in artificial vision like target recognition.

References

1. J.J. Hopfield. Pattern recognition computation using action potential timing for stimulus representation. *Nature* **(376)**:33–36, 1995
2. S.J. Thorpe. Spike arrival times: a highly efficient coding scheme for neural networks in *"Parallel processing in neural system"* R.Eckmiller, G.Hartman and G.Hauske (Eds.) North Holland:Elsevier. 91–94, 1990
3. S.J. Thorpe, D. Fize, C. Marlot. Speed of processing in the human visual system. *Nature* **(381)**:520–522, 1996
4. S.J. Thorpe, J. Gautrais. How can the visual system process a natural scene in under150 ms? On the role of asynchronous spike propagation. In *Proceeding of the 5th European Symposium on Artificial Neural Networks* M. Verleysen (Eds.), Bruxelles: De Facto.(in press)79–84, 1997
5. M.V. Tsodyks, M.V. Markram. The neural code between neocortical pyramidal neurons depends on neurotransmitter release probability. *Proc Natl Acad Sci USA*(**94**):719–723, 1997
6. F. Worgotter, R. Opara, K.Funke, U. Eysel. Using latency for object recognition in real and artificial neural networks. *Neuroreport* **(7)**:741–744, 1996

A Model of Clipped Hebbian Learning in a Neocortical Pyramidal Cell

Bruce Graham and David Willshaw

Centre for Cognitive Science, University of Edinburgh
2 Buccleuch Place, Edinburgh EH8 9LW, Scotland, UK
E-mail: *B.Graham@ed.ac.uk* & *D.Willshaw@ed.ac.uk*
Phone: +44 131 650 4408; Fax: +44 131 650 6626

Abstract. A detailed compartmental model of a cortical pyramidal cell is used to determine the effect of the spatial distribution of synapses across a dendritic tree on the pattern recognition capability of the neuron. By setting synaptic strengths according to the *clipped Hebbian* learning rule used in the associative net neural network model, the cell is able to recognise input patterns, but with a one to two order of magnitude decrease in performance compared to the computing units in the network model. Performance of the cell is optimised by particular forms of input signal, but is not altered by different pattern recognition criteria.

1 Introduction

The experimental phenomena of long-term potentiation (LTP) and long-term depression (LTD) suggest certain synaptic pathways within the mammalian central nervous system are modified in a *Hebbian* fashion (for a review see [4]). Neural network models of associative memory use Hebbian learning for pattern storage. Such memories are often proposed as models of cortical networks and their capacities are used to indicate the potential storage capacity of cortical regions, such as the hippocampus [1, 8, 11, 13]. However, the computing units in such models are idealised versions of biological neurons and do not include many potential sources of noise inherent in real neurons.

Here we compare the memory performance of an output unit in an associative net model [14, 15] with that of a realistic compartmental model of a cortical pyramidal cell. The same set of input patterns is associated with the output unit and the pyramidal cell by setting synaptic strengths according to a *clipped Hebbian* learning rule. During pattern recall input patterns are presented to the output unit, or the cell, as cues, and the unit, or cell, should respond preferentially to the patterns with which it was coactive during pattern storage. The decision to respond is based on the weighted sum of the inputs, or *dendritic sum*. The output unit receives a dendritic sum that is simply the number of active inputs to which the unit is connected. On the other hand, the neuron receives input signals distributed across its dendritic tree. These signals vary over time and must travel to the cell body (soma) to be integrated into the final dendritic sum. The temporal nature of the signals and their spatial distribution introduces noise into the dendritic sums that is not present in the idealised associative net model. Measuring performance by the signal-to-noise ratio (S/N) between the dendritic sums due to the patterns that should be recognised and the sums due to those that should not be recognised reveals a one to two order of magnitude reduction in the S/N in the biological case. Thus current model estimates of cortical capacity are likely to be inflated.

2 The Models

In the associative net [14, 15], input and output units with binary activity are connected by synapses that are also binary. Pairs of patterns are stored in the net using a *clipped Hebbian* learning rule that changes the weight of a synaptic connection from 0 to 1 if both input and output units are active for the same pattern pair. Though extremely simple, this learning rule provides both high memory capacity and information efficiency if the patterns are sparse (few active inputs per pattern) [8, 15]. We use a net with 4096 input units, which is of the order of the number of excitatory inputs to a cortical pyramidal cell. Input patterns contain 400 randomly chosen active units. Output activity is set so that an output unit is associated with about 10% of the input patterns. Following pattern storage the weights from a representative output unit are used as the weights of the synapses of the pyramidal neuron. Thus this output unit and the neuron both should recognise the same input patterns during recall.

The neuron model is based on the morphology of a layer 5 pyramidal cell from the visual cortex of a cat [7], as shown in Figure 1. Model parameters are derived from those used by Bernander [2, 3] for the same cell (see figure legend). The neuron is divided into 441 isopotential compartments with the lengths and diameters of each cylindrical compartment being determined by the cell morphology (using the computer programme *ntscable*, written by J.C. Wathey at the Salk Institute). The length of each compartment is less than one-tenth of the space constant. Simulation of the cell is performed using the simulator NEURON [9], which uses backward Euler integration to solve equations for the membrane potential in each compartment at each time step. In most experiments the entire cell is passive, but when action potentials are required as output signals, fast sodium channels and delayed-rectifier potassium channels are added to the soma, with the Hodgkin-Huxley-style kinetics and densities of Bernander et al. [3].

The 4096 inputs to the cell are in the form of excitatory synapses distributed uniformly across the apical dendritic tree in layers 1 to 3 (Fig. 1). During pattern recall the binary activity of an input pattern is represented by trains of action potentials at those synapses corresponding to active inputs. All the spike trains are independent, Poisson-distributed and of the same mean frequency. Only those synapses given a weight of 1 by the *clipped Hebbian* learning during pattern storage will produce a postsynaptic response. The postsynaptic conductance due to a presynaptic action potential is modelled by an alpha function with a maximum conductance of $0.5\text{n}S$ and a time to peak of $1.5\text{m}secs$ [2]. This conductance gives rise to an excitatory postsynaptic potential (EPSP) which

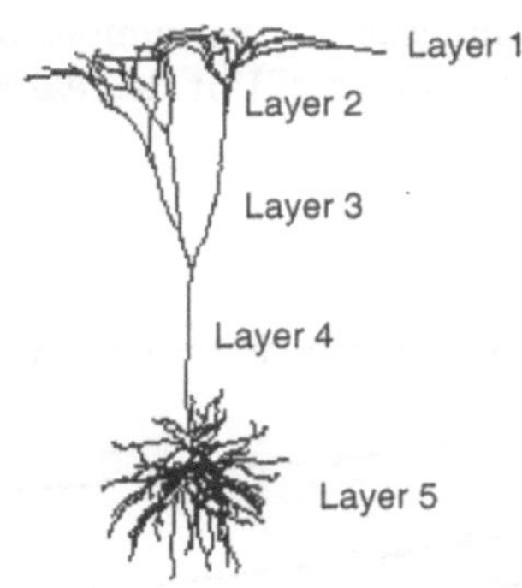

Fig. 1. Morphology of the layer 5 pyramidal cell. The basic passive membrane properties are: capacitance $C_m = 1\mu F/cm^2$; axial resistance $R_a = 200\Omega cm$; membrane resistance $R_m = 20,000\Omega cm^2$ in the soma and $R_m = 50,000\Omega cm^2$ in the dendrites (to approximate the effect of low frequency background excitatory and inhibitory activity on the cell); reversal potential is -66mV.

travels passively along the dendrite until it reaches the soma. The depolarization of the membrane at the soma due to all the EPSPs is the weighted sum of the inputs, or dendritic sum.

The ability of either the output unit or neuron to correctly discriminate the input patterns during pattern retrieval is determined by how much overlap there is between the dendritic sums due to the patterns that should be recognised (*high* patterns) and the sums due to the patterns that should not be recognised (*low* patterns). This is conveniently measured by the signal-to-noise (S/N) ratio between these two classes of dendritic sums [6]:

$$\rho = \frac{(\mu_h - \mu_l)^2}{1/2(\sigma_l^2 + \sigma_h^2)} \tag{1}$$

where μ_h (μ_l) is the mean and σ_h^2 (σ_l^2) is the variance of the dendritic sums from the *high* (*low*) input patterns. If the mean values of the two classes of dendritic sum are well separated and their variances are small, the S/N will be high and the output unit or neuron is likely to discriminate the *high* and *low* patterns correctly.

3 Results

For a given number of stored patterns, the S/N of the output unit, or the cell, was obtained by presenting each stored input pattern as a cue and recording the resultant dendritic sum. The S/N was then calculated from the sample mean and variance of the dendritic sums due to the *high* and *low* input patterns. The S/N of the output unit as a function of the number of stored patterns is shown by the solid line in Figure 2 (plot "net"). It decreases exponentially as the number of stored patterns increases.

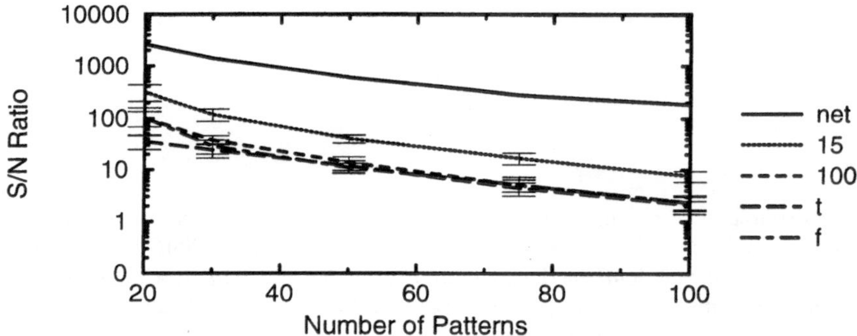

Fig. 2. S/N of the associative net and the pyramidal cell. Legend: net - associative net; 15 - random spikes at 15Hz; 100 - random spikes at 100Hz; t - time of first emitted spike; f - number of emitted spikes. Each pyramidal cell plot is the mean data from 10 runs at each number of stored patterns, with a different spatial distribution of synapses for each run. Error bars show 95% confidence limits.

Figure 2 also shows the S/N of the pyramidal cell when the dendritic sum is the soma depolarization recorded 100m*secs* after the initiation of input activity for input spike trains at 15H*z* (plot "15") or 100H*z* (plot "100"). The S/N of the output unit varies between 8 times and 90 times the pyramidal cell S/N for a given number of stored patterns. Conversely, a S/N of 320, say, is produced by at most 20 patterns in the pyramidal cell, but by 72 patterns in the associative net.

The random inputs at 15H*z* provide approximately 3 times the S/N of inputs at 100H*z*. The higher S/N of lower input frequencies is further demonstrated in Figure 3, where the S/N of 30 stored patterns is plotted against input frequency. This is due to saturation of the pyramidal cell synapses at high frequencies. At 100H*z*, 400 active synapses is enough to depolarize the entire apical dendritic tree to near the synaptic reversal potential.

For actual pattern recognition an output signal must be generated that indicates whether or not the input pattern has been recognised. When sodium and potassium channels are added to the soma, random 100H*z* input triggers action potentials. Figure 2 shows the S/N when the dendritic sum is the first emitted spike time (plot "t") or the number of spikes emitted in the first 200m*secs* of a simulation (plot "f"). This generation of action potentials converts the soma depolarization into an output signal without affecting the S/N. The S/N obtained from just a single spike is as good as that due to the average frequency of firing

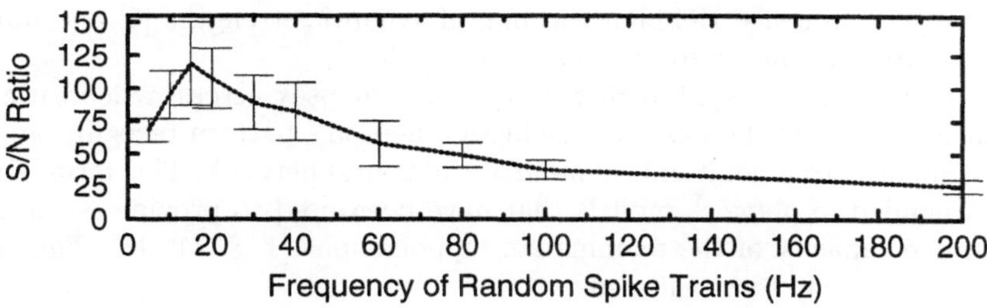

Fig. 3. S/N of soma depolarization for 30 stored patterns with different frequencies of random input. Each plot is the mean data from 10 runs at each number of stored patterns, with a different spatial distribution of synapses for each run. Error bars show 95% confidence limits.

over time. The first spike due to a *high* input pattern is generated about 40m*secs* after the onset of input activity at 100H*z*. This is the minimum time for the cell to recognise an input pattern.

4 Discussion

We have used computer simulations to estimate the signal-to-noise (S/N) ratio of a cortical pyramidal neuron when excitatory synaptic strengths are set according to a *clipped Hebbian* learning rule. For a variety of measurements of the weighted sum of the inputs, the S/N of the neuron is one to two orders of magnitude less than that of output units in the associative net model.

The performance of the neuron is influenced by the form of input signal used to represent an input pattern. For the continuous spike trains used here, lower firing frequencies are better than high frequencies due to synaptic saturation. This is a possible reason for the low average firing rates observed in the mammalian cortex. A single synchronous spike on each input line provides the same performance as the optimum continuous firing frequency (results not shown).

A number of different output signals provide the same recognition capability. Somatic depolarization is converted into action potentials without loss of S/N, and the time of the first emitted spike is as good a recognition indicator as the average firing frequency over a period of time. Thus recognition can be fast.

The use of S/N as a performance measure allows us to ignore issues of threshold setting across multiple neurons and the effect of partial connectivity. Variability in connectivity and in dendritic morphology across cells will introduce further noise into the dendritic sums and make threshold setting difficult.

The synapses in our model are distributed randomly and uniformly across a passive dendritic tree. Pyramidal dendrites are not passive [10] and the effect of active processes on the S/N needs to be explored. Some work has shown how nonlinear processes such as NMDA receptors and calcium channels in the dendrites can influence Hebbian learning of synaptic strengths [5] and amplify the signal from localised groups of synapses [12].

In conclusion, the spatial distribution of synapses across a dendritic tree introduces considerable noise that reduces a neuron's pattern recognition capability compared to a point unit in an artificial neural network. This form of noise is not included in network models that have been used to provide estimates of the storage capacity of the mammalian hippocampus [1, 8, 11, 13]. Thus these estimates are likely to be inflated.

Acknowledgement: to the MRC for financial support under PG 9119632.

References

1. M. Bennett, W. Gibson, and J. Robinson. Dynamics of the CA3 pyramidal neuron autoassociative memory network in the hippocampus. *Phil. Trans. Roy. Soc. Lond. B*, 343:167–187, 1994.
2. O. Bernander. *Synaptic Integration and Its Control in Neocortical Pyramidal Cells.* PhD thesis, California Institute of Technology, 1993.
3. O. Bernander, C. Koch, and R. Douglas. Amplification and linearization of distal synaptic input to cortical pyramidal cells. *J. Neurophys.*, 72:2743–2753, 1994.
4. T. Brown, E. Kairiss, and C. Keenan. Hebbian synapses: biophysical mechanisms and algorithms. *Ann. Rev. Neurosci.*, 13:475–511, 1990.
5. T. Brown, Z. Mainen, A. Zador, and B. Claiborne. Self-organization of Hebbian synapses in hippocampal neurons. In R. Lippmann, J. Moody, and D. Touretzky, editors, *Neural Information Processing Systems 3*, pages 39–45, San Mateo, California, 1991. Morgan Kaufmann.
6. P. Dayan and D. Willshaw. Optimising synaptic learning rules in linear associative memories. *Biol. Cybern.*, 65:253–265, 1991.
7. R. Douglas, K. Martin, and D. Whitteridge. An intracellular analysis of the visual responses of neurones in cat visual cortex. *J. Physiol.*, 440:659–696, 1991.
8. B. Graham and D. Willshaw. Capacity and information efficiency of the associative net. *Network*, 8:35–54, 1997.
9. M. Hines. A program for simulation of nerve equations with branching geometries. *Int. J. Biomed. Comput.*, 24:55–68, 1989.
10. D. Johnston, J. Magee, C. Colbert, and B. Christie. Active properties of neuronal dendrites. *Ann. Rev. Neurosci.*, 19:165–186, 1996.
11. D. Marr. Simple memory: a theory for archicortex. *Phil. Trans. Roy. Soc. Lond. B*, 262:23–81, 1971.
12. B. Mel. Synaptic integration in an excitable dendritic tree. *J. Neurophys.*, 70:1086–1101, 1993.
13. A. Treves and E. Rolls. Computational analysis of the role of the hippocampus in memory. *Hippocampus*, 4:374–391, 1994.
14. D. Willshaw. *Models of distributed associative memory.* PhD thesis, University of Edinburgh, 1971.
15. D. Willshaw, O. Buneman, and H. Longuet-Higgins. Non-holographic associative memory. *Nature*, 222:960–962, 1969.

Hebbian Delay Adaptation in a Network of Integrate-and-Fire Neurons

Christian W. Eurich[1], Jack D. Cowan[2], and John G. Milton[3]

[1] Institut für Theoretische Physik, Universität Bremen
D-28334 Bremen, Germany
e-mail: eurich@physik.uni-bremen.de
[2] Departments of Mathematics and Neurology
The University of Chicago, Chicago, IL 60637, USA
[3] Department of Neurology and Committee on Neurobiology
The University of Chicago, Chicago, IL 60637, USA

Abstract. We study the synchronization properties of a neural network which incorporates time delays. Two layers of integrate-and-fire neurons are connected by delay lines and a Hebbian-type learning rule is applied to allow a self-organizing, adaptive modification of the delays. It is shown that when the network synchronizes to a periodic input of period T, the delays differ by multiples of T. The delay dynamics possess an $(N+1)$-parameter set of fixed points which is locally attracting. Neural networks with delay adaptation may have applications as noise reduction algorithms and for the control of time-delayed dynamical systems.

1 Introduction

Delay lines are essential features of neural circuits for temporal resolution [3,2]. Recently there has been great interest in the ability of networks of spiking neurons with delay lines to synchronize in phase to a periodic input even though the response times of the individual neurons are finite [6,4]. It is observed that the values of the individual time delays in the synchronized network differ by integer multiples of the period, T, of the input.

Here we consider a network of integrate and fire neurons in which there is a Hebbian-type learning rule for delay adaptation, rather than the usual adaptation of synaptic efficacies. Hebb's famous postulate on activity-dependent plasticity in the nervous system [8, p. 62] does not specify the synapse as the site where learning takes place: although Hebb considers synaptic adaptivity to be the most likely mechanism, he also mentions the possibility of an activity-dependent neurobiotaxis [8, p. 63]. Such a growth process would result in a modification of the temporal structure of information processing, and we use a delay learning mechanism to model this behavior. In our network, the fixed points of the delay dynamcis correspond to synchronous network states in which individual delay lines differ by integer multiples of T. The set of fixed points is shown to be locally attracting.

2 The Model

A complete description of the network of integrate and fire neurons with a Hebbian-type delay adaptation rule is given elsewhere [4]. Briefly, consider a neural network which has two layers with a periodic input (period T) to layer 1 (Fig. 1). Layer 1 consists of N spiking neurons which are assumed to fire exactly when the amplitude of the input signal is maximal, i.e. the k-th presynaptic spike occurs at kT where k is an integer. The presynaptic neurons are connected to the postsynaptic neuron by axons which serve as delay lines. The presynaptic spikes arrive at the postsynaptic neuron at times

$$t_{k,n} = kT + \tau_n \ , \tag{1}$$

where τ_n is the time delay introduced by the n-th axon. The postsynaptic neuron is modeled as an integrate-and-fire neuron, i. e. after re-scaling, the dimensionless membrane potential, U, satisfies $dU/dt + U = J(t)$, where the time is measured in units of the membrane time constant, and the dimensionless input current J is given by

$$J(t) = \sum_{k=-\infty}^{\infty} \sum_{n=1}^{N} w_0 \delta(t - t_{k,n}) \ , \tag{2}$$

where w_0 is the synaptic weight (herein assumed to be the same constant for all synapses). The neuron has a firing threshold, θ, and an absolute refractory period, t_{R}.

According to Hebb's proposition, learning takes place if the activity of the presynaptic and postsynaptic neurons is positively correlated within a certain time window. In our case the transmission delays, τ_n, are adaptively modified. Let $s^p_{j,n} \equiv t^p - t_{j,n}$ where j denotes the presynaptic spike closest in time to the p-th postsynaptic spike, t^p. Now we define a window function, $W(s)$, with the following properties: $W(s) < 0$ for $s < 0$, $W(s) = 0$ when $s = 0$, $W(s) > 0$ for $s > 0$ and $W(s) \to 0$ as $s \to \pm\infty$. We take

$$W(s) = \frac{2s/T}{1 + (2s/T)^2} \tag{3}$$

(left inset in Fig. 1). After the p-th postsynaptic spike at time t^p, learning takes place in each of the presynaptic neurons. The delay of the n-th neuron is assumed to change according to the Hebbian rule

$$\tau_n(p+1) = \tau_n(p) + \gamma W(s^p_{j,n}) \tag{4}$$

where $\gamma > 0$ is the learning rate. Thus, when a postsynaptic spike is preceeded by a presynaptic one, i.e. $s^p_{j,n} > 0$, the learning rule leads to an increase in the delay such that successive presynaptic spikes arrive a little later. When a postsynaptic spike is followed by a presynaptic spike, $s^p_{j,n} < 0$, and the reverse happens. In order to account for the slowness of the delay adaptation process, we assume that γ is small, or equivalently, that $\gamma|W(s)| \leq |s|$.

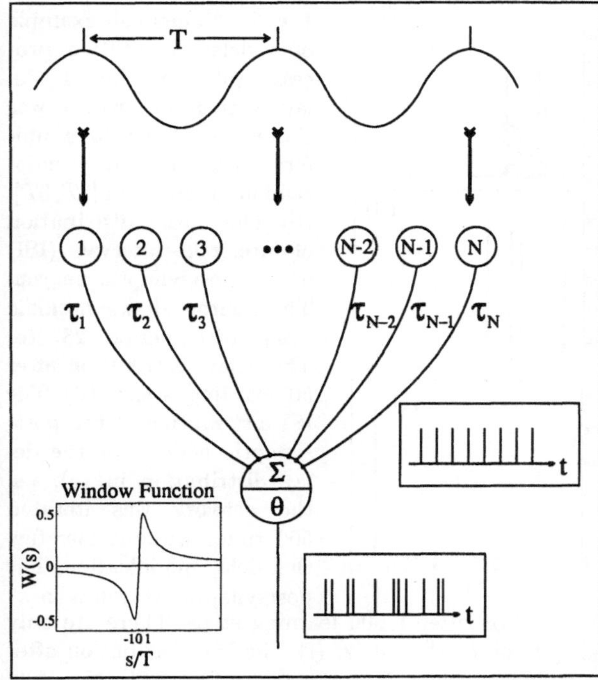

Fig. 1. Schematic diagram of the model. At the top, the periodic input signal is depicted. The peaks at the maxima symbolize the release of spikes in the presynaptic neurons, which are shown in the middle. The upper inset on the right hand side indicates that each presynaptic neuron is exactly periodic. $\tau_1 \ldots \tau_N$ designate the presynaptic delays. The inputs are summed up by the postsynaptic integrate-and-fire-neuron (bottom) which initially shows a more complicated yet periodic firing pattern (lower inset on the right hand side). The window function for the delay dynamics is shown in the left inset.

Numerical simulations show that under the delay dynamics (4), random initial delays eventually reach a fixed point where they differ by multiples of the input period, T. Accordingly, the firing pattern of the postsynaptic neuron changes from a complicated periodic firing to a period-1 orbit with period T. (This is true for $t_{\mathrm{R}} \leq T$. For $t_{\mathrm{R}} > T$, the postsynaptic neuron will fire with a period kT, where $k = \min\{i \in \mathbb{N}^0 : iT > t_{\mathrm{R}}\}$.) An example for the learning dynamics of the network is shown in Fig. 2. Thus it is possible for a network with a finite response time to fire in phase with an external periodic forcing.

3 Theorem

Theorem 1. *The delays given by*

$$\tau_n = \tau^0 + j_n T \quad (\tau^0 \in [0;T[,\ j_n \in \mathbb{N}^0,\ n = 1,\ldots,N) \tag{5}$$

form an $N+1$-parameter set of fixed points of the dynamics (4). In each of these states, the postsynaptic neuron fires periodically with period T if $t_{\mathrm{R}} \leq T$. The set of fixed points is attracting, i. e., a small disturbation in the delays leads to a convergence towards one of the fixed points.

Proof. In the frequency domain the effect of a delay line is to introduce a phase shift which is 2π when the delays of the presynaptic neurons differ by multiples

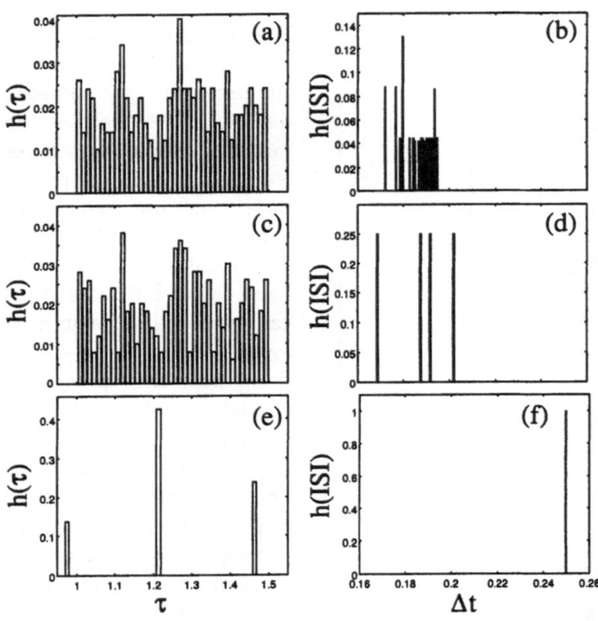

Fig. 2. Numerical example of a delay adaptation process. (a) The initial delay distribution which was chosen according to a uniform distribution of pseudo-random numbers in $[2T, 3T]$. (b) The initial distribution of interspike intervals (ISI) of the postsynaptic neuron. The neuron shows periodic firing with period 23. (c) The delay distribution after 50 learning steps. (d) The ISI distribution of the postsynaptic neuron for the delay distribution in (c), i.e. the network was iterated 500 times without any further delay modification. The postsynaptic neuron is in a period-4 orbit. (e) The delay distribution after 1,600 learning steps. There are only three delays which differ by multiples of T; $\tau^0 = 0.22$. (f) The ISI distribution after 1.600 learning steps. A period-1 orbit is reached. Parameters are $T = 0.25$, $N = 400$, $w_0 = 0.005$, $\theta = 1$, $t_R = 0.05$, $\gamma = 0.001$.

of T. Therefore, the postsynaptic neuron gets a synchronous input which leads to a periodic firing with period T. For each presynaptic neuron n ($n = 1, \ldots, N$), $s^p_{j,n} \equiv 0$, and the learning rule (4) leads to $\tau_n(p+1) = \tau_n(p)$, because $W(0) = 0$. Thus, (5) are fixed points of the delay dynamics.

For $N = 2$, the set of fixed points (5) is shown in Fig. 3a. Without loss of generality, we assume that $j_n \equiv 0$ for $n = 1, \ldots, N$. Consider a slightly disturbed delay distribution, $\tau_n = \tau^0 + \Delta\tau_n$, where

$$2|\Delta\tau_n| \le \epsilon < \min\{T/2, t_R\} \quad (n = 1 \ldots, N) \ . \tag{6}$$

Figure 3b shows such a delay distribution as a continuous function of the delay, τ. At the same time, the graphics shows the time course of the presynaptic spike arrivals: spikes with the shortest delays (left side of the distribution) arrive first, those with the longest delays (right side of the distribution) arrive last. This input sequence is repeated with period T. The dotted vertical line represents the postsynaptic spikes at the original fixed point: spikes occur at times $t^p = pT + \tau^0$. Due to the disturbation, the spike time may shift to a new value, τ^1 (solid vertical line), where $|\tau^1 - \tau^0| \le \epsilon$. Due to the condition that $\epsilon < t_R$, only one postsynaptic spike can occur during each period, T.

After the release of the spike in the postsynaptic neuron, the learning rule (4) is applied. The condition $\epsilon < T/2$ leads to a contraction of the interval

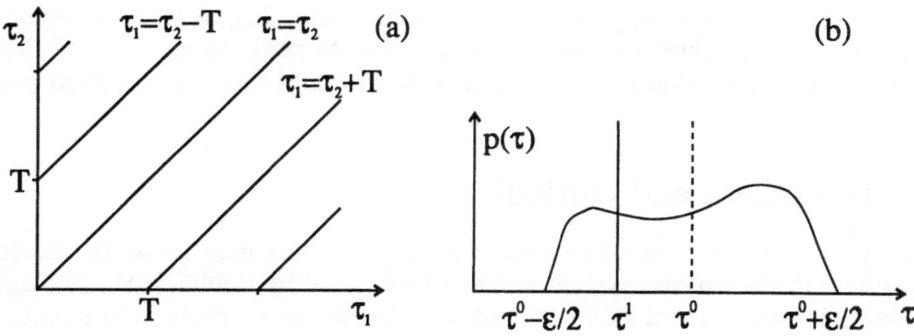

Fig. 3. (a) The set of fixed points of the delay dynamics for $N = 2$; delays differ by multiples of the input period, T. (b) A disturbed delay distribution, $p(\tau)$, depicted as a continuous function of τ. The vertical lines represent spikes of the postsynaptic neuron. For further explanations, see text.

$[\tau^0 - \epsilon/2; \tau^0 + \epsilon/2]$, whereby all delays are shifted towards τ^1. (If ϵ exceeded $T/2$, some of the delays would shift towards $\tau^1 \pm kT$.) This can be seen as follows: If, for a presynaptic neuron n, $\tau_n > \tau^1$, then $s_n = \tau^1 - \tau_n < 0$. This leads to a decrease of the delay, because $W(s) < 0$ for $s < 0$. The condition $\gamma|W(s)| \le |s|$ implies that the new delay still exceeds τ^1, i. e., the new delay is shifted towards τ^1. A similar argument holds for presynaptic neurons n where $\tau_n < \tau^1$. Delays $\tau_n = \tau^1$ are not changed. The contraction can be written as follows: Let $\tau^1_{\min}$, $\tau^1_{\max}$ and $\tau^2_{\min}$, $\tau^2_{\max}$ designate the old and new interval boundaries, respectively. Then, the learning rule implies

$$\tau^2_{\min} \ge \tau^1_{\min}\,, \quad \tau^2_{\max} \le \tau^1_{\max}\,, \quad \text{and} \quad |\tau^2_{\max} - \tau^2_{\min}| < |\tau^1_{\max} - \tau^1_{\min}|\,. \tag{7}$$

(The equalities $\tau^2_{\min} = \tau^1_{\min}$, $\tau^2_{\max} = \tau^1_{\max}$ hold if the postsynaptic spike is shifted to the left or right boundary of the interval, respectively.) The changes in the delays may result in another shift of the postsynaptic spike, $\tau^1 \mapsto \tau^2 \in [\tau^2_{\min}; \tau^2_{\max}]$. A subsequent application of the learning rule leads to another contraction of the delay interval.

Thus, an iteration of the learning rule (4) gives a sequence of intervals $[\tau^m_{\min}; \tau^m_{\max}]$ ($m \in \mathbb{N}$) such that

$$\tau^{m+1}_{\min} \ge \tau^m_{\min}\,, \quad \tau^{m+1}_{\max} \le \tau^m_{\max}\,, \quad \text{and} \quad |\tau^{m+1}_{\max} - \tau^{m+1}_{\min}| < |\tau^m_{\max} - \tau^m_{\min}|\,. \tag{8}$$

Let $l^m \equiv |\tau^m_{\max} - \tau^m_{\min}|$ be the width of the delay distribution at the m-th step. We have shown that the sequence l^m is monotonically decreasing. It has the lower boundary 0. Thus, l^m converges. Now assume that the limiting value exceeds zero, $\lim_{m\to\infty} l^m \equiv l > 0$. Let $\tau^\infty_{\min}$ and $\tau^\infty_{\max}$ designate the minimal and maximal boundary of the corresponding interval, respectively, and let the postsynaptic spike occur at a time $pT + \tau^\infty$, where $\tau^\infty \in [\tau^\infty_{\min}; \tau^\infty_{\max}]$. The contraction property requires that the interval $[\tau^\infty_{\min}; \tau^\infty_{\max}]$ be mapped onto itself (i. e., the interval cannot shift on the τ-axis). Especially, the minimal delay and the maximal delay remain unchanged. However, only those delays for which $s = 0$ do not change, i.e., $\tau_{\min}^{\infty} = \tau^{\infty}$ and $\tau_{\max}^{\infty} = \tau^{\infty}$. This means that $l \not> 0$ which contradicts our assumption. Thus, the interval length goes to zero. As a result, all delays converge to some value τ^{∞}; this set of delays belongs to the set of fixed points (5), q.e.d.

4 Discussion and Outlook

The possibility that activity-dependent neurobiotaxis may lie at the basis of plasticity in the nervous system was first recognized by Hebb [8]. Morphological changes in neurons and axons [3] and activity-dependent changes in membrane conductances [12,9] would be anticipated to be reflected by changes in, respectively, conductive and synaptic delays. Although these observations strongly favor considerations of neural networks which incorporate some form of adaptive delay modification, such networks have only recently begun to attract attention [1,10,7,4]. We anticipate that neural networks which incorporate delay adaptation will be very useful. For example, such networks are capable of noise reduction by utilizing both temporal and spatial averaging; this noise reduction scales as $\sim N^{-1/2}$, where N is the number of presynaptic neurons [4]. Delay adaptation is also likely to find applications in the control of dynamical systems containing retarded variables [11,5].

References

1. Baldi, P., Atiya, A. F.: How delays affect neural dynamics and learning. IEEE Trans. Neural Networks **5** (1994) 612–621
2. Carr, C. E.: Processing of temporal information in the brain. Annu. Rev. Neurosci. **16** (1993) 223-243
3. Carr, C. E., Konishi, M.: A circuit for detection of interaural time differences in the brain stem of the barn owl. J. Neurosci. **10** (1990) 3227-3246
4. Eurich, C. W., Cowan, J. D., Milton, J. G.: A Hebbian learning rule for delay adaptation in integrate-and-fire neural networks. Preprint, submitted
5. Eurich, C. W., Milton, J. G.: Noise-induced transitions in human postural sway. Phys. Rev. E **54** (1996) 6681–6684
6. Gerstner, W., Kempter, R., van Hemmen, J. L., Wagner, H.: A neuronal learning rule for sub-millisecond temporal coding. Nature **383** (1996) 76–78
7. Glünder, H., Hüning, H.: Detection of spatio-temporal spike patterns by unsupervised synaptic delay learning. In: Elsner, N, Schnitzler, H.-U. (eds): Brain and Evolution. Thieme, Stuttgart (1996) 800
8. Hebb, D. O.: The Organization of Behavior. Wiley, New York (1949)
9. Markham, H., Tsodyks, M.: Redistribution of synaptic efficacy between neocortical pyramidal neurons. Nature **382** (1996) 807–810
10. Napp-Zinn, H., Jansen, M., Eckmiller, R.: Recognition and tracking of impulse patterns with delay adaptation in biology-inspired pulse-processing neural net (BNP) hardware. Biol. Cybern. **74** (1996) 449-453
11. Pyragas, K.: Continuous control of chaos in self-controlling feedback. Phys. Lett. A **170** (1992) 421–428
12. Turrigiano, G., Abbott, L. F., Marder, E.: Activity-dependent changes in the intrinsic properties of cultured neurons. Science **264** (1994) 974–977

Hippocampal Formation Trains Independent Components via Forcing Input Reconstruction

András Lőrincz[1]

[1] Department of Chemical Physics, Institute of Isotopes of the Hungarian Academy of Sciences Budapest, P.O.B. 77, Hungary, H-1525

[2] Department of Adaptive Systems, Attila József University, Szeged, Dóm Square 9, Hungary, H-6720

email:lorincz@iserv.iki.kfki.hu

Abstract. It is assumed that higher order concept formation utilizes independent components (ICs). It is argued that ICs require dynamic input reconstruction networks (RNs) to form a reliable internal representation. Input reconstruction, however, can be slow and poor with ICs on substrates with lossy dynamics. A model of the hippocampal formation is proposed that develops the ICs on lossy RNs by means of locking inputs to the internal representation and thus forcing fast reconstruction and cancelling losses. It is assumed that upon training ICs can lock themselves, thus hippocampal lesion mostly affects anterograde memories.

1 Introduction

Studies on adaptive goal oriented systems utilizing episodic memories and reinforcement learning [1] led to the conclusion that higher order concepts (HOCs) can be formed by means of an appropriate representation. The concepts are developed by peeling off unimportant components of the representation. A component can be peeled off if no matter what value that component has the given episode can happen, with the proviso that the episode is independent of the component value. This suggests that independent components (ICs) [2, 3, 4] should be developed for HOC formation. However, ICs form a poor internal representation owing to the fact that ICs may overlap to a considerable extent. Let us consider a 'grandmother cell' and a 'grandfather cell': the internal activities of both may be large if either the 'grandmother' or the 'grandfather' is the actual input. Internal representation can be improved by means of reconstruction networks (RNs) [5] that undergo relaxation by combining the memory traces on the basis of the internal activities, form an input vector, compare the latter with the actual input, and modify the internal representation to decrease the difference. This method corresponds to Wittmeyer's pseudoinverse algorithm [5]. Input reconstruction becomes (1) slow if memories, such as ICs, overlap to a considerable extent and also (2) poor if the error integration of the scheme is lossy. Here, a control scheme is proposed as a model of the trisynaptic loop of the hippocampus (HC) that compensates losses, locks the input and the reconstructed input, and develops ICs.

2 Locked lossy reconstruction networks

Assume that we have an input vector $\mathbf{x}$ ($\mathbf{x} \in \mathbf{R}^N$) and a 'direct internal representation' $\mathbf{a_d}$ ($\mathbf{a_d} \in \mathbf{R}^n$) connected by memory matrix Q ($Q \in \mathbf{R}^n \times \mathbf{R}^N$) formed by n memory vectors $\mathbf{q}_i$, ($i = 1, \ldots, n$) of dimension N with $n < N$. The memory vector (or memory trace) $\mathbf{q}_i$ represents the feedforward connection structure to neuron i, and the jth element of memory vector $\mathbf{q}_i$, that is q_{ij}, is the ijth element of memory matrix Q. The relation between the input vector and the 'direct internal representation' is: $\mathbf{a_d} = Q^T\mathbf{x}$. This low dimensional internal representation is then used to reconstruct the input by linearly superimposing the memory traces in accordance with the internal representation. However, direct internal representation needs correction in order to optimize reconstruction; the correcting term is computed by means of a relaxation equation. The reconstructed input $\mathbf{y}$ is compared to the original input and the difference is evaluated (filtered) by the memory matrix thereby giving rise to a correcting term to the internal representation. Relaxation of internal representation $\mathbf{a}$ stops when the internal representation of the input and that of the reconstructed input become identical [5]. The scheme gives rise to the following equations:

$$\mathbf{a} = \mathbf{a_d} + Q^T\mathbf{c} \tag{1}$$

$$\dot{\mathbf{c}} = -\mathbf{c} + \lambda(\mathbf{x} - \mathbf{y}) \tag{2}$$

where $\mathbf{y} = Q\mathbf{a}$, and a lossy term $-\mathbf{c}$ was introduced tha spoils the temporal integration and thus the aimed reconstruction. The loss can be compensated by means of a control architecture by noting that the 'desired' rate of change of $\mathbf{c}$ is equal to $\dot{\mathbf{c}}_\mathbf{d} = \lambda(\mathbf{x} - \mathbf{y})$ instead of the 'experienced' rate of change $\dot{\mathbf{c}}$. This is a speed-field tracking problem that can be solved to arbitrary precision by means of the SDS control architecture [6, 7]. The control vector may be written as

$$\mathbf{u}(\mathbf{c}, \dot{\mathbf{c}}, \dot{\mathbf{c}}_\mathbf{d}) = \mathbf{A}(\dot{\mathbf{c}}_\mathbf{d} - \dot{\mathbf{c}}) + \Lambda \int \mathbf{A}(\dot{\mathbf{c}}_\mathbf{d} - \dot{\mathbf{c}})dt \tag{3}$$

where $\mathbf{A}$ is the input to control vector association. It has been mentioned that this control equation and the reconstruction dynamics are closely related [5]. The original SDS equations can be simplified if the control equation is fast and thus $\mathbf{A}$ can be pulled out of the integral. In this way the control equation corresponds to the computed torque method leading to an architecture with a direct and a temporally integrated channel followed by an associative stage. Let us use the control vector to contribute to the internal representation in a linear fashion ($\mathbf{a} = \mathbf{a_d} + Q^T\mathbf{c} + \mathbf{u}$). The control equation (Eq. (3)) cannot set all of the components of $\dot{\mathbf{c}}_\mathbf{d}$ arbitrarily close to $\dot{\mathbf{c}}$ unless matrix Q is square and invertible. We do not need this property, however. For our purposes it is satisfactory if the controller can set $Q^T\dot{\mathbf{c}}_\mathbf{d}$ arbitrarily close to $Q^T\dot{\mathbf{c}}$. The control scheme provides the desired result that $Q^T\mathbf{c}$ is arbitrarily close to zero. This involves that if $Q^T\dot{\mathbf{c}}$ is averaged over some short time window the averaged value will also be close to zero and thus $Q^T\mathbf{x}$ will be approximately equal to $Q^T\mathbf{y}(= Q^TQ\mathbf{a})$. Thus up to the extent of the short time window (determined by Λ) simultaneity also

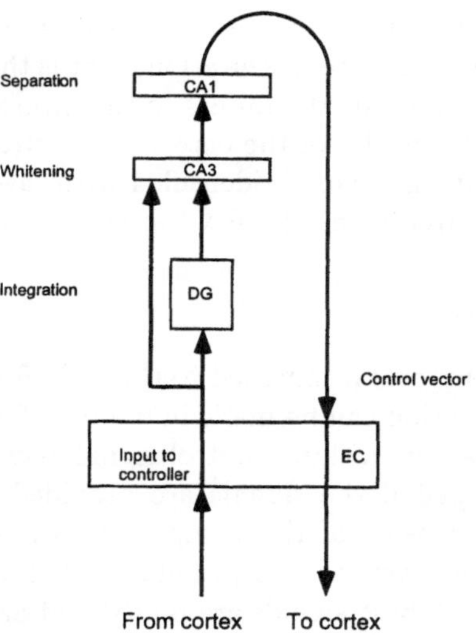

Fig. 1. Architecture of the temporal locking control scheme
The architecture comprises the entorhinal cortex (EC), which is a reconstruction network, and the hippocampus (HC), this being a temporal locking control architecture that targets the output layer of EC. The controller's input propagates via two independent channels: the first directly excites the CA3 subfield, the other is temporally integrated at the dentate gyrus (DG) and then excites the CA3 subfield. The CA3 subfield is a whitening stage followed by a separation stage – the CA1 subfield. The output of the HC serves a double purpose: it locks the input and the reconstructed input of the EC together and also trains the EC (and possibly the associative areas) to hold the estimates of the independent components.

applies to the output layer. Condition $Q^T\mathbf{x} = Q^TQ\mathbf{a}$ is the exact condition of reconstruction dynamics [5] and thus the control architecture makes feasible the pseudoinverse computation of $\mathbf{a}$ despite of the losses.

The architecture is depicted in Fig. 1. The different computational stages are named according to the biological substrate that we are modelling. The entorhinal cortex, which is an RN, receives sensory information from other regions in the cortex. This information is processed by the entorhinal cortex and $\dot{\mathbf{c}}_\mathbf{d}$ and $\dot{\mathbf{c}}$ are also computed. The difference of these terms is the input of the controller, i.e., the HC. According to Eq. (3) the difference reaches a summing stage in two ways: directly, and after temporal integration. The direct pathway can be identified with the direct excitatory inputs reaching the CA3 area whereas the integrating unit and the integrated excitatory signals can respectively be identified with the dentate gyrus and the mossy fiber efferents of the dentate gyrus that innervate the CA3 area.

The summed inputs need to be transformed into a control vector by means of the associative matrix $\mathbf{A}$ and that control vector should return to the same

RN that gave rise to the input of the controller. According to the setting the controller controls the internal representation, that is the output of the entorhinal cortex. The layers that provide the inputs to the hippocampal formation can be identified with layers II and III of the entorhinal cortex. The layers that receive the outputs of the controller can be identified with layers V and VI of the same area. The trisynaptic circuit *and* the RN together form a loop.

2.1 *Associative stage*

The requirement of temporal locking led to the need for an associative stage in a natural fashion. Association can be made in many different ways that we restrict by the assumption that the locking control signal also forms ICs.

ICs can be developed if the signals are provided by independent sources: the joint probability density of the components of the original signal vector $\mathbf{s} = (s_1, ..., s_k)$ can be given as the product of the marginal densities of the individual components. These signals are mixed and are covered with noise, i.e., the input to the system is

$$\mathbf{x} = M^T \mathbf{s} + \mathbf{n} \tag{4}$$

where $\mathbf{n}$ denotes the noise, M is a constant mixing matrix with full column rank, and the presentation index (time) has been dropped for simplicity. The task is to develop an internal representation, or memory traces that correspond to the components of vector $\mathbf{s}$.

Association that leads to independent components (IC) can be built up from three stages [4]. There is a first stage that whitens, i.e., decorrelates and rescales the input giving rise to outputs of equal variance. The second stage makes use of the whitened information and does "blind source separation" [2, 3, 4]. The outputs of this stage are then used at a third stage to train a memory matrix. The columns of that matrix represent the individual independent components.

In accordance with the anatomical constraints these three stages correspond to the CA3 and the CA1 subfields of the hippocampal formation and the entorhinal cortex itself with the entorhinal cortex holding the ICs.

Whitening with reconstruction architecture. Whitening in the RN can be achieved simply by making use of 'recurrent collaterals'. Denoting the matrix representing the recurrent collaterals by P, Eqs. (1) and (2) can be modified accordingly and the temporal derivative of $\mathbf{a}_{\mathrm{CA3}}$ can be written as follows:

$$\dot{\mathbf{a}}_{\mathrm{CA3}} = -\mathbf{a}_{\mathrm{CA3}} + \lambda_1 Q^T_{\mathrm{CA3}}(\mathbf{x}_{\mathrm{CA3}} - \mathbf{y}_{\mathrm{CA3}}) + \lambda_2 Q^T_{\mathrm{CA3}} P_{\mathrm{CA3}} \mathbf{a}_{\mathrm{CA3}} \tag{5}$$

By setting the two λs equal to each other and adjusting (training) P_{CA3} so that $Q^T_{\mathrm{CA3}} Q_{\mathrm{CA3}} = Q^T_{\mathrm{CA3}} P_{\mathrm{CA3}}$, the relaxed equation is

$$\mathbf{a}_{\mathrm{CA3}} = \lambda Q^T_{\mathrm{CA3}} \mathbf{x}_{\mathrm{CA3}} \tag{6}$$

Expressing $\mathbf{y}_{\mathrm{CA3}}$ at the same time yields

$$\mathbf{y}_{\mathrm{CA3}} = \lambda Q_{\mathrm{CA3}} Q^T_{\mathrm{CA3}} \mathbf{x}_{\mathrm{CA3}} \tag{7}$$

If one utilizes Hebbian learning between $\mathbf{a}_{\mathrm{CA3}}$ and $\mathbf{y}_{\mathrm{CA3}}$ and sets $\lambda = 1.0$ the following learning equation can be achieved:

$$\Delta Q_{\mathrm{CA3}} = \eta Q_{\mathrm{CA3}}(I - Q_{\mathrm{CA3}}^T \mathbf{x}_{\mathrm{CA3}} \mathbf{x}_{\mathrm{CA3}}^T Q_{\mathrm{CA3}}) \tag{8}$$

where I denotes the identity matrix. This equation is exactly the whitening equation of Laheld and Cardoso [8]. This whitening model of the CA3 subfield gives rise to orthonormal memories, i.e. (fast) feedforward processing for the recurrent RN of the CA3 subfield.

Separation with reconstruction architecture. The CA1 subfield exhibits limited within layer excitatory interconnectivity, this being a contrasting difference between the CA3 and the CA1 subfields. Thus we assume no recurrent collaterals in this case. Also, we are looking for a separation stage that gives rise to fast processing, i.e., the model should develop orthonormal memory components that gives rise to feedforward processing in RNs. One of the options is Oja's nonlinear learning rule that provides close to orthonormal components over whitened inputs:

$$\Delta Q_{\mathrm{CA1}} = \eta(\mathbf{x}_{\mathrm{CA1}} \mathbf{g}(\mathbf{a}_{\mathrm{CA1}}^T) - Q_{\mathrm{CA1}} \mathbf{g}(\mathbf{a}_{\mathrm{CA1}}) \mathbf{g}(\mathbf{a}_{\mathrm{CA1}}^T)) \tag{9}$$

where $\mathbf{g}(.)$ denotes the same nonlinear function for all of the components.

Learning the independent components. The last stage of independent component analysis is the stage where the ICs are to be estimated, i.e., learnt. According to the anatomical constraints of the model this stage is either the entorhinal cortex itself, or the association cortices with entorhinal afferents. The learning rule of this stage can be given as [4]:

$$\Delta Q = \eta(\mathbf{x} - Q\mathbf{u})\mathbf{u}^T \tag{10}$$

where the unindexed vector $\mathbf{x}$ and matrix Q refer to components of the entorhinal cortex. Here, $\mathbf{x}$ denotes the input of this stage, i.e., the cortical afferents of the entorhinal cortex, whereas $\mathbf{u}$ denotes the output of the controller, i.e., the output of the separating stage ($\mathbf{u} \cong Q_{\mathrm{CA1}}^T \mathbf{x}_{\mathrm{CA1}}$). This training rule minimizes the mean square error of expression $\| \mathbf{x} - Q\mathbf{u} \|^2$ so that the rows of matrix Q become independent components. When the reconstruction process is in effect, the net making use of matrix Q relaxes to vector $\mathbf{a}$ that minimizes the expression $\| \mathbf{x} - Q\mathbf{a} \|^2$. In other words, the RN always minimizes expression $\| \mathbf{x} - Q\mathbf{a} \|^2$, whereas the HC tunes the components of the net to form independent components: matrix Q approximates matrix M^T of Eq. (4) and the reconstruction equations correspond to the computation of multiplying the input with the left pseudoinverse of matrix Q.

3 Conclusions

A model of the hippocampus (HC) was proposed that develops independent components (ICs) via temporal locking of reconstruction architectures. The model utilizes a speed-field tracking control architecture. The locking control architecture that locks and trains long term memories (LTMs) nicely fits the anatomical contraints of the trisynaptic loop of HC including the direct and indirect (via the dentate gyrus) excitatory paths between the entorhinal cortex and the CA3 subfield, the recurrent collaterals of the CA3 subfield, the limited interconnectivity of the CA1 subfield and the input-output organization of the entorhinal-hippocampal system. The model fits physiological properties of HC too: At different levels of computational complexity IC analyis gives rise to components that are quasi-independent but are temporally correlated like the edges of natural scenes [9]. In case of a labyrinth problem thus 'place cells' can emerge in an 'agent' of longer computational history. Another physiological property of HC is its putative double computational role [10] that the model supports by saying that temporal locking is important during exploration whereas LTM consolidation can be performed during sleep. According to the model LTMs are made of ICs. Since hippocampal subjects have almost intact retrograde memories the model suggests that ICs of long term memories can lock themselves upon formation.

References

1. Kalmár, Z., Szepesvári, C., Lőrincz, A.: Generalized dynamic concept model as a route to construct adaptive autonomous agents. Neural Network World **5** (1995) 353–360
2. Jutten, C., Herault, J.: Blind separation of sources, Part I: An adaptive algorithm based on neuromimetic architecture. Signal Processing **24** (1991) 1–10
3. Comon, C.: Independent component analysis - A new concept?. Signal Processing **36** (1994) 287–314
4. Karhunen, J., Oja, E., Wang, L., Vigário, R., Joutsensalo, J.: A class of neural networks for independent component analysis. IEEE Trans. on Neural Networks (1997) In press
5. Lőrincz, A.: Towards a unified model of cortical computation II: From control architecture to a model of consciousness. Neural Network World **7** (1997) 137-152
6. Szepesvári, C., Cimmer, S., Lőrincz, A.: Dynamic state feedback neurocontroller for compensatory control. Neural Networks (1997) In press
7. Szepesvári, C., Lőrincz, A.: Approximate inverse-dynamics based robust control using static and dynamic state feedback. Neural Adaptive Control Theory, World Sci. Singapore, **2** In press
8. Laheld, B., Cardoso, J.F.: Adaptive source separation with uniform performance. Proc. EUSIPCO-94 **2** (1994) 183–186
9. Bell, A.J., Sejnowski, T.J.: Edges are the independent components of natural scenes. Advances in Neural Information Processing Systems **9** (1997) 831-837
10. Buzsáki, Gy.: Two-stage model of memory trace formation: A role for "noisy" brain states. Neuroscience **31** (1989) 551-570

Part II:
Cortical Maps and Receptive Fields

Nature vs. Nurture in the Development of Tangential Connections and Functional Maps in the Visual Cortex

Siegrid Löwel, Kerstin E. Schmidt and Wolf Singer

Max-Planck Institut für Hirnforschung, Deutschordenstr. 46,
D-60528 Frankfurt am Main
e-mail: schmidt@mpih-frankfurt.mpg.de / loewel@ifn-magdeburg.de

Abstract. A series of experiments concerned with mechanisms underlying the development of the visual cortex revealed that long-range tangential connections display at least the following three characteristics: i) in strabismic but not in normally raised cats, intracortical fibers preferentially connect cell groups activated by the same eye ('ocular dominance selectivity'), ii) within the subsystems of the left and right eye domains, they extend primarily between neurons activated by similar stimulus orientations ('orientation selectivity') and iii) they exhibit an anisotropy with respect to the cortical axes by preferentially linking neurons with colinearly aligned receptive fields ('axial specificity'). These results are compatible with the idea of a selective stabilization of tangential fibers between coactive neurons (the "fire together, wire together"-hypothesis). Optical imaging of functional maps in area 17 of strabismic cats further revealed that iso-orientation domains are continuous across the borders between adjacent ocular dominance columns. This rather supports an experience-independent initial development of orientation preference maps. To what extent spontaneous versus visually driven activity patterns might be involved both in the development of tangential connections and in functional maps is discussed.

1 Intrinsic connections

The specific capabilities of our brains depend critically on the interactions of a large number of neurons. For a thorough understanding of the working of brains it is therefore indispensable to know the connectivity patterns and the functional interactions between these groups of neurons. A characteristic feature of the visual cortex of ferrets, cats and monkeys is the existence of long-ranging tangentially oriented axon collaterals that interconnect regularly spaced clusters of cells. In kitten visual cortex, tangential connectivity patterns develop mainly after birth from an initially rather homogeneous distribution of interconnected neurons. Evidence accumulated in the last years indicates that the adult specificity is attained in an experience-dependent way. A very fruitful working hypothesis to elucidate the mechanisms underlying the development of these connections turned out to be the "wire together, fire together"-hypothesis.

According to this hypothesis (based on ideas developed by [1]), connections between two neurons are stabilized when the cells are active in synchrony, while connections are weakened and finally get lost in out-of-synchrony constellations. We were able to gather evidence in favor of this hypothesis in a series of experiments in the visual cortex of strabismic cats. In these animals, the optical axes of the two eyes are no longer aligned so that the images on the two retinae cannot be brought into register. As a consequence, the responses mediated by anatomically corresponding retinal loci in the two eyes are no longer correlated.

If there is selective stabilization of connections between those cells that exhibit correlated activity while the animals grow up, intracortical connections in strabismic animals should display at least the following three characteristics: i) they should extend primarily between neurons driven by the same eye (experiment I), ii) within the subsystems of the left and right eye domains, they should extend primarily between neurons preferring similar stimulus orientations (experiment II). Since in our visual world contrast borders are mostly elongated (at least over short distances), neurons with colinearly aligned receptive fields should have a very high probability of being coactivated. Therefore, iii) intracortical connections should extend primarily between neurons whose receptive fields are aligned colinearly (experiment III). These predictions were tested by an approach combining the following methods: we first visualized the functional architecture of primary visual cortex (area 17) of cats with optical imaging of intrinsic signals [2] and then injected retrograde tracers into functionally identified cortical domains. Finally the resulting connectivity patterns were superimposed on and quantitatively compared with 2-deoxyglucose (2-DG) labeled activity maps of the same cortical regions [3-5].

Experiment I: In area 17 of strabismic but not normally raised cats, tangential intracortical fibers preferentially connected cell groups activated by the same eye [3]. Experiment II: 50-70% of the retrogradely labeled neurons were located in the same eye/same orientation (OR-) domains visualized with 2-DG while these occupied only about 30% of the cortical surface (see also [6,7]). Experiment III: After injections of beads into horizontal OR-columns, the distribution of retrogradely labeled cells was elongated along the cortical representation of the horizontal meridian (Figure 1). Injections into vertical OR-columns revealed anisotropies of the cell plots along the vertical meridian [4]. In both cases, neurons with receptive fields that are aligned colinearly in visual space had more numerous tangential connections (see also [8]).

Taken together, these anatomical results are all compatible with the idea of a selective stabilization of tangential fibers (during early postnatal development) between coactive groups of neurons. They thus support the hypothesis that the strength of intrinsic connections in the primary visual cortex of adult cats reflects the frequency of previous correlated activation. The experimental evidence appears convincing for the 'ocular dominance selectivity' of the tangential fibers, since these fibers connected cell groups activated by the same eye only in strabismic but not in normally raised cats [3]. The situation is less clear for the 'orientation selectivity' and 'axial anisotropy' of the tangential fibers. There

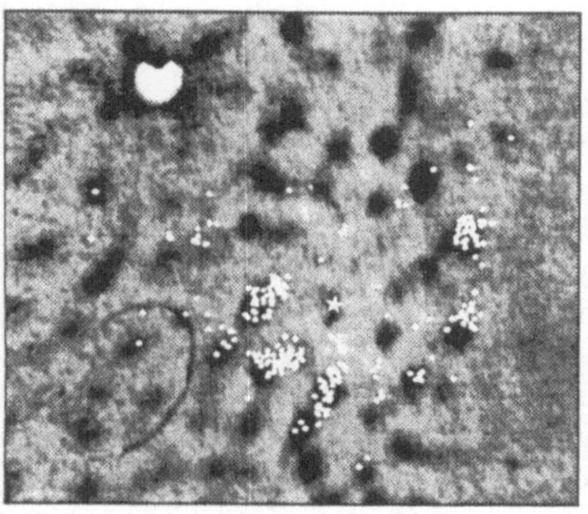

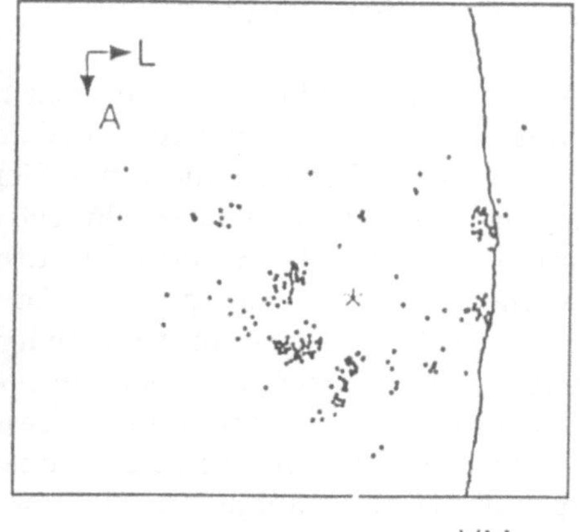

Fig. 1. Functional specificity of tangential connections in the primary visual cortex (area 17) of cats. (A) Autoradiograph of a flat-mount section demonstrating the layout of 2-DG labeled monocular OR-columns of a strabismic cat whose right eye had been stimulated with moving horizontal gratings. After an injection of green beads into a column preferring these gratings (marked with a white asterisk), retrogradely labeled neurons (white dots) are in register with the 2-DG labeled OR-domains. (B) Schematic drawing of the labeled neurons (black dots) relative to the cortical axes (VM=vertical meridian, continuous line). The distribution of labeled cells is elongated along the horizontal meridian (orthogonal to the VM). Note that tangential connections display both modular (A) and axial (B) specificity. L, lateral, A, anterior.

is no systematic investigation to date on parameters or rearing conditions that abolish or change one or the other component as was the case for the strabismic cats. Recently, modular and axial selectivity of tangential fibers was even reported in the visual cortex of very young tree shrews as early as 1-3 weeks after eye opening [9]. It thus seems that either a small amount of visual experience is already sufficient or that visual experience is not needed at all for establishing the modular and axial specificity of horizontal connections as seen in adult animals. On the other hand, the recent evidence that blockade of cortical but not of retinal activity prevents the initial development of clustered horizontal connections in area 17 of ferrets clearly demonstrated that neuronal activity is necessary for the development of tangential connections [10]. The role of spontaneous activity waves in this process is a highly debated issue at the moment but its exact role for both the development of intrinsic connections and cortical maps remains to be determined. The close relationship between intrinsic connections and OR-columns indicates that experience-independent mechanisms might also play a role in the development of the latter system.

2 Functional maps

Accumulating evidence suggests that maps of OR-preference might develop according to different rules compared to the system of OD-columns: while it is clearly established that the segregation of thalamo-cortical afferents into alternating OD-columns is driven by activity-dependent competition between the two sets of afferents for cortical territory, and that correlated activity seems to be the major organizing principle for the development of OD-columns (for review see [11]), visually driven activity seems to play a less important role for OR-preference maps to develop. In both area 17 of ferrets and area 18 of cats, OR-preference maps are present already one week after eye opening and their layout does not seem to change dramatically in the following two weeks [12,13]. In addition, OR-maps are identical for the two eyes in cats raised without binocular visual experience [14,15]. While the latter two studies were concerned with the development of cortical maps in area 18, we focused our attention on area 17 in strabismic cats. If visually driven correlated activity played an equally important role for the development of OR-columns as it does for the development of OD-columns, then OR-domains activated by different eyes should distribute independently, i.e. they should not be continuous across the boundaries between different OD-domains. A continuous course of OR-domains across OD-borders, however, would support an experience-independent initial development of OR-columns as suggested for area 18.

To distinguish between these two possibilities, we visualized the layout of iso-OR- and OD-columns in area 17 of strabismic cats using optical imaging of intrinsic signals, and analyzed the topographic relationship between the two functional systems [16,17]. Optical imaging revealed segregated OD-columns since monocular iso-OR-domains were different for the left and right eye. Most interestingly, iso-OR-columns were continuous across OD-borders (Figure 2A). Quantitative analysis indicated a preponderance of steep intersection angles between iso-OR-contours and OD-columns (Figure 2B) [16,17].

Imaging of the functional architecture of area 17 in strabismic cats revealed that iso-OR-columns are continuous across the borders between different OD-columns. In light of classical experiments about the development of OR-selectivity in single neurons (for a review see [18]) and the more recent imaging experiments in cat area 18 [14,15], the most likely explanation for this observation is that the basic layout of OR-preference maps is specified before the age at which OD-columns start to segregate (about three weeks postnatally) and thus before experience (and hence visually driven activity) begins to influence the development of cortical architecture (the "critical" period; e.g. [19]). This does not imply that neuronal activity plays no role in organizing OR-maps. As briefly discussed above for intrinsic connectivity patterns spontaneous activity waves are also ascribed a role in the expression of functional maps in the visual cortex. Traveling waves of both cortical [20] and thalamic [21,22] origin have been observed so far and thus could instruct the initial layout of OR-maps and the early clustering of tangential connections.

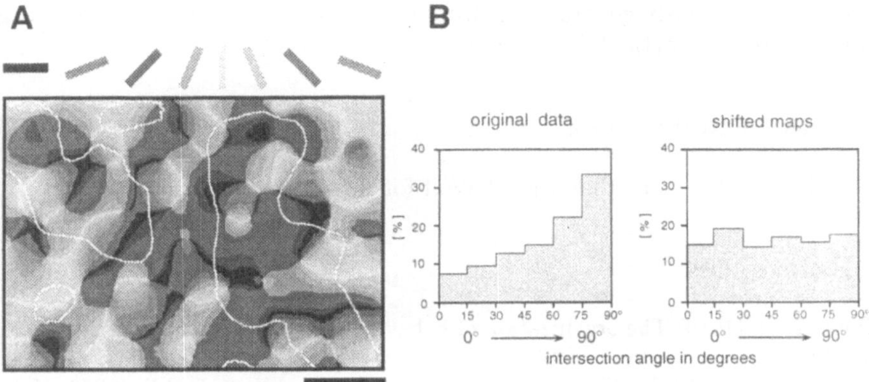

Fig. 2. Topographic relationship between iso-OR- and OD-columns in the visual cortex of a strabismic cat. (A) Superposition of the OR-preference ('angle') map and the outlined borders of adjacent OD-columns (white contours). The preferred OR for every region in the imaged cortex is coded by gray levels according to the scheme on the top of the figure. Scale bar 1mm. (B) Histograms of intersection angles between iso-OR- and OD-columns. x-axis, intersection angle in degrees from 0° to 90° (divided into 6 classes); y-axis, percentage of intersection angles in the respective class. Left histogram, original data; right histogram, shifted maps: iso-OR-contours of one animal superimposed with the OD-borders of another animal. Note that domains of like OR-preference labeled by the same gray in the angle map are continuous across the borders of adjacent OD-domains and that intersection angles between 75° and 90° are most abundant in the original data.

According to the proposed scenario, the development of OR-maps precedes that of OD-columns (at least in cats). Therefore the spatial arrangement of OD-columns should be influenced by the layout of the preexisting OR-map. The fact that iso-OR-contours tend to intersect the OD-borders at steep angles indeed indicates a systematic topological relation between the two maps as originally suggested by [23], and previously observed in macaque monkey striate cortex [24,25]. To what extent there are differences in the layout of OR-columns between strabismic and normally raised cats awaits the detailed analysis of these maps in normal animals.

3 Conclusions

Studies of both the development of tangential connections and functional maps demonstrated that experience and hence visually driven activity play a major role in the structuring of the visual cortex. However, recent experiments also revealed limits of the "experience-dependent" ("nurture") concept and pointed towards spontaneously generated neuronal firing patterns as an additional organizing principle of cortical development.

4 Acknowledgements

It is a pleasure for us to thank Fred Wolf for inspiring discussions.

5 References

[1] Hebb, D.O. (1949) The organization of behavior. A neuropsychological theory. New York: Wiley.
[2] Grinvald, A., Lieke, E., Frostig, R.D., Gilbert, C.D. and Wiesel, T.N. (1986) Nature 324: 361-364.
[3] Löwel, S. and Singer, W. (1992) Science 255: 209-212.
[4] Schmidt, K.E., Goebel, R., Löwel, S. and Singer, W. (1997a) Europ. J. Neurosci. 9:1083-1089
[5] Schmidt, K.E., Kim, D.-S., Singer, W., Bonhoeffer, T. and Löwel, S. (1997b) J. Neurosci., 15:5480-5492
[6] Gilbert, C.D. and Wiesel, T.N (1989) J. Neurosci. 9: 2432-2442.
[7] Malach, R., Amir, Y., Harel, M. and Grinvald, A. (1993) Proc. Natl. Acad. Sci. USA 90: 10469-10473.
[8] Bosking, W.H., Zhang, Y., Schofield, B. and Fitzpatrick, D. (1997) J. Neurosci. 17: 2112-2127.
[9] Crowley, J.C., Bosking, W.H., Foster, M. and Fitzpatrick, D. (1996) Soc. Neurosci. Abstr. 22: 404.10.
[10] Ruthazer, E.S. and Stryker, M.P. (1996) J. Neurosci. 16: 7253-7269.
[11] Stryker, M.P. (1991) In: Development of the visual system (Lam, D.M.-K. and Shatz, C.J., eds.), pp 267-287. Cambridge, MA: MIT Press.
[12] Chapman, B., Stryker, M.P, and Bonhoeffer, T. (1996) J. Neurosci. 16: 6443-6453.
[13] Gödecke, I., Kim, D.-S., Bonhoeffer, T. and Singer, W. (1997) Europ. J. Neurosci. 17:in press.
[14] Gödecke, I. and Bonhoeffer, T. (1996) Nature 379: 251-254.
[15] Kim, D.-S. and Bonhoeffer, T. (1994) Nature 370: 370-372.
[16] Löwel, S., Schmidt, K., Kim, D.-S., Singer, W. and Bonhoeffer, T. (1994) Eur. J. Neurosci. Suppl. 7: 48.06.
[17] Schmidt, K.E., Kim, D.-S., Singer, W., Bonhoeffer, T. and Löwel, S. (1994) Soc. Neurosci. Abstr. 20: 137.7.
[18] Henry, G.H., Michalski, A., Wimborne, B.M. and McCart, R.J. (1994) Prog. Neurobiol. 43: 381-437.
[19] Wiesel, T.N. (1982) Nature 299: 583-591.
[20] Huttenlocher, P.R. (1967) Exp. Neurol. 17: 247-262.
[21] Kim, U., Bal, T. and McCormick, D.A. (1995) J. Neurophysiol. 74: 1301-1323.
[22] McCormick, D.A., Trent, F. and Ramoa, A.S. (1995) J. Neurosci. 15: 5739-5752.
[23] Hubel, D.H. and Wiesel, T.N. (1977) Proc. R. Soc. Lond. B 198: 1-59.
[24] Bartfeld, E. and Grinvald, A. (1992) Proc. Natl. Acad. Sci. USA 89: 11905-11909.
[25] Obermayer, K. and Blasdel, G.G. (1993) J. Neurosci. 13: 4114-4129.

Geometric Relationships Between Feature Maps in Cat Visual Cortex

M. Hübener[1], D. Shoham[2], S. Schulze[1], G. Brändle[1], A. Grinvald[2], and T. Bonhoeffer[1]

[1]Max-Planck-Institut für Psychiatrie, Am Klopferspitz 18a, 82152 Martinsried, Germany
[2]The Weizmann Institute of Science, Rehovot 76100, Israel

Abstract. We used optical imaging of intrinsic signals to simultaneously visualize the orientation, ocular dominance, and spatial frequency maps in cat visual cortex. The analysis of the geometric relationships between these three columnar systems revealed that they are spatially related such that in visual cortex all stimulus features are represented at least once for every point in visual space.

1 Introduction

A characteristic feature of the visual cortex is that neurons with similar response properties are clustered together in columns. Apart from the well known orientation and ocular dominance columns [1, 2] evidence has accumulated in recent years that other receptive field properties like direction and spatial frequency preference are also organized in a columnar fashion [3, 4]. This multitude of columnar systems raises the question of how these different systems are arranged within the cortex and whether they have a consistent geometrical relationship. Optical imaging of intrinsic signals [5, 6] is well suited to address these issues because it allows the mapping of large regions of the visual cortex. In this study we analyze the geometric relationships between the orientation, ocular dominance, and spatial frequency domains in cat primary visual cortex.

2 Methods

Cortical maps in area 17 of 8–13 week old cats were visualized using optical imaging of intrinsic signals. The surface of the cortex was illuminated with red light (707 nm) and images were acquired while the animal was visually stimulated, using either a slow-scan CCD-camera or an enhanced video imaging system. Visual stimuli consisted of moving, high-contrast gratings presented binocularly or monocularly at four orientations and at different spatial and temporal frequencies.

Orientation preference maps were calculated by vectorially summing, for each pixel, the responses to all orientations and assigning the angle of the resulting vector to this pixel. Differential ocular dominance maps were computed by dividing the sum of the single-condition maps obtained with stimulation of the contralateral eye by the sum of the ipsilateral eye's single-condition maps. Similarly, spatial frequency maps were calculated by adding all orientation maps of one spatial frequency and dividing the result by the sum of the orientation maps of the second spatial frequency.

Geometric relationships between the columnar systems were assessed with a detailed quantitative analysis. Intersection angles of iso-orientation lines with borders between ocular dominance columns were determined by computing, for both maps, the gradient fields. The angular difference between the two gradient fields corresponds to the intersection angles. This calculation was applied only to pixels located on borders between ocular dominance columns. In the same way we computed the intersection angles of iso-orientation lines with borders between spatial frequency domains. The precise positions of orientation centers were determined as follows: first, for each point in the vectorial orientation preference map, we calculated the sum of the absolute values of divergence and curl. Next we searched for local maxima in the resulting array and sorted these maxima in descending order of value. Then we applied a threshold to discard, under visual control, all maxima below a certain value that were not located on orientation centers. With this standardized procedure we were able to find the exact positions of all orientation centers in the orientation preference maps. The center and border regions of ocular dominance columns were defined by way of the pixel values in the differential maps: each map was divided into 10 regions of equal area, with those parts encompassing the pixels with the lowest, or highest values corresponding to the centers of the contralateral, or ipsilateral eye columns. Thus, the combined center regions made up 20% of the area of an ocular dominance map. Accordingly, the border regions were defined as 20% of the area of a map with intermediate pixel values. The center and border regions of the spatial frequency domains were defined in the same manner. To determine the centers of low spatial frequency domains we used an algorithm searching for local minima in the low pass filtered spatial frequency map.

In some cases the cortical tissue was processed for cytochrome oxidase after the optical imaging experiment in order to relate features of cortical maps to the pattern of the cytochrome oxidase blobs.

3 Results

Clear orientation and ocular dominance maps were obtained in most experiments. Iso-orientation domains appeared as round or elongated patches with a center to center spacing of roughly one millimeter. In contrast, individual ocular dominance columns had the shape of 0.5 millimeter wide, curved bands running across the cortex over a distance of a few millimeters. As described previously [7] orientation domains were found to be arranged in a pinwheel-like fashion around orientation centers. In monkey visual cortex it has been found that the orientation centers are usually located in the centers of ocular dominance columns and that iso-orientation lines cross the borders between ocular dominance columns preferentially at right angles [8, 9, 10]. Since ocular dominance columns are arranged less orderly in cat compared to monkey visual cortex, we wondered whether such a relationship might also be present in cat visual cortex. Indeed we found the very same geometric relationships in the visual cortex of the cat: iso-orientation lines tended to cross the borders between adjacent ocular dominance columns at right angles and many orientation centers were located in the middle of the ocular dominance columns. To quantify these observations we

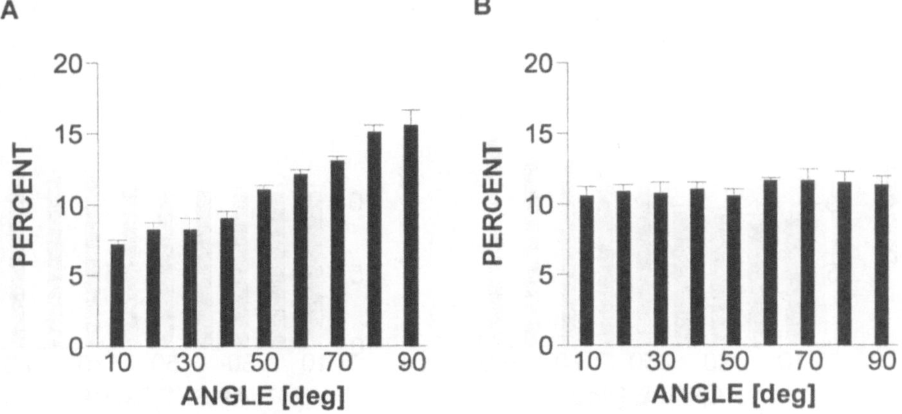

Fig. 1. Quantitative analysis of intersection angles. A: averaged distribution (from 6 maps, error bars are given as SEM) of intersection angles between iso-orientation lines and ocular dominance borders. The histogram reveals a clear predominance of large intersection angles. B: control, orientation preference maps from one cat were related to ocular dominance maps from a different cat, the angle distribution is flat (n = 6 maps).

determined the crossing angles of iso-orientation lines with borders between ocular dominance columns for every point on this border. Figure 1a shows the averaged distribution of these angles from 6 maps analyzed in this way: there is a clear preponderance of right angles. Since we were concerned that this skewed distribution might not result from a specific spatial relationship between the maps but could rather be the consequence of the geometry of each map itself, we also measured the intersection angles for maps originating from different animals. To this end we overlaid ocular dominance maps from one cat with orientation preference maps from a different cat. In these control cases the angle distribution is essentially flat (Fig. 1b). The difference between the two distributions is significant ($p<0.05$, Wilcoxon signed-rank test, n = 6 maps). We also determined the frequency of orientation centers in different regions of the ocular dominance map (Fig. 2). We found a significantly

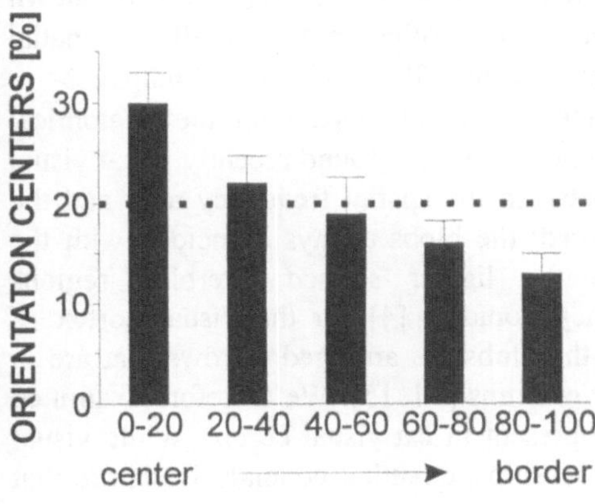

Fig. 2. Relative frequency of orientation centers in different regions of ocular dominance maps (n = 6, error bars are SEM). The maps were divided into 5 regions of equal area, with the 0-20 percentile denoting the center and the 80-100 percentile the border regions of the ocular dominance columns. The dotted line indicates the expected value (20 %) if the orientation centers were distributed randomly. Note the high incidence of orientation centers in the center regions of the ocular dominance columns.

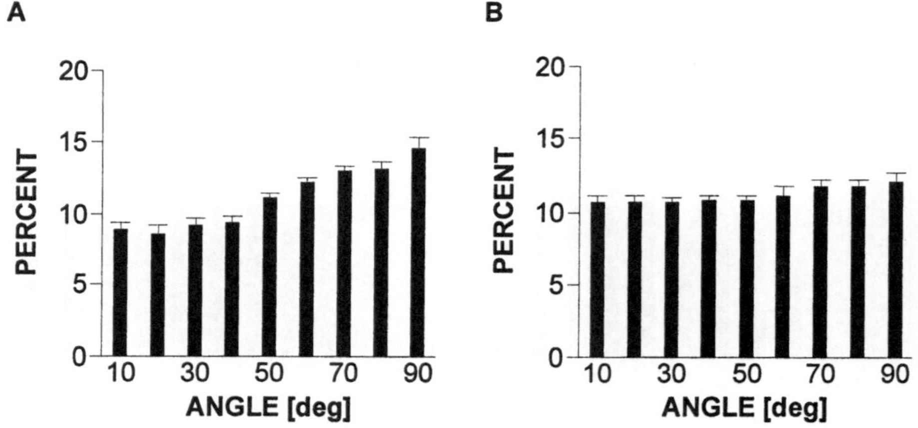

Fig. 3. Intersection angles between iso-orientation lines and borders between spatial frequency domains. A: averaged distribution from 13 maps, large angles are clearly over-represented. B: in the control cases (n = 13 maps) the distribution is again almost flat.

higher proportion of orientation centers in the middle of ocular dominance domains (30.3%) than expected if the orientation centers were distributed randomly (20%; $p<0.01$, χ^2-test, n = 6 maps).

Stimulation with gratings of different spatial frequencies revealed the presence of a spatial frequency map in cat visual cortex. In many cases these maps had a characteristic layout: regions with a preference for low spatial frequencies formed isolated patches that were embedded in a contiguous "matrix" of high spatial frequency preference. Neurons in the spatial frequency domains also responded differentially to variations of the temporal frequency of the stimuli: low spatial frequency domains were activated best by high temporal frequencies while the high spatial frequency domains preferred low temporal frequencies. The spatial frequency domains too were not independent from the orientation map. Rather, we found similar, albeit somewhat weaker spatial relationships again: iso-orientation lines frequently crossed the borders between neighboring spatial frequency domains at right angles (Fig. 3; $p<0.05$, Wilcoxon signed-rank test, n = 13 maps) and, as shown in Figure 4, orientation centers were found more often in the middle of spatial frequency domains (27.3%, expected value: 20%; $p<0.05$, χ^2-test, n = 13 maps).

In some experiments we compared the functional maps with the anatomical pattern of the cytochrome oxidase blobs, which have been found recently in cat visual cortex [11, 12, 13]. A clear correlation between the spatial frequency map and the cytochrome oxidase staining pattern emerged: the blobs always coincided with the low spatial frequency domains, while the lighter stained interblob regions corresponded to the high spatial frequency domains [4]. In the visual cortex of macaque monkeys it has been found that the blobs are arranged in rows that are in precise register with the ocular dominance columns [14, 15]. We therefore examined whether such a relationship might also be present in cat visual cortex. While visual inspection did not reveal any obvious relationships, quantitative analysis proved that

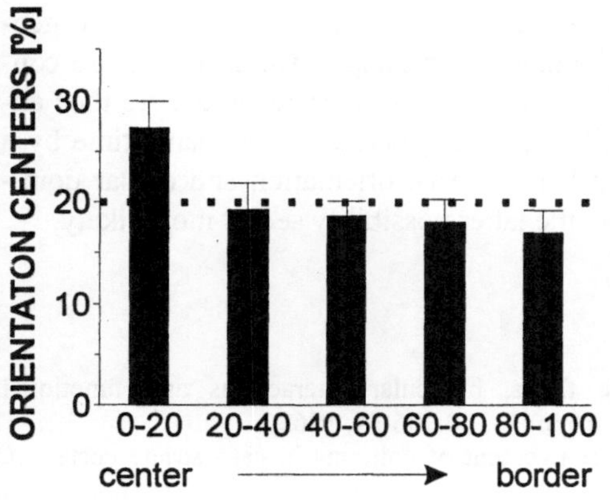

Fig. 4. Relative frequency of orientation centers in different regions of spatial frequency maps (n = 13). Same conventions as in Figure 2. Orientation centers are found more often in the center regions than near the borders of spatial frequency domains.

both systems are not totally independent from each other: there is a tendency for the low spatial frequency domains to rather lie in the centers of ocular dominance columns than on their borders. This tendency becomes apparent when counting the centers of low spatial frequency domains near the borders between ocular dominance columns: on average only 4.1% of these centers are located in the border region of ocular dominance columns (defined as 20% of the total map area; $p<0.05$, χ^2- test, n = 6 maps). Thus, since the low spatial frequency domains coincide with the cytochrome oxidase blobs, this supports the notion that in cat visual cortex the blobs are spatially related to ocular dominance columns [13, but see 12]. Clearly, however, this relationship is less pronounced than in the visual cortex of macaque monkeys. Interestingly, a recent study on squirrel monkey visual cortex reported that the blobs in this species are not aligned with ocular dominance columns [16], indicating that differences in the relationship between blobs and ocular dominance columns do not simply reflect a dichotomy between primates and carnivores.

4 Conclusion

The results of this study show that specific rules govern the geometric relationships between the ocular dominance, spatial frequency, and orientation domains in cat primary visual cortex. These rules, however, are not rigidly followed and we therefore do not find that the visual cortex is a "crystalline" structure built from identical modules that are repeated over and over. However, geometric relationships between columnar systems clearly exist thus raising the question whether such a layout might be advantageous for visual information processing. One functional reason for this arrangement might be that the predominance of right angle crossings between the different columnar systems ensures that all possible combinations of stimulus features are represented at least once in a given region of the visual cortex. Such a design would avoid the occurrence of functional blind spots for a particular stimulus attribute in the visual field.

Intimately related to their possible functional role is the question of how these geometric relationships emerge during development. In principle two scenarios are conceivable: either one map is established first and the maps later generated are then adjusted to the first one, or, alternatively all maps are generated at the same time by a coupled developmental process. Since, at least in cats, orientation and ocular dominance maps appear at about the same time, the latter possibility seems more likely.

5 References

1. D.H. Hubel, T.N. Wiesel. Receptive fields, binocular interactions and functional architecture in the cat's visual cortex. *J. Physiol.,* 160:106-54, 1962.
2. D.H. Hubel, T.N. Wiesel. Shape and arrangement of columns in cat's striate cortex. *J. Physiol.,* 165:559-68, 1963.
3. N.V. Swindale, J.A. Matsubara, M.S. Cynader. Surface organization of orientation and direction selectivity in cat area 18. *J. Neurosci.,* 7:1414-1427, 1987.
4. D. Shoham, M. Hübener, S. Schulze, A. Grinvald, T. Bonhoeffer. Spatio-temporal frequency domains and their relation to cytochrome oxidase staining in cat visual cortex. *Nature,* 385:529-33, 1997.
5. A. Grinvald, E.E. Lieke, R.D. Frostig, C.D. Gilbert, T.N. Wiesel. Functional architecture of cortex revealed by optical imaging of intrinsic signals. *Nature,* 324:361-4, 1986.
6. T. Bonhoeffer, A. Grinvald. Optical imaging based on intrinsic signals: The Methodology. In: *Brain Mapping: The Methods*, A. Toga, J.C. Mazziotta, eds. San Diego, CA: Academic Press, Inc., pp. 55-97, 1996.
7. T. Bonhoeffer, A. Grinvald. Iso-orientation domains in cat visual cortex are arranged in pinwheel-like patterns. *Nature,* 353:429-31, 1991.
8. E. Bartfeld, A. Grinvald. Relationships between orientation preference pinwheels, cytochrome oxidase blobs and ocular dominance columns in primate striate cortex. *Proc. Natl. Acad. Sci. USA,* 89:11905-9, 1992.
9. K. Obermayer, G.G. Blasdel. Geometry of orientation and ocular dominance columns in monkey striate cortex. *J. Neurosci.,* 13:4114-29, 1993.
10. G.G. Blasdel, K. Obermayer, L. Kiorpes. Organization of ocular dominance and orientation columns in the striate cortex of neonatal macaque monkeys. *Visual Neurosci.,* 12:589-603, 1995.
11. K.M. Murphy, R.C. Van Sluyters, D.G. Jones. Cytochrome-oxidase activity in cat visual cortex: is it periodic? *Soc. Neurosci. Abstr.,* 16:292, 1990
12. R.H. Dyck, M.S. Cynader. An interdigitated columnar mosaic of cytochrome oxidase, zinc, and neurotransmitter-related molecules in cat and monkey visual cortex. *Proc. Natl. Acad. Sci. USA,* 90:9066-9, 1993.
13. K.M. Murphy, D.G. Jones, R.C. Van Sluyters. Cytochrome-oxidase blobs in cat primary visual cortex. *J. Neurosci.,* 15:4196-208, 1995.
14. J.C. Horton, D.H. Hubel. Regular patchy distribution of cytochrome oxidase staining in primary visual cortex of macaque monkey. *Nature,* 292:762-4, 1981.
15. J.C. Horton. Cytochrome oxidase patches: a new cytoarchitectonic feature of monkey visual cortex. *Philos. Trans. R. Soc. Lond. [Biol.],* 304:199-253, 1984.
16. J.C. Horton, D.R. Hocking. Anatomical demonstration of ocular dominance columns in striate cortex of the squirrel monkey. *J. Neurosci.,* 16:5510-22, 1996.

Supported by the Max-Planck-Gesellschaft, the ISF/Israel Academy, and the EC Biotech Program.

A Linear Hebbian Model for the Development of Spatiotemporal Receptive Fields of Simple Cells

S. Wimbauer, O. Wenisch, and J.L. van Hemmen

Physik-Department, Technische Universität München,
D-85747 Garching bei München, Germany

Abstract. We propose a linear Hebbian model that describes the development of spatiotemporal receptive fields as a competition between four types of input onto a cortical cell, viz., non-lagged ON and OFF inputs and lagged ON and OFF inputs. The outcome of the development is determined mainly by the spatial and the temporal correlations between the different inputs. We indicate what type of input correlation leads to a wide range of direction selectivity indices as found experimentally.

1 The model

Most simple cells in the visual cortex of cats and many other mammals respond selectively to spatial and spatiotemporal stimulus parameters such as orientation [6] and the direction of stimulus motion [7]. The combined spatial and temporal response properties of cortical neurons are summarized by the notion of a spatiotemporal receptive field (RF).

Two types of spatiotemporal RFs can be observed experimentally, viz., spatiotemporally separable and non-separable RFs [2,3]. In the former case the response function can be written as the product of a spatial and a temporal part whereas in the latter case this is not possible. Only spatiotemporally inseparable RFs give rise to a direction-selective response of a cell [1,12]. An example is shown in Figs 1.b and c. A light bar with an orientation perpendicular to the image plain moving into positive x-direction will evoke a large response, if, in a subregion of homogeneous polarity (ON or OFF), the latency for small x exceeds that for larger x. In this way input signals arrive simultaneously. If the bar moves to the right at a speed v that corresponds to the slope of the ON subregion in the x-t plane, the latency difference is compensated and a short but strong response is evoked at the cell.

What are the physiological mechanisms that underlie such a spatiotemporal RF structure? Motivated by experimental results of Saul and Humphrey [11] we assume that the convergence of four different types of spatiotemporal channels from the LGN onto a cortical cell is responsible for the structure of a RF. These are ON- and OFF-type channels that come in two different temporal "flavors", namely, non-lagged and lagged ones. Lagged channels show a delayed excitation as compared to non-lagged channels in response to a stimulus. The spatiotemporal channels are modeled by the product of a spatial linear response function

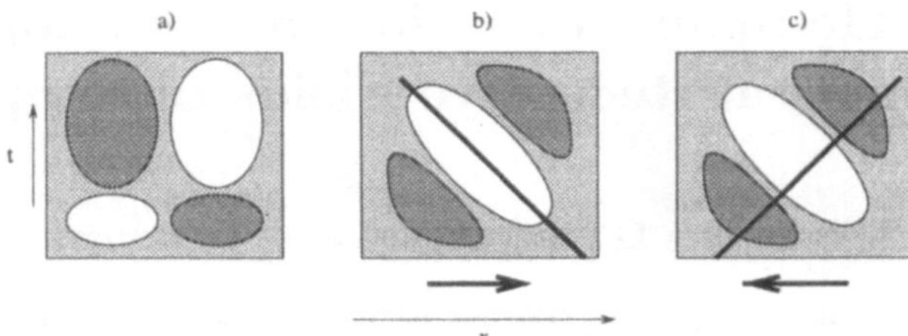

Fig. 1. Spatiotemporal RFs for one spatial dimension (axis of motion). A larger value of t corresponds to a larger time difference between stimulus and response. White and dark patches specify areas of predominant ON and OFF inputs, respectively. a) A spatiotemporally *separable* RF. b) and c) A spatiotemporally *inseparable* RF that is sensitive to rightward but *not* to leftward motion.

$R^c(\boldsymbol{\alpha}, \boldsymbol{\alpha}')$ given by a Mexican-hat function and a temporal response function $L^\tau(t - t', \boldsymbol{\alpha})$, where the index c stands for ON and OFF inputs, and the index τ labels lagged and non-lagged inputs. Typical shapes of non-lagged and lagged response-functions can be found in the second line of Fig 2. The product $R^c(\boldsymbol{\alpha}, \boldsymbol{\alpha}')L^\tau(t - t', \boldsymbol{\alpha})$ describes the response of a cell in the lateral geniculate nucleus (LGN) to the activity $S(\boldsymbol{\alpha}', t')$ of the photoreceptors. On the level of the cortex, these channels are summed and weighted by the feedforward synapses $J^{c,\tau}(\mathbf{x}, \boldsymbol{\alpha})$ from the LGN to a simple cell. The input from other cortical cells is modeled by an intracortical interaction function $I(\mathbf{x}, \mathbf{x}')$. Altogether, we obtain the following functional form for the local potential of a simple cell,

$$h(\mathbf{x}, t) = \sum_{\mathbf{x}'} I(\mathbf{x}, \mathbf{x}') \sum_{c=\mathrm{ON,OFF}} \sum_{\tau=\mathrm{nl,l}} \sum_{\boldsymbol{\alpha}} J^{c,\tau}(\mathbf{x}', \boldsymbol{\alpha}) \int_{-\infty}^{\infty} d\boldsymbol{\alpha}' \int_{-\infty}^{\infty} dt' R^c(\boldsymbol{\alpha}, \boldsymbol{\alpha}')L^\tau(t - t', \boldsymbol{\alpha})S(\boldsymbol{\alpha}', t'), \tag{1}$$

where $\mathbf{x}$ and $\boldsymbol{\alpha}$ denote retinotopical positions in the cortex and the LGN.

It is commonly assumed that an activity-driven modification of synapses during the critical period shortly before and after birth takes place according to some sort of Hebbian learning rule [5]. In this article, we combine the spatial models of [9] with the spatiotemporal model of [14] and study the emergence of direction selectivity through a competition between lagged and non-lagged inputs, cf. Wimbauer [13]. A related model has concurrently been explored by Feidler et al. [4]. In our model only thalamocortical synapses are modified by Hebbian learning, whereas intracortical synapses are kept fixed. Through the linear response model (1) the correlation between pre- and postsynaptic activity, that drives the development in a Hebbian way, is transformed into a spatiotemporal correlation function $C^{c,c';\tau,\tau'}$ for the inputs from the LGN. The following developmental equation is obtained,

$$\frac{dJ^{c,\tau}(\mathbf{x}, \boldsymbol{\alpha}, t)}{dt} = \lambda A(\mathbf{x} - \boldsymbol{\alpha}) \sum_{\mathbf{x}'} I(\mathbf{x}, \mathbf{x}') \sum_{c'=\mathrm{ON,OFF}} \sum_{\tau'=\mathrm{nl,l}} \sum_{\boldsymbol{\alpha}'} C^{c,c';\tau,\tau'}(\boldsymbol{\alpha}, \boldsymbol{\alpha}')J^{c',\tau'}(\mathbf{x}', \boldsymbol{\alpha}', t). \tag{2}$$

In the above equation $A(\mathbf{x} - \boldsymbol{\alpha})$ denotes an arbor function that restricts the possible synaptic wirings that might emerge from the learning process. During the simulations which we describe in the following a subtractive constraint term is added to Eq. 2 to model competition between the different types of input [10].

2 Results

Orientation and direction selectivity already develop before eye opening [8]. We therefore assume that uncorrelated noise in the photoreceptors drives the development during this period so that the correlation function for the inputs from the LGN in (2) factorizes into a spatial and a temporal part $C^{c,c';\tau,\tau'}(\boldsymbol{\alpha}, \boldsymbol{\alpha}', t) = C^{\tau,\tau'} C^{c,c'}(\boldsymbol{\alpha}, \boldsymbol{\alpha}')$. The spatial correlation function is given by a Mexican-hat function, whereas the temporal one reduces to a 2×2 matrix

$$C^{\tau,\tau'} = \begin{pmatrix} C^{\mathrm{nl,nl}} & C^{\mathrm{nl,l}} \\ C^{\mathrm{l,nl}} & C^{\mathrm{l,l}} \end{pmatrix} = \begin{pmatrix} 1 & \mathrm{corr} \\ \mathrm{corr} & 1 \end{pmatrix} . \tag{3}$$

The value of corr depends on the form of the non-lagged and lagged response functions, in particular, on the additional delay introduced by a lagged channel as compared to a non-lagged one. If one chooses the parameter values for the two temporal response functions in agreement with measurements of Saul and Humphrey [11], the resulting corr is small and the synaptic weights for non-lagged and lagged inputs develop nearly independently. The outcome of a typical simulation run for this case is shown in Fig. 2. The delay of the lagged response has been chosen in such a way that corr $= -0.05$.

Fig. 2 also exhibits the way in which spatiotemporal RFs arise. The preferred orientation can be integrated out so that we are left with one spatial dimension, the axis of motion. We have two input channels, viz., non-lagged (top left) and lagged (top right) cells in the LGN. The strength of their synaptic connections to cortical neurons has been plotted. After multiplication ($\times$) with a (non)-lagged temporal response function and subsequent addition ($+$) one obtains the spatiotemporal RFs of these neurons. We now turn to the details.

In the top line of Fig. 2, the difference between ON and OFF inputs is displayed for a 5×5 sublattice of a 32×32 grid for both non-lagged and lagged channels. Each of the small squares corresponds to one cortical cell and shows the distribution of synaptic weights that link this cell with cells in the LGN that are located in a 13×13 grid projecting onto this cortical cell. A single square thus corresponds to the cell's spatial RF split into non-lagged and lagged inputs. We have used a grey-scale coding to indicate the synaptic strength with white denoting positive values or predominant ON inputs and black specifying negative values or predominant OFF inputs.

The spatial RFs formed by non-lagged and lagged inputs alone form an orientation map. Since the correlation between the two types of temporal input is very weak and due to different (random) initial conditions, the two maps develop nearly independently. The orientation and phase of the non-lagged and lagged spatial RF of *one* cortical cell will therefore be nearly uncorrelated.

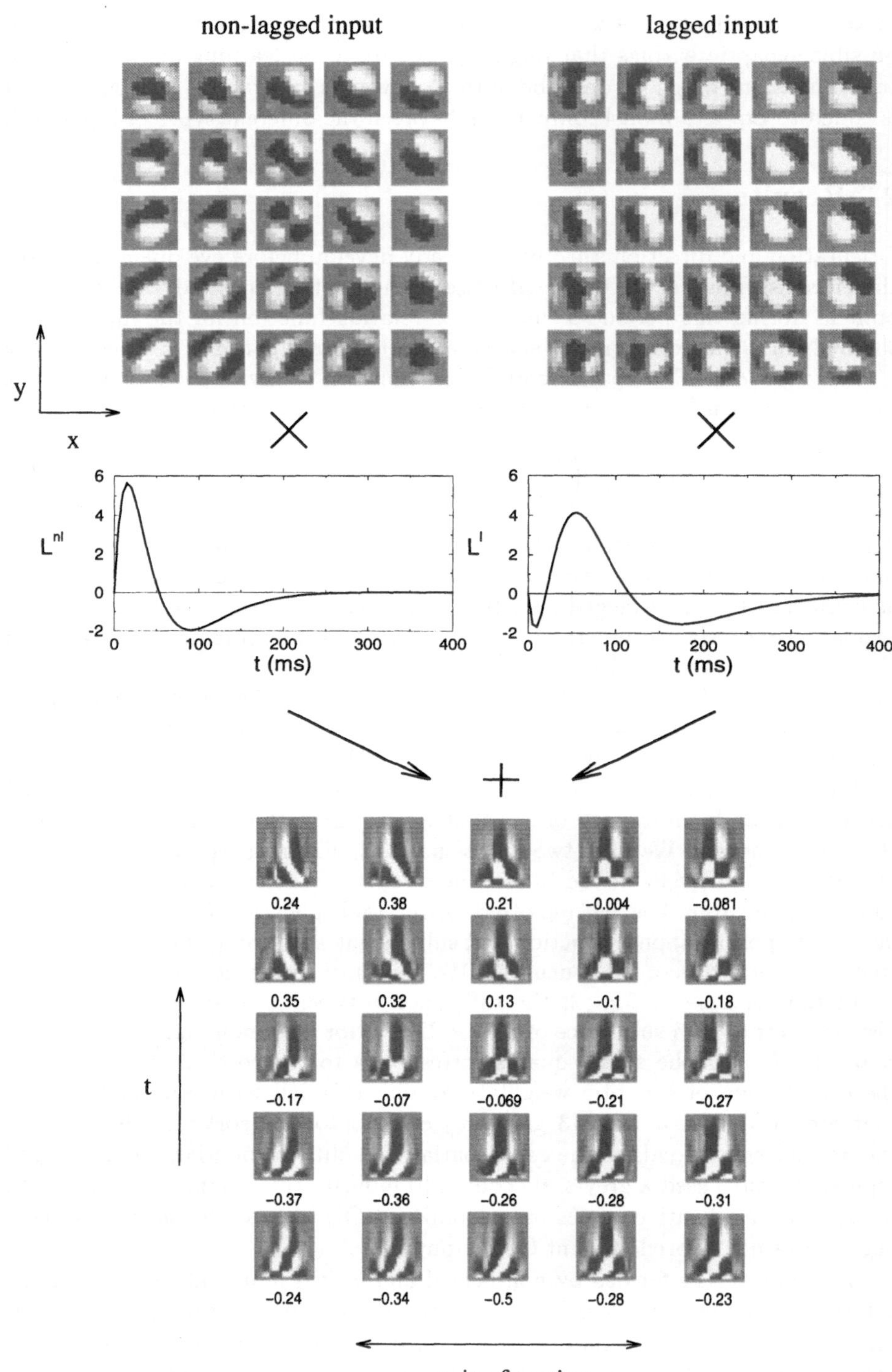

Fig. 2. Simulation run for weak correlations between non-lagged and lagged inputs (corr = −0.05). The numbers below the spatiotemporal RFs (bottom) indicate their direction selectivity index.

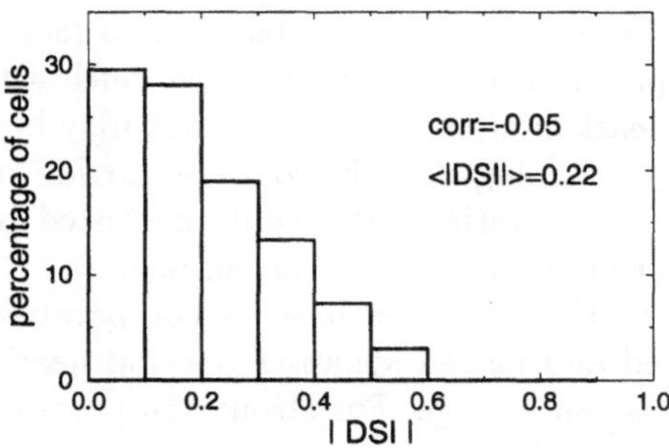

Fig. 3. Distribution of the absolute value of the direction selectivity index (DSI) for the simulation run of Fig. 2.

To obtain a spatiotemporal RF the spatial RFs for the two temporal input types are multiplied by their respective temporal response function and added. In the third row of Fig 2 the resulting spatiotemporal RFs are displayed after an integration along the preferred orientation. The direction selectivity index of each cell can be found beneath the RF and is defined as $DSI = (R_p - R_n)/(R_p + R_n)$ where R_n and R_p denote the maximum response of a cell to a sine wave grating for right- and leftward motion.

A good example of a cell that shows strong direction selectivity is the one in the center of the last row at the bottom of Fig. 1. The ON and OFF subregions are tilted clockwise in the x-t plane. The cell therefore responds better to leftward than to rightward motion. Looking at the respective spatial RFs of the lagged and non-lagged inputs, one notices that the non-lagged input has a spatial phase of about $\pi/2$, whereas the lagged input has approximately zero phase. It turns out that a difference in spatial phase between the lagged and non-lagged inputs is an *essential* prerequisite for the emergence of a direction-selective response to an elongated stimulus.

In contrast, the spatiotemporal RF in the top right corner is nearly spatiotemporally separable. That is, it can be described by a product of a spatial and a temporal response function and, thus, responds equally well to rightward and leftward motion, as is indicated by the low direction selectivity index -0.081. The corresponding spatial RFs for non-lagged and lagged inputs both have a phase near $\pi/2$. This example demonstrates that a whole range of spatiotemporal RFs with different degrees of direction selectivity emerges once there is only a weak correlation between the two types of temporal input channels.

We have plotted the corresponding distribution of the absolute value of the direction selectivity index in Fig. 3. This histogram can be compared with the distribution of the very same quantity which one obtains from reverse correlation measurements. As compared to Fig. 17.c of [2], the good agreement of the two plots, i.e., between experiment and model prediction, is evident.

Due to the nearly independent growth of non-lagged and lagged synapses both types of temporal input may not only develop different spatial phases, which finally results in a direction-selective RF, but also different preferred orientations. The second cell from the left in the first row of Fig. 2 may serve as an example of this case. Different orientations of non-lagged and lagged inputs correspond to a rotation of the preferred orientation in time. Such a RF property appears to be uncommon, at least in adult animals [2], but may be present during early development. After eye opening the RFs could be further modified by vision, in particular, moving lines or gratings; this could be tested by studying temporal invariance of preferred orientation in young animals.

For subtractive constraints there exists a comparatively broad region for corr, where non-lagged and lagged synapses develop nearly independently and direction-selective RFs can emerge. For strong temporal correlations, however, the maps of non-lagged and lagged inputs are nearly identical and agree both in phase and orientation for each cortical cell. Hence in this case, the resulting spatiotemporal RFs are all spatiotemporally separable and show only very weak direction selectivity.

References

1. E. H. Adelson and J. R. Bergen. Spatiotemporal energy models for the perception of motion. *J. Opt. Soc. Am. A*, 2:284–299, 1985.
2. G. C. DeAngelis, I. Ohzawa, and R. D. Freeman. Spatiotemporal organization of simple-cell receptive fields in the cat's striate cortex. I. general characteristics and postnatal development. *J. Neurophysiol.*, 69:1091–1117, 1993.
3. G. C. DeAngelis, I. Ohzawa, and R. D. Freeman. Receptive-field dynamics in the central visual pathways. *TINS*, 18:451–458, 1995.
4. J. C. Feidler, A. B. Saul, A. Murthy, and A. L. Humphrey. Hebbian learning an the development of direction selectivity: the role of geniculate response timings. *Network*, 8:195–214, 1997.
5. D. O. Hebb. *The Organization of Behavior*. Wiley, New York, 1949.
6. D. H. Hubel and T. N. Wiesel. Receptive fields of single neurons in the cat's striate cortex. *J. Physiol.*, 148:574–591, 1959.
7. D. H. Hubel and T. N. Wiesel. Receptive fields, binocular interaction and functional architecture in the cat's visual cortex. *J. Physiol.*, 160:106–154, 1962.
8. D. H. Hubel and T. N. Wiesel. Receptive fields of cells in striate cortex of very young, visually inexperienced kittens. *J. Neurophysiol.*, 26:994–1002, 1963.
9. K. D. Miller. A model for the development of simple cell receptive fields and the ordered arrangement of orientation columns through activitydependent competition between ON- and OFF-center inputs. *J. Neurosci.*, 14:409–441, 1994.
10. K. D. Miller and D. J. C. MacKay. The role of constraints in Hebbian learning. *Neural Comput.*, 6:100–126, 1994.
11. A. B. Saul and A. L. Humphrey. Spatial and temporal response properties of lagged and nonlagged cells in cat lateral geniculate nucleus. *J. Neurophysiol*, 64:206–224, 90.
12. A. B. Watson and A. J. Ahumada. Model of human visual motion sensing. *J. Opt. Soc. Am. A*, 2:322–342, 1985.
13. S. Wimbauer. *Raumzeitliche rezeptive Felder — Modellierung von Antworteigenschaften und Entwicklung einfacher Zellen im visuellen Cortex*. Verlag Harri Deutsch, Frankfurt a. M., 1996.
14. S. Wimbauer, W. Gerstner, and J. L. van Hemmen. Emergence of spatiotemporal receptive fields and its application to motion detection. *Biol. Cybern.*, 72:81–92, 1994.

Synapse Clustering Can Drive Simultaneous ON-OFF and Ocular-Dominance Segregation in a Model of Area 17

Martin Stetter[1], Elmar W. Lang[2] and Klaus Obermayer[1]

[1]Dept. of Computer Science, Technische Universität Berlin, Germany, and
[2]Dept. of Biophysics, Universität Regensburg, Germany
E-mail: moatl@cs.tu-berlin.de

Abstract In the primary visual cortex of monkeys, the development of ocular dominance and orientation selectivity is at least partially driven by neural activity. We propose a modified Hebb-type learning mechanism, which takes into account non-specific components of activity-dependent synaptic modification. It is shown analytically, that ocular dominance and ON-OFF-segregation occur simultaneously in a linear network as soon as left-eye and right-eye synapses tend to cluster on the surface of the postsynaptic neuron. Simulations show, that this mechanism is robust against the introduction of network nonlinearities such as rectifying transfer functions and intracortical recurrency. The results imply, that details of single cell properties can have considerable influence on the behaviour of high level developmental models.

1 Introduction

In the input layer 4C of monkey primary visual cortex, most of the neurons are nearly monocular, are arranged in ocular dominance stripes and in addition show receptive fields with distinct ON-response and OFF-response subfields. While the development of ocular dominance requires both pre- and postsynaptic activity [1], the activity dependence of ON-OFF-segregation mechanisms is still under discussion. However, in monkeys both ocular dominance and orientation selectivity (which is often suggested to be initialized by elongated ON-OFF-subfields) are known to emerge, at least in part, before birth [2].

Linear correlation based learning (CBL) models for the activity-dependent development of the functional cortical architecture have been found to show either ON-OFF-segregation of simple receptive fields [3] and their arrangement into orientation maps [4,5], or the development of ocular dominance patterns [6] after Hebb-training with uncorrelated noise. However, it was shown recently [7], that ON-OFF-subfields and ocular dominance cannot develop simultaneously in linear Hebb-trained CBL-models, but require nonlinearities to be present.

In this work we formulate a higher order synaptic modification rule, which includes a term proportional to the proximity of the synapses onto a given postsynaptic neuron. We show analytically that as soon as left-eye and right-eye driven synapses tend to cluster on the postsynaptic neuron, the CBL-model shows both ON-OFF- and ocular dominance segregation. It is concluded that the geometry of single neurons can considerably influence self-organization processes rendering the investigation of developmental mechanisms with more detailed neuron models an important future task.

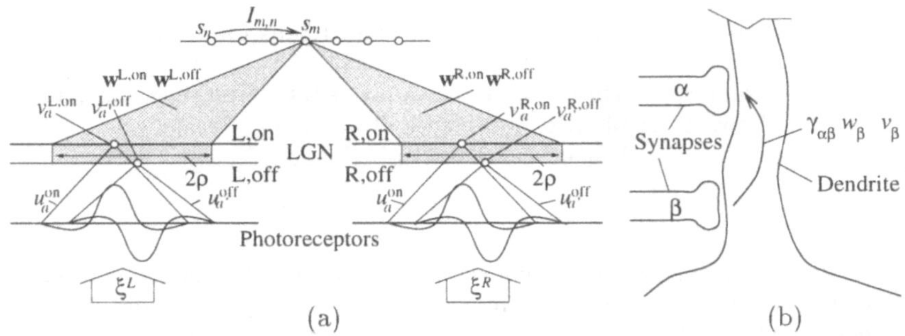

Fig. 1 (a) Neural network architecture. Four sublayers of LGN-neurons with DOG-shaped receptive fields project to a layer of cortical neurons, which are mutually coupled by lateral connections. **(b)** The influence of higher order synaptic plasticity and geometry effects on learning. Synaptic modification of a synapse α is influenced by nearby synapses β proportional to their presynaptic activities.

2 Model Setup and Synaptic Modification Rule

We consider a feed-forward neural network with four 2-dimensional input layers which correspond to the L,on- L,off-, R,on-, and R,off-populations of LGN-cells (L,R abreviate left-eye and right-eye driven neurons), and a two-dimensional output-layer modeling layer 4C of striate cortex (Fig. 1a). Each output neuron m receives non-negative afferent connections $w_a^{M,\mu}$ from LGN-cells a of each layer ($M \in \{\mathrm{L}, \mathrm{R}\}$, $\mu \in \{\mathrm{on}, \mathrm{off}\}$) within the circular afferent projection area (radius ρ, see fig. 1a), as well as intracortical synaptic connections $I_{m,n}$ from other cortical cells n. The four input layers in turn are driven via spatially opponent receptive fields $u_{a,i}^{\mu}$ by two photoreceptor layers. Presentation of a binocular input pattern ξ_a^M thus leads to LGN-activities $v_a^{M,\mu} = f(\sum_{i \in \mathrm{rf}(a)} u_{a,i}^{\mu}\, \xi_i^M)$, where rf$(a)$ runs over all photoreceptors connected to LGN-cell a. The activities s_m of the output units become

$$s_m = F\left(\sum_n I_{m,n} s_n + \sum_{M=\mathrm{L,R}} \sum_{\mu=\mathrm{on,off}} \sum_{a \in \mathrm{rf}(m)} w_{m,a}^{M,\mu}\, v_a^{M,\mu}\right). \tag{1}$$

$F(h)$ and $f(h)$ denote transfer functions, which are rectifying, saturating and piecewise linear.

Recent experiments on hippocampal slices indicate, that activity-dependent modification of a synapse causes a similar modification of nearby synapses (up to 100 μm on the postsynaptic surface), irrespective of their activation states

[8]. Hence, we consider a synaptic modification rule

$$\delta w_{m,a}^{M,\mu} = \eta \left(s_m \, v_a^{M,\mu} + \sum_{N,\nu,b} \gamma_{a,b}^{M,\mu;N,\nu} \, w_{m,b}^{N,\nu} \, v_b^{N,\nu} \right) - D_m \tag{2}$$

for a given synapse that is composed of a standard Hebb-term (pre- and postsynaptic coincidence) and a nonspecific contribution proportional to the presynaptic activities of nearby synapses (fig. 1b). The weights are subject to the constraints $0 \leq w_{m,a}^{M,\mu} \leq w_{\max}$. The coefficients $\gamma_{a,b}^{M,\mu;N,\nu}$ contain information about the geometric relationship between synapses on the postsynaptic neuron m, and D_m is a subtractive decay-term which keeps the sum of the synaptic strengths to a postsynaptic neuron constant [4]. We now assume that synapses driven by one eye tend to lie closer to each other than to synapses driven by the contralateral eye. Therefore, the γ-values in eq. (2) will on average be larger for $M = N$ than for the other cases. In the simplest case, which is analytically tractable, the geometry factors take the same nonzero value γ_0 for same-eye driven synapses and vanish for opposite-eye driven synapses, i.e. $\gamma_{a,b}^{M,\mu;N,\nu} \equiv \gamma_0 \, \delta_{M,N}$. This leads to the simplified learning rule

$$\delta w_{m,a}^{M,\mu} = \eta \left(s_m \, v_a^{M,\mu} + \gamma_0 \sum_b (w_{m,b}^{M,\text{on}} \, v_b^{M,\text{on}} + w_{m,b}^{M,\text{off}} \, v_b^{M,\text{off}}) \right) - D_m. \tag{3}$$

3 Analytic Solution for Linear Neurons

In this subsection we perform an analytical treatment of the emerging afferent synaptic patterns in the special case of linear transfer functions, unconstrained synaptic development (best fulfilled during the initial stage of the development) and (for simplicity) vanishing lateral interactions. Insertion of eq. (1) into (3) and short time averaging (denoted by $<>_t$) yields the learning rule

$$\Delta \mathbf{w}^{M,\mu} = \sum_{N,\nu} \mathbf{G}^{M,\mu;N,\nu} \cdot \mathbf{w}^{N,\nu} + \gamma \sum_{\nu} \mathbf{w}^{M,\nu} \tag{4}$$

where $dw_a^{M,\mu}/dt =< \delta w_a^{M,\mu} >_t$, $\gamma = \gamma_0 < v_b^{M,\mu} >_t$ (assuming stationary first order statistics), the column vectors $\mathbf{w}^{M,\mu}$ contain the individual weights $w_a^{M,\mu}$ as components and $\mathbf{G}^{M,\mu;N,\nu}$ are input correlation matrices with elements $G_{a,b}^{M,\mu;N,\nu} = < v_a^{M,\mu} \, v_b^{N,\nu} >_t$. The learning rate, the decay term and the index m of the postsynaptic neuron have been omitted for simplicity. If one assumes identical statistics within each of the four input layers ($\mathbf{G}^{M,\mu;M,\mu} = \mathbf{G}^{N,\nu;N,\nu} \equiv \mathbf{G}^{\text{ss}} \; \forall M, N, \mu, \nu$) and symmetry of ON- and OFF- as well as left-eye and right-eye statistics, i.e. $\mathbf{G}^{\text{L,on;L,off}} = \mathbf{G}^{\text{R,on;R,off}} \equiv \mathbf{G}^{\text{so}}$, $\mathbf{G}^{\text{L,on;R,on}} = \mathbf{G}^{\text{L,off;R,off}} \equiv \mathbf{G}^{\text{os}}$, and $\mathbf{G}^{\text{L,on;R,off}} = \mathbf{G}^{\text{L,off;R,on}} \equiv \mathbf{G}^{\text{oo}}$, one obtains

$$\frac{d}{dt} \begin{pmatrix} \mathbf{w}^{\text{L,on}} \\ \mathbf{w}^{\text{L,off}} \\ \mathbf{w}^{\text{R,on}} \\ \mathbf{w}^{\text{R,off}} \end{pmatrix} = \left(\begin{pmatrix} \mathbf{G}^{\text{ss}} & \mathbf{G}^{\text{so}} & \mathbf{G}^{\text{os}} & \mathbf{G}^{\text{oo}} \\ \mathbf{G}^{\text{so}} & \mathbf{G}^{\text{ss}} & \mathbf{G}^{\text{oo}} & \mathbf{G}^{\text{os}} \\ \mathbf{G}^{\text{os}} & \mathbf{G}^{\text{oo}} & \mathbf{G}^{\text{ss}} & \mathbf{G}^{\text{so}} \\ \mathbf{G}^{\text{oo}} & \mathbf{G}^{\text{os}} & \mathbf{G}^{\text{so}} & \mathbf{G}^{\text{ss}} \end{pmatrix} + \gamma \begin{pmatrix} 1 & 1 & 0 & 0 \\ 1 & 1 & 0 & 0 \\ 0 & 0 & 1 & 1 \\ 0 & 0 & 1 & 1 \end{pmatrix} \right) \cdot \begin{pmatrix} \mathbf{w}^{\text{L,on}} \\ \mathbf{w}^{\text{L,off}} \\ \mathbf{w}^{\text{R,on}} \\ \mathbf{w}^{\text{R,off}} \end{pmatrix}, \tag{5}$$

where **1** and **0** are matrices which contain 1 an 0 in each element respectively. Eq. (5) can be diagonalized and then describes the decoupled time development of a DC-mode, an ocular dominance mode (OD) and two ON-OFF-segregation modes (OR+) and (OR-) [7],

$$\mathbf{w}^{\mathrm{DC}} = (\mathbf{w}^{\mathrm{L,on}} + \mathbf{w}^{\mathrm{L,off}} + \mathbf{w}^{\mathrm{R,on}} + \mathbf{w}^{\mathrm{R,off}}), \tag{6}$$

$$\mathbf{w}^{\mathrm{OD}} = (\mathbf{w}^{\mathrm{L,on}} + \mathbf{w}^{\mathrm{L,off}} - \mathbf{w}^{\mathrm{R,on}} - \mathbf{w}^{\mathrm{R,off}}), \tag{7}$$

$$\mathbf{w}^{\mathrm{OR+}} = (\mathbf{w}^{\mathrm{L,on}} - \mathbf{w}^{\mathrm{L,off}} + \mathbf{w}^{\mathrm{R,on}} - \mathbf{w}^{\mathrm{R,off}}), \tag{8}$$

$$\mathbf{w}^{\mathrm{OR-}} = (\mathbf{w}^{\mathrm{L,on}} - \mathbf{w}^{\mathrm{L,off}} - \mathbf{w}^{\mathrm{R,on}} + \mathbf{w}^{\mathrm{R,off}}). \tag{9}$$

Their time development is given by

$$\dot{\mathbf{w}}^{\mathrm{DC}} = \mathbf{G}^{\mathrm{DC}} \cdot \mathbf{w}^{\mathrm{DC}} = 4(\mathbf{G}^{\mathrm{ss}} + \mathbf{G}^{\mathrm{so}} + \mathbf{G}^{\mathrm{os}} + \mathbf{G}^{\mathrm{oo}} + 2\gamma\mathbf{1}) \cdot \mathbf{w}^{\mathrm{DC}}, \tag{10}$$

$$\dot{\mathbf{w}}^{\mathrm{OD}} = \mathbf{G}^{\mathrm{OD}} \cdot \mathbf{w}^{\mathrm{OD}} = 4(\mathbf{G}^{\mathrm{ss}} + \mathbf{G}^{\mathrm{so}} - \mathbf{G}^{\mathrm{os}} - \mathbf{G}^{\mathrm{oo}} + 2\gamma\mathbf{1}) \cdot \mathbf{w}^{\mathrm{OD}}, \tag{11}$$

$$\dot{\mathbf{w}}^{\mathrm{OR+}} = \mathbf{G}^{\mathrm{OR+}} \cdot \mathbf{w}^{\mathrm{OR+}} = 4(\mathbf{G}^{\mathrm{ss}} - \mathbf{G}^{\mathrm{so}} + \mathbf{G}^{\mathrm{os}} - \mathbf{G}^{\mathrm{oo}}) \cdot \mathbf{w}^{\mathrm{OR+}}, \tag{12}$$

$$\dot{\mathbf{w}}^{\mathrm{OR-}} = \mathbf{G}^{\mathrm{OR-}} \cdot \mathbf{w}^{\mathrm{OR-}} = 4(\mathbf{G}^{\mathrm{ss}} - \mathbf{G}^{\mathrm{so}} - \mathbf{G}^{\mathrm{os}} + \mathbf{G}^{\mathrm{oo}}) \cdot \mathbf{w}^{\mathrm{OR-}}. \tag{13}$$

The growth of the DC-component, Eq. (10) is compensated by the subtractive decay term, therefore the remaining equations determine, which receptive field profile will emerge. In order to discuss eqs. (11) - (13), it is convenient to interpret the correlation matrices and input vectors as discretized functions of space, i.e. $G_{a,b} = G(\mathbf{r}_a - \mathbf{r}_b)$ and $w_a = w(\mathbf{r}_a)$.

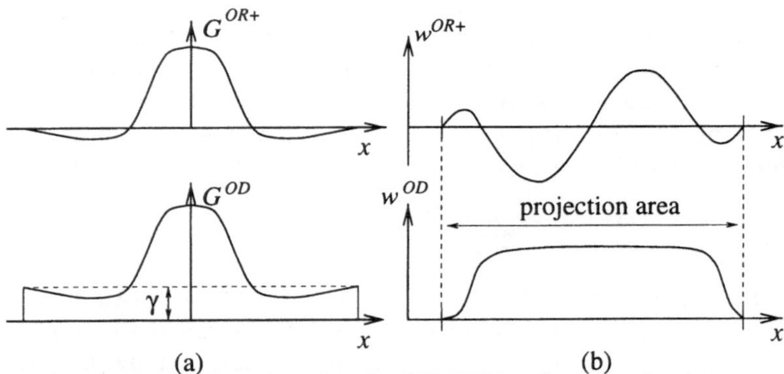

Fig. 2 (a) Correlation functions for the ON-OFF and ocular dominance modes. **(b)** Corresponding eigenmode to the maximal eigenvalue. The proposed algorithm elevates the OD correlation function which becomes non-negative and shows a non-oscillatory leading eigenmode.

On- and OFF-segregation within the afferent projection areas requires spatially oscillating OR-eigenfunctions to emerge, while ocular dominance evolves only with a non-oscillating OD-eigenmode (i.e. the whole projection area must be either left- or right dominated). Consequently, G^{OD} must be non-negative within the projection area while $G^{\mathrm{OR+}}$ and $G^{\mathrm{OR-}}$ must show spatial oscillation at the same time [3]. As long as intrageniculate interactions do not change the second order spatial correlations, the correlation functions $G^{M,\mu;N,\nu}(\mathbf{r})$ derive as a convolution of the LGN receptive field profiles and thus (provided the LGN filters are spatially located) cause oscillations within *all* correlation functions. Therefore, standard Hebb-training in linear CBL-networks cannot produce both ON-OFF-segregation and ocular dominance. The presently considered learning rule adds a constant DC-component to the ocular dominance correlation function thus elevating it to a non-negative state (fig. 2a). In this case, both oscillatory OR-eigenvectors and non-oscillatory OD-eigenvectors evolve simultaneously. For a parameter regime, where the OD and OR- correlation functions have similar maximum eigenvalues, a linear combination of OR- and OD-eigenmodes emerges showing both ocular dominance and ON-OFF-subfields. It is worth mentioning, however, that this occurs robustly only in the presence of decay terms in the learning rule [9].

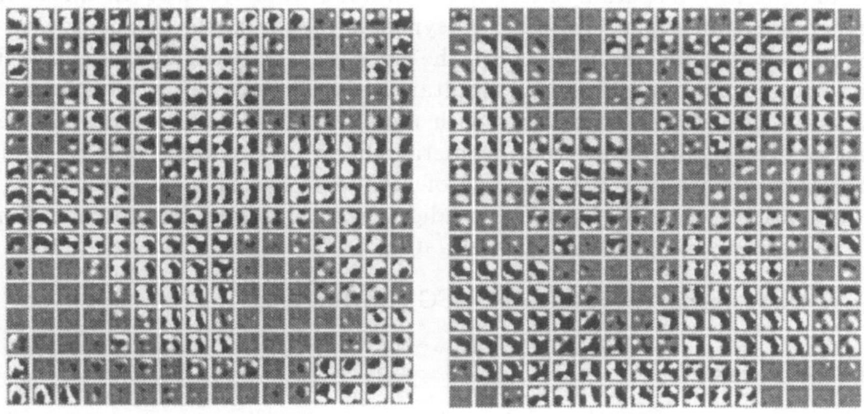

ON-OFF connections (left eye) ON-OFF connections (right eye)

Fig. 3 Binocular afferent weight distributions for a patch of 16x16 output neurons out of a network with 32x32 units. Bright values mark ON-dominated regions, dark values OFF-dominated regions of the afferent projection area. Synaptic strengths vanish at grey regions. Individual receptive fields show elongated ON-OFF subfields and ocular dominance, which is arranged in patchy clusters over the model cortex. Parameters (lengths in units of grid constants): Projection radius $\rho = 6$, DOG-filter radii: $R_c = 1.1$ and $R_s = 3.3$; lateral interaction radii: $R_c = 1.5$, $R_s = 3.0$, strength: $I_0 = 15$; afferent connection strengths were randomly initialized between 0.1 and 0.4, cutoff values were 0 and 1; $\gamma = 0.4$.

4 Simulation Results

The learning rule was tested numerically with sequences of uncorrelated random patterns as inputs and a subtractive decay term that kept the sum over all synapses to a postsynaptic neuron constant [4]. For each training step, the network was allowed to stabilize for 7 iterations before the afferent weights were updated according to Eq. (2). The learning rate was $\eta = 0.005$ for which 40000 iterations were needed for a complete training run. Previous tests showed, that this large number of iterations is necessary to make sure that the learning dynamics is slow compared to single pattern presentations (as used in eq. (4)). However, increased learning rates leave the present mechanism largely unaffected and affect the detailed structure of the emerging orientation map only. Fig. 3 shows the binocular receptive fields of a 16x16 patch out of a 32x32 network trained with $\gamma = 0.4$. One observes both elongated ON-OFF subfields and prominent ocular dominance patches. Similar simulation results can also be achieved by adding positive between-eye correlations corresponding to animals fixating objects after eye-opening. In this postnatal situation the model predicts more frequent binocular receptive fields in accord with experimental findings for cats. Moderate positive between-eye correlations do not destroy ocular dominance, however, because they loose the competition with the dc-term in eq. (11).

In summary, a spatially non-specific component of synaptic modification together with functional clustering of synapses on the postsynaptic neuron was found to be capable of driving both the emergence of structured receptive fields and ocular dominance patterns simultaneously. The mechanism works in linear networks as well as in networks with nonlinear units and intracortical recurrencies. The model shows, that geometric effects on the single neuron level can considerably influence the outcome of high-level neural models rendering the combination of detailed learning paradigms with high-level developmental models an exciting task for future investigation.

Acknowledgements: Supported by DFG (Ob 102/2-1).

References

1. M. P. Stryker and W. Harris. Binocular impuls blockade prevents the formation of ocular dominance columns in cat visual cortex. *J. Neurosci.*, 6:2117–2133, 1986.
2. T. N. Wiesel and D. H. Hubel. Ordered arrangement of orientation columns in monkeys lacking visual experience. *J. Comp. Neurol*, 158:307–318, 1974.
3. M. Stetter, E. W. Lang, and A. Müller. Emergence of orientation selective simple cells simulated in deterministic and stochastic neural networks. *Biol. Cybern.*, 68:465–476, 1993.
4. K. D. Miller. A model for the development of simple cell receptive fields and the ordered arrangement of orientation columns through activity-dependent competition between on- and off-center inputs. *J. Neurosci.*, 14:409–441, 1994.
5. M. Stetter, A. Müller, and E. W. Lang. Neural network model for the coordinated formation of orientation preference and orientation selectivity maps. *Phys. Rev. E*, 50:4167–4181, 1994.
6. K. D. Miller, J. B. Keller, and M. P. Stryker. Ocular dominance column development: Analysis and simulation. *Science*, 245:605–615, 1989.
7. C. Piepenbrock, H. Ritter, and K. Obermayer. Linear correlation-based learning models require a two-stage process for the development of orientation and ocular dominance. *Neural Proc. Lett.*, 3:31–37, 1996.
8. F. Engert and T. Bonhoeffer. Synapse specifity of long-term potentiation breaks down at short distances. *Nature*, page submitted, 1997.
9. C. Piepenbrock, H. Ritter, and K. Obermayer. The joint development of orientation and ocular dominance: Role of constraints. *Neural Comp.*, 9:in press, 1997.

Must Pinwheels Move During Visual Development ?

Fred Wolf and Theo Geisel

Max-Planck Institut für Strömungsforschung, Bunsenstr.10 , 37073 Göttingen and SFB Nichtlineare Dynamik, Universität Frankfurt, Germany.

Abstract. The pinwheel–like arrangement of iso–orientation domains around orientation centers is a ubiquitous structural element of orientation preference maps in primary visual cortex. Here we investigate how activity–dependent mechanisms constrain the way in which orientation centers can form during visual development. We consider the dynamics of a large class of models for the activity–dependent self–organization of orientation preference maps. We prove for this class of models that the density of orientation centers which proliferate as orientation selectivity arises from an unselective state exhibits a universal lower bound. At least π/Λ^2 pinwheels must form initially, where Λ is the characteristic wavelength of iso–orientation domains. Due to topological constraints the density of orientation centers can only change by discrete creation and annihilation events. Consequently densities lower than π/Λ^2 must develop through an initial overproduction and subsequent annihilation of pinwheels. Monitoring the density of orientation centers during development therefore offers a powerful novel approach to test whether orientation preference arises by activity–dependent mechanisms or is genetically predetermined.

1 The development of orientation maps

In the visual cortex orientation selectivity and an initial pattern of orientation columns arises over a period of several days around the time of eye opening [1, 2]. It is an attractive hypothesis that orientation selectivity and the pattern of iso–orientation domains arise via activity–dependent refinement of cortical circuitry during early life [3, 4]. Furthermore, since the underlying connectivity remains in a plastic state over a period of months rather than weeks [5], it is conceivable that the pattern orientation columns reorganizes after its initial emergence as a result of activity–dependent changes in the underlying connections.

Pinwheel–like patterns of iso–orientation domains are a ubiquitous structural element of orientation preference maps in the visual cortex [6]. Here we study the proliferation and dynamics of pinwheels in a large class of models for the activity–dependent development of orientation columns. The spatial pattern of orientation preferences and selectivities can be described by a single complex order parameter field $z(\mathbf{x}) = |z(\mathbf{x})|e^{i2\vartheta(\mathbf{x})}$ where $|z(\mathbf{x})|$ measures the degree of orientation selectivity, $\vartheta(\mathbf{x})$ is the orientation preference map, and $\mathbf{x}$ denotes the tangential position in the cortical sheet [6]. It has been hypothesized that the development of orientation columns can be mathematically described by a dynamics of the order parameter field $z(\mathbf{x})$ [6]. In this paper we show that quantitative predictions on the proliferation and dynamics of pinwheels can be derived from this hypothesis without explicitly specifying the dynamics of $z(\mathbf{x})$.

2 Topological considerations

In any species investigated so far $\vartheta(\mathbf{x})$ contains topological point defects [6]. Their positions $\mathbf{x}_j$, the pinwheel centers, are the zeros of $z(\mathbf{x})$. Close to this points iso–orientation domains are arranged in pinwheels–like patterns. These pinwheels come in two varieties distinguished by their topological charges $q_j = \frac{1}{2\pi} \oint_{C_j} \nabla\vartheta(\mathbf{x}) d\mathbf{s} = \pm\frac{1}{2}$.

For elementary topological reasons, continuous changes of $z(\mathbf{x})$ that might occur during development can affect the defect configuration $\{\mathbf{x}_j, q_j\}$ only in three ways: (1) *motion* of defects, (2) *creation* and (3) *annihilation* of pairs of defects of opposite charge. The number of pinwheels therefore can only change by discrete creation and annihilation events.

Reflecting the roughly repetitive arrangement of orientation columns the Fourier representation of $z(\mathbf{x})$ exhibits a single finite band centered around modes of characteristic wavelength Λ. The density of orientation centers ρ quantifies the deviation of a map from a defect–free state. Λ and ρ are independent quantities characterizing a given orientation map. The scaled density $\hat{\rho} = \rho\,\Lambda^2$ gives the mean number of orientation centers in a region of size Λ^2 and measures the relative abundance of pinwheels in a particular system.

3 Emergence of orientation columns

On the level of individual neurons and synapses, the self–organization hypothesis assumes that afferent and cortical activity patterns guide a continuous refinement of geniculocortical connections. Many theoretical studies (for reviews see [6, 7]) demonstrated that Hebb–type remodeling can lead to the emergence of orientation columns by amplifying weak orientation biases induced by random fluctuations in connectivity or activity patterns. During this process intracortical interactions force neurons to develop orientation preferences similar to those of their neighbors. Consequently the way in which the pattern of orientation selectivities and preferences is modified at a particular point in time depends strongly on the pattern of orientation selectivities and preferences already present.

As an idealization of this picture we assume that the effective modification of $z(\mathbf{x})$ under the stream of sensory activity can be predicted to a large extend from the knowledge of $z(\mathbf{x})$ itself. I.e. it is assumed that $z(\mathbf{x})$ follows a dynamics

$$\frac{\partial}{\partial t} z(\mathbf{x}) = F[z(\cdot)] + \xi\,, \qquad (1)$$

where $F[\cdot]$ is an unknown nonlinear operator and $\xi(\mathbf{x}, t)$ is a spatio–temporal random process modeling intrinsic and stimulus induced fluctuations. To ensure

that there are no other sources of spatial structure but random fluctuations Eq.(1) must exhibit three basic symmetries.
(i) Equivariance under orientation shifts: $F[e^{i\phi}\, z] = e^{i\phi}\, F[z]$.
(ii) Equivariance under translations of the cortical sheet:
$F[\check{T}_y\, z] = \check{T}_y\, F[z]$ with $\check{T}_y\, z(\mathbf{x}) = z(\mathbf{x}+\mathbf{y})$.
(iii) Equivariance under rotations of the cortical sheet:
$F[\check{R}_\beta\, z] = \check{R}_\beta\, F[z]$ with $\check{R}_\beta\, z(\mathbf{x}) = z\left(\begin{bmatrix} \cos(\beta) & \sin(\beta) \\ -\sin(\beta) & \cos(\beta) \end{bmatrix} \mathbf{x}\right)$.
(i) enforces that all orientations are treated equally. (ii) and (iii) ensure that the cortex does not prefer a particular pattern of iso–orientation domains.

Many properties of the dynamics (1) are determined by this set of symmetries. The state $z(\mathbf{x}) = 0$ is a stationary solution as long as the dynamics is invariant under orientation shifts ($F[0] = e^{i\phi}\, F[0] \Rightarrow F[0] = 0$). In its vicinity the dynamics is governed by

$$\frac{\partial}{\partial t}\, z(\mathbf{x}) = \check{L}\, z(\mathbf{x}) + \xi(\mathbf{x}, t)\,, \tag{2}$$

where $\check{L}$ is a linear operator. Like $F[\cdot]$, the operator $\check{L}$ must be invariant under translations and rotations of the cortical plane. This implies that $\check{L}$ is diagonal in Fourier representation and its eigenvalues $\lambda(k)$ depend only on the modulus of the wave vector $k = |\mathbf{k}|$. Eq.(1) and (2) give rise to a columnar pattern if there is one interval (k_l, k_h) of wave numbers with positive eigenvalues. While Eq.(2) only describes the primary emergence of an orientation map, Eq.(1) also captures the saturation of orientation selectivity and possible rearrangements of orientation columns in a subsequent nonlinear phase of development.

4 Pinwheel proliferation

Most importantly these properties already impose a lower bound on the density of pinwheels which form as the orientation map arises from the unselective state $z = 0$. Eq.(2) is solved in terms of the Green's function $G(\mathbf{x}, t) = \frac{1}{2\pi} \int d^2k\, e^{-i\mathbf{k}\mathbf{x} + \lambda(|\mathbf{k}|)t}$ by

$$z(\mathbf{x}, t) = \int d^2y \int_0^t dt'\, G(\mathbf{y} - \mathbf{x}, t - t')\, \xi(\mathbf{y}, t')\,.$$

$z(\mathbf{x})$ thus is the mean of a set of random variables. According to the central limit theorem the field $z(\mathbf{x})$, established during the early (linear) phase of development, therefore realizes a Gaussian random field(GRF). This enables one to calculate the density of pinwheels which proliferate during the symmetry breaking phase. The density of zeros in a GRF is determined by its power spectrum $P(|\mathbf{k}|) = |\int d^2x\, z(\mathbf{x})\, e^{i\mathbf{k}\mathbf{x}}|^2$

$$\rho = \frac{1}{4\pi} \frac{\int d^2k\, |\mathbf{k}|^2 P(|\mathbf{k}|)}{\int d^2k\, P(|\mathbf{k}|)}$$

[8]. Without loss of generality, we assume $\int_0^\infty dk\, P(k) = 1$. For convenience the characteristic wavelength Λ is defined as $\Lambda \equiv 2\pi/\bar{k}$ where $\bar{k} = \int_0^\infty dk\, k\, P(k)$ is the mean wave number. The density of zeros can then be rewritten as

$$\begin{aligned}
\rho &= \frac{1}{4\pi\,\bar{k}} \int_0^\infty dk\, k^3 P(k) \\
&= \frac{\pi}{\Lambda^2}\left(1 + 3\int_0^\infty dk\, \frac{(k-\bar{k})^2}{\bar{k}^2} P(k) + \int_0^\infty dk\, \frac{(k-\bar{k})^3}{\bar{k}^3} P(k)\right)
\end{aligned}$$

Only the third term of this expression can become negative. If one assumes that $P(k)$ vanishes in the range $k > 4\bar{k}$, which holds for spectra containing a single peak centered at $\bar{k}$, the modulus of the third term is bounded by the second

$$\begin{aligned}
&\left|\int_0^\infty dk\, (k-\bar{k})^3\, P(k)\right| \le \int_0^\infty dk\, \left|(k-\bar{k})^3\right| P(k) \\
&\le 3\bar{k}\int_0^\infty dk\, \left|(k-\bar{k})^2\right| P(k) = \left|3\bar{k}\int_0^\infty dk\, (k-\bar{k})^2\, P(k)\right|
\end{aligned}$$

so that

$$\rho = \frac{\pi}{\Lambda^2}\,(1 + \alpha) \tag{3}$$

with $\alpha \ge 0$. Consequently π/Λ^2 forms a lower bound for the density of pinwheels in such a GRF. At least π/Λ^2 pinwheels must therefore proliferate during the primary establishment of orientation selectivity in any model that conforms to the symmetries (i-iii) and can be cast in to the form of or can be approximated by Eq.(1). Even if no further rearrangement would occur, the observation of more than π/Λ^2 pinwheels early in development would reveal an important fingerprint of the activity–dependent self–organization of orientation preference.

5 Pinwheel annihilation

The development of lower pinwheel densities is not prohibited by Eq.(1). However, states of lower density can only form through an early high density state and subsequent annihilation of pinwheel pairs during the nonlinear phase of development. Thus the considered class of models predicts that pinwheel annihilation must occur in those species in which a pinwheel density lower than π/Λ^2 is observed in the adult.

To determine whether pinwheel annihilation is indeed an intrinsic property of the activity–dependent refinement of orientation maps we have investigated different biologically plausible models for the development of orientation columns [9, 10]. In these models pinwheels initially proliferate in large numbers. In agreement with the theory developed above, pinwheel creation leads to the formation of at least π/Λ^2 orientation centers. All investigated models exhibit spontaneous pinwheel annihilation in the nonlinear phase of development. The degree and the velocity of pinwheel annihilation vary for different initial patterns, models,

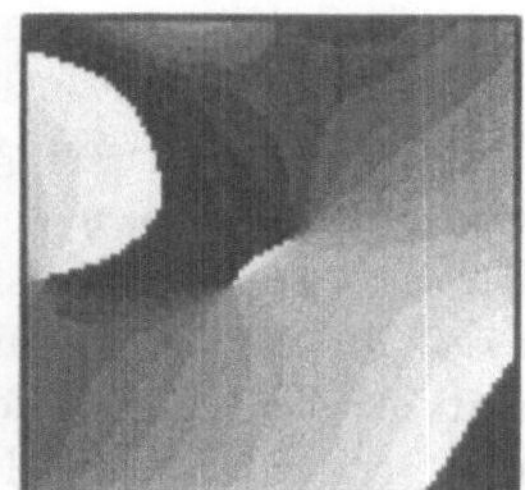

Fig. 1. Detail from the development of an orientation preference map in a simulation of the elastic network. The figure shows the annihilation (from left to right) of a pair of pinwheels. Pinwheel annihilation replaces a pair of pinwheels by a linear zone, in which iso–orientation domains form a system of parallel stripes. The panels display the patterns of preferred orientations $\vartheta(\mathbf{x}) = \frac{1}{2}arg(z(\mathbf{x}))$ coded by grey values at three successive times during the simulation.

and choice of parameters. For every individual pattern, however, pinwheel annihilation proceeds fastest for strong lateral interactions. Fig.1 displays a typical example from a numerical simulation of the elastic network [9]. This behavior conforms with the intuitive notion that lateral cooperation favors smooth mappings and hence should tend to remove discontinuities from the orientation preference map. Pinwheel annihilation breaks the statistical isotropy of the initial pattern and leads to growing regions of stripe–like iso–orientation domains.

Predicting pinwheel annihilation in principle provides a highly significant means to test the self–organization hypothesis. The preferred orientation of the set of columns between annihilating orientation centers changes by 90° during pinwheel annihilation (see Fig.1). Since this requires a radical reshaping of receptive fields it would be incompatible with the assumption that orientation preference is genetically prespecified [11, 12] as well as with the notion that activity–dependent mechanisms only mediate the "selective stabilization" of a subset of exuberant initial connections[13]. Pinwheel annihilation is only possible if the functional architecture of visual cortex remains in a state of flux beyond the primary establishment of orientation selectivity. Direct observation of pinwheel annihilation would therefore provide unequivocal evidence for the self–organization of orientation preference.

6 Discussion

How does the theoretically predicted bound compare to experimental observations? Obermayer and Blasdel have recently quantified the density of pinwheels in adult and juvenile macaque monkeys [14]. In adult animals they observe a relatively high scaled density of about $\hat{\rho} \approx 3.7$. Consistent with our theory this relatively high density is already present in the youngest animals studied.

Relatively low densities of pinwheels have been observed in different species. In area 17 of adult cats Bonhoeffer et al. report a density of $2.1/mm^2$ [15]. Since the characteristic wavelength of iso–orientation domains in this species is about $1.1mm$ this implies a scaled density of $\hat{\rho} = 2.5$ significantly below $\pi \approx 3.1$. Also in tree shrews Bosking et al. have recently observed regions of predominantly stripe–like iso–orientation domains and consequently relatively low densities of pinwheels [16]. It is therefore predicted that pinwheels should in fact move and annihilation in these species during early live, if the development of orientation columns can be described by a model that conforms to the symmetries (i–iii).

We conclude that the density and dynamics of pinwheels during development may provide essential clues to the basic mechanisms of cortical development. Our results underline the importance of careful quantification and interspecies comparison of the geometrical properties of cortical orientation maps. They also strongly suggest that chronic optical imaging studies of the development of orientation maps should extend well beyond the primary emergence of orientation selectivity.

Acknowledgements It is a great pleasure to thank Siegrid Löwel for many open ended and inspiring discussions.

References

1. Chapman, B. and Stryker, M. P. *J. Neuroscience* **13**, 5251–5262 (1993).
2. Chapman, B., Stryker, M. P., and Bonhoeffer, T. *J. Neuroscience* **16**, 6443–6453 (1996).
3. von der Malsburg, C. *Kybernetik* **14**, 85–100 (1973).
4. Nass, M. M. and Cooper, L. N. *Biol. Cybernetics* **19**, 1–18 (1975).
5. Daw, N. W. *Visual Development.* Plenum Press, New York, (1995).
6. Swindale, N. *Network: Computation in Neural Systems* **7**, 161–247 (1996).
7. Miller, K. D. In *Models of Neural Networks III,* Domany, E., van Hemmen, J., and Schulten, K., editors. Springer–Verlag, NY (1995).
8. Halperin, B. I. In *Physics of Defects, Les Houches, Session XXXV, 1980,* Balian, R., Kléman, M., and Poirier, J.-P., editors (North–Holland, Amsterdam, 1981).
9. Durbin, R. and Mitchinson, G. *Nature* **343**, 644–647 (1990).
10. Obermayer, K., Blasdel, G. G., and Schulten, K. *Phys. Rev. A* **45**, 7568–7589 (1992).
11. Wiesel, T. N. and Hubel, D. H. *J. Comp. Neurol.* **158**, 307–318 (1974).
12. Stryker, M. P. In *Neuroscience Res. Prog. Bull., Vol. 15, No.3,* chapter VII, 454–462. MIT Press (1977).
13. Changeux, J.-P. and Danchin, A. *Nature* **264**, 705–712 (1976).
14. Obermayer, K. and Blasdel, G. G. *Neural Computation* **9**, 555–575 (1997).
15. Bonhoeffer, T., Kim, D.-S., Malonek, D., Shoham, D., and Grinvald, A. *Europ. J. Neuroscience* **7**, 1973–1988 (1995).
16. Bosking, W. H., Zhang, Y., Schofield, B. R., and Fitzpatrick, D. *J. Neuroscience* **17**, 2112–2127 (1997).

Extending the TRN Model in a Biologically Plausible Way

Francesco Frisone, Luca Perico, and Pietro G. Morasso

Department of Informatics, Systems, Telecommunications
University of Genova, Italy
Via Opera Pia 13, I-16145 Genova, Italy - E: friso@dist.unige.it

Abstract. The *Topology Representing Network* (TRN) model is extended by using an activation dynamics which implicitly orders the neurons according to the distance from the input pattern. This allows to apply the same Hebbian learning method to the thalamo-cortical and cortico-cortical connections. The model proposed combines a process of diffusion (via the excitatory topologically organized connections) and a process of competitive distribution of activation which tends to sharpen the active map region. The dynamics is analyzed taking into account the excitatory nature of the majority of cortical synapses and the puzzling presence of long-range competition without long-range inhibition. The model is shown to be more consistent than TRN or other self-organizing paradigms with a number of neurophysiological facts.

1 Biological plausibility of cortical map models

The somatotopic layout of many cortical areas has long suggested a kind of topologic organization, associated with a dimensionality reduction of the representational space, thus motivating the development of a (large) family of self-organizing network models.

From its beginning, however, the effort has been affected by a number of misconceptions, partly due to the over-emphasis in the neurophysiological community on the distribution of receptive fields in cortical areas. Only recently, however, a new understanding of the cortex is emerging as a continuously adapting dynamical system, shaped by competitive and cooperative lateral connections [15] and such organization is not static but changes with ontogenetic development together with patterns of thalamocortical connections [4]. Shortly, it has been suggested that cortical areas can be seen as a massively interconnected set of elementary processing elements, which constitute a computational map [5]. The misconceptions about cortical functionality from the modeling point of view can be reduced to the following three items:

- *flatness* of cortical maps (related to the locality of lateral connections);
- *fixed* lateral connections (versus plastic thalamo-cortical connections, which determine receptive-field properties);
- *Mexican-hat* function of lateral interactions (it implies a significant amount of recurrent inhibition for the formation of localized responses by lateral feedback);

The flatness assumption which characterizes the classic map models [1, 6] is contradicted by the fact that the structure of lateral connections is not genetically determined but depends mostly on electrical activity during development. More precisely, they have been observed to grow exuberantly after birth[4] and reach their full extent within a short period; during the subsequent development, a *pruning* process takes place so that the mature cortex is characterized by a well defined pattern of connectivity, which includes a large amount of non-local connections: this rules out all the models limited to a purely 2-D circuitry. The superficial connections to non-neighboring columns are organized into characteristic patterns: a collateral of a pyramidal axon typically travels a *characteristic lateral distance* without giving off terminal branches and then it produces tightly terminal clusters (possibly repeating the process several times over a total distance of several millimeters). Such characteristic distance is not a universal cortical parameter and is not distributed in a purely random fashion but is different in different cortical areas: 0.43 mm in the primary visual area, 0.65 mm in the secondary visual area, 0.73 mm in the primary somatosensory cortex, 0.85 mm in the primary motor cortex, and up to several mm. in the infero-temporal cortex (area 7a) [3, 13, 2]. Thus, the development of lateral and afferent connections depends on the cortical activity caused by the external inflow, in such a way to capture and represent the (hidden) correlations in the input channels. Each individual lateral connection is "weak" enough to go virtually unnoticed while mapping the receptive fields of cortical neurons but the total effect on the overall dynamics of cortical maps can be substantial, as is revealed by cross-correlation studies [14]. The TRN model [7] is attractive because it allows the formation of patterns of lateral connections which are adapted to the input distribution and capture its underlying topological structure. However, the corresponding learning process is rather artificial and un-naturally separated from the learning of input connections.

Lateral connections from superficial pyramids tend to be recurrent (and excitatory) because 80% of synapses are with other pyramids and only 20% with inhibitory inter neurons, most of them acting intra-columnarly [10]. Recurrent excitation is likely to be the underlying mechanism which produces the synchronized firing which has been observed in distant columns [14]. The existence (and preponderance) of massive recurrent excitation in the cortex is in contrast with what could be expected, at least in primary sensory areas, considering the ubiquitous presence of peristimulus competition (or "Mexican-hat pattern") which has been observed time ago in many pathways, as the primary somatosensory cortex and has been confirmed by direct excitation of cortical areas as well as correlation studies; in other words, in the cortex there is a significantly larger amount of long-range inhibition than expected from the density of inhibitory synapses.

In general, "recurrent competition" has been assumed to be the same as "recurrent inhibition", for providing an antagonistic organization that sharpens responsiveness to an area smaller than would be predicted from the anatomical funneling of inputs. Thus, an intriguing question is how long-range competition can arise without long-range inhibition and a possible solution is the mechanism of *gating inhibition* based on a *competitive distribution of activation*, proposed by [11] and further investigated by [9, 8, 12].

2 Cortical dynamics

For simplicity, we lump the generic i-th cortical column into a single processing element, characterized by an activity level V_i, a receptive field center $\mathbf{w}_i$ (or vector prototype), and two kinds of inputs (h_i^{lat} and h_i^{ext}):

$$\frac{dV_i^x}{dt} + \gamma_i V_i^x = f(h_i^{lat} + h_i^{ext}) \tag{1}$$

The equation simply says that V_i evolves under the action of three competing influences: (1) a self-inhibition (weighted by γ_i), (2) a net input h_i^{lat} coming from the set of lateral connections, (3) a net input h_i^{ext} determined by thalamo-cortical connections (or cortico-cortical connections across different cortical areas). The h_i^{lat} term is a recurrent input, intended to express the massive lateral excitatory connections:

$$h_i^{lat} = \sum_{j \in \mathcal{N}_i} C_{ij} \, V_j \tag{2}$$

where $\mathcal{N}_i$ is the set of columns laterally connected to the given element and the lateral connection weights C_{ij} are positive and symmetric. For the external input we simply used a Gaussian distribution, $G_i(\mathbf{x}) = e^{-\|\mathbf{x}-\mathbf{w}_i\|^2/2\sigma^2}$, where $\mathbf{w}$ identifies the thalamo-cortical connections and $\mathbf{x}$ the corresponding thalamic input pattern.

The non-linearity on the right-hand side of equation 1 is intended to provide a mechanism which combines diffusion (of the input Gaussian distribution) and competitive activation, which sharpens the response pattern and centers it around the peak of the input distribution. The proposed implementation is given by the following model

$$\frac{dV_i}{dt} + \gamma_i U_i = \sum_{j \in \mathcal{N}_i} C_{ij} U_j + G_i(\mathbf{x}) U_i \tag{3}$$

with $U_j = V_j / \sum_k V_k$ and $\gamma > 1$, which implements the the idea of *competitive distribution of activation* [11], modified by a gating action on the input signal. At steady state, the dynamics of this model yields stable population codes like that in fig. 1 (for the simple case of a 2-D cortical map) without any need of lateral inhibition. Particularly interesting is the transient behavior which is determined by the sudden shift of the input variable x (say the selection of a new target): it is a characterized by a *diffusion process* (which initially flattens the population code spreading the activity pattern over the whole network) and a *re-sharpening process* (which builds up faster and faster as the diffused wavefront reaches the new designated target). The combination of the two processes is the *propagation* of the population code toward the new target following a geodesic in the characteristic manifold of the map.

3 Learning

In the TRN model the lateral connections do not influence the network dynamics and there are different learning rules for the thalamo-cortical connections and the lateral connections: the former rule is Hebbian but the latter one uses an explicit ordering procedure which is biologically quite non-plausible and contradicts the basic requirement of all the neural models (parallelism and locality). In the proposed extension, on the contrary, the same Hebbian rule is utilized uniformly for adapting the thalamo-cortical W_{ij} and cortico-cortical C_{kl} connection weights. In the former case the weights are increased proportionally to the correlation between the thalamic input and the cortical activation $(X_i V_J)$ and, in the latter, to the correlation between the corresponding cortical activities $(V_k V_l)$. A normalization factor is applied in both cases. The ordering procedure in the TRN model is implicitly performed by the cortical dynamics of equation 1, which attributes to the units, at steady state, an activation level U_i which is higher the closer the receptive field center $\mathbf{w}_i$ to the input pattern $\mathbf{x}$. Learning should only be applied at steady state, after the application of a new input pattern. Alternatively, the learning rate could be modulated by a signal which globally measures the global resting state of the map and depresses the rate during transients similar to the saccadic suppression of visual input.

Figure 2 shows that the result of the learning process for this model is equivalent to the TRN model: a "honeycomb" pattern for the simple case of a map excited by a 2-D input uniformly distributed in a circular domain. The map was initialized by choosing randomly the receptive field centers and the cross-connections (90% of the total full-connection pattern). As learning proceeds both kinds of connection weights are modulated, thus shifting the receptive field centers and, at the same time, pruning the lateral connection weights which falls under a given threshold (10% of the normalized range of values). As can be seen in the figure the pruning process is quite radical and adapted to the topology of the input distribution.

4 Conclusions

A computational models is described of cortical maps as dissipative structures of non-equilibrium generated by an irreversible learning process which builds internal representations of the external world. The same Hebbian learning is applied to both thalamo-cortical and cortico-cortical connections, using the cortical dynamics as a mechanism of "credit assignment". However, the thalamic input patterns are still modeled in a lumped way. The next step will be to consider a totally distributed model in which also the tuning of the receptive field centers is performed with the same pruning mechanism proposed for the cortico-cortical connections. [1]

[1] **Acknowledgments.** This work was partly supported by EU projects SPEECH-MAPS, MIAMI, MOVAID, by a MURST 40% project and an ISS project.

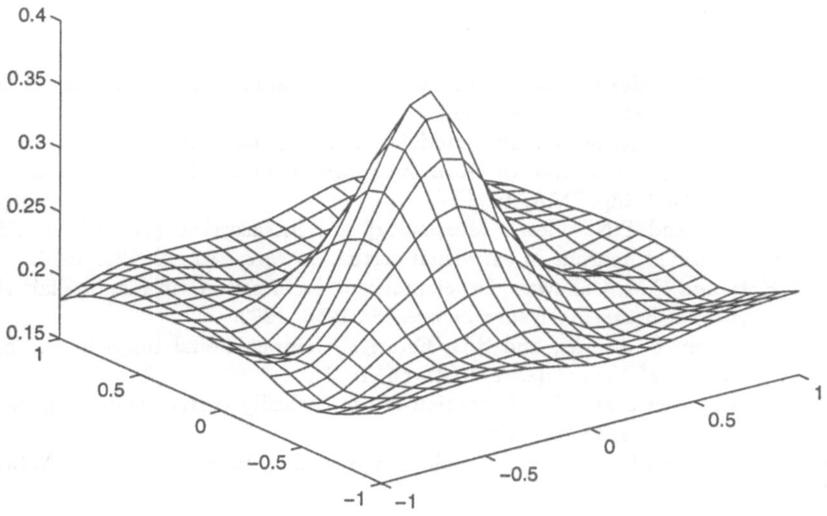

Figure 1: Distribution of activity on a cortical map at steady-state.

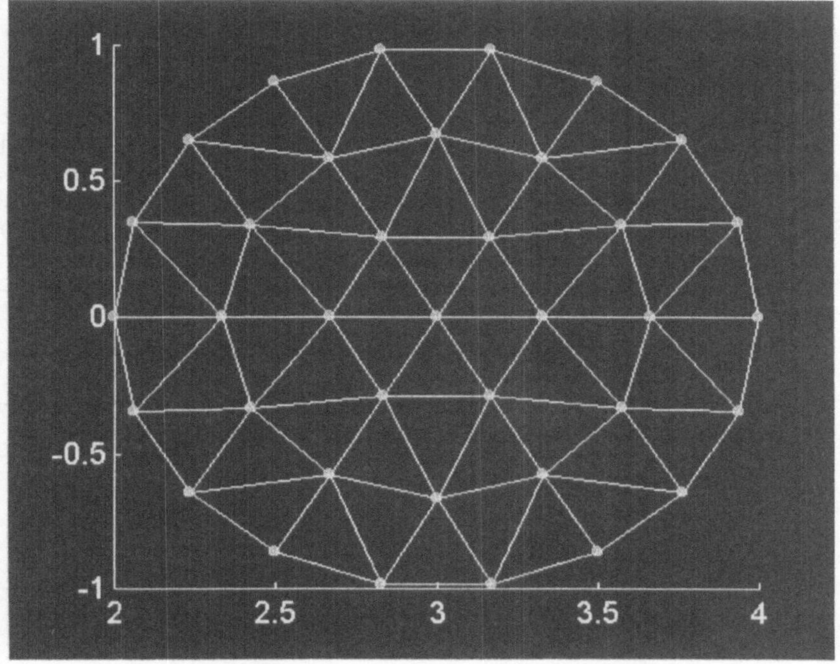

Figure 2: Distribution of receptive field centers and lateral connections of a trained cortical map: a "honeycomb" pattern for the simple case of a map excited by a 2-D input uniformly distributed in a circular domain.

References

1. S. Amari. Dynamics of pattern formation in lateral-inhibition type neural fields. *Biological Cybernetics*, 27:77–87, 1977.
2. W.H. Calvin. Cortical columns, modules, and hebbian cell assembles. In M.A. Arbib, editor, *The handbook of brain theory and neural networks*, pages 269–272. MIT Press, Cambridge, MA, 1995.
3. C.D. Gilbert and T.N. Wiesel. Morphology and intracortical projections of functionally identified neurons in cat visual cortex. *Nature*, 280:120–125, 1979.
4. L.C. Katz and E.M. Callaway. Development of local circuits in mammalian visual cortex. *Annual Review of Neuroscience*, 15:31–56, 1992.
5. E. I. Knudsen, S. du Lac, and S.D. Esterly. Computational maps in the brain. *Annual Review of Neuroscience*, 10:41–65, 1987.
6. T. Kohonen. Self organizing formation of topologically correct feature maps. *Biological Cybernetics*, 43:59–69, 1982.
7. T. Martinetz and K. Schulten. Topology representing networks. *Neural Networks*, 7:507–522, 1994.
8. P. Morasso and V. Sanguineti. How the brain can discover the existence of external egocentric space. *Neurocomputing*, 12:289–310, 1996.
9. P. Morasso, V. Sanguineti, and G. Spada. Neocortical dynamics in sensorimotor control. In J. Mira Mira, editor, *Proceedings of the International Conference on Brain Processes, Theories and Models*, pages 503–512. MIT Press, Cambridge, Mass, USA, 1995.
10. A. Nicoll and C. Blakemore. Patterns of local connectivity in the neocortex. *Neural Computation*, 5:665–680, 1993.
11. J.A. Reggia, C.L. D'Autrechy, G.G. Sutton III, and M. Weinrich. A competitive distribution theory of neocortical dynamics. *Neural Computation*, 4:287–317, 1992.
12. V. Sanguineti, P. Morasso, and F. Frisone. Cortical maps of sensorimotor spaces. In P. Morasso and V. Sanguineti, editors, *Self-organization, Computational Maps, and Motor Control*, pages 1–36. Elsevier Science Publishers, Amsterdam, 1997.
13. H.D. Schwark and E.G. Jones. The distribution of intrinsic cortical axons in area 3b of cat primary somatosensory cortex. *Experimental Brain Research*, 78:501–513, 1989.
14. W. Singer. Development and plasticity of cortical processing architectures. *Science*, 270:758–764, 1995.
15. J. Sirosh, R. Mikkulainen, and Y. Choe. *Lateral interactions in the cortex*. Hypertext Book, http://www.cs.utexas.edu/users/nn/web-pubs/htmlbook96, 1996.

SOM-Model for the Development of Oriented Receptive Fields and Orientation Maps from Non-Oriented ON-center OFF-center Inputs

D. Brockmann[1], H.-U. Bauer[1], M. Riesenhuber[2], and T. Geisel[1]

[1] Max-Planck-Institut für Strömungsforschung, Postfach 28 53, 37018 Göttingen, Germany
[2] Department of Brain and Cognitive Sciences and Center for Biological and Computational Learning, Massachusetts Institute of Technology, E25-221, Cambridge, MA 02139, USA

Abstract. We investigate the development of oriented receptive fields and an orientation map in a SOM-model which is driven by rotationally symmetric stimuli in ON-center and OFF-center input layers. To this end we use the high-dimensional variant of the SOM-algorithm which allows to develop the internal structure of receptive fields as well as the layout of a neural map. We calculate a state diagram for this model, identify parameter regimes in which the rotational symmetry of the stimuli is broken, corroborate the analytical results by simulations, and investigate an extended version of the model which includes ocular dominance development.

1 Introduction

Many aspects of the development of individual receptive fields, and of topographic neural maps have been explained by activity-driven self-organization processes. Assumptions about underlying mechanisms can be cast in mathematical form in such a general fashion that many phenomena, even in different sensory modalities, can be explained within the same modeling framework. One such general framework is Kohonen's Self-Organizing Map (SOM [4]) which has successfully accounted for various aspects of visual, auditory and somatosensory maps.

The development of oriented receptive fields and orientation maps in the visual cortex has been hypothesized to result from a competition of correlated ON-center and OFF-center inputs to the map neurons which could be driven by prenatal spontaneous retinal activity [5, 6]. It is an interesting question whether or not SOM-models can also exhibit this pattern formation behavior. If they do, consistency with other modeling frameworks is maintained, and we come closer to a unified view on map formation. If they do not, the reason for the discrepancy to the other models must be identified; possible conclusions about underlying mechanisms can be drawn. To support this hypothesis in the SOM-framework we need:

a) a "high-dimensional" SOM-model which describes receptive fields as a distribution of synaptic weights over some input layer (as opposed to receptive fields as points in some feature space)
b) a break of rotational symmetry from the unoriented stimuli driving the system to the oriented receptive fields forming the map.

We present here mathematical and numerical results for such a SOM-model with ON-center and OFF-center inputs. The mathematical results were obtained using a recently described analysis technique for high-dimensional SOMs. In the final section of the paper we also describe results for an extended version of the model which includes the development of ocular dominance columns. A more detailed description of this work has been submitted elsewhere [9].

2 Analysis of Orientation Map Model

Neurons in a SOM are characterized by positions $\mathbf{r}$ in a map lattice $\mathcal{A}$, and receptive fields $\mathbf{w_r}$ in a map input space $\mathcal{V}$. The input space is assumed to consist of one (or several) layer(s) of input channels. A stimulus $\mathbf{v}$ consists of a distribution of activity over these input channels, normalized to a constant sum. The stimulus is mapped onto that neuron $\mathbf{s} \in \mathcal{A}$, the receptive field $\mathbf{w_s}$ of which matches $\mathbf{v}$ best, e.g. $\mathbf{s} = \arg\max_{\mathbf{r}\in\mathcal{A}}\{\mathbf{w_r}\cdot\mathbf{v}\}$. Presenting a random sequence of stimuli and performing adaptation steps,

$$\Delta\mathbf{w_r} = \epsilon h(\mathbf{r}-\mathbf{s})\,(\mathbf{v}-\mathbf{w_r})\,, \tag{1}$$

the internal shape of individual receptive fields as well as the map layout self-organize simultaneously. The neighborhood function $h(\mathbf{r}-\mathbf{s}) = \exp\left(-||\mathbf{r}-\mathbf{s}||^2/2\sigma^2\right)$ ensures that neighboring neurons align their receptive fields, and ϵ determines the magnitude of the adaptation step.

We consider a projection geometry analogous to that proposed in [5, 6]. Layers of ON-center and OFF-center cells serve as a map input space and project to the map layer (Fig. 1). We assume our stimuli to consist of an activity peak in one layer, complemented by an activity annulus in the other layer.

Stimuli are represented as difference-of-Gaussians (DOG), the positive part of which resembles the activity in one input-layer (i.e. ON or OFF), the negative part in the other. Stimulus parameters are the relative widths of the gaussians comprising the DOG, and the relative amplitude k of the annulus-shaped negative part of the DOG. In the simulations, the center positions and polarity (i.e. peak-activity in the ON- or OFF-layer) are chosen at random.

The receptive fields which result after the self-organization process belong to one out of three qualitatively different possible states. The states can be distinguished by the way the set of all stimuli is tesselated, i.e. distributed among the map neurons. **System** $\mathcal{B}$: Each neuron responds to both an ON- and an OFF-stimulus, each located at the same retinal position. This tesselation yields neurons with orientation insensitive receptive fields. **System** $\mathcal{S}$: As in system $\mathcal{B}$, each neuron responds to stimuli of both polarities, but now displaced one step

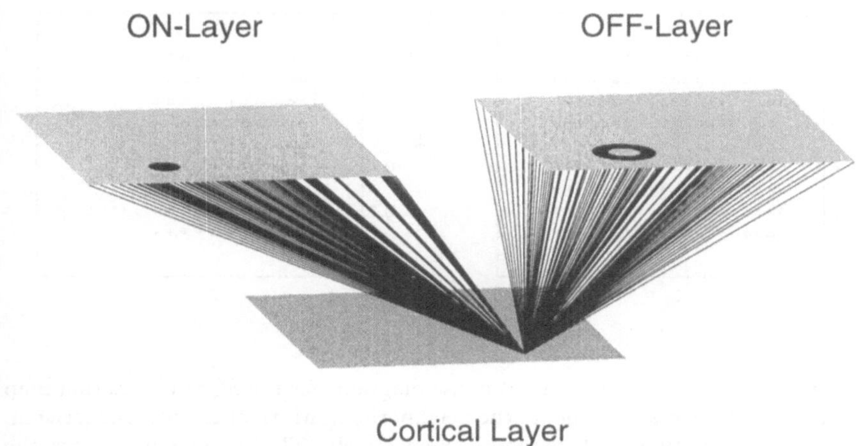

Fig. 1. Architecture of the SOM model of orientation map development.

along one retinal coordinate. The displacement breaks isotropy. It causes the receptive fields to exhibit internal ON- and OFF-center structure, with orientation specificity. **System $\mathcal{O}$:** Here each neuron responds to two retinally neighboring stimuli of identical polarity. Although this tesselation induces an orientation specificity, it also breaks the symmetry between ON-center and OFF-center inputs to each neuron. Neurons segregate into ON-center and OFF-center dominated populations, analogous to an ocular dominance map.

For each of these tesselations we can evaluate the distortion measure

$$E = \sum_{\mathbf{r},\mathbf{r}'} h(\mathbf{r}-\mathbf{r}') \sum_{\mathbf{v}\in\Omega_{\mathbf{r}}} \sum_{\mathbf{v}'\in\Omega_{\mathbf{r}'}} (\mathbf{v}-\mathbf{v}')^2, \tag{2}$$

where $\Omega_{\mathbf{r}}$ denotes all stimuli which are mapped to neuron $\mathbf{r}$. Each term in E consists of the mean squared difference between stimuli within the same, or between neighboring $\Omega_{\mathbf{r}}$'s, weighted by the neighborhood function $h(\mathbf{r}-\mathbf{r}')$. Assuming that the SOM-algorithm leads to a minimization of E, the final state of the receptive field vectors and the map can be determined by comparing the values of $E^{\mathcal{B}}$, $E^{\mathcal{S}}$ and $E^{\mathcal{O}}$ (For a motivation and a more detailed description of this method see [8]). Fig. 2a shows a state diagram in the parameter-plane which is calculated in this way. All three possible map states are predicted to occur in some region of parameter space.

To corroborate the mathematical analysis above and to actually obtain orientation maps we also investigated the model numerically. We ran simulations with 16×16 neuron maps, at various parameters. Classifying the resulting receptive

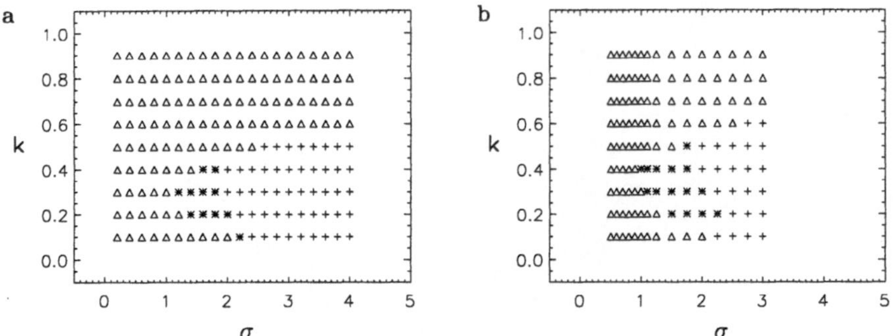

Fig. 2. Analytical (**a**) and numerical (**b**) phase diagrams for the SOM-orientation map model. The parameters σ and k denote the neighborhood width of the SOM-algorithm, and the annulus amplitude of the stimuli, respectively. The $+$ symbol denotes the non-oriented state $\mathcal{B}$, $\star$ denotes the oriented state $\mathcal{S}$, and $\triangle$ the (non-biological) state $\mathcal{O}$.

fields with regard to the states $\mathcal{B}, \mathcal{S}, \mathcal{O}$, we obtained the state diagram depicted in Fig. 2b. The correspondence to the mathematically obtained diagram (Fig. 2a) is strikingly good, underlining the value of the distortion measure method for the analysis of high-dimensional SOMs. Fig. 3 shows exemplary receptive fields of neurons in a segment of the map. The receptive fields show a multi-lobed structure and are clearly oriented.

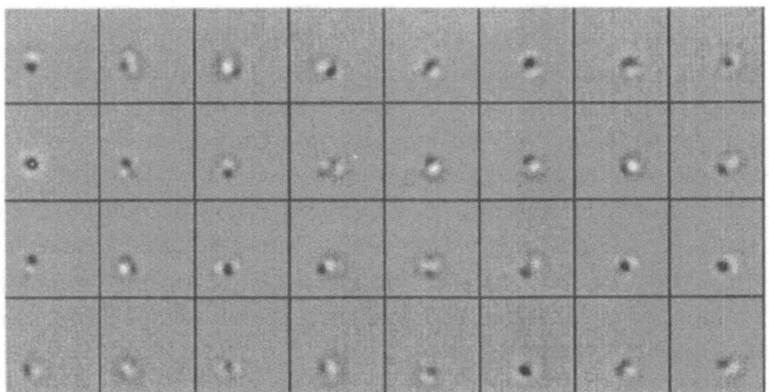

Fig. 3. Sample receptive fields of an orientation map with parameters chosen to yield map state $\mathcal{S}$.

3 Orientation and Ocular Dominance

Finally, we complement the two ON-center and OFF-center input cell layers for one eye by two further ON-center and OFF-center cell layers for the other eye. The repertoire of possible patterns in this extended model should go beyond merely oriented receptive fields in an orientation map, it should also include monocular receptive fields and ocular dominance maps, and combinations of the two types of patterns.

Stimuli in the extended model consist of activity distributions in all four input layers. While the difference in shape of the activity distributions between ON-center and OFF-center layers is the same as before, the partial stimuli are assumed to be of identical shape in the corresponding layers for either eye, but attenuated by a factor of c, $0 \leq c \leq 1$, in one of the eyes (analogous to the assumptions underlying a recently analyzed SOM-based model for ocular dominance formation [3, 2]. The analysis technique introduced above can be applied to this more complicated case as well, considering the different tesselation possibilities for four stimuli per neuron.

The parameter space in this extended model is rather large, which renders simulating the SOM for every set of parameters computationally unfeasible. Evaluation of the distortion measure for different map-layouts, however, enabled us to obtain phase-diagrams analogous to those depicted in fig. 2., and lead us directly to a parameter-regime in which to expect the combined development of orientation and ocular dominance. Simulating in this promising parameter region indeed produced maps with monocular, oriented receptive fields. Fig. 4 shows one such combined map in a plot that displays the boundaries of the isoocularity domains superimposed on the orientation map. Determining the transition lines between isoocularity regions in the simulated map, and computing the intersection angles with the isoorientation lines at these locations, one finds that isoorientation lines intersect the boundaries between isoocularity regions preferably at large angles, consistent with experimental observations [1, 7]. This effect cannot be accounted for in linear models.

Acknowledgement

This work has been supported by the Deutsche Forschungsgemeinschaft through SFB 185 "Nichtlineare Dynamik", TP E6.

References

1. Bartfeld, E., Grinvald A.: Relationship between orientation preference pinwheels, cytochrome oxidase blobs and ocular-dominance columns in primate striate cortex. *Proc. Nat. Acad. Sci. USA* **89** (1992) 11905–11909
2. Bauer, H.-U., Brockmann, D., Geisel, T.: Analysis of ocular dominance pattern formation in a high-dimensional self-organizing-map-model. *Network* **8** (1997) 17–33

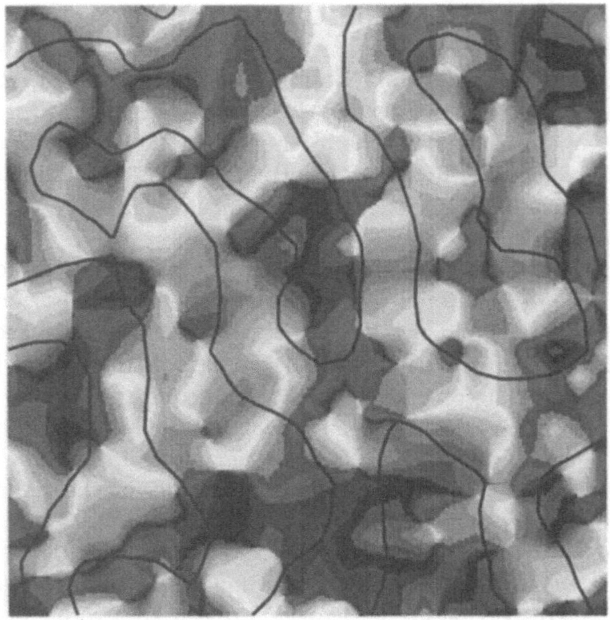

Fig. 4. Combined ocular dominance and orientation map. Isoocularity domain boundaries are superimposed on the orientation map as black lines.

3. Goodhill, G. J.: Topography and ocular dominance: a model exploring positive correlations. *Biol. Cyb.* **69** (1993) 109–118
4. Kohonen, T.: The Self-Organizing Map, *Springer Berlin* (1995)
5. Miller, K. D.: A model for the development of simple-cell receptive fields and the ordered arrangement of orientation columns through activity dependent competition between On- and Off-center inputs. *J. Neurosci.* **14** (1994) 409–441
6. Miyashita, M., Tanaka, S.: A mathematical model for the self-organization of orientation columns in visual cortex. *NeuroRep.* **3** (1992) 69–72
7. Obermayer, K., and Blasdel, G. G.: Geometry of orientation and ocular dominance columns in monkey striate cortex. *J. Neurosci.* **13** (1993) 4114–4129
8. Riesenhuber, M., Bauer, H.-U., Geisel, T.: Analyzing phase transitions in high-dimensional self-organizing maps. *Biol. Cyb.* **75** (1996) 397–407
9. Riesenhuber M., Bauer H.-U., Brockmann D., and T. Geisel: Breaking rotational symmetry in a self-organizing map-model for orientation map development. *submitted to Neur. Comp.* (1997)

On the Anatomical Basis of Field Size, Contrast Sensitivity, and Orientation Selectivity in Macaque Striate Cortex: A Model Study

Ute Bauer[1], Péter Adorján[2], Michael Scholz[2], Jonathan B. Levitt[3], Jennifer S. Lund[3], and Klaus Obermayer[2]

[1] Techn. Fak., Universität Bielefeld, P.O.-Box 10 01 31, D-33501 Bielefeld, Germany
[2] FB Inf., Technische Universität Berlin, Franklinstr.28-29, D-10587 Berlin, Germany
[3] Dept. of Visual Science, Inst. of Ophthalmology, UCL, London EC1V 9EL, U.K.

Abstract. Neurons in layer 4C in macaque striate cortex show a differential change in receptive field size and achromatic contrast sensitivity with depth, and exhibit orientation selective responses in the upper 4Cα sublayer. Using a computational model we first demonstrate that the observed change in receptive field size and contrast sensitivity can arise from a differential convergence of afferents from the P and M subdivisions of the lateral geniculate nucleus onto layer 4C spiny stellate cells - if one postulates that the two anatomically identified M1 and M2 subpopulations of the M afferents differentially project to different depth in the 4Cα subdivision. Number ratios and response properties of both M subpopulations are predicted and may now be tested experimentally. We then show that realistic orientation selective responses in upper 4Cα can emerge intracortically as a result of local lateral interactions, which are anisotropic, between spiny stellate cells and inhibitory interneurons. The model assumes that orientation bias and tuning are generated by the same cortical circuits and predicts a receptive field dynamics with an initial non orientation specific response.

1 Introduction

In macaque monkeys, layer 4C in striate cortex can be either divided into the two subdivisions 4Cα and 4Cβ which correspond to the termination zones of P and M thalamic input, or into the three output subdivisions lower 4Cβ, which projects primarily to 4A, mid-4C, which projects to the interblob territories in 2/3, and upper 4Cα, which projects to the direction selective layer 4B [13]. Thus, layer 4C seems to convert two input streams with different physiological properties into three output streams carrying different information; the question arises as to how information transfer occurs within layer 4C.

Blasdel and Fitzpatrick [1] and Hawken and Parker [4] reported that receptive field size and achromatic contrast sensitivity of neurons show a gradient from P-like to M-like properties from the bottom to the top of the layer 4C, this is the first evidence for an interaction between the incoming P- and M-streams. Although P- and M-afferents terminate in 4C in a segregated manner, the dendritic arbor of a spiny stellate cell in mid-4C intrudes into both termination zones and may form the anatomical substrate for this interaction. In the following we explore this hypothesis and we demonstrate that a good fit with experimental data can be obtained.

Blasdel and Fitzpatrick [1] and Hawken and Parker [4] furthermore showed that orientation selectivity emerges for the first time in layer 4C and is most prominent in upper 4Cα within 4C; it is coincident with the emergence of long (up to 1.5 mm) anisotropic, laterally spreading intralayer connections of excitatory and inhibitory stellate cells [13,5]. This has motivated us to investigate the hypothesis that anisotropic lateral connections between layer 4C neurons form the anatomical substrate of an **intracortical** origin of orientation preference and tuning.

The results presented at this conference are the beginnings of a detailed study of the anatomical, physiological and computational properties of layer 4C in macaque striate cortex.

2 The Connectionist Model

Our computational model consists of the visual field layer, the LGN layers with the P- and M-populations and the eight 4C "sublayers" representing neurons at eight different depths D (Fig. 1a). Only monocular ON-cells are considered, binocular and ON-OFF interactions are neglected. Model parameters correspond to $5° - 8°$ eccentricity (for references see [6]).

If the stimulus $l(\boldsymbol{X})$ is presented to the visual field layer, the afferent input I_t to a cell at location $\boldsymbol{U} = (u_1, u_2)$ in a geniculate layer is calculated by

$$I_t(\boldsymbol{U}) = \int_{\Re^2} d\boldsymbol{X}\, l(\boldsymbol{X}) \left\{ k_c^t \exp\left(-\frac{(\boldsymbol{U}-\boldsymbol{X})^2}{(r_c^t)^2}\right) - k_s^t \exp\left(-\frac{(\boldsymbol{U}-\boldsymbol{X})^2}{(r_s^t)^2}\right)\right\} \quad (1)$$

with center and surround radii r_c^t, r_s^t, peak sensitivities k_c^t, k_s^t, LGN cell type $t \in \{P, M\}$, and $\boldsymbol{X} = (x_1, x_2)$ are the visual field coordinates. The output $O_t = f(I_t)$ of a geniculate cell is then calculated via a transfer function f which is determined by fitting the contrast-response function of the model-neurons to a Michaelis-Menten function $\mathrm{M}_t\, c/(\mathrm{M}_t\, \mathrm{G}_t^{-1} + c)$ with maximum response M_t, contrast gain G_t and stimulus contrast c. In section 3, parameters were taken from a Gaussian distribution with mean values and standard deviations given in [12]; in section 4 only the mean values were used.

The afferent input to a cortical cell at location $\boldsymbol{U}$ and depth D is given by

$$A(\boldsymbol{U}, D) = \sum_{t \in \{M1, M2, P\}} P_t(D) \int_{\Re^2} d\boldsymbol{X}\; w_t\,(\boldsymbol{U}-\boldsymbol{X}) O_t(\boldsymbol{X}), \quad (2)$$

where $P_t(D)$ is the proportion of the geniculate P and M input to a cortical cell at depth D (Fig. 1b) and $w_t(\boldsymbol{X})$, $\int d\boldsymbol{X} w_t(\boldsymbol{X}) = 1$, is a normalized circular symmetric weight kernel which is proportional to the areal overlap between an afferent axonal arbor at position $\boldsymbol{X}$ and a spiny stellate dendritic field at position $\boldsymbol{U}$ (cylinders in Fig. 1a).

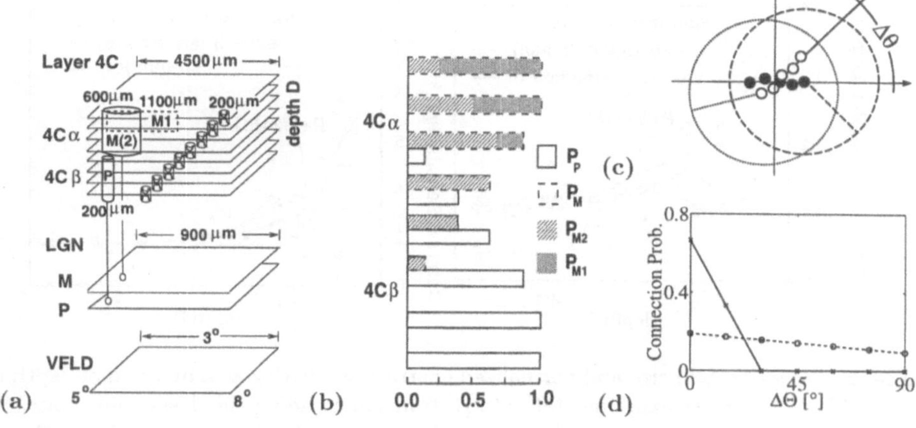

Fig. 1. **(a)** Network architecture. The large cylinders on the left represent the axonal arbors for P-, M1- and M2-afferents; the small cylinders on the right represent the dendritic arbors of layer 4C spiny stellate cells. **(b)** Proportion of P- and M- (M1- and M2-) input to a cortical cell as a function of depth D in layer 4C. The numbers correspond to the best fit against the experimental data (see Fig. 2). **(c)** Scheme of an orientation hypercolumn. The filled and empty circles indicate the centers of receptive fields of cells from two orientation columns with orientation difference $\Delta\Theta$ of preferred orientation. The biggest distance between field centers in one orientation column is $\approx 0.2°$. The diameter of the receptive fields (dotted and dashed circles) is $\approx 0.25°$. **(d)** The probability of excitatory (solid line) and inhibitory (dotted line) lateral connections assumed for the model in upper $4C\alpha$ as a function of the difference of preferred orientations.

For the numerical simulations of section 3 the neurons in the cortical layers D were not connected, and their output was taken to be proportional to their total input (2). For the numerical simulations of section 4 we assume two populations of cells, excitatory cells e which correspond to spiny stellate neurons and inhibitory cells i which correspond to local interneurons. Cells whose receptive fields lie along one axis in visual space belong to one "orientation column" (Fig. 1c). Cells are connected via excitatory ($e \to e$, $e \to i$) and inhibitory ($i \to e$, $i \to i$) connections which fall off with the difference in axis-angle as shown in Fig. 1d. Given this wiring scheme, the orientation of the axis across visual space determines the preferred orientation of "its" cortical cells.

Cortical neurons are modelled as connectionist units, whose state m is interpreted as a "membrane potential". Let θ denote the preferred orientation of a cell as given by the orientation of a ray in Fig. 1c, index i the position in the ray, $p = \{\text{exc, inh}\}$ the type, and t the time. Then we obtain

$$\frac{d}{dt} m_{\theta,i}^{(p)}(t) = -m_{\theta,i}^{(p)}(t) + I \,, \tag{3}$$

where I is the sum of the afferent input A (eq. 2) and the lateral input L,

$$L(\Theta, q, i, t) = \sum_{p,j,\Theta'} N(p, q, |\Theta - \Theta'|)\, W(p, q)\, g(m(\Theta', p, j, t)). \tag{4}$$

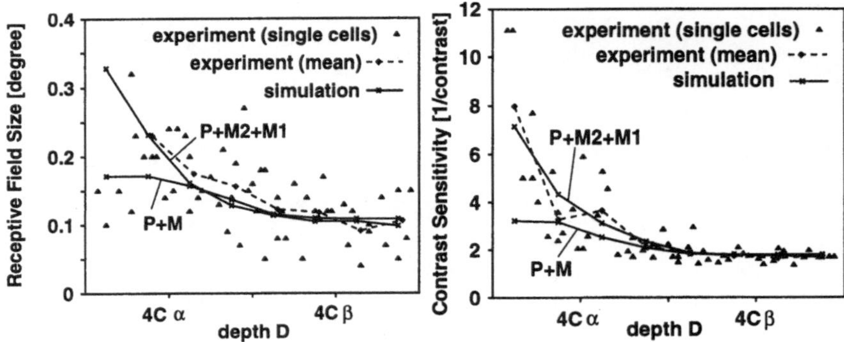

Fig. 2. Receptive field size and normalized contrast sensitivity as a function of depth in layer 4C. Dots denote experimental data [1] from single units; the dashed line connects their mean values for eight equally spaced depth intervals. Solid lines indicate "best" model predictions, the corresponding proportion of P- vs. M1 and M2-inputs are shown in Fig. 1b. Receptive field size and contrast sensitivity of model cells were determined as described in [1].

$N(p, q, |\Theta - \Theta'|)$ is the number of synapses made by a cortical cell of type p and orientation Θ with its target cell of type q and orientation Θ'. $W(p, q)$ is the strength of a single synapse between cell types p and q. The cortical transfer function $g(x)$ describes the transformation between the membrane potential and the output firing frequency and is taken to be semilinear. Since orientation selective cells predominantly appear in upper 4Cα, only the M-pathway was implemented for the simulations of section 4.

3 Receptive Field Size and Contrast Sensitivity in 4C

In order to match model predictions to data we have explored in detail how response properties of our model cortical cells change with model parameters, in particular with the ratios of P- vs. M- and M1- vs. M2-inputs to cortical cells. Figure 2 shows the "best predictions" for one (P+M) and two (P+M1+M2) kinds of M-input. The P+M model provides a good match for lower 4Cβ and mid-4C but fails to predict the increase in contrast sensitivity and receptive field size in upper 4Cα. A good fit can be obtained everywhere for the P+M1+M2 model, i.e. under the assumption that a class of M-cells (M1) with values for receptive field size and contrast sensitivity from the upper part of the M-distribution selectively projects to upper 4Cα [2]. The model predicts that approximately 12% of M cells belong to M1, that their physiological parameters ($r_c = 0.13^o$, $r_s = 1.4^o$, $G =$ 3 spikes sec^{-1} %contrast^{-1}, M = 80 spikes sec^{-1}) are roughly one standard deviation above the mean values for the M population and that it is the afferent input arising from the M1 group that drives the cortical neurons at the top of 4Cα.

4 Intracortical Origin of Orientation Selectivity

Figure 3 shows simulation results for laterally connected cortical units. If a grating is presented, the sum of all afferent inputs to an orientation column is biggest if the orientation of the grating matches the orientation of the connectional axis, hence an orientation bias is generated. The bias is subsequently amplified by the recurrent excitatory and inhibitory connections, resulting in a contrast invariant tuning width (Fig. 3a) as reported by [10]. The model predicts a non-oriented initial response (Fig. 3b), similar to what has been reported by [8] for cat area 17, but different from the predictions of models assuming a Hubel- and Wiesel style afferent orientation bias (cf. [11]). Blocking inhibition in the whole orientation hypercolumn (cf. [9]), i.e. reducing the strength of the $i \to e$ and $i \to i$ connections, broadens orientation tuning and finally eliminates orientation selectivity, but the broadening predicted by the model is mainly due to saturation in the cortical firing rate. Selective blockade of cross-orientation inhibition (cf. [3])

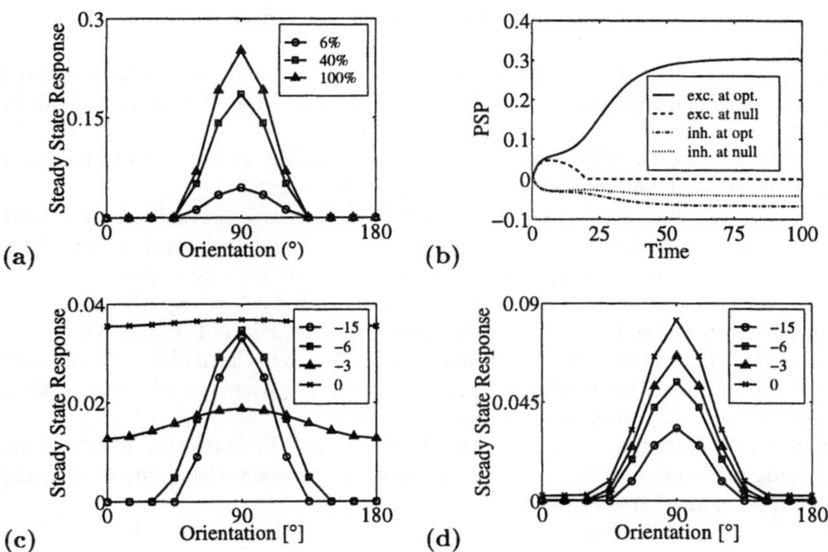

Fig. 3. **(a)** Orientation tuning curves of simulated cortical cells at different levels of contrast. **(b)** Total excitatory and inhibitory input to a simulated cortical cell at null and at optimal stimulus orientation. For nonoptimal stimuli there is a fast decay of excitation due to cross-orientation inhibition after a nonspecific initial phase of excitatory input. For optimal stimuli the initial excitation is amplified by recurrent excitation and controlled by iso-orientation inhibition. **(c)** Orientation tuning curves as a function of the strength of cross-orientation inhibition ($\Delta\Theta > 30°$). **(d)** Orientation tuning curves for a simulated blockade of inhibition in a single cell. In (b), (c), and (d) stimuli were grating with optimal spatial frequency and $c = 5\%$ contrast.

also broadens orientation tuning (Fig. 3c), but activity remains below saturation level. Intracellular blockade of inhibition (cf. [7]) does not affect orientation selectivity in the model because its response is driven by sharply tuned excitatory inputs arriving at the cortical cell. Iso-orientation inhibition is important for controlling runaway excitation; cross-orientation inhibition is necessary to establish sharp orientation tuning but can be weaker. In summary, realistic orientation selective responses can indeed be established intracortically.

Acknowledgments Supported by DFG (Ob 102/2-1), HFSPO (RG-98/94), MRC G9409137, and a DFG scholarship (GK 231) to U.B.

References

1. G.G. Blasdel and D. Fitzpatrick. Physiological organization of layer 4 in macaque striate cortex. *J. Neurosci.*, 4(3):880–895, 1984.
2. G.G. Blasdel and J.S. Lund. Termination of afferent axons in macaque striate cortex. *J. Neurosci.*, 3:1389–1413, 1983.
3. U. T. Eysel, J. M. Crook, and H. F. Machemer. GABA-induced remote inactivation reveals cross-orientation inhibition in the cat striate cortex. *Exp. Brain Res.*, 80:626–630, 1990.
4. M. J. Hawken and A. J. Parker. Contrast sensitivity and orientation selectivity in lamina iv of the striate cortex of old world monkeys. *Exp. Brain. Res.*, 54:367–372, 1984.
5. J. S. Lund. Local circuit neurons of macaque monkey striate cortex: I. neurons of laminae 4C and 5A. *J. Comp. Neurol.*, 257:60–92, 1987.
6. J. S. Lund, Q. Wu, P. T. Hadingham, and J. B. Levitt. Cells and circuits contributing to functional properties in area V1 macaque monkey cerebral cortex: Bases for neuroanatomically realistic models. *J. Anatomy*, 187:563–581, 1995.
7. S. Nelson, L. Toth, B. Sheth, and M. Sur. Orientation selectivity of cortical neurons during intracellular blockade of inhibition. *Science*, 265:774–777, 1994.
8. X. Pei, T. R. Vidyasagar, M. Volgushev, and O. D. Creutzfeld. Receptive field analysis and orientation selectivity of postsynaptic potentials of simple cells in cat visual cortex. *J. Neurosci.*, 14:7130–7140, 1994.
9. H. Sato, N. Katsuyama, H. Tamura, Y. Hata, and T. Tsumoto. Mechanisms underlying orientation selectivity of neurons in the primary visual cortex of macaque. *J. Physiol.*, 494:757–771, 1996.
10. G. Sclar and R. D. Freeman. Invariance of orientation tuning with stimulus contrast. *Exp. Brain Res.*, 46:457–461, 1982.
11. D. C. Somers, S. B. Nelson, and M. Sur. An emergent model of orientation selectivity in cat visual cortical simple cells. *J. Neurosci.*, 15:5448–65, 1995.
12. P.D. Spear, R.J. Moore, C.B.Y. Kim, J. Xue, and N. Tumosa. Effects on aging on primaten visual system: Spatial and temporal processing by lateral geniculate neurons in young adult and old rhesus monkeys. *J. Neurophysiology*, 72:402–420, 1994.
13. T. Yoshioka, J. B. Levitt, and J. S. Lund. Independence and merger of thalamocortical channels within macaque monkey primary visual cortex: anatomy of interlaminar projections. *Vis. Neurosci.*, 11:467–489, 1994.

Statistics of Natural and Urban Images

Christian Ziegaus and Elmar W. Lang

Institute of Biophysics, University of Regensburg, D–93040 Regensburg, Germany
christian.ziegaus@rphs1.physik.uni-regensburg.de

Abstract. We investigated ensembles of artificial and real–world grey-scale images to find different invariance properties: translation invariance, scale invariance and a new hierarchical invariance recently proposed by Ruderman [1]. We found that the assumption of translational invariance can be taken for granted. Our results concerning the scale invariance are qualitatively the same as those found by Ruderman [1] and others. The deviations of the distributions of the logarithmically transformed images from a Gaussian distribution cannot be seen as clearly as stated by Ruderman [1]. Depending on the preprocessing of the images the results concerning the hierarchical invariance differed widely. It seems that this new invariance can be confirmed only for logarithmically transformed images.

1 Introduction

The visual system is expected to be optimally adapted to the statistics of natural images it deals with. Hence the statistical properties of visual input patterns are of primary interest. Since there is no way to collect enough data to fully characterize an image environment, recent investigations [1–4] seek to identify a simple underlying structure or invariance property in the image probability distribution. One such symmetry frequently assumed is translational invariance. Undoubtedly their most robust statistical property is an invariance to scale [1]. Recently, evidence has been presented supporting the notion of a hierarchical invariance in natural scenes. It relates to the conversion of exponential histograms to Gaussian distributions via local non–linear transformations.

2 Ensembles

2.1 The natural and the urban ensemble

Various scenes for the natural ensemble NAT have been taken within a forest with a CCD video camera. Scenes for the urban ensemble ZIV have been filmed around the campus of the University of Regensburg. From individual frames an average image has been constructed and then about 100 smaller sized images (256×256 pixels each) have been extracted at random. To evaluate the image statistics raw pixel intensities $\Phi(\boldsymbol{x})$ have been transformed either linearly or non–linearly to difference intensities $\Phi(\boldsymbol{x}) \equiv D(\boldsymbol{x})$ or log–contrast intensities $\Phi(\boldsymbol{x}) \equiv L(\boldsymbol{x})$ with zero mean distributions.

2.2 The artificial ensemble K2

Fourier images with $1/|\boldsymbol{k}|$–amplitude spectra and random phases were generated on a computer and inverse Fourier–transformed into the spatial domain to generate an image which is perfectly scale invariant, hence is characterized by a power spectrum $S(|\boldsymbol{k}|) \propto 1/|\boldsymbol{k}|^2$. After determining for each Fourier–transformed image the corresponding maximal and minimal pixel intensities I_{max} and I_{min} the latter have been rescaled linearly to span the intervall $[0, \ldots, 255]$.

The actual spread of the rescaled pixel intensities can be characterized by the relation $\langle I_{max} \rangle \approx \frac{\sigma_I^2}{\sqrt{2}M}$. The distribution of pixel intensities $I(\boldsymbol{r})$ after rescaling with I_{max} is thus given by

$$P(I) = \frac{\sigma_I}{255M\sqrt{\pi}} \exp\left\{ -\left(\frac{\sigma_I}{2M}\right)^2 \left(\frac{I}{127.5} - 1\right)^2 \right\}, \tag{1}$$

with M the number of pixels in each spacial direction and σ_I the standard deviation of the distribution.

3 Translational Invariance

If images are indeed translationally invariant then any two–point–correlation function $A_{\boldsymbol{x}}(\boldsymbol{y}) = \langle \Phi(\boldsymbol{x}) \Phi(\boldsymbol{x}+\boldsymbol{y}) \rangle$ of pixel intensities at points $\boldsymbol{x}$ and $\boldsymbol{y}$ will depend on their relative distance only and not on $\boldsymbol{x}$ itself. Besides calculating, for different starting points $\boldsymbol{x}$, the second order correlation function $A_{\boldsymbol{x}}(\boldsymbol{y})$ of pixel intensities an average correlation function has been determined also according to the relation

$$\bar{A}(\boldsymbol{y}) = \frac{1}{N_{\boldsymbol{x}}} \sum_{\boldsymbol{x}} A_{\boldsymbol{x}}(\boldsymbol{y}) = \frac{1}{N_{\boldsymbol{x}}} \sum_{\boldsymbol{x}} \langle \Phi(\boldsymbol{x}) \Phi(\boldsymbol{x}+\boldsymbol{y}) \rangle \tag{2}$$

which is the mean over the number of pixels contained in any given image.

The average correlation functions of the image ensembles are compared in Fig. 1. Though obviously translationally invariant, rotational isotropy is lost as vertical structures (stemming from trees largely) dominate the natural images, whereas the average correlation function from images of the urban environment reflect the predominance of horizontal contours.

4 Scale Invariance

If scale invariance prevails, the power spectrum $S_\Phi(\boldsymbol{k})$ of the intensity distribution should, after averaging over all orientations, scale like

$$S_\Phi(|\boldsymbol{k}|) \propto \frac{1}{|\boldsymbol{k}|^{2-\eta}}, \tag{3}$$

with $|\boldsymbol{k}|$ the modulus of the spatial frequency and $\eta \neq 0$ an anomalous exponent. Scale invariance does not tell the form of the stationary distribution from

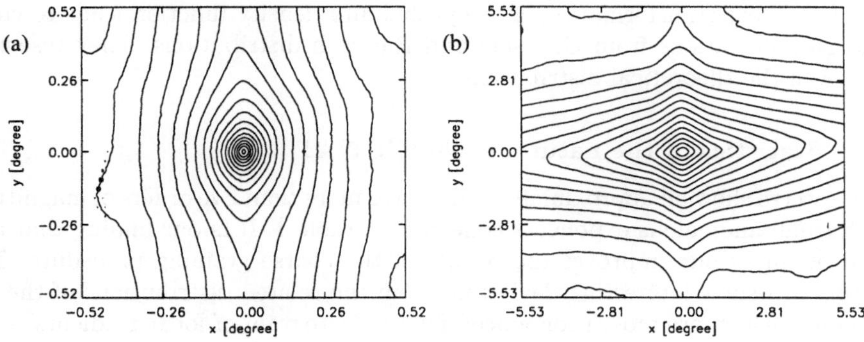

Fig. 1. Average correlation function $\bar{A}(\boldsymbol{y})$ for the ensembles NAT (a) and ZIV (b) for the difference intensities. Contours are shown at equal intervals of correlation.

which the $\Phi(\boldsymbol{x})$ are drawn. Both aspects may be further investigated through the process of *coarse graining* [5], which replaces an $N \times N$ block of pixel intensities by their average. If the probability distribution $P_N(\Phi)$ is scale invariant (scaling variable N), normally distributed difference intensities $\Phi(\boldsymbol{x}) \equiv D(\boldsymbol{x})$ would result in a Gaussian distribution. But in case of non–linearly transformed log–contrast intensities $\Phi(\boldsymbol{x}) \equiv L(\boldsymbol{x})$ the correspondingly transformed and normalized distribution with zero average becomes

$$P(L) = \frac{\sigma_I \sigma_L}{255M\sqrt{\pi}} \exp\left(\sigma_L L + \bar{L}\right) \exp\left\{ - \left(\frac{\sigma_I}{2M}\right)^2 \left(\frac{\exp\left(\sigma_L L + \bar{L}\right)}{127.5} - 1\right)^2 \right\}$$

$$\sigma_L^2 = \int L^2 P(L)\, dL \quad \bar{L} = \int_0^{255} \ln(I)\, P(I)\, dI = \int_{-\infty}^{\ln(255)} L P(L)\, dL. \qquad (4)$$

Except for investigating the distribution of difference intensities or the logarithmic contrast values it is advantages also to consider the distribution of local gradients $G \approx |\nabla \Phi|$ in the images. If the $\Phi(\boldsymbol{x}) \equiv D(\boldsymbol{x})$ are normally distributed and scale invariant, a Rayleigh distribution of local gradients would result [6]. The corresponding distribution of local gradients $G = \sqrt{G_x^2 + G_y^2}$ in case of logarithmically transformed pixel intensities $\Phi(\boldsymbol{x}) \equiv L(\boldsymbol{x})$ as well as the average local gradient $\bar{G} = \int dG_x \int dG_y G P(G_x, G_y)$ must be obtained numerically assuming factorization of the joint probability density $P(G_x, G_y) = P(G_x)\, P(G_y)$ with $P(G_{x,y})$ chosen according to (4).

4.1 Statistics of the artificial ensemble K2

For the ensemble K2 $S(|\boldsymbol{k}|) \propto |\boldsymbol{k}|^{\alpha}$ holds with $\alpha = -1.969 \pm 0.002$ the best estimate in the least squares sense. A coarse graining of the pixel intensities has been performed for the scales $N = 1, 2, 4, 8, 16$, and 32. It is to be noted that the non–linear transformation of the probability density function leads to characteristic deviations from Gaussian and Rayleigh distributions. They resemble very well the theoretical distributions.

4.2 Statistics of the natural ensemble NAT

The spectra show an approximate scaling over more than five orders of magnitude with small anomalous exponents collected in Table 1. It is our finding that any kind of smoothing improves the results of the coarse graining procedure. The histograms show interesting deviations from the related distributions of the K2 ensemble which are most pronounced for the histogram of local gradients.

Table 1. Summary of the anomalous exponent η according to (3).

ensemble	transformation	slope $-2+\eta$	η
K2	linear	-1.969 ± 0.002	0.031
K2	non–linear	-1.933 ± 0.002	0.067
NAT	linear	-2.319 ± 0.054	-0.319
NAT	non–linear	-2.369 ± 0.062	-0.369
ZIV	linear	-2.634 ± 0.052	-0.634
ZIV	non–linear	-2.665 ± 0.054	-0.665

4.3 Statistics of the urban ensemble ZIV

The images of this ensemble contain man–made structures only with prominent horizontal and vertical contours. This is reflected in the power spectrum which is very asymmetric. It is also of interest that the anomalous scaling exponent η of the orientationally averaged power spectrum shows the largest anomalous exponent of all image ensembles considered in this study (cf. Table 1).

4.4 Comparison of logarithmical contrast distributions

Figure 2 shows the distributions of the logaritmically transformed pixel intensities for all ensembles together with the theoretically derived distribution and the histograms of local gradients. The distributions of the logarithmically transformed pixel intensities show for all ensembles only small differences at high log contrast values when compared to the theoretically distribution. The comparison of the experimentally obtained distributions of local gradients of the ensembles NAT and ZIV with the theoretical histogram shows an increasing probability of very small and very large gradients in going from the theoretical distribution to NAT and ZIV. This is due to the increasingly pronounced edges and large unstructured areas in the latter images. Related observations have been made with histograms of difference intensities and corresponding local gradients.

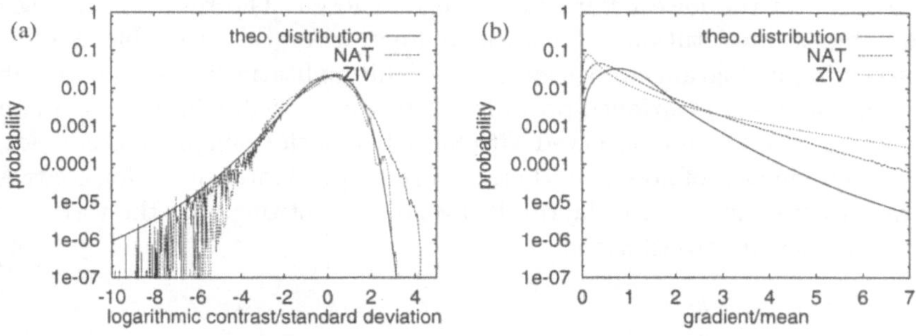

Fig. 2. Scaling of the distributions of logarithmic contrast values (a) and gradients (b) for the real–world ensemble NAT/ZIV and the theoretically derived histograms.

5 Hierarchical Invariance

Recently the possibility of a hierarchical invariance of natural images has been discussed [1]. It is related to the observation that simple linear filtering of logarithmic contrasts produces exponential histograms much like those observed experimentally. If these exponential tails are due to a superposition of many distributions with largly varying variances, one may try to find a local non–linear transformation which can turn the distributions to Gaussians. The transformation proposed by Ruderman [1] amounts to calculating

$$\Psi(\boldsymbol{x}) = \frac{\Phi(\boldsymbol{x}) - \bar{\Phi}(\boldsymbol{x})}{\sigma(\boldsymbol{x})} \tag{5}$$

with $\bar{\Phi}(\boldsymbol{x})$ the average pixel intensity (whether linearly or non–linearly transformed) within a block of size $N \times N$ pixels and $\sigma(\boldsymbol{x})$ the standard deviation of pixel intensity fluctuations within the block. Besides these *variance modified images* one may as well consider so called *variance images*. These may be constructed either by substituting any pixel intensity by the related variance of pixel intensity fluctuations within block $N \times N$ or by substituting the whole block of pixel intensities by the variance of their intensity fluctuations thereby changing the image size. We will refer to both procedures as variance images *without* and *with block substitution*, respectively. The pixel intensities of these variance images may be transformed in the same way as the original images. The interesting thing is that these variance images seem to exhibit similar histograms as do the original images. This observation asks for the possibility to iterate this procedure and to its possible outcome. In doing so one may select the block size according to the smallest possible kurtosis of the distribution of pixel intensities as Gaussian distributions are characterized by vanishing kurtosis.

5.1 Results for the ensembles K2 and NAT

Ten iterations with 100 images have been performed to obtain variance modified images of block sizes 3×3 to 19×19 pixels. For every iteration the block size resulting in the lowest kurtosis has been chosen. The resulting histograms without block substitution are shown in Figs. 3 for the ensemble NAT. The corresponding histograms of ensemble K2 yield qualitatively the same results.

If a hierarchical invariance prevails, all histograms should exhibit a similar shape which is clearly not observed with linearly transformed pixel intensities but seems to hold in case of non–linearly transformed pixel intensities (log contrasts). Contrary to scale invariance the result does depend strongly on the way the raw pixel intensities are transformed.

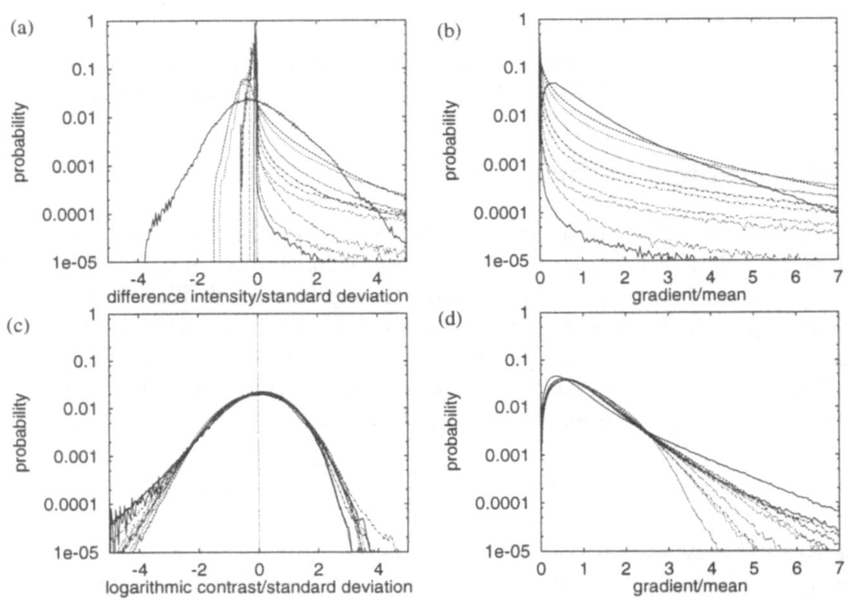

Fig. 3. Distributions of the linearly (a, b) and non–linearly (c, d) transformed pixel intensities (a, c) and the corresponding local gradients (b, d) for ensemble NAT and the iteration procedure without block substitution (10 steps with 100 pictures).

References

1. Ruderman, D. L.: The statistics of natural images Network **5** (1994) 517–548
2. Ruderman, D. L.: Designing receptive fields for highest fidelity Network **5** (1994) 147–155
3. Field, D. J.: Relations between the statistics of natural images and the response properties of cortical cells J. Opt. Soc. Am. **4** (1987) 2379–2394
4. Field, D. J.: What is the goal of sensory coding? Neural Comput. **6** (1994) 559–601
5. Kadanoff, L. P.: Scaling laws for ising models near T_c Physics **2** (1966) 263–273
6. Gardiner, C. W.: Handbook of Stochastic Methods (Berlin, Heidelberg: Springer) 1983

A CBL Network Model with Intracortical Plasticity and Natural Image Stimuli

Thomas Burger and Elmar W. Lang

Institut für Biophysik und physikalische Biochemie
Universität Regensburg, D-93040 Regensburg, Germany
thomas.burger@rphs1.physik.uni-regensburg.de

Abstract. We present a simplified binocular neural network model of the primary visual cortex with separate ON/OFF-pathways and modifiable afferent as well as intracortical synaptic couplings. Natural image stimuli drive the weight adaptation which follows Hebbian learning rules stabilized with constant norm and constant sum constraints. The simulations consider the development of orientation selective cortical cells and orientation maps under different conditions concerning stimulus patterns and lateral couplings. Strong short range excitatory lateral connections emerge between individual cortical neurons with inhibitory couplings being less specific and rather diffuse.

1 Introduction

In the past several computational models have been deviced to investigate the mechanisms controlling the activity-dependent self-organization of cortical feature maps in the primary visual cortex [7–10,13–15]. However, almost all previous network models used non-modifiable lateral couplings in cortex. But recent experimental findings demonstrate a rather substantial plasticity of lateral connections in the primary visual cortex of monkeys and cats [2,4,3,6,5]. Exploration of CBL (Correlation Based Learning) models with modifiable lateral couplings is thus essential to an understanding of how such plastic lateral interactions may affect the development and structures of cortical receptive fields and cortical feature maps. Sirosh and Miikkulainen investigated lateral plasticity within their LISSOM-model, which is an extension of competitive SOM (Self Organizing Map) models [11,12]. Receptive field profiles and cortical feature maps have not been examined so far, however. Furthermore the difficulties concerning a neurobiological justification of SOM-type models remain. A correlation-based Hebbian learning model with lateral plasticity has not been described in the literature so far.

As visual stimuli random input patterns were used in the majority of these developmental neural network models. Also idealized correlation functions of input activities were substituted for a direct training with real input patterns, whether random or structured. As cortical maps seem to emerge prenatally but mature to adult levels postnatally only, the respective role of structured versus random input patterns is still obscure and needs to be clarified.

In this study we investigate a correlation-based neural network model with modifiable afferent and lateral synaptic couplings trained with natural image stimuli.

2 The model

2.1 The network architecture

The binocular model possesses two retinal input layers, the neurons of which project their non-negative output from within localized receptive fields to separate cell layers with fixed ON- and OFF-center-surround contrast filters (LGN cells). They are implemented as differences of Gaussians (DOG). The total center radius of the mexican hat profile together with the cortical receptive field size determine the type of RF-structure obtained. They have been chosen to yield bilobed RF-structures during prenatal development [13,1]. The afferent activity of these preprocessing units converge onto neurons of the output layer which may correspond to cortical simple cells.

The connections between input neurons and LGN cells are fixed. However, the afferent connections converging onto cortical neurons are plastic. Their non-negative weights, corresponding to exclusively excitatory afferents, are denoted as $w^{\alpha,\beta}(\boldsymbol{r}_y, \boldsymbol{r}_x)$, where $\alpha \in \{\text{left, right}\}$, $\beta \in \{\text{on, off}\}$, $\boldsymbol{r}_y$ denoting a cortical cell and $\boldsymbol{r}_x$ denoting a LGN cell. The finite size of the afferent arbor is characterized by a radially symmetric Fermi-function $A(|\boldsymbol{r}_x - \boldsymbol{r}_y|)$:

$$A(|\boldsymbol{r}_x - \boldsymbol{r}_y|) = \frac{1}{1 + \exp \frac{|\boldsymbol{r}_x - \boldsymbol{r}_y| - 4.0}{0.3}}. \tag{1}$$

Besides these afferent connections there are also plastic cortical lateral connections in the model. The variable $v_E(\boldsymbol{r}_y, \boldsymbol{r}_{y'})$ denotes the excitatory synaptic coupling strength between a postsynaptic cortical neuron at $\boldsymbol{r}_y$ and a presynaptic cortical neuron at $\boldsymbol{r}_{y'}$ and $v_I(\boldsymbol{r}_y, \boldsymbol{r}_{y'})$ denotes a corresponding inhibitory connection. These coupling strengths cannot change sign during adaptation. They are both non-negative, for convenience.

The excitatory and inhibitory lateral couplings exist within circular regions with radii r_{LE} and r_{LI} respectively. Also $r_{LE} \leq r_{LI}$ will be assumed throughout. Furthermore the variable v_{EI} is introduced allowing a more convenient evaluation of the lateral coupling structure: $v_{EI} = v_E - v_I$.

All neurons are assumed to operate in the linear regime of their activation function, i.e. the postsynaptic activity of a neuron between its threshold value and its saturation value rises linearly with the summed presynaptic activities.

2.2 The learning rules

For the afferent synaptic weights w a Hebbian learning rule with a multiplicative constraint (Yuille decay) is used [16,13,14].

$$\Delta w^{\alpha,\beta}(\boldsymbol{r}_y, \boldsymbol{r}_x, t \to t+1) = \eta_{heb}\, y(\boldsymbol{r}_y, t)\, x^{\alpha,\beta}(\boldsymbol{r}_x, t) -$$

$$-\eta_{dec}\left[\sum_{\boldsymbol{r}_x,\alpha,\beta}\left(w^{\alpha,\beta}(\boldsymbol{r}_y,\boldsymbol{r}_x,t)\right)^2\right] w^{\alpha,\beta}(\boldsymbol{r}_y,\boldsymbol{r}_x,t), \tag{2}$$

where $t = 0, 1, 2, \ldots$ denotes sequential update cycles and η_{heb} and η_{dec} represent appropriately chosen learning rates.

The lateral synaptic weights are modified according to the following version of stabilized Hebb-type learning rules denoted as a constant sum (CS) rule. The CS rule is given by

$$v_X(\boldsymbol{r}_y,\boldsymbol{r}_{y'},t+1) = \Sigma_X \cdot \frac{v_X(\boldsymbol{r}_y,\boldsymbol{r}_{y'},t) + \mu_X \cdot \eta_{lat,heb}\, y(\boldsymbol{r}_y,t)\, y(\boldsymbol{r}_{y'},t)}{\sum_{\boldsymbol{r}_{y'}}\left[v_X(\boldsymbol{r}_y,\boldsymbol{r}_{y'},t) + \mu_X\, \eta_{lat,heb}\, y(\boldsymbol{r}_y,t)\, y(\boldsymbol{r}_{y'},t)\right]}. \tag{3}$$

where $X \in \{E, I\}$, $\mu_E \equiv 1$ and $\mu_I \in \{-1, 1\}$. The parameter μ_X accounts for an anti-Hebbian learning of the inhibitory lateral synaptic weights considered later. The factors Σ_E and Σ_I denote the sum of excitatory and inhibitory intracortical synaptic coupling strengths of any cortical neuron, respectively.

$$\Sigma_X = \sum_{\boldsymbol{r}_{y'}}\left(v_X(\boldsymbol{r}_y,\boldsymbol{r}_{y'})\right). \tag{4}$$

These sums are independently kept constant during a simulation.

3 Results

The simulations intended to investigate a postnatal development of synaptic couplings driven by structured stimulus patterns. For that purpose a greyscale picture showing a natural scene with thicket and foliage was used. The picture had a size of (256 × 256) pixels out of which smaller images matching the size of the input layers of the model network were extracted at random and presented to the network as stimulus patterns before each update cycle. Each stimulus pattern was presented to both input layers of the network to account for correlations resulting from fixation during binocular vision. A parameter Δ_x was further introduced to effect a horizontal shift of both input patterns and takes account of the fact that the two eyes of mammals are not looking at exactly the same scene. In addition this parameter allows to simulate misalignment of both monocular receptive fields as occurs in case of strabismic vision. Note that Δ_x describes a constant disparity value for the entire image. A disparity variation across image regions seems not to be necessary, because the input layers represent a small part of the visual field only.

Lateral synaptic weights were implemented with the excitatory connection radius being shorter than the radius of the inhibitory interactions. Further $\Sigma_E = \Sigma_I$ was employed.

With natural image stimuli no bilobed monocular $(0,1)$ profiles developed anymore (for notation details see [8]). Instead unstructured $(0,0)$-like profiles emerged, which usually were a little bit deformed, however. These structural features were robust against any variation of the model parameters [13].

A horizontal shift $\Delta_x = 10$ of one retinal image against the other led to a complementary arrangement of the monocular $(0,0)$ receptive field profiles or ON- and OFF-patches, respectively. The patches formed relatively small clusters.

In contrast to the monocular receptive field structures within the patches, most of the binocular neurons located at the borders between the clusters were orientation-selective and had bilobed $(0,1)$-like receptive fields. These orientation specific neurons gave rise to a recognizable orientation map (Fig. 1(a)). But this map exhibited no prominent pinwheel vortices as obtained with random stimuli [1].

The excitatory lateral synaptic weight strengths were relatively stable and inhibitory weights formed diffuse fluctuating clusters. Averaging over many input patterns resulted in strong excitatory connections between individual cortical neurons only, whereby connections to other cells were very weak or negligible (Fig. 2(a, b)). As a result no mexican hat type lateral coupling structure appeared as in the case of random input patterns [1]. Moreover most of the excitatory connections now were reciprocal, but any distinct specificity with regard to the orientation preference and selectivity of the coupled neurons was still absent.

Using equal radii for both inhibitory and excitatory interactions led to monocular $(0,0)$ receptive field profiles and binocular bilobed $(0,1)$-like receptive fields, too. The binocular orientation map now contained $\pm 1/2$ vortices exclusively, which, furthermore, occurred in pairs of opposing polarity (Fig. 1(b)).

The lateral coupling structure was characterized by stable clusters of excitatory and inhibitory connections (Fig. 2(c)). The lateral connections were reciprocal and self-couplings were generally excitatory. Neurons with either equal or orthogonal orientation preferences were connected by inhibitory lateral couplings in most cases. On the other hand, the majority of cells with a nearly 45° difference in their orientation preference were linked with excitatory couplings. The individual lateral coupling structures of any two cortical neurons were identical, if they were connected excitatorily, and were complementary, if they were connected inhibitorily. Further these corresponding lateral coupling structures were shifted according to the different topographic locations of both cortical neurons.

Until now natural images only, which do not contain any man-made objects, have been used as visual stimuli. In contrast, now an image from an urban environment was used, which included many sharp edges and straight lines oriented horizontally and vertically in most cases. This led to an emergence of unstructured $(0,0)$-like as well as bilobed $(0,1)$-like monocular and binocular receptive fields. All cortical cells exhibited orientation preferences of 90° only corresponding to the predominance of vertical contours in the input images. The lateral couplings were structured in the same way as before.

3.1 Sequential combination of random and structured input pattern phases

The simulations started with random input patterns and after corresponding structures have stabilized structured input patterns were presented to the input layers of the model network. Again $r_{LE} < r_{LI}$ and$\Sigma_E = \Sigma_I$ were used. During the postnatal developmental phase a horizontal shift $\Delta_x = 10$ of both retinal images was employed.

The prenatal development led to bilobed receptive fields and realistic orientation maps [1]. During the postnatal development a reorganization of the receptive fields started immediately. The monocular bilobed $(0, 1)$ profiles rearranged into deformed $(0, 0)$ structures. However, most of the binocular receptive fields remained orientation selective and still formed $(0, 1)$-like profiles. These orientation specific cortical cells were organized into a binocular orientation map containing $\pm 1/2$ vortices exclusively.

Furthermore strong selective excitatory lateral connections disappeared and inhibitory lateral synaptic weights formed extended fluctuating clusters. Averaging over many input patterns the v_{EI} couplings corresponded to mexican-hat-like structures approximately, even though these structures were less well developed than in case of random input patterns only.

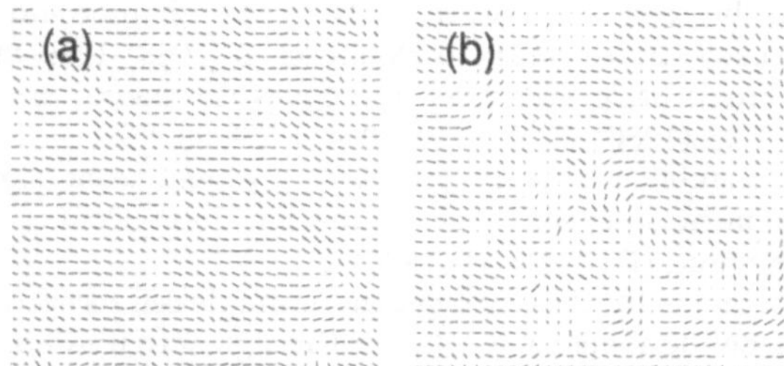

Fig. 1. Binocular orientation maps using either (a) $r_{LE} < r_{LI}$ or (b) $r_{LE} = r_{LI}$ and natural images . Each line in the map belongs to a cortical neuron. The line orientation denotes the direction of spatial frequency vector of that sinus wave, which causes the maximum response, and the line length denotes the extent of selectivity (for details see [14]. Note, that the vector direction is perpendicular to the optimal stimulus orientation.

References

1. Burger, T., Kussinger, M., Ziegaus, C., Lang, E. W.: Emergence of orientation maps in area 17 of the cerebral cortex: A correlation-based model with afferent and lateral plasticity of synaptical weights and real input patterns. Proceedings of the 25th Göttingen Neurobiology Conference 1997, Ed. Elsner, N., Wässle, H. (Stuttgart, Germany: Georg Thieme Verlag) (1997)

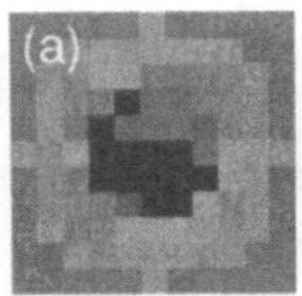

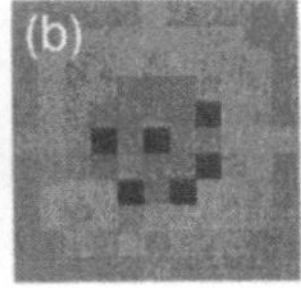

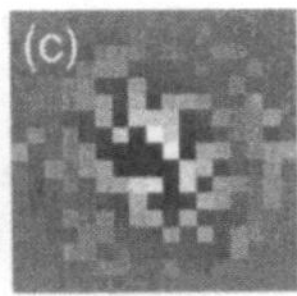

Fig. 2. Lateral coupling structures of individual cortical cells obtained by using natural images and (a), (b) $r_{LE} < r_{LI}$ or (c) $r_{LE} = r_{LI}$ and averaging over many input patterns. White denotes the strongest inhibitory, black the strongest excitatory formal synaptic weights v_{EI}. A vanishing lateral synaptic weight is denoted by a intermediate grey value. The central pixel grey value denotes the self-coupling of the neuron considered.

2. Callaway, E. M., Katz, L. C.: Emergence and refinement of clustered horizontal connections in cat striate cortex. J. Neurosci. **10** (1990) 1134–53
3. Gilbert, C. D., Hirsch, J. A., Wiesel T. N.: 1990 Lateral interactions in visual cortex. Cold Spring Harbor Symposia on Quantitative Biology **55** 663–77
4. Gilbert, C. D., Wiesel, T. N.: Columnar specificity of intrinsic horizontal and corticocortical connections in cat visual cortex. J. Neurosci. **9** (1989) 2432–42
5. Katz, L. C., Callaway, E. M.: Development of local circuits in mammalian visual cortex. Ann. Rev. Neurosci. **15** (1992) 31–56
6. Katz, L. C., Gilbert, C. D., Wiesel, T. N.: Local circuits and ocular dominance columns in monkey striate cortex. J. Neurosci. **9** (1989) 1389–99
7. Linsker, R.: From basic networks principles to neural architecture (series). Proc. Natl. Sci. **83** (1986) 7508–12, 8390–4, 8779–83
8. Linsker, R.: Designing a sensory processing system: what can be learned from principal component analysis? Proc. of the Int. Joint Conf. on Neural Networks (IJCNN, Washington (DC), USA) (1990)
9. Miller, K. D.: A model for the development of simple cell receptive fields and the ordered arrangement of orientation columns through activity-dependent competition between ON- and OFF-center inputs. J. Neurosci. **14** (1994) 409–41
10. Olshausen, B. A., Field, D. J.: Emergence of simple-cell receptive field properties by learning a sparse code for natural images. Nature **381** (1996) 607–609
11. Sirosh, J., Miikkulainen, R.: Cooperative self-organization of afferent and lateral connections in cortical maps. Biol. Cybern. **71** (1994) 65–78
12. Sirosh, J., Miikkulainen, R.: Topographic receptive fields and patterned lateral interaction in a self-organization model of the primary visual cortex. Neural Computation **9** (1997) 577–94
13. Stetter, M., Lang, E. W., Müller, A.: Emergence of orientation selective simple cells simulated in deterministic and stochastic neural networks. Biol. Cybern. **68** (1993) 465–476
14. Stetter, M., Müller, A., Lang, E. W.: Neural network model for the coordinated formation of orientation preference and orientation selectivity maps. Phys. Rev. E **50** (1994) 4167–81
15. Stetter, M., Kussinger, M., Schels, A., Seeger E., Lang, E. W.: Self-organization of cortical receptive fields and columnar structures in a Hebb-trained neural network. Lecture Notes in Computer Science **930** (1995) 37–44
16. Yuille, A. L., Kammen, D. M., Cohen, D. S.: Quadrature and the development of orientation selective cortical cells by Hebb rules. Biol. Cybern. **61** (1989) 183–94

Geometry of Orientation Preference Map Determines Nonclassical Receptive Field Properties

U. Ernst, K. Pawelzik, F. Wolf, and T. Geisel

MPI für Strömungsforschung
D-37073 Göttingen

Abstract. We propose a simple mechanism for the nonclassical receptive field property of orientation contrast sensitivity. Our model includes the long-range lateral connections linking cell populations of similar orientation preference, and the dynamics of local microcircuits which introduce a differential interaction whose sign depends on the post- and presynaptic activation. We demonstrate that the geometry of an orientation preference map determines the positions of cells sensitive for orientation contrasts, and we propose a simple statistical method to check the predictions of our model for experimentally given maps.

1 Introduction

Recently Sillito and coworkers [1] demonstrated that stimulation beyond the classical receptive field can not only modulate, but radically change a neuron's response to oriented stimuli. They revealed that patch-suppressed cells when stimulated with contrasting orientations inside and outside their classical receptive field can strongly respond to stimuli oriented orthogonal to their nominal preferred orientation.
Here we analyze the emergence of such complex response patterns in a simple model of primary visual cortex. We show that the observed sensitivity for orientation contrast can be explained by the differential interaction between the local lateral microcircuitry and the long-range lateral connections between iso-oriented domains. The simplicity of our model allows for a rigorous analytical investigation of the mechanisms underlying cross-orientation enhancement in a one-dimensional model. In particular we demonstrate that the observed properties might arise without specific connections between sites with cross-oriented classical receptive fields.
To be more realistic, we also simulate the visual cortex as a two-dimensional rather than a one-dimensional system, thus gaining an additional degree of freedom. Surprisingly, the existence of contrast-sensitive cells seems now to be restricted to specific regions in this artificial cortex. We interpret this phenomenon in terms of a conflict between the pattern formation process of the neuronal dynamics and the boundary conditions induced by the orientation preference of each cell population. We show that the formation of orientation contrast sensitive cells depends on the existence of large patches of cells having the same

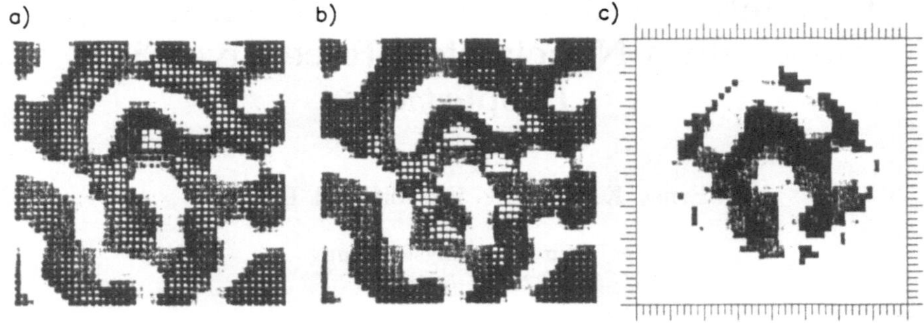

Fig. 1. Structure and response properties of the model network. a) Coupling structure from one neuron on a grid of $N = 1600$ elements projected on the orientation preference map which was used for stimulation. Inhibitory and excitatory couplings are marked with black and white squares, respectively, the sizes of which represent the coupling strength. b) Activation patterns of the network driven by a central stimulus of radius $r_c = 11$ and horizontal orientation. c) Self-consistent orientation map calculated from the activation patterns for all stimulus orientations. Note that this map matches the input orientation preference map shown in a) and b).

orientation preference. Regions with a broad distribution of orientations (e.g. pinwheels) are not suitable for the development of contrast-sensitive cells. These inhomogeneities also lead to the emergence of direction selective cells under some stimulation conditions, where the dynamics shows strong effects of hysteresis.

2 Model

Abstracting from biophysical details, our cortex model consists of a layer of $N \cdot N$ rate coded units, each of them representing a pool of biological neurons (Fig.1). The stimulus consists of a center patch and a surround annulus of oriented moving gratings. Input from the stimulus is projected onto the cortex via the LGN. The divergence of inputs from one specific location on the retina, and the orientation tuning of one cortical unit, are modeled by a convolution with Gaussian functions in cortical and orientation space, respectively. The lateral interaction in the cortical layer has two contributions. Short-range connections with typical length scales less than the width of one hypercolumn are Mexican-hat shaped and homogeneous in orientation space. Long-range connections were established preferentially between cells of similar orientation.
The long-range coupling strength ω_l depends on the postsynaptic and presynaptic activation. ω_l is positive, if either the postsynaptic activation or presynaptic activation is low, and negative otherwise

$$\omega_l \approx c_1 - tanh(0.55 * (c_2 - I)/c_3) \tag{1}$$

with varying constants c_1, c_2, and c_3.

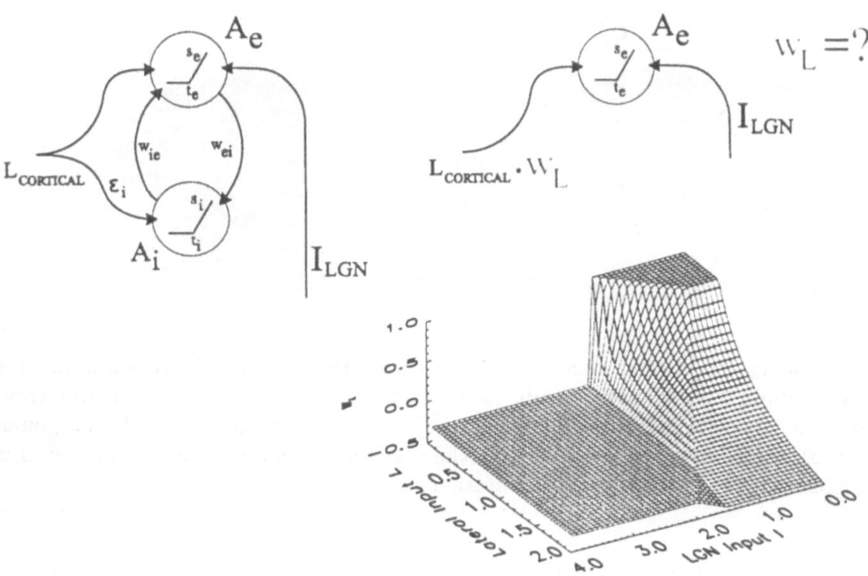

Fig. 2. Cortical microcircuit (upper left) and simplified cortical circuit (upper right). The behaviour of inhibitory interneurons and disynaptic connections shown left is modeled by a differential coupling weight ω_l in the simplified circuit shown right. ω_l can be extracted analytically (lower). If the input I from the LGN and the input L from the lateral connections is low, $\omega_l > 0$. If either I or L is large, $\omega_l < 0$. The shape of ω_l for constant L or I matches the shape of ω_l chosen heuristically in Eq.1.

This differential interaction allows us to model the effects of inhibitory interneurons or other disynaptic cells by modulating the synaptic strength, while using only a single rate-coded unit at each cortical position (see Fig.2, upper). ω_l can be derived analytically from a cortical microcircuit where an inhibitory and an excitatory cell population are connected reciprocally, both of them receiving input from the LGN and from lateral cortical connections. If the inhibitory unit has a higher gain than the excitatory unit, ω_l appears to have qualtitatively the same shape as mentioned above (Fig.2, lower). This differential interaction, whose sign changes with the degree of postsynaptic and/or presynaptic activation, yields the key for the understanding of the mechanisms for surround-suppression and orientation contrast enhancement.

3 Results

We assumed a rather weak selectivity of the afferent connections and a restricted contrast, which implies that every stimulus provides some input also to orthogonally tuned cells. This means that long-range excitatory connections, while not effective when only the surround is stimulated, can very well be sufficient for driving cells if the stimulus to the center is orthogonal to their preferred orientation (Contrast enhancement, positive differential interaction $\omega_l > 0$). Similarily, the center being stimulated with the preferred orientation, sub-threshold input

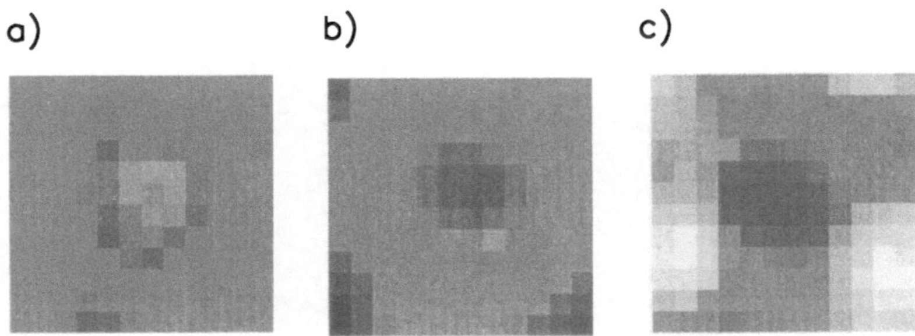

Fig. 3. Changes in pattern of activity induced by the additional presentation of a surround stimulus. Grey levels encode increase (darker grey) or decrease (lighter grey) in activation. a) center and surround parallel, b) and c) center and surround orthogonal. While in b), the center is stimulated with the preferred orientation, in c), the center is stimulated with the non-preferred orientation.

from outside the classical receptive field can also enhance the response to an orthgonal surround. But as soon as a **parallel** surround is presented, long-range lateral connections act inhibitory on the center unit ($\omega_l < 0$, negative differential interaction), leading to a strong suppression of the cell's response (Patch suppression)(Fig.3).
These results show a surprising agreement with previous findings on non-classical receptive field properties which culminated in the paper by Sillito et al. [1], and they clearly demonstrate that the known intracortical interactions might lead to surprising effects on receptive fields.

The mexican-hat shaped coupling structure induces blob-like patches of firing activity. Two blobs can not coexist within a radius being roughly of the same size as the inhibitory surround of one cell population. In one spatial dimension, this interaction leads to a strong competition between blobs, while in two spatial dimensions, the blobs have a much higher ability to shift away from each other. To imagine the consequences of the geometry, suppose the presentation of a stimulus with horizontal center and vertical surround. Depending on the geometry of the orientation preference map, we can have one of three different cell configurations (Fig.4). The center consists either of a majority of 'horizontal' cells (A), a majority of 'vertical' cells (B), or a more or less homogeneous mixture of both and other cell types (C). Only in (B), the 'vertical' cells will be activated by additional input from long-range interactions, responding to the non-preferred orientation. In (A) and (C), the total input to 'vertical' cells is also above threshold, but the response is suppressed by the much higher activation of 'horizontal' cells which can be found nearby.
To quantify the inhomogeneity of the map's geometry, we introduce the function $H(\mathbf{r}_0, \Phi_0)$ which is 1, if all cortical cells at their positions $\mathbf{r}$ have the same orientation $\Phi(\mathbf{r}) = \Phi_0$, and -1, if all cells in the cortex $\mathbf{C}$ have the opposite orientation $\Phi = \Phi_0 + \pi/2$ of the cell at position $\mathbf{r}_0$

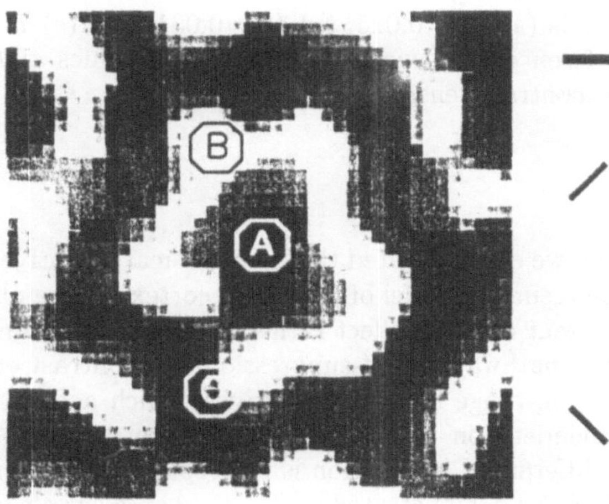

Fig. 4. Orientation preference map. Depending on the exact position, we sometimes find locally a majority of 'horizontal' cells (A), 'vertical' cells (B) or a mixture of all possible orientations (C, typically at pinwheel positions). The existence of orientation contrast sensitive cells is strongly affected by the geometry of the orientation map.

$$H(\mathbf{r_0}, \Phi_0) = \left(\frac{1}{\sqrt{2\pi\sigma_r^2}}\right)^d \int_{\mathbf{C}} \exp\left(-\frac{|\,\mathbf{r}-\mathbf{r_0}\,|^2}{2\sigma_r}\right) \cdot cos(2 \cdot (\Phi(\mathbf{r}) - \Phi_0)) \cdot d\mathbf{r} \quad . \quad (2)$$

$d = 2$ denotes the dimension of the cortical surface, and σ_r controls the width of the spatial averaging. For a finite number of neurons, H can also be written as a sum. For one-dimensional cortices with regular columnar structure, $\Phi(\mathbf{r}) = c \mid \mathbf{r} \mid$ and $H = const.$ (translational invariance), but in real cortical maps, Φ does not depend linearly on $\mid \mathbf{r} \mid$, and differs from -1 to +1. From the statistical properties of orientation preference maps, we are able to give an approximation for the probability density $\rho(\Delta\Phi \mid \Delta r)$ of the orientation difference $\Delta\Phi$ of two cells, given their distance Δr

$$\rho(\Delta\Phi \mid \Delta r) \approx \frac{1}{\sqrt{2\pi\sigma(\Delta r)^2}} \sum_{n=-\infty}^{\infty} exp\left(-\frac{(\Delta r - \mid R(\Delta\Phi, n) \mid)^2}{2\sigma(\Delta r)^2}\right) \quad , \quad (3)$$

where $\sigma(r)$ is a monotonically increasing function of r, and

$$R(\Delta\Phi, n) = \lambda \cdot \mid n + \frac{\Delta\Phi}{\pi} \mid \quad (4)$$

is the average distance to cells with orientation difference $\Delta\Phi$ in the n-th next orientation column (λ is the average orientation column width).
From our simulations, we calculated the average for H for all cells showing (a) patch suppression, (b) patch-suppression and cross-orientation enhancement, and (c) patch-suppression and **no** cross-orientation enhancement. For e.g. $\sigma_r =$

5.0 the analysis yields (a) $H = 0.025$, (b) $H = 0.031$, and (c) $H = 0.019$. Given an orientation preference map, we assume that high values of H signal a high probability to find contrast sensitive cells.

4 Summary

In this contribution, we demonstrated that nonclassical receptive field properties can be explained in a simple model of the visual cortex. These phenomenons appear to be an emergent network effect rather than a special connection scheme in the afferent input pathway. Patch-suppression and contrast enhancement are consequences of a long-range lateral connectivity which acts either excitatorily or inhibitorily, dependent on the degree of activation of the post- and presynaptic cells. This differential interaction is the typical behaviour of the cortical microcircuitry, which we demonstrated analytically on a simple cortical subunit. As can be seen in our simulations, the geometry of orientation preference maps puts some restrictions on the existence of cells sensitive for cross-oriented stimuli. We expect that the statistical measure H provides a tool to detect locations, where such cells may be found with a high probability. At the same time, this framework may offer a test if the mechanisms revealed in the simulations are consistent with the biological nature of nonclassical receptive fields.

References

1. A.M. Sillito, K.L. Grieve, H.E. Jones, J. Cudeiro, and J. Davis, *Visual cortical mechanisms detecting focal orientation discontinuities*, Nature **378**, 492-496 (1995).
2. D. Ts'o, C.D. Gilbert, and T.N. Wiesel, J. Neurosci **6**, 1160-1170 (1986).
3. C.D. Gilbert and T.N. Wiesel, J. Neurosci. **9**, 2432-2442 (1989).
4. S. Löwel and W. Singer, Science **255**, 209 (1992).
5. R. Malach, Y. Amir, M. Harel, and A. Grinvald, PNAS **90**, 10469-10473 (1993).
6. J.A. Hirsch and C.D. Gilbert, J. Neurosci. **6**, 1800-1809 (1991).
7. M. Weliky, K. Kandler, D. Fitzpatrick, and L.C. Katz, Neuron **15**, 541-552 (1995).
8. M. Stemmler, M. Usher, and E. Niebur, Science **269**, 1877-1880 (1995).
9. L.J. Toth, D.C. Somers, S.C. Rao, E.V. Todorov, D.-S. Kim, S.B. Nelson, A.G. Siapas, and M. Sur, preprint 1995.
10. U. Polat, D. Sagi, Vision Res. **7**, 993-999 (1993).
11. J.J. Knierim and D.C. van Essen, J. Neurophys. **67**, 961-980 (1992).
12. H.R. Wilson and J. Cowan, Biol. Cyb. **13**, 55-80 (1973).
13. R. Ben-Yishai, R.L. Bar-Or, and H. Sompolinsky, Proc. Nat. Acad. Sci. **92**, 3844-3848 (1995).
14. D. Somers, S.B. Nelson, and M. Sur, J. Neurosci. **15**, 5448-5465 (1995).

A Model for Orientation Tuning and Contextual Effects of Orientation Selective Receptive Fields

Hauke Bartsch, Martin Stetter, and Klaus Obermayer

Dept. of Computer Science, Technische Universität Berlin, Germany,
E-mail: hauke@cs.tu-berlin.de

Abstract. We investigate a mean-field model which has previously been used to explain the response properties of orientation selective neurons in the primary visual cortex of monkeys and cats [2]. Two mutually coupled orientation hypercolumns are setup as local amplifiers based on local recurrent excitation and inhibition. We first investigate the individual hypercolumns. The model correctly predicts contrast invariant tuning, but analytical and numerical results show that the contrast response functions of individual orientation columns do not saturate. We therefore hypothesize that the cortical saturation effects found experimentally may be a consequence of the non-linear properties of single neurons rather than being an effect of different gains for inhibitory and excitatory cells [13]. We then extend this model to cover non-classical receptive fields and contextual effects. The model correctly predicts effective iso-orientation inhibition between hypercolumns. As long as parameters are chosen to ensure contrast invariant orientation tuning, however, net cross-orientation facilitation emerges only, if cells of different orientation preference are connected across hypercolumns. These results hint at deficiencies of this simple approach and suggest that contextual effects are mediated by populations of neurons, which are not take part of the local gain control.

1 Introduction

In the primary visual cortex of monkeys and cats, orientation selective neurons show tuning curves, whose widths are roughly independent of contrast [11,12]. Their contrast response curves (CRF) saturate well below their maximum possible firing rate and the CRF of cells stimulated with optimally oriented bars and gratings is suppressed by additional nonoptimal stimuli within its classical receptive field [4]. The orientation tuning curves are modulated by stimuli outside the classical receptive field (referred to as surround stimuli), which generate iso-orientation inhibition and cross-orientation facilitation for all center orientations [3,10] whose magnitude is dependent on contrast. Also shifts of orientation tuning curves as a result of surround stimulation have been reported [5].

Several models have been setup to explore subgroups of these phenomena [9,8]. Ben-Yishai et al. [2] suggested a mean-field model for orientation selectivity in the primary visual cortex, which is based on within-hypercolumn recurrent excitation local in orientation space and within-hypercolumn inhibition which is less specific. They showed that the model exhibits a "marginal phase", for which cortical orientation tuning is independent of stimulus contrast and of the magnitude of the afferent orientation bias. Motivated by their results we explore

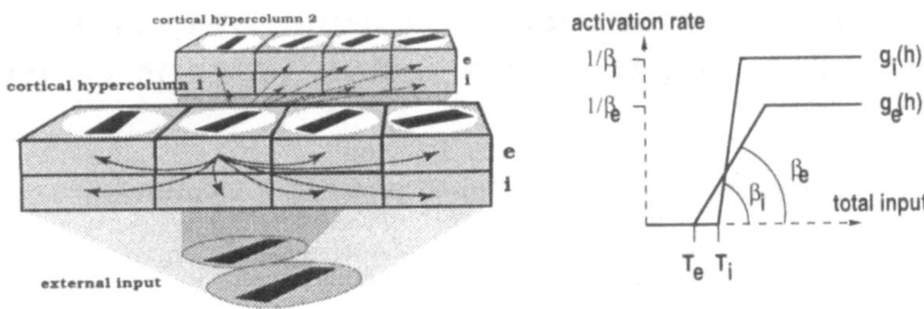

Fig. 1. (Left) Cartoon of the computational model. e and i denote excitatory and inhibitory cells. Only 4 out of 50 "orientation columns" are shown. **(Right)** Transfer functions for the excitatory,$g_e(h)$, and inhibitory, $g_i(h)$, cells.

their ansatz in more detail in order to provide evidence against or in favor of the idea of the "marginal phase" and in order to relate contextual effects to intracortical wiring patterns. We first analytically and numerically investigate the tuning properties of single hypercolumns. We then extend the model to two mutually coupled hypercolumns in order to compare prediction with experimental data about contextual effects. We conclude, that the assumptions underlying the model are not sufficient and suggest (i) that non-linear properties of single neurons are important and (ii) that at least some contextual effects are mediated by neurons which are not part of local gain control.

2 The Model

The model consists of two coupled cortical hypercolumns $a = 1, 2$ with excitatory and inhibitory neurons $\alpha = e, i$ (Fig. 1, left). Neurons are grouped into 50 "orientation columns" which receive orientationally biased input from the LGN and which are characterized by the corresponding preferred orientation $\theta \in [-\pi/2, \pi/2]$. Neurons are binary stochastic units with activation rates $g_\alpha(h)$, h being their total synaptic input after stimulation with external orientations θ^a,

$$h_\alpha^a(\theta,t) = \frac{1}{\pi} \sum_{b=1,2} \sum_{\beta=e,i} \int_{-\pi/2}^{\pi/2} K_{\alpha,\beta}^{a,b}(\theta-\theta')\, m_\beta^b(\theta',t)\, d\theta' + h_{ext}(\theta-\theta^a), \quad (1)$$

where $m_\alpha^a(\theta,t)$ is the average activation of population (a, α, θ) at time t. Input h is composed of an intracortical component from all other neurons mediated by sixteen interaction functions $K_{\alpha,\beta}^{a,b}(\theta-\theta')$, and two external orientation biased inputs, $h_{ext}^a = h_{ext}(\theta-\theta^a)$, which are roughly proportional to the logarithm of stimulus contrast [1]. Following Ben-Yishai et al. [2] the dynamics of the average activations is given by

$$\tau \frac{d}{dt} m_\alpha^a(\theta,t) = -m_\alpha^a(\theta,t) + g_\alpha(h_\alpha^a(\theta,t)), \quad (2)$$

where $g_\alpha(h)$ are rectifying, piecewise linear functions with thresholds T_α and slopes β_α (Fig. 1, right).

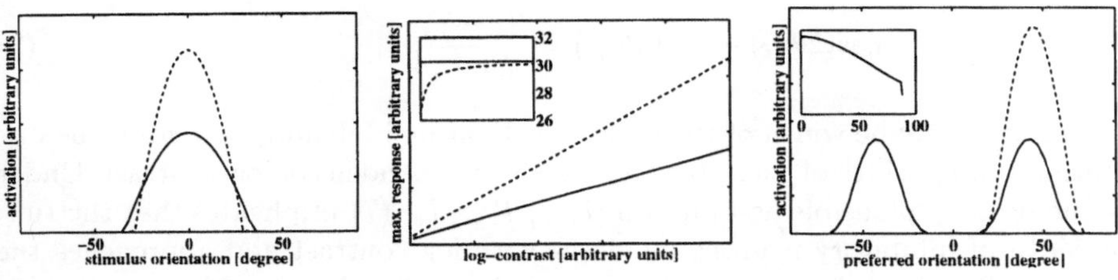

Fig. 2. Results for a single hypercolumn. **(Left)** Orientation tuning curves for excitatory (solid) and inhibitory (dashed) neurons (numerical results). **(Center)** Maximum response and orientation tuning widths (inset) as a function of log-contrast for excitatory (solid) and inhibitory (dashed) neurons (analytical results). Parameters were $E = 10, \beta_e = T_e = 1, \beta_i = T_i = 2$. **(Right)** Numerical results for cortical activity patterns for an oriented stimulus (dashed) and for an additionally superimposed cross-oriented stimulus (solid). The inset shows peak activity as a function of orientation difference between the superimposed stimuli. Activation patterns were broad gaussian profiles with widths $\pi/6$. Parameters were: $E_2 = 10,\ I_0 = 3$.

3 Results

3.1 Analytical Results for a Single Hypercolumn

In order to simplify calculations we parametrize the external input by $h_{ext}(\theta) = c(1 - \varepsilon + \varepsilon\cos(2\theta))$, where c is the log-contrast of a grating or bar stimulus and ε, $0 \leq \varepsilon \leq 0.5$, is the orientation bias, and the intra-hypercolumn interactions K,$K_{e,e}(\theta) = K_{i,e}(\theta) \equiv E(\theta) = E_0 + E_2\cos(2\theta), K_{e,i}(\theta) = K_{i,i}(\theta) \equiv I(\theta) = -I_0 - I_2\cos(2\theta)$ by their zeroth and second fourier coefficients $0 \leq E_2 \leq E_0$ and $0 \leq I_2 \leq I_0$. Setting $\theta^a = 0$, the stationary solutions $m_\alpha(\theta)$ of (1, 2) are fully described by their zeroth and second fourier components $m_{0,\alpha}$ and $m_{2,\alpha}$. In the regime of weak afferent orientation bias, $\varepsilon \ll 1$, and strong intracortical interactions, $\beta_e E_2 - \beta_i I_2 > 2$, we obtain

$$m_{0,\alpha} = \frac{\beta_\alpha}{\pi}(E_2 m_{2,e} - I_2 m_{2,i})(\sin(2\theta_{c,\alpha}) - 2\theta_{c,\alpha}\cos(2\theta_{c,\alpha})) \tag{3}$$

$$m_{2,\alpha} = \frac{\beta_\alpha}{\pi}(E_2 m_{2,e} - I_2 m_{2,i})(\theta_{c,\alpha} - \sin(4\theta_{c,\alpha})/4), \tag{4}$$

where $\theta_{c,\alpha}$ $(\alpha = e, i)$,

$$\cos(2\theta_{c,\alpha}) = \frac{T_\alpha - c - E_0 m_{0,e} + I_0 m_{0,i}}{E_2 m_{2,e} - I_2 m_{2,i}}, \tag{5}$$

are the widths of the activation blobs in orientation space for the cell populations α. Equations (4, 5) are defined for spatially modulated solutions only.

For $E_0 = I_0 = E_2 = E$ and $I_2 = 0$, i.e. Iso-orientation excitation and unspecific inhibition, (4, 5) yield

$$\theta_{c,e} - \frac{1}{4}\sin(4\theta_{c,e}) = \frac{\pi}{\beta_e E}, \qquad \beta_e E > 2 \tag{6}$$

$$\cos(2\theta_{c,i}) - \cos(2\theta_{c,e}) = \frac{T_i - T_e}{E m_{2,e}}, \tag{7}$$

which show that the width of the cortical activation blob and, therefore, the orientation tuning width of excitatory neurons is independent of log-contrast. Under the biologically plausible assumption [7,13] $T_i > T_e$, (7) implicates that the tuning width of inhibitory neurons increases with log-contrast and approaches the tuning width of excitatory neurons from below (Fig. 2, left and center inset). Numerical simulations show that for $I_2 \neq 0$ both tuning widths depend on c, the magnitude of variations being related to I_2/E_2 but being less than 10%. Inserting $m_{0,e} - m_{0,i}$ from (3) into (4,5) we obtain for the second Fourier coefficient $m_{2,e}$,

$$m_{2,e} = (c - T_e)\left(\frac{E^2}{\pi}(\beta_i f(2\theta_{c,i}(c)) - \beta_e f(2\theta_{c,e})) - E\cos(2\theta_{c,e})\right)^{-1}. \tag{8}$$

Since $\theta_{c,i} \leq \theta_{c,e}$, $m_{2,e}$ as a function of c remains above a straight line with slope $((E^2 f(2\theta_{c,e})/\pi)(\beta_i - \beta_e) - E\cos(2\theta_{c,e}))^{-1}$ and the CRF does not saturate (Fig. 2, center). Consequently, the saturation of the CRF observed experimentally does not derive from intracortical interactions coupled with different contrast gains for inhibitory and excitatory neurons as has previously been suggested [12].

3.2 Numerical Results

In order to characterize the response of the model to two superimposed stimuli with different orientations presented within the classical receptive field, we applied a bimodal stimulus $h_{ext}(\theta) = c(\exp(-(\theta - \Delta\theta)^2/2(\pi/6)^2) + \exp(-(\theta + \Delta\theta)^2/2(\pi/6)^2))$ to a single hypercolumn. If excitatory connections connect cells with similar preferred orientation while inhibitory connections are unspecific, $E(\theta) = E_2\cos^{12}(\theta)$, $I(\theta) = I_0$, the model predicts cross-orientation suppression. Fig. 2 (right) shows the corresponding activation patterns for one oriented stimulus (dashed) and for an additionally superimposed cross-oriented stimulus (solid). In the latter case, the activity is reduced by about 50%. The inset shows maximum activity as a function of orientation difference between the superimposed stimuli. Activity decreases monotonically with increasing orientation difference which agrees with data from cells in cat area 17 [4]. We now consider two identical hypercolumns with local circuitry as described above which are coupled via long-range connections from excitatory cells in one hypercolumn onto excitatory and inhibitory targets in the other hypercolumn. This configuration serves as a model for contextual effects which arise when one hypercolumn is stimulated in its classical receptive field while the second hypercolumn processes a stimulus surrounding the classical receptive field of the first hypercolumn, influencing its activity via long-range connections. Under the assumption that lateral connections between hypercolumns are localized in orientation space [14], the model correctly predicts iso-orientation suppression (graph not shown), but fails to show cross-orientation facilitation which has been found in cats and monkeys [3,10]. This is a consequence of the "marginal phase", i.e. the parameter regime which ensures contrast invariant orientation tuning independent of

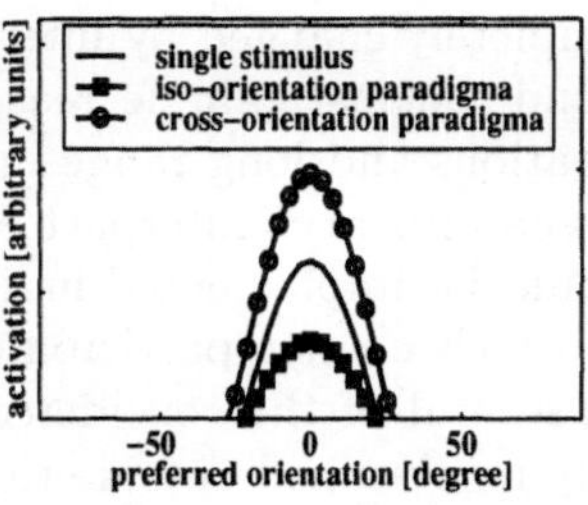

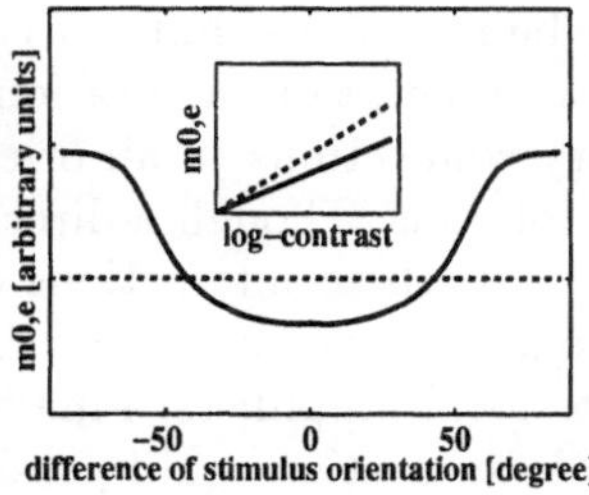

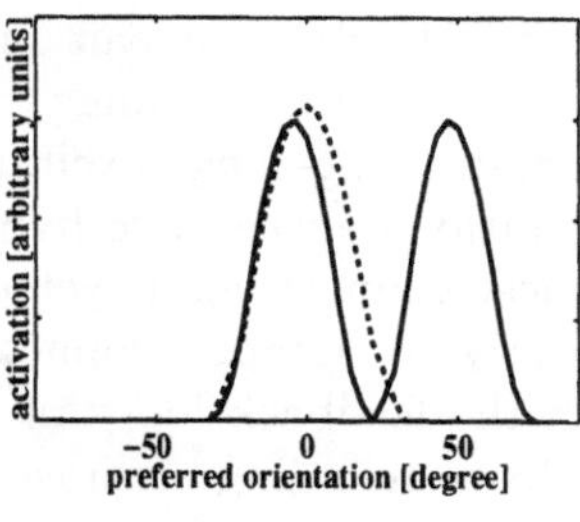

Fig. 3. (Left) Activation pattern of hypercolumn 1 stimulated with 0°, once with a silent hypercolumn 2 (solid), with an iso-oriented stimulation (squares) and with a cross-oriented stimulation (circles) of the second hypercolumn. **(Center)** Peak activation as a function of orientation difference between center and surround stimuli (solid). The dashed line indicates the activation level for stimulation of one hypercolumn only. Inset: Contrast-response functions for stimulation of hypercolumn 1 (dashed) and for iso-orientation stimulation of both hypercolumns (iso-orientation surround, solid). Parameters as in Fig. 2, and nonzero fourier components $K^{i,j\neq i}_{e,e,0} = K^{i,j\neq i}_{e,e,2} = 1, K^{i,j\neq i}_{i,e,2} = 6$, $j = 1, 2$. **(Right)** Activation with center stimulus alone (dashed) and with an additional surround stimulus with orientation difference 45° (solid). Parameters as before, but inclusion of long-range inhibition $K^{i,j\neq i}_{e,i,2} = 1, K^{i,j\neq i}_{i,i,2} = 6$, $j = 1, 2$.

the strength of the afferent bias: For the cross-orientation paradigm, active neurons in each hypercolumn project to neurons in the other hypercolumn which are silent and thus have no effect. Only when excitatory connections between hypercolumns are less specific within orientation space and also project to "cross-orientated" columns, the model predicts both iso-orientation suppression **and** cross-orientation facilitation (Fig. 3 left, center). Under those conditions, an iso-oriented stimulus in the surround also decreases the gain of the contrast response function, a result which has recently been reported from macaque striate cortex [6]. If taken literally, however, the necessary circuitry contradicts the anatomical data on stepped connections [14]. At intermediate orientation differences, activation blobs attract each other by a small amount for purely excitatory couplings between hypercolumns but repel each other if inhibitory connections between hypercolumns with similar angular dependence are present (Fig. 3, right), the size of the effect increasing with the strength of the coupling.

4 Discussion

We have investigated, the emergence of classical receptive field properties and contextual effects from different recurrent connection paradigms in a model of mutually coupled hypercolumns. For strong local interactions, i.e. in the "marginal phase", the model correctly predicts contrast-insensitive orientation tuning widths, cross-orientation suppression within the classical receptive field, iso-orientation suppression from a surrounding stimulus and contrast gain reduction. In the marginal phase, however, non-saturating CRFs are an inherent feature. Since an alternative approach, which uses detailed compartmental model neurons [12], shows saturating CRFs, our model suggests nonlinear summation on single neurons to be the origin of saturation. Because in the marginal phase cortical activation can only be modulated but not completely changed by inter-hypercolumn couplings, cross-orientation facilitation and contrast gain decrease require long-range excitatory connections to all orientations and long range inhibition between two hypercolumns. Though a direct anatomical counterpart of these connections is yet unclear, these interactions could be implemented indirectly by stepped connections to local interneurons, which do not participate in the feedback loops of the local amplifiers. Finally, several of the considered effects including contrast gain saturation could possibly be obtained outside the marginal phase. Since this is at the expense of contrast-independent orientation tuning, no single wiring paradigm seems to be suitable to account for all observed effects. It seems therefore necessary to consider more than one type of mutually coupled local circuits in future studies.

Acknowledgements: Supported by DFG (Ob 102/2-1) and HFSPO (RG-98/94).

References

1. D. G. Albrecht and D. B. Hamilton. Striate cortex of monkey and cat: contrast response function. *J. Neurophysiol.*, 48:217–237, 1982.
2. R. Ben-Yishai, R. Lev Bar-Or, and H. Sompolinski. Theory of orientation tuning in visual cortex. *Proc. Natl. Acad. Sci. USA*, 92:3844–3848, 1995.
3. C. Blakemore and E. A. Tobin. Lateral inhibition between orientation detectors in the cat's visual cortex. *Exp. Brain Res.*, 15:439–440, 1972.
4. G. C. DeAngelis, J. G. Robson, I. Ohzawa, and R. D. Freeman. Organization of suppression in receptive fields of neurons in cat visual cortex. *J. Neurophysiol.*, 68:144–163, 1992.
5. C. D. Gilbert and T. N. Wiesel. The influence of contextual stimuli on the orientation selectivity of cells in primary visual cortex of the cat. *Vision Res.*, 30:1689–1701, 1990.
6. J. B. Levitt and J. S. Lund. Contrast dependence of contexual effects in primate visual cortex. *Nature*, 387:73–76, 1997.
7. D. A. McCormick, B. W. Connors, J. E. Lighthall, and D. A. Prince. Comparative electrophysiology of pyramidal and sparsely spiny stellate neurons of the neocortex. *J. Neurophysiol.*, 54(4):782–806, 1985.
8. T. Mundel, A. Dimitrov, and J. Cowan. A simple model for cortical orientation selectivity. In G. Tesauro et al., editors, *NIPS.* MIT Press, 1996. in press.
9. K. Pawelzik, U. Ernst, F. Wolf, and T. Geisel. Orientation contrast sensitivity from long-range interactions in visual cortex. In G. Tesauro et al., editors, *NIPS.* MIT Press, 1996. in press.
10. A. M. Sillito, K. L. Grieve, H. E. Jones, J. Cudeiro, and J. Davis. Visual cortical mechanisms detecting focal discontinuities. *Nature*, 378:492–496, 1995.
11. B. C. Skottun, A. Bradley, G. Sclar, I. Ohzawa, and R. D. Freeman. The effects of contrast on visual orientation and spatial frequency discrimination: a comparison of single cells and behaviour. *J. Neurophysiol.*, 57:773–786, 1987.
12. D. C. Somers, S. B. Nelson, and M. Sur. An emergent model of orientation selectivity in cat visual cortical simple cells. *J. Neurosci.*, 15:5448–5465, 1995.
13. E. Todorov, A. Siapas, and D. Somers. A model of recurrent interactions in primary visual cortex. In T. Leen G. Tesauro, D. Touretzky, editor, *Advances in Neural Information Processing Systems 8.* MIT Press Cambridge, Massachusetts, 1996.
14. T. Yoshioka, G. G. Blasdel, J. B. Levitt, and J. S. Lund. Relation between patterns of intrinsic lateral connectivity, ocular dominance and cytochrome oxidase-reactive regions in macaque monkey striate cortex. *Cereb. Cortex*, page in press., 1997.

Objective Functions for Neural Map Formation

Laurenz Wiskott[1] and Terrence Sejnowski[123]

[1]Computational Neurobiology Laboratory
[2]Howard Hughes Medical Institute
The Salk Institute for Biological Studies, San Diego, CA 92186-5800
{wiskott,terry}@salk.edu, http://www.cnl.salk.edu/CNL

[3]Department of Biology
University of California, San Diego
La Jolla, CA 92093

Abstract

A unifying framework for analyzing models of neural map formation is presented based on growth rules derived from objective functions and normalization rules derived from constraint functions. Coordinate transformations play an important role in deriving various rules from the same function. Ten different models from the literature are classified within the objective function framework presented here. Though models may look different, they may actually be equivalent in terms of their stable solutions. The techniques used in this analysis may also be useful in investigating other types of neural dynamics.

1 Introduction

Computational models of neural map formation can be considered on at least three different levels of abstraction: detailed neural dynamics, abstract weight dynamics, and objective functions from which weight dynamics may be derived as gradient flows. Objective functions provide many advantages in analyzing systems analytically and in finding stable solutions by numerical simulations. The goal here is to provide a unifying objective function framework for a wide variety of models and to provide means by which analysis becomes easier.

2 Correlations

The architecture considered here consists of an input layer all-to-all connected to an output layer without feed-back connections. Input neurons are indicated by ρ (retina), and output neurons by τ (tectum). The dynamics in the input layer is described by neural activities a_ρ, which yield mean activities $\langle a_\rho \rangle$ and correlations $\langle a_\rho, a_{\rho'} \rangle$. Assume these activities propagate in a linear fashion through feed-forward connections $w_{\tau\rho}$ from input to output neurons and *effective* lateral connections $D_{\tau\tau'}$ among output neurons. $D_{\tau\tau'}$ is assumed to be symmetrical and represents functional aspects of the lateral connectivity rather than the connectivity itself. We also assume a linear correlation function $\langle a_{\rho'}, a_\rho \rangle$ and $\langle a_{\rho'} \rangle =$ constant. The activity of output neurons then is $a_\tau = \sum_{\tau'\rho'} D_{\tau\tau'} w_{\tau'\rho'} a_{\rho'}$. With $i = \{\rho, \tau\}$, $j = \{\rho', \tau'\}$, $D_{ij} = D_{ji} = D_{\tau\tau'} D_{\rho'\rho} = D_{\tau\tau'} \langle a_{\rho'}, a_\rho \rangle$, and

$A_{ij} = A_{ji} = D_{\tau\tau'}\langle a_{\rho'}\rangle$ we obtain mean activity and correlation

$$\langle a_\tau \rangle = \sum_{ij} A_{ij} w_j \, , \tag{1}$$

$$\langle a_\tau, a_\rho \rangle = \sum_{ij} D_{ij} w_j \, . \tag{2}$$

Since the right hand sides of Equations (1) and (2) are formally equivalent, we will discuss only the latter, which contains the former as a special case. This correlation model is accurate for linear models [e.g. 2, 5, 7, 8] and is an approximation for non-linear models [e.g. 3, 6, 10, 11, 12, 13].

3 Objective Functions

With Equation (2) a linear Hebbian growth rule can be written as $\dot{w}_i = \sum_j D_{ij} w_j$. This dynamics is curl-free, i.e. $\partial \dot{w}_i / \partial w_j = \partial \dot{w}_j / \partial w_i$, and thus can be generated as a gradient flow. A suitable objective function is $H(\mathbf{w}) = \frac{1}{2}\sum_{ij} w_i D_{ij} w_j$ since it yields $\dot{w}_i = \partial H(\mathbf{w})/\partial w_i$.

A dynamics that cannot be generated by an objective function directly is $\dot{w}_i = w_i \sum_j D_{ij} w_j$ [e.g. 5], because it is not curl-free. However, it is sometimes possible to convert a dynamics with curl into a curl-free dynamics by a coordinate transformation. Applying the transformation $w_i = \frac{1}{4}v_i^2$ yields $\dot{v}_i = \frac{1}{2} v_i \sum_j D_{ij} \frac{1}{4} v_j^2$, which is curl free and can be generated as a gradient flow. A suitable objective function is $H(\mathbf{v}) = \frac{1}{2}\sum_{ij} \frac{1}{4} v_i^2 D_{ij} \frac{1}{4} v_j^2$. Transforming the dynamics of $\mathbf{v}$ back into the original coordinate system, of course, yields the original dynamics for $\mathbf{w}$. Coordinate transformations thus can provide objective functions for dynamics that are not curl-free. Notice that $H(\mathbf{v})$ is the same objective function as $H(\mathbf{w})$ evaluated in a different coordinate system. Thus $H(\mathbf{v}) = H(\mathbf{w}(\mathbf{v}))$ and H is a Lyapunov function for both dynamics.

More generally, for an objective function H and a coordinate transformation $w_i = w_i(v_i)$

$$\dot{w}_i = \frac{\mathrm{d}}{\mathrm{d}t}\left[w_i(v_i)\right] = \frac{\mathrm{d}w_i}{\mathrm{d}v_i}\dot{v}_i = \frac{\mathrm{d}w_i}{\mathrm{d}v_i}\frac{\partial H}{\partial v_i} = \left(\frac{\mathrm{d}w_i}{\mathrm{d}v_i}\right)^2 \frac{\partial H}{\partial w_i} \, , \tag{3}$$

which implies that the coordinate transformation simply adds a factor $(\mathrm{d}w_i/\mathrm{d}v_i)^2$ to the original growth term obtained in the original coordinate system. Equation (3) shows that fixed points are preserved under the coordinate transformation in the region where $\mathrm{d}w_i/\mathrm{d}v_i$ is defined and finite but that additional fixed points may be introduced if $\mathrm{d}w_i/\mathrm{d}v_i = 0$.

Table 1 shows two objective functions and the corresponding induced dynamics terms they induce under different coordinate transformations. The first objective function, L, is linear in the weights and induces constant weight growth (or decay) under coordinate transformation $\mathcal{C}^1$. The growth of one weight does not depend on other weights. L can be used to differentially bias individual links, as required in dynamic link matching. The second objective function,

Q, is a quadratic form. The induced growth rule for one weight includes other weights and is usually based on correlations between input and output neurons $\langle a_\tau, a_\rho \rangle = \sum_j D_{ij} w_j$, in which case it induces topography. Q may also be induced by the mean activities of output neurons $\langle a_\tau \rangle = \sum_j A_{ij} w_j$.

		Coordinate Transformations	
		$\mathcal{C}^1$	$\mathcal{C}^w$
		$w_i = v_i$ $\left(\frac{dw_i}{dv_i}\right)^2 = 1$	$w_i = \frac{1}{4} v_i^2$ $\left(\frac{dw_i}{dv_i}\right)^2 = w_i$
Objective Functions $H(\mathbf{w})$		Growth Terms: $\dot{w}_i = ... + ...$	
L	$\sum_i \beta_i w_i$	β_i	$\beta_i w_i$
Q	$\frac{1}{2} \sum_{ij} w_i D_{ij} w_j$	$\sum_j D_{ij} w_j$	$w_i \sum_j D_{ij} w_j$
Constraint Functions $g(\mathbf{w})$		Normalization Rules: $w_i = ...$	$\forall i \in I_n$
$I_=$, $I_\geq$	$\theta_i - w_i$	θ_i	θ_i
$N_=$, $N_\geq$	$\theta_n - \sum_{j \in I_n} \beta_j w_j$	$\tilde{w}_i + \lambda_n \beta_i$	$\tilde{w}_i + \lambda_n \beta_i \tilde{w}_i$
$Z_=$, $Z_\geq$	$\theta_n - \sum_{j \in I_n} \beta_j w_j^2$	$\tilde{w}_i + \lambda_n \beta_i \tilde{w}_i$	$\tilde{w}_i + \lambda_n \beta_i \tilde{w}_i^2$

Table 1: Objective functions, constraint functions, and the dynamics terms they induce in two different coordinate systems $\mathcal{C}^1$ and $\mathcal{C}^w$. Specific terms are indicated by the symbols in the left column plus a superscript taken from the first row representing the coordinate transformation. For instance, the growth term $\beta_i w_i$ is indicated by L^w and the subtractive normalization rule $w_i = \tilde{w}_i + \lambda_n \beta_i$ is indicated by $N^1_=$ (or $N^1_\geq$). $N^w_=$ and $Z^1_=$ are multiplicative normalization rules. For the classifications in Table 2 this table has to be extended by two additional coordinate transformations and two other methods of deriving normalization rules from constraints, leading to terms such as $N^{\alpha w}_{\approx}$.

4 Constraints

A constraint is either an inequality describing a surface between valid and invalid region, e.g. $g(\mathbf{w}) = w_i \geq 0$, or an equality describing the valid region as a surface, e.g. $g(\mathbf{w}) = 1 - \sum_{j \in I} w_j = 0$. A normalization rule is a particular prescription for how the constraint has to be enforced. Thus constraints can be uniquely derived from normalization rules but not vice versa. Normalization rules can be *orthogonal* to the constraint surface or *non-orthogonal*. Only the orthogonal normalization rules are compatible with an objective function. A non-orthogonal normalization rule can lead to a combined dynamics (growth rule plus normalization rule) that decreases the objective function value, such that the objective function would not even be a Lyapunov function.

Reference		Classification			
Bienenstock & von der Malsburg	[2]		Q^1	$\mathrm{I}^1_{\geq}$	$\mathrm{N}^1_{\approx}$
Goodhill	[3]		Q^1	$\mathrm{I}^{\bar{1}}_{\geq}$	$\mathrm{N}^1_{=}\ \mathrm{N}^w_{=}$
Häussler & von der Malsburg	[5]	L^w	Q^w	$\mathrm{I}^w_{>}$	$\mathrm{N}^w_{\simeq}$
Konen & von der Malsburg	[6]		$\mathrm{Q}^{\alpha w}$		$\mathrm{N}^{\alpha w}_{=}$
Linsker	[7]	L^1	Q^1	$\mathrm{I}^1_{\geq}$	
Miller et al.	[8]		Q^α	$\mathrm{I}^\alpha_{\approx}\ \mathrm{I}^\alpha_{\geq}$	$\mathrm{N}^\alpha_{=}\ (\mathrm{N}^w_{=})$
Obermayer et al.	[10]		Q^1		$\mathrm{Z}^1_{=}$
Tanaka	[11]		Q^w	$\mathrm{I}^w_{>}$	$\mathrm{N}^{\alpha w}_{\approx}\ (= \mathrm{N}^w_{\approx})$
von der Malsburg	[12]		Q^1		$\mathrm{N}^w_{=}$
Whitelaw & Cowan	[13]		$\mathrm{Q}^1\ \mathrm{Q}^\alpha$		$\mathrm{N}^?_{=}$

Table 2: Classification of weight dynamics in previous models.

The method of Lagrangian multipliers can be used to derive orthogonal normalization rules from constraints. If the constraint $g(\mathbf{w}) \geq 0$ is violated for $\tilde{\mathbf{w}}$, the weight vector has to be corrected along the gradient of the constraint function g, which is orthogonal to the constraint surface, $w_i = \tilde{w}_i + \lambda \partial g / \partial \tilde{w}_i$. The Lagrangian multiplier λ is determined such that $g(\mathbf{w}) = 0$ is obtained. If no constraint is violated, the weights are simply taken to be $w_i = \tilde{w}_i$.

Consider the effect of a coordinate transformation $w_i = w_i(v_i)$. An orthogonal normalization rule can be derived from a constraint function $g(\mathbf{v})$ in a new coordinate system $\mathcal{V}$. If transformed back into the original coordinate system $\mathcal{W}$ one obtains an in general non-orthogonal normalization rule:

$$\text{if constraint is violated :} \qquad w_i = \tilde{w}_i + \lambda \left(\frac{\mathrm{d}w_i}{\mathrm{d}\tilde{v}_i} \right)^2 \frac{\partial g}{\partial \tilde{w}_i} \,. \tag{4}$$

This has an effect similar to the coordinate transformation in Equation (3). These normalization rules are indicated by a subscript $=$ (for an equality) or $\geq$ (for an inequality), because the constraints are enforced immediately and exactly. Table 1 shows several constraint functions and their corresponding normalization rules as derived in different coordinate systems by the method of Lagrangian multipliers. There are only two types of constraints. The first type is a *limitation constraint* I that limits the range of individual weights. The second type is a *normalization constraint* N that affects a group of weights, usually the sum, very rarely the sum of squares as indicated by Z.

5 Classification of Existing Models

Table 1 summarizes the different objective functions and derived growth terms as well as the constraint functions and derived normalization rules discussed in this paper. Since the dynamics needs to be curl-free and the normalization rules orthogonal in the same coordinate system, only entries in the same column may be combined to obtain a consistent objective function framework for a system. Classifications of ten different models are shown in Table 2. The models [2, 5, 6, 7, 10] can be directly classified under one coordinate transformation. The models [3, 8, 11, 12] can probably be made consistent with minor modifications. The applicability of our objective function framework to model [13] is unclear. Another model [1] is not listed because it can clearly not be described within our objective function framework.

6 Discussion

A unifying framework for analyzing models of neural map formation has been presented. Objective functions and constraints provide a formulation of the models as constraint optimization problems. From these, weight dynamics, i.e. growth rule and normalization rules, can be derived in a systematic way. Different coordinate transformations lead to different weight dynamics, which are closely related because they usually have the same set of stable solutions. We have analyzed ten different models from the literature and find that the typical system contains the quadratic term Q, a limitation constraint I, and a normalization constraint N (or Z).

In addition to the unifying formalism, the objective function framework provides deeper inside into several aspects of neural map formation:

• Functional aspects of the quadratic term Q can be easily analyzed. For instance, if $D_{\rho\rho'}$ is a negative constant and $D_{\tau\tau'}$ is a positive Gaussian and in combination with a positive linear term L, topography is ignored and the map is *expanding*, i.e. even without normalization rules, each output neuron eventually receives the same total sum of weights. More complicated effective lateral connectivities can be superimposed from simpler ones.

• Because of the possible expansion effect of L + Q it should be possible to define a model without any constraints.

• The same objective functions and constraints evaluated under different coordinate transformations provide different weight dynamics that may be equivalent with respect to the stable solutions they can converge to. This is because stable fixed points are preserved under coordinate transformations with finite derivatives.

• In [9] a clear distinction between multiplicative and subtractive normalization was made. However, the concept of equivalent models shows that normalization rules have to be judged in combination with growth rules, e.g. $N^w + I^w + Q^w$ (multiplicative normalization) is equivalent to $N^1 + I^1 + Q^1$ (subtractive normalization).

• Models of dynamic link matching [2, 6] introduced similarity values rather implicitly. A more direct formulation of dynamic link matching can be derived from the objective function L + Q.

• Objective functions provide a link between neural dynamics and algorithmic systems. For instance, the C-measure proposed in [4] as a unifying objective function for many different map formation algorithms is a one-to-one mapping version of the quadratic term Q.

The objective function framework provides a basis on which many models of neural map formation can be analyzed and understood in a unified fashion.

Furthermore, coordinate transformations as a tool to derive objective functions for dynamics with curl, to derive non-orthogonal normalization rules, and to unify a wide range of models might also be applicable to other types of models, such as unsupervised learning rules, and provide deeper insight there as well.

Acknowledgment

We are grateful to Geoffrey J. Goodhill, Thomas Maurer, and Jozsef Fiser for carefully reading the manuscript and useful comments. Laurenz Wiskott has been supported by a Feodor-Lynen fellowship by the Alexander von Humboldt-Foundation, Bonn, Germany.

References

[1] Amari, S. (1980). Topographic organization of nerve fields. *Bulletin of Mathematical Biology*, 42:339–364.

[2] Bienenstock, E. and von der Malsburg, C. (1987). A neural network for invariant pattern recognition. *Europhysics Letters*, 4(1):121–126.

[3] Goodhill, G. J. (1993). Topography and ocular dominance: A model exploring positive correlations. *Biol. Cybern.*, 69:109–118.

[4] Goodhill, G. J., Finch, S., and Sejnowski, T. J. (1996). Optimizing cortical mappings. In Touretzky, D., Mozer, M., and Hasselmo, M., editors, *Advances in Neural Information Processing Systems*, volume 8, pages 330–336, Cambridge, MA. MIT Press.

[5] Häussler, A. F. and von der Malsburg, C. (1983). Development of retinotopic projections — An analytical treatment. *J. Theor. Neurobiol.*, 2:47–73.

[6] Konen, W. and von der Malsburg, C. (1993). Learning to generalize from single examples in the dynamic link architecture. *Neural Computation*, 5(5):719–735.

[7] Linsker, R. (1986). From basic network principles to neural architecture: Emergence of orientation columns. *Ntl. Acad. Sci. USA*, 83:8779–8783.

[8] Miller, K. D., Keller, J. B., and Stryker, M. P. (1989). Ocular dominance column development: Analysis and simulation. *Science*, 245:605–245.

[9] Miller, K. D. and MacKay, D. J. C. (1994). The role of constraints in Hebbian learning. *Neural Computation*, 6:100–126.

[10] Obermayer, K., Ritter, H., and Schulten, K. (1990). Large-scale simulations of self-organizing neural networks on parallel computers: Application to biological modelling. *Parallel Computing*, 14:381–404.

[11] Tanaka, S. (1990). Theory of self-organization of cortical maps: Mathematical framework. *Neural Networks*, 3:625–640.

[12] von der Malsburg, C. (1973). Self-organization of orientation sensitive cells in the striate cortex. *Kybernetik*, 14:85–100.

[13] Whitelaw, D. J. and Cowan, J. D. (1981). Specificity and plasticity of retinotectal connections: A computational model. *J. Neuroscience*, 1(12):1369–1387.

Relative Time Scales in the Self-Organization of Pattern Classification: From "One-Shot" to Statistical Learning

Klaus Kopecz and Karim Mohraz

University of Marburg, Dept. of Neurophysics, 35032 Marburg, Germany

Abstract. We propose a biologically plausible learning scheme which enables a system to classify patterns based on the presentation of one single example. During a learning mode, the system recognizes whether a category for a presented pattern has been instantiated before, or whether it must be classified as unknown. In this case a new category is created autonomously. The proposed "one-shot" learning rules are characterized by certain time scale relations between system parameter dynamics and input dynamics. We show that reversing these relations (leading to a statistical learning regime), the learning dynamics can be reduced to a Kohonen learning scheme. Our results show that both "one-shot" and statistical learning in biological systems might be governed by identical laws.

1 Introduction

The visual system of higher mammals is able to categorize patterns based on single or few examples. When classifying a pattern as "unknown", a new category can be created which is used to classify similar patterns during succeeding presentations. Here, we adopt the view that this process of learning from a single example is reflected in the adaptation of parameters characterizing the neural system (e.g. strength of interneural connections) and is driven by neural activity. Further, we must require that the time scale of learning, τ_{learn}, is smaller than the characteristic time, $\tau_{pattern}$, with which pattern information varies. In a biological neural system we must also consider (at least) one time constant τ_{act} of the neural activity dynamics, which must be chosen small enough to organize the learning process. Thus, in summary we require that in the "one-shot" learning regime the following relation holds:

$$\tau_{act} \ll \tau_{learn} \ll \tau_{pattern} \qquad (1)$$

Note, that these relations are different from the relations which define the usually considered regime of statistical learning (e.g. Kohonen-learning), where we have

$$\tau_{act} \ll \tau_{pattern} \ll \tau_{learn} \qquad (2)$$

so that the learning system gathers statistics about the pattern environment.

Following the idea of "competitive Hebbian learning" [3], we postulate biologically plausible neural activity and learning dynamics which implement a "one-shot" learning system for the purpose of pattern classification. The system recognizes previously shown patterns whereas for a new pattern a new set of neurons is recruited to represent the new category. Thus, the system exhibits basically the functionality shown by ART networks [2].

The neural activity dynamics is chosen to be of a competitive nature to select a unique group of classificator neurons which took part in learning. Neural weights are changed according to the product of presynaptic and postsynaptic activity (Hebbian modification). Further, we introduce additional dynamics of neural thresholds, which desensitize neurons once they took part in learning and at the same time increase their pattern specificity. We show that when considering the same learning equations in the regime of statistical learning, an algorithm similar to those used for the formation of Kohonen maps can be derived.

2 "One-shot" pattern classification

The layout of our system is depicted in Fig. 1. A pixel k with value I_k of a normalized input pattern is coupled to a classificator neuron i by adaptable weights w_{ik}. Here, classificator neurons are arranged in a one-dimensional chain (henceforth called the "neural field") for simplicity. A two-dimensional arrangement could have been used similarly. Neurons within this chain are laterally coupled by fixed weights c_{ij}. The dynamics of the neural field is given by

$$\tau_{act} \frac{d}{dt} u_i(t) = -u_i(t) + h_i(t) + \sum_j c_{ij} S[u_j(t)] + f_i(t) \tag{3}$$

$$f_i(t) = \sum_k w_{ik}(t) I_k(t) \tag{4}$$

Here, $u_i(t)$ is the membran potential of neuron i and $S(u_i)$ its activity, where

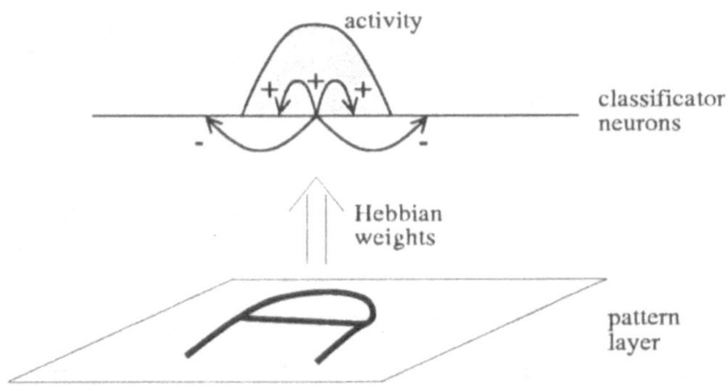

Fig. 1. Model architecture

$S(\cdot)$ is a positive threshold function with threshold zero. We use a piecewise linear function, but any similar variant, like a sigmoid, can be used. $h_i < 0$ is the subthreshold resting potential of neuron i in the absence of activity and input. Alternatively, $-h_i$ can be identified with the threshold of $S(\cdot)$. Obviously, this threshold affects the sensitivity of a neuron to become activated by input signals (in the following we refer to h_i simply as *sensitivity* of neuron i). This input $f_i(t)$ into the neural field is given by the projection of the pattern $\mathbf{I}(t)$ through the weights w_{ik}. The lateral couplings c_{ij} are chosen such that in the presence of weak input, stationary solutions of the neural field dynamics are unique localized clusters of activity located around the maximum of the input distribution f_i. Through the competitve interaction, other neurons become inhibited. Learning will take place only for the activated subset (competitive learning). Following [1], we choose the coupling to be composed of a local excitatory part and a global inhibitory contribution:

$$c_{ij} = c(|i-j|) = k_0 \exp\left[-\frac{(i-j)^2}{2\sigma^2}\right] - H_0 \tag{5}$$

The Hebbian weights w_{ik} are initialized with small random positive values. In the following we use the notation $\mathbf{w}_i$ to refer to the weight vector attached to neuron i. The sensitivities h_i are set such that all neurons can easily be activated by arbitrary input. When a certain pattern is presented, the neural field dynamics will select a subset of neurons at a random location which then take part in the following learning dynamics:

$$\tau_{learn} \frac{d}{dt}\mathbf{w}_i = \left[-\mathbf{w}_i + S(u_i)\mathbf{I}\right] S(u_i) \tag{6}$$

If we consider the condition $\tau_{act} \ll \tau_{learn} \ll \tau_{pattern}$, then we can regard $S(u_i)$ as being relaxed to a localized cluster, and $\mathbf{I}$ as being constant. Thus, in this limit, it is reasonable to assume that the learning dynamics converge to a stationary state before input information changes. This state will be given by

$$\mathbf{w}_i = S(u_i)\mathbf{I} \tag{7}$$

which embodies the idea of Hebbian learning. When presenting a second pattern the selected location on the neural field should differ from the previous one (except when this second pattern is similar to the first one). This can be achieved by decreasing the sensitivity h_i of those neurons, which already had been activated. At the same time, this increases the specificity of the neurons to these patterns which led to their activation during training. Thus, we postulate the additional learning dynamics:

$$\tau_{learn} \frac{d}{dt} h_i = -(h_i - h_{final}) S(u_i) \tag{8}$$

with the stationary solution $h_i = h_{final}$.

Fig. 2 shows the response of the neural field to a sequence of patterns. In

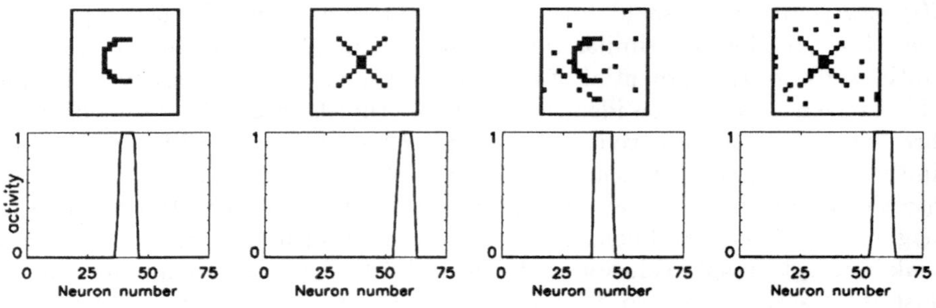

Fig. 2. System response to a sequence of patterns.

each column, the presented pattern together with the resulting stationary activity is depicted. Presentation order is from left to right. When presenting the first pattern "**C**", the neural field selects a random location according to the random initial weight distribution. At these locations, weight vectors are adjusted according to the correlation of pattern and activity and simultaneously, the sensitivity h_i is reduced. When presenting the second pattern "**X**", the overlap with pattern "**C**" is not large enough to reactivate the same neurons again. Instead, a different population is recruited which then adapt their weight vectors and sensitivities. However, the system is capable of generalization as depicted in the last two columns of Fig. 2, where disturbed versions of the previously learned patterns are presented. These noisy patterns are correctly classified as long as there is sufficient overlap with a stored pattern. Whenever this overlap is below a certain threshold, a pattern is detected as being "new", and a new category (a new set of neurons) is instantiated. The parameter which affects the critical overlap separating "classified as known pattern" and "recruit a new set of neurons" is the sensitivity h_{final} which is approached during fast learning (cf. eqn. 8). If this sensitivity is very low, a pattern must match exactly the prototype used for training to provide maximal input for the corresponding classificator neurons. Then it is very easy to activate a different region on the neural field with a noisy or similar pattern (note that all neurons which have not been activated before have an initial high sensitivity). If the final sensitivity h_{final} is higher, it will be easier to activate neurons with patterns not exactly matching their learned weight vectors.

3 The limit of statistical learning

In the statistical limit the self-organizing process becomes very similar to the scheme used for the formation of Kohonen maps with time varying neighborhood function and learning rate (cf. e.g. [5]). This is seen as follows: For better comparison, we transform the learning differential equations into numerical one-step Euler approximations with discretization interval $dt \ll \tau_{learn}$ and change the unit of time to iteration steps. Further defining the learning rate $\eta = dt/\tau_{learn} \ll 1$

results in:

$$\mathbf{w}_i(t+1) = \mathbf{w}_i(t) + \eta S(u_i)\left[-\mathbf{w}_i(t) + S(u_i)\mathbf{I}(t)\right] \tag{9}$$

$$h_i(t+1) = h_i(t) + \eta S(u_i)\left[-h_i(t) + h_{final}\right] \tag{10}$$

Now, $\mathbf{I}(t)$ changes every learning step, which means that patterns change fast compared to the learning dynamics ($\eta \ll 1$). Still, the activity $S(u_i)$ evolves fast compared to the input so that it is fully relaxed to define a neighborhood for learning. Eqn. (9) is similar to the Kohonen learning rule, where $S(u_i)$ is the neighborhood function. The difference is that $S(u_i)$ appears again as an additional factor in front of $\mathbf{I}(t)$. The second learning equation (10) leads to a decrease in the width and height of the neighborhood function over time. This can be seen by plotting the width and maximal amplitude of stationary activity distributions of the neural field dynamics (3) versus the sensitivity h (which can be considered as the statistical average of all h_i). Fig. 3 shows that the width decreases monotonically with decreasing h (continuous line). Hence, the decay of h to its minimum h_{final} by the learning dynamics (10) is equivalent to reducing the width of the neighborhood function $S(u_i)$ over time.

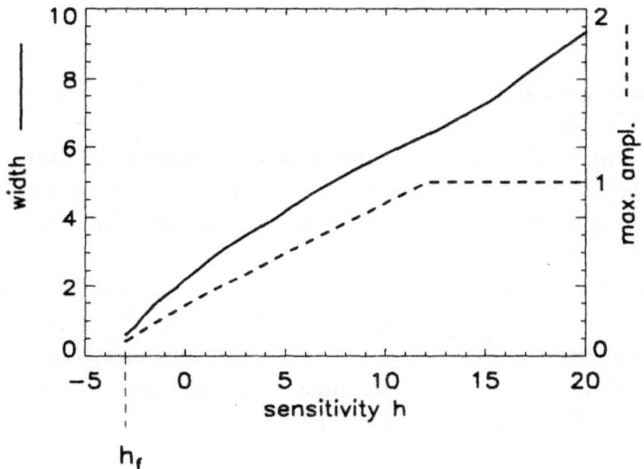

Fig. 3. Decrease of width (continuous line) and height (dashed line) of the neighborhood function with decreasing mean sensitivity h.

As obvious from Fig. 3, dashed line, the reduction of h has the additional effect of decreasing the maximal amplitude A of the neighborhood function. Due to the appearance of the product $\eta S(u_i)$ in eqn. (9), this defines an effective learning rate $\eta_{eff} = \eta\, A$, which decreases over time. Thus, learning will be global and fast for early periods and local and slow for late periods of time.

In summary, the equations (3,6,8) implement a learning scheme which has been shown to produce topographically ordered stationary states with small fluctuations (cf. e.g. [5]).

4 Summary and Conclusion

We proposed activity and learning dynamics which produce a competitive Hebbian learning scenario. We studied the behavior of these dynamics under varying time scale relations between learning and pattern dynamics. When learning is fast compared to pattern presentation times, the system works in a "one-shot" learning mode where categories are defined autonomously from single examples. Whereas this is basically the same functionality shown by ART networks [2], our learning scheme is considerably more simple.

In the statistical regime, where learning proceeds slow compared to pattern dynamics, the system realizes a Kohonen learning scheme including a shrinking neighborhood function as learning proceeds and a decaying learning rate. As the here proposed dynamics are plausible to be realized in the biological system, our study supports the biological motivation of competitive learning algorithms (see also [4] for a somewhat similar biological implementation of Kohonen learning). Further, we can postulate that biological nervous systems use the same learning principles for different types of knowledge aquisition by changing time scale relations between neural processes.

References

1. S. Amari. Dynamics of pattern formation in lateral–inhibition type neural fields. *Biol. Cybern.*, 27:77–87, 1977.
2. G.A. Carpenter and S. Grossberg. Art2: Stable self-organization of pattern recognition codes for analogue input patterns. *Appl. Opt.*, 26:4919–4930, 1987.
3. J Hertz, A Krogh, and RG Palmer. *Introduction to the Theory of Neural Computation*. Addison-Wesley, 1991.
4. R. Miikkulainen. Self–organizing process based on lateral inhibition and synaptic resource distribution. In T. Kohonen, K. Mäkisara, O. Simula, and J. Kangas, editors, *Artificial Neural Networks*, pages 415–420. Elsevier, North-Holland, 1991.
5. H Ritter and K Schulten. On the stationary state of kohonen's self-organizing sensory mapping. *Biol. Cybern.*, 54:99–106, 1986.

Acknowledgements:
This work was supported by the German Ministry of Research and Technology (BMBF) through grant no. 01 M 3013D. We thank Reinhard Eckhorn for fruitful comments.

Realization of Geometric Illusions and Geometry of Visual Space with Neural Networks

Jinhui Chao[1], Miyata Yasuhiko[1], Shinich, Yoshida[2]

[1] Dept. of Electrical and Electronic Eng., Chuo University, Tokyo, Japan
[2] Dept. of Computational Intelligence and Systems Science, Tokyo Institute of Technology

Abstract. A novel neural network model is shown to realize various geometric illusion phenomena in human vision, which also makes it possible to quantitatively analyze the geometrical deformation of visual space. In sequel, we can predict the geometric illusions based on the proposed model.

1 Introduction

Illusions in human vision, especially geometric illusions, have been a theme for century-lasting efforts in order to explore deep and mysterious mechanisms of visual information processing in human brain. It is believed that any unifying theory on geometric illusions will greatly magnify understanding on 2D visual perception then bring up new paradigm in image processing techniques.
Most researches on the subject so far have been in the perceptional psychology field and intended to explain geometric illusion phenomena under various ad-hoc psychological hypotheses for each particular phenomenon. e.g. the Ebbinghaus illusion is interpreted by a hypothesis that human vision system is liable to amplify an acute-angel; the Müller-Lyer illusion is explained using a hypothesis that visual perception amplifies the distant objects (the perspective hypothesis). One of the reason for so many different psychological theories is that only practical approach to verify the hypotheses is psychological experiments, which are known to be lack of stability and repeatability and hard to control. Therefore, verification of these psychological hypotheses and associated illusion theories is a very difficult task. e.g. the question about where the illusion occurs remained unsolved: is it in the retina or in the cortex? Psychologists preferred and tried hard to prove the second answer, since it may introduce some interesting mental process. But it is difficult to cancel out the other possibilities. Presently no unifying psychological or physiological model is known.
In this paper, we attribute the geometrical illusions to result of the deformation or curvature of the visual space in the certain visual environments, specifically due to particular arrangement of the inductive lines. Theory based on the similar assumption does exist, e.g. the theory by Luneburg on classical illusions such as alleys or horopters[5]. However, the visual space in his theory is constantly and uniformly curved. On the other hand, it is known that the visual spaces in

geometric illusions are much complicated than in the above classical illusions and are non-uniform. These features make the modeling and explicit calculation of the spatial deformation in geometric illusion very hard.

In this paper, we assume the geometric illusions are results of a very basic physiological mechanism, the lateral inhibition dynamics between locally connected neurons. This is the only assumption we needed in this model.

The lateral inhibition modelhave been known for quite long time[1] [6], [4][7]. But it has never been widely accepted since there are no convincing and verifiable evidences to show that illusions can be produced without other causes. It also met serious difficulty such as the paradox of displacement [2]. More seriously, this model express figures with the extreme points of the potential and inhibitions from each neurons are added to form these extreme or peaks. In fact, it does not work in computer simulations since it requires too strong inhibition so that it attenuates or even erases the figures and often scatters a connected figure to isolated parts.

Our aim is to build an engineering model which makes possible quantitative, constructive and verifiable research of the geometric illusions.

In the first place, we propose a novel "shift" model for lateral inhibition in which the relative distances between positions of firing neurons are changed under action of inhibition. This model not only always produces stable figures easily, it naturally introduces the Riemannian metric tensor of visual space and the "tension force" which is an important attribute of geodesics in Riemannian space. In particular, we show that we can realize various illusion phenomena with this unifying computational model.

In the second place, this model is then applied for quantitative analysis of the intrinsic Riemannian geometric properties such as twist or bending of the visual space in which geometric illusions are observed. In particular, the distributions of Gaussian curvature of visual space in various geometric illusion phenomena are calculated from the neural network model. Furthermore, we calculate also the geodesics in the visual space based on the above curvature tensors. These geodesics coincide with the so-called negative-illusion figures which show the opposite distortion comparing with the ordinary illusion figures.

It is very interesting that, we started from a totally neuro-physiological model but eventually proved the celebrated hypothesis of Luneburg in psychophysics: the subjective straight lines are geodesics in the visual space as Riemannian space. At the same time, we also disprove Luneburg's theory that the visual space is a Riemannian space with constant-negative-curvature.

As a result of these studies, we can in the sequel to predict the occurrence of geometric illusions in any particular visual environments.

2 Shift Model of Lateral Inhibition For Illusion

We assume the lateral inhibitive connection in our neural network such that each visual neuron has a Mexican-hat shaped activation or feedback function. Instead of calculating the peak values, we choose in this discrete model to shift the positions of excitatory points (or excitatory neurons) in the visual field according to the excitatory-inhibitory interaction dynamics. (In a continuous neural distribution, this could be occured in the form of exchanges between firing neurons and neighboring unfired neurons). Specifically, for each neuron, the activities of the excitatory neurons around it are accumulated on four directions (each separated by 45°), then by the vectorial sum of these actions, one determines the shift amount of the neuron.

Another interesting feature of this model is that if the lateral interaction function has sharp peak near the origin, then a strong tension force will be observed between the neighboring excitatory neurons lying on the figure lines. It is known that in Luneburg's theory, which assume that subjectively straight lines are the geodesics in the curved visual space. In the sense of calculus of variation, this is equivalent to introduction of a (elastic) potential energy on each figures in visual space, which means that each of lines is regarded as an elastic string restrained on the curved visual space. Thus, this totally physiology-based model naturally coincides with the mathematical theory of psycho-physiological vision by Luneburg[5].

3 Intrinsic Geometry of Visual Space

Luneburg's model contained an important and well-accepted theme that illusions are caused by curvature of visual space and the most stable visual perceptive objects —the subjective straight-lines are geodesics in the curved space.

It is known that since Luneburg's work [5] , the geometry of the visual space has become a central theme for classic illusion phenomena. Arguments have continued on Luneburg's assumption, based on acute theoretical consideration, that the visual space is of constant-negative-curvature Riemannian. Unfortunately, this attractive theory has not been verified. It was even impossible for such study on those important but more complicated illusion phenomena such as the geometric illusions.

Our neural network model proved to be a very useful and convenient tool for study of the geometry of visual space in various visual phenomena. Besides, it provides an opportunity to verify by constructive engineering approach the psychophysical theory by Luneburg.

In particular, the illusions is regarded to be caused by nonuniform deformation in visual space, which is a result of the local lateral excitatory-inhibitory dynamics of retina visual neurons. The relative displacements between the firing neurons in the visual field caused by the lateral inhibition dynamics naturally determine the Riemannian metric tensor at each point in visual space such that the metric is extended between separate excitatory neurons but shrunken between close ones.

After the neural network converged to a stable figure which is a reproduction of geometric illusion in human vision, we can analyze the deformation of visual

space in this case by calculation of the Riemannian metric and curvature tensor. Specifically, the metric tensor of visual space in geometric illusions can be calculated from the neural network model either directly or through estimation of the Jacobian matrices point-wisely.
When a point (x, y) moves to (x', y'), one can estimate the Jacobian $J = \begin{pmatrix} j_{11} & j_{12} \\ j_{21} & j_{22} \end{pmatrix}$ by solving linear equations $\begin{pmatrix} x' \\ y' \end{pmatrix}_{i,j} = J \begin{pmatrix} x \\ y \end{pmatrix}_{i,j}$ for certain neighborhood. Then to derive metric tensor by

$$G_{i,j} = J^T J \equiv \begin{pmatrix} g_{11}(i,j) & g_{12}(i,j) \\ g_{21}(i,j) & g_{22}(i,j) \end{pmatrix}_{i,j} \tag{1}$$

Or estimate the metric tensor G by solving the point-wise system of linear equations of

$$\|(x, y) - ((x + \Delta x)', (y + \Delta y)')\|^2_{i,j} = (\,\Delta x, \Delta y\,)_{i,j}\, G \begin{pmatrix} \Delta x \\ \Delta y \end{pmatrix}_{i,j} \tag{2}$$

The Gaussian curvature K can then be obtained by Gauss' Theorema Egregium

$$K = k_1 k_2 = \frac{\partial^2 g_{12}}{\partial x \partial y} - \frac{1}{2}\frac{\partial^2 g_{22}}{\partial x^2} - \frac{1}{2}\frac{\partial^2 g_{11}}{\partial y^2} \tag{3}$$

where the partial differentiations are calculated as

$$\frac{\partial^2 g_{12}}{\partial x \partial y} = \frac{1}{\Delta x_1 \Delta y_1}\{(g_{12}(i+1, j+1) - g_{12}(i+1, j)) - (g_{12}(i, j+1) - g_{12}(i, j))\} \tag{4}$$

It turned out that the deformation of visual space in geometric illusion is much involved comparing with the classic illusions considered by Luneburg. e.g., no constant-curvature can be observed any longer. The visual space shows certain complex patterns of distribution of curvature which are enviroment dependent. This also makes study of deeper geometric properties such as geodesics (as negative illusions) and connection possible.
Below, we calculate the geodesics in the visual space whose Gaussian curvature distribution is obtained from the neural network model.
A geodesic $\boldsymbol{u}(t)$ is defined as the integral curve of the differential equation

$$0 = \ddot{\boldsymbol{u}}^i + \sum_{j,k} \Gamma^i_{jk} \dot{\boldsymbol{u}}^j \dot{\boldsymbol{u}}^k. \tag{5}$$

Here t is arc length as natural parameter. Christoffel symbol Γ^i_{jk}

$$\Gamma^i_{jk} = \frac{1}{2} \sum_l g^{il}(g_{lj,k} + g_{lk,j} - g_{jk,l}) \tag{6}$$

is defined by partial derivatives of g_{ij} and inverse matrix g^{ij}
Since the geodesics represent negative geometric illusions in the visual space with given curvature distribution , one can then predict the positive illusion by geodesics which lie in the Riemannian space whose Riemannian metric is defined as the inverse matrix of the metric matrix in the visual space in which potential illusion is expected.

4 Simulations

Various illusions are realized by our model, which are illusions seen by computers in the similar way to human vision. Besides, cuvature of visual space provides a quantitative description of subjective characteristics in geometric illusions, which makes it possible to study these phenomena in a systematic way. The two Müller-Lyer illusions are realized by the proposed model in fig. 1, 3. While the Gaussian curvature distributions in these two visual spaces are shown in fig. 2, 4. Fig. 5-7 shown realization of the Wundt illusion, the Gaussian curvature distribution in the visual space. Fig. 6 demonstrates that the central of the space has positive curvature but negative one around. It is interesting to notice the existence of the positive curvature in the distribution, which proved to play a very important role in the generation of illusion. This phenomena, however, was not predicted by both psychological explanations and even neurophysiological theory on geometrical illusion. Fig. 7 shows the geodesic derived from the visual space with the Gaussian curvature distribution in fig. 6 which represents the negative Wundt illusion. The simulation results of the Hering illusion are shown in figures 8-10.

References

1. F.Ratliff:Holden-Day ,San Francisco(1965)
2. S. Morinaga, H. Ikeda, Jap. J. Psycho., 36, p.231-238 (1965)
3. Howard, L. Resnikoff, Springer-Verlag, 1989
4. R.B.Howard : Psychonomic Monograph Supplements, 4, 3, p.57(1971)
5. R.K.Luneburg : Princeton Univer. Press, 1947
6. L.Ganz,Psycho. Review, 73, p.128(1966)
7. T. Morita, K. Fujii, Jap. J. Inst. Eletr. Com Eng. pp.1857-1863, 1966.
8. Blackmore, R.H.S.Carpenter, M.A.Georgeson, Nature, 228, p.37(1970)

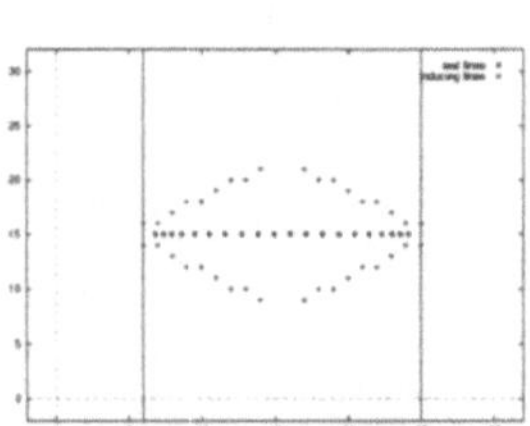

Fig. 1 Realization of Müller-Lyer illusion (1)

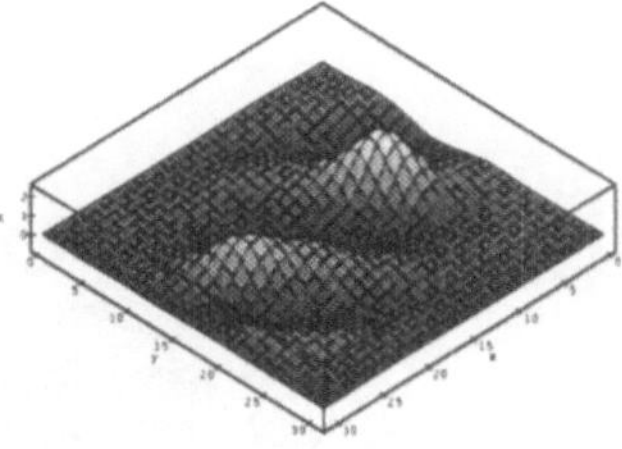

Fig. 2 Curvature in (1)

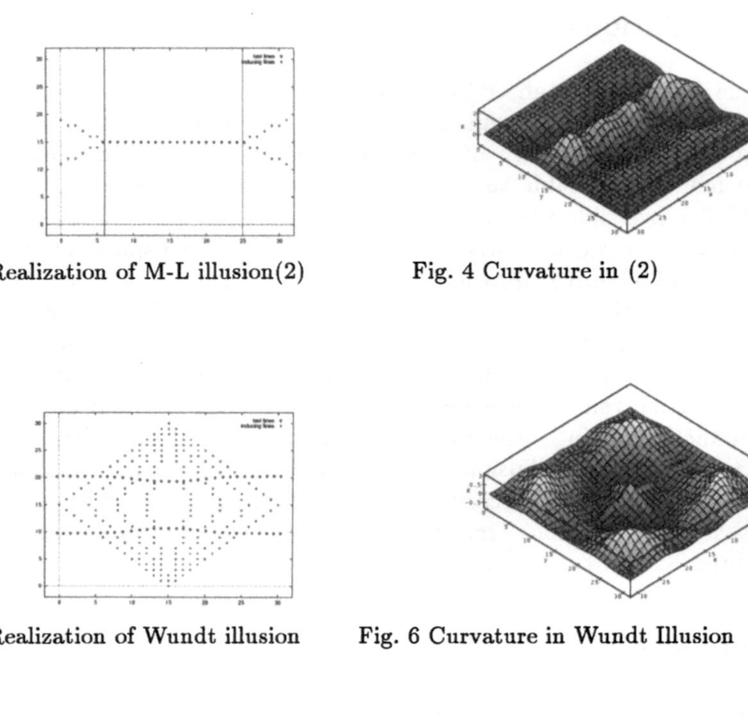

Fig. 3 Realization of M-L illusion(2)

Fig. 4 Curvature in (2)

Fig. 5 Realization of Wundt illusion

Fig. 6 Curvature in Wundt Illusion

Fig.7 Geodesic as negative Wundt Illusion

Fig.8 Realization of Hering illusion

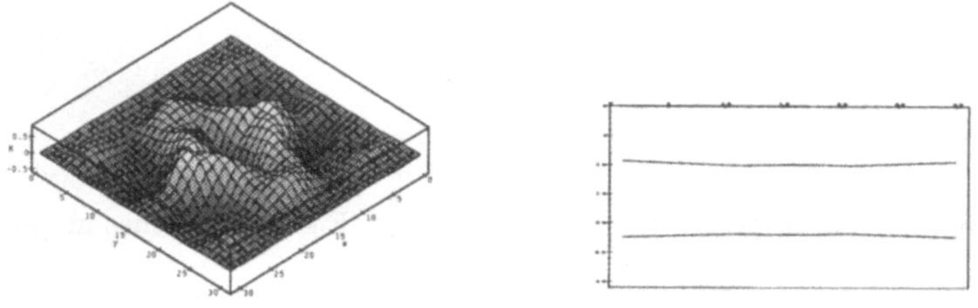

Fig.9 Curvature in Hering Illusion

Fig.10 Geodesic as negative Hering Illusion

Part III:
Learning: Theory and Algorithms

The Support Vector Method

Vladimir N. Vapnik

AT&T Labs-Research
100 Schultz Dr. Red Bank, NJ 07701, USA
vlad@research.att.com

Abstract. The Support Vector (SV) method is a new general method of function estimation which does not depend explicitly on the dimensionality of input space. It was applied for pattern recognition, regression estimation, and density estimation problems as well as for problems of solving linear operator equations. In this article we describe the general idea of the SV method and present theorems demonstrating that the generalization ability of the SV method is based on factors which classical statistics do not take into account. We also describe the SV method for density estimation in a set of functions defined by a mixture of an infinite number of Gaussians.

1. Introduction

The Support Vector (SV) method is a general approach to function estimation problems, i.e. to problems of finding the function $y = f(x)$ given by its measurements y_i (with noise) at some (usually random) vectors x

$$(y_1, x_1), ..., (y_\ell, x_\ell). \tag{1}$$

It can be applied for pattern recognition (to estimate indicator functions), for regression (to estimate real-valued functions), for estimation of densities, and for solving linear operator equations.

The idea behind this method is very simple: one maps n-dimensional vectors x of the input space X into high-dimensional (even infinite dimensional) vectors z of the feature space Z, where using the samples

$$(y_1, z_1), ..., (y_\ell, z_\ell) \tag{2}$$

corresponding to (1) one constructs a linear function

$$y = (w * z) + b. \tag{3}$$

In (3) we denote by $(w * z)$ an inner product between two vectors.

Two problems arise with this approach:

1. Which properties should the linear function (3) satisfy in order to approximate well the desired function .
2. How to construct such a hyperplane in a high-dimensional space.

Note that we are interested in the case where the dimensionality of a feature space is much larger than the number of available observations. In classical statistics this case refers to the situation called "the curse of dimensionality". One can consider the SV machine as a tool to overcome this curse.

Two theoretical results made construction of SV machines possible:

1. The generalization ability of the learning machine depends on the capacity of a set of functions that the machine implements (on the VC dimension of a set of functions) rather than on the dimensionality of a space. Therefore a function that describes the data well and belongs to a set with low VC dimension will generalize well regardless of the dimensionality of the space [4].
2. To construct a hyperplane one only needs to evaluate an inner product between two vectors of the training data. In Hilbert space inner products have simple kernel representation and therefore can be easily evaluated.

These two singularities make the SV approach an efficient and powerful method for function estimation.

The SV method was discovered in 1965 for constructing separating hyperplanes in the pattern recognition problem (see [5]). Then in 1992 – 1995 it was generalized for constructing non-linear separating functions (but linear in the feature space) [2], [3]. In 1995 it was generalized for estimation of real-valued functions [4]. Lastly in 1996 it was applied for solving linear operator equations [6].

In section 2 we describe the SV method for constructing linear functions, then in section 3 we describe theorem on generalization ability of this method, in section 4 we introduce the general form of the SV method, and in section 5 we will describe the method for solving linear operator equations where we consider as an example the problem of density estimation.

2. Problem of constructing the optimal hyperplane

Consider the pattern recognition problem, i.e. the problem of estimating indicator function. Let (1) be i.i.d. data where y takes only two values $y \in \{-1, 1\}$. Suppose that data (2) can be separated by a hyperplane. That is, there exists a hyperplane

$$u = (w * z) + b$$

such that $\mathrm{sign}(u) = y$.

Our goal is to construct a separating hyperplane possessing the maximal margin. That is, the hyperplane that correctly separates the data and has maximal distance to the closest vector z from (2). One can show that this problem has the following mathematical expression: minimize the quadratic form $(w * w)$ subject to constraints

$$y_i[(w * z_i) + b] \geq 1, \quad i = 1, ..., \ell. \tag{4}$$

For the non-separable case we generalize the setting of the problem as follows [2]: we introduce nonnegative slack variables $\xi_i \geq 0$ and minimize the functional

$$\Phi(a) = C \sum_{i=1}^{\ell} F(\xi_i) + (w * w) \tag{5}$$

subject to constraints

$$y_i[(w * z_i) + b] \geq 1 - \xi_i, \quad i = 1, ..., \ell. \tag{6}$$

In (5) $F(\xi)$, $F(0) = 0$ is a given monotonically increasing function and C is a constant defined a priori. For the non-separable case, if $F(\xi) = \xi^\sigma$, $\sigma > 0$, is sufficiently small, and C in (5) is sufficiently large, the solution to our optimization problem defines the hyperplane that minimizes the number of training errors. If, however, the data are separable then the solution is the optimal hyperplane.

To solve this optimization problem one has to find the saddle point of the Lagrangian

$$L = C \sum_{i=1}^{\ell} F(\xi_i) + (w * w) - \sum_{i=1}^{\ell} \alpha_i [y_i[(w * z_i) + b] - 1 + \xi_i] - \sum_{i=1}^{\ell} \lambda_i \xi_i, \tag{7}$$

(minimum over w, b, and ξ and maximum over $\alpha_i \geq 0, \lambda_i > 0$).

Let us choose $F(\xi) = \xi$, the smallest σ that leads to a simple quadratic optimization problem. With this loss function we obtain the following solution: the optimal hyperplane has the form

$$y = \sum_{i=1}^{\ell} \alpha_i^0 (z * z_i) + b_0, \tag{8}$$

where vector $\alpha_0 = (\alpha_1^0, ..., \alpha_\ell^0)$ provides the maximum to the functional

$$W(\alpha) = \sum_{i=1}^{\ell} \alpha_i - \frac{1}{2} \sum_{i,j=1}^{\ell} y_i y_j \alpha_i \alpha_j (z_i * z_j) \tag{9}$$

subject to constraints

$$\sum_{i=1}^{\ell} \alpha_i y_i = 0, \qquad 0 \leq \alpha_i \leq C, \ \ i = 1, .., \ell. \tag{10}$$

Since the optimal solution must satisfy the Kuhn-Tucker conditions

$$\alpha_i^0\{y_i[(w * z_i) + b_0] - 1 + \xi_i\} = 0, \tag{11}$$

the number of nonzero coefficients α in the expansion (8) is usually small. Let us call vectors z_i corresponding to nonzero α_i the support vectors.

To construct an approximation to a linear real-valued function we introduce the so called loss function with ε-insensitive zone:

$$\xi_i = |y_i - (w * z_i)|_{\varepsilon_i} = \begin{cases} 0 & \text{if } |y_i - (w * z)|_{\varepsilon_i} < \varepsilon_i \\ |y_i - (w * x_i)|_{\varepsilon_i} - \varepsilon_i. & \text{otherwise.} \end{cases} \tag{12}$$

To estimate linear real-valued functions we minimize the functional (5) where values of ξ_i are defined by the expression (12). In Eq. (5) one can use different functions $F(u)$. However three of them: linear, quadratic and Huber lead to simple quadratic optimization problems.

In particular for the linear function $F(u) = u$ repeating the transformations of the Lagrange functional we obtain the estimator

$$y = \sum_{i=1}^{\ell}(\alpha_i^* - \alpha_i)(z * z_i) + b, \tag{13}$$

where to find coefficients $\alpha_i^*, \alpha_i,\ i = 1, ..., \ell$ we maximize the functional

$$W(\alpha^*, \alpha) = -\sum_{i=1}^{\ell}(\alpha_i^* + \alpha_i)\varepsilon_i + \sum_{i=1}^{\ell} y_i(\alpha_i^* - \alpha_i) - \frac{1}{2}\sum_{i,j=1}^{\ell}(\alpha_i^* - \alpha_i)(\alpha_j^* - \alpha_j)(z_i * z_j) \tag{14}$$

subject to constraints

$$\sum_{i=1}^{\ell}(\alpha_i^* - \alpha_i) = 0, \quad 0 \leq \alpha, \alpha_i^* \leq C. \tag{15}$$

We also consider a slightly different form of the optimal hyperplane which we call the optimal hyperplane with a fixed margin. Consider the hyperplane (8) (or (13)) with coefficients that minimize the functional

$$\Phi(\xi) = \sum_{i=1}^{\ell} F(\xi_i) \tag{16}$$

subject to constraint $(w * w) \leq A^2$. To construct a separating hyperplane we maximize the functional

$$W(\alpha) = \sum_{i=1}^{\ell}\alpha_i - \frac{1}{2d}\sum_{i,j=1}^{\ell} y_i y_j \alpha_i \alpha_j (z_i * z_j) - \frac{dA^2}{2} \tag{17}$$

subject to constraints

$$\sum_{i=1}^{\ell} \alpha_i y_i = 0, \qquad 0 \le \alpha_i \le 1, \;\; d \ge 0, \;\; i = 1,..,\ell. \tag{18}$$

There exists an analogous solution for estimating real-valued functions.

3. Statistical properties of the optimal hyperplane.

Optimal hyperplanes possess some remarkable statistical properties.

Consider the pattern recognition problem. Suppose that the training data can be separated by a hyperplane without error. Suppose that the optimal hyperplane is expanded on N support vectors and is such that $A = (w * w)$. One can show that the optimal separating hyperplane is unique however its expansion on the SV is not. Let us call the SVs that are in all expansions *the essential SVs.* We denote the number of essential SVs by $\mathcal{N}$. Suppose that the maximal norm of the SVs is bounded by D ($||z|| \le D$).

Consider the following random variable h that specifies the smallest of the following three values: two random values A^2D^2 and $\mathcal{N}$ and the dimensionality of space n

$$h = \min(A^2D^2, \; \mathcal{N}, \; n) + 1. \tag{19}$$

The following theorem is true [5].

Theorem. *If any training data of size ℓ can be separated without error then the expectation of the probability of test error for the optimal hyperplane has the bound*

$$EP \le \frac{Eh}{\ell}. \tag{20}$$

The important message from this theorem is that small dimensionality is not the only reason for good generalization. There are two other reasons (number of support vectors or a small value of D^2A^2). Classical statistics ignored these opportunities and considered only one reason for generalizations: small dimensionality of space. Therefore with increasing dimensionality of space the curse of dimensionality appears.

The SV approach uses new opportunities. It maps input vectors into a very high-dimensional feature space and constructs the optimal hyperplane in this space. The motivation is: the SV method relies on other factors than dimensionality to attain good generalization.

4. Kernel representations of inner products

Note the following important fact: both the expression for the optimal hyperplane (8), (13) and the objective function for estimating the parameters of the optimal hyperplane (9), (14) depend only on the inner product $(z * z_i)$ between two vectors. It is known that in Hilbert space the inner product has a compact representation. This allows us to construct hyperplanes in high-dimensional spaces.

Suppose that we map an input vector $x = (x^1, ..., x^n)$ into Hilbert space $z(x) = (\phi_1(x), ..., \phi_k(x), ...,)$ where we construct the optimal hyperplane. The following equivalent representation of the inner products in Hilbert spaces holds

$$\sum_{i=1}^{\infty} a_i \phi_i(x_m) \phi_i(x_k) \Longleftrightarrow K(x_m, x_k)$$

To guarantee that the symmetric function $K(x_m, x_i)$ is an inner product in some feature space it is necessary and sufficient that the condition

$$\int K(u, v) g(u) g(v) du dv > 0 \tag{21}$$

is valid for any nonzero function from L_2 (Mercer's theorem).

Therefore if the symmetric kernel function $K(x_i, x)$ satisfies Mercer's condition then one can replace the inner product $(z * z_i)$ in all equations of Section 2 by the kernel $K(x_i, x)$, i.e. using the technique of Section 2 one can estimate a function of the form

$$f(x) = \sum_{i=1}^{\ell} \beta_i K(x_i, x) + b, \tag{22}$$

where $\beta = y_i \alpha_i$ for pattern recognition problems and $\beta = (\alpha_i^* - \alpha_i)$ for real function estimation problems.

The following kernels satisfy Mercer's conditions [1], [4]: polynomial kernels

$$K(x, x_i) = [(x * x_i) + 1]^k \quad k \geq 1$$

radial basis function kernels, for example,

$$K(x, x_i) = \exp\{-a|x - x_i|^2\}.$$

With these kernels we obtained a good performance in pattern recognition, function approximation and regression estimation problems [3], [4].

For signal processing one can use kernels that perform spline expansion, Fourier series expansion and so on [6]. To construct expansion on spline of order d with an infinite number of nodes one uses the kernel

$$K(x, x_i) = \sum_{r=0}^{d} \frac{C_d^r}{2d - r + 1} [\min(x, x_i)]^{2d-r+1} |x - x_i|^r + \sum_{r=0}^{d} x^r x_i^r.$$

To construct one-dimensional Fourier series expansion on N terms

$$f(x) = a_0 + \sum_{k=1}^{N} (a_k \sin kx + b_k \cos kx)$$

one can use the Dirichlet kernel

$$K(x, x_i) = \frac{\sin(N + 1/2)(x - x_i)}{\sin 1/2(x - x_i)}.$$

It is known, however, that non-regularized Fourier expansion does not approximate functions well. Therefore one uses the expansion

$$f(x) = a_0 + \sum_{k=1}^{N} q^k (a_k \sin kx + b_k \cos kx), \quad 0 < q < 1,$$

This regularized Fourier expansion can be obtained using the kernel

$$K_q(x_i, x_j) = \frac{1 - q^2}{1 - 2q\cos(x_i - x_j) + q^2},$$

To construct multidimensional expansions it is sufficient to use kernels that are coordinate-wise products of one-dimensional kernels.

5. Solving operator equation. Density estimation

One can use the SV technique for solving the linear operator equation [6]

$$Af(t) = F(x) \tag{23}$$

on the basis of measurements F_i of the right hand side at the points x_i with accuracy ε_i. In other words one has to solve the equation on the basis of triplets

$$(x_1, F_1, \varepsilon_1), ..., (x_\ell, F_\ell, \varepsilon_\ell).$$

Consider the solution of the equation in the set of functions

$$f(t, g) = \int_{-\infty}^{\infty} g(\tau)\psi(t, \tau)d\tau, \tag{24}$$

where $\psi(t, \tau)$ is some known function and $g(\tau)$ is an unknown function from L_2. To find the solution $f(x, g)$ means to specify function $g(u)$ in (24).

Suppose that operator A maps this set of functions into another set

$$F(x, g) = \int_{-\infty}^{\infty} g(\tau)\phi(x, \tau)d\tau \tag{25}$$

in a one-to-one manner where $\phi(x,\tau) = A\psi(t,\tau)$. For any fixed x one can consider the function (25) as an inner product in Hilbert space. Therefore one can construct the kernel

$$K(x_i, x_j) = \int_{-\infty}^{\infty} \phi(x_i,\tau)\phi(x_j,\tau)d\tau. \tag{26}$$

Using this kernel one estimates the function $g(\tau)$ on the basis of the technique described in Section 2.

$$g(\tau) = \sum_{i=1}^{\ell}(\alpha_i^* - \alpha_i)\phi(x_i,\tau).$$

Putting the expression for $g(\tau)$ back into (24) one obtains

$$f(t) = \sum_{i=1}^{\ell}(\alpha_i^* - \alpha_i)\mathcal{K}(x_i,t), \tag{27}$$

where by $\mathcal{K}(x_i,t)$ we denote the cross-kernel function

$$\mathcal{K}(x_i,t) = \int_{-\infty}^{\infty} \phi(x_i,\tau)\psi(t,\tau)d\tau. \tag{28}$$

The main problem in solving the linear operator equation using the SV technique is to calculate both the kernel function and the cross-kernel function.

In [6] we calculated these functions for the PET problem. Below we present another example: the SV method of density estimation on bounded support.

Let us start with the one dimensional case. According to the definition estimating a density on $[0,1]$ means solving the integral equation

$$\int_0^x p(t)dt = F(x),$$

where $F(x)$ is a distribution function. We are given the data $x_1,...,x_\ell$ instead of a distribution function. It is known that that for any x^* the frequency $F_{emp}(x^*)$ of event $\mathcal{A} = \{x_i \leq x^*\}$ approximates the distribution function $F(x)$ well: for any fixed x^* the estimate is unbiased and has the variance

$$\sigma(x^*) = \sqrt{\ell^{-1}F(x^*)(1-F(x^*))}.$$

Therefore one can solve the equation using triplets

$$(x_1, F_1, \varepsilon_1), ..., (x_1, F_1, \varepsilon_1)$$

where $F_i = F_{emp}(x_i)$, $\varepsilon_i = \delta\sigma(x_i)$, and $\delta > 0$ is a chosen constant.

Let us solve this problem in the set of Gaussian mixtures

$$p(t) = \frac{1}{\sqrt{2\pi}\sigma} \int_{-\infty}^{\infty} g(\tau) \exp\left\{-\frac{(t-\tau)^2}{2\sigma^2}\right\} d\tau.$$

For this set of functions one can calculate the kernel and cross-kernel functions using the erf-function

$$erf(x) = \frac{2}{\sqrt{\pi}} \int_0^x e^{-t^2} dt.$$

One obtains the kernel

$$K(x_i, x_j) = \sigma \left(\int_0^{\frac{x_i}{2\sigma}} \mathrm{erf}(u) du + \int_0^{\frac{x_j}{2\sigma}} \mathrm{erf}(u) du - \int_0^{\frac{|x_i - x_j|}{2\sigma}} \mathrm{erf}(u) du \right)$$

and the cross-kernel

$$\mathcal{K}(x_i, t) = \frac{1}{2} \left[\mathrm{erf}\left(\frac{x_i - t}{2\sigma}\right) + \mathrm{erf}\left(\frac{t}{2\sigma}\right) \right].$$

With these functions one obtains the SV solution of the density estimation problem: using the kernel function one finds the SVs and the expansion coefficients $\beta_i = \alpha_i^* - \alpha_i$ and then using the cross-kernel function, the SVs, and this coefficients one defines an approximation of the density (27).

To estimate multidimensional density one needs to construct corresponding kernels that are simple coordinate-wise products of the corresponding one-dimensional kernel and cross-kernel functions.

References

1. B.E. Boser, I.M. Guyon, and V.N. Vapnik. A training algorithm for optimal margin classifier. *Proceedings of the 5th Annual ACM Workshop on Computational Learning Theory.* pp 144-152, Pittsburgh, PA, 1992.

2. C. Cortes and V.Vapnik. Support Vector Network. *Machine Learning.* 20:273-297, 1995.

3. H. Drucker, C.J. Burges, L. Kaufman, A. Smola, and V. Vapnik. Support vector regression machines. *In Advances in Neural Information Processing Systems 9*, 1997, MIT Press.

4. V. Vapnik. *The Nature of Statistical Learning Theory.* Springer Verlag, 1995, New-York.

5. V.N. Vapnik and A. Ya. Chervonenkis. *Theory of Pattern Recognition* (in Russian) Nauka, Moscow, 1974. German translation: W.N. Wapnik, A Ja. Tscherwonenkis *Teorie der Zeichenerkennung*, Akademia, Berlin, 1979.

6. V. Vapnik, S. Golowich, and A. Smola. Support vector method for function approximation, regression estimation and signal processing. *In Advances in Neural Information Processing Systems, 9.*, 1997 MIT Press.

On the Significance of Markov Decision Processes

Richard S. Sutton

Department of Computer Science
University of Massachusetts, Amherst, MA USA
http://www.cs.umass.edu/~rich

Abstract. Formulating the problem facing an intelligent agent as a Markov decision process (MDP) is increasingly common in artificial intelligence, reinforcement learning, artificial life, and artificial neural networks. In this short paper we examine some of the reasons for the appeal of this framework. Foremost among these are its generality, simplicity, and emphasis on goal-directed interaction between the agent and its environment. MDPs may be becoming a common focal point for different approaches to understanding the mind. Finally, we speculate that this focus may be an enduring one insofar as many of the efforts to extend the MDP framework end up bringing a wider class of problems back within it.

Sometimes the establishment of a problem is a major step in the development of a field, more important than discovery of solution methods. For example, the problem of supervised learning has played a central role as it has developed through pattern recognition, statistics, machine learning and artificial neural networks. Regulation of linear systems has practically defined the field of control theory for decades. To understand what has happened in these and other fields it is essential to track the origins, development, and range of acceptance of particular problem classes. Major points of change are marked sometimes by a new solution to an existing problem, but just as often by the promulgation and recognition of the significance of a new problem. Now may be one such time of transition in the study of mental processes, with Markov decision processes being the newly accepted problem.

Markov decision processes (MDPs) originated in the study of stochastic optimal control (Bellman, 1957) and have remained the key problem in that area ever since. In the 1980s and 1990s, incompletely known MDPs were gradually recognized as a natural problem formulation for reinforcement learning (e.g., Witten, 1977; Watkins, 1989; Sutton and Barto, 1998). Recognizing the common problem led to the discovery of a wealth of common algorithmic ideas and theoretical analyses. MDPs have also come to be widely studied within AI as a new, particularly suitable kind of planning problem, e.g., as in decision-theoretic planning (e.g., Dean et al., 1995), and in conjunction with structured Bayes nets (e.g., Boutilier et al., 1995). In robotics, artificial life, and evolutionary methods it is less common to use the language and mathematics of MDPs, but again the problems considered are well expressed in MDP terms. Recognition of this common problem is likely to lead to greater understanding and cross fertilization among these fields.

MDPs provide a simple, precise, general, and relatively neutral way of talking about a learning or planning agent interacting with its environment to achieve a goal. As such, MDPs are starting to provide a bridge to biological efforts to understand the mind. Analyses in MDP-like terms can be made in neuroscience (e.g., Schultz, et al., 1997; Houk et al., 1995) and in psychology (e.g., Sutton and Barto, 1990; Barto et al., 1990). Of course, the modern drive for an interdisciplinary understanding of mind is larger than the interest in MDPs; the interest in MDPs is a product of the search for an interdisciplinary understanding. But MDPs are an important conceptual tool contributing to a common understanding of intelligence in animals and machines.

1 The MDP Framework

A Markov decision process comprises an *agent* and its *environment*, interacting as in Figure 1. At each of a sequence of discrete time steps, $t = 1, 2, 3, \ldots$, the agent perceives the state of the environment, s_t, and selects an action, a_t. In response to the action, the environment changes to a new state, s_{t+1} and emits a scalar reward, $r_{t+1} \in \Re$. The dynamics of the environment are stationary and Markov, but are otherwise unconstrained. In *finite MDPs* the states and actions are chosen from finite sets. In this case the environment is characterized by arbitrary probabilities and expected rewards, $P^a_{ss'}$ and $R^a_{ss'}$, for each possible transition from a state, s, to a next state, s', given an action, a.

At each time step, the agent implements a mapping from states to probabilities of selecting each possible action. This mapping is called the agent's *policy*, denoted π_t, where $\pi_t(s, a)$ is the probability that $a_t = a$ if $s_t = s$. The agent's goal is to maximize the total amount of reward it receives over the long run. More formally, in the simplest case, the agent should choose each action a_t so as to maximize the expected discounted return:

$$V^\pi(s) = E\left\{r_{t+1} + \gamma r_{t+2} + \gamma^2 r_{t+3} + \cdots \mid s_t = s, \pi\right\}, \tag{1}$$

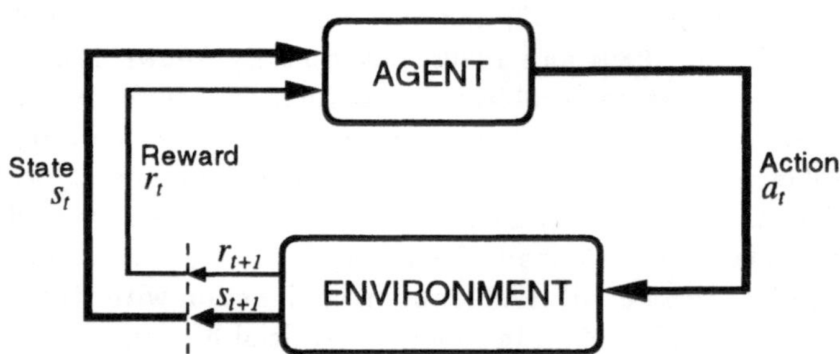

Fig. 1. The agent-environment interaction in Markov decision processes.

where γ, $0 \leq \gamma < 1$, is a discount-rate parameter akin to an interest rate in economics. $V^\pi(s)$ is called the *value* of state s under policy π, and the function V^π is called a *state-value* function. An *optimal policy*, denoted π^*, is a policy whose values are greater than or equal to that of all other policies at all states.

MDPs were originally studied under the assumption that the dynamics of the environment—$P^a_{ss'}$ and $R^a_{ss'}$—is completely known. The issue in this case is just the relative efficiency of various ways of computing optimal policies. In reinforcement learning the same problem has been studied under the assumption that the dynamics is completely unknown. A wide variety of intermediate cases have also been studied in reinforcement learning and optimal control. Other extensions include various kinds of approximation of optimal solutions, non-Markov dynamics, and undiscounted goals.

The MDP framework is abstract and very flexible, allowing it be applied to many different problems and in many different ways. For example, the time steps need not refer to fixed intervals of real time; they can refer to arbitrary successive stages of decision making and acting. The actions can be low-level controls such as the voltages applied to the motors of a robot arm, or high-level decisions such as whether or not to have lunch or to go to graduate school. Similarly, the states can take a wide variety of forms. They can be completely determined by low-level sensations, such as direct sensor readings, or they can be more high-level and abstract, such as symbolic descriptions of objects in a room. Some of what makes up a state could be based on memory of past sensations or even be entirely mental or subjective. For example, an agent could be in "the state" of not being sure where an object is, or of having just been surprised in some clearly defined sense. Similarly, some actions might be totally mental or computational. For example, some actions might control what an agent chooses to think about, or where it focuses its attention. In general, actions can be any decisions we want to learn how to make, and the state can be anything that might be useful in making them.

It is revealing to contrast the MDP framework with other problem formulations. In adaptive control, the environment ("plant") is taken to be a linear or nonlinear system. The critical difference from MDPs is that the objective is to track a given desired trajectory. In addition, it is assumed that errors can be determined at each time and smoothly reduced without greatly degrading performance at other times. These differences make adaptive control suitable for a wide class of regulation problems, but not as a model of an animal or artificial agent as a whole. In the context of an overall agent we need to allow more general environment dynamics and more general goals. Adaptive control may be well suited to guiding a robot arm along a given trajectory, but not for choosing the trajectory, or for deciding whether to reach for one object rather than another.

The MDP framework differs from control theoretic and other older frameworks primarily in that it is more general. This makes the MDP framework more widely applicable, but harder to solve exactly. Today's increased interest in autonomous agents (requiring greater generality) and greater computational resources (enabling search for better approximations) shifts the tradeoff in favor of the MDP framework.

2 Implications

The MDP framework with completely known dynamics has recently been studied within artificial intelligence (AI) as a *planning* problem. This formulation of planning is more general than that previously considered in AI, which has classically assumed deterministic state-transitions and a set of goal states all equally desirable and all paths to which were equally desirable. AI planning thus reduced to finding a single action sequence taking the agent from a start state to a goal state. The sequence of actions could be executed *open-loop*, that is, without sensing, because the environment was deterministic (and completely known) and thus sensing added no new information. In the last decade or two AI has come to realize that this conception of planning and execution was too limited, that execution has to be sensitive to unforeseen events and that planning has to take into account their probabilities of occurrence. This seems inevitably to lead to plans that are more like policies ("universal plans") than they are like action sequences. Today a significant segment of AI planning research has embraced the view that all behavior should be closed loop, that stochastic or incompletely foreseen events should be considered normal rather than the exception. This growing segment of the AI community is also using the MDP framework (e.g., Dean et al., 1995; Barto, Bradtke and Singh, 1995).

I should emphasize at this point that using the MDP framework does not mean that AI researchers, or reinforcement learning researchers, or others, are reverting to, or doing no more than, the prior work using MDPs. The new research, where it is significant, always brings in new issues and new challenges that were not the focus of prior research in MDPs. For example, AI planning research using MDPs may consider much larger and more richly structured state spaces. Other important extensions are approximation methods, for example, anytime planning issues (Dean et al., 1995), incomplete dynamics knowledge (as in reinforcement learning), and sampling methods (e.g., Tesauro and Galperin, 1997).

The most important implication of MDPs for planning methods is the importance that it suggests should be placed on the concepts of policies and value functions as long-term memory structures that accumulate planning results. Much of the planning literature in both AI and control assumes planning is a one-shot computation in which a planning problem is presented and completely solved before taking action. In an MDP framework, planning is typically too complex to expect it ever to be solved completely; certainly one does not want to hold up action waiting for a complete solution. Instead one seeks planning algorithms that form good approximate plans (policies) quickly and then gradually improves their quality the more time passes. This is the heart of MDP-based planning methods such in dynamic programming and reinforcement learning. The results of planning accumulate in the improving policy function and, even more so, in the approximate value function.

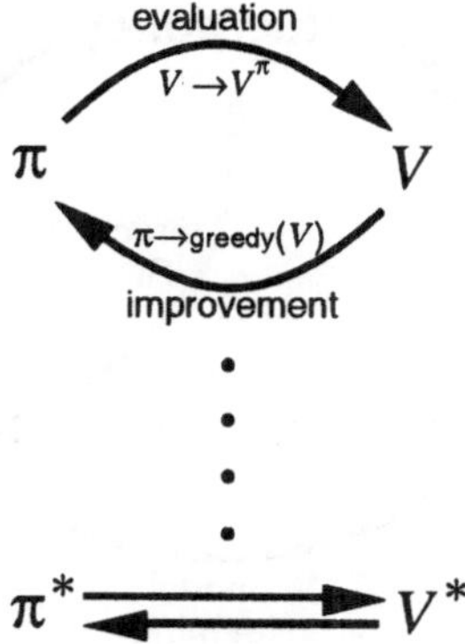

Fig. 2. In generalized policy iteration, a general schema for solving MDPs, approximate policy and value functions continually interact until settling at their optimal values.

In our recent book (Sutton and Barto, 1998), Andy Barto and I suggest that a wide range of methods for solving MDPs can be understood as a recursive ratcheting interplay between the approximate policy and value functions. We call the interplay *Generalized Policy Iteration* (GPI) after a related dynamic programming method, "policy iteration." The overall idea of GPI is suggested by Figure 2. At any moment in time the agent maintains both a policy, π, and a value function, V. The policy influences the value function via an "evaluation" process that modifies the value function to more accurately predict the rewards actually received under the policy, that is, to approximate V^π as given by (1). Simultaneously, the value function influences the policy to improve in a local, greedy fashion based on V. For example, the policy may be moved toward the greedy policy that deterministically selects the action, a, in each state, s, which maximizes $\sum_{s'} P^a_{ss'}[R^a_{ss'} + \gamma V(s')]$.

The policy and value function influence each other but, as they change, the targets that they provide for each other also change. The two provide moving targets for each other, as suggested by the curved lines in the top of Figure 2, causing a joint evolution of both functions toward a solution that satisfies both processes, as suggested by the bottom of figure. When a policy and value function are found that are undisturbed by either process, then they are guaranteed to be the optimal policy and its value function, π^* and V^{π^*}. Moreover, in many cases one can guarantee steady improvement to near this stable point.

Another way of expressing the important role of value and policy functions in incremental planning is suggested by Figure 3. Here these functions are seen as one of three fundamental data structures interrelated by planning, acting and learning processes. The policy determines actions and thus experience, experience feeds into learning processes for improving the agent's model of its world,

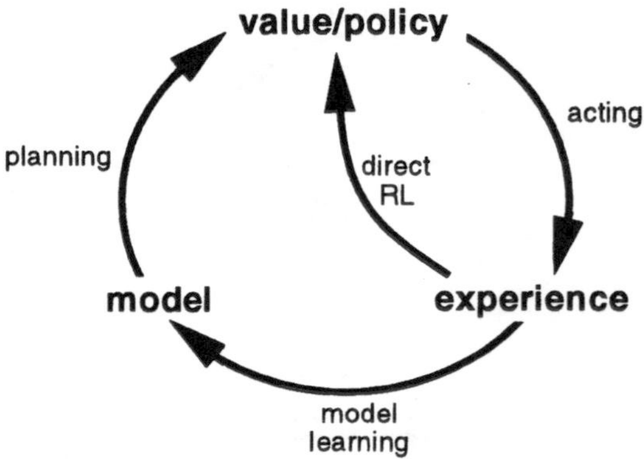

Fig. 3. The MDP framework suggests a continual incremental cycle of interaction and improvement among learning, planning and acting processes.

and the model in turn feeds into planning processes which improve the value and policy functions, completing the cycle. Experience may also influence the value and policy functions direct through direct reinforcement-learning methods.

3 Extensions and Reductions

Part of what makes the MDP framework appealing is that, even when it needs to be extended, the extensions are often done in such a way as to reduce the extended problem back to an MDP. The most prominent example of this is the extension of MDPs to the non-Markov case, in which the state is not fully observable on each time step. The classical approach to partially-observable MDPs (POMDPs) is to estimate the probability with which the environment is in each of its possible states at each time. It turns out that this probability distribution can itself be treated as the state of the whole POMDP process, and then all the classical MDP methods can be applied, albeit in a larger and more complex state space. Others have proposed simply adding memories of earlier observations to the state representation (e.g., McCallum, 1995) and then proceeding with the same methods as approximations. In some cases, convergence results can still be obtained for such ad hoc approaches (e.g., Singh et al., 1994, 1995).

Another important but simple extension is that from finite MDPs to general MDPs with continuous state and action variables. In this case, the state space can be arbitrarily large, even infinite. It is no longer possible to represent policy and value function exhaustively, in tables. Instead, they must be represented by parameterized function approximators such as artificial neural networks. Not all methods for the finite/tabular case extend reliably to the use of function approximators, but many do (e.g., see Sutton, 1996; Santamaria et al., 1996). The ability to use function approximators in this way is largely responsible for most of the modern successes of reinforcement learning (e.g., Tesauro, 1995; Crites and Barto, 1996; Van Roy et al., 1996). Again, the extension to a larger and more difficult case is handled by a little additional machinery, but remaining within the same overall framework.

Finally, it has also been proposed that even hierarchical and modular approaches to decision-making and action generation can be incorporated within the nominally low-level MDP framework (Singh, 1992; Sutton, 1995; Precup and Sutton, tion). The idea here is to reason about whole complexes of action—whole subpolicies—as if they were a single action. For example, one might have low-level actions to activate individual muscles, and also subpolicies for reaching, picking up objects, making a phone call, driving to work, or flying to London. In recent work, Precup and I have shown that, in principle, all of these can be immediately incorporated into the MDP framework, including Bellman equations and many of the solution methods, with essentially no changes.

An example is shown in Figure 4. In this gridworld, the primitive actions are steps up, down, right, and left, which usually cause the agent to move to the corresponding cell (but a third of the time they cause it to move to one of the other three neighbors). Two subpolicies have been previously learned for each room that bring the agent efficiently to each hallway state between rooms. Figure 5 shows what happens during planning (via the value iteration algorithm of dynamic programming) when the subpolicies are used as abstract actions in parallel with the primitive actions. In this planning problem, the agent is told that reward can be obtained only at the goal state indicated by the single disk in the first panel of the figure. The value function is 1 in this state and zero elsewhere on this, the first iteration. On the next two iterations, the region of accurate valuation spreads out slowly, by one neighboring cell per iteration. In normal value iteration, this process would continue, cell by cell, until the whole state space was correctly valued. In this example it would take approximately 18 iterations. Instead, starting with the fourth iteration we see the correct valuations suddenly jumping back and being filled in a room at a time rather than a state at a time. This is due to the inclusion of the subpolicies for room-to-room abstract actions, and their associated models, in the planning process. The system is able to plan about moving from room-to-room over an indefinite sequence of actions in exactly the same way as it plans about moving from state-to-state in one time step. The result is much faster planning and determination of an appropriate policy.

4 Conclusion

The MDP framework is general and simple, with an elegant and compact theory. It seems to capture something essential about using cause-and-effect to achieve goals. Even as we consider more complex cases, we are able to usefully bring them back to the base case of MDPs. MDPs are widely used in reinforcement learning, but they may also be relevant much more widely. The concepts of policy and value function, Bellman equation and generalized policy iteration, may be useful and intuitively relevant throughout cognitive science.

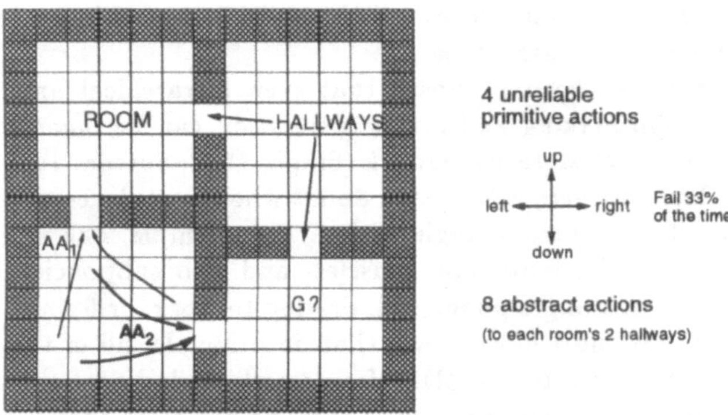

Fig. 4. An environment with cell-to-cell actions and learned room-to-room abstract actions (subpolicies).

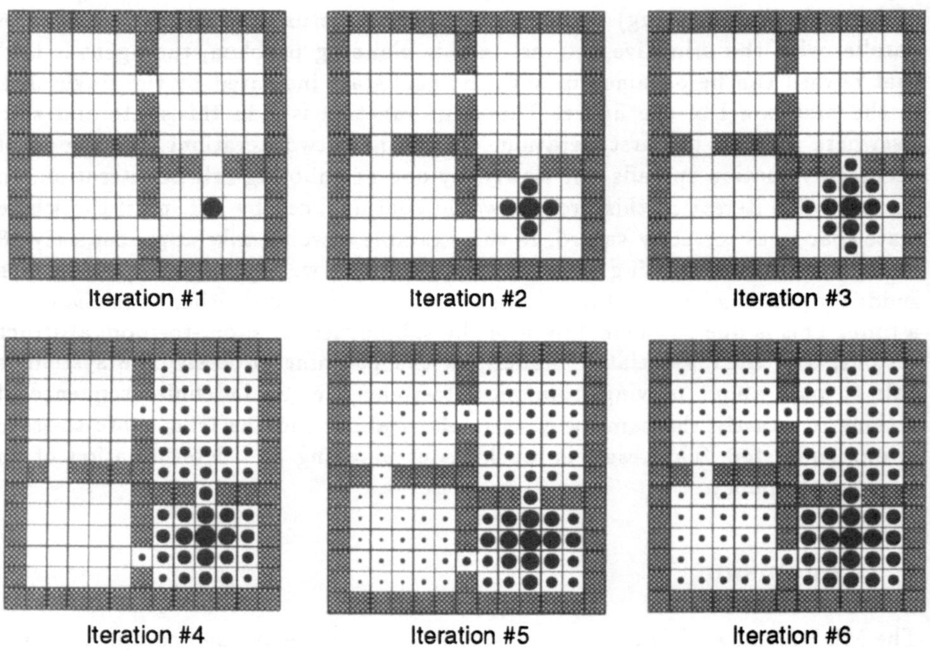

Fig. 5. Value iteration using abstract actions. After the third iteration, room-to-room planning dominates and quickly finds the optimal value function and policy to the goal state indicated in the first panel.

References

Barto, A. G., Bradtke, S. J., and Singh, S. P. (1995). Learning to act using real-time dynamic programming. *Artificial Intelligence*, 72:81–138.

Barto, A. G., Sutton, R. S., and Watkins, C. J. C. H. (1990). Learning and sequential decision making. In Gabriel, M. and Moore, J., editors, *Learning and Computational Neuroscience: Foundations of Adaptive Networks*, pages 539–602. MIT Press, Cambridge, MA.

Bellman, R. E. (1957). A Markov decision process. *Journal of Mathematical Mech.*, 6:679–684.

Boutilier, C., Dearden, R., and Goldszmidt, M. (1995). Exploiting structure in policy construction. In *Proceedings of the Fourteenth International Joint Conference on Artificial Intelligence.*

Crites, R. H. and Barto, A. G. (1996). Improving elevator performance using reinforcement learning. In D. S. Touretzky, M. C. Mozer, M. E. H., editor, *Advances in Neural Information Processing Systems: Proceedings of the 1995 Conference*, pages 1017–1023, Cambridge, MA. MIT Press.

Dean, T. L., Kaelbling, L. P., Kirman, J., and Nicholson, A. (1995). Planning under time constraints in stochastic domains. *Artificial Intelligence*, 76(1-2):35–74.

Houk, J. C., Adams, J. L., and Barto, A. G. (1995). A model of how the basal ganglia generates and uses neural signals that predict reinforcement. In Houk, J. C., Davis, J. L., and Beiser, D. G., editors, *Models of Information Processing in the Basal Ganglia*, pages 249–270. MIT Press, Cambridge, MA.

McCallum, A. K. (1995). *Reinforcement Learning with Selective Perception and Hidden State.* PhD thesis, University of Rochester, Rochester.

Precup, D. and Sutton, R. S. (in preparation). Multi-time models for temporally abstract planning.

Santamaria, J. C., Sutton, R. S., and Ram, A. (1996). Experiments with reinforcement learning in problems with continuous state and action spaces. Technical Report UM-CS-1996-088, Department of Computer Science, University of Massachusetts, Amherst, MA 01003.

Schultz, W., Dayan, P., and Montague, P. R. (1997). A neural substrate of prediction and reward. *Science*, 275:1593–1598.

Singh, S. P. (1992). Transfer of learning by composing solutions of elemental sequential tasks. *Machine Learning*, 8:323–339.

Singh, S. P., Jaakkola, T., and Jordan, M. I. (1994). Learning without state-estimation in partially observable Markovian decision problems. In Cohen, W. W. and Hirsch, H., editors, *Proceedings of the Eleventh International Conference on Machine Learning*, pages 284–292, San Francisco, CA. Morgan Kaufmann.

Singh, S. P., Jaakkola, T., and Jordan, M. I. (1995). Reinforcement learing with soft state aggregation. In G. Tesauro, D. Touretzky, T. L., editor, *Advances in Neural Information Processing Systems: Proceedings of the 1994 Conference*, pages 359–368, Cambridge, MA. MIT Press.

Sutton, R. S. (1995). TD models: Modeling the world at a mixture of time scales. In Prieditis, A. and Russell, S., editors, *Proceedings of the Twelfth International Conference on Machine Learning*, pages 531–539, San Francisco, CA. Morgan Kaufmann.

Sutton, R. S. (1996). Generalization in reinforcement learning: Successful examples using sparse coarse coding. In Touretzky, D. S., Mozer, M. C., and Hasselmo, M. E., editors, *Advances in Neural Information Processing Systems: Proceedings of the 1995 Conference*, pages 1038–1044, Cambridge, MA. MIT Press.

Sutton, R. S. and Barto, A. G. (1990). Time-derivative models of Pavlovian reinforcement. In Gabriel, M. and Moore, J., editors, *Learning and Computational Neuroscience: Foundations of Adaptive Networks*, pages 497–537. MIT Press, Cambridge, MA.

Sutton, R. S. and Barto, A. G. (1998). *Introduction to Reinforcement Learning.* MIT Press/Bradford Books, Cambridge, MA.

Tesauro, G. J. (1995). Temporal difference learning and TD-Gammon. *Communications of the ACM*, 38:58–68.

Tesauro, G. J. and Galperin, G. R. (1997). On-line policy improvement using monte-carlo search. In *Advances in Neural Information Processing Systems: Proceedings of the 1996 Conference*, Cambridge, MA. MIT Press.

Van Roy, B., Bertsekas, D. P., Lee, Y., and Tsitsiklis, J. N. (1996). A neuro-dynamic programming approach to retailer inventory management. Technical Report LIDS-P-?, Laboratory for Information and Decision Systems, Massachusetts Institute of Technology.

Watkins, C. J. C. H. (1989). *Learning from Delayed Rewards.* PhD thesis, Cambridge University, Cambridge, England.

Witten, I. H. (1977). Exploring, modelling and controlling discrete sequential environments. *International Journal of Man-Machine Studies*, 9:715–735.

Economical Reinforcement Learning for Non Stationary Problems

Nathalie Chatenet[1] and Hugues Bersini[2]

[1] Cemagref de Bordeaux (division ORH)
50, Avenue de Verdun, Gazinet - 33 612 Cestas Cedex - FRANCE

[2] Université libre de Bruxelles - IRIDIA - CP194/6
50, avenue Franklin Roosevelt - 1050 BRUXELLES - BELGIUM

Abstract. In a lot of reinforcement learning applications and solutions, time information is implicitly contained in state information. Although time sequential decomposition is inherent to dynamic programming, this aspect has been simply omitted in usual Q_Learning applications. A non stationary environment, a non stationary reward/punishment or a time dependent cost to minimize will naturally lead to non stationary optimal solutions in which time has to be explicitly accounted in the search for the optimal solution. Although largely neglected so far, non stationarity and the computational cost it is likely to rapidly induce, should instead becomes of concern to the reinforcement learning community. The particular nature of time entails a dedicated processing when attempting to develop economical and heuristic solutions. In this paper two of such possible heuristics are proposed, justified and illustrated on a simple application.

1 Introduction

An optimal policy for optimal control problems is said to be non stationary when the control policy is explicitly dependent on time i.e. when the optimal action to execute in a given state will be different according to the instant. First of all, reinforcement learning is viewed as an approximate, asynchronous and online version of Dynamic Programming and it is worth noticing that the largest part of reinforcement learning or Q_learning applications assumes the solutions to be stationary. The robot in the labyrinth [6] is one of the most prototypical reinforcement learning problem and indeed, the optimal path is time independent i.e. optimal move only depends on the current robot position. As a matter of fact, the original presentation of Q_learning [2] considers the optimal action to be only dependent on the current state and accordingly the Q_values to be only dependent on the current state-action: $Q = Q(x, a)$. Although inherently present in dynamic programming formulation, time simply disappears in a lot of reinforcement learning theoretical and practical developments. $Q(x, a)$ is the expected infinite horizon discounted return if action a is taken in state x, without any concern for the actual instant this action is executed. Nevertheless, real world is non stationary and a good move at time t could simply be disastrous later on when turning in this same state again. A non stationary environment, a non stationary reward/punishment or a time dependent cost to minimize will naturally lead to non stationary optimal solutions in which time has to be explicitly accounted in the search for the optimal solution. Although largely neglected, non stationarity should instead becomes of concern to the reinforcement learning community. Solving non stationary problems corresponds in adding time as another variable in whatever reinforcement learning application. Apparently, the turning of $Q(x,a)$ into $Q(x,a,t)$ is nothing really complicated and, accordingly, the Q_learning well-known recurrent equation would simply turn out to be:

$$\Delta Q(x(t),a,t) = \alpha[r(t) + \gamma Max_a Q(x(t+1),a,t+1) - Q(x(t),a,t)]$$

Every development both at theoretical and practical levels should easily extend to this updated non stationary version of Q_learning. However, a major worry for reinforcement learning and Q_learning, even more than for dynamic programming, is the slowness to converge and the exponential scalability with the dimension of the state space. Attempts to maintain minimal state space in Q_learning applications, either by incremental, decremental or clustering strategies, have become quite popular [3, 4, 1]. Among all variables, time is a crucial one. First, it can easily take a huge amount of values, then its special nature makes possible the use of dedicated heuristic approaches. In this paper we propose various systematic and economical processings of time while running reinforcement learning algorithms for non stationary problems. These processings are based on the perception of non stationary solutions as approximately composed of successive pieces of stationary solutions. Non stationary optimal solution can be gradually approached by decomposing more and more the control temporal horizon into time intervals in which the solution remains stationary. In other words, the optimal action $a(x,t)$ can be replaced by a vector of actions $[a_{T_1}(x), a_{T_2}(x), ..., a_{T_n}(x)]$ and accordingly $Q(x,a,t)$ replaced by a vector $[QT_1(x,a), QT_2(x,a), ...QT_n(x,a)]$, where $a_{T_i}(x)$ is the optimal action to be executed in state x during the time interval $[T_{i-1}, T_i]$ and $QT_i(x,a)$ is its associated Q-value during this same interval.

The quality of the solution can be progressively improved, if necessary, by increasing the number of decompositions. Such a gradual increase of the number of states is a classical approach in the reinforcement learning community but, here time is concerned, which by no means is a variable which rapidly tends to explode. In the next section a simple optimal control problem presenting a non stationary solution will be described. The running of the classical time-independent Q_learning would lead to sub-optimal solution. This simple application will serve as a benchmark for comparing three approaches: first, the obvious and not economical extension of Q_learning obtained by adding time as just another variable, then two non stationary heuristic Q_learning. By gradually adding interval of stationary solutions in various ways, these heuristic approaches can progressively reach the optimal solution but at different convergence speed.

2 A Simple Numerical Application

Suppose a one dimensional process given by the following deterministic difference equation: $x_{k+1} = x_k + u_k$; with :

- $k = 0, 1...., T-1$ with $T = 24$ which designates the problem temporal horizon,
- u_k can restrictively takes one among the 10 values {-5,-4,-3,-2,-1,1,2,3,4,5},
- x_k is an integer, which can only be comprised in the interval [0,10].

The target must be reached at $k+1 = 24$ and be one of the two values {8,9}. At each action u is associated a cost given by :

Actions	-5	-4	-3	-2	-1	1	2	3	4	5
Costs	6	6.8	7.6	8.4	9.2	1	2	3	4	5

The reinforcement associated to the target is :

$$r(x_T) = \begin{cases} 9 & \text{if } x_T \in \{8,9\} \\ -1 & \text{otherwise} \end{cases}$$

The optimal control objective is to minimize the following discounted cost:

$$E = \sum_{k=0}^{k=23} 0.9^k c_k(x_{k+1}, u_k) \text{ with } c_k(x_{k+1}, u_k) = \begin{cases} c(u_k) & \text{if } k < 23 \\ c(u_{23}) - r(x_{24}) & \text{otherwise} \end{cases}$$

The initial condition is x(0)=10.

By using dynamic programming, it is easy to compute the optimal non stationary solution as shown in Figure 1.

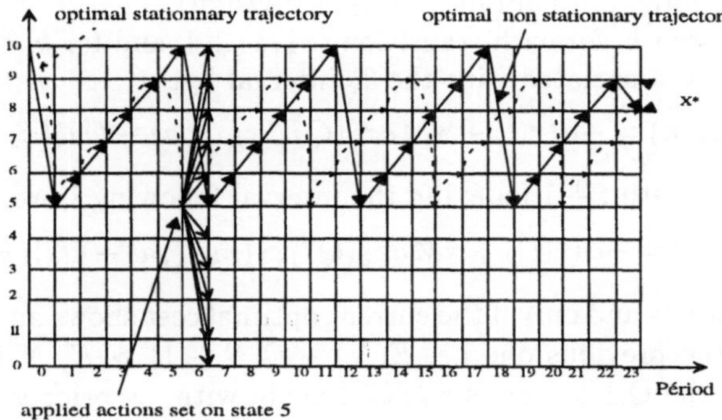

Fig. 1. Stationary and Non Stationary Optimal Solutions of the simple Benchmark.

The solution is non stationary since in position 9 the optimal action is 1 for k<23 and -1 for k=23. Its corresponding cost is 10.75. The stationary suboptimal solution equally found by dynamic programming is also indicated in Figure 1. Its corresponding cost is 12.55. The figure helps to understand the natural tendency of non stationary optimal solutions to cluster in successive periods of stationary solutions and why a promising approach for non stationary problems indeed consists in progressively adding time intervals characterized by stationary solutions.

3 Two Heuristic Non Stationary Q-Learning Algorithms

The Turnpike theorem, more deeply discussed in [7, 5], tells the following: For certain categories of optimal control problem, it is always possible to find a time interval in which the optimal non stationary solution can be closely approximated by a stationary solution, the quality of the approximation being dependent on the cost difference between the optimal non stationary solution and the optimal stationary solution, and thus on the length of the interval. The discovery of the right stationary intervals so as to minimize the approximation error will be the aim of the two heuristic approaches presented now. These methods both consist in gradually adding new time intervals with stationary solutions. Whenever a new time interval is added or the temporal decomposition of the problem is refined, the initial Q_values for this new temporal structure is the one obtained for the previous structure.

3.1 Algorithm H1

The first algorithm is the most intuitive and the most general one. It does not assume any prior knowledge of the temporal decomposition of the problem and simply consists in progressively dividing the temporal horizon into more intervals of equal size. A certain temporal decomposition is given by $QT_i(x, a)$ with $i = 1..N$ where N is the current number of time intervals. N takes initial value 1 and, following a certain number of learning iterations (called `ITER1`), takes value 2 then 3 then 4 ... The length of each time interval is given by $\frac{T}{N}$ (adjusting the length of the last interval to obtain the right division).

At each time step k, for each transition (x_k, x_{k+1}), and for $k < T_{i+1} - 1$, we adjust the Q_values associated with the T_i interval by:

$$\Delta QT_i(x_k, a) = -\alpha(r(t) + \gamma Max_a(QT_i(x_{k+1}, a) - QT_i(x_k, a))$$

and for $k = T_i - 1$ with $i + 1$ indexing the interval following i, we have:

$$\Delta QT_i(x_k, a) = -\alpha(r(t)) + \gamma Max_a(QT_{i+1}(x_{k+1}, a) - QT_i(x_k, a))$$

After `ITER1` iterations and only if the current optimal cost shows an improvement with respect to the previous one i.e. $E(N) - E(N-1) > E$, N is once again incremented and the Q_learning is released again with, as prior value, the ones obtained by the previous decomposition i.e.

$$\text{Initial } QT_i(x_k, a) \text{ for } N + 1 = \text{Final } QT_j(x_k, a) \text{ for } N, \text{ for } k \in (Ti \cap Tj).$$

The temporal decomposition terminates when no improvement is allowed by the last decomposition. Other very similar time incremental decompositions can be easily imagined.

3.2 Algorithm H2

This second algorithm performs a backward decomposition in time. The first division in two time intervals $\{T_1, T_2\}$ is done in the following way:

$T_1 = [1, T-2]$ and $T_2 = [T-1]$.

After ITER1, the second more refined division is:

$T_1 = [1, T-3], T_2 = [T-2], T_3 = [T-1]$ and so forth.

The recurrent updated equations of Q_learning are the same as in the previous chapter. With a view to the small benchmark described in the previous chapter, adding new time interval backward in time seems to be a very natural thing to do, since the precise time for the non stationarity to occur is likely to be close to the end.

4 Experimental Comparisons

Three non stationary Q_learning versions have been tested on the simple application described in section 2. For all versions, whenever one action in a state makes the process to exceed the interval [0,10], it is suppressed from the list of possible actions in that state and another action is selected. The whole control process runs during 24 time steps before to repeat again. Such a run amounts to one iteration. The first method is the simple extension of Q_learning, including time as a new variable, like explained in the introduction. For this particular application, for every action/state there are 24 Q_values to maintain. The emphasis in this article deals with an economical processing of time so that we do not expect to have tackled whatever other classical and important issues of reinforcement learning in any interesting way. The exploration/exploitation dilemma is faced in the following way. Actions are always chosen randomly. After ITER2 initial iterations, the best cost achieved so far (selecting the best action in each state) is computed. Then every ITER3 iterations, whenever the best cost reached so far has decreased of 10% with respect to the previous one, the worst action in each state is deleted from the list of the possible actions.

In the experimental results shown here, we have ITER1=600, ITER2=2500 and ITER3=500.

For the two non stationary heuristic approaches, every ITER1 iterations, the temporal decomposition is changed adding one new time interval. In Figure 2, the convergence of the minimal cost is plotted as a function of the number of iterations.

Not surprisingly, the H2 version of the Q_learning is the one which converges faster on this problem. In fact the perfect temporal decomposition in two time steps is the first to be tried by this algorithm. In general, there is no reason for such an ideal situation to be always the case and a lot of other alternative gradual decompositions could be equally good candidates to try.

5 Conclusions

The work presented here first aims at increasing the awareness of the reinforcement learning community for the importance of the so far neglected non stationary optimal solutions. Time is often hard to omit and is computationally

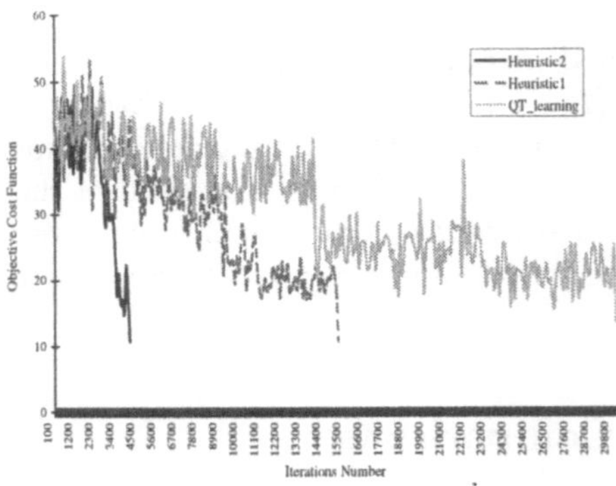

Fig. 2. Convergence speed of the three non stationary Q_Learning approaches.

expensive. Secondly, rather than treating time as any other state variable, its particular nature entails a dedicated processing when attempting to develop economical and heuristic solutions. For instance, the backward decomposition at the basis of one of the heuristics would make no sense for any other variable. Also the existence of stationary parts hidden somewhere in the complete solution seems here again a particular property inherent to time and justifying the approaches proposed in the paper. We believe this paper to be a first step when the reinforcement learning solutions have to explicitly take time into account, and we suspect that some of the ideas developed in its content could also be of interest for reducing the computational cost of some dynamic programming applications.

References

1. A.W. Moore. Variable Resolution Dynamic Programming : Efficiently Learning Action Map in Multivariate Real-valued State-spaces. In *Proceedings of the Eighth International Workshop on Machine Learning*, pages 333–337, 1991.
2. C. Watkins. Technical Note Q_learning. *Machine Learning*, 8:279–297, 1992.
3. D. Chapman and L. Kaebling. Input Generalization in Delayed Reinforcement Learning: An Algorithm and Performance Comparisons, 1991.
4. M. Dorigo and H. Bersini. A Comparison of Q_Learning and Classifier Systems. In *Proceedings of SAB III*, pages 248–255. MIT Press, 1994.
5. N. Chatenet. Adaptation of Turnpike Theorem in a Variant of Q_learning. Technical Report 96016, University of Bordeaux I, 1996.
6. R. Sutton. Integrated Modeling and Control Based on Reinforcement Learning and Dynamic Programming. In *Advances in Neural Information Processing 3*, pages 471–478, 1991.
7. S. Achmanov. *Programmation Linéaire*. MIR, 1984.

A Double Gradient Algorithm to Optimize Regularization

Thomas Czernichow
Thomas.Czernichow@iit.upco.es
Instituto de Investigación Tecnológica, Universidad Pontificia Comillas
Santa Cruz de Marcenado, 26.
28015 Madrid, Spain

Abstract. We present in this article a new technique dedicated to optimise the regularization parameter of a cost function. On the one hand the derivatives of the cost function with regards to the weights permits to optimise the network. On the other the derivatives of the cost function with regards to the regularization parameter permits to optimize the smoothness of the function achieved by the network. We show that by oscillating between these two gradient descent optimisations we achieve the task of regulating the smoothness of a neural network. We present the results of this algorithm on a task design to clearly express the network's level of smoothness.

1. Introduction

It is well known that when the training database is finite (of size N_l), the solution to the problem of finding the function which minimizes the Mean Squared Error (MSE) is ill posed :

$$\hat{f}(x) = \underset{h}{Argmin}\left[\sum_{t=1}^{N_l}\left(y_t - h(x_t)\right)^2\right] \tag{1}$$

The problem is ill posed in at least two terms. Firstly the solution is not unique, and secondly the variation from one solution to another with regards to the variation to the database might be high. To rectify these defaults many paths exist, like the choice of an appropriate learning algorithm or to impose restrictions on the type of the functions h. In the case of neural neworks one could use for example an early stopping algorithm to smooth the achieved function ([1]). Another solution would be to diminish the number of weights, or the number of hidden units ([2, 3]) for a given architecture. One could also include a second term to the cost function (eq. 1) penalizing parts of the space function in which we search h ([4]). This second term Φ, the stabilizer, is multiplied by a regularization parameter λ, which modulates the strengh of the penalty :

$$\hat{f}(x) = \underset{h}{Argmin}\left[\sum_{t=1}^{N_l}\left(y_t - h(x_t)\right)^2 + \lambda\Phi\left(h(x)\right)\right] \tag{2}$$

When the function h is entirely defined by its parameters it is possible that Φ would be a function of these ones ([5, 6]). In one of the very popular methods, the "ridge" regression ([7]), the penalty term is the sum of the squared parameters :

$$\hat{f}(x) = \underset{h}{Argmin}\left[\sum_{t \in L}\left(y_t - h(x_t)\right)^2 + \lambda \sum_{p \in I} w_p^2\right]$$

Where L is the set of index of the learning database, I the weights' index and h the function achieved by the network. We will use here one hidden layer networks to expose the algorithm, although it is not linked with any particular architecture.

2. The algorithm

The ridge regression tends to favorize small weights by it corresponds to assuming a gaussian prior distribution on the weights which would be centered at zero. The smoothing parameter corresponds to an estimation of this variance under this assumption. One of the problems with all regularization techniques is to choose the smoothing parameter λ ([8]). The usual way to estimate its value is to use the validation base. Namely, estimate the network for some values of λ, and elect the one that gives the smallest MSE on the validation base. This method is both unprecise (the choice of the values to test is made blindly), and long (for each value a whole estimation process has to be run). Some methods, however, keep the previous weights as starting points for the following estimation. Our method is a gradient descent variation of this technique.

As said before, we will use to estimate the network the regularized MSE cost function, $C_L = \sum_{t \in L} \|\hat{y}_t - y_t\|^2 + \lambda \sum_{p \in I} w_p^2$, where $\hat{y}_t = h(x_t)$ is the estimation made by the network. Once the optimization process (using $\frac{\partial C_L}{\partial w_p}$, $\forall p \in I$) as reached a minimum ($w^* = \underset{w}{Argmin}(C_L)$), we would then like to find, at w^*, which λ would minimise the error on the validation base ($C'_V = \sum_{t \in V} \|\hat{y}_t - y_t\|^2$). To achieve this goal we need to compute the following derivatives : $\left.\frac{\partial C'_V}{\partial \lambda}\right|_{w^*}$. The appendix presents the computation of this gradient and the linked assumptions. Our approach consists in starting with a value λ_0. We then find the optimal weights w^{0*} with regards to this value and compute $\left.\frac{\partial C'_V}{\partial \lambda}\right|_{w^{0*}}$. Using the delta rule we compute a second value $\lambda_1 = \lambda_0 - \varepsilon_0 \left.\frac{\partial C'_V}{\partial \lambda}\right|_{w^{0*}}$.

Fig. 1. Scheme of the algorithm

More generally, the Double Gradient Algorithm (DGA) alternates in between finding the optimal weights w^{i*} at λ_i set constant, and computing λ_i from $w^{(i-1)*}$ using the delta rule. The gradient step (ε_i) follows a decreasing curve on 1/i to fit the so called Robbins-Monro converging conditions. The search algorithm used to find w^{i*} takes $w^{(i-1)*}$ as starting point, and stops when the weights have a relative change below 1E-2, assuming that the inputs have been normalised. We will now present the results on a time series specially built to enlight the control of the regularization.

3. Increasing frequency function

The following function is interesting because the frequency of the signal increases with *x*, although we keep picking the points with a uniform distribution.

$$g(x) = cos(1/exp(-x))$$

It is then very easy to see where is the cutoff frequency of the estimator, as it then stays straight at the expectancy. We have compared different optimisation methods to estimate their precisions. All the tries are made with 200 points, taken uniformly from [0; 3.5] in each of the three databases, with an added noise of variance 0.1, and 100 hidden neurones to be sure that the network will be over-parametrized.

Methods	Learning base	Validation base	Test base
Best weights on the Validation base	0.08 (0.04)	0.22 (0.05)	0.22 (0.06)
Fixed λ (best)	0.09 (0.02)	0.16 (0.02)	0.17 (0.04)
Linearly Increasing λ	0.10 (0.04)	0.24 (0.06)	0.22 (0.05)
DGA λ_0=1E–6	0.06 (0.01)	0.15 (0.01)	0.20 (0.01)
DGA λ_0=1E–3	0.08 (0.02)	0.14 (0.008)	0.18 (0.02)
DGA λ_0=1	0.07 (0.005)	0.14 (0.009)	0.16 (0.03)

Tab. 1. mean and (std) of the MSE on the three databases for 4 different methods.

The first method used consists in choosing the best weights on the validation base, with no regularisation parameter (λ=0). Table 1 shows the mean and the standard deviation (std) of 100 tries each using differents initialisations of the weights. The second method includes a fixed value of λ from 1E-6 to 3 (see fig. 2 (a) for a zoom on the minimum point of the validation set). Each point is made of 20 tries and we have put the error values of the minimum (for λ=0.001) in table 1. The third method comes from the idea that at the beginning the smoothing parameter should be small, because the function achieved by the network needs to "bend" to modelize the desired output. During the learning λ should increase so as to prevent over-fitting, to prevent the network to "bend" too much. For this method, the parameter λ starts at 1E-6 and increased linearly by steps of 1E-6. The algorithm stops when the error on the learning base stays still. The three other results of table 1 are from the DGA algorithm for three different initialisations of λ.

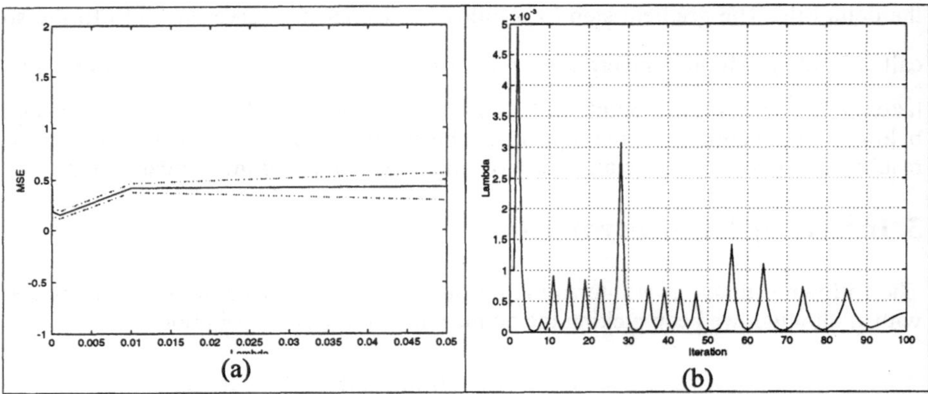

Fig. 2. (a) mean of the RMSE and confidence interval (±2.std) for each of the 20 tries at fixed value of Lambda and (b) evolution of the lambda parameter starting at 1e-3 for one estimation.

The mean number of oscillations between the optimisation of λ and w^* is about 12 for the three initialisations. The first search of the minimum takes about 200 steps, but all the other (after the first change of λ) are in less than 10, which enforces the assumption that we stay near the optimum. When λ is too small at the beginning the algorithm starts to increase it until the networks is sufficiently smoothed (fig. 2.a). When its value is too big it will continualy reduce it until reaching a bottom value at which it will start to slowly oscillate to keep the correct smoothing factor. From the point of view of the DGA, at least, it seems preferable to starts with a high value of λ. The intuition is that it seems more easy to 'bend' little by little an over-parametrized model than to unfold it when it has caught all the variance of a database. The numerical reason is that when choosing a high value of λ the matrix $\left(\lambda Id + \Delta^T \Delta\right)$ will be, at the beginning, conditionned in a better way for inversion (see appendix).

4. Conclusion

We have presented in this article an optimizing method of the regularization parameter in the case of the ridge regression. The approach is based on the computation of the gradient of the error on the validation base with regards to this parameter. The results shows that this double gradient descent based algorithm is very rapid because each iterations starts very close from the next minimum, and very efficient in controling the complexity of the estimator. We insist on the fact that we have used a worst case situation with really over-parametrized models. In cases where it is not the algorithm, although less needed, functions better because we are closer to the assumptions.

Appendix

We use the regularized MSE cost function, $C_L = \sum_{t \in L} \left\| \hat{y}_t - y_t \right\|^2 + \lambda \sum_{p \in I} w_p^2$, and assume that the optimization of the network has returned a set of optimal weights :

$w^* = \underset{w}{Argmin}(C_L)$. At this minimum (we'll omit the star exponent from now on) we have $\frac{\partial C_L}{\partial w_i} \approx 0, \forall i \in I$. Our first hypothesis is that for the error on the validation base, $C_V = \sum_{t \in V} \|\hat{y}_t - y_t\|^2 + \lambda \sum_{i \in I} w_i^2$, this is still true. Deriving this cost function with regards to any weight w_{i_0} leads to our first equation :

$$\forall i_0 \in I, \quad \sum_{t \in V} (\hat{y}_t - y_t) \frac{\partial \hat{y}_t}{\partial w_{i_0}} + \lambda w_{i_0} = 0 \tag{3}$$

Nevertheless, what we really would like is to minimize the MSE on the validation base : $C'_V = \sum_{t \in V} \|\hat{y}_t - y_t\|^2$. We want to find, at w^*, which λ would minimise this error. To achieve this goal we need to compute the following derivatives :

$$\frac{\partial C'_V}{\partial \lambda} = 2 \sum_{t \in V} (\hat{y}_t - y_t) \frac{\partial \hat{y}_t}{\partial \lambda} \tag{4}$$

This calculation will be made in two steps. Firstly in decomposing the derivatives of $\hat{y}_{t_0}$ (for any pattern t_0) with regards to λ :

$$\forall t_0 \in V, \quad \frac{\partial \hat{y}_{t_0}}{\partial \lambda} = \sum_{k \in I} \frac{\partial \hat{y}_{t_0}}{\partial w_k} \frac{\partial w_k}{\partial \lambda} \tag{5}$$

And secondly by deriving the equation (3) with regards to λ :

$$(3) \Rightarrow \forall i_0 \in I, \quad \sum_{t \in V} (\hat{y}_t - y_t) \frac{\partial^2 \hat{y}_t}{\partial \lambda \partial w_{i_0}} + \sum_{t \in V} \frac{\partial \hat{y}_t}{\partial \lambda} \frac{\partial \hat{y}_t}{\partial w_{i_0}} + \lambda \frac{\partial w_{i_0}}{\partial \lambda} + w_{i_0} = 0$$

We will assume here that the first term of the equation is neglectable for it is a second order term, ponderated by what should also be small, the error (please note that we are at a local minimum of the cost function). By doing this we can easily extract :

$$\forall i_0 \in I, \quad \frac{\partial w_{i_0}}{\partial \lambda} = -\frac{1}{\lambda} \left[\sum_{t \in V} \frac{\partial \hat{y}_t}{\partial \lambda} \frac{\partial \hat{y}_t}{\partial w_{i_0}} + w_{i_0} \right]$$

Which using (5) and interverting the sum signs lead to :

$$\forall t_0 \in V, \quad -\lambda \frac{\partial \hat{y}_{t_0}}{\partial \lambda} = \sum_{k \in I} \frac{\partial \hat{y}_{t_0}}{\partial w_k} w_k + \sum_{t \in V} \frac{\partial \hat{y}_t}{\partial \lambda} \sum_{k \in I} \frac{\partial \hat{y}_{t_0}}{\partial w_k} \frac{\partial \hat{y}_t}{\partial w_k} \tag{6}$$

The matrix $\Delta = \left(\frac{\partial \hat{y}_t}{\partial w_k} \right)_{\substack{t=1,\ldots,N \\ k=1,\ldots,K}}$ has as many lines as the number of weights (renumbered from 1 to K), and as many columns as the size of V (renumbered from 1 to N). We will also use its column vector $\Delta_{t_0} = \left[\frac{\partial \hat{y}_{t_0}}{\partial w_1} \cdots \frac{\partial \hat{y}_{t_0}}{\partial w_K} \right]^T$, the derivative vector of $\hat{y}$ with regards to lambda $D_\lambda \hat{y} = \left[\frac{\partial \hat{y}_1}{\partial \lambda} \cdots \frac{\partial \hat{y}_N}{\partial \lambda} \right]^T$, and the weight vector $W = \left[w_1 \cdots w_K \right]^T$. *Id* is the identity of the matrix space. Using these notations lead to :

$$(6) \Rightarrow \forall t_0 \in V, \quad -\lambda \frac{\partial \hat{y}_{t_0}}{\partial \lambda} = \left(\Delta_{t_0}^T W \right) + \sum_{t \in V} \frac{\partial \hat{y}_t}{\partial \lambda} \left(\Delta_{t_0}^T \Delta_t \right)$$

which in matrix form is : $-\lambda D_\lambda \hat{y} = \Delta^T W + \Delta^T \Delta D_\lambda \hat{y}$. This last equation gives the desired derivatives :

$$D_\lambda \hat{y} = -\left(\lambda Id + \Delta^T \Delta \right)^{-1} \Delta^T W .$$

It is no surprise that the ridge regularization leads to ridge regression. Indeed, if the matrix $\Delta^T \Delta$ is not of rank *N*, it is possible to find a λ which makes it inversible.

Acknowledgment

My work at the IIT has been enabled by the Lavoisier grant of the french ministry of foreign affairs. A part of this work has been made at the Institut National des Télécommunications. I would like to thank Stephane Canu for fruitfull discussions and for advising the function used as test in this article.

References

[1] C. M. Bishop, “Chap. 9 Learning and Generalization,” in *Neural Networks for Pattern Recognition*: Oxford University Press, ISBN 0-19-853864-2, 1995, pp. 332-384.

[2] T. Czernichow, “Architecture Selection through Statistical Sensitivity Analysis,” presented at ICANN'96, Bochum, 1996.

[3] T. Czernichow, “Apport des réseaux récurrents à la prévision de séries temporelles, application à la prévision de consommation d'électricité,” PhD thesis, Intelligence Artificielle et Reconnaissance de Formes. Université Pierre et Marie Curie (Paris 6), Lab. Laforia/INT-SIM,1996.

[4] A. N. Tikhonov, V.Y. Arsenin, *Solution of ill-posed problems*. Washington, D.C.: W.H. Wilson, 1977.

[5] A. Weigend, D. Rumelhart, B. Huberman, “Generalization by weight elimination with application to forecasting,” presented at Neural Information Processings 3, pp. 875-882, 1991.

[6] Y. Chauvin, “Dynamic behavior of constrained back-propagation networks,” presented at Advances in Neural Information Processings 2, pp. 643-649, 1990.

[7] S. Hanson, D. Burr, “Minkowski-r back propagation : Learning connectionist models with non-euclidian error signals,” presented at Neural Information Processing Systems, American Institute of Physics, New York, pp. 348-357, 1988.

[8] V. Morosov, *Methods for solving incorrectly posed problems*. Berlin: Springer Verlag, 1984.

Global Least-Squares vs. EM Training for the Gaussian Mixture of Experts

N.P.Bradshaw, A.Duchâteau and H.Bersini

IRIDIA - ULB (CP 194/6), 50, av. F.Roosevelt, 1050-Brussels.

Abstract. Since the introduction of the mixture of experts models and the EM algorithm for training them, maximum likelihood training of such networks has been shown to be a very useful and powerful tool for function estimation and prediction. A similar architecture is derived by other researchers from the application of fuzzy rules. Such systems are often trained by a straightforward global error minimisation procedure. This paper argues that in certain situations global optimisation is the most appropriate approach to take despite its apparent lack of statistical justification compared to the maximum likelihood approach. Moreover a composition of the two approaches often gives the minimal error on both the training and validation sets.

1 Introduction

The large number of local model architectures currently being used for supervised learning can be divided into two different approaches, namely the statistically-based – for example the MOE (mixture of experts) as introduced by Jordan and Jacobs in [4] – and the fuzzy-based – such as the fuzzy-rule models of Takagi and Sugeno [8]. Both approaches use similar architectures, that of simple local models which are weighted or gated by "higher-level" components with output

$$y(x) = \sum_{j=1}^{K} \alpha_j \mathcal{N}(c_j; v_j) w_j^T x \tag{1}$$

(where $\mathcal{N}(mean; variance)$ represents the gaussian density function). The major difference lies in the training method applied. The preferred method for training MOE systems is to apply a statistical analysis of the system and the assumed data distribution, and then to define a likelihood function which may be maximised directly by a gradient method or, often, by the EM algorithm. Fuzzy (or "neuro-fuzzy") techniques are usually not justified so formally. They are used to model non-linearities in functions without trying to model precisely the underlying function or process. In this sense their underlying assumptions are more general but consequently less well-informed.

There is an a priori distinction between the two approaches. The ML (maximum likelihood) approach makes an assumption of separability of classes while the fuzzy or global optimisation approach does not. Thus in the case where the assumption is valid we would expect the ML/EM approach to be more successful (since it is making use of more information). What is not so clear is the relative performance of the two systems when the separation is less complete. That is the subject of this paper.

2 Theoretical Assumptions

The EM approach to training the architecture makes use of an underlying statistical model in order that a likelihood function can be defined. Let us consider here the two-layer model with one layer of local experts and one layer of gaussian gating functions, as this is isomorphic to the Takagi-Sugeno (TS) fuzzy model. In order to pursue an ML approach to selecting the model parameters, it is necessary to interpret the model statistically. That is, to interpret the deterministic output of the model as the expectation of the statistical model. So that the experts can be optimised individually, each data point is assumed to be generated by a single linear expert, and the gating function represents the probability, conditional on the input, that a particular expert generated that data point. Indeed it is hard to devise another meaningful probabilistic interpretation of the system. The EM algorithm can be applied to this system to produce the maximally likely parameters given a training sample.

This is not, however, the inspiration of fuzzy approach, in which the gaussians are not introduced as a probabilistic models, but simply to represent uncertainty in the system in an approximate form. The fuzzy approach seeks to model the underlying process functionally. It does not seek some "real" underlying parameters.

2.1 Mixture of Experts

Within the framework of the MOE architecture and the EM algorithm we must specify the exact optimisation technique to use. A recurring problem with the optimisation has been with the M-step. The function to be maximised in most cases does not have a maximum that can be found analytically. In this case it has been common to use an iterative method [5] or else to improve, but not maximise, the objective function [6]. However Xu et al. showed in [10] that if the gating functions are gaussian as in the case considered here, then by maximising the joint likelihood of the output and input values (rather than the more usual conditional likelihood of the outputs given the inputs) a closed form could be obtained for the M-step maximisation. They also showed that this approximation gave good results on their test data as well as considerably speeding up the algorithm. This approach has been used in the paper.

E-Step

$$h_j(t) = \frac{\alpha_j \mathcal{N}(c_j; v_j)\mathcal{N}(x_t - w_j; \sigma_t)}{\sum_i \alpha_i \mathcal{N}(c_i; v_i)\mathcal{N}(x_t - w_i; \sigma_t)}.$$

M-Step

For the gating

$$\alpha_j' = \frac{1}{N}\sum_t h_j(t),\ c_j' = \frac{1}{\sum_t h_j(t)}\sum_t h_j(t)x_t$$

$$\Sigma_j' = \frac{1}{\sum_t h_j(t)}\sum_t h_j(t)[x_t - c_j'][x_t - c_j']^T.$$

The new linears, w_j', are found by a pseudo-inverse method, and their associated covariances are given by

$$\sigma_j' = \frac{1}{\sum_t h_j(t)}\sum_t h_j(t)[w_j'^T x_t - y_t][w_j'^T x_t - y_t]^T.$$

2.2 The Fuzzy Model

The same architecture was described by Takagi and Sugeno as an algorithmic interpretation of a system of fuzzy rules. A gaussian gating function represents the domain of validity of each rule, and also its "strength" at each point. It is hoped that such a system should be able to model systems which are not describable by more traditional "crisp" rules. The effectiveness of such a system is measured by its error (usually its mean squared error) on the problem domain. So-called neuro-fuzzy techniques can be used [2] to optimise this error over a training set and thus give a set of optimal rules based on the examples and the prior knowledge used to set up the rules.

For the purposes of the experimental work below, a two-stage optimisation was used for global minimisation of the least squared error of expression (1) on the trainnig data. First the linear parameters were found by a least squares optimisation (in this case recursive least squares) and then the parameters of the gaussian functions were found using the Levenberg-Marquardt algorithm [7]. This two-stage process was continued until the squared error failed to decrease by a factor of more than $\frac{1}{0.99}$.

2.3 A Theoretical Comparison

The practical, qualitative distinction between the two models lies in their treatment of the relationship between the linears and the gaussians. The mixture of experts model tries to fit a piecewise linear to the model in which the gaussians should only represent the uncertainty between two or more candidate linears. This is essentially a competition between the linears. The global error minimisation approach makes no such distinction between the sets of functions. The architecture simply defines a set of candidate functions from which we must select the best fit. In this sense the algorithm is co-operative. Neither set of functions is assumed to have a precise interpretation and the form of the solution could as easily be interpreted as gaussian functions gated by linears as vice versa. Thus the optimisation is, a priori, less constrained. Hence we might expect the possibility both for accurate solutions and over-fitted solution to be higher.

3 Experimental Comparison

In order to test this reasoning a comparative experiment was set up. Two models were used to analyse three artificial data sets. The first model, sometimes referred to here as "EM", was the mixture of experts model with gaussian gating functions as described by Xu et al. in [10]. The second, "LM" for short, was the same architecture with the parameters trained to minimise mean squared error on the training data. The worry that unrepresentative linear functions might be generated by an unconstrained optimisation scheme led to the suggestion of a hybrid approach. This consists of taking the output of a MOE/EM optimisation – which will hopefully find sensible linear functions – and feeding it into the initial condition of a global algorithm to make full use of the modelling capability of the gaussian functions. This is referred to as "EM2LM".

The training sets were constructed to test the expected strengths and weaknesses of each algorithm. Set *xu1* is generated by the same mixture of linears as used by Xu et al. in [10] while *xu2* is the same with gaussian noise (mean 0, $variance^2$ 0.3) added to the inputs (the x_t). The base model is piecewise linear with a prior of 0.25 for the left-hand linear and 0.75 for the right. In this case the ML approach may be expected to deal better with the noise than a straightforward error minimisation approach. Set *aab1* consists of a quadratic part, $y = x^2$, and a linear part, $y = x + 4$, and should be less easily modelled under an assumption of piecewise linearity. The priors for each part are 0.5. The function *aab1* is shown in figure 1. The results of the experiments are tabulated in table 1.
It can be seen from these results that the training error is always lower with the direct minimisation approach, as one might have expected. The error-minimising approach has not proved to be best in validation in all cases. In the case of noisy linear data *xu2*, the EM algorithm showed better performance in validation. This is a case in which the correct assumption of an underlying linear model allows the EM algorithm to see through the noise on the inputs and find the most appropriate linear model. As regards running time, both methods showed a large amount of variability, but they seemed broadly comparable, with neither being clearly quicker in the simulations.
In one example of the failure possible with global optimisation two linears became very closely associated and by means of very large – and opposite – displacements yielded a very close approximation to the generating function, albeit one in which the linears bear no relationship to the underlying model. This feature of global optimisation may lead to poor results in the case where the underlying structure *is* linear.

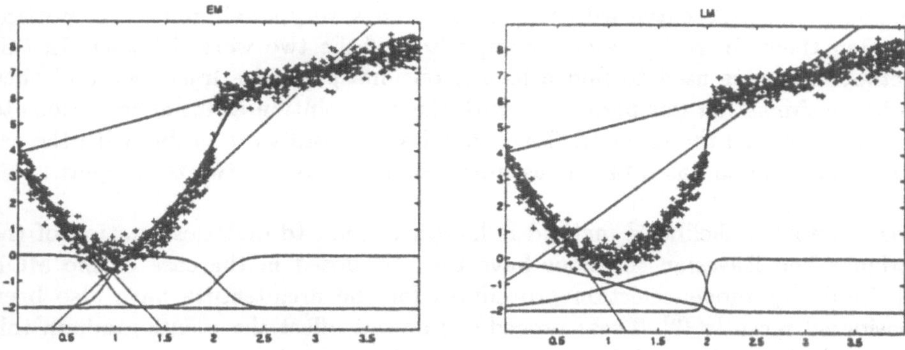

Fig. 1. (1) Non-linear example *aab1* with a solution found by EM with 3 linear experts. (2) Non-linear example *aab1* with a solution found by Levenberg-Marquardt error minimisation with 3 linear functions. The linears and normalised gaussians (re-scaled for ease of reading) are also shown.

Training method	Number of linears	Test set *xu1* RMS error	Test set *xu1* X-Val error	Test set *xu2* RMS error	Test set *xu2* X-Val error	Test set *aab1* RMS RMS	Test set *aab1* X-Val X-Val
EM	2	0.3677	0.3811	0.5324	0.5257	0.3057	0.3152
LM	2	0.3606	0.3724	0.5289	0.5257	0.1423	0.1438
EM2LM	2	0.3610	0.3732	0.5287	0.5249	0.1554	0.1583
EM	3	0.3613	0.3733	0.5320	0.5251	0.1536	0.1552
LM	3	0.3546	0.3670	0.5249	0.5346	0.1312	0.1350
EM2LM	3	0.3601	0.3713	0.5283	0.5247	0.1129	0.1148
EM	4	0.3614	0.3733	0.5320	0.5252	0.1302	0.130
LM	4	0.3539	0.3683	0.5285	0.5314	0.1213	0.1223
EM2LM	4	0.3599	0.3710	0.5282	0.5246	0.1081	0.1114

Table 1. Experimental results for *xu1*, *xu2* and *aab1*. Training was performed on a thousand points and validation on a further ten thousand drawn from the relevant model. Results were averaged over ten runs. EM2LM refers to the method in which the output of an EM optimisation is used as the intial parameter estimate for an LM optimisation.

4 Conclusions

A significant difference in both training and validation error was found between the two models, and also their composition. This was especially so when noise was low and the underlying model non-linear. Qualitatively it could be seen that the MOE model gave sharper transitions between linear components even when, as in the case of the quadratic function, these were not justified. By contrast the fuzzy model was able to smooth the corners to give a better approximation to the underlying function. This can be best seen in figure 1 where the fuzzy approximation adapts more closely to the quadratic curve while the MOE/EM version models the curve with two linears and a very small non-linear portion between them. In many cases a compositon of the two worked better. In this method, EM was used to find a locally optimal piecewise linear fit, and then global optimisation was performed with the EM solution as intial conditions to accurately model the function. These qualitative results are in line with the expectations we had based on the assumptions made by the two techniques.

The maximum likelihood method is known to lead to under-estimation of the variance and Bayesian solutions have been proposed in the case of the MOE [9]. Similarly, model-selection procedures for the architecture have also been considered recently [3]. These procedures do not affect the points made in this paper which consider minimising a given risk on the training set. Rather they constitute new ways of defining the risk while still remaining in the probability-based model. The incremental addition of linears has been used a a model-selection technique for the global minimsiation model [1] while the addition of a regularisation term to the error function has also been suggested.

Acknowledgments This work was carried out with the support of Honeywell Inc..

References

1. H Bersini, A. Duchateau, and N. Bradshaw. Using incremental learning algorithms in the search for minimal and effective fuzzy models. In *Proceeding of FUZZ-IEEE*. IEEE, 1997.
2. M. Brown and C.J. Harris. *Neurofuzzy adaptive modelling and control*. Prentice-Hall, Hemel Hempstead, 1994.
3. R.A. Jacobs. Bias/variance analyses of mixtures-of-experts architectures. *Neural Computation*, 9(2):369–384, 1997.
4. M.I. Jordan and R.A. Jacobs. Hierarchies of adaptive experts. *NIPS*, 4:985–993, 1992.
5. M.I. Jordan and R.A. Jacobs. Hierarchical mixtures of experts and the EM algorithm. *Neural Computation*, 6(181-214), 1994.
6. M.I. Jordan and L. Xu. Convergence results for the EM approach to mixtures of experts architectures. *Neural Networks*, 8(9):1409–1431, 1995.
7. W.H. Press, S.A. Teukolsky, W.T. Vetterling, and B.F. Flannery. *Numerical Recipes in C*. CUP, 1988.
8. T. Takagi and M. Sugeno. Fuzzy identification of systems and its applications to modeling and control. *IEEE Transactions on Systems, Man and Cybernetics*, 15(1):116–132, 1985.
9. S. Waterhouse, D. MacKay, and T. Robinson. Bayesian methods for mixtures of experts. *NIPS*, 8:351–357, 1996.
10. L. Xu, M.I. Jordan, and G.E. Hinton. An alternative model for mixtures of experts. In *NIPS* 7, pages 633–640, 1995.

Accelerated Learning in Boltzmann Machines Using Mean Field Theory

H.J. Kappen[1] and F. B. Rodríguez[2]

[1] RWCP SNN Laboratory, Department of Biophysics, University of Nijmegen, Geert Grooteplein 21, NL 6525 EZ Nijmegen, The Netherlands
[2] Instituto de Ingeniería del Conocimiento & Departamento de Ingeniería Informática, Universidad Autónoma de Madrid, Canto Blanco,28049 Madrid, Spain

Abstract. The learning process in Boltzmann Machines is computationally intractible. We present a new approximate learning algorithm for Boltzmann Machines, which is based on mean field theory and the linear response theorem. The computational complexity of the algorithm is cubic in the number of neurons.

In the absence of hidden units, we show how the weights can be directly computed from the fixed point equation of the learning rules. We show that the solutions of this method are close to the optimal and give a significant improvement over the naive mean field approach.

1 Introduction

Boltzmann Machines (BMs) [1], are networks of binary neurons with a stochastic neuron dynamics, known as Glauber dynamics. Assuming symmetric connections between neurons, the probability distribution over neuron states $\mathbf{s}$ will become stationary and will be given by the Boltzmann-Gibbs distribution $P(\mathbf{s})$. The Boltzmann distribution is a known function of the weights and thresholds of the network. However, computation of $P(\mathbf{s})$ or any statistics involving $P(\mathbf{s})$, such as mean firing rates or correlations, requires exponential time in the number of neurons. This is due to the fact that $P(\mathbf{s})$ contains a normalization term Z, which involves a sum over all states in the network, of which there are exponentially many. This problem is particularly important for BM learning.

Using statistical sampling techiques [2], learning can be significantly improved [1]. However, the method has rather poor convergence and can only be applied to small networks.

In [3, 4], an acceleration method for learning in BMs is proposed using mean field theory by replacing $\langle s_i s_j \rangle$ by $m_i m_j$ in the learning rule. It can be shown [5] that such a naive mean field approximation of the learning rules does not converge in general. Furthermore, we argue that in the correct treatment of mean field, the correlations can be computed using the linear response theorem [6].

2 Boltzmann Machine learning

The Boltzmann Machine is defined as follows. The possible configurations of the network can be characterized by a vector $\mathbf{s} = (s_1, .., s_i, .., s_n)$, where $s_i = \pm 1$ is the state of the neuron i, and n the total number of the neurons. Neurons are updated using Glauber dynamics.

Let us define the energy of a configuration s as

$$-E(\mathbf{s}) = \frac{1}{2}\sum_{i,j} w_{ij}s_i s_j + \sum_i s_i\theta_i. \tag{1}$$

After long times, the probability to find the network in a state s becomes independent of time (thermal equilibrium) and is given by the Boltzmann distribution

$$p(\mathbf{s}) = \frac{1}{Z}\exp\{-\beta E(\mathbf{s})\}. \tag{2}$$

$Z = \sum_{\mathbf{s}} \exp\{-\beta E(\mathbf{s})\}$ is the partition function which normalizes the probability distribution.

Learning [1] consists of adjusting the weighs and thresholds in such a way that the Boltzmann distribution approximates a target distribution $q(\mathbf{s})$ as closely as possible.

A suitable measure of the difference between the distributions $p(\mathbf{s})$ and $q(\mathbf{s})$ is the Kullback divergence [7]

$$K = \sum_{\mathbf{s}} q(\mathbf{s}) \log \frac{q(\mathbf{s})}{p(\mathbf{s})}. \tag{3}$$

Therefore, learning consists of minimizing K using gradient descent [1]

$$\Delta w_{ij} = \eta\Big(\langle s_i s_j\rangle_c - \langle s_i s_j\rangle\Big), \quad \Delta\theta_i = \eta\Big(\langle s_i\rangle_c - \langle s_i\rangle\Big). \tag{4}$$

The parameter η is the learning rate. The brackets $\langle\cdot\rangle$ and $\langle\cdot\rangle_c$ denote the 'free' and 'clamped' expectation values, respectively.

The computation of both the free and the clamped expectation values is intractible, because it consists of a sum over all unclamped states. As a result, the BM learning algorithm can not be applied to practical problems.

3 The mean field method and the linear response correction

The mean field approximation consists of introducing n mean fields m_i that minimize the mean field free energy [2]:

$$F = -\log Z' = -\frac{1}{2}\sum_{i,j} w_{ij}m_i m_j - \sum_i \theta_i m_i$$
$$+ \frac{1}{2}\sum_i \left((1+m_i)\log(1+m_i) + (1-m_i)\log(1-m_i)\right). \tag{5}$$

Z' is the mean field approximation of the partition function Z. The mean fields are given by the coupled set of equations:

$$m_i = \tanh(\sum_j w_{ij} m_j + \theta_i) \tag{6}$$

We can now compute the mean firing rates and correlations in the mean field approximation:

$$\langle s_i \rangle = \frac{1}{Z}\frac{dZ}{d\theta_j} \approx \frac{1}{Z'}\frac{dZ'}{d\theta_j} = m_i, \tag{7}$$

$$\langle s_i s_j \rangle \approx \frac{1}{Z'}\frac{d^2 Z'}{d\theta_i d\theta_j} = \frac{1}{Z'}\frac{d}{d\theta_j}(Z' m_i) = m_i m_j + A_{ij} \tag{8}$$

with $A_{ij} = \frac{dm_i}{d\theta_j}$. To arive at Eqs. 7 and 8 we have used the mean field equations Eq. 6. Eq. 8 is known as the linear response theorem [6]. The inverse of the matrix A can be directly obtained by differentiating Eq. 6 with respect to θ_i. The result is:

$$(A^{-1})_{ij} = \frac{\delta_{ij}}{1 - m_i^2} - w_{ij} \tag{9}$$

Thus, our approximation consists of replacing the free expectation values in Eqs. 4 by their linear response approximations Eqs. 6, 7-9. The clamped quantities are directly computed from the data. The inclusion of hidden units is straigthforward and is discussed elsewhere [5]. The complexity of the method is dominated by the computations in the free phase. The computation of the linear response correlations involves the inversion of the matrix A, which requires $\mathcal{O}(n^3)$ operations. The computation of the mean firing rates through fixed point iteration of Eq. 6 requires $\mathcal{O}(n^2)$ or $\mathcal{O}(n^2 \log n)$ operations, depending on whether fixed precision in the components of m_i or in the vector norm $\sum_i m_i^2$ is required. Thus, the full mean field approximation, including the linear response correction, computes the gradients in $\mathcal{O}(n^3)$ operations.

3.1 No hidden units

For the special case of a network without hidden units we can make significant simplifications. In this case, the gradients Eqs. 4 can be set equal to zero and can be solved directly in terms of the weights and thresholds, i.e. no 'gradient based learning' is required. First note, that $\langle s_i \rangle_c$ and $\langle s_i s_j \rangle_c$ can be computed exactly from the data for all i and j. Let us define $C_{ij} = \langle s_i s_j \rangle_c - \langle s_i \rangle_c \langle s_j \rangle_c$.

The fixed point equation for $\Delta\theta_i$ gives

$$\Delta\theta_i = 0 \Leftrightarrow m_i = \langle s_i \rangle_c . \tag{10}$$

The fixed point equation for Δw_{ij}, using Eq. 10, gives

$$\Delta w_{ij} = 0 \Leftrightarrow A_{ij} = C_{ij}, i \neq j. \tag{11}$$

The fixed point equations are only imposed for the off-diagonal elements of Δw_{ij} because the Boltzmann distribution Eq. 2 does not depend on the diagonal elements w_{ii}. The condition $\Delta w_{ii} = 0$ is automatically satified in the exact method. However, in the approximate method things are different. The solution depends on w_{ii} in Eq. 6 and the condition $\Delta w_{ii} = 0$ must be enforced explicitly to ensure that $1 = \langle s_i^2 \rangle = 1 - m_i^2 - A_{ii}$. Thus one must impose instead of Eq. 11 the stronger condition $A_{ij} = C_{ij}$ for all i, j, which is equivalent to $(A^{-1})_{ij} = (C^{-1})_{ij}$ for all i, j. Using Eq. 9 we obtain

$$w_{ij} = \frac{\delta_{ij}}{1 - m_i^2} - (C^{-1})_{ij} \tag{12}$$

In this way we have solved m_i and w_{ij} directly from the fixed point equations. The thresholds θ_i can now be computed from Eq. 6:

$$\theta_i = \tanh^{-1}(m_i) - \sum_j w_{ij} m_j \tag{13}$$

Note, that this method does not require fixed point iterations to obtain mean firing rates m_i in terms of w_{ij} and θ_i. Instead, the 'inverse' computation of θ_i given m_i and w_{ij} is required in Eq. 13.

4 Results

In this Section we will compare the accuracy of the linear response correction with the exact method and with the naive mean field approximation. We restrict ourselves to networks without hidden units. Of course, there are many probability estimation problems, for which the BM without hidden units is a poor model. The optimal solution can be found using the exact gradient descent method. Our main concern is whether the linear response approximation will give a solution which is sufficiently close to the optimal solution, and not whether the optimal solution is good or bad.

The correct way to compare our method to the exact method is by means of the Kullback divergence. However, this comparison can only be done for small networks. The reason is that the computation of the Kullback divergence requires the computation of the Boltzmann distribution, Eq. 2, which requires exponential time due to the partition function Z. In addition, the exact learning method requires exponential time.

In order to show the performance of the linear response correction, we have compared it with the results obtained with the exact method and with a 'mean field' method that ignores correlations. For the exact method (K_{ex}), we have used a gradient descent method with a momentum term. The mean firing rates and correlations are computed exactly using the probability distribution and summing over all states. For the linear response method (K_{lr}), we obtain the weights and thresholds from Eq. 12 and Eq. 13. In the case of the naive mean field approximation (K_{mf}), we assume a factorized model:

$$P_{mf}(\mathbf{s}) = \Pi_i \frac{1}{2}(1 + s_i m_i). \tag{14}$$

The mean firing rates are given by $m_i = \langle s_i \rangle_c$.

We compared the methods on a number of typical examples in Fig. 1. Each neuron value $s_i^\mu = \pm 1, i = 1, \ldots, n, \mu = 1, \ldots, p$ is generated randomly and independently with equal probability. The three methods are compared by computing the Kullback divergence, using Eq. 3, that we obtain for each method on each of the data sets. The network size was varied from 3 to 10 neurons. For each data set we compute $K_{lr} - K_{ex}$ and $K_{mf} - K_{ex}$. In the Figure, we show these values averaged over all data sets, as well as their variances.

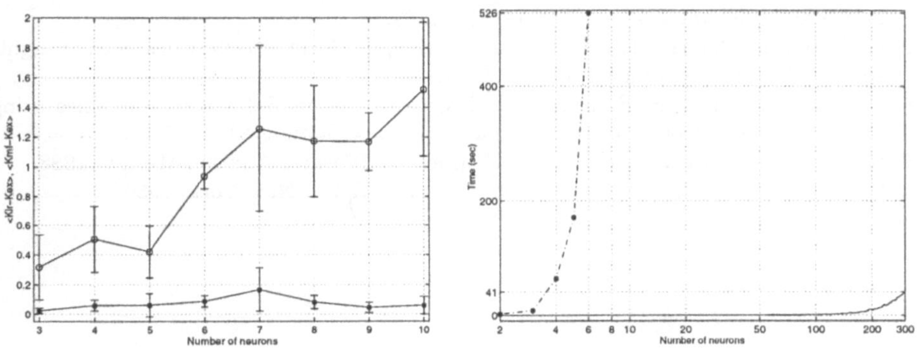

Fig. 1. (Left) Kullback divergence relative to exact method, for mean field approximation (open symbols) and linear response method (close symbols). The number of patterns $p = 2n$. Results are averaged over 4 data sets. The error bars indicate the variance over the data sets. (Right) CPU required for the linear response method and the exact method as a function of the network size. For each number of neurons, one data set was generated consisting of $2n$ patterns. In addition to the simulation data, we show the best fits using the expected $\mathcal{O}(n^3)$ and $\mathcal{O}(2^n)$ time complexity.

We conclude that the linear response correction gives a good approximation to the exact results. The naive mean field approximation that ignores the correlations is much worse, as should be expected. It indicates that correlations play a significant role in these learning problems.

For networks with a large number of units, one can show the quality of the linear response method by means of a pattern completion task i.e. the network must be able to generate the rest of a pattern, when part of the pattern is shown. This comparison is reported in [5]. It is concluded that the use of the linear response correction give important improvements over the naive mean field approach also for large networks.

We stated before, that the time required for the exact method is $\mathcal{O}(2^n)$ and $\mathcal{O}(n^3)$ for the linear response method. We plot the computation time for both methods in Fig. 1 for networks up to 300 neurons and observe run-times in accordance with this complexity.

References

1. D. Ackley, G. Hinton, and T. Sejnowski. A learning algorithm for Boltzmann Machines. *Cognitive Science*, 9:147–169, 1985.
2. C. Itzykson and J-M. Drouffe. *Statistical field theory.* Cambridge monographs on mathematical physics. Cambridge University Press, Cambridge, UK, 1989.
3. C. Peterson and J.R. Anderson. A mean field theory learning algorithm for neural networks. *Complex systems*, 1:995–1019, 1987.
4. G.E. Hinton. Deterministic Boltzmann learning performs steepest descent in weight-space. *Neural Computation*, 1:143–150, 1989.
5. H.J. Kappen and F.B. Rodríguez. Efficient learning in Boltzmann Machines using linear response theory. *Neural Computation*, page Submitted, 1997.
6. G. Parisi. *Statistical Field Theory.* Frontiers in Physics. Addison-Wesley, 1988.
7. S. Kullback. *Information theory and statistics.* Wiley, New York, 1959.

Adaptive Online Learning for Nonstationary Problems

Siegfried Bös

Information Representation Lab, FRP, RIKEN
Wako–shi, Saitama 351–01, Japan
email: boes@fugu.riken.go.jp

Abstract. An adaptation algorithm for online training is examined. For stationary tasks it can reduce the learning rate to reach the best convergence. Instead of simple annealing, it keeps the learning rate flexible, such that it can also adapt to nonstationary tasks. Different tasks, abrupt or gradual changes, and different guidance measures are discussed.

1 Introduction

Online training is an effective training algorithm and it is the first choice when the number of data is large or time for training is limited. It assumes that an example is used only once to update all the weights and then thrown away, such that a memory to store the training data is not necessary. In supervised training, using example–pairs of inputs $\mathbf{x}_\mu$ and correct outputs $\mathbf{z}^*_\mu$, an update requires only information available at time t, such that $t = \mu$, i.e.

$$W_i(t+1) = W_i(t) + \eta_t\, \mathcal{F}[\,\mathbf{z}^*(t), \mathbf{W}(t), \mathbf{x}(t)]\,. \tag{1}$$

Different update–functions $\mathcal{F}$ can be used, the most common one is the negative local gradient, i.e. $-\partial\text{loss}/\partial W_i$.

The choice of the learning rate η_t is rather crucial in online training. Since each example is presented only once, the learning rate determines how much emphasis is given to each example. Most publications deal with *time–independent* learning rates, i.e. $\eta_t = \eta_0$, which are fixed during the whole training process. In this case, the following dilemma is typical. A larger learning rate increases the convergence speed, however the fluctuations of the final solution around the optimum also become stronger. In other words, a fast convergence requires a large learning rate and a good solution, a small learning rate.

To overcome this dilemma, the learning rate can be made *explicitly time–dependent*, i.e. $\eta_t = \eta(t)$; it is annealed from a large value at the beginning of training to a small value at the end. General proofs by Amari [1] and Opper [2] have indicated that with a suitably annealed learning rate online learning can become asymptotically efficient, i.e. it convergences as fast as the optimal batch solution. Unfortunately, these proofs require a learning rate of matrix–form, i.e. $\eta_t \sim \eta(t)\,\mathbf{G}^{-1}$. The Fischer information $\mathbf{G}$ is the averaged second derivative of the loss–function and can take the curvature of the problem into account. Until now, no study exists where the matrix online learning is applied to a specific model. We will therefore circumvent this open problem by using a simple model for illustration, in which the Fischer information is just a multiple of the identity matrix.

Online learning has also other favorable properties; it can follow nonstationary rules and it is not that prone to be trapped in a local minimum. However, these properties are lost if the learning rate is simply annealed. Therefore, an adaptation scheme has been proposed [3], in which the learning rate is chosen as *implicitly time-dependent.* The dependence is determined by a differential equation, i.e. $\eta_t \rightarrow \dot{\eta}(t) = f[\eta(t)]$.

In this paper we study the adaptation algorithm more closely. The algorithm should adapt properly to tasks, which change either abruptly or gradually. On the other hand, it should not adapt if the new input is only a corrupted or very noisy data-point.

2 The Model

First we recall briefly a few properties of the model, which we want to use. We consider a supervised learning problem, where the task is provided by a teacher network. Both networks are single-layer perceptrons, with $z = g(h)$ and $h := N^{-\frac{1}{2}}\mathbf{W}\,\mathbf{x}$. The student has a linear output, i.e. $g(h) = h$. The teacher variables will always be indicated by stars, i.e. $\mathbf{W}^*$, z^* and $g^*(h^*)$, which can be a nonlinear function, or a linear one with some random noise. The loss-function is the usual quadratic deviation. Training error E_{T} and generalization error E_{G} are averages of the loss-function over all examples respectively over all possible inputs. The time evolution of the system can be described by the two order parameters, $R(t) := N^{-1}\mathbf{W}^*\,\mathbf{W}$ and $Q(t) := N^{-1}(\mathbf{W})^2$. If the inputs $\mathbf{x}$ are drawn randomly from a normal distribution $\mathcal{N}(0,1)$, then the typical generalization error is given by $E_{\mathrm{G}}(\alpha) = \frac{1}{2}\,[\,G - 2HR(\alpha) + Q(\alpha)\,]\,/2$, with $\alpha := P/N$, and $G :=< [g^*(h^*)]^2 >_{h^*}$ and $H :=< g^*(h^*)\,h^* >_{h^*}$.

Batch training, which was studied in [4], exhibits overtraining, a very typical phenomena of unrealizable tasks. How overtraining can be avoided by early stopping or weight decay to receive a optimal solution, can be found in the reference. Here we will need only the asymptotical behavior of the two errors, $E_{\mathrm{G}}(\alpha) \simeq E_\infty(1 + 1/\alpha)$ and $E_{\mathrm{T}}(\alpha) = E_\infty(1 - 1/\alpha)$.

Online training on the other hand can be described by differential equations for $\dot{R}(\tau)$ and $\dot{Q}(\tau)$, again with $\tau := \mu/N$. In [5] the analytical solution of these differential equations for a fixed learning rate η_0 was discussed. The typical behavior was found, clearly illustrating the trade-off between fast convergence and good final results. In the extended work [6], the annealing of the learning rate was studied. It can be shown, that the optimal asymptotical annealing is $\eta(t \rightarrow \infty) = b/t^a$ with $a = 1$ and $b = N$. In the range $\alpha = \mathcal{O}(10^3)$, online training reaches the same convergence speed as batch training.

This solution is consistent with the general results of [1] and [2]. Furthermore, in [7] it is shown, that the optimal annealing rate for general networks using $\eta(t)\,\mathbf{G}^{-1}$ is $\eta(t) = 1/t$. Only for the linear single-layer perceptron, where the inverse of the Fischer information is just $(\mathbf{G})^{-1} = N\mathbb{1}$, it simplifies to a scalar learning rate. The application of a matrix learning rate to a multilayer network is currently under consideration. After this brief update, we can now discuss the schemes to adapt the learning rate.

3 Adaptive Learning Rate

In order to preserve the ability to follow nonstationarities and to avoid local minima, Sompolinsky *et al.* [3] proposed the following adaptation scheme,

$$\dot{\eta}(t) = -c_1\, c_2\, \mathcal{M}(t)\, \eta(t) + c_1 \eta^2(t)\,, \tag{2}$$

with the parameters c_1 and c_2 and the guidance measure $\mathcal{M}(t)$.

They haven chosen the generalization error as the guidance, $\mathcal{M}(t) = E_{\mathrm{G}}(t)$, which makes it easy to understand the mechanism of the algorithm. In the undisturbed case, the error converges proportional to t^{-1} to zero, such that the first term guiding the adaptation behaves like the second term. The solution is $\eta(t) = t^{-1}$, as in simple annealing. If the rule changes, the error increases, as does the learning rate. The higher learning rate allows the network to follow the new rule rapidly. In the case of a local minimum, the generalization does not converge to zero, leaving the learning rate also on a finite value. The probability to escape from the local minimum is not zero, as only the total gradient is zero, however not all of the local gradients.

Unfortunately, the generalization error is only useful for realizable tasks, a a nice illustration, or for the calculation of optimal behavior. In unrealizable tasks, the best asymptotical solution has a finite asymptotical error E_∞, which has to be subtracted, leading to the rest–error $\Delta E_{\mathrm{G}} := E_{\mathrm{G}} - E_\infty$. Practically, neither generalization error E_{G} nor asymptotical error E_∞ can be measured, such that measurable approximations or other empirical measures need to be found.

First we will try to find a measurable empirical rest–error. The asymptotic error E_∞ is unknown for every learning task. It is, however, hidden in the two formulas for generalization error and training error of batch learning. If we assume efficiency, then the online results will converge against the batch results and we can use the difference between the two errors, which is exactly what we are looking for, i.e. $E_{\mathrm{G}}(t) - E_{\mathrm{T}}(t) \rightarrow 2\,\Delta E_{\mathrm{G}}(t)$. Note, that similar equations hold quite generally also for other networks. Generalization error and training error have to be estimated empirically on two different finite test–sets, one consisting of non–learned examples, the other of learned examples.

The generalization error is not the only measure, which can be used for guidance of the adaptation algorithm. Other measures with similar behavior can also be used. In [8], it was proposed to use the average local gradient, i.e.

$$\mathcal{M}(t) \quad \rightarrow \quad \left\langle \frac{\partial\, \mathrm{loss}[z^*(t), \mathbf{W}(t), \mathbf{x}(t)]}{\partial W_i} \right\rangle_i . \tag{3}$$

Both empirical measures show fluctuations, which can be reduced using a *moving average*, i.e.

$$\overline{f_t} := (1-\epsilon)\,\overline{f_{t-1}} + \epsilon f_t = \epsilon \sum_{i=0}^{t-1} (1-\epsilon)^i\, f_{t-i}\,, \tag{4}$$

with a small $\epsilon \in [0,1]$ determining the range of the moving average. Now we will test the adaptation scheme on two nonstationary tasks. We will also examine, how the empirical measures perform relative to the solution guided by the theoretical rest–error.

4 Tests and Discussion

First we check, whether the adaptive algorithm can follow a task, which is abruptly changing. Up to a time T_0, the task is stationary, then an abrupt change to an uncorrelated task occurs. The abrupt change results in a large increase of the guidance measure and is therefore rather easy to detect. Fig. 1 shows, how the different schemes, fixed, annealed and adaptive learning rate, handle this task.

More difficult are gradually drifting tasks. There again the task is stationary until time T_0, then it starts to change gradually. We assume that the drift is linear in the direction to another uncorrelated task; other drifts are possible. After T_1 time–steps, the task is defined by another orthogonal teacher vector. Fig. 2 shows the results for this layout.

With the theoretical guidance, the adaptation algorithm works in both tasks very good. It is able to choose the most appropriate learning rate for the problem. The parameters c_1 and c_2 from (2) can be chosen from a wide range ($c_1 > 1$ and $c_2 > 0$). For the abrupt change, $c_2 \simeq 1.5$ is a good choice. For the drift, a higher value for c_2 of about 5 is better, giving more weight to the adaptation term (2).

However, the theoretical measure is so sensible to changes, that it would even increase the learning rate, if an example with very high noise is presented. This is definitely not a desired feature. By the use of the moving average this oversensitivity is automatically reduced.

The detection of abrupt changes is hereby not effected, however, gradual drifts are then detected only with a certain time lag. In Fig. 2 we can see, that the empirical local gradient is able to adapt within a time lag. The empirical rest–error has more problems to follow drifts, it is more suited for asymptotical adaptation. Due to limited space, a detailed discussion of drifting tasks can not be given here. It should be noted, that there is certainly a tradeoff between strong smoothing and sensitive detection of slight drifts.

In this paper a powerful algorithm for adaptive online learning was illustrated. The results are easily transferable to more general and more complicated systems. We have chosen the linear single–layer perceptron as a model, because it allows a more extensive study. Furthermore, it is right now the only model, for which the optimal annealing of the learning rate is known. Of the two empirical

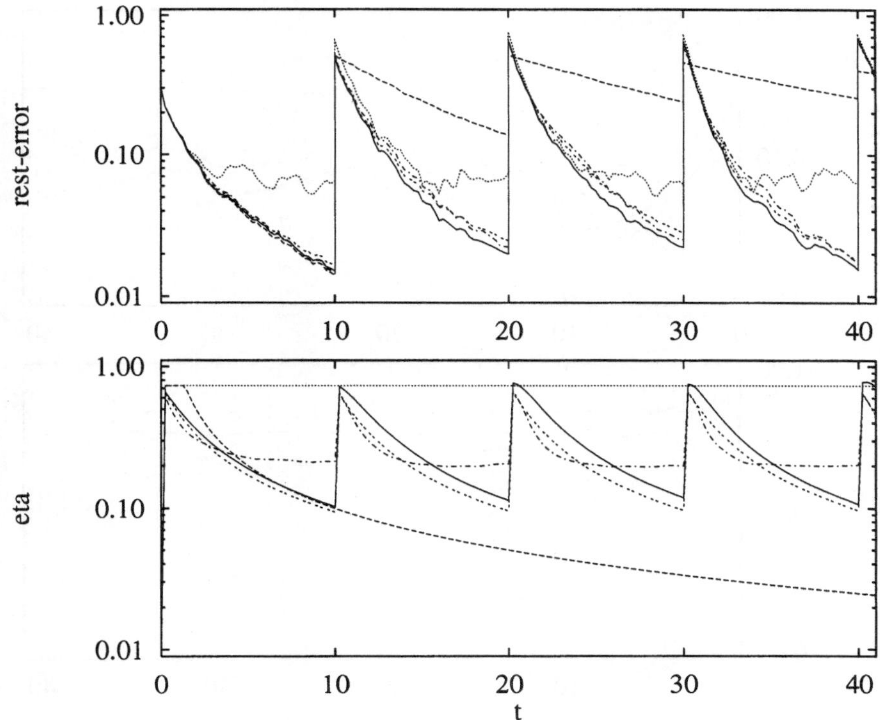

Fig. 1. Sudden change: (a) Performance measured by the rest–error $\Delta E_G = E_G - E_\infty$. A fixed learning rate (*dotted line*) is able to follow all changes, however not always optimal, since the optimal η_0 is unknown and depends on T_0; here $\eta_0 = 0.7$. An annealed learning rate (*dashed line*) is better for the stationary task, however its adaptation ability decreases with the decreasing η. The adaptation schemes combines the advantages of both schemes. The best results are achieved with the theoretical rest–error (*solid line*). The empirical guides, empirical rest–error (*dash–dotted line*) and averaged local gradient (*double–dashed line*), perform slightly worse. (b) Behavior of the corresponding learning rate shown by the same line types. (Parameters: $N = 100$, 1 trial, $G = 0.84$, $H = 0.78$, $T_0 = T_1 = 10N$).

guidance measures, the averaged local gradient, is very promising. It can be calculated easily during the update and with a suitable moving average it provides a reliable measure.

Acknowledgment. I would like to thank Shun–ichi Amari, Noburu Murata, Klaus–Robert Müller, Manfred Opper and Bruno Orsier for valuable discussions. Emily Helle I want to thank for her advise concerning the presentation.

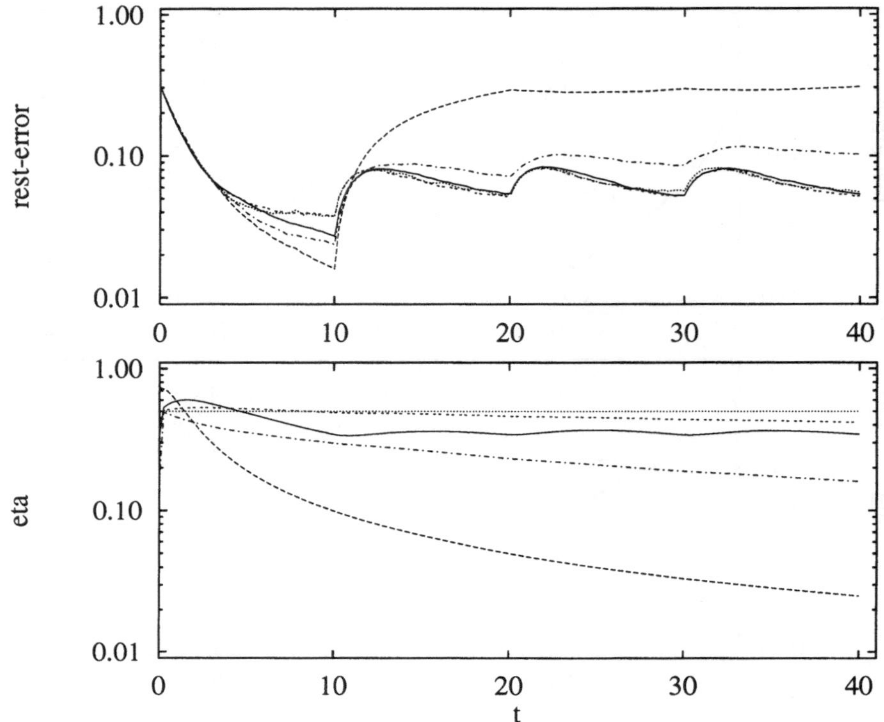

Fig. 2. Continuous drift: (a) Performance measured by the rest–error. A fixed learning rate (*dotted line*) is also able to follow the drift. The best choice for η_0 is unknown, here $\eta_0 = 0.5$. An annealed learning rate (*dashed line*) is not able to follow. The adaptive learning rate is superior for the stationary task and also able to follow task drifting task. Good results are achieved with the theoretical rest–error (*solid line*). The empirical guides, empirical rest–error (*dash–dotted line*) and averaged local gradient (*double–dashed line*), have more problems with this task, as discussed in the text. (b) Behavior of the corresponding learning rates. (Parameters: as above, 100 trials).

References

1. Amari S. (1997), in *NIPS 9*, MIT Press, in press.
2. Opper M. (1996), *Phys. Rev. Lett.* **77**, p. 4671–4674.
3. Sompolinsky H., Barkai N. & Seung H.S. (1995), in *Neural Networks: The Statistical Mechanics Perspective*, World Scientific, p. 105–130.
4. Bös S. (1995), in *ICANN'95*, p. 111–116, and to be submitted.
5. Bös S. (1996), in *ICANN'96*, Springer LNCS 1112, p. 89–94.
6. Bös S., Murata N., Amari S., & Müller K.-R. (1997), submitted.
7. Murata N., Bös S., Amari S., & Müller K.-R. (1997), in preparation.
8. Murata N., Müller K.-R., Ziehe A., & Amari S. (1997), in *NIPS 9*, MIT Press, in press.

Weight Discretization Due to Optical Constraints and Its Influence on the Generalization Abilities of a Simple Perceptron

Maissa Aboukassem, Steffen Schwember, Steffen Noehte, Reinhard Männer

Lehrstuhl für Informatik V der Universität Mannheim
Mannheim, Germany**

Abstract. Motivated from the optical implementation of NN which can be realized by storing the weights in holograms with a limited number of gray values, we focus our investigation on the dependence of the generalization and training errors of a simple perceptron with discrete weights, on the number of allowed discrete values (there are 2^p allowed values for a bit precision of p) and on the training set size. Our starting point is the teacher pupil paradigm. The teacher is defined by fixing its continuous weights to random values. The pupil network that is only allowed to have discrete values was trained to learn the rule produced by the teacher with simulated annealing. For $\alpha < \alpha_s$, where α encodes the training set size, weight configurations exist so that the training set can be reproduced without error whereas the generalization error is nonzero. For $\alpha > \alpha_s$ there is no weight configuration of the pupil which can reproduce the training set without error and for $\alpha \to \infty$ both training and generalization errors asymptotically converge to an ϵ_{min}. We found that between a precision of 5 bit and 8 bit there was no remarkable improvement in the generalization ablitity of the pupil perceptron. This result is very useful for the optical implementation since optical constraints for storing weights in holograms restrict precision to a maximum value of 6 bit.

1 Background

1.1 The Simple Perceptron

The perceptron consists of N input nodes and one output node which is a threshold function of the input.
The task is to learn an input output mapping subjected to the inequalities

$$\forall \mu : \qquad \lambda(\xi^\mu, J) = \frac{1}{\sqrt{N}} \sigma^o(\xi^\mu) \sum_{i=1}^{N} J_i \xi_i^\mu > \kappa \qquad \mu = 1, ..., p \qquad (1)$$

** Special thanks to Prof.Dr.H.Horner and Priv.Doz.Dr.R.Kühn from the Institute of Theoretical Physics at Heidelberg University for supporting our work with very helpful discussions and hints.

where $\xi_i^\mu \in \pm 1$ is an input pattern at node i, $\sigma^0(\xi^\mu)$ denotes the teacher's answer to pattern ξ^μ and J_i are the weights of the perceptron. The patterns are chosen to be random and uncorrelated. For normalized weights κ in equation 1 is called the stability, a finite value $\kappa \ll O(\frac{1}{\sqrt{N}})$ enables discrimination between stored and random input and gives error correction ability.

1.2 The Teacher Pupil Problem

The teacher pupil method is a natural approach to study the generalization ability of neural networks, if we want to get general information about their behavior independent of a special training set.
For a given NN architecture (a perceptron with continuous weights in our case), a teacher NN is defined by fixing its weights to random values according to some distribution. This teacher NN realizes a function which represents the concept to be learned. The training set is generated by propagating patterns drawn from another distribution as input (in our case the questions are binary random patterns) through the teacher network. This output is the answer to be recorded together with the question, as one sample of the training set.
A pupil network with the same architecture as the teacher (in our case a perceptron with discrete weights) is trained with the training set, and the training errors ϵ_T and generalization errors ϵ_G are studied:

$$\epsilon_T = \frac{1}{\alpha N} \sum_\mu \Theta(-\lambda(\xi^\mu, J)). \tag{2}$$

$$\epsilon_G = \frac{1}{\pi} arccos(\frac{1}{||J^o|| ||J||} \sum_i J_i^o J_i). \tag{3}$$

where J_i^o denotes the teacher's weights. In order to get results independent of the special realization of the random variables in the previous steps (questions in the training set, pupil weights), averages over the underlying distributions are performed.

2 Implementation and Parameters of the Simulated Annealing Algorithm

We implemented the simulated annealing using the metropolis method of [6]. First the costs (energy) of a network with randomly initialized weights is calculated using the following equation:

$$E = \sum_{\mu=1}^{\alpha N} \Theta(-\lambda(\xi^\mu, J))(-\lambda(\xi^\mu, J))^n \tag{4}$$

Then weight changes are applied to find the network configuration with the lowest cost. Changes that lower the costs are always accepted whereas those that raise them are accepted with a certain probability dependent on the temperature

$$\text{acceptance probabiblity} = exp(\frac{-\Delta E}{T}). \tag{5}$$

where $\Delta E = (E_{aft} - E_{bef})$ describes the costs differences between the new and the old weight configuration. When accepting weight configurations with higher costs the system is given the chance to leave local minima and when lowering the temperature the probability of their acceptance decreases causing the system to strive for a minimum.
The following parameters of the simulated annealing algorithm had to be adjusted:

- parameter n of cost function
- initial temperature
- cooling schedule
- step size of temperature changes
- number of weight changes per connection and temperature step
- mechanism of weight changes
- termination criteria

The results represented in this paper have been obtained by setting $n = 1$ in the cost function of equation 4. To define an appropriate initial temperature we started with a fairly high value and observed the acceptance rate during the annealing process. The cost function for the simple perceptron showed a constant movement towards a minimum despite temporary increases. We infered from these results that the starting temperature could be chosen as low as $T_0 = 0.3$ resulting in an accpetance rate of 0.1. Previous investigations showed that using a reciprocal schedule

$$\frac{1}{T} \leftarrow \frac{1}{T} + a \tag{6}$$

resulted only in more computational expenditure compared to the linear schedule

$$T \leftarrow T - \Delta T \tag{7}$$

with $\Delta T = 0.01$ used in our simulations. The weights are real numbers and their values were calculated by equally dividing the interval [-1,1] by the number of approved discrete weight values 2^p which varies with precision p. After initializing the connection strengths with random values drawn from the pool of allowed weight values, changes were applied in cyclic order. The number of Monte Carlo Steps (MCS) denotes the number of weight changes per temperature and connection. Its value depends on the number of input nodes and was fixed to 20MCS for 100 input nodes. The algorithm terminated when

- the problem has been learned with zero training error.
- zero temperature has been reached.
- the acceptance rate dropped below 0.01.

The binary learning patterns were generated by creating a real valued random number using the internal generator. If smaller than 0.5 the value were set to -1 otherwise to 1.
To exclude influence of coincidence we averaged over 50 randomly chosen pupil configurations and newly initialized training patterns with every pupil. We additionally established a pocket mechanism where the best temporary solution were stored. The costs of the final solution were compared to those of the pocket solution and the cheaper one stated the actual solution.

3 Results and Conclusions

In the case of a teacher network with continuous weights and a pupil with continuous weights conventional training algorithms are able to find the teacher configuration in finite time. The situation is quite different if the allowed weights of the pupil are constrained to be discrete and the teacher is still allowed to have continuous weights. We see in figure 1 that for $\alpha < \alpha_s$ weight configurations still exist such that the training set can be reproduced without errors. For $\alpha > \alpha_s$ both training and generalization errors are finite. α_s depends on the bit precision of the pupil and a theoretical calculation of α_s (Aboukassem; to be published) by means of statistical mechanics (replica) methods qualitatively agrees with our experimental results.
When storing the weights in holograms - the heart of an optical vector matrix multiplier - optical constraints restrict the number of realizable gray values to 6bit. Comparison of the fits for various precisions in figure 2 shows that an increase of precision from 5 to 8bit results only in gradual improvements of the generalization abilities. That means an optical implementation would be possible without remarkable loss of performance with even 5bit.
We also studied the influence of reducing the training expense on the performance of the pupil by training the pupil network with Monte Carlo simulations at $T = 0$. The learning and generalization errors obtained by the simulated annealing algorithm and Monte Carlo simulations at $T = 0$ for various bit precisions of the weights and for networks with 100 input nodes are shown in figure 1 below.

Simulated annealing gives better results for all α than the Monte Carlo at $T = 0$. But computational savings of the Monte Carlo simulations justify the use of this method which can find a weight configuration with the same generalization ability when the training set size coded in α is sufficiently larger.
We additionally investigated the behavior at larger training set sizes (α), fitted the results with a power function (see following tables) and found that $\epsilon_G, \epsilon_T \sim \frac{1}{\alpha}$ is a good fit for large α. The same asymptotic behavior has been obtained for a pupil perceptron with continuous weights by Engel in [1]. Figure 2 also shows a fit for 3bit precision. It can also be infered that both generalization error and learning error strive for a value ϵ_{min} and that simulated annealing achieves better generalization abilities than the direct discretization.

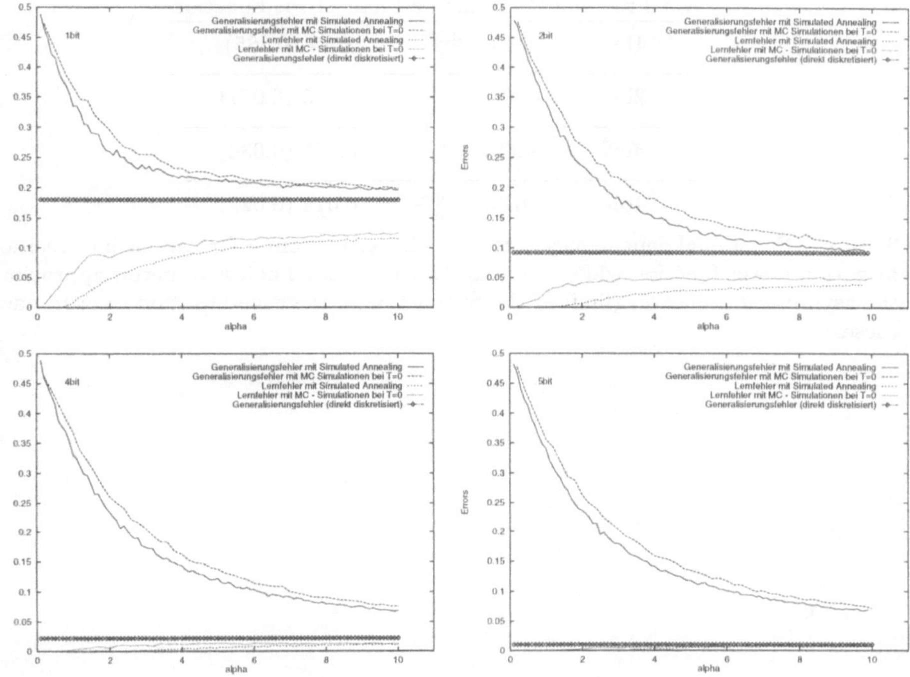

Fig. 1. Comparison of learning and generalization errors obtained by Simulated Annealing and Monte Carlo simulations at $T = 0$ for various bit precisions

Bit Precision	Fit Function	Limit of Fit Function
1bit	$0.181 + \frac{0.155}{\alpha}$	0.181 (0.18)
2bit	$0.072 + \frac{0.246}{\alpha}$	0.072 (0.09)
3bit	$0.034 + \frac{0.398}{\alpha}$	0.034 (0.044)
4bit	$0.016 + \frac{0.515}{\alpha}$	0.016 (0.022)
5bit	$0.012 + \frac{0.541}{\alpha}$	0.012 (0.01)

Table 1. Functional dependency of the generalization error from α for various bit precisions; fit with a power function while including only the last 30 points to better approximate the asymptotic behavior. (generalization error gained by direct discretization in parantheses)

Bit Precision	Fit Function	Limit of Fit Function
1bit	0.14 - $\frac{0.23}{\alpha}$	0.14 (0.185)
2bit	0.055 - $\frac{0.17}{\alpha}$	0.055 (0.071)
3bit	0.027 - $\frac{0.1}{\alpha}$	0.027 (0.034)
4bit	0.014 - $\frac{0.024}{\alpha}$	0.014 (0.022)

Table 2. Functional dependency of the learning error from α for various bit precisions; fit with a power function while including only the last 30 points to better approximate the asymptotic behavior. (limit of the fit function of the generalization error in parantheses)

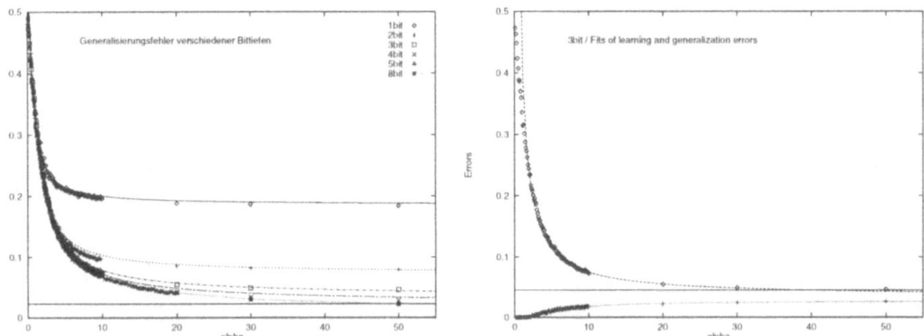

Fig. 2. Generalization error vs. α for various bit precisions, N=100, averaged over 50 pupils (left). Learning and generalization errors obtained by Simulated Annealing for 3 bit precision (right).

References

1. Engel, A.: Uniform convergence bounds for learning from examples. Modern Physics Letters B bf 8 (1994) 1683-1708
2. Biehl, M. Watkin, T.L.H. Rau, A.: The statistical mechanics of learning a rule. Review of Modern Physics bf 65 (1993)
3. Horner, H.: Dynamics of learning and generalization in a binary perceptron model. Zeitschrift für Physik B - Condensed Matter bf 87 (1992) 371-376
4. Horner, H.: Dynamics of learning and generalization in perceptrons with constraints. Physica A bf 200 (1993) 552-562
5. Lange, R.: Perfect learning in neural networks. PhD thesis Ruprecht-Karls-Universität Heidelberg (1995)
6. Metropolis, M. Rosenbluth,A.W. Rosenbluth, M.N. Teller, A.H. Teller, E. Equation of state calculations by fast computing machines. Journal of Chemical Physics bf 21 (1953) 1087-1092
7. Patel, H.-K.: Computational complexity, learning rules and storage capacities: a monte carlo study for the binary perceptron. Zeitschrift für Physik B - Condensed Matter bf 91 (1993) 257-266
8. Schwember, St.: Untersuchungen zur Generalisierungsfähigkeit des Simple Perzeptrons mit diskreten Gewichten. Diplomarbeit Ruprecht-Karls-Universität Heidelberg (1997)

Wavelet Frames Based Estimator

Skander Soltani[1], Stéphane Canu[1], Daniel Boichu[2], Yves Grandvalet[1]

[1] Heuristique & Diagnostic des Systèmes Complexes
U.M.R. C.N.R.S. 6599
[2] Division de Mathématiques Appliquées
Université de Technologie de Compiègne
B.P. 20529, 60205 Compiègne Cedex, France.
e-mail: Skander.Soltani@hds.utc.fr

Abstract. This paper introduces a new wavelet frames-based functional estimation method (*i.e.* a wavelet-based neural network) which works for more than one dimension functions. The use of frames and wavelets in our approach yields to robust decomposition with an interesting parsimonious property: compression of information in few coefficients. This approach is illustrated using the problem of estimating radioactivity in Chernobyl area.

1 Introduction

Our problem is to estimate a functional dependency between an input $\mathbf{x}$ and an output $\mathbf{y}$ of a system. For this purpose, we use a set of observations $\{(x_i, y_i),$ $x_i \in \mathbb{R}^d, y_i \in \mathbb{R},\ i = 1 : N\}$ supposed to be taken according to an unknown density law $\mathbb{P}(\mathbf{x}, \mathbf{y})$.
In the general case, the nature of this dependency is unknown, so we use flexible estimators, these ones can approximate large function sets. The performance of the estimation is measured by a generalization criterion or cost function. We will restrain ourselves to the quadratic one given by

$$C_{\text{gen}}(f) = \mathbb{E}_{\mathbf{xy}}\{(\mathbf{y} - f(\mathbf{x}))^2\}. \tag{1}$$

The minimum of $C_{\text{gen}}(f)$ on measurable functions space $\mathcal{F}$ is given by

$$r(x) = \mathbb{E}\{\mathbf{y}|\mathbf{x} = x\}. \tag{2}$$

This is the regression function, thus, our problem becomes the estimation of the regression function using the set of observations (called training set).
The probability law $\mathbb{P}(\mathbf{x}, \mathbf{y})$ is unknown, several estimators use the empirical distribution, this leads to minimization of the empirical criterion

$$C_{\text{emp}}(f) = \frac{1}{N}\sum_{i=1}^{N}(y_i - f(x_i))^2. \tag{3}$$

Unfortunately, this minimization problem is ill-posed because in $\mathcal{F}$ there exists an infinity of functions for which $C_{\text{emp}}(f) = 0$. One method to well pose the

problem is to regularize it by adding a high frequency penalty term, the solution is called spline estimator [6]. Nevertheless, neural networks minimize the empirical criterion within a parametric function family [1]. By using the empirical criterion, one can estimate function decomposition on a basis (for instance Fourier series [7]). The use of these estimators is limited because of parameter estimation problems.
Another method consists on searching within sets $\mathcal{F}_m$ of $\mathcal{F}$. The set $\mathcal{F}_m$ contains all the subspaces (noted $\mathcal{G}_m^k$, $k \in \mathbb{N}$) which are generated by m functions. The estimator related to each of the subspaces is obtained by minimization of the empirical criterion. We look for the set $\mathcal{F}_{m^*}$ which satisfies:

$$\mathcal{F}_{m^*} = \arg\min_{\mathcal{F}_m \in \mathcal{F}} \mathbb{E}_{\mathbf{xy}}\{(y - [\arg\min_{\widehat{f} \in \mathcal{F}_m} C_{emp}](x))^2\}. \tag{4}$$

So we search for the set within which we can find an estimator with minimal cost function, the estimator itself is determined using the empirical criterion. We have so a dilemma between bias and variance. This kind of methods have also been developed by Vapnik in the framework of structural risk minimization [8]. In this paper, we introduce a functional estimation method based on the criterion (4) and on wavelet frames decompositions. This kind of decomposition includes redundancy and compression of information.

2 Wavelet Frames Based Estimator

The estimator presented in this section is defined by a finite linear combination of wavelets extracted from a frame [4]. Wavelets and frames are concepts with interesting properties exploited in designing the estimator. In order to introduce the details concerning the estimator, some fundamental notions are recalled in the first part of this section.

2.1 Wavelet Frames

A frame is a set of non-independent and complete functions $\varphi_\lambda(x)$[3] which satisfies some conditions given in [4] such that every function $f \in \mathbb{L}^2(\mathbb{R}^d)$ can be written as follows:

$$f(x) = \sum_\lambda c_\lambda \varphi_\lambda(x). \tag{5}$$

The coefficients c_λ are given by $c_\lambda = \langle f, \varphi_\lambda^* \rangle$ where functions $\varphi_\lambda^*(x)$ constitute a dual frame. For a given frame, there exists an infinity of possible dual frames, this means that the decomposition is not unique [3][4]. For instance, one can search for the dual frame which provides the decomposition with minimal $\sum_\lambda |c_\lambda|^2$.
In the context of multidimensional wavelet frames theory, one can define projective or radial frames, we restrain ourselves to the first case:

[3] in the general case, λ is a multi-index of integers.

projective frames: functions $\varphi_\lambda(x)$ (called ridgelets in [3]) are defined by:

$$\varphi_\lambda(x) = a_0^{k/2}\varphi(a_0^k(\langle u_{l\theta_k}, x\rangle - mt_0)) \qquad \lambda = (k,l,m), \tag{6}$$

where $a_0 > 1$, $t_0 \in \mathbb{R}$ are called elementary dilatation and translation, vectors $u_{l\theta_k}$ are directive unitary vectors, they belong to the unitary sphere $S^{d-1}(\mathbb{R})$, $l \in \mathbb{Z}^d$ and θ_k is the elementary $(d-1)$-tuple directions angles at scale k.
In both cases (projective and radial), elementary parameters must satisfy conditions given by Calderón-Zygmund theory [3][4]. The wavelets must also satisfy the admissibility condition which is an oscillation constraint, so neither the hyperbolic tangent nor the Gaussian functions, widely used for neural networks and RBF estimators respectively, can constitute a frame. Among coefficients c_λ, a lot are close to zero because of spatio-frequential properties[4] of the function $\varphi(x)$, the idea of wavelet frames-based estimator is precisely to use for the estimation a finite number of functions $\varphi_\lambda(x)$.

2.2 Wavelet Frames Based Estimator

The estimator $\widehat{f}_W(x)$ is written as a linear combination of the wavelets which generate a subspace $\mathcal{G}_{m^*}^{k^*} \subset \mathcal{F}_{m^*}$. Recall that $\mathcal{G}_{m^*}^{k^*}$ is a subspace generated by m^* wavelets, $\widehat{f}_W(x)$ is written as follows

$$\widehat{f}_W(x) = \sum_{\lambda \in \mathcal{G}_{m^*}^{k^*}} \widehat{c}_\lambda \varphi_\lambda(x). \tag{7}$$

The first problem is to find the best size of the estimator using the criterion (4) (i.e. m^*), then, we use the training data to look for the best m^* wavelets in the frame to be used for the estimation (i.e. space $\mathcal{G}_{m^*}^{k^*}$ or simply k^*).
In our case, the space $\mathcal{F}$ is equal to $\mathbb{L}^2(\mathbb{R}^d)$. A solution of (4) is practically impossible because $\mathbb{L}^2(\mathbb{R}^d)$ is of infinite dimension. $\mathcal{F}$ is so restricted to become a finite dimension space within which we can find an acceptable solution. This restriction is based on the explanation power of each wavelet in the frame.
Actually, only some low dilatations are considered, higher ones are eliminated because the related wavelets:

- don't include a sufficient number of observations in their supports,
- correspond to higher frequencies[5] included in the function, this can be seen as a regularity assumption on the nature of the solution.

For each of the considered dilatation, we remove wavelets with few observations in their support, thus we have constructed the set $\mathcal{L}$ of cardinality $L < \infty$, so instead of searching in $\mathbb{L}^2(\mathbb{R}^d)$ because of constraints imposed by the data, we use only the elements of $\mathcal{L}$ (also called library). This modification gives

$$\mathcal{F}_{m^*} = \arg\min_{\mathcal{F}_m \in \mathcal{L}} \mathbb{E}_{\mathbf{xy}}\{(y - [\arg\min_{\widehat{f} \in \mathcal{F}_m} C_{emp}](x))^2\}. \tag{8}$$

[4] wavelets are rapidly vanishing both in space and frequency domains.
[5] wavelets are band-pass filters.

The problem can now be solved in finite time, but it is combinatorial because we must consider all possible spaces with elements of $\mathcal{L}$. One can avoid this by using subset selection methods like forward subset selection method, this one uses sequential procedure for selecting in the library $\mathcal{L}$ best approximating functions [2][9], the solution is of course not optimal but quite satisfying. The generalization cost C_{gen} is also unknown because the probability law $\mathbb{P}(\mathbf{x}, \mathbf{y})$ is unknown, we use resampling methods on the raw data to estimate it, we can use for instance cross-validation or one of different variants of the bootstrap (sophisticated, 0.632.,$\cdots$) [5]. The subset selection is then applied on each of the resulting samples. Thus, we find the optimal wavelets to be used (i.e. m^*), but this is insufficient because we have no information about the wavelets to be used, so we apply again subset selection on the original training set to determine them (i.e. k^*).

3 Application

The algorithm previously described have been tested on a problem of spatial interpolation described next section. In the following section we will give results found by the projective wavelet frames only.

3.1 Nature of the data

The data represents Cs^{137} concentration measures in Gomel area (Belarus)[6], the contamination is due to Chernobyl fallouts. The available set of data is composed of empirical average values of several measures taken at 594 different positions (latitude and longitude) in the area. If we take into account the geographical characteristics of the area (mountains, lakes,...), we can suppose that position of measures follows a probability law $\mathbb{P}(\mathbf{x})$. If we suppose that measurement equipment perturbations has zero expectation, the problem of Cs^{137} concentration estimation becomes a regression estimation one.

3.2 Results

The first thing to do is to choose elementary dilatation a_0, translation t_0 and direction u_0 using Calderón-Zygmund theory. Actually, it is very hard to check these conditions, but by taking elementary parameters relatively small, we are quite sure that they are satisfied. We have taken $a_0 = 2$, $t_0 = 1$ and $\theta_0 = \frac{\pi}{4}$.
For complexity control, we have tested three variants of the bootstrap, in this case the averaging, sophisticated and 632. The number of bootstrap samples was 20 in all cases. The results do not differ too much, the optimal number of wavelets found by each of the methods is practically the same, this shows the stability of the used procedure. We can conclude that bootstrap stabilizes the forward subset selection method which is known by its instability, the following

[6] they were provided in the context of the European project INTAS N^{O} 942361.

table shows the minimal, the maximal and average number of selected wavelets for 10 experiments.

Table 1. number of selected wavelets for 10 experiments

	mean number	minimal number	maximal number
Averaging	56	45	62
Sophisticated	51	43	62
632.	47	38	53

On figures (1) and (2), we show some typical results, we have the estimation of the regression surface and its contours, on the first we can see how the surface fits the data. On the second, we can see the different levels of the surface.
We can remark that the results are satisfying, the surfaces fit well the data and we have variations of the levels near abrupt changes of the measures. In some cases, we remark some variations of the surface where no data is available, this means that the method has some difficulties to extrapolate outside the domain of the observations.

4 Conclusion

The method gives promising results for the Cs^{137} concentration estimation problem. It has two important qualities: redundancy and compression of information due to wavelet frames properties. The fact that we have used bootstrap for complexity control avoids us the instability of forward subset selection method. Other methods of complexity control like ridge or lasso regression will be tested in the next future. Many other aspects like effect of the choice of the elementary parameters and estimator convergence have to be prospected.

References

1. Christopher M. Bishop. *Neural Networks for Pattern Recognition*. Oxford University Press, 1995.
2. Leo Breiman. Statistics and nets: Understanding nonlinear models from their linear relatives. NIPS 94, Tutorial 6, November 1994.
3. Emmanuel J. Candès. Harmonic analysis of neural networks. Technical report, Department of Statistics, University of Stanford, October 1996.
4. Ingrid Daubechies. *Ten Lectures on Wavelets*. CBMS-NSF Regional Conference Series on Applied Mathematics, No **61**, SIAM, 1992.
5. Bradley Efron and Robert J. Tibshirani. *An Introduction to the Bootstrap*, volume 57 of *Monographs on statistics and Applied probability*. Chapman & Hall, 1993.
6. Frederico Girosi, Michael Jones, and Tomaso Poggio. Regularization theory and neural networks architectures. *Neural Computation*, 7(2):219– 269, 1995.
7. Wolfgang Härdle. *Applied Nonparametric Regression*, volume 19. Cambridge University Press, 1990.

8. Vladimir N. Vapnik. *The Nature of Statistical Learning Theory.* Springer-Verlag, 1995.
9. Qinghua Zhang. Wavelets and regression analysis. In Anestis Antoniadis and George Oppenheim, editors, *Wavelets and Statistics*, volume 103 of *Lectures in Statistics*, pages 397–407. Springer Verlag, 1995.

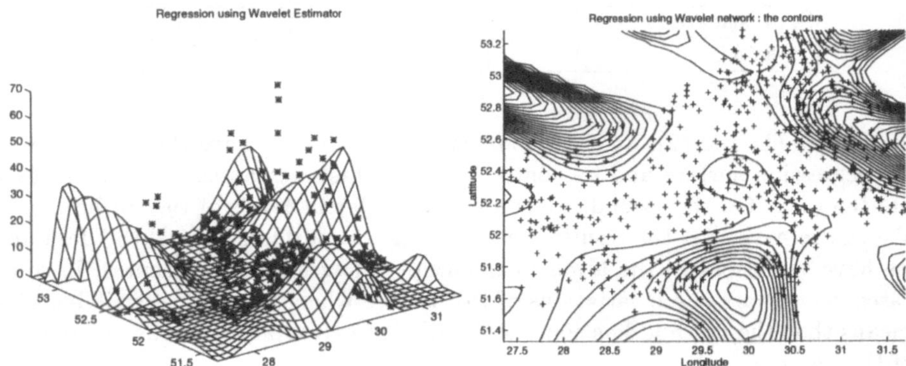

Fig. 1. Surface and contours when using the sophisticated bootstrap, we have used 'o' for measures less than 5, '+' for measures between 5 and 20 and '*' for measures higher than 20.

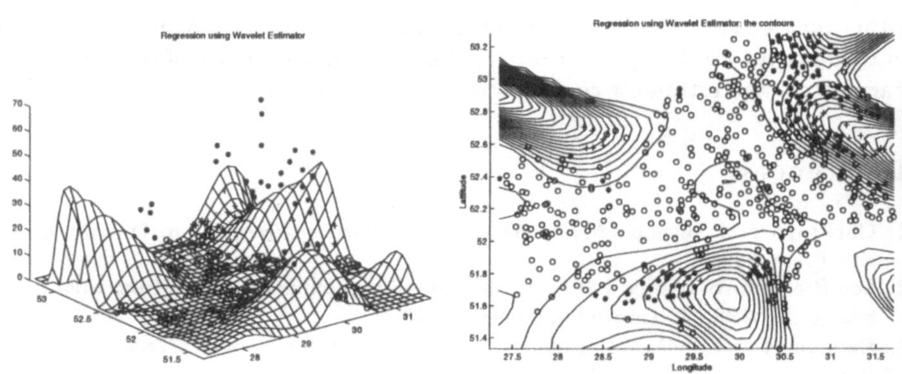

Fig. 2. Surface and contours when using the 632. bootstrap.

A Spatio-Temporal Perceptron for on-Line Handwritten Character Recognition

Nasser Mozayyani and Gilles Vaucher

SUPÉLEC
B.P. 28, 35511 Cesson Sévigné Cedex, France
E-mail: Nasser.Mozayyani@supelec.fr
Gilles.Vaucher@supelec.fr

Abstract. The objective of this work is the application of the spatio-temporal multilayer perceptron (ST-MLP) developed in our laboratory to the recognition of on-line handwritten characters. The ST-MLP integrates a spatio-temporal data coding defined in the complex domain.

Starting from the stroke of a character produced by a digitizing tablet, we conduct the recognition process in two steps. This procedure which is classic in this domain, consist of a preprocessing step and a recognition one. The first step (segmentation step), identifies some elementary (basic) lines, called primitives, from the stroke of the character. Then we utilise the ST-MLP to recognize the traced character from the primitives provided.

1 Introduction

As a followup of a broadbased effort to introduce time in neural networks by spatio-temporal (ST) coding, we hereby take up the case of multilayer perceptron. Previously we have presented a successful integration of the same coding in the Kohonen maps [1].

The ST coding, defined in the complex domain, is specially conceived for the data of ST nature. The second dimension, obtained by utilising the complex values, permit us to integrate the time factor in the data handled by the network at the level of each neural unit [1]. A basic, first version of ST multilayer perceptron applied on a simple example is presented in [3]. Here we present a more concrete application as well as a network with better characteristics.

In numerous industrial fields the processing of ST patterns has to be faced. Some examples of these domains are : the movement analysis, lipreading and handprinted character recognition on which our lab is working currently. Here we will present the case of on-line recognition of isolated handwritten characters. First we will present briefly the ST coding of data and the ST-MLP utilised. Later on, we will elaborate on the utilisation of the ST-MLP for the character recognition.

[1] It is worth mentioning that there is a lot of works in the literature concerning the integration of time in the neural networks. For an overview see [2] and [1].

2 Spatio-temporal multilayer perceptron

2.1 Coding of time

For introducing time in neural networks, as explained in [1], we use the ST representation given in [4, 5] to code the input events. By this coding, each event having an amplitude ρ and being produced at a date τ is associated to a complex number x in the input of each neuron as shown in Fig.1 :

Fig. 1. Complex representation of an event

Taking a trigonometric semi-circle $]-\frac{\pi}{2},+\frac{\pi}{2}[$, as the domain of variation of ϕ, the date τ can vary by $]-\infty,+\infty[$ in a one-to-one relation. The amplitude of x is constrained to $\mathbb{R}_+$ to avoid the problems of discontinuity posed by the the imaginary axis as explained in [6].

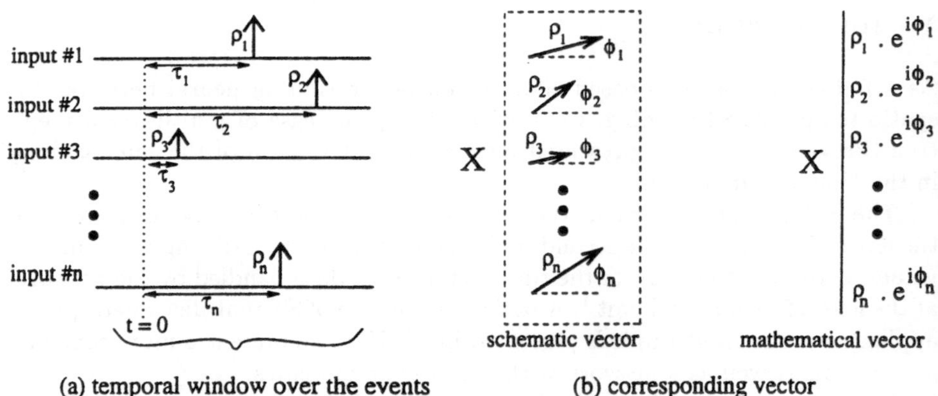

Fig. 2. Example of vectorial representation of events

The sequence of events coded in this way are presented to the network in vectors. Each vector's component corresponds to an input data which is an event defined by an amplitude and a time delay relative to a common time origin (see example of Fig. 2).

2.2 Extension of perceptron to complex domain

When we extend one classic artificial neuron to the complex domain, all parameters can be transposed in the new space as follows :

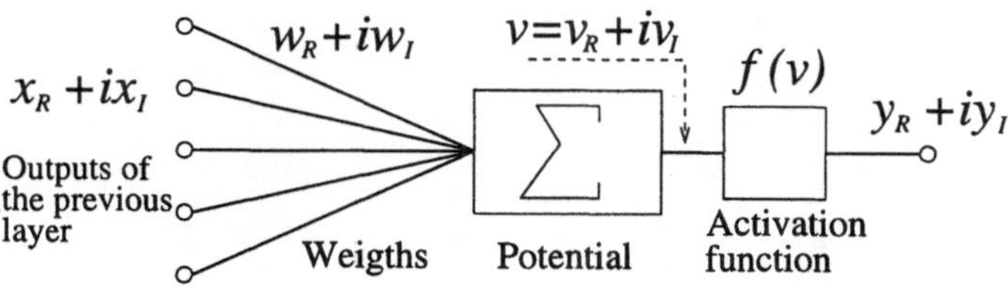

Fig. 3. An artificial neuron extended to complex domain

But for extending the multilayer perceptron to the complex domain, the problem of derivability for calculating the back propagation error has to be solved. For bypassing this difficulty, the complex model of neurons is presented in the form of a real neuron and an imaginary neuron strongly coupled. The problem is thereby taken into real domain [7, 8, 3]. At the end, we take back the result in the complex domain. This triple transposition (from spatio-temporal to complex) then (from complex to real) and again (from real to complex) is resumed as :

$$(\rho, \tau)^n \quad \rightarrow \mathbb{C}^n \quad \rightarrow \mathbb{R}^{2n} \quad - - \rightarrowtail \mathbb{C}^n \tag{1}$$

As far as the activation function is concerned, we have chosen to adopt the one proposed by T. Masters [8] :

$$f(x+iy) = p(x+iy), \qquad \text{with} \quad p = \frac{tanh(1.5\sqrt{x^2+y^2})}{\sqrt{x^2+y^2}} \tag{2}$$

This function applies a hyperbolic tangent on the module of the complex input value while transferring the same phase at the output.

3 Recognition of isolated handwritten characters

In the domain of handprinted manuscript recognition, there is usually a set of preprocessings before passing to the recognition step [9]. The preprocessing can be : filtering, normalization, resampling, segmentation or a group of them. Here we carry out just a segmentation step by identifying some basic primitives.

3.1 Preprocessing

The digitizing tablet provide us with (X_t, Y_t) coordinates of movement of the stylus at each instant t. Some authors retrieve from these values, the orientation of displacement at the instant t along with certain hypothesis for the amplitude. For example, Castaing et al. [10] have chosen to decompose the stroke of a character into basic bits (lines) each one characterized by a direction according to the 8-topology of Freeman [11] (Fig. 4.a dotted lines).

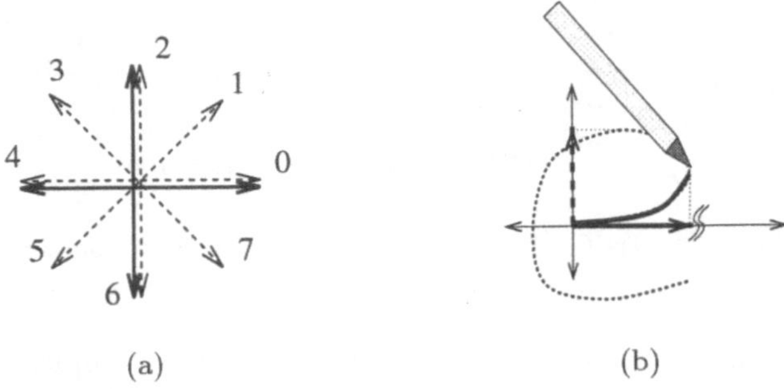

Fig. 4. (a) 8-topology of Freeman in dotted lines, and our choice of four orientation in solid line (b) Movement of the stylus translated in oriented lines

We follow the same notion but in a different way with a simpler choice concerning the length of lines as well as their temporal order, by supposing :

- four basic orientation instead of eight (Fig. 4.a bold lines).
- to quantify the length of the elementary lines, a choice of six possible lengths in each orientation for covering all displacement of the stylus from very small movement while tracing small curls, eg. in 'v', upto more bigger lines, eg. in 'l' (the lengths of lines are equally distributed between a maximum and a minimum),
- a temporal tag representing the date at which the line has been detected.

The strategy for translating the movement of the stylus to basic lines, which we will from now on call *primitives*, is as follows : as soon as the stylus moves in one of the four orientations, a detector of lines is turned on, upto an interruption of movement in this direction (Fig.4.b). This is done independently for all orientations. Hence we produce at the instant 't' (where 't' is the time at the end of detection) a primitve classified according to its orientation and its length. In this manner, the stroke of a character is composed of sequences of different primitives.

3.2 Recognition

For adapting these primitives to the ST-MLP input format, we have to apply the ST coding on them. Then we associate one input with each primitive, so that the same primitives appear as a sequence of spikes to a neuron (Fig. 5). The set of these sequences represents a segmented character for the network.

The ST-MLP, receives these sequences of spikes and is supposed to recognize the 26 letters of alphabet c.f. Fig. 5. The structure of the ST-MLP as well as its parameters are the same as what presented in §2.2. The only difference is that the network used here requires only one layer for such a task.

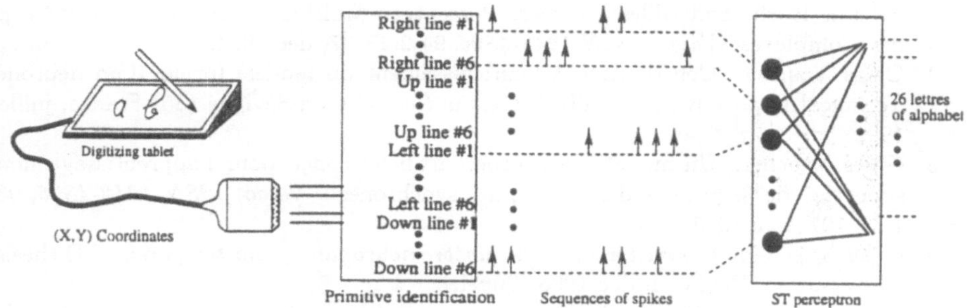

Fig. 5. Recognition system

The training on this network was done on a database of 260 examples : 10 examples for each character. The letters were simples : lowercases, mono-scripter and carefully written.

The test base comprised of the same number of examples constructed in the same manner. We obtain a rather good recognition of 95% of all letters after two thousand iterations.

4 Conclusion

The introduction of a spatio-temporal coding in the classic architecture of neural networks permits the processing of spatio-temporal patterns. As we have seen in [1] with a ST Kohonen map and here with a ST perceptron, one can envisage to use this coding in other neural models for treating the ST applications.

As for the domain of handwritten character recognition, we are presently working to obtain a better and comparable performance by using a standard data base. The next step will be the extension of the model to the recognition of words.

5 Acknowledgements

This work was carried out at *Supélec, Rennes Campus*, France. We would like to thank the people who helped us in revising this paper (A.R. Baig and L. Mé).

References

1. Nasser Mozayyani, Véronique Alanou, Jean-François Dreyfus, and Gilles Vaucher. A spatio-temporal data-coding applied to kohonen maps. In *International Conference on Artificial Neural Networks*, volume 2, pages 75–79. EC2 et Compagnie, Oct. 1995.
2. Jean-Cédric Chappelier and Alain Grumbach. Time in neural networks. *SIGART Bulletin*, 5(3):3–11, Jul 1994.
3. Étienne Roché and Gilles Vaucher. Perceptron Multi-Couches Étendu aux Corps des Complexes. *Valgo, ISSN 1243-4825*, 94-2:23–37, déc 1994.
4. Gilles Vaucher. Contribution à l'enrichissement du modèle formel d'un neurone. Technical Report R-SI-GV-TH-A 1.1, Supélec, Cesson-Sévigné (35) France, juillet 1992.
5. Gilles Vaucher. Un modèle de neurone artificiel conçu pour l'apprentissage non-supervisé de séquences d'événements asynchrones. *Valgo, ISSN 1243-4825*, 93-1:66–107, mai 1993.
6. Gilles Vaucher. *A la recherche d'une algèbre neuronale spatio-temporel.* PhD thesis, Université de Rennes I, dec 1996. Supélec.
7. Gordon R. Little, Steven C. Gustafson, and Robert A. Senn. Generalization of the backpropagation neural network learning algorithm to permit complex weights. *Applied Optics*, 29(11):1591–2, April 1990.
8. Timothy Masters. *Signal and image processing with neural networks.* John Wiley & Sons, Inc., 1994.
9. S. Bercu, B. Delyon, and G. Lorette. Segmentation pour une méthode de reconnaissance d'écriture cursive "en-ligne". In CRIN-CNRS GRCE Afcet, editor, *Actes du Colloque CNED'92, Traitement de l'écriture et des documents*, pages 144–151, Juillet 1992.
10. Jacqueline Castaing, Pierre Brézellec, and Henry Soldano. Une méthode symbolique pour la reconnaissance de l'écriture manuelle en ligne. In AFIA AFCET, editor, 10e *congrès RFIA, Reconnaissance des Formes et Intelligence Artificielle*, pages 580–587, Jan 1996.
11. A. Belaïd and Y. Belaïd. *Reconnaissance des formes: Méthodes et Application.* InterEdition, 1992.

Learning Oscillations Using Adaptive Control

Martin Georg Weiß

Graduiertenkolleg Technomathematik, Universität Kaiserslautern,
Erwin-Schrödinger-Straße, 67663 Kaiserslautern, Germany

Abstract. We study a model for learning periodic signals in continouus time Hopfield networks proposed by Doya and Yoshizawa ([3]) that can be considered as a model for temporal pattern memory in animal motor systems. A network receives an external oscillatory input and adjusts its weights so that this signal can be reproduced approximately as the network output after some time. We use tools from adaptive control theory to derive an algorithm for weight matrices with special structure. If the input is generated by a network of the same structure the algorithm converges globally and does not exhibit the deficiencies of the back-propagation based approach of Doya and Yoshizawa.

1 Introduction

In [3] the following model for learning of motions was proposed: Trajectories of motion are assumed to be stored in some parts of the motor nervous system. Whenever we try to memorize a new motion, e. g. riding a bicycle or swimming, we achieve our goal by conscious repetition in a way of supervised learning. During this process some high level components of the nervous systems adjust or influence lower level components such that the difference between the trajectory that should be generated and the one that is actually produces is minimized. This constitutes an on-line learning method. After a while we can repeat the desired trajectory seemingly without thinking. The lower level components can autonomously generate the signals necessary for the control of the muscles. No supervision is needed any more.

Based on these assumptions a neural network was suggested that operates in two modes: In the first mode the network receives a periodic signal to be learned. The weights are adapted by recurrent backpropagation with the usual squared difference between the reference signal and the network output as the minimization criterion. After some learning period, when the error is sufficiently small, the weight adaptation is cut off and the network operates in the second mode: The output replaces the reference signal as the input so that the network is running autonomously now. After successful learning the network should be able to sustain an output oscillation that approximates the reference signal.

The results in [3] show that simple wave forms can be learned and sustained. Signals containing more than one period can not be sustained. However, no theoretical explanation could be given whether this was due to the well-known deficiencies of the recurrent back-propagation algorithm (like the use of approximate partial derivatives, see e. g. [6]) or to some intrinsic properties of the problem. Accordingly no proof of convergence could be given.

In this paper we propose a learning algorithm based on techniques from adaptive control theory that sheds more light on the problem and allows for easy convergence speedup. Nevertheless, the control law uses only structures that can be interpreted as neural networks. No linear filters are required. In Section 2 we introduce the model and mathematical problem formulation. Then our learning rule is formulated with the proof of convergence in Section 3 and implications for the learning problem are given. Numerical results are shown in Section 4.

2 Problem Formulation

We consider additive Hopfield network with state $x \in \mathbb{R}^m$, scalar input u, scalar output y and adjustable weights $W(t) \in \mathbb{R}^{m \times m}$:

$$\begin{aligned} \dot{x} &= -\tau x + W(t)\sigma(x) + b\sigma(u) \\ y &= c^t x \end{aligned} \tag{1}$$

The function σ is assumed sufficiently smooth with the following properties like $\sigma = \tanh$ or $\sigma = \arctan$: $\|\sigma\|_\infty < \infty$, $\|\sigma'\|_\infty < \infty$, $\sigma'(x) > 0$ and $\sigma(-x) = -\sigma(x)$ for all $x \in \mathbb{R}$. Only the output can be observed.

The system receives a periodic input signal u with period T that it should learn in the following sense: The weights W are adjusted by a differential equation so that the output y asymptotically tracks the reference signal u.

In order to be able to replicate u exactly we assume that u is generated by a network of the same structure running in closed loop:

$$\begin{aligned} \dot{x}^\star &= -\tau x^\star + W^\star\sigma(x^\star) + b\sigma(y^\star) \\ y^\star &= c^t x^\star \end{aligned} \tag{2}$$

We choose the following structure in dimension $n + 1 = m$

$$W = A(w) = \begin{bmatrix} J_n(-\alpha) & 0_{n\times 1} \\ w^t & -\alpha \end{bmatrix}, \quad b = \mathbf{e}_n, \quad c = \mathbf{e}_{n+1} \tag{3}$$

where $J_n(-\alpha)$ denotes a Jordan block of size n and $w \in \mathbb{R}^n$ corresponding to self-inhibitory connections for $\alpha > 0$. We assume that α is fixed and known; only $w_1, \ldots, w_n$ can be adjusted. The choice of this structure is motivated by the following: The unit vectors b and c enable us to write the closed loop system (2) as an autonomous Hopfield system without inputs

$$\dot{x}^\star = -\tau x^\star + (A(w^\star) + bc^t)\sigma(x^\star) \ . \tag{4}$$

It is easily seen that we can choose n eigenvalues of $A(w^\star) + bc^t$ symmetric to the real axis by appropriate choice of $w^\star$. Eigenvalue conditions are a main ingredient in bifurcation theory which we use to generate periodic orbits. For suitable values of $w^\star$ a Hopf bifurcation occurs around the fixed point 0 of (4) yielding an asymptotically stable periodic orbit, see [9]. A similar structure has also been used in [7] where the linear case $\sigma = \mathrm{id}$ was studied. On the other hand it is well known that - unlike the linear case - this structure reminiscent of normal forms in linear control theory cannot produce all dynamics which are possible for fully variable weight matrix W (see [1]).

Now our learning problem can be stated as follows: Find an adaptation rule f for w such that for all initial values $w(0) \in \mathbb{R}^n$, $x(0) \in \mathbb{R}^{n+1}$ the system

$$\begin{aligned} \dot{x}^\star &= -\tau x^\star + (A(w^\star) + bc^t)\sigma(x^\star) \\ \dot{x} &= -\tau x + A(w)\sigma(x) + b\sigma(y^\star) \\ \dot{w} &= f(w, y^\star, x) \end{aligned} \tag{5}$$

achieves state and parameter convergence, provided that $x^\star$ is periodic:

$$\lim_{t\to\infty} (y(t) - y^\star(t)) = 0 \tag{6}$$

$$\lim_{t\to\infty} (w(t) - w^\star) = 0 \tag{7}$$

3 Learning Rule

Consider the first mode of the model, the learning phase. Define the state error $e := x - x^\star$ and the parameter error $p := w - w^\star$. In analogy to [7] we propose the parameter adaptation law

$$\dot{w} = \dot{p} = -(y - y^\star)Q\sigma(\underline{x}) \tag{8}$$

where we abbreviate $\underline{x} = [x_1, \ldots, x_n]^t$ and $Q \in \mathbb{R}^{n,n}$ is an arbitrary positive definite matrix. For $Q = \eta I$ the rule is local. Similarly to [7] this rule can be motivated by a time-varying quadratic Lyapunov function. We have

Theorem 1. *Consider (5) with $\sigma \in C^1$. Let $x^\star$ be nontrivially periodic and $\tau > -\alpha\|\sigma'\|_\infty$. Then (8) solves the output tracking problem (6). If $\sigma(x_1^\star), \ldots, \sigma(x_n^\star)$ are linearly independent then the state and parameter error converge to zero exponentially. Otherwise the weight vector converges to an affine subspace containing $w^\star$:*

$$w(t) \;\to\; w^\star + \{\sigma(\underline{x}^\star(t)) : t \in \mathbb{R}\}^\perp$$

Proof. We only sketch the ideas. The right hand side of (5) is linearly bounded so solutions exist on $\mathbb{R}^+$. The key idea is to define an differential equation for the state error, given solutions x and $x^\star$:

$$\dot{e} = (-\tau I + A(w^\star)\mathrm{diag}(\sigma'(\xi_i(t))))\, e + cp^t\sigma(\underline{x}) \quad =: \tilde{A}(t)e + cp^t\sigma(\underline{x})$$

The functions ξ_i are determined by the mean value theorem for x_i and $x_i^\star$. Neglecting an exponentially decaying inhomogeneity that does not influence convergence one can write the combined error system in the form

$$\begin{bmatrix} \dot{e} \\ \dot{p} \end{bmatrix} = \begin{bmatrix} & \tilde{A} & \begin{matrix} 0_{n\times n} \\ \sigma(\underline{x}^\star)^t \end{matrix} \\ 0_{n\times n} & -Q\sigma(\underline{x}^\star) & 0 \end{bmatrix} \begin{bmatrix} e \\ p \end{bmatrix} \tag{9}$$

and apply techniques similar to well-known facts from adaptive control of time-invariant systems that can be found in [5] and [8]. The linear independence condition on the $\sigma(x_i^\star)$ constitutes the persistency of excitation condition required for parameter convergence.

Remark. The conditions $\sigma(0) = 0$ and $\sigma(-x) = -\sigma(x)$ are not used in the proof. These properties make life easier when dealing with bifurcation theory. We do not use the fact that the periodic orbit results from a Hopf bifurcation, anyway. Neither is it necessary to have $\sigma(x) > \epsilon > 0$ for all $x \in \mathbb{R}$ as in [4].

Remark. The gain matrix Q can be made arbitrarily large (in the sense $Q > \eta I$, η large) but numerical simulations show that the convergence speed saturates for large η. The speed of convergence can be increased by using time-varying gains $R(t)^{-1}$ instead of Q in a learning rule

$$\dot{w} = \dot{p} = -(y - y^\star)R(t)^{-1}\sigma(\underline{x}) \tag{10}$$

where R is adapted on-line according to the differential equation

$$\dot{R} = \begin{cases} 0, & 0 \leq t \leq t_0 \\ \frac{1}{t}(-R(t) + \sigma(\underline{x})\sigma(\underline{x})^t), & t \geq t_0 \end{cases} \tag{11}$$

Here $t_0 > 0$ is necessary to avoid the singularity in the right hand side. It can be shown ([9]) that for arbitrary positive definite initial values $\tilde{Q}(0)$ the solutions of (11) approach the autocovariance of $\sigma(\underline{x}^\star)$

$$R^\star := \lim_{t\to\infty} R(t) = \lim_{T\to\infty} \frac{1}{T}\int_0^T \sigma(\underline{x}^\star(s))\sigma(\underline{x}^\star(s))^t ds \tag{12}$$

which is positive definite iff $\sigma(x_1^\star), \ldots, \sigma(x_n^\star)$ are linearly independent. This matrix gives an optimal weighting (in a sense explained in [9]) for the convergence of the components of w.

We stop the learning process after some time $\bar{t}$ when state and parameter error are sufficiently small, thus entering the second mode of the model, the replication phase. That is, we keep $w(t) = w(\bar{t})$ fixed and let the learning system (1) run in closed loop mode: $u = y$. It is known from the theory of Hopf bifurcation (see [2]) that the existence of the periodic solution is stable under parameter variation and that the resulting trajectories depend continuously in Hausdorff metric on the parameters. Therefore the solutions $x^\star(t)$ and $x(t)$ will not stay close (as the periods will generally be different but the qualitative behaviour will be similar.

4 Computer Simulations

Let us consider a system with $n = 2$ where $\sigma(x) = \tanh(1.8x)$, $\tau = 1$, $\alpha = 0.1$, $w^\star{}_1 = -7.2$, $w^\star{}_2 = 6.6$. System (5) was started with $x(0) = 0$, $w(0) = 0$ and a point $x^\star(0)$ on the systems periodic orbit. The learning gains were $Q = I$. Fig. 1 shows the reference system output $y^\star$ on the left hand side and the output of the reference (dotted) and learning (solid) system when learning was stopped at $\bar{t} = 50$ with $w(t) = w(\bar{t})$ for all $t > \bar{t}$. The learning system produces an output very similar in shape but with greater period. In Fig. 2 both the tracking error $y - y^\star$ and the parameter error $w - w^\star$ clearly can be seen to converge exponentially. In another simulation we consider the above system

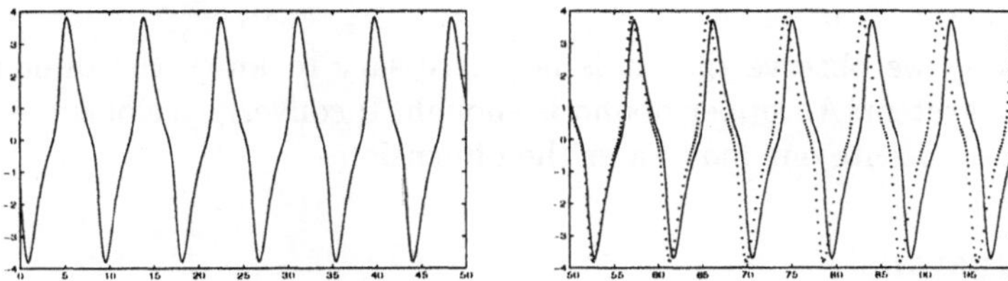

Fig. 1. Output of the systems

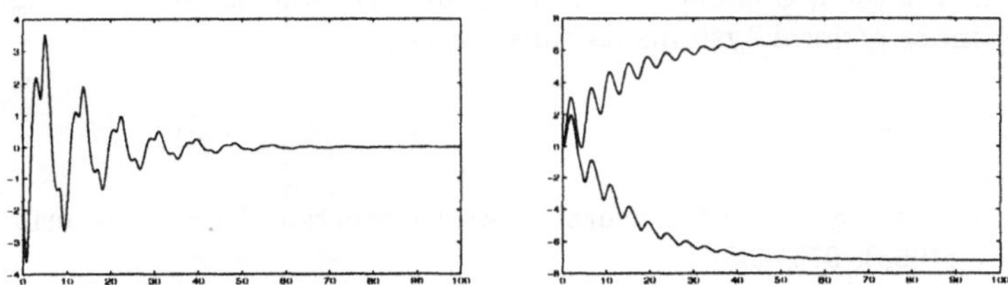

Fig. 2. Tracking error and parameter error

in $n + 1 = 3$ dimensions as a system in 5 dimensions by trivially extending the reference weight vector to $w^\star = [0\,0\,-7.2\,6.6]^t$. Obviously this system can generate the same output as above. Due to the higher dimension convergence is much slower (Fig. 3, left). However, if we apply adatation of the gain matrix (10) we can achieve convergence in much shorter time (Fig. 3, right; $Q = 38I_4$). Note the different periods of time plotted in the two cases! One could argue that the adaptation of R simply increases the gain to speed up convergence. Indeed, $||R^\star||_2 \approx 38$ in our case. So this claim is refuted by the choice of Q. The convergence with $Q = 38I_4$ is not much faster than with $Q = 38I_4$. This shows that it is necessary to treat different directions in parameter error space differently.

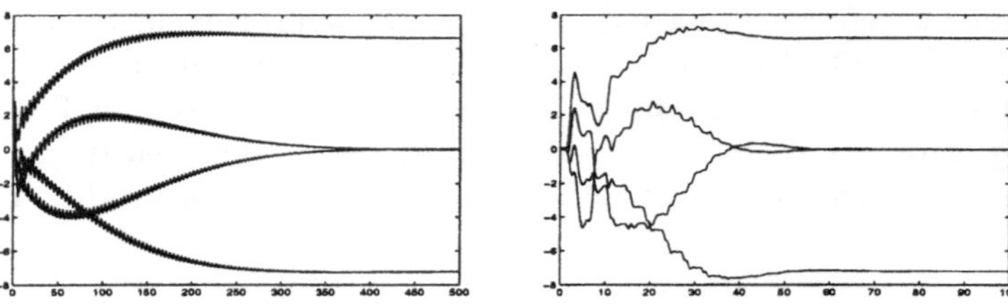

Fig. 3. Parameter error with and without adaptation of gains

By the way we observe that it is not necessary to know the dimension of the reference system: An upper bound is enough. If convergence of some leading weights to zero occurs one may lower the dimension.

5 Conclusions

For a special type of network we have presented an simple and fast learning rule working with neural network dynamics only. The condition of linear independence remains to be investigated. Similar results can be obtained for discrete time systems with more technical arguments. The full weight matrix can be identified using an open loop version of the algorithm under the quite restrictive assumption $\|W^{\star}\|_2 < 1$ when all of the state is accessible. These results will be reported elsewhere ([9]). The algorithm heavily relies on the fact that the parameters enter the network dynamics linearly and is therefore not easily extended to more general systems as those in [1].

References

1. Albertini, F., Sontag, E. D.: For neural networks, function determines form. Neural Networks **7** (1993), 975–990
2. Amann, H.: Gewöhnliche Differentialgleichungen. de Gruyter, Berlin 1983
3. Doya, K., Yoshizawa, S.: Adaptive neural oscillator using continuous time backpropagation. Neural Networks **2** (1989), 375–385
4. Kosmatopolous, E. B., Christodoulou, M. A., Ioannou, P. A.: Dynamical neural network that ensure exponential identification error convergence. Neural Networks **10** (1997), 299–314
5. Narendra, K., Annaswamy, A.: Stable Adaptive Systems. Prentice Hall, Englewood Cliffs, New Jersey 1989
6. Pearlmutter, B. A.: Gradient calculations for dynamic recurrent neural networks: a survey. IEEE Trans. Neural Networks **6** (1995), 1212–1228
7. Reinke, R.: Adaptive Regeln zum Lernen und Reproduzieren von periodischen Signalen mit dynamischen Netzwerken. Dissertation, Universität Kaiserslautern, 1994
8. Sastry, S., Bodson, M.: Adaptive Control. Prentice Hall, Englewood Cliffs, New Jersey 1989
9. Weiß, M. G.: Learning periodic signals with neural networks. in preparation

Creation of Neural Networks Based on Developmental and Evolutionary Principles

Peter Eggenberger

AILab and Software Engineering
Department of Computer Science
University of Zurich
Winterthurerstrasse190, 8057 Zurich, Switzerland
FAX: +41-1-363 00 35; Email: eggen@ifi.unizh.ch

Abstract. In this paper we propose a biological inspired model to develop the structure of artificial neural networks. The model is based on an artificial genetic regualtory system, which controls the development of the neural network. The model allows for different cell types which are the result of different intercellular communication processes. Different cell types will also lead to different connection patterns of the neural networks. The goal of the proposed model is to investigate the question how the local genetic processes are able to construct the structure of a neural network.

1 Introduction

This paper reports on a computer model which is able to evolve artificial neural networks (ANN) in three dimensions. The proposed model is based on an artificial genetic regulatory system (AGRS) which controls the epigenetic processes of the development of an ANN by means of strictly local interactions between the cells and the genes. Since every cell contains the same genome, the differences between the cells are due to different signals which the cells receive from other cells. The main developmental processes as cell division, cell death, cell differentiation and development of the neural connections are under the dynamic control of the AGRS. Variations in the artificial genomes will lead to different intercellular communication patterns and different developmental possiblities of the cells and, in consequence, will change the structure and functions of the ANNs. As the biological neural networks are created by means of nonlinear and local genetic dynamics, I propose a model in which the relationship between the genome, development and an ANN can be studied in a rather simplified, but yet interesting way.

Kitano [12] reduced the size of the genome using a graph generation grammar to encode neural network topologies which has a better scaling behavior than direct encoding schemes. Belew [1] used a grammar to simulate developmental processes. His scheme is context sensitive, but it is restricted to pre-specified neural network topologies. Gruau and Whitley [9] encoded the developmental process as a grammar tree. In this approach the cells inherit their connections and no context sensitive development is possible. Fleischer and Barr [8] used a genetic encoding (hand coded) to specify the developmental processes by means of ordinary differential equations which were coupled with if-clauses to allow for differential gene expression. Vaario [17] has proposed a grammar-based simulation tool, in which the developmental process is described by a set of rules. Nolfi [15] proposed a developmental model for neural networks based on cell division and cell migration. The major flaw of this approach is that the number of the genes in the genome grows with the number of neurons which leads to a bad scaling behavior. Dellaert and Beer [4, 3, 5] proposed a model based on Boolean networks to evolve autonomous agents with developmental processes. With the proposed method he was able to get the simplest Braitenberg vehicles.

2 Implementation of Developmental Processes

As in biological neural systems the whole topology and the structure develop, I implemented a model which is a simple abstraction of those developmental processes. The artificial cells are enabled by the genome to communicate, to change their genetic states and to respond by producing different artificial chemicals. As a result the cells become different and their abilities, for instance, to connect to a partner cells will also be different between the cell types. In other words, the dynamics of the ARGRS determine the topology of the ANN.

2.1 Gene Regulation and Gene Products

To simulate cell differentiation, I introduced regulatory mechanisms of gene expression in the genome. In contrast to usual genetic algorithms a structured genome is used with two classes of genes: regulatory units and structural genes. The regulatory units are some kind of switches to turn on or off the structural genes they control. Structural genes encode for specific substances, which are described below. Among these substances, a special one, the transcription factors (TF), can switch on or off the regulatory units. The activity of a structural gene is regulated in the following way. Every cell contains a list of transcription factors (TF) which influence its genome. The TF's as well as the regulatory units are implemented as strings of integers. The two strings are then compared and an artificial affinity is assigned to them. In a second step, also the concentration of the TF's is taken into account. Depending on the different sources and the diffusion constants, the concentration is calculated for every TF in the cell. The product of the affinity and the concentration of every TF at a regulatory unit is calculated and the products summed. The same is repeated for every regulatory unit of a gene. The total sum is then put in a sigmoidal function and if a fixed threshold is exceeded, the gene is activated or inhibited (See equations 1,2,3). The activity of structural gene depend on the states of its regulatory units and is calculated as follows:

$$r_j = \sum_{i=1}^{n} \mathrm{aff}_i * conc_i \tag{1}$$

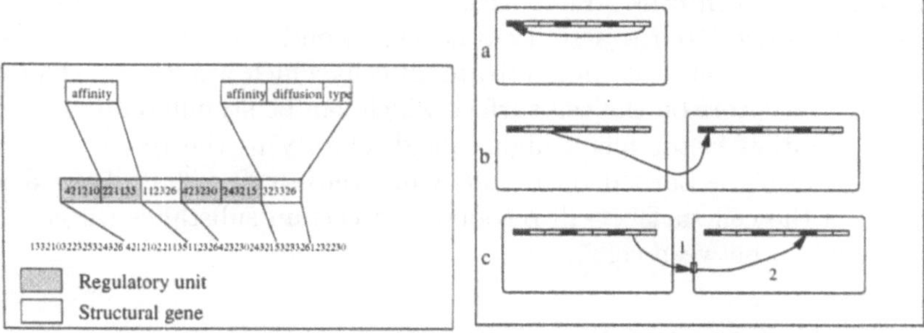

Fig. 1. Left: The genes are represented as integers and they are used to specify substance classes and properties of the produced artificial substances. Right: a) intracellular (within cell) communication. b) intercellular (between cells) communication. c) receptor-based, intercellular communication.(These mechanisms are the base of cell differentiation).

$$a_k = \frac{1}{1 + exp^{-}(\sum_{j=1} r_j)} \tag{2}$$

$$g_k = \begin{cases} -1.0 & : \quad a_k < 0.2 \\ 1.0 & : \quad a_k > 0.8 \\ 0.0 & : \quad otherwise \end{cases} \tag{3}$$

- aff_i = affinity of the ith TF with the jth regulatory unit gene
- conc_i = concentration of the ith TF
- r_j = activity of the jth regulatory unit of a structural gene
- a_k = total sum of the activities of all regulatory units of the kth gene
- g_k = activity of kth gene

(More details about how the genes are regulated can be found in [6, 7] or can be posted from the web.) If a structural gene is expressed different artificial chemicals are produced which I describe below. A TF has the function to regulate the gene activities and these substances are used to modulate the development processes. Cell adhesion molecule (CAM) are able to connect cells to each other, if the affinity between the CAMs of different cells are high enough affinity Receptor are used to regulate the communication between the cells by filtering out unspecific TFs. Artificial functions as cell division, cell death or searching for partners are called, if the corresponding structural gene has the right markers. The product classes are encoded in the genome by the first three integers of the structural genes (see fig 1).

2.2 Cell Differentiation

Cell induction is a fundamental mechanism of cell differentiation. To simulate cell induction we implemented three different possible pathways to exchange information between cells . First, there are substances which do not leave the cell and which regulate the activity of genes. Second, there are substances which can penetrate the cell wall and activate all cells which are nearby. Third, there are specific receptors on the cell surface which can be stimulated by substances. If a transcription factor has a high enough affinity to the receptor, a gene or a group of genes is turned on or off. Only those cells which have a specific receptor on the cell surface will respond to a certain substance (filtering of the communication between cells).

2.3 The Neurons and their Connections

As cells can become different, they will produce also different substances. To connect two neurons, there are two different classes of CAMs. These are stored in lists in the cell and are used differently. The members of the first list of one neuron are compared to the second list of another neuron. If two adhesion molecules of the two different lists have a high affinity to each other, a link from the first cell to the second cell is established. Note, that the direction of the link is given by the two types of lists. But not all cells will look through all the lists of all cells and will decide to whom a link should be established. The code of the CAM is used to define a search range. This range determines the maximal distance of two cells in the grid which can connect with each other. As explained in figure 1 the gene has a part named range, which is used to determine how long the search range for a neuron is. Minimally, the cell looks only at the nearest neighbors and maximally, a cell tries with every cell to build a contact. The weights are randomly assigned and can be changed by a Hebbian rule. To determine if a synapse will be inhibitory or excitatory is determined by the type of the connections of the CAMs. In figure 2 some examples of evolved neural network structures are shown. We use standard artificial neurons which obey the following equations:

$$n_i(t+1) = \sigma(\sum_{i=1}^{n} \omega_{i,j} n_j(t) - \theta_i) \tag{4}$$

The σ-Function is defined as

$$\sigma(x) = \frac{1}{1 + exp^{-\alpha x}} \tag{5}$$

The output of the neuron i is represented as n_i. Time t is taken as discrete. The synaptic weight $\omega_{i,j}$ represents the strength of the connection between neuron j and neuron i. During a time step the neuron sums up all active synapses and if a fixed threshold θ is exceeded the output of the neuron will fire.

The resulting neural nets are quite naturally heterogenous and not all neurons do the same thing anymore. These neurons have different possibilities to connect to each other. Also the communication between the neurons is different because communication is receptor dependent and not all neurons have the same receptors. The genetic regulatory network coupled with the possiblities of communication is the bases of an heterogenous, self-modulating neural network.

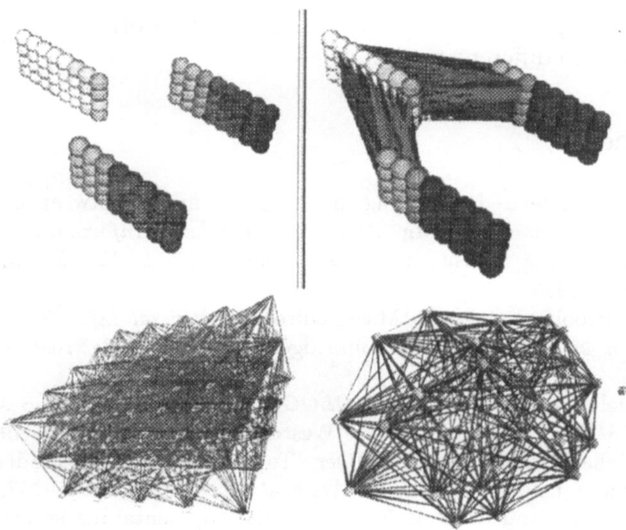

Fig. 2. In the upper figure one can see that depending on the different state of a cell, it will connect to different partner cells. In the lower part two further examples of developed structures of artificial neural networks are illustrated.

3 Discussion

As the number of genes in the genome is insufficient to specify all synapses, epigenetic processes with their combinatorial expression of sets of genes are used in Nature to specify the connections between neurons [11].

The used concepts contains the following biological ideas which reduce the data which has to be specified in an artificial genome:

- as in biology every artificial cell contains the same genetic information
- developmental processes as cell division, cell death, cell migration and cell differentiation are implemented
- no explicit encoding of the structure of the neural network, no information of a specific cell with whom it should be connected
- especially important is the fact that the genome will not necessarily grow, if the number of neurons or cells will grow.
- no direct encoding of genetic information for cell types, cell position or links to other cells, because these things depend on the dynamics of the epigenetic processes

In this work we showed that artificial neural networks can also be investigated on the genetic level. The question how the genetic regulatory network in Nature is able to develop the structure of a neural network is in my opinion an important one and which can now be investigated by the proposed computational model. In the future I will use the model to build automatically artificial neural networks (p.e. reentry maps [16] between different neural network layers) which can adapt automatically to different tasks.

References

1. Richard K. Belew. Interposing an ontogenic model between genetic algorithms and neural networks. In *Advances in neural Information Processing Systems (NIPS),S.J. Hanson and J.D. Cowan and C.L.Giles,Morgan Kauffman:San Mate,1993*, 1993.
2. Rodney Brooks and Pattie Maes, editors. *Artificial Life IV: Proceedings of the Workshop on Artificial Life*, Cambridge, MA, 1994. MIT Press. Workshop held at the MIT.
3. Frank Dellaert. *TOWARD A BIOLOGICALLY DEFENSIBLE MODEL OF DEVELOPMENT*. PhD thesis, Case Western Reserve University, 1995.
4. Frank Dellaert and Randall D. Beer. Toward an evolvable model of development for autonomous agent synthesis. *Artifical Life IV*, pages 246–257, 1994.
5. Frank Dellaert and Randall D. Beer. A developmental model for the evolution of complete autonomous agents. In [14], pages 393–401, 1996.
6. Peter Eggenberger. Cell interactions for development in evolutionary robotics. In [14], pages 440–448, 1996.
7. Peter Eggenberger. Evolving morphologies of simulated 3d organisms based on differential gene expression. In [10], page in press, 1997.
8. Kurt Fleischer and Alan H. Barr. A simulation testbed for the study of multicellular development: The multiple mechanisms of morphogenesis. In [13], pages 389–416, 1992.
9. Frederic Gruau and D. Whitley. The cellular developmental of neural networks: the interaction of learning and evolution. Technical Report 93-04, Laboratoire de l'Informatique du Parallélisme, Ecole Normale Supérieure de Lyon, France, 1993.
10. Phil Husbands and Inman Harvey, editors. *Fourth European Conference of Artificial Life*. MIT Press, 1997.
11. Eric R. Kandel, James H. Schwartz, and Thomas M. Jessell. *Essentials of Neural Science and Behavior*. Appleton & Lange, 1995.
12. Hiroaki Kitano. Designing neural networks using genetic algorithms with graph generation system. *Complex Systems*, 4:461–476, 1990.
13. Christopher G. Langton, editor. *Artificial Life III: Proceedings of the Workshop on Artificial Life*, Reading, MA, 1994. Addison-Wesley. Workshop held June, 1992 in Santa Fe, New Mexico.
14. Pattie Maes, Maja J. Mataric, Jean-Arcady Meyer, Jordan Pollack, and Stewart W. Wilson, editors. *From animals to animats 4: Proceedings of the fourth international conference on simulation of adaptive behavior*. MIT Press, 1996.
15. Stefano Nolfi, Dario Floreano, Orazio Miglino, and Francesco Mondada. How to evolve autonomous robots: Different approaches in evolutionary robotics. In P. Meas R. A. Brooks, editor, *Artificial Life IV*. Cambridge, MA: MIT Press, 1994.
16. Olaf Sporns, Giulio Tononi, and Gerald M. Edelman. Modeling perceputal grouping and figure-ground segregation by means of active reentrant connections. *Proceedings of the National Academy of Science, USA*, 88:129–133, 1991.
17. Jari Vaario. Modelling adaptive self-organization. In [2], 1994.

A Boosting Algorithm for Regression*

A. Bertoni, P. Campadelli, M. Parodi

Dipartimento di Scienze dell'Informazione
Università degli Studi di Milano
via Comelico 39/40, I-20135 Milano (Italy)

Abstract. A new boosting algorithm ADABOOST-RΔ for regression problems is presented and upper bound on the error is obtained. Experimental results to compare ADABOOST-RΔ and other learning algorithms are given.

1 Introduction

Boosting refers to the general problem of producing a very accurate prediction algorithm by appropriately combining rough and moderately inaccurate ones. It works by calling repeatedly a given "weak" learning algorithm on various distributions on the training set, and combining the hypotheses obtained with a linearly separable boolean function.

The boosting algorithm ADABOOST proposed by Freund and Schapire [1] and presented in Section 2 has been successfully applied to improve the performance of different learning algorithms used for classification problems both binary and multiclass [2]. The first extension for regression problems $f : X \to [0,1]$ is the algorithm ADABOOST-R [1]. In this paper we present ADABOOST-RΔ, a different and more general extension for problems $f : X \to [0,1]^m$. The main theoretical result is an upper bound on the error.

To analyse the performance of ADABOOST-RΔ we have done experiments using backpropagation as "weak" learning algorithm. Preliminary results show good convergence properties of ADABOOST-RΔ; notably it is able to lower both the mean and the maximum error on either the training and the test set. For regression problems $f : X \to [0,1]$ it works better than ADABOOST-R.

2 Preliminary definitions and results

Given a set X and $Y = \{0,1\}$, let $\mathcal{P}$ be a probability distribution on $X \times Y$. An $N-$sample is a sequence $\langle(x_1,y_1),\ldots,(x_N,y_N)\rangle$ with $(x_k,y_k) \in X \times Y$; we call $Samp_N$ the set of all the $N-$samples and $Samp = \cup_N Samp_N$.

Given a class of functions $H \subseteq \{f \mid f : X \to \{0,1\}\}$, a learning algorithm $\mathcal{A}$ on H is a function $\mathcal{A} : Samp \to H$.

* Supported by grants CT 9305230.ST 74, CT 90.021.56.74

Let $S = \langle(x_1, y_1), \ldots, (x_N, y_N)\rangle$ be a N−sample and let $h = \mathcal{A}_S$ be the hypothesis output by $\mathcal{A}$ on input S, the empirical error $\widehat{\epsilon}$ of $\mathcal{A}$ on S is

$$\widehat{\epsilon} = \frac{\#\{(x_k, y_k) \mid y_k \neq h(x_k)\}}{N}$$

while the generalization error is

$$\epsilon_g = \mathcal{P}\{y \neq h(x)\}.$$

Under weak conditions on H (that is the Vapnik-Chervonenkis dimension [3] of H is finite), choosing elements of $X \times Y$ randomly and independently according to $\mathcal{P}$, for sufficiently large samples, the empirical error is close to the generalization error with high probability [4]. For this reason a good learning algorithm should minimize the empirical error. Often this is a difficult task because of the large amount of computational resources required and computationally efficient algorithms are usually moderately accurate.

Boosting is a general method for improving the accuracy of a learning algorithm. In particular we refer to the algorithm ADABOOST presented in [1] and described below. It has in input a learning algorithm $\mathcal{A}$, a N−sample $S = \langle(x_1, y_1), \ldots, (x_N, y_N)\rangle$, a distribution D on the elements of S, an integer T and gives in output a final hypothesis $h_f = HS(\sum_{k=1}^{T} w_k h_k - \lambda)$ with $h_k \in H \quad (k = 1, T)$; HS denotes the function $HS(x) =$ if $x \geq 0$ then 1 else 0. Even whether $\mathcal{A}$ is moderately inaccurate, for a sufficiently large T the error made by the final hypothesis h_f on the sample S can be made close to 0. Besides, the generalization ability of h_f is good since the Vapnik-Chervonenkis dimension of the family of the final hypothesis does not grow too much [1].

Algorithm ADABOOST

Input: a N−sample $S = \langle(x_1, y_1), \ldots, (x_N, y_N)\rangle$, a distribution D on S, a learning algorithm $\mathcal{A}$, an integer T.
Initialize the weight vector $w_i^1 = D(i)$ for $i = 1, \ldots, N$
Do for $t = 1, 2, \ldots, T$
1. Set $\mathbf{p}^{(t)} = \frac{\mathbf{w}^{(t)}}{\sum_{i=1}^{N} w_i^{(t)}}$
2. Choose randomly with distribution $\mathbf{p}^{(t)}$ the sample $S^{(t)}$ from S; call the learning algorithm $\mathcal{A}$ and get the hypothesis $h_t = \mathcal{A}_{S^{(t)}}$
3. Calculate the error $\epsilon_t = \sum_{i=1}^{N} p_i^{(t)} \mid h_t(x_i) - y_i \mid$
4. Calculate $\beta_t = \frac{\epsilon_t}{(1-\epsilon_t)}$
5. Set the new weights vector to be: $w_i^{(t+1)} = w_i^{(t)} \beta_t^{1-|h^{(t)}(x_i)-y_i)|}$
Output the hypothesis $h_f = HS\left(\sum_{k=1}^{T}(\log \frac{1}{\beta_t}) h_t(x) - 1/2 \sum_{k=1}^{T}(\log \frac{1}{\beta_t})\right)$

An upper bound to the error $\epsilon = \sum_{k=1}^{N} D(k) \cdot \mid h_f(x_k) - y_k) \mid$ is given by the following

Theorem 1 (Freund-Schapire). *Suppose the learning algorithm $\mathcal{A}$, when called by ADABOOST, generates hypotheses with errors $\epsilon_1, \ldots, \epsilon_T$. Then the error $\epsilon = \sum_{k=1}^{N} D(k) \cdot \mid h_f(x_k) - y_k \mid$ of the final hypothesis h_f output by ADABOOST is bounded by*

$$\epsilon \leq 2^T \prod_{t=1}^{T} \sqrt{\epsilon_t(1-\epsilon_t)}.$$

ADABOOST can be applied to classification problems with 2 classes. It has been generalized to multiclass problems ($Y = \{1, \ldots, K\}$) and to regression problems ($Y = [0,1]$) [1]. Roughly speaking, the main idea of the algorithm ADABOOST.**R** designed by Freund and Shapire for regression problems, is that of transforming the regression problem into a classification one using the total order relation $\leq$ on the real numbers. For example, every hypothesis $h : X \to [0,1]$ is transformed into the boolean function $\widehat{h} : X \times [0,1] \to \{0,1\}$ with

$$\widehat{h}(x,y) = \begin{cases} 1 & y \geq h(x) \\ 0 & \text{otherwise} \end{cases}$$

In the next paragraph we present a different way of transforming a regression problem for functions with values in $[0,1]^m$, into a classification problem. It develops an idea presented in [5] and it is based on the notion of norm in R^m.

3 The algorithm ADABOOST-RΔ

In this section we show a boosting algorithm ADABOOST-RΔ for regression problems. In this setting, Y is $[0,1]^m$ instead of $\{0,1\}$; as before, a sample S, chosen at random according to a probability distribution $\mathcal{P}$ on $X^m \times Y$, is given to a learning algorithm $\mathcal{A}$, that outputs a hypothesis $h : X \to [0,1]^m$.

Given a norm $\| \;\|$ on R^m, fixed $\Delta > 0$, we say that x and $\tilde{x}$ "Δ- agree" if $\|x - \tilde{x}\| \leq \Delta$. We consider as generalization error ϵ_g^Δ of a hypothesis h the probability that y and $h(x)$ does not "Δ- agree", that is $\epsilon_g^\Delta = \int HS(\|h_(x) - y\| - \Delta) dP$. Analogously, the empirical error ϵ^Δ on a sample $S = \langle (x_1, y_1), \ldots, (x_N, y_N) \rangle$ is $\epsilon^\Delta = \sum (HS(\|h(x_k - y_k\| - \Delta))$.

Algorithm ADABOOST-RΔ

Input: a $N-$sample $S = \langle (x_1, y_1), \ldots, (x_N, y_N) \rangle$, a distribution D on S, a learning algorithm $\mathcal{A}$, an integer T, a real numberΔ.

Initialize the weight vector $w_i^1 = D(i)$ for $i = 1, \ldots, N$

Do for $t = 1, 2, \ldots, T$

1. Set $\mathbf{p}^{(t)} = \frac{\mathbf{w}^{(t)}}{\sum_{i=1}^{N} w_i^{(t)}}$

2. Choose randomly with distribution $\mathbf{p}^{(t)}$ the sample $S^{(t)}$ from S; call the learning algorithm $\mathcal{A}$ and get the hypothesis $h_t = \mathcal{A}_{S^{(t)}}$

3. Calculate the error $\epsilon_t = \sum_{i=1}^{N} p_i^{(t)} HS(\|h_t(x_i) - y_i\| - \Delta)$
if $\epsilon_t > 1/2$ then $T = t - 1$ and abort loop

4. Calculate $\beta_t = \frac{\epsilon_t}{(1-\epsilon_t)}$
5. Set the new weights vector to be $w_i^{(t+1)} = w_i^{(t)} \beta_t^{1-HS(\|h_t(x_i)-y_i\|-\Delta)}$
Output the hypothesis $h_f = argmax_{y\in[0,1]^m} \sum \alpha_t HS(\Delta - \|h_t(x_i) - y_i\|)$ where $\alpha_t = \log \frac{1}{\beta}_t$

An upper bound to the error $\epsilon^{2\Delta} = \sum(HS(\|h_f(x_k - y_k\| - 2\Delta))$ is given by following

Theorem 2. *Suppose the learning algorithm $\mathcal{A}$, when called by ADABOOST-RΔ, generates hypotheses h_t with errors $\epsilon_1, \ldots, \epsilon_T < 1/2$ and let h_f be the final hypothesis output by ADABOOST-RΔ. Then*

$$\sum_{\|h_f(x_k)-y_k\|>2\Delta} D(k) \leq 2^T \prod_{t=1}^{T} \sqrt{\epsilon_t(1-\epsilon_t)}$$

Proof. (Outline) We transform the regression problem $X \to [0,1]^m$ into a classification problem $X \times [0,1]^m \to \{0,1\}$

- the sample $S = \langle (x_i, y_i) \mid i = 1, N\rangle$ is transformed into the sample $\widehat{S} = \langle ((x_i, y_i), 0) \mid i = 1, N\rangle$
- the distribution $D(i)$ is transformed into $\widehat{D}(i) = D(i)$ on the sample $\widehat{S}$
- the algorithm $\mathcal{A}$ with input S is transformed into the algorithm $\widehat{\mathcal{A}}$ with input $\widehat{S}$ with the rule: $\widehat{\mathcal{A}}_{\widehat{S}}(x_i, y_i) = HS(\|\mathcal{A}_S(x_i) - y_i\| - \Delta)$

Let $w^{(t)}$ and $\epsilon^{(t)}$ be respectively the weights vector and the error at step t of the algorithm ADABOOST-RΔ and let $\widehat{w}^{(t)}$ and $\widehat{\epsilon}^{(t)}$ $\epsilon^{(t)}$ be respectively the weight vector and the error at step t of the algorithm ADABOOST applied to the associated classification problem. By induction it can be proved that

$$\widehat{w}^{(t)} = w^{(t)}, \quad \widehat{\epsilon}^{(t)} = \epsilon^{(t)} \quad (1 \leq t \leq T)$$

Let $\widehat{h}_f(x, y)$ be the final hypothesis given by ADABOOST and $h_f(x)$ be the final hypothesis given by ADABOOST-RΔ. If $\|h_f(x_i) - y_i\| \geq 2\Delta$ then $\widehat{h}_f(x_i, y_i) = 1$. Let us suppose, on the contrary, that $\widehat{h}_f(x_i, y_i) = 0$, then

$$\sum_t \alpha_t HS(\|h_t(x_i) - y_i\| - \Delta) < 1/2 \sum_t \alpha_t \tag{1}$$

Let $I = \{t \mid \|h_t(x_i) - y_i\| < \Delta\}$, the inequality (1) becomes $\sum_{t\notin I} \alpha_t < 1/2 \sum_t \alpha_t$ and the following relation hold

$$\sum_{t\in I} \alpha_t > \sum_{t\notin I} \alpha_t \tag{2}$$

Since $h_f = argmax_{y\in[0,1]^m} \sum \alpha_t HS(\Delta - \|h_t(x_i) - y_i\|)$ then

$$\sum_t \alpha_t HS(\Delta - \|h_t(x_i) - h_f(x_i)\|) \geq \sum_t \alpha_t HS(\Delta - \|h_t(x_i) - y_i\|) \quad (3)$$

From (2) and (3) follows

$$\sum_{t\in I} \alpha_t HS(\Delta - \|h_t(x_i) - h_f(x_i)\|) > \sum_{t\notin I} \alpha_t (1 - HS(\Delta - \|h_t(x_i) - h_f(x_i)\|) \geq 0 \quad (4)$$

From the inequality (4) one can assert that there exists $\tilde{t} \in I$ such that $HS(\Delta - \|h_{\tilde{t}}(x_i) - h_f(x_i)\|) = 1$; for such $\tilde{t}$ it holds that $\|h_f(x_i) - h_{\tilde{t}}(x_i)\| < \Delta$ and $\|h_{\tilde{t}}(x_i) - y_i\| < \Delta$. Hence by the triangular inequality we obtain: $\|h_f(x_i) - y_i\| < 2\Delta$ but this is against the hypothesis.

Since $\|\widehat{h}_f(x_i) - y_i)\| > 2\Delta$ implies $\widehat{h}_f(x_i, y_i) = 1$, using Theorem 1 we conclude

$$\sum_{\|\widehat{h}_f(x_i) - y_i)\| > 2\Delta} D(i) \leq \sum_{\widehat{h}_f(x_i, y_i) = 1} D(i) \leq 2^T \prod_{t=1}^{T} \sqrt{\epsilon_t(1 - \epsilon_t)}$$

□

4 Experimental results

In this section we present some preliminary results to evaluate the performance of ADABOOST-RΔ in terms of learning accuracy and computational efficiency. The experiments have been done as follows:

- the functions to be learned are functions $f : R \to R$ or $g : R^2 \to R^2$
- the "weak algorithm" $\mathcal{A}$, given in input to ADABOOST-RΔ, is backpropagation on neural networks of fixed architecture
- the parameter Δ has been set at 1.5δ, where δ is the error on the training set made by backpropagation and preliminarly computed.

The experiments show that ADABOOST-RΔ exhibits better convergence properties than backpropagation: after the same number of epochs the accuracy on either the training and the test sets is higher. A qualitative example of this behaviour is shown in Figure 1. Figure 1a and 1b show respectively the interpolation of the function $f(x) = (\sin(10\cdot x) + 2)/4 + \sin(50\cdot(x + 0.5)^2)/15 + 0.1N(0, 0.1)$ made by backpropagation and by ADABOOST-RΔ.

Quantitative results for a function $g : R^2 \to R^2$, described by the the expression $g(x, y) = \sin(\frac{6}{5\cdot(x+0.5)}) \cdot \sin(\frac{6}{5\cdot(y+0.5)})/3 + 0.5$ with gaussian noise $0.2 \cdot N(0, 0.2)$ added, are given in Table 1 (left). Two different network architectures (a denotes the architecture 2-10-10-2, b denotes the architecture 2-15-15-2) have been trained for different numbers of epochs.

For functions $f : R \to R$ we have also compared the performance of ADABOOST-RΔ with the algorithm ADABOOST-R. Table 1 (right) shows the results relative to the function $f(x) = \sin(\frac{1}{(0.03\cdot x)})/5 + 0.2 \cdot N(0, 0.2)$

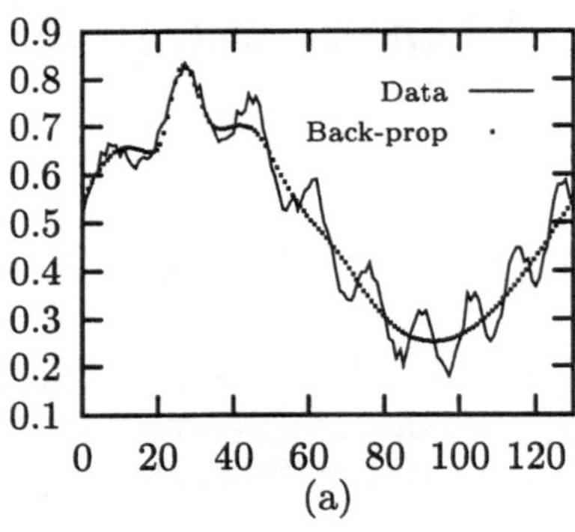

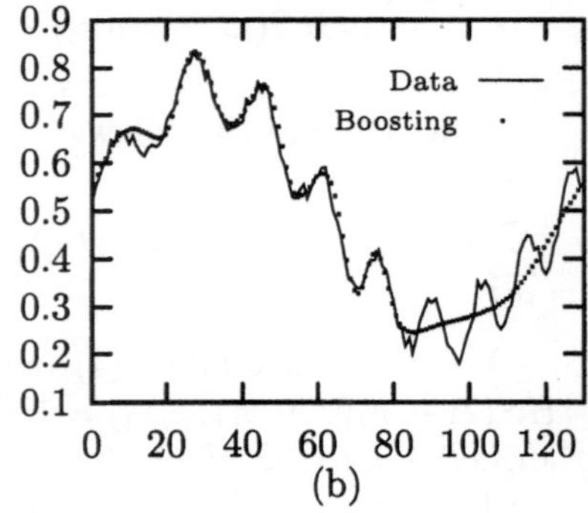

Fig. 1. Comparison between back-propagation and ADABOOST-RΔ.

str	data	epochs	T	mean err.	max err.
a	tr	1000	5	0.0428	0.1607
a	test	1000	5	0.0433	0.1545
a	tr	5000	1	0.0471	0.1739
a	test	5000	1	0.0475	0.1664
b	tr	1500	4	0.0417	0.1687
b	test	1500	4	0.0425	0.2101
b	tr	6000	1	0.0434	0.2054
b	test	6000	1	0.044 0	0.2404

alg.	data	epochs	T	mean err.	max err.
F.S.	tr	1000	6	0.021	0.111
F.S.	test	1000	6	0.020	0.128
Δ	tr	1000	5	0.017	0.058
Δ	test	1000	5	0.018	0.073
b.pr.	tr	6000	1	0.031	0.204
b.pr.	test	6000	1	0.036	0.228

Table 1.

References

[1] Freund, Y., Schapire, R.E.: A decision-theoretic generalization of on-line learning and an application to boosting. Internal Report of AT & T, September (1995)

[2] Freund, Y., Schapire, R.E.: Experiments with a new boosting algorithm. Machine Learning: Proc. of Thirteenth Int. Conf. (1996) 148–156

[4] V.N. Vapnik (1982) Estimation of Dependences Based on Empirical Data. Springer-Verlag.

[3] V.N. Vapnik, A.Y. Chervonenkis (1971) On the uniform convergence of relative frequencies of events to their probabilities. *Theory of Probability and its Applications*, 16(2): 264-280.

[5] Freund, Y.: Boosting a weak learning algorithm by majority. Information and Computation **121 (2)** (1995) 256–285

Combining Regularized Neural Networks

Michiaki Taniguchi and Volker Tresp

Siemens AG, Corporate Technology
Otto-Hahn-Ring 6
81739 München, Germany

Abstract. In this paper we show that the improvement in performance which can be achieved by averaging depends critically on the degree of regularization which is used in training the individual neural networks. We compare four different averaging approaches: simple averaging, bagging, variance-based weighting and variance-based bagging. Bagging and variance-based bagging seem to be the overall best combining methods over a wide range of degrees of regularization.

1 Introduction

Several authors have noted that averaging neural networks which were trained either on identical training data ([6]) or on bootstrap samples of the training data (a procedure termed "bagging predictors" by [1]) can improve performance considerably. Theory and experiments both show that averaging helps most if the errors in the individual neural networks are not positively correlated and if the neural networks have only small bias. On the other hand, it is well known from theory and experiment that best performance of a single predictor is typically achieved if some form of regularization (weight decay), early stopping or pruning are used. All three methods tend to decrease the variance and increase the bias of the neural network. Therefore, we expect that if we use a single neural network we would employ a higher degree of regularization than if we employ averaging. In this paper we investigate the effect of regularization on averaging. In addition to simple averaging and bagging we also perform experiments using combining principles where the weighting functions are dependent on the input (variance-based weighting, variance-based bagging).

2 Averaging Regularized Estimators

Our goal is to estimate the conditional expected value $t(x) = E(y|x)$ using neural networks. Each neural network was trained on the training set L, consisting of K samples, to minimize the cost function

$$C_i = \sum_{k=1}^{K}(y^k - f_i(x^k))^2 + \lambda \sum_{j=1}^{J} w_{ij}^2, \quad i = 1, \dots, M \tag{1}$$

where $\{w_{ij}\}_{j=1}^{J}$ are the weights in the $i-$th neural network and J is the number of weights in each network. The first term is the squared error between the prediction of the neural network and the target in the training data L and the second term is a weight-decay penalty weighted by the regularization parameter $\lambda \geq 0$. Let $f_i^{C,\lambda}(x)$ denote the i-th optimized network if weight decay parameter λ is used.

Averaging is only useful if the individual neural networks differ in their prediction. Neural networks trained on an identical data set only vary because the optimization was initialized with different random initial weights and the optimization procedure terminates in different local minima. Alternatively, Breiman suggested to train the neural networks using bootstrap replicates $L_1^B, \ldots, L_M^B$ to decorrelate the individual neural networks, a procedure Breiman calls bagging predictors ([1]). The bootstrap replicate L_i^B is generated by randomly sampling K-times from the original training data set L with replacement ([2]) Let $f_i^{B,\lambda}(x)$ denote the i-th optimized network if weight decay parameter λ is used and if bootstrap replicate L_i^B is used for training.

In the experiment we considered combined neural networks which can be written as

$$\hat{t}(x) = \sum_{i=1}^{M} g_i(x) f_i(x) = \frac{1}{n(x)} \sum_{i=1}^{M} h_i(x) f_i(x), \tag{2}$$

where $n(x) = \sum_{j=1}^{M} h_j(x)$ is the normalizing factor and $h_i(x) \geq 0, \forall i = 1, \ldots, M$. Note, that we allow for the possibility that the weighting functions $g_i(x)$ depend on the input x.

Simple Averaging (AV): In simple averaging, we set $h_i(x) = 1, \quad i = 1, \ldots, M$. Past experiments have shown that simple averaging can improve performance considerably. For experimental results and theoretical background of simple averaging see [3], [4], [6] and [10].

Bagging (BA): The only difference to AV is that the individual networks are trained on bootstrap replicates. Although the individual neural networks are less correlated, we expect that each individual neural network contains more variance (and possibly more bias) because the training set contains fewer distinct data if compared to the case where each neural network is trained on the complete data set.

Variance-based Weighting (VW): In variance-based weighting ([9]) we set

$$h_i(x) = \frac{1}{var(f_i(x))}, \quad i = 1, \ldots, M.$$

The intuitive idea of the variance-based approach is that when an neural network is uncertain about its own prediction for a certain input then this neural network is not competent for this input and obtains a low weight. For calculation of the variance it is important to be clear about which random process we average. If the expected value is taken over all random initial weights the variances of all neural networks are identical and we would obtain simple averaging. On the other hand if we consider that each neural network has found a different local minimum we can consider each neural network to be a distinct model. Now, we only average over the noise on the targets. We consider that the targets were generated by the random process $y^k = t(x^k) + \gamma$ where $t(x^k) = E(y^k|x^k)$ and γ is independent zero-mean noise with variance σ^2. The variance of an individual predictor for input x can be estimated as

$$var(f_i(x)) \approx \sigma^2 \ \theta_i(x)^T H_i^{-1} \theta_i(x) \quad H_i \approx \sum_{k=1}^{K} \frac{\partial f_i(x^k)}{\partial w_i} \frac{\partial f_i(x^k)}{\partial w_i}^T \tag{3}$$

where $\theta_i(x) = \frac{\partial f_i(x)}{\partial w_i}$ is the output sensitivity of the neural networks $f_i(x)$ w.r.t the weights w_i at the input x and where $w_i = (w_{i1}, \ldots, w_{iJ})'$ are the weights in the i-th neural network ([8]) and H_i is an approximation of the Hessian.
Note that the computational cost of calculating H_i is $O(KJ^2)$. The cost of the inversion of H_i amounts to $O(J^3)$.

Variance-based Bagging (VB): The neural networks are trained on the bootstrap replicates $\{L_i^B\}_{i=1}^M$ of the original training set L. The Hessian matrix in Equation 3 is now calculated using the bootstrap samples ([7]).

3 Experiment

In this section, we present experimental results using the Breast Cancer data set[1]. In the experiment $M = 25$ neural networks were combined. All neural networks had the same fixed architecture, i.e. a multilayer perceptrons with a single hidden layer of 10 hidden units. We divided the data bases randomly in two independent sets: the training set L and the test set T. To obtain statistically significant results we repeated the experiment 10 times ($R = 10$ runs).

The data set contains 699 samples with 9 input variables consisting of cellular characteristics and one binary output with 458 benign and 241 malignant cases. All input variables were normalized to zero mean and a standard deviation of one. For every run we divided the data base randomly in $K = 599$ training samples and $P = 100$ test samples.

The Figure 1 (top) shows the test set error of the individual neural networks as a function of the regularization parameter λ. The large values for $\lambda = 0$ indicates that neural networks without regularization extremely overfit the data. For $\lambda \approx 1.25$ the networks trained on complete data obtain best performance and for $\lambda = 2$ the networks trained on bootstrap data. As expected, networks trained on bootstrap samples perform worse since the networks trained on bootstrap replicates have seen fewer distinct data. The Figure 1 (second) shows the averaged standard deviation of the prediction of the neural networks trained with complete data. Clearly visible is the monotonous decrease with growing λ. Note, that the variance for bootstrap networks is always larger. The figure clearly

[1] The data base has been recorded at the University of Wisconsin Hospitals, Madison.

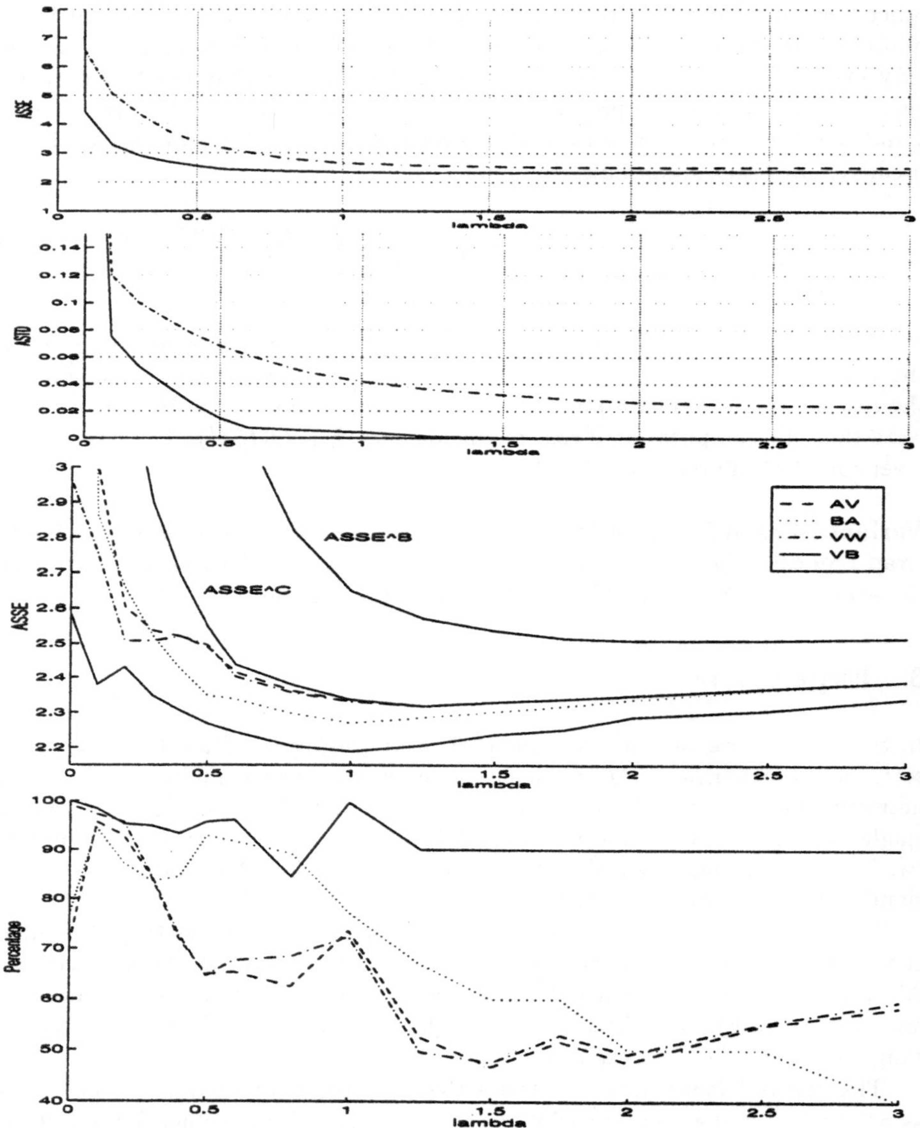

Fig. 1. Breast Cancer Data. Top: Test set error of the individual neural networks trained on complete data (continuous) and on bootstrap data (dash-dotted). Second: Standard deviation of the prediction of the neural networks trained on complete data (continuous) and on booststrap data (dash-dotted). Third: Test set error of the different averaging approaches. Displayed is AV (dashed), BA (dotted), VW (dash-dotted), and VB (lower continuous line). The highest continuous line shows the average performance of the individual bagging networks and the second highest line shows the average performance of the networks trained on the complete data. Bottom: The figure shows the number of single neural networks trained on complete data which are worse than the averaging methods (in percent).

demonstrates that bagging increases variance in the prediction. The test set performances of the different combination methods is plotted in Figure 1 (third). With no regularization, all averaging methods show dramatically better performance if compared to the individual neural networks. The variance-based approaches VW and VB are both better than the other averaging methods if λ is close to zero. This result seems to indicate that the variance-based methods successfully recognize the large variance in networks due to local overtraining. With increasing λ the individual networks as well as all averaging methods improve performance. In particular for $0.3 < \lambda < 1$ the performance of bagging is impressive and is superior to AV and VW. Most strikingly however, VB shows very good performance for any λ and seems to combine the advantages of both VW and BA. Figure 1 (bottom) shows the number of single neural networks which were worse than the averaging methods in percent. The impressive performance of VB is also apparent: in the large majority of settings for λ, VB is better than all individual networks! Bagging shows excellent performance up to $\lambda \approx 1$ and AV and VW are best for small values of $\lambda <\approx 0.5$.

4 Discussion

As predicted, the variance in the networks decreases rapidly with increasing weight-decay parameter λ if networks are trained on complete data. On the other hand, if networks are trained on bootstrap replicates, we obtain large variance in the networks even at relatively large values of λ. The relative improvement in performance by averaging increases with increasing variance in the estimates and bias hurts. Therefore for all averaging methods the relative improvement is maximum if no weight decay is used. On the other hand the performances of the networks improve with weight decay. So —as confirmed by experiment— regularization also improves the averaged systems. Simple averaging (AV) shows good performance at small values of λ. Even better at small λ is variance-based weighting (VW). The reason is that local overtraining is reflected in large variance. The corresponding neural network obtains consequently a small weight. With increasing λ, the performance of both AV and VW become comparable and both approach the performance of an average individual neural network trained on complete data for large λ. This confirms that all neural networks are highly correlated for large λ as already noted in Figure 1. Bagging (BA) displays better performance than AV and VW up to intermediate values of λ except when λ is extremely small or zero. This can be explained by the fact that training on bootstrap samples results in considerable variance in the networks even for large λ. Variance-based bagging (VB) seems to combine the advantages of both variance-based weighting and bagging. If networks are overtrained they locally have large variance and obtain a small weight locally. Training on bootstrap replicates introduces additional variance in the networks which is particularly useful for large λ. In our experiment variance-based bagging was the overall best combining method over a wide range of degrees of regularization. Similar results were obtained for further data sets.

5 Conclusions

Based on our experiments we can conclude that — in comparison with the individual neural networks — averaging improves performance at all levels of regularization. In particular we also obtain improvements with respect to optimally regularized neural networks, although the degree of improvement is application specific. Averaging is less sensitive with respect to the regularization parameter λ if compared to the individual neural networks. Especially if the individual neural networks overfit, averaging still gives excellent performance. Overall, bagging and variance-based bagging which both use networks trained with bootstrap replicates work well for a wide range of values of λ. At extremely small values of λ, variance-based weighting and variance-based bagging are clearly superior to the other averaging approaches.

Acknowledgements

This research was partially supported by the Bundesministerium für Bildung, Wissenschaft, Forschung und Technologie, grant number 01 IN 505 A.

References

1. Breiman, L.: Bagging Predictors. TR No. **421**, Dept. of Statist., Berkeley (1994)
2. Efron, B. and Tibshirani, R.: *An Introduction to the Bootstrap.* Chapman/Hall (1993).
3. Jacobs, R. A.: Methods for Combining Experts' Probability Assessment. *Neural Computation,* **7** (1995) 867–888.
4. Krogh, A. and Vedelsby, J.: Neural Network Ensembles, Cross Validation, and Active Learning. *Adv. in Neural Inf. Proc. Systems* **7**. Cambridge MA: MIT Press (1995).
5. Meir, R.:. Bias, Variance and the Combination of Estimators: The Case of Linear Least Squares. TR: Dept. of Electrical Engineering, Technion, Haifa (1994)
6. Perrone, M. P.: *Improving Regression Estimates: Averaging Methods for Variance Reduction with Extensions to General Convex Measure Optimization.* PhD thesis. Brown University (1993)
7. Taniguchi, M., Tresp, V.: Variance-based Combination of Estimators trained by Bootstrap Replicates. Proc. Inter. Symp. on Artificial Neural Networks, Taiwan (1995)
8. Tibshirani, R.: A Comparison of Some Error Estimates for Neural Network Models. TR Dep. of Stat, Univ. of Toronto (1994)
9. Tresp, V. and Taniguchi, M.: Combining Estimators Using Non-Constant Weighting Functions. *Adv. in Neural Inf. Proc. Systems* **7** Cambridge MA: MIT Press (1995)
10. Wolpert, D. H.: Stacked Generalization. *Neural Networks,* **5** (1992) 241–159

Making Stochastic Networks Deterministic

Stefan M. Rüger

Department of Computing
Imperial College of Science, Technology and Medicine
180 Queen's Gate, London SW7 2BZ, England
s.rueger@doc.ic.ac.uk

Abstract. Graphical models are considered more and more as a key technique for describing the dependency relations of random variables. Various learning and inference algorithms have been described and analysed. This article demonstrates how an important subclass of graphical models can be treated by transforming the underlying model into a regular feedforward network with special, yet deterministic, activation functions. Inference and the relevant quantities for learning can be calculated exactly in these networks. Moreover, all the known techniques for feedforward networks can be exploited and applied here.

1 Introduction

Graphical models [9] aim at describing conditional independence relations of random variables. Given a specific graphical model, *inference* is the task of calculating the joint posterior probability of a set of variables, given the observations for some other variables, while the value of still other variables may be unknown. *Learning* means the computing of the parameters of the graphical model from data. Thus, stochastic relations that are given solely by examples can be stored in a parameter vector. Inference and learning in general graphical models are NP-hard problems [1, e. g.]. This suggests that research should be directed away from the search for efficient inference and learning rules in general graphical models, and towards the design of special-case algorithms.

One interesting class of graphical models are decimatable multivalued Boltzmann machines [8], as they contain hidden Markov models as a special case [2]. Section 2 reviews these Boltzmann machines.

This article shows how inference and learning in a decimatable Boltzmann machine, which is a stochastic and recurrent network, can equivalently be performed in a specific deterministic feedforward network, which we will call *the corresponding mapping network*. Though this is certainly not the only possible algorithm for inference and learning (and indeed [9] discusses more general algorithms), our method can easily be implemented and lends itself to parallelisation.

2 Decimatable Multivalued Boltzmann Machines

The underlying structure of a multivalued Boltzmann machine is given by a set $N = \{0, \dots, n\}$ of nodes and a set $E \subset \{(i, j) \in N \times N | i < j\}$ of edges. Each node $a \in N$ can assume one of m_a activations $1, 2, \ldots, m_a$. A $m_a \times m_b$ matrix v_{ab} is associated to every edge $(a, b) \in E$, and v_{ba} shall be understood as the transposed matrix of v_{ab}. The component $v_{ab}(s_a, s_b) \in \mathbb{R}$ denotes the interaction of node a having activation s_a with node b having activation s_b. Usually, N is partitioned into four mutually disjoint sets I, H, O and $\{0\}$, denoting the input, hidden and output nodes, and the bias node, respectively. At every time-step, the Boltzmann machine produces a random activation vector s with $s_o := 1$ and $s_a \in \{1, m_a\}$ for $a \in I \cup H \cup O$. The sequence of the observed activation vectors is a Markov chain, and, once in equilibrium, s obeys a Boltzmann-Gibbs distribution

$$P_v(s) := \frac{\exp(-h(v,s))}{Z_v}, \quad \text{where} \quad h(v,s) := -\sum_{(a,b)\in E} v_{ab}(s_a, s_b) \tag{1}$$

is the defining quantity of a Boltzmann machine[1]. Z_v is defined to be the normalisation term and is also known as the partition sum.

Clamping values to input nodes (and sometimes to output nodes) changes the Boltzmann-Gibbs distribution in an obvious way: the clamped node acts as a bias node, and its effect can be subsumed under the existing bias node 0. Technically, clamping a value s_i to the node i corresponds to changing the bias weights $v_{oj}(1, s_j)$ by $v_{ij}(s_i, s_j)$ of all those nodes j with $(i, j) \in E$ or $(j, i) \in E$. At the same time, the node i, together with all edges connected to i, can be disregarded. If a complete pattern ξ is clamped to a subset X of N, a new Boltzmann machine is created whose Boltzmann-Gibbs distribution is the conditional Boltzmann-Gibbs distribution of the original Boltzmann machine, given ξ. We denote all quantities of the new Boltzmann machine with Greek superscripts, e.g., $N^\xi := N \setminus X$, $P^\xi_{v^\xi}(s^\xi) := P_v(s|\xi)$, etc.

One possible application of Boltzmann machines is in marginalising the Boltzmann-Gibbs distribution over all possible activations of the hidden nodes — thus obtaining a distribution p_v from P_v — to approximate a desired distribution r of activations of nodes in $I \cup O$. That is, either it is known that all the activations s_a, $a \in I \cup O$, shall follow a joint distribution r, or we want that the empirical distribution r of activation subvector samples is approximated by p_v. Let α be an activation vector of input nodes, β be an activation vector of hidden nodes and γ be an activation vector of output nodes. They are all subvectors of the activation vector $s = \alpha\beta\gamma$. Let $q(\alpha) = \sum_\gamma r(\alpha\gamma)$ be the probability of an input pattern α. Assuming that the distribution of input patterns is *not* to be learned, the information gain

$$\mathrm{IG}(r, p_v) = \sum_\alpha q(\alpha) \sum_\gamma r(\gamma|\alpha) \log\left(\frac{r(\gamma|\alpha)}{p_v(\gamma|\alpha)}\right) \tag{2}$$

is a commonly used error measure for the deviation of r with respect to p_v. Gradient descent for (2) results in the learning rule

$$\Delta v_{ij}(s_i, s_j) = -\eta \frac{\partial\, \mathrm{IG}(r, p_v)}{\partial v_{ij}(s_i, s_j)} = \eta \overline{\frac{1}{p_v(\gamma|\alpha)} \cdot \frac{\partial p_v(\gamma|\alpha)}{\partial v_{ij}(s_i, s_j)}}^{r(\alpha\gamma)}, \tag{3}$$

[1] we abandon the usual factor T that only serves to scale the weight vector v

where $\eta \in \mathbb{R}^+$ is the learning rate. $\overline{f(\cdot)}^{r(\cdot)}$ denotes the expectation value w. r. t. r or, if r is given by samples, the average over the samples.

(3) reduces the learning problem to the inference problem, i. e. the computation of the conditional probability

$$p_v(\gamma|\alpha) = \sum_{\beta} P_v(\beta\gamma|\alpha). \tag{4}$$

Inserting the definition of conditional probabilities, it holds that for every β

$$p_v(\gamma|\alpha) = \frac{P_v(\beta\gamma|\alpha)}{P_v(\beta|\alpha\gamma)} =$$
$$\exp\Big(\sum_{(0,c)\in E} v_{oc}(1,\gamma_c) + \sum_{\substack{(a,c)\in E\\ a\in I, c\in O}} v_{ac}(\alpha_a,\gamma_c) + \sum_{\substack{(c,d)\in E\\ c,d\in O}} v_{cd}(\gamma_c,\gamma_d)\Big)\frac{Z_{v\alpha\gamma}^{\alpha\gamma}}{Z_{v\alpha}^{\alpha}}. \tag{5}$$

Equation (5) links inference (4) and learning (3) to the partition sum, and this transformation is polynomial in $|N|$. This result and all other results of the preceding paragraphs are valid for general multivalued Boltzmann machines. In principle, Z_v can be calculated by summing all the $m_1 m_2 \ldots m_{|N|}$ terms $\exp(-h(v,s))$, which is intractable in $|N|$.

An attractive special case are networks, where one node after the other can be replaced by edges and weights in such a way that in each step the probability distribution of the remaining network's activation vector coincides with the marginal distribution of the original network's activation vector. Figure 1a and Figure 1b show the removal of a node $c \neq 0$ with all corresponding edges and the insertion of a new edge between neighbour nodes of c. Figure 1c visualises the fact that any new weight matrix associated with a new edge has to be added to an existing weight matrix owing to the linearity of h in v.

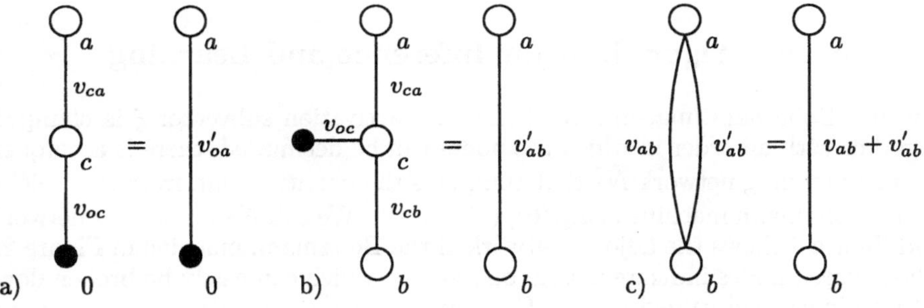

Fig. 1. Decimation rules for multivalued Boltzmann machines

Straightforward calculation [8] yields for v'_{oa} (Figure 1a)

$$v'_{oa}(1,s_a) = \log\left(\sum_{s_c=1}^{m_c} \exp(v_{oc}(1,s_c) + v_{ac}(s_a,s_c))\right) \tag{6}$$

and for v'_{ab} (Figure 1b)

$$v'_{ab}(s_a, s_b) = \log \left(\sum_{s_c=1}^{m_c} \exp(v_{oc}(1, s_c) + v_{ac}(s_a, s_c) + v_{bc}(s_b, s_c)) \right) . \tag{7}$$

The partition sum remains unaffected in both examples.

Figure 2a shows an example of a Boltzmann machine whose partition sum can be computed by decimating nodes step by step until a simple Boltzmann machine with one nontrivial node z and the weight v'_{oz} is reached. Since decimation in multivalued Boltzmann machines does not change the partition sum Z, it holds that

$$Z = \sum_{s_z=1}^{m_z} \exp(v'_{oz}(1, s_z)). \tag{8}$$

Boltzmann machines whose partition sums can be calculated with decimation (after clamping input, or input and output nodes) seem to be ideal candidates for a sufficiently interesting class of tractable networks — we call them *decimatable.*

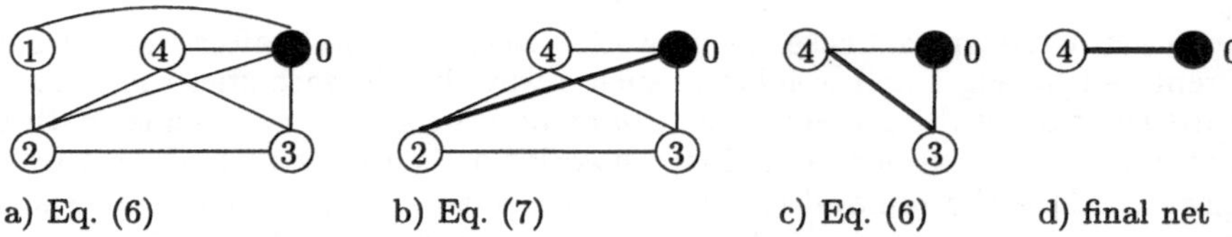

Fig. 2. Decimation of a multivalued Boltzmann machine

3 Efficient Algorithms for Inference and Learning

Given a Boltzmann machine N^ξ (where an activation subvector ξ is clamped) together with an order in which the nodes can be decimated, there is a uniquely defined mapping network $\tilde{N}^\xi$ that computes the partition sum from the weights of the Boltzmann machine using (6), (7) and (8). We call $\tilde{N}^\xi$ an *adjoint* network, and Figure 3 shows the adjoint network of the Boltzmann machine in Figure 2a. The squared nodes indicate matrix operations, which can easily be broken down into ordinary real-valued nodes. The empty nodes are identity nodes (copying input to output). They are not necessary and only shown for the sake of visualisation.

Usually, an adjoint network has a clamped weight vector v^ξ as input and calculates $Z^\xi_{v^\xi}$. It is easy to construct a mapping network that computes the inference terms $p_v(\gamma|\alpha)$ combining the adjoint networks $\tilde{N}^\alpha$, $\tilde{N}^{\alpha\gamma}$ and the process of clamping with Equation (5). This *corresponding mapping network* outputs $p_v(\gamma|\alpha)$ and has v, α and γ as input.

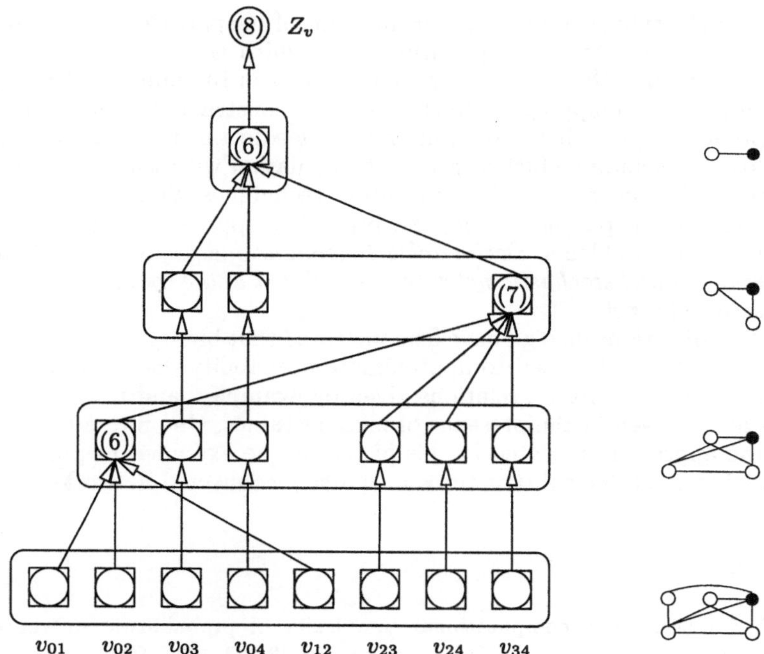

Fig. 3. An adjoint network for the computation of the partition sum Z_v

A generalisation of the backpropagation algorithm allows computation of all the partial derivatives $\partial p_v(\gamma|\alpha)/\partial v_{ij}(s_i, s_j)$ — they are needed for the learning rule (3) — in a simple forward-backward pass [7]. Thus, the gradient of the information gain (2) can be computed very efficiently.

The learning rule (3) was given by gradient descent in the cost function (2). It is well-known that pure gradient descent is not optimal for learning. Either certain minima of the cost function cannot be found if the learning rate is too large, or learning takes a long time if η is too small, cf. [5, trade-off-theorem]. Virtually every convex-optimization method utilizes cost function values for efficient optimization.

The knowledge of $p_v(\gamma|\alpha)$ can be used for an efficient information gain computation. Hence, all established convex-optimisation methods like conjugate gradient, the Broyden-Fletcher-Goldfarb-Shanno algorithm and other second-order methods [4] can be exploited or, e. g., stable dynamic learning-rate adaptation [5], which uses comparatively few but well-chosen evaluations of the information gain. What is more, the decimation-based algorithms of this section are exact, as opposed to algorithms based on Gibbs sampling.

4 Discussion

This article emphasises the importance of the partition sum of decimatable Boltzmann machines. One result is that both inference and improved convex-optimisation learning can be expressed in terms of the partition sum. In decimatable Boltzmann machines the partition sum (which is, in general, intractable) can be calculated with a complexity that is linear in the number of nodes, once the corresponding mapping network has been constructed. The advantage of this approach is that all usual techniques of feedforward neural networks can be employed - including efficient gradient calculation with the backpropagation algorithm and more advanced optimisation techniques. Since neural networks lend themselves to parallelisation, the presented approach seems to be easily scalable to large problems. *Decimatable Boltzmann machines offer all the benefits of recurrent and stochastic networks — and that at the cost of a deterministic feedforward network!*

Small applications [6, e.g.] have demonstrated that binary decimatable Boltzmann machines can be used to approximate probability distributions, to store knowledge about stochastic relations given by examples, and to do inference in such databases even in the presence of missing values. One interesting question is how the approach presented in this paper can be extended to previous work [3] exploiting tractable substructures in sparsely connected networks.

References

1. G. F. Cooper. The computational complexity of probabilistic inference using Bayesian belief networks. *Artificial Intelligence*, 42:393–405, 1990.
2. D. J. C. MacKay. Equivalence of Boltzmann chains and hidden Markov models. *Neural Computation*, 8(1):178–181, 1996.
3. M. J. Nijman and H. J. Kappen. Efficient learning in sparsely connected Boltzmann machines. In C. von der Malsburg, W. von Seelen, J. C. Vorbrüggen and B. Sendhoff, editors, *Artificial Neural Networks - ICANN 96*, pages 41–46. Springer-Verlag, 1996.
4. W. H. Press, S. A. Teukolsky, W. T. Vetterling and B. P. Flannery. *Numerical Recipes in C.* Cambridge University Press, 1988.
5. S. M. Rüger. Stable dynamic parameter adaptation. In D. Touretzky, M. Mozer and M. Hasselmo, editors, *Advances in Neural Information Processing Systems 8*, pages 225–231. MIT Press, 1996.
6. S. M. Rüger. Decimatable boltzmann machines for diagnosis: Efficient learning and inference. In *Proceedings of IMACS'97, Berlin (accepted)*, 1997.
7. S. M. Rüger. *Zur Theorie künstlicher neuronaler Netze.* Verlag Harri Deutsch, Thun, Frankfurt/Main, 1997.
8. L. K. Saul and M. I. Jordan. Boltzmann chains and hidden Markov models. In G. Tesauro, D. S. Touretzky and T. K. Leen, editors, *Advances in Neural Information Processing Systems 7*, pages 435–442. MIT Press, 1995.
9. P. Smyth, D. Heckerman and M. I. Jordan. Probabilistic independence networks for hidden Markov probability models. *Neural Computation*, 9(2), 1997.

Acknowledgment. This work was partly supported by the Fujitsu European Centre for Information Technology.

Unsupervised Learning in Networks of Spiking Neurons Using Temporal Coding

Berthold Ruf and Michael Schmitt*

Institute for Theoretical Computer Science, Technische Universität Graz
Klosterwiesgasse 32/2, A-8010 Graz, Austria
E-mail: {bruf, mschmitt}@igi.tu-graz.ac.at

Abstract. We propose a mechanism for unsupervised learning in networks of spiking neurons which is based on the timing of single firing events. Our results show that a topology preserving behaviour quite similar to that of Kohonen's self-organizing map can be achieved using temporal coding. In contrast to previous approaches, which use rate coding, the winner among competing neurons can be determined fast and locally. Hence our model is a further step towards a more realistic description of unsupervised learning in biological neural systems.

1 Introduction

In the area of modelling information processing in biological neural systems, there is an ongoing debate about which essentials have to be taken into account (see e.g. [3,13,11,9]). Discrete models, such as threshold gates or McCulloch-Pitts neurons, are undoubtedly very simplistic descriptions of biological neurons. Models with real-valued output, such as the sigmoidal gate, where analogue values are interpreted as firing rates of biological neurons, are more suitable for the modelling of neural processes in terms of analogue computations. However, both types of models do not capture a phenomenon which is widely believed to be the basis of fast analogue computations in biological neural networks: the timing of single action potentials. Models of spiking neurons that use temporal coding are a first attempt to explore the computational significance of this phenomenon [1,7,11]. Furthermore, <u>s</u>piking <u>n</u>euron <u>n</u>etworks (SNNs), where the computations are based on this coding scheme, have recently been shown to be computationally more powerful than networks consisting of threshold or sigmoidal gates with respect to time and network complexity [9].

In this paper we investigate unsupervised learning processes in SNNs. On the basis of a construction introduced in [8] we show how competitive learning can be performed by SNNs using temporal coding. We extend this idea to a learning mechanism that is closely related to one of the most successful paradigms of unsupervised learning: the <u>s</u>elf-<u>o</u>rganizing <u>m</u>ap (SOM) by Kohonen [6].

Topology preserving maps have been found in many regions of the brain, e.g. in the visual, auditory, or somatosensory cortex [1]. The SOM provides a

* To whom correspondence should be addressed.

possible explanation how such maps can develop. Previous versions of the SOM assume that the output of a neuron is characterized by its firing rate and not by the timing of single firing events. Using lateral connections the procedure for detecting the so-called winner neuron is usually implemented as a recurrent network. This has the consequence that the winner neuron is detected not before the network has settled down into an equilibrium state. However, since the computation relies on the convergence of the network, this approach disregards the benefits that temporal coding offers to fast information processing in SNNs.

In addition to these conventional implementations there has also been some research on biologically more realistic models of self-organizing map algorithms, e.g. by Kohonen [5], Sirosh and Miikkulainen [14], Choe and Miikkulainen[2]. Also in these approaches the output of a neuron is assumed to correspond to its firing rate, and learning takes place in terms of this rate after the network has reached a stable state of firing.

In the following we propose a mechanism for unsupervised learning in SNNs where computing and learning are based on the timing of single firing events. In contrast to the standard formulation of the SOM, our construction has the additional advantage that the winner among the competing neurons can be determined fast and locally by using lateral excitation and inhibition. These lateral connections also constitute the neighbourhood relationship among the neurons. We assume that initially neurons which are topologically close together have strong excitatory lateral connections whereas remote neurons have strong inhibitory connections. During the learning process the lateral weights are decreased, thus reducing the size of the neighbourhood.

In a series of computer simulations we have investigated the capability of the model to form topology preserving mappings. For the evaluation of these mappings, instead of relying on visual inspection, we used a measure for quantifying the neighbourhood preservation which was recently studied [4]. Our results show that the model exhibits the same characteristic behaviour as the SOM. The typical emergence of topology preserving behaviour could be observed for a wide range of parameters.

We also studied the effect of weight normalization. It is used in several implementations of the SOM (see e.g. [2,6,14]) but its biological relevance is controversial. We performed simulations with and without weight normalization after each learning cycle. Comparing the results we discovered that after a certain number of learning cycles approximately the same degree of topology preservation could be achieved regardless whether the weights were normalized or not. The model of unsupervised learning that we propose in this paper is therefore a candidate for a more realistic description of fast analogue computation in biological neural systems. Moreover, it also provides a link to possible industrial applications via silicon implementations in pulse coded VLSI [10].

The paper is organized as follows: In Section 2 we introduce the formal model of a spiking neuron. In Section 3 we propose the mechanism for unsupervised learning in networks of these model neurons. The simulation results are presented in Section 4.

2 The Spiking Neuron Model

Recently Maass has shown in [8] how leaky integrate-and-fire neurons can compute weighted sums in temporal coding, where the firing time of a neuron encodes a value in the sense that an early firing of the neuron represents a large value. More precisely, one considers a neuron v which receives excitatory input from m neurons $u_1, \ldots, u_m$, where the corresponding weights are denoted by $w_1, \ldots, w_m$. Each u_i fires exactly once within a sufficiently small time interval at a time t_i with $t_i = T_1 - s_i$, where s_i is the ith input to v and T_1 some constant. Usually T_1 is given by the time when a reference spike is fired by some additional input neuron to v. Under certain weak assumptions (basically the initial rising segments of the excitatory postsynaptic potentials have to be linear) one can guarantee that v fires at a time determined by $T_2 - \sum w_i s_i$ with T_2 being some constant.

This computation can also be performed on the basis of "competitive temporal coding", such that no explicit reference times T_1 and T_2 are necessary (see [8] for details).

3 Unsupervised Learning

On the basis of this construction it is now possible to implement various types of unsupervised learning in the context of SNNs as follows: Given a set S of m-dimensional input vectors $\mathbf{s}^l = (s_1^l, \ldots, s_m^l)$ and an SNN with m input neurons and n output neurons, where each output neuron v_j receives synaptic "feedforward" input from each input neuron u_i with weight w_{ij} and "lateral" synaptic input from each output neuron $v_k, k \neq j$, with weight $\tilde{w}_{kj}$. At every cycle of the learning procedure one $\mathbf{s}^l \in S$ is randomly chosen and the input neurons are made fire such that they temporally encode $\mathbf{s}^l$. Each v_j then starts to compute $\sum_i w_{ij} s_i^l$ as described in Section 2. If we assume that the input vector and the weight vector for each neuron are normalized, then this weighted sum represents the similarity between the two vectors with respect to the euclidean distance. Hence the earlier v_j fires, the more similar is its weight vector to the input vector and the output neurons form a competitive layer, where the winner is the neuron which fires first. Note that the firing time of the winner is not influenced by the firing of the other output neurons.

If the lateral connections are strongly inhibitory, such that the firing of the winner neuron, say v_k, inhibits all other neurons in the competitive layer from firing, one can implement in a straightforward way competitive learning: One simply has to apply the standard competitive learning rule (also known as instar learning rule) to the winner neuron v_k. This rule is given for v_k by:

$$\Delta w_{ik} = \eta(s_i^l - w_{ik}), \qquad i \in \{1, \ldots, m\}$$

where $\mathbf{s}^l$ is the current pattern and η the learning rate.

For realizing the SOM in our model, one has to find a way to implement the neighbourhood function on the basis of locally available information. We use the lateral connections among the output neurons to reflect the structure of the neighbourhood by assuming that initially neurons which are topologically close together have strong excitatory lateral connections whereas remote neurons have strong inhibitory connections. This means that the firing of the winner neuron, say v_k, at time t_k drives the firing times of neurons in the neighbourhood of v_k towards t_k, thus increasing the values they encode. The firing of remote neurons is postponed by the lateral inhibition. We suggest the following learning rule:

$$\Delta w_{ij} = \eta \frac{T_{out} - t_j}{T_{out}} (s_i^l - w_{ij}) \tag{1}$$

where t_j is the firing time of the jth output neuron. The learning rule applies only to neurons that have fired before a certain time T_{out} (which has to be chosen sufficiently large). The factor $(T_{out} - t_j)/T_{out}$ realizes the neighbourhood function, which is largest for the winner neuron and decreases for neurons which fire at later times.

During the learning process the lateral weights $\tilde{w}_{kj}$ are decreased, thus reducing the size of the neighbourhood.[1] As in the standard formulation of the SOM, η is slowly decreased during learning.

4 Simulations

We tested our approach with one of the standard examples for the SOM, where two-dimensional input patterns are chosen randomly from a square and the competitive units are expected to organize themselves in a topology preserving grid. For the tests we used an array of 5×5 competitive units. We normalized the inputs by adding a third input component, which was chosen such that all input vectors have the same norm. The feedforward weights were initialized from random values around the midpoint of the patterns. The lateral weights for the immediate neighbours were chosen slightly positive, for the second neighbours zero and for all other neurons negative. In order to examine the effect of normalization we performed two series of experiments. In the first the feedforward weights were normalized after each application of (1), in the second they remained unnormalized.

In order to quantify the neighbourhood preservation we used the method of "metric multidimensional scaling" (see e.g. [4]) which is based on the measure

$$M_{\text{MDS}} = \sum_{i=1}^{N} \sum_{j<i} (F(i,j) - G(M(i), M(j)))^2. \tag{2}$$

where $N = |S|$. The function M represents the mapping of the network, i.e. $M(i)$ is the index of the winner neuron in the competitive layer for input $\mathbf{s}^i$. The

[1] This requires that the lateral weights change their sign during learning, which is not very realistic in the context of biological networks. However, one can consider instead two connections, one excitatory, which decreases and one inhibitory which increases during learning.

function F specifies for any given pair of patterns how dissimilar they are, and G is the measure of dissimilarity between pairs of weight vectors. Obviously, a value of $M_{\mathrm{MDS}} = 0$ indicates a perfectly topology preserving mapping M. In order to take the random initialization of the weights into account we computed the relative neighbourhood distortion which is the actual value of M_{MDS} divided by the maximum initial value of M_{MDS}. This yields an initial value from the interval $[0,1]$ and makes the results for different initializations comparable. Figure 1 shows that – regardless whether the weights are normalized after each cycle or not – the relative M_{MDS} stabilizes for our example indeed at approximately the same small value.

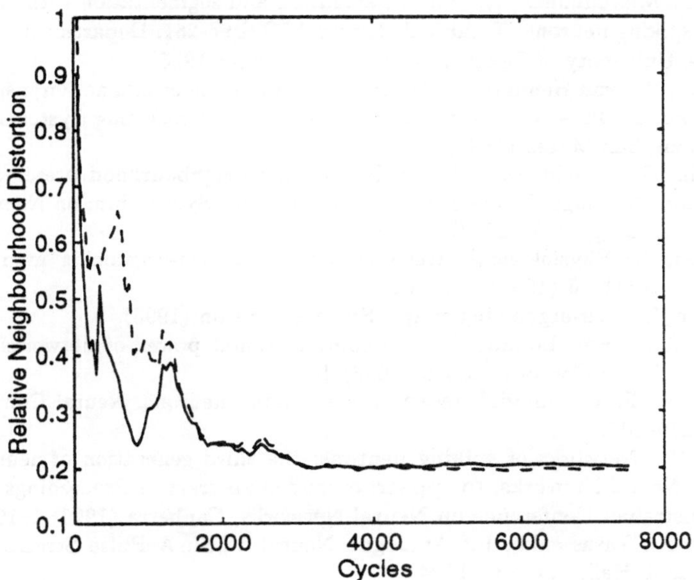

Fig. 1. Development of the relative neighbourhood distortion for the example described in Section 4, where the solid (dashed) line shows the case with (without) weight normalization after each cycle.

5 Conclusions

To the best of our knowledge, this is the first implementation of Kohonen's learning algorithm for SNNs using temporal coding. This work is a further step towards showing that biological neurons can indeed achieve a topology preserving behaviour using a similar learning procedure like the one suggested by Kohonen.

As mentioned above we have assumed as in [8] that the neurons are of the leaky integrate-and-fire type, where the initial segment of the postsynaptic potentials rises linearly. Recent simulations [12] have shown that even when using a more detailed neural model which includes nonlinear effects and more realistic shapes (e.g. α-functions) for the postsynaptic potentials, spiking neurons can still compute weighted sums in the way described in Section 2. This indicates that these simplifying assumptions can be dropped.

References

1. Arbib, M. A.: The Handbook of Brain Theory and Neural Networks. MIT Press, Cambridge (1995).
2. Choe, Y., Miikkulainen, R.: Self-organization and segmentation with laterally connected spiking neurons. Technical Report AI TR 96-251, Department of Computer Science, University of Texas at Austin, September 1996.
3. Gerstner, W., van Hemmen, L. H.: How to describe neuronal activity: spikes, rates, or assemblies? In Advances in Neural Information Processing Systems 6, Morgan Kaufmann, San Mateo (1994) 463–470.
4. Goodhill, G. J., Sejnowski, T. J.: Quantifying neighbourhood preservation in topographic mappings. Proceedings of the 3rd Joint Symposium on Neural Computation, San Diego, CA (1996) 61–82.
5. Kohonen, T.: Physiological interpretation of the self-organizing map algorithm. Neural Networks **6** (1993) 895–905.
6. Kohonen, T.: Self-organizing maps. Springer, Berlin (1995).
7. Maass, W.: Lower bounds for the computational power of networks of spiking neurons. Neural Computation **8** (1996) 1–40.
8. Maass, W.: Fast sigmoidal networks via spiking neurons. Neural Computation **9** (1997) 279–304.
9. Maass, W.: Networks of spiking neurons: the third generation of neural network models. Neural Networks, to appear; extended abstract in Proceedings of the Seventh Australian Conference on Neural Networks, Canberra (1996) 1–10.
10. Murray, A., Tarassenko, L.: Analogue Neural VLSI: A Pulse Stream Approach. Chapman & Hall, London (1994).
11. Rieke, F., Warland, D., de Ruyter van Steveninck, R., and Bialek, W.: SPIKES: Exploring the Neural Code. MIT Press, Cambridge (1996).
12. Ruf, B.: Computing functions with spiking neurons in temporal coding. In J. Mira, R. Moreno-Díaz and J. Cabestany (eds.). Biological and Artificial Computation: From Neuroscience to Technology. Proceedings of the International Work-Conference on Artificial and Natural Neural Networks IWANN'97, Lecture Notes in Computer Science, vol. 1240, Springer, Berlin (1997) 265–272.
13. Sejnowski, T.: Time for a new neural code? Nature **376** (1995) 21–22.
14. Sirosh, J., Miikkulainen, R.: Topographic receptive fields and patterned lateral interaction in a self-organizing model of the primary visual cortex. Neural Computation **9** (1997) 577–594.

Experiments on Regularizing MLP Models with Background Knowledge

Arto Selonen and Jouko Lampinen

Laboratory of Computational Engineering,
Helsinki University of Technology, Finland.

Abstract. In this contribution we present results of using possibly inaccurate knowledge of model derivatives as part of the training data for a multilayer perceptron network (MLP). Even simple constraints offer significant improvements and the resulting models give better prediction performance than traditional data driven MLP models.

1 Introduction

An increasingly important application of neural networks is in modeling nonlinear plants for simulation and control purposes. Often the training must be done with measurements from normal operating situations. Then, depending on the operating statistics during the data collection, many important features of the process behavior may be lacking from the data.

An important type of knowledge from any process to be controlled is the effect of control variables to the output variables. Typically the process experts can tell the direction and coarse magnitude of the change in product variables as function of the control variables, even though they could not estimate the actual output variables. Thus the knowledge is based on the model derivatives, $\frac{\partial o_k}{\partial x_i}$, where x_i is an input variable and o_k is an output variable. The expert knowledge is in general numerically inaccurate, since it is impossible to take into account all the process parameters.

Using MLP-type networks for process modeling has been very popular, since the same network architecture can learn to solve rather different problems, solely depending on the training data. However, this is also one of the most serious weaknesses of neural networks in real world applications. With the general network architecture and training it is rather difficult to incorporate existing domain knowledge, such as partial derivatives, into the models. Also, the generic network structure tends to produce over-fitted models. For industrial applications this means models that predict poorly in new situations, models that may change drastically after each new training, and models that typically have incorrect mappings from inputs to outputs.

In general, the generalization error (that is, the error with an independent test set) can be reduced by adding background knowledge from the problem into the network [5]. More specifically, the generic learning capacity of the network must be reduced, so that the variations in the training data have less effect on the training results. This requires that known characteristics of the solution must be somehow encoded into the training procedure.

Various methods have been presented for training different order partial derivatives along with measurement data, including [2], [9] and [13]. Especially useful method for adding knowledge as hints was proposed by Abu-Mustafa [1]. Unfortunately, precise values for the partial derivatives are usually not known, but rather the information available is inaccurate, or fuzzy in nature. Other approaches include enforcing certain smoothness to the network mappings by limiting the number of the hidden units or connection weights [6]. A popular and very efficient regularization method is weight decay [7], where large weights are penalized. A related, widely used regularization method is 'early stopping', where the training is stopped when the error on validation set is minimum.

A practical problem with regularization methods is that they are generic in nature, and often can not utilize process specific knowledge. Similarly, the parameters that control the regularization usually have no physical interpretation. Therefore, finding the optimal parameters is based on properties like validation set error, instead of available process information.

In this contribution we show how using a set of rules describing the model derivatives can improve the network performance through good regularization. The rules need to be provided by the experts, but analysis of the data-based models can suggest useful candidate rules for expert screening. Resulting models have good generalization capabilities, their behaviour is more easily understood by other experts, and typically the models do not differ much from one training to another.

2 Training algorithm for rules

The proposed approach of adding prior knowledge to the MLP model is based on defining a generic cost function augmented to the data fit error, that can encode expert knowledge represented as fuzzy rules. Note that we are not using any fuzzy methods here, but because the knowledge that is available in many process modeling problems can be expressed in fuzzy terms we are pursuing cost function based methods where this knowledge can be used to enhance the MLP model.

The training database for the network thus consist of data samples and rules. The data samples consist of input vectors and desired output vectors $\{x^p, y^p\}$, where p denotes the index of the sample, while the rules consist of sets $\{x^p, y^p, R_{ik}^p, F_{ik}^p\}$ for any inputs i, outputs k and samples p, where R_{ik}^p is the a priori mean value of the partial derivative $\frac{\partial o_k}{\partial x_i}\big|_{x_i^p}$ and F_{ik}^p is the additional cost function. The total sum squared error of the network for the training data becomes

$$E = E_{data} + \alpha E_{rule} = \sum_{k,p} \frac{1}{2}(y_k^p - o_k^p)^2 + \alpha \sum_{i,k,p} \frac{1}{2} F_{ik}^p (R_{ik}^p, D_{ik}^p)^2 \qquad (1)$$

where D^p_{ik} is the partial derivative produced by the network. To support knowledge expressed in fuzzy terms we define the rules by a 'membership function', shown in Fig. 1, where the values have similar interpretation as fuzzy membership functions. The 'membership function' μ^p_{ik} is mapped to cost function according to Eq. 2.

$$F^p_{ik}(R_{ik}, D_{ik}) = (\frac{1}{\mu^p_{ik}} - 1)(R_{ik} - D_{ik}) \tag{2}$$

The interpretation of the values of μ are following:

$\mu = 1$: The cost function goes to zero, and the rule provides a non-informative prior for D_{ik}, that is, D_{ik} can have any value where μ is one. In natural language or fuzzy terms, the rule states that the correct value is somewhere in the area where $\mu = 1$ but there is no more accurate knowledge (such as the mean etc.). In Fig. 1 the rule says that the derivative may be anything between 1 and 2.5.

$0 < \mu < 1$: Gaussian distribution for D_{ik} is assumed, with mean value given by R^p_{ik}. The variance of the prior distribution is proportional to $\frac{\mu}{1-\mu}$. Small value of μ biases the solution towards the prior R_{ik}, so that μ should be set according to assumptions of the accuracy of the expectation for R_{ik}. This is a statistical prior that, to our knowledge, has no fuzzy interpretation. However, as will be shown by the experiments, it can be used to efficiently regularize the solution.

With trapezoidal form for μ, such as in Fig. 1, different hard and soft limits for the values D_{ik} can be specified. With small μ the solution can be forced towards R_{ik}, indicating forbidden values. The training algorithms for minimizing the error in Eq.1 have been reported in [9] and [12].

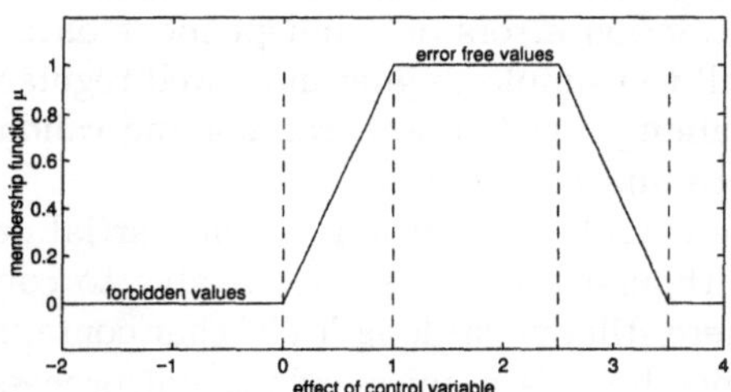

Fig. 1. Example of the 'membership function' for one input-output relation.

3 Background knowledge as regularization method

A real world process was used for the experiments. The goal was to predict one quality measurement of a paper manufacturing process. This was done using 19 selected on-line measurements chosen by the experts. A total of 2240 measurements were available after removal of false data. This included only typical errors such as single, large spikes, or areas that had only constant values (indicating the sensor was off-line). The network architecture was two-hidden-layer MLP (19-9-5-1) with sigmoidal hidden layers and linear output layer, and the data was scaled to [-1,1].

A 7 fold cross validation scheme was used to analyze the network performance, that is, the data was divided into seven blocks of equal size (320 points each), and one block at a time was chosen to be an independent test set. Rest of the data was used to train the network with either data, or data and predefined rules. The training was repeated ten times for each configuration using different initial weights. In addition, the models from all the 10 runs were also used in combination as a 'committee' ie. the ten outputs were averaged to get yet another model. This was done to reduce the sensitivity of the 'early stopping' regularization to initial weights [4].

In all the experiments the networks were trained using the RPROP-algorithm [11] with early stopping regularization: the modeling data (six blocks) was divided into training and validation sets of equal size, and the training was stopped at the minimum of the validation set error.

3.1 Rule set R1

For the rule based training eight of the model inputs were chosen by the experts as the most effective ones. Since only the sign of the effect was known, all the eight rules were of the form $\frac{\partial y}{\partial x} > 0$ (later referred as rule set R1). This was implemented by setting all the target values to $R_{ik} = 1$ and the respective 'membership function' so that $\mu(D) = 1$ when $D > 0$ and $\mu(D) = 0.01$ otherwise.

As can be seen from Table 1, using rules in addition to data gives a clear improvement in prediction errors over independent data sets. However, the rules in R1 were still too simple to guarantee well regularized models. Some of the individual training runs had low training and validation set errors, yet the test set error was unacceptable.

Fig. 2 shows the overall distributions of the partial derivative values for three inputs. Even though the rule set R1 is able to constrain values onto the positive side, there still appear long 'tails' that contain unrealistic values. Finding proper upper bounds requires additional process information that was not available at the time.

3.2 Rule set R2

Linear models are known to be robust predictors with easily explained behaviour. Combining linear models with, e.g. MLP based models could solve some of the problems that follow the nonlinear methods. Using partial derivative values as rules gives two simple ways of implementing this: either require the second order derivatives to be small [2] or require the first order derivatives $\frac{\partial o}{\partial x_i} = c_i$ to have compact distribution around the mean. This way the rules constrain linear relationships for selected inputs while still allowing for nonlinearity for other inputs.

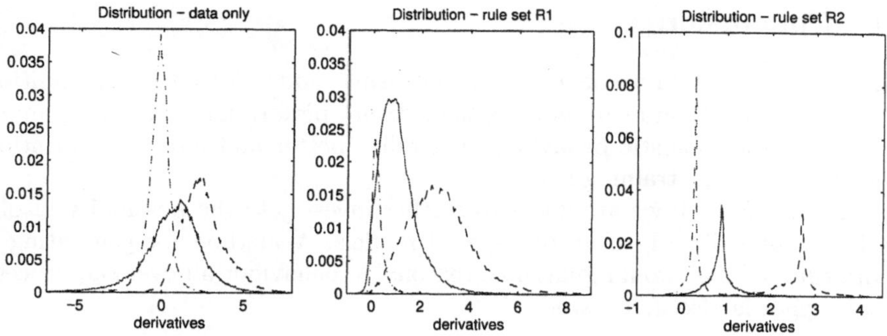

Fig. 2. Distributions of the partial derivative values $\frac{\partial o}{\partial x_i}$ for three inputs x_i having associated rules. The values were calculated using all the data samples and all the 70 models built resulting in 156800 values per rule.

For the R2 rule set gaussian distribution around the mean was assumed for each derivative. The mean values were estimated using the respective distributions in the data-based models, and refined by the rule set R1 distributions (shown in Fig. 2). In some cases a reasonable target value could only be obtained from the rule-based models. Using these we collected estimates $R_{ik} = c_i$ for all the eight 'rule inputs' and the second rule set R2 was made. This time the 'membership functions' were set to $\mu(D) = 0.5$ for all the rules.

As was expected from a semi-linear model, the prediction errors are reasonably small. From Fig. 2 we also see that the estimates c_i were fairly good and the resulting models actually have mainly linear relationships with wanted partial derivative values. Note, however, that one of the resulting distributions has a lower 'plain' before its actual peak, implying that maybe the target value c_i should have been lower, or that there are other, more non-linear dependencies present. Again, consulting process experts would be in place.

Models obtained with rule set R2 give robust predictions, are easier to explain and can be produced quite easily. Because of the good regularization, the standard deviation of the prediction errors between training runs is also considerably smaller than with models based solely on measurement data.

Table 1. Sum squared errors of the models (average of 10 runs) for each test block.

Mean square error of model output from data								
Training method	I	II	III	IV	V	VI	VII	Total
Data, early stop	36.3	20.1	3.75	2.95	5.91	4.01	5.9	79.0
Data (committee)	14.2	10.1	2.14	1.22	3.35	3.20	3.5	37.7
Rule set R1	16.3	12.0	2.15	2.92	4.08	3.74	4.2	45.3
R1 (committee)	9.3	7.5	1.42	1.45	3.10	3.04	2.5	28.3
Rule set R2	9.8	9.6	1.48	1.67	3.03	3.08	3.1	31.7
R2 (committee)	9.6	4.5	1.12	1.51	2.71	2.89	2.8	25.1

4 Conclusion

In this contribution we have shown that using model derivatives in addition to measurement data can improve the network performance. Resulting models have more realistic behaviour, generalize better and have less variation between different training runs.

In future work we are going to further investigate the possibility of obtaining rules directly from data-based models. A method for generating a core rule set that would guarantee reasonable behaviour all over the process state space is also under study.

References

1. Y. S. Abu-Mostafa, Hints and the VC dimension, Neural Computation, Vol. 5, No. 2, 1993, pp. 278-288.
2. C. Bishop, Curvature-Driven Smoothing: A Learning Algorithm for Feedforward Networks, IEEE Tr. on Neural Networks, Vol. 4, No. 5, Sep 1993, pp. 882-884.
3. C. Bishop, Regularization and Complexity Control in Feed-forward Networks, Proc. ICANN'95, Vol. 1., 1995, pp. 141-148.
4. C. Bishop, Neural Networks for Pattern Recognition, Oxford University Press, 1995.
5. S. Geman, E.Bienenstock and R.Doursat, Neural Networks and the Bias/Variance Dilemma, Neural Networks 4, 1992, pp. 1-58.
6. S. Haykin, Neural Networks, A Comprehensive Foundation, Macmillan, New York, NY, 1994.
7. G. Hinton, Connectionist Learning Procedures, Artificial Intelligence, 40, 1989, pp. 185-234.
8. J. Lampinen and O.Taipale, Optimization and Simulation of Quality Properties in Paper Machine with Neural Networks, *Proc. of IEEE International Conference on Neural Networks*, Orlando, FL, June 28 - July 2, 1994, pp. 3812-3815.
9. J. Lampinen and A. Selonen, Multilayer Perceptron Training with Inaccurate Derivative Information, Proc. 1995 IEEE International Conference on Neural Networks ICNN'95, Perth, WA, Vol. 5, pp. 2811-2815, 1995.
10. Y. Le Cun, B. Boser, J.S. Denker, D. Henderson, R.E. Howard, W. Hubbard, L.D. Jackel Backpropagation applied to hand written zip code recognition, *Neural Computation*, Vol. 1, 1989, pp. 541-551.
11. M. Riedmiller and H. Braun, A direct adaptive method for faster backpropagation learning: The RPROP algorithm, in H. Ruspini, (Ed), Proc. IEEE Internation Conf. on Neural Networks, San Francisco, 1993, pp. 586-591.
12. A. Selonen, J. Lampinen, L. Ikonen, Using External Knowledge in Neural Network Models, in Proc. SPIE on Intelligent Robots and Computer Vision XV: Algorithms, Techniques, Active Vision, and Materials Handling, Vol. 2904, Boston, MA, 1996, pp. 239-249.
13. P. Y. Simard, B. Victorri, Y. LeCun and J. Denker, Tangent prop - a formalism for specifying selected invariances in an adaptive network, In *Neural Information Processing Systems*, Vol. 4, San Mateo, CA, 1992.

Elliptical Basis Function Networks for Classification Tasks

Steffen Gutjahr[1] and Joachim Feist[2]

[1] University of Karlsruhe, Institute of Logic, Complexity and Deduction Systems, 76128 Karlsruhe, Germany, Email: gutjahr@ira.uka.de

[2] Neurotec Hochtechnologie GmbH, 88046 Friedrichshafen, Germany, Email: jf@neurotec.de

Abstract. In this paper we compare variants of elliptical basis function networks for classification tasks. The networks are introduced as density estimators and then modified towards RBF networks. Node reduction is accomplished by a genetic algorithm(GA). Two different kinds of node connections are compared. As a second degree of freedom different types of basis functions are investigated. On an artificial test set of the time series domain the impact of dimensionality is considered.

1 Introduction

The PNN (Probabilistic Neural Network, [11]) is a network formulation of density estimation. In order to classify a feature vector x one estimates the density $p(x|C_k)$ of each class C_k and combines these estimates by the rule of Bayes (eq. 1) to yield a posteriori class probabilities $P(C_k|x)$. These allow to make an optimal decision. In addition the probabilities tell us about the certainty of a classification.

$$P(C_k|x) = \frac{p(x|C_k)P(C_k)}{\sum_{i=0}^{K-1} p(x|C_i)P(C_i)}. \tag{1}$$

Density estimation is accomplished by using the Parzen Window ([8]) technique. One possible way of looking at this technique is to build a sphere of influence $p(s,x)$ around each training sample s and to add them up for each of the K classes.

$$p(x|C_k) = \frac{1}{\#C_k} \sum_{s \in C_k} p(s,x) \tag{2}$$

Exactly this is done by the PNN. Eq. 3 gives the details, the only parameter is the width σ of the gaussians, M denotes the dimension of the input space:

$$p_{PNN}(s,x) = \frac{1}{\sigma^M \sqrt{(2\pi)^M}} \exp\left(-\frac{1}{2}\frac{\|x-s\|}{\sigma}^2\right) \tag{3}$$

To put this into a network is straightforward. In figure 2a we have one node in layer one for each of the N training samples. The weights leading from the input to the layer one node are the co-ordinates of the sample. The node computes the distance $d(s,x)$ from the test vector x to the training sample s and outputs the value of the gaussian as in eq. 3. Each node in layer one performs part of the density estimation for the class it belongs to. The outcome of each of the layer one cells is added separately for the different classes (eq. 2) by the connections to the output cells with weight one.

Density Estimation in higher dimensions is a very difficult task. [10] states that density estimation in more than five dimensions is nearly impossible from the theoretical point of view. In [1] it could be shown that the PNN does not estimate densitys properly already in two dimensions. Improvements have been proposed that use more general basis functions.

The MNN (Minimal Error Neural Network, [7, 6]) and the EBF (Elliptical Basis Function Network, [2]) use a multivariate gaussian with different covariance matrices Σ_s for each sample s. The formula is a generalization of eq. 3:

$$p_{EBF|MNN}(s,x) = \frac{1}{\sqrt{(2\pi)^M |\Sigma_s|}} e^{-\frac{1}{2}(x-s)^T \Sigma_s^{-1}(x-s)} \tag{4}$$

The network structure stays the same, only the computations in the nodes are different. Each node s has the $M \times M$ matrix as parameter. Figure 1 gives an idea of how the construction of the sphere of influence of each training sample is accomplished. Shape and size have to be determined. The ellipses show areas of the input space for which the node for training sample s will output the same activation.

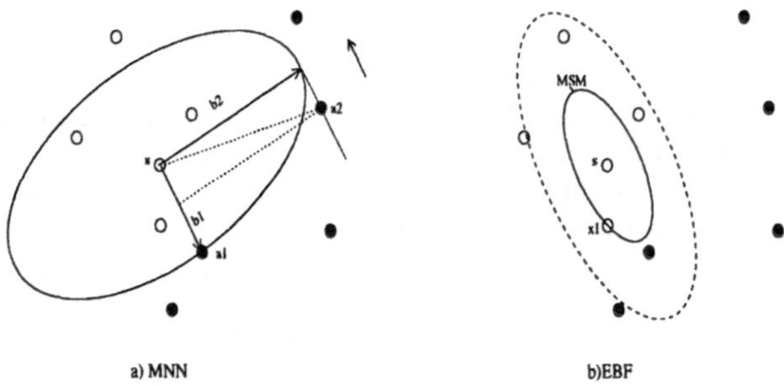

Fig. 1. Sphere of Influence of EBF and MNN

The MNN trys to separate the densities of the different classes. It looks for samples of other classes, building a rectangular coordinate system of the ellipsoid. The shape is determined by extending the sphere of influence just as far as samples of other classes are reached. The size is determined by making sure that densities of different classes do overlap only by five percent.

The EBF just considers samples of the same class. The shape is determined by computing the local covariance of the members of the training set. The size can be defined in terms of the Q nearest samples having an average mahalanobis distance (expression in the exponent of eq. 4) of one. The details for the version used in this paper can be found in [1].

2 Reduction of Basis Functions

The models described so far need to keep the complete training data. To reduce the number of basis functions, different methods have been proposed. For the EBF [2] and the MNN [6] it has been suggested to cluster the training data. Instead of this often used method we use a GA that selects the basis functions according to the performance on a validation set. This method has the following advantages:

- we do not need to decide about the final number of basis functions,
- we do not need a measure of distance in the input space for clustering,
- the node reduction is optimized for the classifier used.

The last point is a big advantage, since clustering the input space has a good intuitive justification, but an optimization according to the network performance on a validation set might come up with solutions that are better suited for the classifier used.

The genetic optimization uses a binary gene string representing the presence or absence of the training examples. Fitness is measured by the error on the evaluation set. Mutation and cross-over operators are used in a natural way as described in [4].

3 Weights

The reduction of neurons makes it computationally feasible to introduce weights from the basis functions to the output layer. As it is generally done in RBF networks we will compute optimal weights by a regression minimizing the mean squared error on the training set. So far there were only connections between nodes that belonged to the same class and the corresponding output neuron. Now we can introduce as well weights to output neurons of other classes. However, we will show later on that leaving these additional weights out may increase generalization properties.

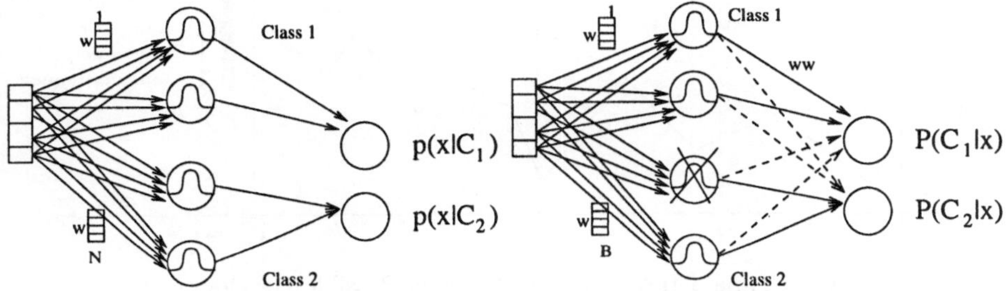

Fig. 2. PNN and Genetically Optimized Regression NN

The regression uses all the training samples and tunes the weights in such a way, that they activate the output neuron with one if it is the appropriate class and zero otherwise (1-out-of-k coding). This makes the outputs model the a posteriori class probability ([9]). Figure 2b illustrates the new architecture. The reduction of nodes is symbolized by the marked out neuron. The new connections are shown by the dashed lines. We call the network we get through this procedure "Genetically Optimized Regression Neural Network". We will use the abbreviation "GREBF" when the nodes are tuned with the EBF algorithm and "GRMNN" when the MNN is used.

4 Different Basis Functions

Instead of using the gaussians as basis functions for density estimation other kernel functions can be used. Some of these are considered superior on theoretical grounds [10].

With the results of [3] we can make use of an even bigger set of basis functions. Lowe shows that even using non-local basis functions at the location of the training feature vectors makes sense if one tunes weights by minimization of the mean squared error. Here we will investigate results for the spline and sigmoid as basis functions. In eq. 5, $d(s,x)$ is the local Mahalanobis distance as it was already used in the exponent of eq. 4. Σ_s is computed by the EBF or MNN algorithm.

$$p_{Spline}(s,x) = d(s,x)^2 log(d(s,x)) \qquad p_{Sigmoid}(s,x) = \frac{1}{1+e^{-d(s,x)}} \tag{5}$$

5 Tests

The classifiers were tested on an artificial time series. EBF and MNN are variants without node reduction. An ARMA-Model (Auto Regressive Moving Average was used with a chaotic time series as the noise process. We get a nonlinear short memory time series. Eq. 6 shows the ARMA(1,5) process used, and eq. 7 the quadratic map chaotic time series.

$$X_t = X_{t-1} + \sum_{j=0}^{4} \beta_j \varepsilon_{t-j} \text{ mit } \beta_0 = 0.3, \beta_1 = 0.2, \beta_2 = 0.1, \beta_3 = 0.2, \beta_4 = 0.3 \tag{6}$$

$$q_t = 4q_{t-1}(1 - q_{t-1}) \qquad \varepsilon_t = q_t - 0.5 \tag{7}$$

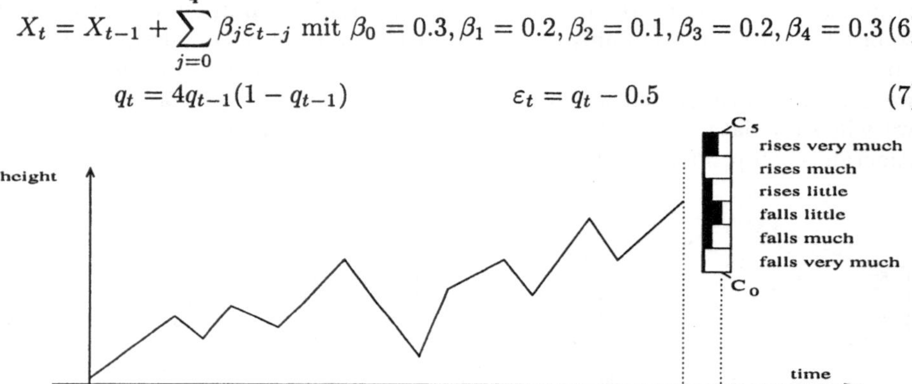

Fig. 3. Time Series with Six Classes of Possible Development

For training 700 samples are used, including 200 samples for the validation set of the GA, testing was performed on 1000 samples. The nature of the classification task is shown in figure 3. We want to find out the class of the change of value of the time series. This gives us a six class problem. The output values of the proposed network estimate the a posteriori class probabilities $P(C_i|x)$. Additionally we can get a classification of the direction of change by using the following equation.

$$\begin{aligned} P(falls|x) &= P(falls\ very\ much|x) + P(falls\ much|x) + P(falls\ little|x) \\ P(rises|x) &= P(rises\ very\ much|x) + P(rises\ much|x) + P(rises\ little|x) \end{aligned} \tag{8}$$

5.1 Different Basis Functions

We used twenty past changes in value as input vector. Table 1 shows the results with the different basis functions and different number of weights for the GREBF. The three best sets of basis functions were used and the results of these three classifiers averaged. The results are best when the gaussian is used. This might be due to the underlying EBF algorithm which is constructed having this basis function in mind. It is better to set the dashed lines in figure 2b a priori to zero. The fewer number of weights makes an over-fitting less likely.

The sigmoid is the best non-local basis function. It is surprising that in this case leaving out the additional weights is superior. Setting them to zero was motivated by the interpretation of the process as a density estimation. But this cannot be the reason for this to function for nonlocal basis functions.

For the second non-local kernel the additional weights are clearly needed, without them classification with the spline as basis function is very poor. But the more complex model is still not able to reach the values of the other models.

The gaussian proved to be the best classifier when looking at network complexity as well. The genetic optimization resulted in an average of 145 basis functions, whereas the models with the other basis functions needed 215.

The network was also tested using the MNN as algorithm to construct the shape of the basis function. But this algorithm separates the densitys of the different classes too much, resulting in output values that are either near zero or one. This makes it difficult for the regression to boost the performance with the weights. The GRMNN was always worse than the GREBF.

		GREBF+Gauss	GREBF+Spline	GREBF+Sigmoid
Few Weights	Six Classes	**37.8**	25.0	**37.5**
	Two Classes	**82.7**	72.3	**82.0**
All Weights	Six Classes	37.1	**36.0**	37.0
	Two Classes	80.7	**80.7**	79.9

Table 1. Hit Rate and Direction Hit Rate for Different GREBF Networks

5.2 Dimension

RBF-type networks and density estimation techniques are known to suffer from high dimensionality. The time series problem was used to investigate for what dimensionality the GREBF with the gaussian as basis function is best suited. Different numbers of past difference values of the time series were used. The results are summarized in table 2. The EBF and MNN are versions that keep all basis functions.

For this problem the MNN is the best classifier. With just the five most recent changes of value of the time series it achieves a very high accuracy in estimating the future change. However, it is also the classifier to suffer most by increased dimensionality.

The GREBF is best for input dimensions 10–20. The backpropagation neural network performs best for very high dimensions. It is surprising that it needs many redundant input dimensions to yield best performance whereas the other classifiers work best with just the absolutely necessary input dimensions.

6 Test on Real Data

Reduction of basis functions works best when used on real data sets where we face outliers and noisy samples that are untypical and are therefore best omitted. The diabetes dataset used is a two class problem consisting of 768 cases and having 8 input dimensions. We use the same test setup as in [5] (12 cross-validation) except that we do not normalize the dimensions of the dataset. The average performance of the three best basis function sets found for the GREBF with few weights and the gaussian as basis function had a hit rate of 76.5%. In order to have a result of a usable classifier rather than an average performance of three classifiers we use the outputs of the three networks by averaging over the output values, which yields an accuracy of 76.7%. This is better than the results for the MNN (72.1%), the EBF (72.1% as well) the results for the Backpropagation Network (75.2%) and the RBF (75.7%) mentioned in [5]. With this result the GREBF would be in 5th place of the 23 classifiers used in [5], the best classifier reaching a performance of 77.7%.

Dimension	MNN	EBF	GREBF	NN
5	**49.5**	41.2	40.1	31.1
10	30.8	32.1	**38.8**	31.4
15	25.4	35.6	**38.3**	31.4
20	25.2	28.0	**37.8**	32.0
40	19.5	27.3	30.0	**31.3**

Dimension	MNN	EBF	GREBF	NN
5	**83.9**	83.2	81.6	78.7
10	74.0	**81.0**	79.3	79.3
15	65.8	79.5	**83.0**	80.4
20	64.2	63.8	**82.7**	82.4
40	57.6	63.9	76.0	**81.1**

Table 2. Hit Rate for the Six Class Problem and the Direction Decision for Different Numbers of Input Dimensions

References

[1] Feist,Joachim; Scott, Paul D. *Uncertainty in Connectionist Learning*, Paper based on M. Sc. Thesis, University of Essex, 1995. `http://i11www.ira.uka.de:80/~jofeist/paper.ps`

[2] Johnston, Lloyd P.M.; Kramer, Mark A. *Probability Density-Estimation Using Elliptic Basis Functions* in AIChE Journal, American Inst. of Chemical Engineers, N.Y., Vol.40 No.10, 1994.

[3] Lowe, D. *On the Use of Nonlocal and Non Positive Definite Basis Functions in Radial Basis Function Networks*, Artificial Neural Networks, Conf. Publication No. 409, 1995.

[4] Masters,T., *Practical Neural Network Recipes in C++*,Academic Press,San Diego,1993.

[5] Michie D., Spiegelhalter D. J., Taylor C. C. *Machine Learning, Neural and Statist. Classification*, Ellis Horwood, Hempstead, 1994.

[6] Musavi, M; Ahmed, W; Chan K.; Fabris K. B; Hummels D. M. *On the Training of Radial Basis Function Classifiers* in Neural Networks, Vol. 5, Pergamon Press, New Yourk, 595–603, 1992.

[7] Musavi, M.; Kalantri, K.; Ahmed, W.; Chan K. H *A Minimum Error Neural Network (MNN)* Neural Networks, Vol. 6, Pergamon Press, New York, 397-407, 1993.

[8] Parzen, E. *On the Estimation of a Probability Density Function and Mode* Annals of Mathematical Statistics 33, 1065-1076, 1962.

[9] Richard, Michael D.; Lippmann, Richard P. *Neural Network Classifiers Estimate Bayesian A Posteriori Probabilities* in Neural Computation v. 3(4),Cambridge, MA: MIT Press, 461-483, 1991.

[10] Scott, D. *Multivariate Density Estimation*, N.Y. Wiley, 1992.

[11] Specht, Donald F. *Probabilistic Neural Networks* in Neural Networks v. 3(1), Pergamon Press, New York, 109-118, 1990.

Probabilistic Neural Networks with Rotated Kernel Functions

Ingo Galleske and Juan Castellanos

Facultad de Informática, Universidad Politecnica de Madrid, Spain

Abstract. For the automatical determination of the "smoothing parameter" and the enhancement of the generalization ability of the standard Probabilistic Neural Network(PNN), a method to construct the covariance matrix of the Gaussian kernel functions of the training pattern is proposed. Based on the minimization of the local error, the constant potential surface of the Gaussian function provides two matrices: a rotation matrix and a matrix of variances, which are both combined to calculate the desired covariance matrix. The new approach was applied to the two-spiral problem, where training was done with a reduced pattern set. The efectiveness is demonstrated in a comparison between the PNN and the new model on the generalization to the entire set.

1 Introduction

Artificial Neural Networks are used in many different types of applications: one main field is the classification problem, which consists in assigning an untrained pattern to the correct class. A Neural Network architecture which has this ability is the Probabilistic Neural Network (PNN) [4], [5]. The main idea is to apply the Bayes decision rule in a multidimensional input space to establish a classification, i.e. to separate the input space into different regions. For that, Specht used a kernel estimator to design a Neural Network to estimate the probability density which underlies the problem. But this solution is not unproblematic: It needs one unit for each training pattern, which is especially tiresome for problems with a large sample. The main advantage over other types of Neural Networks is the rapid training process which only consists in a one-pass learning. A second advantage is the possibility to add new patterns without repeating the complete training process. For these reasons, applications can mainly be found in real-time problems [1].

In the original contribution, Specht [4] used the multidimensional Gaussian function as kernel estimator. He stated that the determination of the "smoothing parameter", the variance of the Gaussian kernel function, does not play an important role; other researchers found out, that a correct value *is* of importance and a wrong choice for this parameter could result in a great misclassification[2],[6]. Therefore, some procedures to enhance the generalization ability were developped [2]. This paper presents a new way to determine automatically the necessary covariance matrices of the kernel functions, freeing the designer of the network from finding the correct values. The disadvantage is the loss of the *very* high training speed of the standard PNN, to the advantage of gaining a network, which still is *much* faster to train than Backpropagation nets, additionally giving a better generalization ability and consequently a better performance on unseen training patterns.

This work starts with a general introduction into the mathematical basis used by the PNN and a description of its architecture. Based on this, a new way for the parameter estimation of the multidimensional kernel functions is explained, followed by the concrete description of the calculation of the necessary parameters. Experiments, which show the better generalization will also be presented.

2 Theoretical foundations

The original Probabilistic Neural Network The general PNN consists of 3 layers of Neurons (see Fig. 1). The n input units indicate the n-dimensional problem space. The next layer of units (pattern layer) consists of p units; p being the number of training patterns of all classes. These two layers are completely connected. The one-pass learning of the PNN sets the connections between the input layer and the pattern layer in the following way: The weights, which lead into a pattern unit p_i, are set to the coordinates of the corresponding training pattern in the n-dimensional input space. The pattern units feed their output without exception only into that class unit of the next layer which represents the class it belongs to. These connections cannot be learned and have their weights fixed to the reciprocal value of the number of units of the corresponding class. Above this layer is a decision network, which selects the class unit with the highest activation stating the class to which the input pattern belongs.

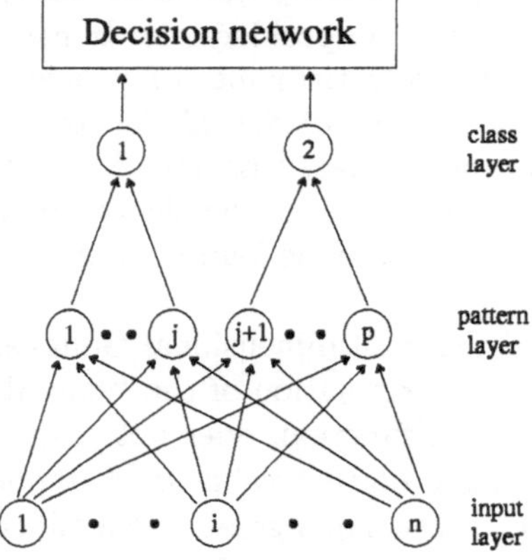

Fig. 1. Architecture of the standard Probabilistic Neural Network.

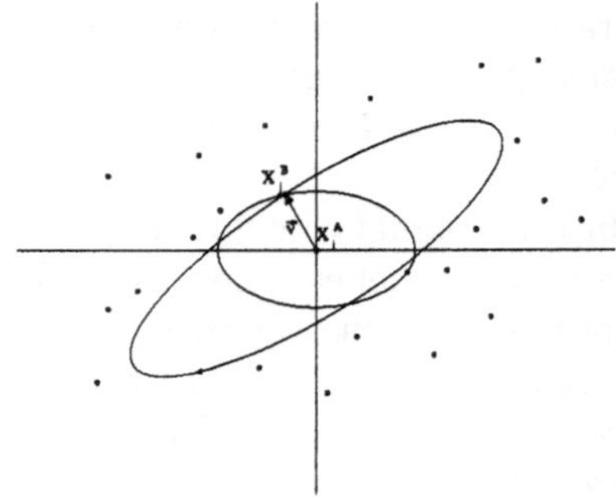

Fig. 2. The rotation of the ellipse.

The most important point in Probabilistic Neural Networks are the activation functions of the pattern units (the class units only have a single summation function $a(h) = \sum_{i=k}^{l} o(i)$, where the pattern units k to l belong to class h). The output functions of the network are equal to the activation, i.e. $o(i) = a(i)$ for all units. The activation functions of the units in the pattern layer are those functions, which finally determine the complete density function of the underlying problem. Generally, multi-dimensional Gaussian functions are used:

$$f(X) = \frac{1}{(2\pi)^{\frac{n}{2}}|\Sigma|^{-\frac{1}{2}}} * \exp\left(-\frac{1}{2}(\boldsymbol{X} - \boldsymbol{X}_p)^T \Sigma^{-1}(\boldsymbol{X} - \boldsymbol{X}_p)\right) \quad (1)$$

with $\boldsymbol{X}$ point under observation, Σ matrix of covariances, $|\Sigma| = \det \Sigma$, $\boldsymbol{X}_p$ center of Gaussian function. The application of these kernel functions is legitimated, as proved in the fundamental article by Parzen [3].

The general problem to resolve by the netword designer, is to find the best estimation of Σ_i for all training pattern i. The original PNN uses the same matrix Σ for all training pattern, leading to an architecture which is very fast to train, but which does not have a very good generalization ability.

The Modified Model The algoritm proposed here, calculates for each pattern i its own matrix Σ_i in a still quiet fast way. For that, each matrix Σ_i will be subdivided into four matrices.

The multi-dimensional Gaussian functions have a Constant Potential Surface (CPS) (points of same probability) which are hyper-ellipsoides. The form of these CPS are determined by the parameter Σ. In a simple two-dimensional case, Σ is a 2x2 matrix of the following form:

$$\Sigma = \begin{bmatrix} \sigma_1^2 & \sigma_{12} \\ \sigma_{12} & \sigma_2^2 \end{bmatrix}$$

σ_1^2 and σ_2^2 are the lengths of the principal axes of the CPS, which in this case is an ellipse. σ_{12}, if not equal to zero, indicates a stretching of the ellipse combined with a rotation.

The idea proposed in this work is to allow the ellipse to rotate around any angle in the hope to achieve a better generalization ability of the Network. In Fig.2 the rotated ellipse has bigger principal axes leading to better generalization. For that, the covariance matrix Σ is subdivided into four matrices $\Sigma = \mathbf{RSSR}^T$. $\mathbf{S}$ is a diagonal matrix with the elements on the diagonal indicating the length of the principal axis of the ellipse. $\mathbf{R}$ is a simple rotation matrix, which rotates the ellipse to the desired position.

The next chapter will explain in detail, how the unknown rotation matrix $\mathbf{R}$ and the matrix of the principal axis $\mathbf{S}$ are calculated using a general recursive formula for n-dimensions.

3 Calculation of the matrices R and S

In this chapter a general procedure for calculating the rotation matrix $\mathbf{R}$ and the matrix of the variances $\mathbf{S}$ is described. These two matrices are the constituents of Σ, the matrix of covariances of the Gaussian kernel estimator of the activation function of the Artificial Neurons.

It will be presented a recursive formula for the estimation of the unknown parameters of this model in a general n-dimensional input space. First, σ_n and the last row vector of $\mathbf{R}$ are calculated. With the knowledge of σ_n the calculation of σ_{n-1}, and with the knowledge of the last row vector of $\mathbf{R}$ the calculation of the second last row vector can be done; and so forth until $\sigma_n, \sigma_{n-1}, \ldots, \sigma_2$ are known, and σ_1, as well as the first row vector of $\mathbf{R}$, can be estimated.

For each of the training pattern of all classes, the following procedure has to be executed. For the pattern X_i^A, the first axis a_n of the ellipsoid is searched. a_n is the length, using simple Euclidean metric, of the shortest connection vector between the pattern under observation and all pattern of the opposite class. The calculation of the other principal axes a_i is done using the following procedure:

The general formula of an hyper-ellipsoid in an n-dimensional space is

$$\frac{x_1^2}{a_1^2} + \frac{x_2^2}{a_2^2} + \cdots + \frac{x_i^2}{a_i^2} + \cdots + \frac{x_n^2}{a_n^2} = 1.$$

a_{n-1} is found setting $a_1 = a_2 = \ldots = a_{n-1}$. This simplification is possible, because the principle axis are not yet located in the $(n-1)$-dimensional subspace, which is perpendicular to a_n, and therefore can be rotated around any angle. Afterwards, the calculation of a_{n-2} with the known variables a_n, a_{n-1} is possible setting $a_{n-2} = a_{n-3} = \cdots = a_1$, until finally the calculation of a_1, the last axis of the ellipsoid, is possible. The variances $\sigma_1, \ldots, \sigma_n$, which constitute the matrix $\mathbf{S}$ are set to half of the length of the principal axis of the calculated hyper-ellipsoid: $\sigma_1 = \frac{1}{2}a_1, \sigma_2 = \frac{1}{2}a_2, \ldots, \sigma_n = \frac{1}{2}a_n$. This decision is legitimated, because the variances of the training pattern of the different classes do not overlap this way and therefore at least 50% of the probability mass do not overlap either.

To calculate the principal axis a_i of the hyper-ellipsoid, the distances of the pattern under observation X_i^A to all pattern of the opposite class X_j^B are calculated using the following recursive formula:

$$d(X_i^A, X_j^B) = a_i = \begin{cases} \left(\dfrac{\prod\limits_{j=i+1}^{n} a_j \left(\sum\limits_{j=1}^{i} x_j \right)}{\prod\limits_{j=i+1}^{n} a_j - \left(\sum\limits_{j=i+1}^{n} \left(\prod\limits_{\substack{k=i+1 \\ k \neq j}}^{n} a_k \right) x_j \right)} \right)^{\frac{1}{2}} & \text{,if denominator} \geq 0 \\ \infty & \text{,if denominator} < 0 \end{cases}$$

where i starts from $n-1$ and decrements to 1. $d(X_i^A, X_j^B)$ could be interpreted as the length of the connection vector from X_i^A to X_j^B. The procedure choses the shortest connection vector of all vectors calculated using the above distance measure.

The connection vectors $\mathbf{v}_1, \ldots, \mathbf{v}_n$ between X_i^A and the respective X_j^B's are used to construct the rotation matrix $\mathbf{R}$, where $\mathbf{v}_1$ is the last row vector, $\mathbf{v}_2$ is the second last row vector, until finally $\boldsymbol{v}_n$ is the first row vector of $\mathbf{R}$. $\boldsymbol{v}_i$ is proyected to the subspace, which is perpendicular to the known vectors $\boldsymbol{v}_{i+1}, \ldots, \boldsymbol{v}_n$. The last vector $\boldsymbol{v}_1$ has to be chosen, that the resulting system is a "right-hand" coordinate system. The orthonormal vectors $\boldsymbol{v}_i$ constitute the row vectors of the matrix $\mathbf{R}$.

As described before, the rotation matrix $\mathbf{R}$ and the matrix of the variances $\mathbf{S}$ are used to construct the unknown matrix Σ of the Gaussian kernel functions:$\Sigma = \mathbf{RSSR}^T$.

As seen in equation 1, the matrix of covariances Σ is not needed directly, only its determinant and the inverse of the covariance matrix Σ^{-1}. The calculation of the determinant of Σ is unproblematic. Σ^{-1} is not determined by the inversion of Σ, but more easily and faster using the constituents $\mathbf{R}$ and $\mathbf{S}$: $\Sigma^{-1} = \mathbf{RS}^{-1}\mathbf{S}^{-1}\mathbf{R}^T$, which can be calculated easily, because $\mathbf{S}$ is a diagonal matrix and its inversion can be done without any problem.

4 Experimental results

The above described method was implemented to compare the generalization ability of this model versus the original Probabilistic Neural Network. The results of the two-spiral problem, known in the area of backpropagation training and proposed by A.Wieland, are used for this purpose.

The task is to classify points in the (x-y)-plane to one of two classes. The classes have the form of a spiral, making it difficult to be learned by a standard backpropagation network, which is "only" a linear classifier. The PNN seemed to offer a better solution, not depending on linear separation. The difficulty in the application of the PNN is the determination of the covariance matrix, which is not known in advance. The wrong choice could result in an error on the training set between 0% and 33%. The reason is the high density of pattern in the center, which affect the output on outer spiral arms. Using the model described in this work, which calculates this matrix separately for all pattern, results in a perfect separation of the two spirals with 0% error.

For testing the generalization ability, the two models had to learn the two spirals with a reduced set of training pattern and to generalize to the complete set. Table 1 shows the results for various steps of reduction. It can easily be seen that for all steps the network model proposed is this article performs better, i.e. has a smaller error, than the original PNN.

5 Conclusion

This work presents a technique to overcome certain limities of the Probabilistic Neural Network. The proposed model calculates automatically the "optimal" covariance matrix observing the local environment of the training pattern. The implementation is relatively simple and experiments with the two-spiral problem were done. In comparison with the standard PNN, the new model exhibited better performance in the training phase and in the generalization ability. The

Reduction	0%	5%	10%	15%	20%	25%
PNN (pattern)	0	7	5	11	18	41
proposed model (pattern)	0	3	3	7	11	15
PNN (in %)	0%	3.6%	2.6%	5.6%	9.2%	20.9%
proposed model (in %)	0%	1.5%	1.5%	3.6%	5.6%	7.7%

Table 1. Misclassification in the experiment with reduced training set.

results are very encouraging, so future work will go into the direction to use the proposed procedure in clustering techniques, which reduce the size of the network without loosing its generalization ability.

References

1. Chen C.H., G.H. You (1992), ISBN Recognition Using a Modified Probabilistic Neural Network (PNN). *Proceedings 11th IAPR International Conference on Pattern Recognition*, Vol.II, 419-421
2. Musavi M.T., Chan K.H., Hummels D.M., Kalantri K (1993), On the Generalization Ability of Neural Network Classifiers. *IEEE Transactions on Pattern Analysis and Machine Intelligence* Vol.16, No.6, 659-663
3. Parzen E. (1962), On estimation of a probability density function and mode. *Annals of Mathematical Statistics*, 33, 1065-1076
4. Specht, D.F. (1988), Probabilistic neural networks for classification mapping, or associative memory. *Proceeding, IEEE International Conference on Neural Networks*, 1, 525-532
5. Specht, D.F. (1990), Probabilistic Neural Networks. *Neural Networks*, 3, 109-118
6. Specht, D.F. (1991), Generalization accuracy of probabilistic neural networks compared with backpropagation networks. *IJCNN-91-Seattle: International Joint Conference on Neural Networks*, Vol.1, 887-892

Statistical Control of RBF-like Networks for Classification

Norbert Jankowski[1] and Visakan Kadirkamanathan[2]

[1] Nicholas Copernicus University, Toruń, Poland, e-mail:norbert@phys.uni.torun.pl
[2] The University of Sheffield, UK, e-mail: visakan@acse.shef.ac.uk

Abstract. *Incremental Net Pro* (IncNet Pro) with local learning feature and statistically controlled growing and pruning of the network is introduced. The architecture of the net is based on RBF networks. *Extended Kalman Filter* algorithm and its new fast version is proposed and used as learning algorithm. IncNet Pro is similar to the *Resource Allocation Network* described by Platt in the main idea of the expanding the network. The statistical *novel criterion* is used to determine the growing point. The Bi-radial functions are used instead of radial basis functions to obtain more flexible network.

1 Introduction

The *Radial Basis Function* (RBF) networks [13,12] were designed as a solution to an approximation problem in multi-dimensional spaces. The typical form of the RBF network can be written as

$$f(\mathbf{x}; \mathbf{w}, \mathbf{p}) = \sum_{i=1}^{M} w_i G_i(||\mathbf{x}||_i, \mathbf{p}_i) \tag{1}$$

where M is the number of the neurons in hidden layer, $G_i(||\mathbf{x}||_i, \mathbf{p}_i)$ is the i-th Radial Basis Function, $\mathbf{p}_i$ are adjustable parameters such as centers, biases, etc., depending on $G_i(||\mathbf{x}||_i, \mathbf{p}_i)$ function which is usually choosed as a Gaussian ($e^{-||\mathbf{x}-\mathbf{t}||^2/b^2}$), multi-quadratics or thin-plate spline function[1]. In contrast to many *artificial neural networks* (ANNs) including well known *multi-leyered perceptrons* (MLPs) networks the RBF networks have well mathematical properties. Girosi and Poggio [6,12] proved the existence and uniqueness of best approximation for regularization and RBF networks. In the 1991 Platt published the article on the *Resource-Allocating Network* [11]. The RAN network is an RBF-like network that grows when two criteria are satisfied:

$$\mathbf{y}_n - f(\mathbf{x}_n) = e_n > e_{min}; \qquad ||\mathbf{x}_n - \mathbf{t}_c|| > \epsilon_{min} \tag{2}$$

e_n is equal the current error, $\mathbf{t}_c$ is the nearest center of a basis function to the vector $\mathbf{x}_n$ and e_{min}, ϵ_{min} are some experimentally choosen constants. The growing network can be described by $f^{(n)}(\mathbf{x},\mathbf{p}) = \sum_{i=1}^{k-1} w_i G_i(\mathbf{x},\mathbf{p}_i) + e_n G_k(\mathbf{x},\mathbf{p}_k) =$

[1] For a interesting review of many other transfer function see [3].

$\sum_{i=1}^{k} w_i G_i(\mathbf{x}, \mathbf{p}_i)$, where $\mathbf{p}_k$ includes centers $\mathbf{x}_n$ and others adaptive parameters which are set up with some initial values. If the growth criteria are not satisfied the RAN network uses the LMS algorithm to estimate free parameters. Although LMS algorithm is faster than *Extended Kalman Filter* (EKF) algorithm [1] we decided to used EKF algorithm because it exhibits fast convergence, use lower number of neurons in hidden layer [9] and gives some *tools* which would be useful in control of the growth and pruning process.

The Goal of IncNet Pro The main goal of our researche was to build a network which would be able to adjust the complexity of the network to complexity of the data shown to the network during the learning time.

The IncNet Pro tries to solve the above task in 4 ways: • ESTIMATION: The typical learning process is based on fast EKF algorithm. • GROWING: If the *novelty criterion* is satisfied then a new neuron is added to the hidden layer. • DIRECT PRUNING: IncNet algorithm checks whether or not a neuron should be pruned. If yes, then the neuron with the smallest saliency is removed. • BI-RADIAL FUNCTIONS: The Bi-radial transfer function estimate more complex density of input data through using the separate biases and separate slopes in each dimension and for each neuron.

Similar work has been done in recent years by several authors, but it is quite rare to combine growing and pruning in one network, which is quite important for optimal generalization of the network. Weigend, Rumelhart & Huberman [16] described weight-decay, pruning neurons with smallest magnitude of weights. LeCun et al. [10] described more effective pruning method, *Optimal Brain Damage*. Hassibi in 1993 [7] published the *Optimal Brain Surgeon* algorith, which works without assumption used by LeCun that the Hessian matrix is near diagonal.

RAN network using EKF learning algorithm (RAN-EKF) was proposed by [9]. The M-RAN net [17] is based on RAN-EKF with pruning based on removing neurons with smallest normalized output from hidden layer. The previous version of the IncNet [8] is a RAN-EKF network with statistically controlled growth criterion. Another very good example, derived from MLP network, is the *Cascade–Correlation* algorithm [4]. *Feature Space Mapping* (FSM) system is the system which joins two strategies: growing and pruning, see [2] for more information. For more exhaustive description of ontogenic neural network see [5].

2 The IncNet Pro Framework

Fast EKF: We introduce new fast version of the EKF learning algorithm, described in [1]. The EKF was chosen because it can estimate not only adaptive parameters, but also some others values which will be used in novelty criterion and in pruning.

Covariance matrix $\mathbf{P}_n$ can be quite large for real data because its size is the square of the total number of adaptive parameters. Assuming that correlations between parameters of different neurons are not very important we can simplify the matrix $\mathbf{P}_n$ assuming block-diagonal structure of $\widetilde{\mathbf{P}}_n$ with $\widetilde{\mathbf{P}}_n^i$, $i = 1 \ldots M$.

Diagonal elements represents correlations of adaptive parameters of the i-th neuron.

Let m be the number of adaptive parameters per neuron and M the number of neurons. The size of matrix $\mathbf{P}_n$ is $m \cdot M \times m \cdot M$, but matrix $\widetilde{\mathbf{P}}_n$ has only $m^2 M$ elements not equal to zero. For a given problem $\mathcal{P}$ the complexity of matrix $\mathbf{P}_n$ is $O(M^2)$, and matrix $\widetilde{\mathbf{P}}_n$ just $O(M)$ (m is constant in $\mathcal{P}$)! Using this approximation the fast version of the EKF algorithm is:

$$\begin{aligned}
e_n &= y_n - f(\mathbf{x}_n; \mathbf{p}_{n-1}) \\
\mathbf{d}_n^i &= \frac{\partial f(\mathbf{x}_n; \mathbf{p}_{n-1})}{\partial \mathbf{p}_{n-1}^i} \\
R_y &= R_n + \mathbf{d}_n^{1\,T} \widetilde{\mathbf{P}}_{n-1}^1 \mathbf{d}_n^1 + \cdots + \mathbf{d}_n^{M\,T} \widetilde{\mathbf{P}}_{n-1}^M \mathbf{d}_n^M \qquad i = 1, \ldots, M \\
\mathbf{k}_n^i &= \widetilde{\mathbf{P}}_{n-1}^i \mathbf{d}_n^i / R_y \\
\mathbf{p}_n^i &= \mathbf{p}_{n-1}^i + e_n \mathbf{k}_n^i \\
\widetilde{\mathbf{P}}_n^i &= [\mathbf{I} - \mathbf{k}_n^i \mathbf{d}_n^{i\,T}] \widetilde{\mathbf{P}}_{n-1}^i + Q_0(n)\mathbf{I}
\end{aligned} \tag{3}$$

the suffixes $n-1$ and n denote the priors and posteriors. $\mathbf{p}_n$ consists of all adaptive parameters: weights, centers, biases, etc. To prevent too quick convergence of the EKF, which leads to data overfitting, the $Q_0\mathbf{I}$ adds a *random* change, where Q_0 is scalar (sometimes decreasing to small values around 10^{-5}) and $\mathbf{I}$ is the identity matrix.

Novelty Criterion: Using methods which estimate during learning covariance of uncertainty of each parameter, the network output uncertainty can be determined and the same criterion as in the previous version of IncNet [8] may be used. Then the hypothesis for the statistical inference of model sufficiency is stated as follows:

$$\mathcal{H}_0: \quad \frac{e^2}{\mathrm{Var}[f(\mathbf{x}; \mathbf{p}) + \eta]} = \frac{e^2}{R_y} < \chi^2_{n,\theta} \tag{4}$$

where $\chi^2_{n,\theta}$ is $\theta\%$ confidence on χ^2 distribution for n degree of freedom, $e = y - f(\mathbf{x}; \mathbf{p})$ is the error and $R_y = \mathrm{Var}[f(\mathbf{x}; \mathbf{p}) + \eta]$ (part of EKF) — see Eq. 3.

If this hypothesis is satisfied the current model is sufficient and the IncNet network continues learning using the fast EKF algorithm. Otherwise, a new neuron $(M+1)$-th should be added with some initial parameters. For Gaussian functions $G_{M+1}(\cdot)$ these parameters are: $w_{M+1} := e_n$, $\mathbf{t}_{M+1} := \mathbf{x}_n$, $b_{M+1} := b_0$, $\mathbf{P}_n := \begin{bmatrix} \mathbf{P}_n & 0 \\ 0 & P_0\mathbf{I} \end{bmatrix}$, where e_n is the error for given input vector $\mathbf{x}_n$, b_0 and P_0 are some initial values for bias (depending on a given problem) and covariance matrix elements (usually 1).

Pruning: As a result of the learning process a neuron can become completely useless and should be pruned. Assume the structure of vector $\mathbf{p}_n$ and the covariance matrix as:

$$\mathbf{p}_n = [w_1, \ldots, w_M, \ldots]^T \qquad \mathbf{P} = \begin{bmatrix} \mathbf{P}_w & \mathbf{P}_{wv} \\ \mathbf{P}_{wv}^T & \mathbf{P}_v \end{bmatrix} \tag{5}$$

where $\mathbf{P}_w$ is a matrix of correlations between weights, $\mathbf{P}_{wv}$ between weights and other parameters, $\mathbf{P}_v$ **only** between others parameters (excluding all weights).

Then by checking the inequality $\mathcal{P}$ presented below we can decide whether to prune or not and find the neuron for which value L has smallest saliency and should be pruned.

$$\mathcal{P}: \; L/R_y < \chi^2_{1,\vartheta} \qquad\qquad L = \min_i w_i^2/[\mathbf{P}_w]_{ii} \tag{6}$$

where $\chi^2_{n,\vartheta}$ is ϑ% confidence on χ^2 distribution for one degree of freedom.

Neurons are pruned if the saliency L is too small and/or the uncertainty of the network output R_y is too big.

Bi-radial Transfer Functions: To obtain greater flexibility the bi-radial transfer functiona [3] are used instead of Gaussians. These functions are build from products of pairs of sigmoidal functions for each variable and produce decision regions for classification of almost arbitrary shapes.

$$Bi(\mathbf{x};\mathbf{t},\mathbf{b},\mathbf{s}) = \prod_{i=1}^{N} \sigma(e^{s_i}\cdot(x_i - t_i + e^{b_i}))(1 - \sigma(e^{s_i}\cdot(x_i - t_i - e^{b_i}))) \tag{7}$$

where $\sigma(x) = 1/(1+e^{-x})$. The first sigmoidal factor in the product is growing for increasing input x_i while the second is decreasing, localizing the function around t_i. Shape adaptation of the density $Bi(\mathbf{x};\mathbf{t},\mathbf{b},\mathbf{s})$ is possible by shifting centers $\mathbf{t}$, rescaling $\mathbf{b}$ and $\mathbf{s}$, see Fig. 1. The number of adjustable parameters per processing unit is in this case (excluding weights w_i) $3N$. Dimensionality reduction is possible as in the *gaussian bar case* [3], but we can obtain more flexible density shapes, thus reducing the number of adaptive units in the network. Exponentials e^{s_i} and e^{b_i} are used instead of s_i and b_i to prevent oscillations during learning procedure (learning becomes more stable).

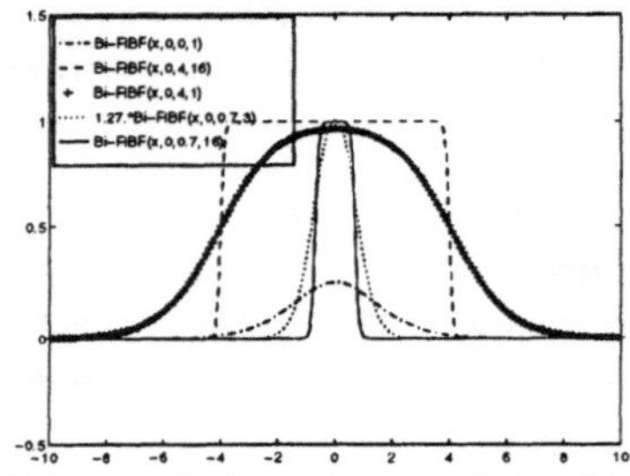

Fig. 1: A few shapes of the bi-radial functions in two dimensions.

Classification using IncNet Pro: k independed IncNet network are used for k-class problem. Each of them receives input vector $\mathbf{x}$ and 1 if index of i-th IncNet is equal to desired number of class, otherwise 0. The output of i-th IncNet Pro network is equal to probability that the vector belongs to i-th class. See figure on the right.

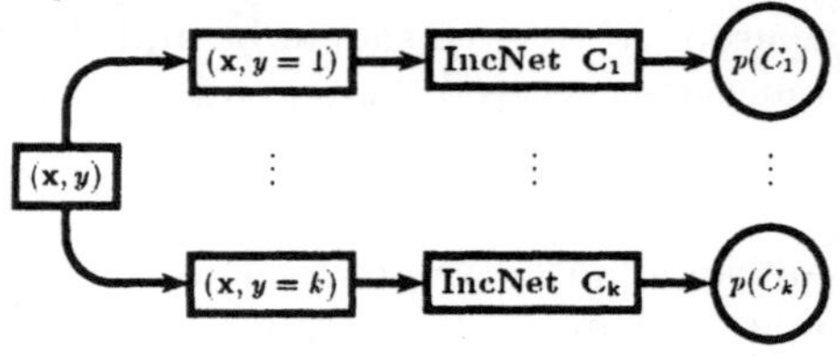

3 Results

The two-spiral problem. The data consists of two sets (training and testing) with 194 patterns each for two spirals. After 10,000 iterations (it took about 35 minutes on PC Pentium 150MHz) we got result which fit 192 points out of 194 (99%) for training set and 191 (98.5%) for the test set. Final net has 79 neurons. The fast version of EKF accelerates computation 50 times in comparision with standard EKF learning. There are other nets which are able to solve the two-spiral problem too, for example one of the best is an MLP using a global optimization algorithm by Shang and Wah [14]. Their network is able to get 100% correct results for the training set but never more than 95.4% for the test set. Although it used only 6 neurons, it takes about 200 minutes to train.

Breast Cancer, Hepatitis, Pima Indians Diabetes, Heart Disease are medical diagnosis benchmarks considered in [15]. Short summary of the data: Breast Cancer – 2 classes, 9 attributes, 699 instances; Hepatitis – 2 classes, 19 attributes, 155 instances; Diabetes – 2 classes, 8 attributes, 768 instances. Heart – 2 classes, 13 attributes, 303 instances.

method	Breast	Hepat.	Diab.	Heart
IncNet	**97.1**	**82.3**	**77.6**	**90.0**
BP	96.7	82.1	76.4	81.3
LVQ	96.6	83.2	75.8	82.9
CART	94.2	82.7	72.8	80.8
Fisher	96.8	84.5	76.5	84.2
LDA	96.0	86.4	77.2	84.5
QDA	34.5	85.8	59.5	75.4
KNN	96.6	85.3	71.9	81.5
LFC	94.4	81.9	75.8	75.1
ASI	95.6	82.0	76.6	74.4

Table 1: Accuracies (%) for medical benchmarks

Breast Cancer problem used 49 neurons and 3000 iterations, the accuracies on training and test sets was very similar: 97.7%, 97.1%, computation time: 5150 sec. Hepatitis data used 97 neurons and 500 iterations, the accuracies on training and test sets was: 98.6%, 82.3%, computation time: 3100 sec. Diabetes data used 100 neurons and 5000 iterations, the test accuracy was better on test set (77.6%) than on the training set (77.2%), computation time: 11200 sec. Heart data used 117 neurons and 1000 iterations, the training accuracy was 92.6% and test was 90.0%, computation time: 7400 sec.

4 Conclusions

The IncNet network is able to control the complexity of its structure by growing and pruning the network. In spite of incremental character of the algorithm, the *pruning time* is determined by theoretical criterion — not in random time moment or by checking the error on the whole training/test data set. Another advantage of the direct pruning is reduction of the time of computation. Nearly all parameters of the network are controlled automatically by EKF algorithm, the other parameters are very similar for different benchmark problems (excluding the biases and slopes, which are defined by the *resolution* of data). Another positive feature of IncNet Pro is the capacity of uniform generalization. In many benchmarks (see section 3) the errors on testing and training sets are much more similar than for other networks.

In some classification problems it would be useful to add the possibility of merging two neurons G_i and G_j which can be replaced by another neuron G_{new}

with a confidence α, for example using the criterion:

$$\int_{d \subseteq \mathcal{D}} |G_i(\mathbf{x}) + G_j(\mathbf{x}) - G_{new}(\mathbf{x})| < \alpha$$

Acknowledgments I'm grateful to prof. W. Duch for many valuable comments and to the Polish Committe for Scientific Research, grant 8T11F00308 for partial support.

References

1. J. V. Candy. *Signal processing: The model based approach.* McGraw-Hill, New York, 1986.
2. W. Duch and G. H. F. Diercksen. Feature space mapping as a universal adaptive system. *Computer Physics Communications*, 87:341–371, 1994.
3. W. Duch and N. Jankowski. New neural transfer functions. *Jour. of Applied Math. and Computer Science.* submitted.
4. S. E. Fahlman and C. Lebiere. The cascade-correlation learning architecture. In D. S. Touretzky, editor, *NIPS.* Morgan Kaufmann, 1990.
5. E. Fiesler. Comparative bibliography of ontogenic neural networks. In *Proceedings of the International Conference on Artificial Neural Networks*, 1994.
6. F. Girosi and T. Poggio. Networks and the best approximation property. AI Lab. Memo, MIT, 1989.
7. B. Hassibi and D. G. Stork. Second order derivatives for network pruning: Optimal brain surgeon. In *NIPS*, 1993.
8. V. Kadirkamanathan. A statistical inference based growth criterion for the RBF network. In *Proc. IEEE. Workshop on Neural Networks for Signal Processing*, 1994.
9. V. Kadirkamanathan and M. Niranjan. A function estimation approach to sequential learning with neural networks. *Neural Computation*, 5(6):954–975, 1993.
10. Y. LeCun, J. Denker, S. Solla, R. E. Howard, and L. D. Jackel. Optimal brain damage. In D. S. Touretzky, editor, *Advances in Neural Information Processing Systems II.* Morgan Kauffman, 1990.
11. J. Platt. A resource-allocating network for function interpolation. *Neural Computation*, 3:213–225, 1991.
12. T. Poggio and F. Girosi. Network for approximation and learning. *Proc. IEEE*, 78:1481–1497, 1990.
13. M. J. D. Powell. Radial basis functions for multivariable interpolation: A review. In J. C. Mason and M. G. Cox, editors, *Algorithms for Approximation of Functions and Data*, pages 143–167. Oxford University Press, 1987.
14. Y. Shang and W. Wah. Global optimization for neural network training. *IEEE Computer*, 29, 1996.
15. B. Šter and A. Dobnikar. Neural networks in medical diagnosis: Comparison with other methods. In A. B. B. et al., editor, *Proceedings of the International Conference EANN '96*, pages 427–430, 1996.
16. A. S. Weigend, D. E. Rumelhart, and B. A. Huberman. Back–propagation, weight elimination and time series prediction. In *Proceedings of the 1990 Connectionist Models Summer School*, pages 65–80. Morgan Kaufmann, 1990.
17. L. Yingwei, N. Sundararajan, and P. Saratchandran. A sequential learning scheme for function approximation using minimal radial basis function neural networks. *Neural Computations*, 9:461–478, 1997.

Improving RBF Networks by the Feature Selection Approach EUBAFES

M. Scherf[1], W. Brauer[2]

[1] GSF - National Research Center for Environment and Health, medis Institute
D-85754 Neuherberg, e-mail: scherf@gsf.de
[2] Institut für Informatik
Technische Universität München, D-80290 München

Abstract. The curse of dimensionality is one of the severest problems concerning the application of RBF networks. The number of RBF nodes and therefore the number of training examples needed grows exponentially with the intrinsic dimensionality of the input space. One way to address this problem is the application of feature selection as a data pre-processing step.
In this paper we propose a two-step approach for the determination of an optimal feature subset: First, all possible feature-subsets are reduced to those with best discrimination properties by the application of the fast and robust filter technique EUBAFES. Secondly we use a wrapper approach to judge, which of the pre-selected feature subsets leads to RBF networks with least complexity and best classification accuracy. Experiments are undertaken to show the improvement for RBF networks by our feature selection approach.

1 Introduction

Neural networks with localised receptive fields (RBF networks) [5] are very popular as classifiers due to their interpretability and fast parameter estimation. Nevertheless a serious disadvantage of RBF networks is the way of addressing the curse of dimensionality. The number of basis functions and the number of training examples which are necessary for accurate parameter estimation increase exponentially with the intrinsic dimensionality of the input space. One way to address this problem is the application of feature selection as a data pre-processing step.
The determination of an optimal feature-subset is based on a search strategy and a criterion function. The search strategy gives candidates of feature subsets while the criterion function decides whether a feature subset is superior to another. Given Q features we have $2^Q - 1$ possible feature subsets thus an exhaustive search is computationally too expensive if Q is large. We address this problem by a feature weighting technique[3], i.e. we assign a continuous weight to every feature and optimise a criterion function with respect to the weights in a

[3] see [7] for an overview of feature weighting approaches

way that the feature weights are either set to one or zero when the optimisation process terminates. Concerning the criterion function, [2] distinguishes wrapper and filter approaches: Wrapper approaches use the accuracy of a classifier to decide the superiority of a feature subset whereas filter approaches use a criterion function which is independent of classifier accuracy. Although the computational cost for training RBF networks is small when compared with multi layer perceptrons, wrapper methods are inapplicable particularly in the presence of many features. On the other hand we have the problem, that filter approaches can not give optimal but only possibly optimal feature subsets. We address this conflict by a procedure which combines filter and wrapper techniques: Applying the fast and robust filter approach EUBAFES, we firstly reduce the number of possible feature subsets to a small number of potentially optimal candidates. Secondly we use the accuracy of an RBF network to pick the optimal feature subset from the pre-selected candidates.

2 The EUBAFES Approach

EUBAFES abbreviates EUclidean BAsed FEature Selection. The approach is motivated by the goal to determine well separated clusters by reinforcing similarities between instances in the same class while deteriorating similarities between instances in different classes.

2.1 Criterion Function

Let $\aleph = \{X^1, X^2, ..., X^N\}$ be a set of N instances, each uniquely labeled to a class. An instance is described by an assignment of Q values to Q features. We will refer to the q'th feature value of the i'th instance by x_q^i. Furthermore let $W = \{w_1, w_2, ..., w_Q | w_i \in \mathbb{R}_0^+\}$ be a set of feature weights. The similarity between instances X^i and X^j is defined as the weighted Euclidean distance:

$$d_{ij}^W = \sqrt{\sum_{q=1}^{Q} w_q \rho(x_q^i, x_q^j)^2} \tag{1}$$

with

$$\rho(x_q^i, x_q^j) = \begin{cases} \mid x_q^i - x_q^j \mid & \text{if feature } q \text{ is continuous} \\ 1 & \text{if feature } q \text{ is discrete and } x_q^i \neq x_q^j \\ 0 & \text{if feature } q \text{ is discrete and } x_q^i = x_q^j \end{cases} \tag{2}$$

According to the demand of reinforcement and deterioration of similarities between instances in the same and in different classes we base the EUBAFES criterion function on the averages of inter- and intraclass-similarities:

$$J = (\sum_{i=1}^{N-1} \sum_{j=i+1}^{N} \delta_{ij} \frac{d_{ij}^W}{N_s} - (1 - \delta_{ij}) \frac{d_{ij}^W}{N_v}), \tag{3}$$

where δ_{ij} is one if instances X^i and X^j are in the same class and zero otherwise, $N_s = \sum_{i=1}^{N-1} \sum_{j=i+1}^{N} \delta_{ij}$ and $N_v = \sum_{i=1}^{N-1} \sum_{j=i+1}^{N} (1 - \delta_{ij})$.
Searching for a local minimum of (3) will tend to reinforce the similarities between *all* instances in the same class, while the similarities between *all* instances in different classes will be deteriorated. Thus feature subsets which cause well separated but multimodal distributed classes would be ignored. We address this problem by introducing a parameter δ_{ij}^{kNN} which is set to one if X^i is one of k nearest neighbours of X^j or vice versa and to zero otherwise and extend (3) to:

$$J = \sum_{i=1}^{N-1} \sum_{j=i+1}^{N} \delta_{ij}^{kNN} (\delta_{ij} \frac{d_{ij}^W}{N_s} - (1 - \delta_{ij}) \frac{d_{ij}^W}{N_v}). \tag{4}$$

In the following sections we will discuss the search for an optimal feature subset.

2.2 Determination of Local Minima with Binary Feature Weights

We apply a gradient descent approach to optimise (4) with respect to the feature weights. To obtain a local minimum with binary feature weights we shrink the searchspace for the feature weights to a hypercube $[0, 1]^Q$, i.e. $w_q \in [0, 1]$ holds for every feature weight. This could either be done by fixing a weight when it reaches a boundary during the optimisation process or by the application of a sigmoid function to the feature weights in (1): $\sigma(w_q) = (1 + exp(-w_q))^{-1}$.
Whether a feature weight will be set to zero or one depends on the sign of its gradient:

$$\frac{\partial J}{\partial w_q} = \sum_{i=1}^{N-1} \sum_{j=i+1}^{N} \delta_{ij}^{kNN} (\delta_{ij} \frac{\rho(x_q^i, x_q^j)^2}{N_s 2 d_{ij}^W} - (1 - \delta_{ij}) \frac{\rho(x_q^i, x_q^j)^2}{N_v 2 d_{ij}^W}). \tag{5}$$

As can be seen from (5), the gradient is based on the relationship between the quadratic distance of features and the Euclidean distance between instances. This is very important because it guarantees that the discrimination property of feature q, i.e. the validity of

$$\sum_{i=1}^{N-1} \sum_{j=i+1}^{N} \delta_{ij}^{kNN} \delta_{ij} \frac{\rho(x_q^i, x_q^j)^2}{N_s} < \sum_{i=1}^{N-1} \sum_{j=i+1}^{N} \delta_{ij}^{kNN} (1 - \delta_{ij}) \frac{\rho(x_q^i, x_q^j)^2}{N_v} \tag{6}$$

is a necessary but not sufficient condition for an increment of w_q if $J < 0$ holds. Consequently EUBAFES will not select all features with discrimination properties according to (6) but a subset of features which together have best discrimination properties.
To obtain a set of potentially optimal feature subsets, we introduce a control-parameter $\tau > 0$ which causes a shift of the square root function arguments:

$$d_{ij}^{\tau,W} = \sqrt{\sum_{q=1}^{Q} w_q \rho(x_q^i, x_q^j)^2 + \tau}. \tag{7}$$

A large τ results in an approximately linear input/output behaviour of the square root function according to weighted sum:

$$\sqrt{\sum_{q=1}^{Q} w_q \rho(x_q^i, x_q^j)^2 + \tau} \approx (\sum_{q=1}^{Q} w_q \rho(x_q^i, x_q^j)^2) * l + t, \tag{8}$$

where l and t are constants. If we replace the square root function in (4) by (8) and calculate the derivative we obtain

$$\frac{\partial J}{\partial w_q} \approx l * (\sum_{i=1}^{N-1} \sum_{j=i+1}^{N} \delta_{ij}^{kNN} (\delta_{ij} \frac{\rho(x_q^i, x_q^j)^2}{N_s} - (1 - \delta_{ij}) \frac{\rho(x_q^i, x_q^j)^2}{N_v})). \tag{9}$$

Thus the gradient will be negative, resulting in an increment of a feature weight, iff (6) holds. Consequently the number of selected features will be the more increased the more τ is increased. Applying EUBAFES with different values for τ will therefore result in different potentially optimal feature subsets. We will give some examples in the next section.

3 Experimental Results

We applied EUBAFES to a number of real world data sets, taken from the UCI Repository of machine learning and used the RBF-DDA neural network approach, introduced by [1] as classifier. The reason for the choice of this approach is its ability to automatically determine the number of RBF units. Thus we are able to compare the complexity of different networks by the number of RBF units. In every experiment we chose $\tau \in]0, 1]$. The number of k-nearest neighbours was set to 100. We initialised the feature weights with $w_q = 0.5$ and did a linear standardisation of the feature values.
We trained the RBF networks with 10-fold crossvalidation except in the case of the vowel data set where a test and training set is donated.

3.1 Vowel Data Set

The problem to solve with this instance base is the speaker independent recognition of eleven steady-state vowels of British English. The instance base is divided in a training set of 528 instances and a test set of 462 instances. Every instance is described by ten continuous valued features. As can be seen from Table 1 the best result is obtained with all features.

3.2 Pima Indians Diabetes Data Set

The Pima Indians Diabetes data set contains 768 instances of eight real-valued features. The underlying task is to decide, whether an at least 21 year old female of Pima Indian heritage shows signs of diabetes according to World Health Organisation criteria. Table 2 shows that best results are obtained with a feature subset which contains not all, but more features than selected with $\tau \approx 0$.

Vowel instance base			
τ	Selected Features	Classification Accuracy	RBF Nodes
$]0.00, 0.02]$	x_1, x_2, x_5, x_6	50%	208
$]0.02, 0.05]$	x_1, x_2, x_5, x_6, x_8	49%	207
$]0.05, 0.07]$	$x_1 - x_6, x_8$	56%	210
$]0.07, 0.08]$	$x_1 - x_8$	57%	208
$]0.08, 0.21]$	$x_1 - x_9$	56%	209
$]0.21, 1.0]$	$x_1 - x_{10}$	**61%**	**210**

Table 1. Results for the vowel training data set.

Pima Indians Diabetes Database				
τ	Selected Features	Sensitivity	Specifity	RBF-Nodes
$]0.00, 0.01]$	$x_1 - x_2, x_5, x_7 - x_8$	82%	50%	530
$]0.01, 0.1]$	$x_1 - x_2, x_5 - x_8$	**86%**	**72%**	**528**
$]0.10, 1.0]$	$x_1 - x_8$	90%	54%	568

Table 2. Results for the Pima Indians Diabetes data set.

3.3 Wisconsin Diagnostic Breast Cancer Data Set

The 30 features of this data set with 569 instances are computed from a digitised image of a fine needle aspirate (FNA) of a breast mass. They describe characteristics of the cell nuclei present in an image. An instance belongs either to the class "benign" or "malign". As can be seen from Table 3 the subset with the least number of features led to best results.

Wisconsin Diagnostic Breast Cancer				
τ	Selected Features	Sensitivity	Specifity	RBF-Nodes
$]0.00, 0.01]$	$x_1, x_3 - x_8, x_{11}, x_{13} - x_{14}, x_{21} - x_{30}$	**96%**	**100%**	**151**
$]0.01, 0.02]$	$x_1, x_3 - x_8, x_{10}, x_{11}, x_{13} - x_{14}, x_{21} - x_{30}$	96%	98%	160
$]0.02, 0.04]$	$x_1 - x_8, x_{10}, x_{11}, x_{13} - x_{14}, x_{21} - x_{30}$	96%	100%	164
$]0.04, 0.06]$	$x_1 - x_{11}, x_{13} - x_{14}, x_{16}, x_{21} - x_{30}$	96%	100%	176
$]0.06, 1.0]$	$x_1 - x_{11}, x_{13} - x_{14}, x_{16}, x_{19}, x_{21} - x_{30}$	96%	100%	184

Table 3. Results for the Wisconsin Diagnosis Breast Cancer data set.

4 Discussion and Related Work

In this paper we introduced the distance-based feature selection technique EUBAFES and showed the improvement of RBF networks by the two-step feature selection approach. An interesting point which underlines our approach is the

fact, that in every example an optimal feature subset was obtained with a different value for parameter τ: In case of the vowel instance base $\tau > 0.21$ led to the selection of every feature which was necessary to obtain best classification accuracy, whereas in case of the wisconsin breast cancer data base $\tau < 0.1$ had to be chosen to obtain a small but optimal feature subset. The reason for this effect is, that distance-based feature selection techniques are not sensitive to class overlap. However due to the flexibility of EUBAFES we are able to address this problem.
We applied a quasi Newton optimisation algorithm to determine a local minimum of the criterion function. To give an example about the computational cost of our method, a Sparc 10 needed about one minute to obtain a feature subset for the vowel training set.
Although the combination of filter and wrapper techniques in context with RBF networks is a new approach, feature selection is a wide and well examined area in statistics and machine learning (see [2] for a detailed overview). A feature selection approach, related to EUBAFES is RELIEF which was originally introduced by [3] and extended to handle noisy, incomplete and multiclass data sets by [4]. RELIEF coincides with EUBAFES regarding the goal to reinforce similarities between instances in the same and deteriorate similarities of instances in different classes. However RELIEF does not give feature subsets but continuous feature weights and uses a different metric and optimisation technique. In [6] we compare both approaches in more detail.

References

1. Berthold M.R., Diamond J., Boosting the Performance of RBF Networks with Dynamic Decay Adjustment, Advances in Neural Information Processing Systems, vol.7 (1995)
2. John G., Kohavi R., Pfleger K., Irrelevant Features and the Subset Selection Problem, Machine Learning: Proceedings of the Eleventh International Conference, (1993) 121-129 William W. Cohne and Haym Hirsh
3. Kira K., Rendell L.: A practical approach to feature selection. Proceedings of the International Conference on Machine Learning, Aberdeen, (1992) 249-256,Sleeman D., Edwards P. Morgan Kaufmann
4. Kononenko I., Estimation attributes: Analysis and extensions of RELIEF in Proceedings of the European Conference on Machine Learning, Catana, Italy (1994) 171-182 Springer Verlag"
5. Moody J. and Darken C.J., Fast learning in networks with locally-tuned processing units Neural Computation, 1, (1989) 281-294
6. Scherf M., Brauer W., Feature Selection by Means of a Feature Weighting Approach. Technical Report No. FKI-221-97, Forschungsberichte künstliche Intelligenz, Institut für Informatik, Technische Universität München (1997)
7. Wettschereck D., Aha D.W., Mohori T., A Review and Empirical Evaluation of Feature Weighting Methods for a Class of Lazy Learning Algorithms. Artificial Intelligence Review (to appear) (1997)

Polynomial Classifiers and Support Vector Machines

Ingo Graf Ulrich Kreßel Jürgen Franke

Daimler–Benz AG, Research and Technology
P.O. Box 2360, 89013 Ulm, Germany
email: Kressel@DBAG.Ulm.DaimlerBenz.Com

Abstract. Polynomial support vector machines have shown a competitive performance for the problem of handwritten digit recognition. However, there is a large gap in performance vs. computing resources between the linear and the quadratic approach. By computing the complete quadratic classifier out of the quadratic support vector machine, a pivot point is found to trade between performance and effort. Different selection strategies are presented to reduce the complete quadratic classifier, which lower the required computing and memory resources by a factor of more than ten without affecting the generalization performance.

1 Introduction

The *support vector machine* is a new and promising approach [1] to adapt classifiers, such as linear and polynomial classifiers, multilayer perceptrons, and radial–basis–functions. Taking into account the finiteness of a given learning set, the support vector machine minimizes the so called *structural risk* in order to reach robust generalization for unseen samples. Since our research group has a long experience in applying *polynomial classifiers* [2], which are usually adapted by pseudo inversion minimizing the *empirical risk* (least mean squares), we are interested in comparing and eventually combining these both approaches.

The *classification of handwritten digits* is used as a test bed for our research activities, which also has a high practical relevance [3]. The learning and test set [4] consists of 10000 digits each (1000 samples per class). The digits are normalized in size (16×16 pixel) and quantized with 8 bits (values 0...255). Some samples are shown in figure 1:

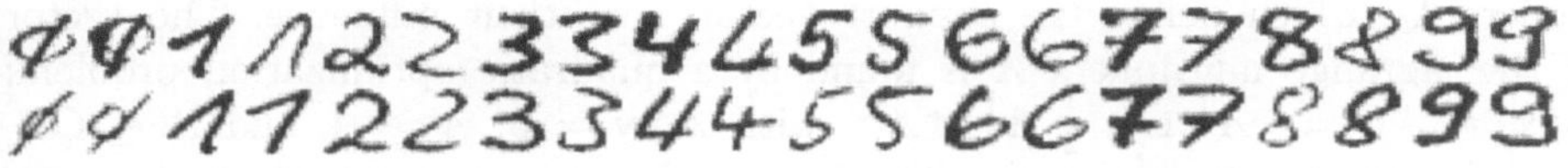

Fig. 1. Samples (size normalized, gray values) of handwritten digits.

The plan of the paper is as follows: In chapter 2 the *polynomial support vector machine* is defined and classification results and computing efforts are given for different polynomial degrees. In chapter 3 it is shown first, how a *complete quadratic classifier* can be calculated out of the quadratic support vector machine. Different *selection strategies* for the complete quadratic classifier are presented next, which reduce the computing effort significantly without affecting the generalization performance. Finally, convincing numerical results are given for the digit classification.

2 Polynomial Support Vector Machines

2.1 Definition

The *support vector machine* (SVM) is a new approach [1] to find the optimum separating hyperplane between two classes (labels are coded by $y_l = +/-1$), taking into account both the minimization of the empirical risk for the given learning samples $\mathbf{x}_l$ and the confidence interval of the adapted classifier for unknown test samples — this comprehensive concept is known as *structural risk minimization*. The name support vector machine is motivated by the fact, that the optimum separating plane turns out to be the weighted superposition of selected samples of the learning set: $\mathbf{w}_{opt} = \sum_l \lambda_l \, y_l \, \mathbf{x}_l$.

By now there exist several extensions, taking care of overlapping classes as well as allowing nonlinear decisions functions [5]. In this paper we are interested in the *polynomial* enhancement of the SVM applied to K eventually overlapping classes, which leads to the following decision functions $\boldsymbol{d}_k$ for each class k:

$$
\begin{aligned}
\boldsymbol{d}_{\mathrm{PSVM}}(\mathbf{x}) &= \mathbf{w}_{opt}^{\mathrm{T}} \, \boldsymbol{\Phi}(\mathbf{x}) + b_{opt} = \\
&= \sum_{\lambda_l > 0} \lambda_l \; y_l \; \boldsymbol{\Phi}(\mathbf{x}_l)^{\mathrm{T}} \, \boldsymbol{\Phi}(\mathbf{x}) + b_{opt} = \\
&= \sum_{\lambda_l > 0} \lambda_l \; y_l \; \boldsymbol{K}_{poly}(\mathbf{x}_l, \mathbf{x}) + b_{opt} = \\
&= \sum_{\lambda_l > 0} \lambda_l \; y_l \; \left(\mathbf{x}_l^{\mathrm{T}} \mathbf{x} + 1\right)^d + b_{opt} \, .
\end{aligned}
\tag{1}
$$

The nonlinear transformation $\boldsymbol{\Phi}(\mathbf{x})$ can either be applied explicitly to the feature vector $\mathbf{x}$ or be included in the computation of the decision function $\boldsymbol{d}_k$ according to *Hilbert–Schmidt theory* and *Mercer's theorem* [6]. For polynomial extensions it is easy to see, that both approaches lead to identical classifier structures. The same idea works also for the computation of the lagrange multipliers λ_l, which define the support vectors and their influence. The vector of the lagrange multipliers $\boldsymbol{\Lambda}$ is found by a quadratic optimization problem: $\max\ F(\boldsymbol{\Lambda}) = \boldsymbol{\Lambda}^{\mathrm{T}}\mathbf{1} - \frac{1}{2}\boldsymbol{\Lambda}^{\mathrm{T}}\mathbf{D}\,\boldsymbol{\Lambda}^{\mathrm{T}}$, with $\boldsymbol{\Lambda}^{\mathrm{T}}\mathbf{y} = 0$, $\boldsymbol{\Lambda} \geq \mathbf{0}$, and $\boldsymbol{\Lambda} \leq C\,\mathbf{1}$. All information about the classification problem is concentrated in the symmetric matrix $\mathbf{D} \in I\!R^{L \times L}$, which elements are defined as follows: $D_{ij} = y_i \, y_j \, \boldsymbol{\Phi}(\mathbf{x}_i)^{\mathrm{T}} \boldsymbol{\Phi}(\mathbf{x}_j) = y_i \, y_j \, (\mathbf{x}_i^{\mathrm{T}} \mathbf{x}_j + 1)^d$. The parameter C in the quadratic problem allows for a trade off in the structural risk minimization between the empirical risk and the confidence interval [7].

The decision functions $\boldsymbol{d}_k$ are adapted for each class k, using $y_l = +1$ as class labels for the samples of class k and $y_l = -1$ for all the other samples. The final class decision is made by a *winner–takes–all strategy* over all $\boldsymbol{d}_k$.

2.2 Comparison

For the already introduced problem of classifying handwritten digits, we varied the degree of the *polynomial SVM* from $d = 1$ up to $d = 4$ (see table 1). The quadratic approach turned out to give the best generalization results (1.51 %). For the linear case ($d = 1$) a large difference between the error rates on the learning and test set can be noticed, which can be reduced by decreasing the value C (in table 1 for all results $C = 1000$ was used; the linear approach showed the best generalization performance for $C = 0.1$ with $L_{error} = 3.40\%$ and $T_{error} = 4.66\%$).

degree of polynomial	error rate learn set	test set	support vectors total	different	effort (coefficients)
linear	1.48 %	8.11 %	3463	2790	2560
quadratic	0.00 %	1.51 %	4324	2720	696320
cubic	0.00 %	1.58 %	5122	2827	723712
$d = 4$	0.00 %	1.99 %	6109	3012	771072

Table 1. Comparison of performance and effort of SVMs for different degrees.

Since the decision functions for the $K = 10$ classes are computed independently, some of the support vectors are identical — reducing the number of different support vectors. Table 1 shows, that the number of different support vectors increases only slightly for larger polynomial degrees.

The *classification effort* for the linear case is very low, since the separating hyperplane can be precomputed out of the support vectors. However, for the quadratic approach (and all higher polynomial degrees) the explicit computation of eq. 1 requires an immense number of operations, which is proportional to the number of support vectors times the number of features.

3 Incomplete Quadratic Classifiers

3.1 Computation of the Quadratic Classifier

Especially for high–dimensional feature spaces (where usually also the number of support vectors is large), the computing and memory requirements of the nonlinear SVM are very high. This problem has already triggered some further research [8]. One result is, that the number of necessary support vectors can be limited in the quadratic case to the feature dimension M, i.e. the classification requires at most M^2 operations (for the 10–digit case we still have 655360 coefficients).

Reconsidering eq. 1, however, we can express the decision function $\boldsymbol{d}_{\mathrm{PSVM}}(\mathbf{x})$ also by applying explicitly the nonlinear transformation $\boldsymbol{\Phi}(\mathbf{x})$. In the quadratic case we get for each class the following structure $\boldsymbol{P}_{qua}(\mathbf{x})$, which has only $\binom{M+2}{2}$ free parameters:

$$\begin{aligned}\boldsymbol{P}_{qua}(\mathbf{x}) \;=\; a_0 \;&+\; a_1\,x_1 + a_2\,x_2 + \ldots + a_M\,x_M \;+ \\ & a_{1,1}\,x_1^2 + a_{1,2}\,x_1x_2 + \ldots + a_{2,2}\,x_2^2 + \ldots + a_{M,M}\,x_M^2\,. \end{aligned} \tag{2}$$

By this step we can already halve the necessary resources for the digit classification without loosing any accuracy: $K \cdot \binom{M+2}{2} = 331530$. The coefficients in eq. 2 can be found by computing the decision functions explicitly:

$$\begin{aligned} \boldsymbol{d}_{\mathrm{PSVM}}(\mathbf{x}) &= \sum_{\lambda_l > 0} \lambda_l \; y_l \; \left(\mathbf{x}_l^{\mathrm{T}} \, \mathbf{x} + 1\right)^2 \; + \; b_{opt} \; = \\ &= \mathbf{x}^{\mathrm{T}} \, \mathbf{Q} \, \mathbf{x} \; + \; \mathbf{p}^{\mathrm{T}} \, \mathbf{x} \; + \; c \; = \; \boldsymbol{P}_{qua}(\mathbf{x}) \, . \end{aligned} \tag{3}$$

It should be mentioned, that this reverse approach allows to compute *complete quadratic classifiers*, which so far could not be adapted by pseudo–inversion (i.e. empirical risk minimization), because the memory requirements (matrix $\binom{M+2}{2} \times \binom{M+2}{2}$) were too high [2].

3.2 Reduction by Polynomial Structure Lists

Having computed the complete quadratic classifier, we search for a way to retain as much recognition performance as possible but to reduce the computing and memory requirements significantly, thus closing the large gap (in performance vs. effort) between the linear and the quadratic approach. In order to do so, we try to identify, which (enhanced) features $\boldsymbol{\Phi}(\mathbf{x})$ have low significance for the classification:

$$\mathbf{d}(\mathbf{x}) \; = \; \mathbf{A}^{\mathrm{T}} \; \boldsymbol{\Phi}(\mathbf{x}) \, , \tag{4}$$

where $\quad \mathbf{d}(\mathbf{x}) \; = \; \Big(d_1(\mathbf{x}) \;\, d_2(\mathbf{x}) \;\, \cdots \;\, d_K(\mathbf{x}) \Big)^{\mathrm{T}} ,$

and $\quad \boldsymbol{\Phi}(\mathbf{x}) \; = \; \Big(1 \;\;\; x_1 \, x_2 \cdots x_M \;\;\; x_1^2 \;\, x_1 x_2 \;\, \cdots \;\, x_2^2 \;\, \cdots \;\, x_M^2 \Big)^{\mathrm{T}} .$

The respective coefficients in $\mathbf{A}$ (complete rows in $\mathbf{A}$) are either set to zero or the corresponding elements in the polynomial structure list $\boldsymbol{\Phi}(\mathbf{x})$ are just deleted. It is also recommended, to recompute the reduced coefficient matrix $\mathbf{A}$ by a linear SVM using the reduced polynomial structure list $\boldsymbol{\Phi}_{red}(\mathbf{x})$ and thus to readapt the remaining coefficients optimally to the classification problem.

Since it is not possible, to exhaustively try out all combinations for the polynomial structure list, we try to estimate the influence of each (enhanced) feature on the classification result and thus get a ranking for the features. This ranking is made independently for the quadratic and linear features, because otherwise the value range for the features must be taken into account.

We defined three different *selection strategies* based on heuristic considerations. A score is assigned to each row m in the coefficient matrix $\mathbf{A}$ depending on the

- euclidean norm: $$S_{euclid}(m) \; = \; \sum_k \; a_{mk}^2 \, , \tag{5}$$
- variance: $$S_{variance}(m) \; = \; \sum_k \; \left(a_{mk} \; - \; \bar{a}_m\right)^2 , \tag{6}$$
- min/max difference: $$S_{min/max}(m) \; = \; \max_k a_{mk} \; - \; \min_k a_{mk} \, . \tag{7}$$

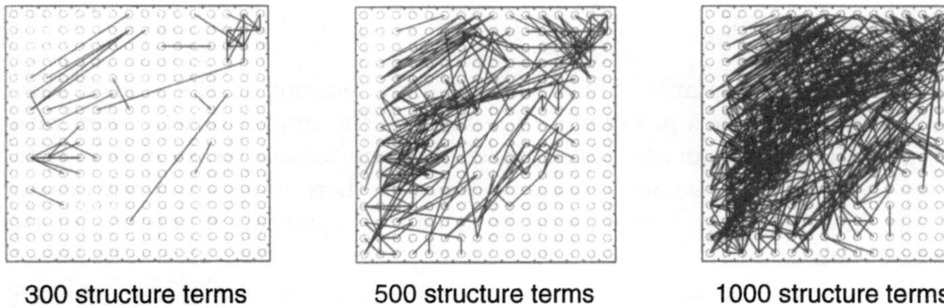

Fig. 2. Visualization of quadratic structure lists for the digit classification.

According to these scores (here only) the quadratic features are ranked and the n–best selected. Figure 2 visualizes the quadratic combinations for different lengths of the polynomial structure list (1 constant value, all 256 linear terms, and 43, 243 resp. 743 quadratic terms) using the min/max–difference strategy.

polynomial structure list (selection strategy)	effort per class (coefficients)	error rate	
		learn set	test set
complete quadratic	33153	0.00 %	1.51 %
min/max difference	500	0.00 %	2.67 %
min/max difference	1000	0.00 %	1.78 %
min/max difference	2500	0.00 %	1.50 %
min/max difference	5000	0.00 %	1.36 %
variance	1000	0.00 %	1.81 %
euclidean norm	1000	0.00 %	1.90 %
four nearest neighbors	1281	0.00 %	2.25 %
randomly	1000	0.00 %	2.54 %

Table 2. Effort and performance for different incomplete quadratic classifiers.

In table 2 the results for the min/max–difference strategy are given for different lengths of the polynomial structure list. A length of 2500 already gives the same generalization performance as the quadratic classifier, but requires less than 1/10 of the computing resources — for $n = 5000$ structure terms the complete quadratic classifier is even outperformed, which, however, might be an artefact of the given problem. The other selection strategies (variance and euclidean norm) had about the same performance.

To cross check our selection strategies, we also choose a systematic approach, which included besides the linear terms (i.e. the pixels directly) the four neighbor combinations of each pixel in the quadratic part of the structure list. Furthermore, we selected the structure terms randomly. Both approaches, the short reaching systematic combinations and also the farer reaching random connections did not reach the performance of the given selection strategies.

4 Conclusion

On the practical example of handwritten digit recognition the *support vector machine* proved to be a powerful (outperforming the empirical risk minimization based on the least mean squares approach) and practicable (reasonable adaption times around one cpu hour) concept. However, there is a gap in performance vs. classification effort between the linear and the quadratic approach for time critical applications.

We succeeded in filling this gap by computing out of the quadratic support vector machine *complete quadratic classifiers*, which so far required for the adaption prohibitively large memory resources. Based on the complete quadratic classifier we presented different *selection strategies*, which reduced for the digit classification the computing and memory resources by more than a factor of ten without loosing generalization performance. We even got a new record (1.36 % errors) on our often used test bed of digit classification [4, 9].

Combining the robust *adaption qualities* of the (linear) support vector machine with known *classification principles* (such as the polynomial structure list [10]) proved to be a strong tool for practical classification problems and opens up a variety of further research in the near future.

References

1. **V. Vapnik**: *The Nature of Statistical Learning Theory*; Springer–Verlag, New York, 1995.
2. **J. Schürmann**: *Pattern Classification: A Unified View of Statistical and Neural Approaches*; John Wiley & Sons, New York, 1996.
3. **R. Wilkinson** et al.: *The First Census Optical Character Recognition Systems Conference*; National Institute of Standards and Technology (NIST), Gaithersburg, 1992.
4. **U. Kreßel**: *The Impact of the Learning–Set Size in Handwritten–Digit Recognition*; p. 1685–1689 in **T. Kohonen** et al. (eds.): *Artificial Neural Networks*; Proceedings of the International Conference on Artificial Neural Networks (ICANN–91), North-Holland, Amsterdam, 1991.
5. **C. Cortes** and **V. Vapnik**: *Support-Vector Networks*; Machine Learning, Nr. 20, p. 273–297, 1995.
6. **R. Courant** and **D. Hilbert**: *Methods of Mathematical Physics*; J. Wiley, New York, 1953.
7. **I. Graf**: *Adaption von Klassifikatoren mit Support–Vektoren*; Master Thesis, Universität Ulm, Abteilung Informationstechnik, Ulm, 1997.
8. **C. Burges**: *Simplified Support Vector Decision Rules*; Proceedings of the 13th International Conference on Machine Learning, p. 71–77, Bari, 1996.
9. **U. Kreßel** and **J. Schürmann**: *Pattern Classification Techniques Based on Function Approximation*; chapter 2 in [11].
10. **J. Franke**: *Isolated Handprinted Digit Recognition*; chapter 4 in [11].
11. **P. Wang** and **H. Bunke** (eds.): *Handbook on Optical Character Recognition and Document Image Analysis*; World Scientific Publishing Company, 1997.

Unique Representations of Dynamical Systems Produced by Recurrent Neural Networks

Masahiro KIMURA and Ryohei NAKANO

NTT Communication Science Laboratories
2-2 Hikaridai, Seika-cho, Kyoto 619-02, Japan

Abstract. This paper considers learning a dynamical system (DS) by a recurrent neural network (RNN). We propose an affine neural dynamical system (A-NDS) as a DS that an RNN actually produces on the output space to approximate a target DS. We present a unique parametric representation of A-NDSs using RNNs and affine sections with the aim of constructing effective learning algorithms.

1 Introduction

Recurrent neural networks (RNNs) are expected to have the capability to model and control various dynamical behaviors and time-series, for example, speech and motion of a physical or biological system [2]. In this paper, we deal with the problem of learning dynamical systems (DSs) by RNNs. Here a DS on $\mathbb{R}^n$ means a flow on $\mathbb{R}^n$ [3, 6]; this is the same notion as a complete vector field on $\mathbb{R}^n$ [3]. As for the learning of a DS by an RNN, we require that the learned RNN must actually produce a DS that approximates the target DS.

An RNN consists of visible units and hidden units. In the problem of learning DSs, all of the visible units are output units as well as input units, and each orbit of a DS to be learned is approximated by an output orbit that the RNN generates on the output space (the state space of the output units). Although supervised gradient descent learning algorithms have been proposed for making an RNN learn several orbits of an arbitrary DS [1], such RNNs have not necessarily learned the target DS. In particular, some could not produce any DSs on the ouput space.

In our previous paper [6], we proposed *neural dynamical systems* (NDSs) as an important class of DSs that RNNs can produce on their output spaces. We also constructed concrete examples of NDSs, and pointed out that these NDSs have relevance to a phase-space learning framework [8] of DSs by RNNs. We call those NDSs *affine neural dynamical systems* (A-NDSs). In Section 2, the definitions of NDSs and A-NDSs are reviewed, and a finite approximation method of an arbitrary DS by an A-NDS is presented.

Accordingly, when the learning of a DS by an RNN is considered, we propose an A-NDS as a DS that the RNN actually produces on the output space to approximate the target DS. An A-NDS on $\mathbb{R}^n$ is represented by a suitable pair of an RNN with n output units and an affine section (see Section 2). However, the representation is not necessarily unique. From the point of view of constructing effective learning algorithms of a DS by an RNN, it is significant to investigate a unique parametric representation of A-NDSs on $\mathbb{R}^n$ using RNNs with n output units and their affine sections, that is, to investigate a nonredundant search set for the learning of DSs by RNNs using A-NDSs. In Section 3, we mathematically construct such a unique parametric representation of A-NDSs by extending Sussmann's work [7] on feedforward neural networks (FNNs).

2 Neural Dynamical Systems

2.1 Recurrent Neural Networks

In this paper, we consider the following widely-used RNNs [2] with n output units; that is, our RNN consists of $n+r$ units (the units 1 to n are visible and the units $n+1$ to $n+r$ are hidden), and the activation state $u_i(t)$ of unit i at time t is governed by a system of ordinary differential equations:

$$\frac{du_i}{dt}(t) = -\frac{1}{\tau}u_i(t) + \sum_{j=1}^{n+r} W_{ij}\tanh(u_j(t)) + W_{i0}, \qquad (i = 1, \cdots, n+r),$$

where $\overline{W} = (W_{ij})_{1\le i\le n+r,\,0\le j\le n+r} \in M_{n+r,n+r+1}(\mathbb{R})$ and τ is a fixed positive constant. Here we denote by $M_{p,q}(\mathbb{R})$ the set of all $p \times q$ real matrices. The adjustable parameters for learning are the number $r \in \mathbb{N}$ of hidden units and the matrix $\overline{W} \in M_{n+r,n+r+1}(\mathbb{R})$ of weights and biases. Let $N_r(\overline{W})$ denote the RNN corresponding to $r \in \mathbb{N}$ and $\overline{W} \in M_{n+r,n+r+1}(\mathbb{R})$, and let $\mathcal{N}(n;r)$ denote the set of all RNNs with n output units and r hidden units. Note that $\mathcal{N}(n;r)$ is identified with $M_{n+r,n+r+1}(\mathbb{R})$.

2.2 Affine Neural Dynamical Systems

We recall the definitions of NDSs and A-NDSs on $\mathbb{R}^n$ [6].

Let $C^\infty(\mathbb{R}^n;\mathbb{R}^r)$ denote the set of all C^∞-maps from $\mathbb{R}^n$ to $\mathbb{R}^r$. For $\bar{\mu} = (N_r(\overline{W}), h) \in \mathcal{N}(n;r) \times C^\infty(\mathbb{R}^n;\mathbb{R}^r)$, we define the C^∞-map $\varphi_{\bar{\mu}} : \mathbb{R} \times \mathbb{R}^n \to \mathbb{R}^n$ by

$$\varphi^t_{\bar{\mu}}(x) = \pi\left(\Phi^t_{\overline{W}}(x, h(x))\right), \qquad (t \in \mathbb{R},\ x \in \mathbb{R}^n),$$

where $\pi : \mathbb{R}^n \times \mathbb{R}^r \to \mathbb{R}^n$ is the natural projection and $\Phi_{\overline{W}} : \mathbb{R} \times \mathbb{R}^{n+r} \to \mathbb{R}^{n+r}$ is the DS that the RNN $N_r(\overline{W})$ produces in the state space $\mathbb{R}^{n+r}$ [5]. The C^∞-map $\varphi_{\bar{\mu}}$ is called the *neural dynamical system* (NDS) produced by the RNN $N_r(\overline{W})$ under section h if it becomes a DS on $\mathbb{R}^n$.

Note that all elements of $\mathcal{N}(n;r) \times C^\infty(\mathbb{R}^n;\mathbb{R}^r)$ cannot necessarily define NDSs on $\mathbb{R}^n$. As a subset of $\mathcal{N}(n;r) \times C^\infty(\mathbb{R}^n;\mathbb{R}^r)$ consisting of elements $\bar{\mu}$ that define NDSs $\varphi_{\bar{\mu}}$ on $\mathbb{R}^n$, we have concretely constructed the set $\mathcal{M}^n_r \subset \mathcal{N}(n;r) \times A(\mathbb{R}^n;\mathbb{R}^r)$ [6]. Here, $A(\mathbb{R}^n;\mathbb{R}^r)$ is the set of all affine maps from $\mathbb{R}^n$ to $\mathbb{R}^r$, and is identified with $M_{r,n+1}(\mathbb{R})$ in the following way: $A(\mathbb{R}^n;\mathbb{R}^r) \ni h$

$\leftrightarrow H = (H_{kj}) \in M_{r,n+1}(\mathbb{R})$; $h_k(x) = \sum_{j=1}^{n} H_{kj}x_j + H_{k0}$, $(k = 1, \cdots, r)$, where $h(x) = (h_1(x), \cdots, h_r(x))$ and $x = (x_1, \cdots, x_n) \in \mathbb{R}^n$. We refer to each element of $A(\mathbb{R}^n; \mathbb{R}^r)$ as an *affine section* of an RNN belonging to $\mathcal{N}(n; r)$. For $\mu \in \mathcal{M}_r^n$, the NDS φ_μ is called the *affine neural dynamical system* (A-NDS) on $\mathbb{R}^n$. For an A-NDS φ on $\mathbb{R}^n$, we call $\mu \in \mathcal{M}_r^n$ a *representation* of the A-NDS φ if $\varphi = \varphi_\mu$. Note that an A-NDS on $\mathbb{R}^n$ is represented by a suitable pair of an RNN with n output units and an affine section.

Using the identification of $\mathcal{N}(n; r) \times A(\mathbb{R}^n; \mathbb{R}^r)$ and $M_{n+r,n+r+1}(\mathbb{R}) \times M_{r,n+1}(\mathbb{R})$, the set $\mathcal{M}_r^n$ is identified with the following set [6]:

$$\{ ((W_{ij}), (H_{kj})) \in M_{n+r,n+r+1}(\mathbb{R}) \times M_{r,n+1}(\mathbb{R})\,;\, W_{n+k\,j} = \textstyle\sum_{i=1}^{n} H_{ki}W_{ij},\\ W_{n+k\,0} = \textstyle\sum_{i=1}^{n} H_{ki}W_{i0} + (H_{k0}/\tau), (j = 1, \cdots, n + r;\ k = 1, \cdots, r) \}.$$

Accordingly, we moreover identify $\mathcal{M}_r^n$ with $M_{n,n+r+1}(\mathbb{R}) \times M_{r,n+1}(\mathbb{R})$ in the following way: $M_{n+r,n+r+1}(\mathbb{R}) \times M_{r,n+1}(\mathbb{R}) \supset \mathcal{M}_r^n \ni (\overline{W}, H) \leftrightarrow (W, H) \in M_{n,n+r+1}(\mathbb{R}) \times M_{r,n+1}(\mathbb{R})$, where W is the submatrix of $\overline{W}$ consisting of the row 1 to the row n of $\overline{W}$.

Let $\mu = (W, H) \in \mathcal{M}_r^n$. Let $\overline{W} \in M_{n+r,n+r+1}(\mathbb{R})$ such that $(\overline{W}, H) \leftrightarrow (W, H)$, and let $h \in A(\mathbb{R}^n; \mathbb{R}^r)$ such that $h \leftrightarrow H$. Notice that μ indicates the pair of the RNN $N_r(\overline{W})$ and the affine section h. Let V_μ be the vector field on $\mathbb{R}^n$ corresponding to the A-NDS φ_μ on $\mathbb{R}^n$. Then we know [6] that

$$\begin{aligned}(V_\mu)_i(x) = &-(1/\tau)x_i + \textstyle\sum_{j=1}^{n} W_{ij}\tanh(x_j)\\ &+ \textstyle\sum_{k=1}^{r} W_{i\,n+k}\tanh(\sum_{j=1}^{n} H_{kj}x_j + H_{k0}) + W_{i0},\ (i = 1, \cdots, n),\end{aligned} \tag{1}$$

where $V_\mu(x) = ((V_\mu)_1(x), \cdots, (V_\mu)_n(x))$ and $x = (x_1, \cdots, x_n) \in \mathbb{R}^n$. Note that the C^∞-map $V_\mu(x)$ is viewed as a three-layer FNN.

2.3 Approximation of Dynamical Systems

A-NDSs are DSs that can actually be produced by RNNs on their output spaces. Below, we give a finite approximation method of an arbitrary DS by an A-NDS, which embodies a phase-space learning framework [8].

Let ψ be a DS on $\mathbb{R}^n$ and X the corresponding vector field on $\mathbb{R}^n$. Given a positive constant ε, a bounded closed interval J containing $0 \in \mathbb{R}$, and a compact subset K of $\mathbb{R}^n$, our purpose is to construct an A-NDS φ_μ, $(\mu \in \mathcal{M}_r^n)$ such that

$$||\psi^t(x) - {\varphi_\mu}^t(x)|| < \varepsilon, \quad (t \in J, x \in K), \tag{2}$$

where $||\cdot||$ is the Euclidean norm on $\mathbb{R}^n$.

We fix $y \in K$. For $\rho > 0$, let B_ρ denote the n-dimensional closed ball with center y and radius ρ. We choose $R > 0$ such that $\psi(J \times K) \subset B_R$, and choose a Lipshitz constant L of the map $X|_{B_{R+\varepsilon}} : B_{R+\varepsilon} \to \mathbb{R}^n$. We also choose $\delta > 0$ such that $\delta < L\varepsilon/(e^{LT} - 1)$, where $T = \max\{|t|; t \in J\}$. By equation (1) and the universal approximation capability of three-layer FNNs [4], we can find $r \in \mathbb{N}$ and $\mu \in \mathcal{M}_r^n$ such that $||X(x) - V_\mu(x)|| \leq \delta$ for all $x \in B_{R+\varepsilon}$, where V_μ is the vector field on $\mathbb{R}^n$ corresponding to the A-NDS φ_μ on $\mathbb{R}^n$. This implies that the A-NDS φ_μ satisfies equation (2) [3]. Note that such a V_μ can be computed by using those methods [2] developed for the learning of FNNs.

3 Representations of Affine Neural Dynamical Systems

As seen in the previous section, A-NDSs constitute an important class of DSs for the learning of DSs by RNNs. In this section, we investigate a unique parametric representation of A-NDSs by extending the concepts of "symmetry" and "irreducibility" in Sussmann's paper [7] for representations of A-NDSs and by proving a similar result to Sussmann's result [7].

3.1 Symmetric Transformations

The set $\mathcal{M}_r^n$ constitutes the representations of A-NDSs on $\mathbb{R}^n$ using RNNs with r hidden units. First, we define some symmetric transformations of $\mathcal{M}_r^n$, where we say that a bijective transformation g of $\mathcal{M}_r^n$ is *symmetic* if for each $\mu \in \mathcal{M}_r^n$, μ and $g(\mu)$ represent the same A-NDS.

We define the bijective transformation θ_ℓ, $(\ell = 1, \cdots, r)$ of $\mathcal{M}_r^n$ as follows: For $\mu = (W, H) \in \mathcal{M}_r^n$, $\theta_\ell(\mu) = (W', H')$;

$$W'_{ij} = W_{ij},\ W'_{i\,n+k} = \begin{cases} -W_{i\,n+k}, & (k = \ell) \\ W_{i\,n+k}, & (k \neq \ell) \end{cases},\ H'_{kj} = \begin{cases} -H_{kj}, & (k = \ell) \\ H_{kj}, & (k \neq \ell) \end{cases},$$

$(i = 1, \cdots, n; j = 0, 1, \cdots, n; k = 1, \cdots, r)$. Note that the transformation θ_ℓ means changing the signs of all the weights associated with the hidden unit $n + \ell$. It is easily seen that θ_ℓ is symmetric.

Let S_r be the set of all permutations on the set $\{1, \cdots, r\}$. For $\sigma \in S_r$, we define a bijective transformation $\tilde{\sigma}$ of $\mathcal{M}_r^n$ as follows: For $\mu = (W, H) \in \mathcal{M}_r^n$, $\tilde{\sigma}(\mu) = (W', H')$;

$$W'_{ij} = W_{ij},\ W'_{i\,n+\sigma(k)} = W_{i\,n+k},\ H'_{\sigma(k)\,j} = H_{kj},$$

$(i = 1, \cdots, n; j = 0, 1, \cdots, n; k = 1, \cdots, r)$. Note that the transformation $\tilde{\sigma}$ means relabeling the hidden units according to the permutation σ. It is easily seen that this transformation is symmetric.

Let $\mathcal{G}_r^n$ be the transformation group of $\mathcal{M}_r^n$ generated by the symmetric transformations θ_ℓ, $(\ell = 1, \cdots, r)$ and $\tilde{\sigma}$, $(\sigma \in S_r)$. Note that $\mathcal{G}_r^n$ is a symmetric transformation group of $\mathcal{M}_r^n$ and a finite group of degree $2^r r!$.

3.2 Irreducible Representations

Next, we consider minimal representations of A-NDSs. Here, $\mu \in \mathcal{M}_r^n$ is called a *minimal representation* of an A-NDS φ on $\mathbb{R}^n$ if $\varphi = \varphi_\mu$ and there does not exist another representation of φ using an RNN with fewer hidden units, i.e., if $\varphi = \varphi_{\mu'}$, $(\mu' \in \mathcal{M}_{r'}^n)$, then $r \leq r'$. In order to express the minimality condition of $\mu \in \mathcal{M}_r^n$ in terms of μ, the concept of irreducibility is proposed. We say that $\mu = (W, H) \in \mathcal{M}_r^n$ is *irreducible* if the following conditions hold:
(A) For $k = 1, \cdots, r$, $(W_{1\,n+k}, \cdots, W_{n\,n+k}) \neq (0, \cdots, 0)$.
(B) For $k = 1, \cdots, r$, $(H_{k1}, \cdots, H_{kn}) \neq (0, \cdots, 0)$.

(C) For $k = 1, \cdots, r$, $(H_{k0}, H_{k1}, \cdots, H_{kn}) \neq (0, 0, \cdots, \overset{i}{\breve{\pm 1}}, 0, \cdots, 0)$, $(i \neq 0)$.
(D) For $k, \ell = 1, \cdots, r$, $k \neq \ell$, $(H_{k0}, H_{k1}, \cdots, H_{kn}) \neq \pm(H_{\ell 0}, H_{\ell 1}, \cdots, H_{\ell n})$.

It is easily seen that if $\mu \in \mathcal{M}_r^n$ is not an irreducible representation of an A-NDS φ, then it is not a minimal representation of φ. Let $\widehat{\mathcal{M}_r^n}$ denote the set of all irreducible representatons in the set $\mathcal{M}_r^n$. Note that $\widehat{\mathcal{M}_r^n}$ is an open subset of $\mathcal{M}_r^n$ ($\simeq \mathbb{R}^{n^2+2nr+n+r}$), and $\mathcal{G}_r^n$ is also a symmetric transformation group of $\widehat{\mathcal{M}_r^n}$ since $g(\widehat{\mathcal{M}_r^n}) = \widehat{\mathcal{M}_r^n}$ for any $g \in \mathcal{G}_r^n$.

We call two functions f_1, f_2 on $\mathbb{R}^n$ *sign-equivalent* if $|f_1(x)| = |f_2(x)|$ for all $x \in \mathbb{R}^n$. For $\mu = (W, H) \in \mathcal{M}_r^n$, let $h_k(x) = \sum_{j=1}^n H_{kj} x_j + H_{k0}$, $(k = 1, \cdots, r)$. Then, the irreducible conditions are described as follows: Condition (A) states that there exists at least one connection from each hidden unit $n + k$ to the output units, condition (B) states that each $h_k(x)$ is not a constant function, condition (C) states that each $h_k(x)$ is not sign-equivalent to the affine functions $x_1, \cdots, x_n$ on $\mathbb{R}^n$, and condition (D) states that no two of $h_1(x), \cdots, h_r(x)$ are sign-equivalent.

3.3 Unique Representations

Now we investigate a unique representation of an A-NDS φ on $\mathbb{R}^n$. First, we prove the following theorem similar to Sussmann's main theorem [7].

Lemma 1 (Sussmann). *Let $f_1, \cdots, f_m$ be nonconstant affine functions on $\mathbb{R}^n$, no two of which are sign-equivalent. Then, the functions* $\tanh \circ f_1, \cdots, \tanh \circ f_m$ *and the constant function* 1 *on $\mathbb{R}^n$ are linearly independent.*

Theorem 2. *Suppose that $\mu = (W, H) \in \widehat{\mathcal{M}_r^n}$ and $\mu' = (W', H') \in \widehat{\mathcal{M}_{r'}^n}$ represent the same A-NDS φ on $\mathbb{R}^n$. Then, $r = r'$, and there exists $g \in \mathcal{G}_r^n$ such that $g(\mu) = \mu'$.*

Sketch of Proof. Let $h_k(x) = \sum_{j=1}^n H_{kj} x_j + H_{k0}$, $(1 \leq k \leq r)$, and $h'_\ell(x) = \sum_{j=1}^n H'_{\ell j} x_j + H'_{\ell 0}$, $(1 \leq \ell \leq r')$. Since μ and μ' are irreducible, and $\varphi_\mu = \varphi_{\mu'}$, we can prove by using Lemma 1 that $r = r'$, $W_{ij} = W'_{ij}$, and there exists a $\sigma \in S_r$ such that $h'_{\sigma(k)}(x) = \rho_k h_k(x)$, and $W'_{i\, n+\sigma(k)} = \rho_k W_{i\, n+k}$, where $\rho_k = 1$ or -1, and $1 \leq i \leq n$, $0 \leq j \leq n$, $1 \leq k \leq r$. We define $g \in \mathcal{G}_r^n$ by $g = {\theta_1}^{(1-\rho_1)/2} \circ \cdots \circ {\theta_r}^{(1-\rho_r)/2} \circ \tilde{\sigma}$. Then, it is easily seen that $g(\mu) = \mu'$. □

Notice that it is much easier to decide that a representation of an A-NDS is irreducible than to decide that it is minimal. We can prove the following corollary from Theorem 2 and the fact that a minimal representation is irreducible [7].

Corollary 3. *Let $\mu \in \mathcal{M}_r^n$ be a representation of an A-NDS φ on $\mathbb{R}^n$. Then μ is a minimal representation of φ if and only if μ is an irreducible representation.*

Let $\mathcal{D}^n$ denote the set of all A-NDSs on $\mathbb{R}^n$. Below, we mathematically construct a unique parametric representation of $\mathcal{D}^n$, a bijection $\widetilde{\mathcal{F}} : \mathcal{A}^n \to \mathcal{D}^n$, based on the representations of A-NDSs by RNNs and affine sections.

The set $\widehat{\mathcal{M}}^n$ of all irreducible representations of A-NDSs on $\mathbb{R}^n$ is expressed by $\bigcup_{r\in\mathbb{N}} \widehat{\mathcal{M}}^n_r$, (disjoint union). Let $\mathcal{F} : \widehat{\mathcal{M}}^n \to \mathcal{D}^n$ be the map indicating the irreducible representations of A-NDSs using RNNs and affine sections; i.e., $\mathcal{F}(\mu) = \varphi_\mu$, $(\mu \in \widehat{\mathcal{M}}^n)$. Corollary 3 implies that the map $\mathcal{F}$ is surjective.

Since the group $\mathcal{G}^n_r$ is a symmetric transformation group of the space $\widehat{\mathcal{M}}^n_r$, the map $\mathcal{F}|_{\widehat{\mathcal{M}}^n_r} : \widehat{\mathcal{M}}^n_r \to \mathcal{D}^n$ induces the map $\widetilde{\mathcal{F}}_r : \widehat{\mathcal{M}}^n_r/\mathcal{G}^n_r \to \mathcal{D}^n$ such that $\mathcal{F}|_{\widehat{\mathcal{M}}^n_r} = \widetilde{\mathcal{F}}_r \circ \pi_r$, where $\widehat{\mathcal{M}}^n_r/\mathcal{G}^n_r$ is the orbit space of the $\mathcal{G}^n_r$-space $\widehat{\mathcal{M}}^n_r$ and π_r: $\widehat{\mathcal{M}}^n_r \to \widehat{\mathcal{M}}^n_r/\mathcal{G}^n_r$ is the natural projection. Theorem 2 implies that the map $\widetilde{\mathcal{F}}_r$ is injective.

We put $\mathcal{A}^n = \bigcup_{r\in\mathbb{N}}(\widehat{\mathcal{M}}^n_r/\mathcal{G}^n_r)$, (disjoint union), and define the map $\widetilde{\mathcal{F}} : \mathcal{A}^n \to \mathcal{D}^n$ by $\widetilde{\mathcal{F}}(\pi_r(\mu)) = \widetilde{\mathcal{F}}_r(\pi_r(\mu))$, $(\mu \in \widehat{\mathcal{M}}^n_r)$. Then, it turns out that the map $\widetilde{\mathcal{F}}$ is a bijection such that $\widetilde{\mathcal{F}}(\pi_r(\mu)) = \varphi_\mu$, $(r \in \mathbb{N}, \mu \in \widehat{\mathcal{M}}^n_r)$. Hence, we have obtained a unique parametric representation of A-NDSs on $\mathbb{R}^n$ based on the representations by RNNs with n output units and their affine sections.

4 Conclusion

We have proposed an A-NDS as a DS that an RNN produces actually to approximate a DS to be learned. For the representations of A-NDSs, we have proposed symmetric transformations and irreducible representations. It is computable to decide whether a representation is irreducible, and the symmetric transformation groups are computable finite groups. Using these concepts, we have constructed a unique parametric representation of A-NDSs by RNNs and affine sections, that is, a nonredundant search set for the learning of DSs by RNNs using A-NDSs.

References

1. Baldi, P., Gradient descent learning algorithm overview: A general dynamical systems perspective, *IEEE Transactions on Neural Networks* **6** (1995), 182–195.
2. Hertz, J., Krogh, A., and Palmer, R. G., *Introduction to the Theory of Neural Computation*. Addison Wesley, 1991.
3. Hirsh, M. W., and Smale, S., *Differential Equations, Dynamical Systems and Linear Algebra*. Academic Press, 1974.
4. Hornik, K., Stinchcombe, M., and White, H., Multilayer feedforward networks are universal approximators, *Neural Networks* **2** (1989), 359–366.
5. Kimura, M., and Nakano, R., Learning dynamical systems from trajectories by continuous time recurrent neural networks, *Proceedings of 1995 IEEE International Conference on Neural Networks* **6** (1995), 2992–2997.
6. Kimura, M., and Nakano, R., Learning dynamical systems produced by recurrent neural networks, *Proceedings of 1996 International Conference on Artificial Neural Networks* (1996), 133–138.
7. Sussmann, H. J., Uniqueness of the weights for minimal feedforward nets with a given input-output map, *Neural Networks* **5** (1992), 589–593.
8. Tsung, F-S., and Cottrell, G. W., Phase-space learning, *Advances in Neural Information Processing Systems* **7** (1995), 481–488.

Generalization of Elman Networks

Barbara Hammer

University of Osnabrück, Albrechtstr. 28, D-49069 Osnabrück, Germany

Abstract. The Vapnik Chervonenkis dimension of Elman networks is infinite. Here, we find constructions leading to lower bounds for the fat shattering dimension that are linear resp. of order $\log^2$ in the input length even in the case of limited weights and inputs. Since finiteness of this magnitude is equivalent to learnability, there is no a priori guarantee for the generalization capability of Elman networks.

1 Introduction

In time series prediction tasks Elman networks compete with simple feedforward networks trained with a moving window technique. Recurrent networks have the advantage that the time context necessary to classify a new event can be handled implicitely. Further, they are in principle more powerful than perceptrons.

There is one advantage of feedforward networks: Since they have finite capacity their worst case generalization error can be limited [9]. Recurrent networks are Turing universal [7]. Therefore it is not surprising that Elman networks with only 2 hidden units have infinite VC dimension as follows from [8, 6].

The same is valid for the fat shattering dimension even if the weights and inputs are restricted. We construct architectures with dimension of order k in the sigmoidal case resp. $\log^2 k$ in the linear case if the input length is at most k. Since finiteness of this dimension is equivalent to learnability there cannot be an a priori guarantee that Elman networks perform correct generalization.

2 Elman Networks

Let $\sigma : \mathbb{R} \to \mathbb{R}$ denote an activation function and $\boldsymbol{\sigma} : \mathbb{R}^n \to \mathbb{R}^n$ the vector function $(x_1 \ldots, x_n) \to (\sigma(x_1), \ldots, \sigma(x_n))$. We will consider three special activation functions: the identity $\mathrm{id}(x) = x$, the perceptron activation $\mathrm{H}(x) = 1$, if $x \geq 0$, $\mathrm{H}(x) = -1$, otherwise, and the sigmoidal function $\mathrm{sgd}(x) = 2/(1+\mathrm{e}^{-x}) - 1$. Let $(\mathbb{R}^m)^*$ denote the set of finite sequences with elements from $\mathbb{R}^m$.

Definition 1. Let $f : \mathbb{R}^{m+n} \to \mathbb{R}^n$ be a function and $x \in \mathbb{R}^n$ a fixed vector. Then the *induced function* $\widetilde{f}_x : (\mathbb{R}^m)^* \to \mathbb{R}^n$ is defined recursively where $(x_1, \ldots, x_t)$ is a sequence of length t and $()$ is the empty sequence:

$$\widetilde{f}_x(()) = x, \quad \widetilde{f}_x((x_1, \ldots, x_t)) = \widetilde{f}_{f(x_1,x)}((x_2, \ldots, x_t))) \ .$$

A function $f : (\mathbb{R}^m)^* \to \mathbb{R}^l$ is computed by an *Elman network* if there exists a natural number n, a context vector $x \in \mathbb{R}^n$, a function $g : \mathbb{R}^{m+n} \to \mathbb{R}^n$

which is the composition of a linear function $\mathbb{R}^{m+n} \to \mathbb{R}^n$ with σ, and a linear function $h : \mathbb{R}^n \to \mathbb{R}^l$, so that $f = h \circ \widetilde{g}_x$.

Note the following technical remarks:
Elman networks as defined above are recurrent neural networks (see e.g. [8]) but with unlimited length of the input sequences. They can be represented as sketched in Fig.1(a): One computation step is implemented using implicit recurrent connections and the actual state is stored in the context units. For time series prediction tasks the inputs often have the special form $(x_1), (x_1, x_2), \ldots$

Elman networks with an additional hidden layers in the recursive part g can be simulated by standard Elman networks: The hidden units are replaced by context units and a delay is introduced (Fig.1(b)). Biases can be simulated by additional weights receiving constant input 1.

We can keep the initial context fixed, e.g. as 0, and simulate the context with an additional weight and input unit. This input is 1 at the beginning and then 0, whereas the other input sequences are extended by an entry 0 at the beginning. Now it is possible to extend any input of length $l \leq k$ to an equivalent input with length exactly k by simply adding $k - l$ zero inputs at the beginning.

By an architecture we mean the set of functions computed by an Elman network with a fixed grouping of neurons but arbitrarily chosen weights.

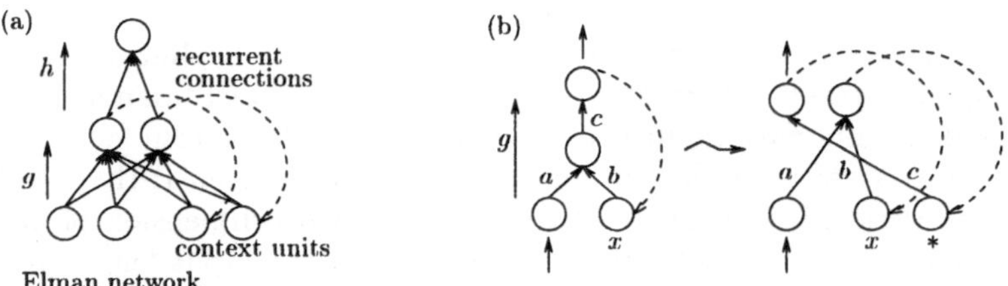

Fig. 1. (a) Standard Elman network; (b) Simulating hidden layers

3 Worst Case Generalization

In a practical learning problem there are given examples $(x_i, f_0(x_i))$ chosen according to a certain probability P of an unknown real function f_0. The task is to find a function f from a set $\mathcal{F}$ so that the risk $E(f) = \int |f(x) - f_0(x)| dP$ becomes minimal. We can e.g. minimize the empirical risk $\hat{E}(f) = \sum_i |f(x_i) - f_0(x_i)|$, but in general $\hat{E}(f)$ underestimates $E(f)$. Assume, $\mathcal{F}$ has finite capacity and the examples are i.i.d. Then there is a theoretical bound for the number of training patterns required so that $\hat{E}(f)$ is representative for $E(f)$ with high probability.

The capacity of a function class mapping to $\{-1, 1\}$ can be measured in terms of the *Vapnik-Chervonenkis dimension* $\mathcal{VC}(\mathcal{F})$. This is the largest size of a set S such that any dichotomy on S can be implemented by a function in $\mathcal{F}$. Finiteness of $\mathcal{VC}(\mathcal{F})$ is equivalent to learnability of $\mathcal{F}$ in the PAC sense.

If the functions map to the entire interval $[-1,1]$, an equivalent condition to learnability is the finiteness of the *fat shattering dimension*: The γ-fat shattering dimension $fat_\gamma(\mathcal{F})$ is the largest size of a set $S=\{x_1,\ldots,x_n\}$ such that reference points r_i and for any dichotomy d on S a function $f\in\mathcal{F}$ exist with $\mathrm{H}(f(x_i)-r_i)=d(x_i)$ and $|f(x_i)-r_i|\geq\gamma$. For $\gamma=0$ the so called *pseudodimension* $\mathcal{PS}(\mathcal{F})$ results. It is $\mathcal{PS}(\mathcal{F})=\mathcal{VC}(\{f_e:X\times\mathbb{R}\to\{-1,1\}\,|\,f_e(x,y)=\mathrm{H}(f(x)-y),f\in\mathcal{F}\})$ and $\mathcal{PS}(\mathcal{F})\geq fat_\gamma(\mathcal{F})$ for all $\gamma>0$. But still the weaker condition $fat_\gamma(\mathcal{F})<\infty$ ensures learnability up to a certain accuracy depending on γ [1].

In contrast to the VC analysis the concept of fat shattering offers e.g. a motivation for weight decay and ensures that boosting techniques can improve generalization whereas the VC dimension is infinite [5, 2].

For a function class parametrized by a set Y the dual class is $\mathcal{F}^\vee=\{f_\cdot(x):Y\to[-1,1],(f_\cdot(x))(y)=f_y(x)\mid x\in X\}$. It is $\mathcal{VC}(\mathcal{F})\geq\log\lfloor\mathcal{VC}(\mathcal{F}^\vee)\rfloor$ [3]. The argumentation in [3] is valid for the uniform pseudodimension and fat shattering dimension, too (i.e. $r_i=r$ for each x_i). This inequality is not valid for the non uniform version of the pseudo- and fat shattering dimension since e.g.

$$F:X\times Y\to[-1,1],\text{ with } X=\{\tfrac{-n+1}{n},\ldots,\tfrac{n-1}{n}\}\text{ and }Y=\{-1,1\}^{2n-1},$$
$$(\tfrac{i}{n},\epsilon_1,\ldots,\epsilon_{n-1})\mapsto(2i+\epsilon_i)/(2n)$$

induces the classes $\mathcal{F}=\{F_y:X\to[-1,1]\mid y\in Y\}$ with $fat_\gamma(\mathcal{F})=2n-1$ and $\mathcal{F}^\vee=\{F_x:Y\to[-1,1]\mid x\in X\}$ with $fat_\gamma(\mathcal{F}^\vee)=1$ for $\gamma<\frac{1}{2n}$.

4 VC Dimension of Elman Networks

Let k denote the maximal length of an input sequence and W the number of weights in an architecture $\mathcal{F}$. Any input of length at most k can be extended to an equivalent input of length exactly k. Therefore the upper bounds for the VC dimension of recurrent networks transfer to this case directly [8]:

$$\mathcal{VC}(\mathcal{F})=\begin{cases}\mathrm{O}(\min\{Wk\log(Wk),W^2+W\log(Wk)\}) & \text{if } \sigma=\mathrm{H}\\ \mathrm{O}(\min\{W\log k,W+k\}) & \text{if } \sigma=\mathrm{id}\\ \mathrm{O}(W^4k^2) & \text{if } \sigma=\mathrm{sgd}\end{cases}$$

Lower bounds follow from [8, 3] resp. Theorem 2:

$$\mathcal{VC}(\mathcal{F})=\begin{cases}\Omega(W\log(Wk)) & \text{if } \sigma=\mathrm{H}\\ \Omega(W\log(k/W)) & \text{if } \sigma=\mathrm{id}\\ \Omega(Wk) & \text{if } \sigma=\mathrm{sgd}\end{cases}$$

Theorem 2. *$\mathcal{VC}(\mathcal{F})=\Omega(W\log(Wk))$ if $\sigma=H$.*

Proof. We write the 2^k different dichotomies of k numbers as rows in a matrix and denote the k different column vectors as $\epsilon_1,\ldots,\epsilon_k$. The k input sequences $y_j=(0,2,4,\ldots,2(2^k-1))+\epsilon_j$ can be shattered by the networks which start with 0 and recursively compute $f_x(y)=(x\vee(2i+\frac{1}{2}\leq y\leq 2i+\frac{3}{2}))$ for $i\leq 2^k-1$.

In W sets of hidden neurons and inputs y_j like before we extend each sequence y_j of set number i to y^i_j by an initial entry w_i. The w_i are W different negative values. The conjunction of $f_x(y) = ((y = w_i) \vee (x = 1))$ with the row extraction computes zero on sets $\neq i$ and else the dichotomy corresponding to the row.

Additionally, according to [9] an Elman architecture with $\mathrm{O}(W)$ weights exists which shatters $W \log W$ input sequences x_j of constant length 3.

We extend the x_j and y^i_j by an extra input which is constantly w resp. $\tilde{w}$ ($w \neq \tilde{w} \in \mathbb{R}$). The conjunction with the test whether this input equals w resp. $\tilde{w}$ is added to the recursive computations shattering the x_j resp. y^i_j. Finally the disjunction of the outputs of these $W+1$ subnetworks is a network with $\mathrm{O}(W)$ weights shattering the x_j and y^i_j. □

Some aspects are worth mentioning:
The network depicted in Fig.2(a) shatters any set of sequences with pairwise different height. Since one can approximate $\mathrm{H}(x)$ by $\mathrm{sgd}(x/\epsilon)$ and $\mathrm{id}(x)$ by $\mathrm{sgd}(\epsilon x)/\epsilon \cdot \mathrm{sgd}'(0)$ $(\epsilon \to 0)$, sigmoidal Elman networks with 3 hidden neurons resp. Jordan networks with 2 hidden neurons shatter a training set that occurs e.g. for inputs with growing context length from only one time series.

The network in Fig.2(b) implements the function $f_a(x) = 1 \Leftrightarrow a$ occurs in x. For inputs of length $\leq k$ it is $\mathcal{VC}(\{f_a \mid a \in \mathbb{R}\}^{\vee}) \geq k$. This shows that the VC dimension of perceptron Elman networks with 4 hidden neurons and recurrent cascade correlation networks [4] with 3 hidden neurons is infinite (Fig.2(c)).

Further any class has infinite VC dimension if it can at least tell whether one of an infinite number of single entries occured in a time series. In the perceptron case a restriction of the number of different possible entries to a finite number r leads to the finite VC dimension $O(W^2 + W \log r)$ [8].

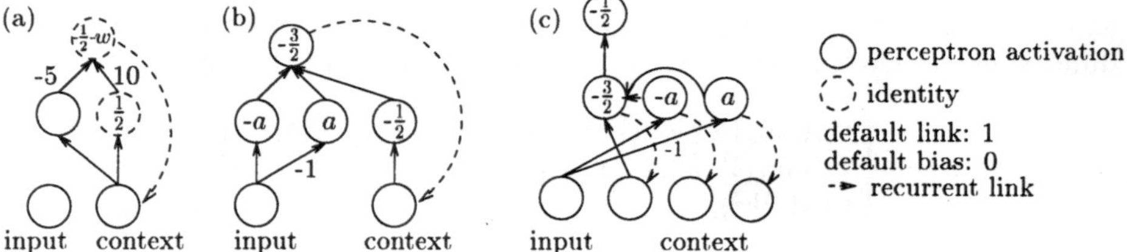

Fig. 2. Networks with infinite VC dimension. (a) Sigmoidal network; (b) Perceptron network; (c) Recurrent cascade correlation network.

5 Fat Shattering Dimension of Elman Networks

Since $fat_\gamma(\mathcal{F}) \leq \mathcal{PS}(\mathcal{F})$, upper bounds of the VC dimension transfer to this case. For unlimited weights the two dimensions coincide. The same is valid for unlimited inputs and linear activation since a division of the weights by s corresponds to a scaling of an input of length k at time i by s^{k-i+1}. Now we assume that the inputs are in $[-1,1]$ and the weights are absolutely bounded by $B \geq 15$. The constructions of [8, 3] leading to the lower bounds for $\mathcal{VC}(\mathcal{F})$ used an approximation process in the sigmoidal case resulting in unlimited weights. In the linear case they used arbitrarily growing inputs. But other architectures with infinite fat shattering dimension can be found. Assume $0 < \gamma \leq 0.37$.

Theorem 3. *The γ-fat shattering dimension of sigmoidal Elman networks with 2 hidden neurons is at least linear in the length of the input sequences.*

Proof. $f(x) : x \to -1 - \text{sgd}(15x - 10) - \text{sgd}(-15x - 10)$ maps each of the intervals $[-0.9, -0.4]$ and $[0.4, 0.9]$ to an interval containing $[-0.9, 0.9]$. Therefore for any sequence $\epsilon_1, \dots, \epsilon_n$ of signs we can find a real value $x \in [-1,1]$ so that $f^i(x) \in [-0.9, -0.4]$ for $\epsilon_i = -$ and $f^i(x) \in [0.4, 0.9]$ for $\epsilon_i = +$. The iterated mapping f can be simulated by an Elman network with 2 hidden neurons and initial context $(-1-x, 0)$ as shown in Fig.3. □

For W weights a lower bound $O(kW)$ can be derived as follows: We simulate W initial contexts by W weights and extended input dimension W. The input sequences are extended by initial entries which are 1 in the place j if the context number j is responsible for this input and 0 in all other places.

Theorem 4. *The γ-fat shattering dimension of Elman networks with linear activation and 2 hidden units is $\Omega(\log^2 k)$ if k denotes the maximal input length.*

Proof. We consider the dual class. The sequence $\sin(i\frac{\pi}{n} + \frac{\pi}{2n})$, $i = 0, 1, \dots$ is the solution of the difference equation

$$x_0 = \sin\left(\frac{\pi}{2n}\right), \; x_1 = \sin\left(\frac{3\pi}{2n}\right), \; x_{i+2} = x_{i+1} - 2\cos\left(\frac{\pi}{n}\right) x_i + 1 \; .$$

The difference equation can be implemented by an Elman network with limited weights, two hidden neurons, and time delay 1. We fix k networks which produce output sequences with alternating n positive and n negative values for $n = 1, 2, 4, 8, \dots, 2^{k-1}$. Any dichotomy of these k networks appears as output at one of the time steps $1, \dots, 2^k$. Therefore these networks can be shattered if one chooses an input of accurate length.

We are interested in the fat shattering dimension. For growing n the value $\sin(\frac{i\pi}{n} + \frac{\pi}{2n})$ becomes small if i equals multiples of n. We requested $|\text{output}| \geq \gamma$. Since $0.37 \leq \sin(\frac{\pi}{8})$, for a sign sequence corresponding to n the first $(2n-8)/16$ values are too small. This holds for the same number of values before and after any sign change, too. For a number k we consider the $3k/2$ networks producing alternating signs for $n = 1, 2, \dots, 2^{3k/2-1}$ as above. We drop the networks $2, 5, \dots, 3k/2 - 1$. In the remaining networks we drop the outputs corresponding to the time steps $1, \dots, 2^{3i-1}/4$ before and after any sign change in network number $3i$ (see Fig.4). In the networks number $1, 4, 7, \dots$ we reverse the signs.

It follows by induction that in the remaining k networks and 2^k not cancelled outputs per sequence still any dichotomy is stored in the outputs at one time step. These outputs have absolute value at least γ. □

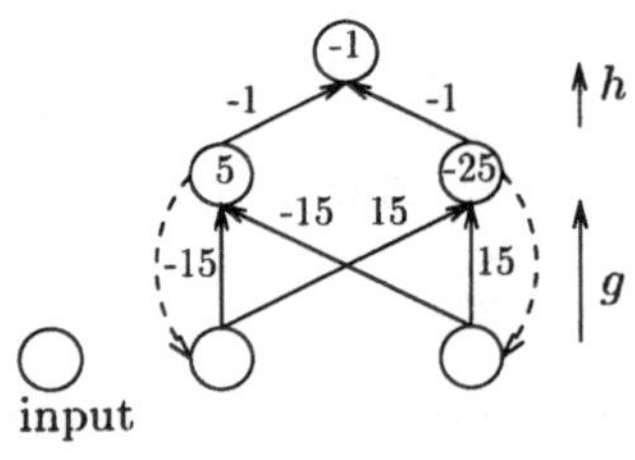

Fig.3. $fat_\gamma(\mathcal{F}) = \infty$

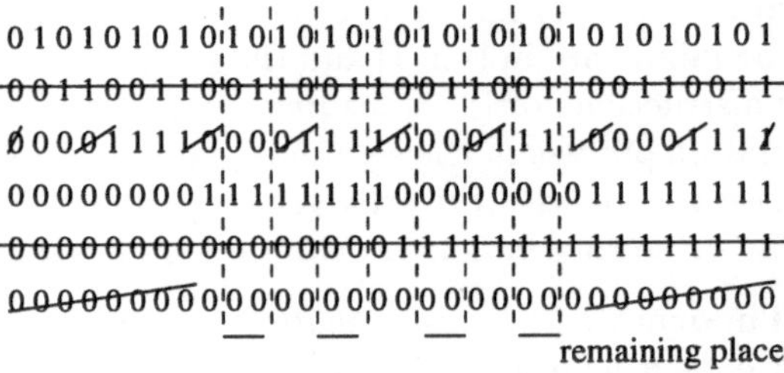

Fig.4. Example for output cancelling

The same argumentation as in the sigmoidal case leads to a bound $\Omega(W \log^2 k)$ if we can use $O(W)$ weights and W additional input neurons.

6 Conclusion

We have shown that Elman networks with restricted weights and inputs and 2 hidden neurons have infinite fat shattering dimension. Since this characterizes learnability, valid generalization cannot be guaranteed unless we possess further information about the learning scenario. The reasons for the infinity are three-fold: The nonlinearity in the sigmoidal case permits chaotic behaviour leading to any desired output; even in the linear case any value can occur at some points of the output. Third, arbitrary information can be encoded in the inputs since the length is not restricted. For example any possible dichotomy can be stored at one time step in the inputs so that the network can easily use this information.

References

1. N. Alon, S. Ben-David, N. Cesa-Bianchi, and D. Haussler. Scale-sensitive dimensions, uniform convergence, and learnability. In *Proc. of 34th IEEE Symp. Foundations Computer Science*, 1993.
2. P. L. Bartlett. The sample complexity of pattern classification with neural networks: the size of the weights is more important than the size of the network. Technical report, Department of Systems engineering, ANU, 1996.
3. B. Dasgupta and E. D. Sontag. Sample complexity for learning recurrent perceptron mappings. *IEEE Transactions Information Theory*, 42, 1996.
4. S. E. Fahlman. The recurrent cascade-correlation architecture. In *Advances in Neural Information Processing Systems*, volume 3, 1991.
5. L. Gurvits and P. Koiran. Approximation and learning of convex superpositions. In *2nd European Conf. Comp. Learning Theorie*, 1995.
6. B. Hammer and V. Sperschneider. Neural networks can approximate mappings on structured objects. In *2nd Int. Conf. Comp. Intelligence and Neuroscience*, 1997.
7. J. Kilian and H. T. Siegelmann. The dynamic universality of sigmoidal neural networks. *Information and Computation*, 128, 1996.
8. P. Koiran and E. D. Sontag. Vapnik-Chervonenkis dimension of recurrent neural networks. In *Proc. of the 3rd European Conf. Comp. Learning Theorie*, 1997.
9. W. Maass. Vapnik-Chervonenkis dimension of neural nets. Technical Report 96-015, NeuroCOLT Technical Report Series, 1996.

Designing Neural Networks by a Combination of Structural Learning and Genetic Algorithms

Masumi Ishikawa and Kazuhiko Nishino

Kyushu Institute of Technology, Iizuka, Fukuoka 820, Japan

Abstract. Kitano proposed to use GA with graph encoding method to have good scalability for hierarchical networks. However, this is not applicable to recurrent networks. The authors propose to use a combination of a structural learning with forgetting(SLF) and GA for designing recurrent neural networks; the former generates quasi-optimal recurrent network structure and the latter prevents local minima by global search. Its applications to two kinds of time series data well demonstrate the superiority to SLF and to a combination of GA and BPTT.

1 Introduction

It is well known that simple structured neural networks are preferable to complex ones from the viewpoint of generalization ability. To this end, various learning algorithms have been proposed, which are roughly classified into destructive learning and constructive learning. Most of them, however, aim at decreasing the computation time and the number of connections and units used in networks.

In contrast to this, the authors claim that to obtain skeletal networks reflecting regularities in data and providing interpretation of hidden units is more important. For this purpose, a structural learning with forgetting(SLF) has been proposed[4], showing its superiority to other methods in terms of the generalization ability, computation time, the structural simplicity of resulting networks, and the ability to extract rules from data[3].

Kitano proposed to use genetic algorithms(GA) with graph encoding method to have good scalability for hierarchical networks[5]. However, resulting network structure is far from optimal, because BP learning is used. Since BP uses all the connections provided, the combination of GA and BP is not effective in the discovery of network structure. Dynamic node creation and elimination are only suggested. Kitano extended this method to accelerate convergence by setting initial connection weights in addition to the graph encoding method [6].

Because the number of possible recurrent network structures is much larger than that of hierarchical ones of the same size, the discovery of recurrent network structure is much more difficult due mainly to local minima. The combination of GA and back propagation through time(BPTT) is not effective due to the same reason as in hierarchical networks. In the present paper the authors propose to design recurrent neural networks by a combination of SLF and GA; SLF generates quasi-optimal recurrent network structure and GA prevents local minima by global search. Since SLF alone has capability to generating structurally simple networks, the combination is expected to have excellent ability.

2 Structural learning with forgetting

The criterion function in the learning with forgetting is,

$$J_f = J + \varepsilon' \sum_{i,j} |w_{ij}| = \sum_k (o_k - t_k)^2 + \varepsilon' \sum_{i,j} |w_{ij}| \tag{1}$$

where the total criterion, J_f, is composed of the mean square error, J, and a regularization term. The regularization term favors small values of connection weights. The weight change, Δw_{ij}, is obtained by differentiating Eq.(1) with respect to the connection weight w_{ij},

$$\Delta w_{ij} = -\eta \frac{\partial J_f}{\partial w_{ij}} = \Delta w'_{ij} - \varepsilon\, sgn(w_{ij}) \tag{2}$$

where $\Delta w'_{ij} (= -\eta \frac{\partial J}{\partial w_{ij}})$ is the weight change due to BP learning, η is a learning rate, $\varepsilon\,(= \eta\varepsilon')$ is the amount of decay at each weight change, and $sgn(w_{ij})$ is a sign function, i.e., 1 when w_{ij} is positive and -1 otherwise.

As shown in Eq.(2), a key idea of SLF is constant decay of connection weights in contrast to ordinary exponential decay. Its extension to recurrent networks, i.e., a combination of SLF and BPTT, is straightforward.

3 Genetic algorithms

In recurrent networks, any connections are allowed in contrast to hierarchical networks. Therefore, graph encoding method is not applicable. In this paper, each chromosome represents a connectivity matrix of a recurrent network. At each generation, the connection weights are modified by a combination of SLF and BPTT during a learning phase. During an evaluation phase, each individual is evaluated by AIC[1] or prediction sum of error squares(PSS). A crossover can entirely change a network structure. To prevent this drastic change, only an exchange of a pair of rows of a connectivity matrix , i.e., incoming connections into a unit, is introduced. Mutation is also introduced, but its occurrence probability is kept small, i.e., 1%, to prevent drastic change of network structure.

The number of individuals at each generation is 20. Candidates of the next generation are the best 10, the best 10 with different initial connection weights, and 10 offsprings generated by crossovers and mutation. From these 30 candidate individuals, best 20 are selected by evaluation using AIC or PSS.

4 Artificial time series data

The external input sequence of length 40 in Figure 1(a) is given to unit 0 of the network in Figure 2(a). The corresponding outputs of visible units, 6 and 7, are used as target output sequences in the learning phase. Figure 1(b) illustrates the output sequence of unit 6. Each individual network has 8 sigmoidal units at

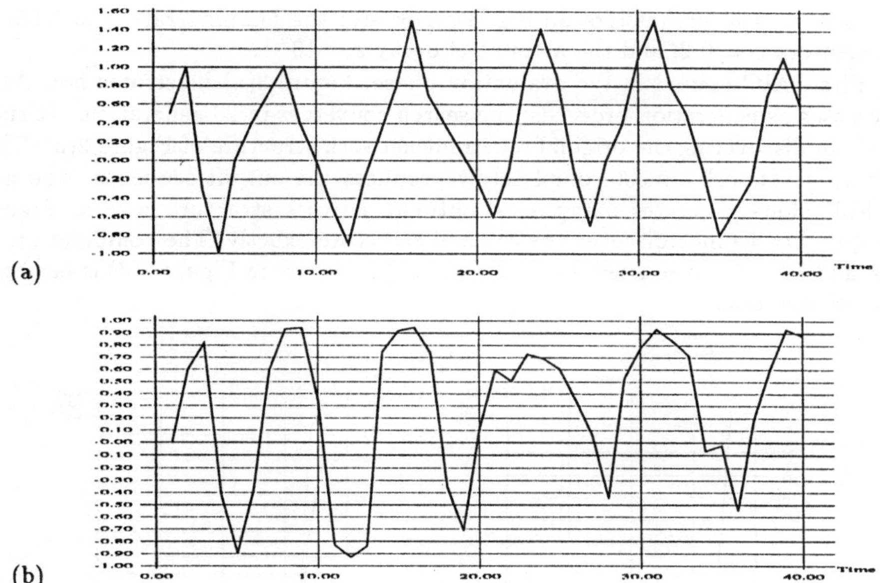

Fig. 1. (a) External input sequence. (b) Output sequence of unit 6.

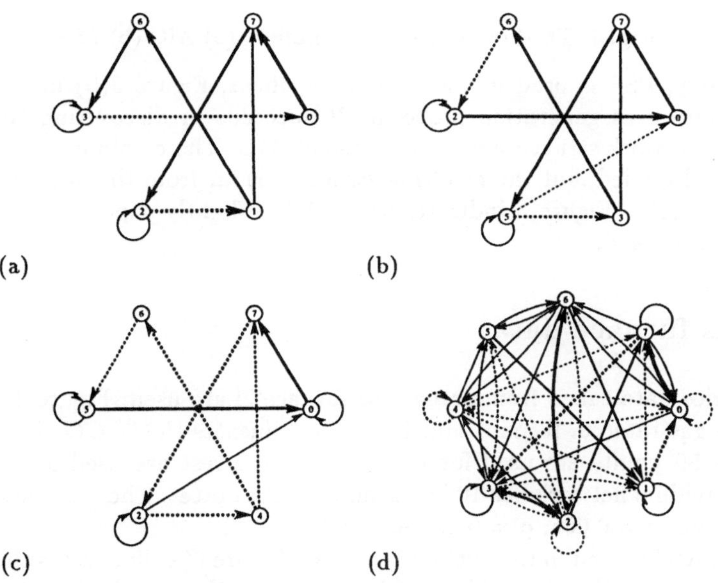

Fig. 2. (a) Original network structure. External input is given to unit 0, and units 6 and 7 are visible, i.e., target outputs are given. (b)(c) Examples of resulting network structure by SLF, (d) An example of network structure by GA and BPTT

the outset. The parameters during learning are: the learning rate $\eta = 0.05$, a momentum $\alpha = 0.2$, and the amount of decay $\varepsilon = 10^{-4}$.

First, AIC is used in the evaluation phase. Figure 3(a) illustrates how AIC decreases as generation proceeds; the search converges in 20 generations. It succeeds in discovering the original recurrent network structure of Figure 2(a). The resulting network almost completely reproduces the output sequence. The use of SLF alone generates many quasi-optimal network structure such as Figure 2(c)(d), but cannot discover the original structure exactly. The combination of GA and BPTT is also tried, but the resulting structure in Figure 2(d) is far from the original one.

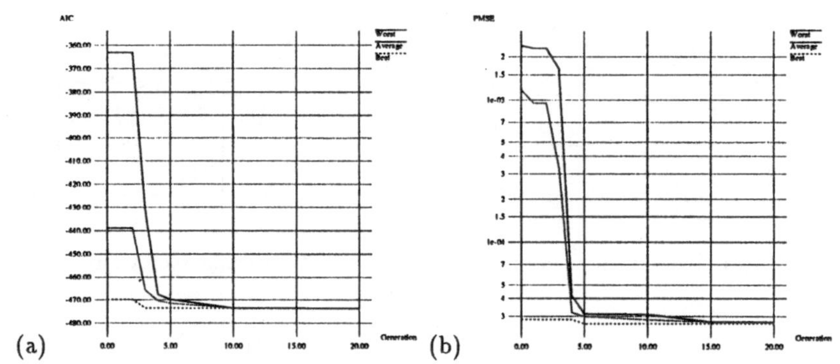

Fig. 3. The fitness at each generation. (a) AIC (b) PSS.

Secondly, PSS is used in the evaluation phase. Figure 3(b) illustrates how PSS decreases as generation proceeds. It succeeds in discovering the original recurrent network structure as in the case of AIC. The combination of GA and BPTT is also tried, but the resulting structure is far from the original one as in the case of AIC. Figure 3 indicates that AIC is slightly superior to PSS in the convergence speed.

5 Gas furnace data

Time series data taken from actual gas furnace[2] are used. Figure 4 illustrates methane gas rate(cu. ft/min) and Figure 5 delineates the % CO_2 in outlet gas. The first 50 points are used for learning and the rest are used for evaluation. Each individual network has 10 linear units at the outset. The parameters during learning are: $\eta = 0.005$, $\alpha = 0.2$, and $\varepsilon = 5 \times 10^{-5}$.

First, AIC is used in the evaluation phase. Figure 6(a) illustrates the decrease of AIC as generation proceeds; in 30 generations the search almost converges. Figure 7(a) illustrates the optimal network structure obtained. Figure 5 also illustrates % CO_2 in outlet gas of the optimal model.

Secondly, PSS is used in the evaluation phase. Figure 6(b) illustrates the decrease of PSS as generation proceeds. Figure 7(b) indicates the optimal network

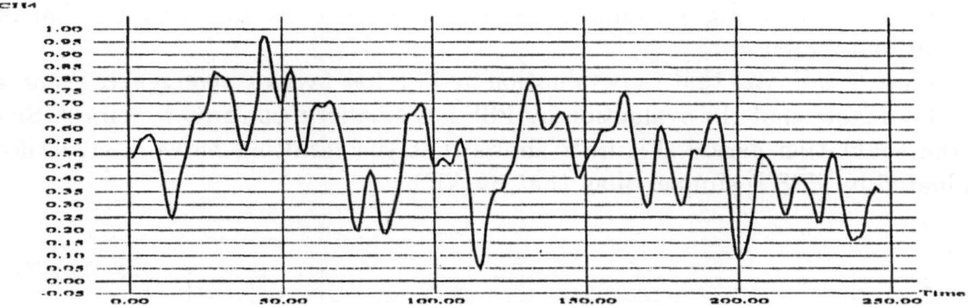

Fig. 4. Normalized methane gas rate

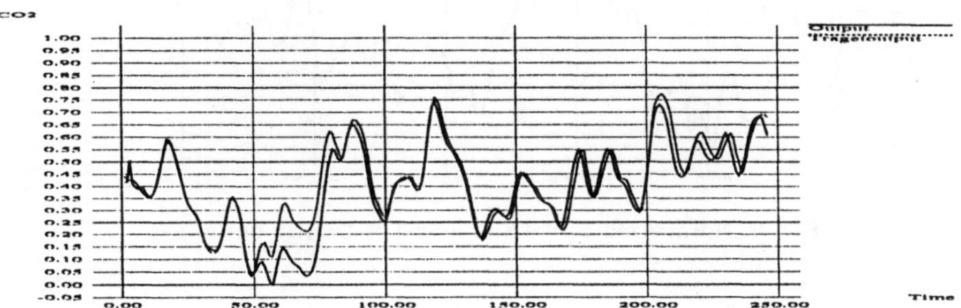

Fig. 5. Normalized actual and model % CO_2 in outlet gas

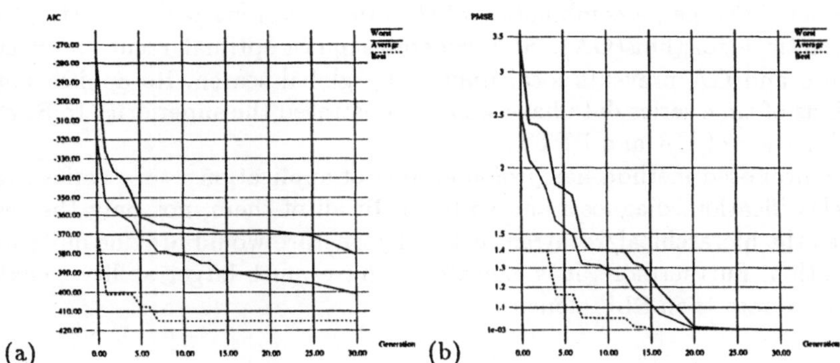

Fig. 6. The fitness at each generation. (a) AIC (b) PSS.

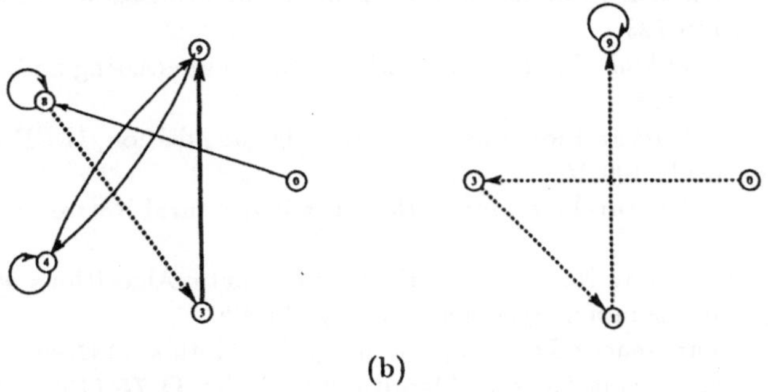

Fig. 7. Optimal network structure. (a) AIC, (b) PSS

structure obtained. Figure 8 illustrates actual % CO_2 in outlet gas and that of the optimal model.

Figure 6 indicates that the evaluation by PSS has faster convergence. Figures 5 and 8 denote that the evaluation by PSS has smaller mean square error(MSE) in the evaluation period. Figure 7 shows that the resulting network structure evaluated by PSS is simpler than that by AIC.

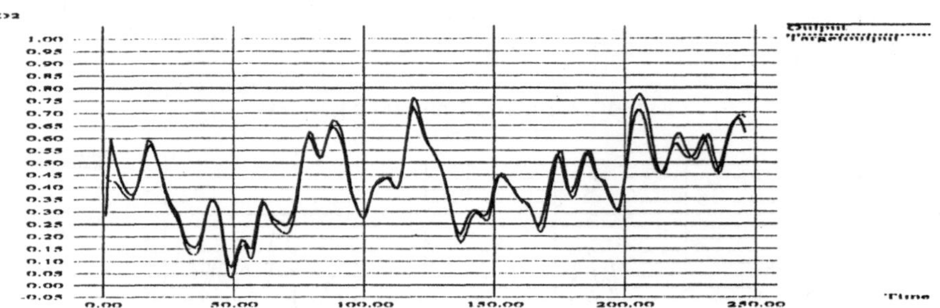

Fig. 8. Normalized actual and model % CO_2 in outlet gas

6 Conclusions

In this paper, the authors have proposed a novel method for designing recurrent neural networks, i.e., a combination of structural learning with forgetting(SLF) and genetic algorithms(GA). SLF generates quasi-optimal recurrent network structure and GA prevents local minima by global search. Its applications to two kinds of time series data have well demonstrated the superiority to SLF and a combination of GA and BPTT.

The proposed method has profound area of applications: time series prediction, classification, diagnosis and so forth. In all of them, not only the results but also the hierarchical/recurrent network structure would provide quite useful information. Furthermore, it is expected to have much larger ability to extract rules from data than SLF alone.

References

1. Akaike, H.: A new look at the statistical model identification, IEEE Trans. on AC **19-6** (1974) 716-723
2. Box, G. E. P., Jenkins, G. M.: Time Series Analysis forecasting and control, Holden-Day (1976) 370-418
3. Ishikawa, M.: Structural learning and knowledge acquisition, IEEE ICNN'96 (1996) Washington D.C., 100-105
4. Ishikawa, M.: Structural learning with forgetting, Neural Networks **9-3** (1996) 509-521
5. Kitano, H.: Designing Neural Networks Using Genetic Algorithms with Graph Generation System, Complex Systems **4** (1990) 461-476
6. Kitano, H.: Neurogenetic Learning : an Integrated Method of Designing and Training Neural Networks Using Genetic Algorithms, Physica **D 75** (1994) 225-238

A Recurrent Self-Organizing Map for Temporal Sequence Processing

Markus Varsta[1], José del R. Millán[2], Jukka Heikkonen[1]

[1]Helsinki University of Technology, Laboratory of Computational Engineering
Miestentie 3, 02150 Espoo, Finland

[2]Joint Research Centre of the European Commission. 21020 Ispra (VA), Italy

Abstract. This paper presents a recurrent self-organizing map (RSOM) for temporal sequence processing. The RSOM uses the history of a pattern (i.e., the previous elements in the sequence) to compute the best matching unit and to adapt the weights of the map. The RSOM is similar to Kohonen's original SOM except that each unit has an associated recursive differential equation. The experimental results show that the RSOM is able to learn and distinguish temporal sequences, and that it can improve EEG-based epileptic activity detection.

1 Introduction

Temporal sequence processing (TSP) is at the heart of a number of applications and different statistical methods have been proposed to deal with (e.g., [2]). Recently, neural networks (NN) have proven to be a promising alternative [11] due to their non-linearity, learning and generalization capabilities. The simplest NNs for TSP rely on feeding the NN with the current as well as several past elements of the sequence. This time-delay NN approach, however, has well-known drawbacks, the most serious being the difficulty to set the proper length of the delay line. Thus, several recurrent NNs have been designed to capture the essential history of the temporal sequence without the need of external time delay mechanisms. In these models the node and learning equations are often described by differential equations and there exist feedback connections [11, 13].

Most recurrent NNs are trained through supervised learning rules. Only a few unsupervised NN models have been proposed for TSP even if they could unfold useful information hidden in the temporal sequences in the same way that they do when dealing with non-sequential input spaces. One of the most interesting unsupervised approaches for TSP is the *Temporal Kohonen Map (TKM)* [1]. In the TKM the activity of a node is defined as a recursive differential equation that takes into account previous elements of the sequence: the current activity of a node is a function of its past activations and its distance to the new input. This paper proposes an enhancement to the TKM, and the resulting algorithm is called the *Recurrent Self-Organizing Map (RSOM)*. Briefly, the RSOM uses the history (or context) of a pattern (by means of recursive differential equations) not only to compute the best matching unit, but also to adapt the weights of the map. The experimental results show RSOM's ability to distinguish temporal sequences and its potential for EEG-based epileptic activity detection.

2 Temporal Sequence Processing by SOM

The Self-Organizing Map (SOM) [7] is a vector quantization method that maps patterns from an input space V_I onto a typically lower dimensional space V_M in such a way that the topological relationships between the input patterns are preserved and the units, which represent points in V_I, approximate closely the probability density function of V_I. The equations of the SOM are

$$\|x(t) - w_b(t)\| = \min_{i \in V_M} \{\|x(t) - w_i(t)\|\}, \quad (1)$$

$$w_i(t+1) = w_i(t) + \gamma(t) N_{ib}(t)(x(t) - w_i(t)), \quad (2)$$

where b is the best matching unit to a given input pattern $x(t)$, $\|\cdot\|$ denotes the Euclidean norm, w_i are the weights of unit $i \in V_M$, $\gamma(t) \in [0,1]$ is the learning rate, and $N_{ib}(t)$ is the "neighborhood kernel" that gives the activation of unit i when the best matching unit is b. Typically, $N_{ib}(t)$ is a Gaussian function.

In TSP the properties of the SOM can be exploited to find a chain of best matching units for sequential patterns that produces a time varying trajectory on the map space. [8] employs this kind of trajectory approach for visualization of speech signals and speech signal variations. There have also been attempts to use contextual information in conjuction with the SOM algorithm. [4] reports that the time-delay approach improves phonemic recognition.

Two of the interesting variations of the SOM algorithm for TSP are the Hypermap [5] and Self-Organizing Operator Map [6] architectures. The Hypermap can use several levels of information to find the best matching unit. For instance, the context in which the pattern occurred is first used to define a subset of units from which the final best matching unit is searched by the pattern itself. The Self-Organizing Operator Map permits to analyze the temporal sequence directly by associating an operator to each unit of the map. These operators, however, are application-dependent and the derivation of their associated learning laws can be rather complicated. [6] proposes a kind of genetic algorithm to adapt operator parameters. If an error function can be defined for an operator, such as in [4] where the operator function is the Linear Prediction Coding, the parameters of the operator can be trained through a gradient optimization approach. Operator maps have been used, for instance, in speech analysis [4].

SOMs with leaky integration of signals have the closest resemblance to our RSOM algorithm to be described in the next section. [3] proposes a three-stage approach to TSP. First, the input pattern $x(t)$ is mapped through the first SOM V_1. Second, the unit responses $r(i,t)$ are collected into a "leaky integrator" memory $s(i,t) = s(i,t-1) + \beta[r(i,t) - s(i,t)]$, where i is a unit of the map V_1, $s(i,t)$ is the memory value for the unit i at time t, and β is the adaptation factor of the memory. Third, the "leaky integrator" memory is then fed to the second SOM V_2. This approach has been shown to improve phoneme recognition.

The Temporal Kohonen Map (TKM) [1] also uses leaky integration. It determines the response of each unit i of the map for the input $x(t)$ in terms of a potential $V_i(t) = d \cdot V_i(t-1) - 1/2\|x(t) - w_i(t)\|^2$, where $0 < d < 1$. The learning rules of TKM are almost identical to the normal SOM, the only difference being that the best matching unit b at time t is searched according to

$V_b(t) = \max_i\{V_i(t)\}$. [1] shows that a TKM of 4×4 units is able to classify 16 possible sequences of length 2 where each pattern is a two-dimensional vector of binary components. It also reports some preliminary results in the formation of semantic word maps from sentences.

3 The Recurrent Self-Organizing Map

The Temporal Kohonen Map [1] does not directly use the history of a given input in the weight updating rule. Indeed, previous inputs are only taken into account in the search of the best matching unit. Our goal, and also the main novelty of the proposed *Recurrent Self-Organizing Map (RSOM)* algorithm, is to utilize the history of a pattern (i.e., the previous elements in the sequence) not only to search the best matching unit, but also to adapt the weights of the map. This is done by associating the following recursive differential equation to each unit i of the map V_M that computes the difference vector $y_i(t)$:

$$y_i(t) = (1-\alpha)y_i(t-1) + \alpha(x(t) - w_i(t)), \tag{3}$$

where α, $0 < \alpha \leq 1$, is a factor determining the influence of earlier difference vectors on the current one. When α is close to 1, the system of Eq. 3 corresponds to a short term memory, whereas when α approaches zero Eq. 3 describes a long term memory. It is possible to state a general solution for the Eq. 3 in the form

$$y_i(t) = \alpha \sum_{k=0}^{n-1} (1-\alpha)^k (x(t-k) - w_i(t-k)) + (1-\alpha)^n y_i(t-n), \tag{4}$$

which explicitly shows the involvement of the earlier inputs. It should also be noted that it is possible to replace Eq. 3 by any other recursive differential equation that might be more suitable for the task at hand.

By replacing $x'_i(t) = x(t) - w_i(t)$, Eq. 3 can be written in a familiar form

$$y_i(t) = (1-\alpha)y_i(t-1) + \alpha x'_i(t), \tag{5}$$

which describes an exponentially weighted linear IIR filter with the impulse response $h(k) = \alpha(1-\alpha)^k$, $k \geq 0$. For further analysis of Eq. 5, see e.g. [12].

Now the equations of the RSOM (search of the best matching unit and adaptation of the weights) are

$$y_b = min_i\{||y_i(t)||\}, \tag{6}$$

$$w_i(t+1) = w_i(t) + \gamma(t)N_{ib}(t)y_i(t), \tag{7}$$

Note that if $\alpha = 1$, no history is used, Eqs. 6 and 7 correspond to the Eqs. 1 and 2 of the SOM. It follows that since RSOM is a direct extension of the normal SOM it "inherits" all the SOM's appealing properties, what it is not necessarily true for the TKM. In addition, while TKM uses just the distance $||x - w_i||^2$ to determine the response of the unit i, RSOM also captures the direction thus being more sensitive to seemingly similar sequences. Furthermore, RSOM is built on principles of digital signal processing what allows the analysis of the functionalities of the unit after learning and also the sound incorporation of more suitable filters. Indeed, the units of the RSOM work upon patterns filtered with a low pass operator, which could be replaced by more complex operators such as band pass filters that are sensitive to periodic phenomena in the sequence.

To end this section, readers should notice some practical details related to the presentation of the input sequence $\{x(1), x(2), \ldots, x, (N-1), x(N)\}$ to the RSOM for learning. At each learning cycle one can randomly select a time t_c, $t_c \geq n$, to build the subsequence $\{x(t_c-n+1), x(t_c-n+2), \ldots, x(t_c)\}$ of n samples to be presented to the RSOM. The value of n has to be sufficiently high so that the sample $x(t_c-n+1)$ has only a minimal effect on the total value of $y_i(t_c)$. Also, at each learning cycle the memory should be reset to zero by setting $y_i(t_c-n) = 0$, $\forall i \in V_M$. Finally, after each weight adaptation cycle t, it is preferable to update the difference vectors according to $y_i(t) = (1-\alpha)y_i(t-1) + \alpha(x(t) - w_i(t+1))$, which corresponds to the updated map. In practice, however, this rule produces only slighty different final maps due to the fine tuning of the weights at the later stages of learning.

4 Experimental Results

We have carried out two sets of experiments to illustrate the performance of the proposed RSOM algorithm. The first one is intended to compare RSOM with TKM and so uses synthetic temporal sequences. The second one works upon real data and shows that RSOM can significantly improve EEG-based epileptic activity detection. The RSOM is implemented using Luttrell's method [9].

[1] shows that a TKM of 4×4 units is able to classify 16 possible sequences of length 2 where each pattern is a two-dimensional vector of binary components. We have repeated the same experiment achieving the same results. We have failed, however, to make the TKM learn longer sequences. On the contrary, our RSOM not only performs as well as the TKM in the reported experiment, but also differentiates correctly longer sequences. We give now the details of one of such experiments, where there are 81 possible sequences of length 2, each pattern being a two-dimensional vector. In other experiments, not reported here, the RSOM is also able to map correctly all possible sequences of length 2 where each pattern is a three-dimensional vector as well as sequences of length 3.

A RSOM of 9×9 units is trained with the following 9 input pairs: $(1,1)$, $(1,3)$, $(1,5)$, $(3,1)$, $(3,3)$, $(3,5)$, $(5,1)$, $(5,3)$, $(5,5)$. If the input pairs $(1,1)$, $(1,3)$, etc. are denoted by the numbers $1, 2, \ldots, 9$, respectively, the RSOM learns to map all possible XY combinations, $X = 1 \ldots 9$ and $Y = 1 \ldots 9$, into separate units, where Y corresponds to the most recently input. In this experiment $\alpha = 0.7$ and the memory is reset to zero only at the beginning of the learning phase. Fig. 1 shows the resulting map. Each XY input sequence is visualized on the location of the two-dimensional weight vector of the best matching unit. Fig. 1 demonstrates that the best matching units are determined by the combination of the two most recent inputs, and that the map performs a perfect classification of all the 81 possible sequences of length 2. Moreover, it can be observed that the last input, Y, determines a block of 3×3 units and that the first input X is used to select the winning unit inside that block.

The second experiment is related to EEG spectral feature clustering for epileptic activity detection. EEG is an important clinical tool for diagnosing, monitoring and managing neurological disorders related to epilepsy. Since epileptiform activity can be noticed in EEG as clearly distinguishable transient waveforms, wavelets have been used to extract suitable features for epilepsy detection. In particular, we have employed two Daubechies' mother wavelets, $Daub_4$ and $Daub_{12}$. The sampling rate of the EEG data was 200 Hz. To extract the spectral features at time t we use a window W^t of 256 samples, $W^t(i) = S_{EEG}(t - 127 + i)$ $i = 0 \ldots 255$, that slides over the EEG sequence S_{EEG}. The discrete wavelet transform for each window W^t is computed by means of Mallat's "with holes" algorithm [10] and a total of 16 energy features f^t are determined for each W^t by calculating the squared sum of the $Daub_4$ and $Daub_{12}$ wavelet coefficients at each scale. The training data contains a total of 150987 16-dimensional feature vectors, of which 5430 patterns correspond to epileptic activity.

The extracted EEG features are first clustered by four SOMs of 3×3, 5×5, 9×9, and 17×17 units. The SOMs are trained using Luttrell's method [9]. After training each unit of the map is labeled according to the plurality rule to belong either to "normal" or "epileptic" activity, and the SOMs are used as classifiers to evaluate the discrimination potentials of the learned feature clusters. Note that for labeling the number of epileptic activity samples mapped into each unit i is multiplied by the factor 145557/5430 for equally weighting both classes. Table 1 shows the confusion matrices of the four SOM classifiers. The values in the matrices have the following meanings, and so the diagonal entries give the number of correctly classified samples:

Right negative	Wrong positive
Wrong negative	Right positive

These results are then compared to the performance of four RSOMs of the same sizes. In this experiment $\alpha = 0.6$ and, in each learning cycle, a sequence of 5 feature vectors f^{t_c-64}, f^{t_c-48}, f^{t_c-32}, f^{t_c-16}, and f^{t_c} is shown to the map with a randomly selected time t_c. Table 1 shows that the labeled RSOMs classifies the training samples better than the corresponding normal SOM. These clustering results suggest that the use of the history can significantly improve EEG-based epileptic activity detection[1]. Indeed, experts seem to recognize epilectic activity from other seemingly similar EEG patterns from the temporal context. Thus, the RSOM may be a valuable tool for automatic analysis and diagnosis.

[1] Readers should keep in mind that the purpose of this experiment is not to classify EEG patterns, but just to compare the discrimination capabilities of the SOM and RSOM algorithms. Of course, better classification results can be achieved if, after presenting an input sample (either a feature vector or a sequence of feature vectors), the activity of the map is fed to a feedforward network (e.g., [14]).

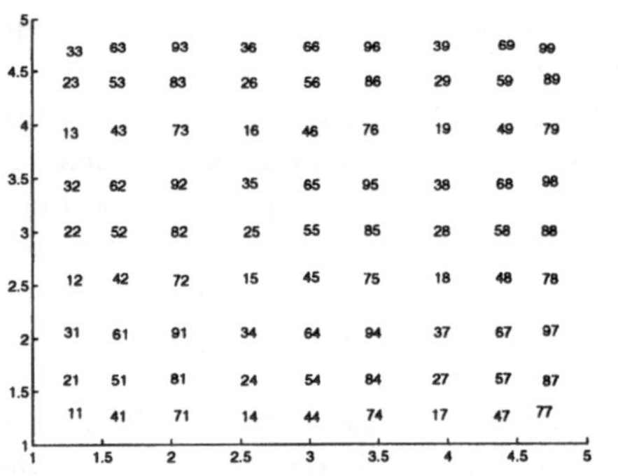

Fig. 1. The learned RSOM of 9 × 9 units.

Table 1. SOM and RSOM confusion matrices obtained in the clustering of EEG spectral features.

# of units	SOM		RSOM	
3 × 3	126966	18591	129269	16224
	405	5025	321	5109
5 × 5	131794	13763	133389	12104
	443	4987	380	5050
9 × 9	133820	11737	134660	10833
	394	5036	317	5113
17 × 17	134465	11092	135935	9558
	338	5092	284	5146

5 Conclusions

We have proposed a Recurrent Self-Organizing Map (RSOM) for temporal sequence processing (TSP) that is a direct extension of the SOM and is built on principles of digital signal processing. The RSOM can be used as a flexible tool to unfold useful information hidden in the temporal data. The experimental results show the potentials of the RSOM, although more systematic experiments in different TSP tasks and with varying parameter values have to be conducted.

Acknowledgements

We thank Dr.Tech. Alpo Värri of Tampere University of Technology in Finland for generously providing us the scored EEG recordings of the ANNDEE project.

References

1. G.J. Chappell & J.G. Taylor (1993). *Neural Networks*, **6**:441–445.
2. N.A. Gershenfeld & A.S. Weigend (1993). In *Times Series Prediction: Forecasting the Future and Understanding the Past*, pp. 1–70, Addison-Wesley.
3. J. Kangas (1991). *ICANN'91*, pp. 1591–1594.
4. J. Kangas (1994). *Dr. Tech. thesis*, Helsinki University of Technology.
5. T. Kohonen (1991). *ICANN'91*, pp. 1357–1360.
6. T. Kohonen (1993). *Proc. IEEE Int. Conf. on Neural Networks*, pp. 1147–1156.
7. T. Kohonen (1995). *Self-Organizing Maps*. Springer-Verlag.
8. L. Leinonen, J. Kangas, K. Torkkola, & A. Juvas (1992). *Journal of Speech and Hearing Research*, **35**:287–295.
9. S.P. Luttrell (1989). *Pattern Recognition Letters*, **10**:1–7.
10. S.G. Mallat (1989). *IEEE Trans. Pattern Anal. Machine Intell.*, **11**:674–693.
11. M.C. Mozer (1993). In *Times Series Prediction: Forecasting the Future and Understanding the Past*, pp. 243–264, Addison-Wesley.
12. J.G. Proakis & D.G. Manolakis (1992). *Digital Signal Processing*. Macmillan.
13. A.C. Tsoi & A.D. Black (1994). *IEEE Trans. on Neural Networks*, **5**:229–239.
14. M. Varsta, J. Heikkonen, & J. del R. Millán (1997). *Int. Conf. on Engineering Applications of Neural Networks*.

An Extended Elman Net for Modeling Time Series

Peter Stagge* Bernhard Sendhoff*

Institut für Neuroinformatik
Ruhr-Universität-Bochum
44780 Bochum, Germany

Abstract The prediction and modeling of dynamical systems, for example chaotic time series, with neural networks remains an interesting and challenging research problem. It seems to be rather natural to employ recurrent neural networks for which we will suggest a new structure based on the Elman net [1]. The major difference to neural networks as proposed by Williams and Zipser [2] is the way we organize the time steps. The dynamic of the network and of the input flow is defined in a way as to guarantee that the information at the input node is available at the output node in one time step, irrespective of the connection matrix. We apply the network to the Lorenz and the Rössler system and comment on the problem of evaluating the quality of a network used as a dynamical model.

1 Introduction

The standard multi layer perceptron is known to be a universal function approximator [3]. The static mapping realized by feed forward neural networks has been frequently applied to the prediction of dynamical systems. Generally the system state is given by a vector $\boldsymbol{q}(t)$ and the neural network model learns a function $\boldsymbol{F}(\boldsymbol{q}(t), \delta t) = \boldsymbol{q}(t + \delta t)$. The identification of the state $\boldsymbol{q}(t)$ itself is a problem, however, we will propose an *internal* method in sec. 2.
The quality of the prediction of nonlinear dynamical systems is usually determined by the deviation between the predicted and the known state at time $t+\delta t$. In addition to the prediction we demand the modeling of a system. The quality of modeling a system is usually determined by iterating the model, i.e. feeding the output at time t back to the input at time $t + \delta t$. The model can then itself be seen as a dynamical system and the deviation between invariants like Liapunov spectrum or entropy based measures determine the modeling quality. Simply observing the attractor structure of the model will also prove useful. For feedforward neural networks it is hard to identify the underlying dynamical process. But it was shown that the modeling capacity of those networks can be improved if iterating the network is incorporated in the learning process, [4] [5].
Since we are looking for dynamical systems as models, it is sensible to use recurrent neural networks. In recurrent structures we have, in addition to the static mapping between input and output, also internal states of the network.

* e-mail: {peter,bs}@neuroinformatik.ruhr-uni-bochum.de

These states work as a short term memory and they are able to represent information about the preceding inputs. Led by this idea that internal states can represent the past, Elman proposed a neural network in which the activations of the hidden layer are used as an additional input in the next time step. Elman strictly kept the concept of layers in the network structure and used this architecture to predict words in simple [1], and more complex [6] sentences. The recurrent network structure proposed in this paper is an extension of the Elman network, which seems to be more suitable for time series modeling.
Williams and Zipser [2] introduced a method to calculate the error gradient with respect to the weights for arbitrarily connected recurrent neural networks. Besides the very high computational cost of their *real time recurrent learning* algorithm, the number of computations increases as $\mathcal{O}(n^4)$, n being the number of neurons, the algorithm has another disadvantage concerning the time series prediction task: When the activations for time step t are computed, it is possible that, depending on the network structure, the input at time t does not reach the output nodes at all. Instead the input information is transformed and kept in some internal activations from where it might influence the output in one of the following time steps. This would implicitly transform a one-step prediction task to a two or more step prediction task.

2 Implicit Embedding

The success of the prediction of a dynamical system strongly depends on the identification of the complete system state. Usually the embedding theorem is used to regroup the scalar measurement values at different times into vectors. Let $\{x(t)\}$, $t \in [0, 1, \cdots, T]$ be the set of measured values. The embedding theorem, [7], then guarantees the existence of a diffeomorphism between the reconstructed vectors

$$\tilde{q}(t) = (x(t),\, x(t-\tau),\, \cdots,\, x(t-(d_E-1)\tau), \tag{1}$$

and the original state space. However, in practice, the determination of the correct values for the *embedding dimension* d_E and for the *time-lag* τ is problematic, [8]. Furthermore, we expect that the correct reconstruction depends on the structure which is used to model the original dynamical system.
Having in mind what the Elman net was originally used for, namely a representation of the past, we will use our recurrent network structure not only to learn the dynamics of the system, but also to identify the correct state space which is needed to achieve optimal representation and prediction. This way we circumvent the problem of model dependent reconstruction. In our system the reconstruction and the modelling problem are handled by the same system simultaneously. Thus, the recurrent networks proposed here, perform implicit reconstruction, which can actually be observed from some single neuron output after successful learning of the system. Of course the more we demand from the network the more flexible we should keep its structure.

3 The Recurrent Network Structure

Following the remarks made in the introduction, we define a recurrent neural network, based upon the work by Elman with the following dynamics and notation:

$$y_i(t+1) = \sigma\left(\sum_{j=1}^{i-1} w_{ij}y_j(t+1) + \sum_{j=1}^{N} r_{ij}y_j(t) + \sum_{k=1}^{K} f_{ik}in_k(t) + \theta_i\right) \quad (2)$$

$y_i(t+1)$	: output of neuron i at time $t+1$
N, K	: Number of Neurons and Inputs
σ	: sigmoidal function, e.g. $tanh()$
w_{ij}	: forward conection matrix, lower triangular, $w_{ii} = 0$
f_{ik}, r_{ij}	: input connection matrix, recurrent connections
θ_i	: threshold weights
$in_k(t)$	: input to the network at time t

We use standard back propagation for the calculation of the error gradients of weights. The Williams & Zipser structure does not contain the first sum on the r.h.s. of equation (4). This makes the r.h.s. a function of t only and seems more natural to physicists, as it looks like a discrete differential equation. However, as we pointed out in the introduction, neglecting this term leads to the problems of insufficient propagation of the input information.

The network, equation (2), itself defines a dynamical system, if we identify the output at time t with the input at time $t+1$:

$$\boldsymbol{y}(t+1) = \boldsymbol{F}(\boldsymbol{y}(t)) \quad (3)$$

This is a typical return map and the dimensionality of its dynamical system equals the number of neurons whose output is fed back to the feedforward calculation in the next step (including the output nodes). Unfortunately, it is quite difficult to make any analytical statements about the behaviour of the dynamics produced by the return map, e.g. chaos, periodic orbits, stable fixpoints or number of attractors and their basins. One result, which is guaranteed by the finitness of the sigmoid function, is the existance of at least one fixpoint.

Figure 1 shows an example of a neural network with the introduced recurrent structure. It has one input and one output node. In terms of layers this network would correspond to a net with five hidden layers consisting of one neuron each, but we believe it is more appropriate to skip the restriction to a layered structure and call figure 1 a *bulk* of neurons instead. This network structure (figure 1) will be used for the simulations in the next section.

4 Experiments with Chaotic Time Series

The system we want to model is the Lorenz attractor [9] which is defined by the three coupled differential equations in (4), which show chaotic behaviour

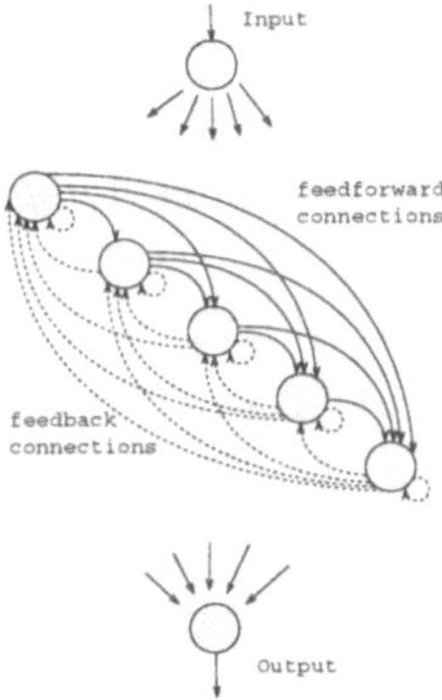

Figure1. Recurrent neural network structure. The lines indicate forward connections and the dashed lines indicate recurrent connections. In one time step all feedforward activations are calculated and propagated through all connections. The feedback connections correspond to the propagation of activations from one time step before.

for the parameter values $\sigma = 16.0$, $r = 45.92$, and $b = 4.0$.

$$\frac{dx(t)}{dt} = -\sigma x + \sigma y, \quad \frac{dy(t)}{dt} = -xz + rx - y, \quad \frac{dz(t)}{dt} = xy - bz. \tag{4}$$

We solved (4) using a 4th order Runge-Kutta method with a time step $dt = 0.01$. The actual time step for the simulations was $\Delta t = 5dt = 0.05$, and we used the data from the x coordinate. We want the network to learn a model of the entire dynamical system. However, the task during the presentation of the training sequence was to predict one time step ahead: $x(t) \to x(t + \Delta t)$. We used a training and a test sequence of 500 data points each, which corresponds to about 26 stretching and folding processes in the attractor. We took standard backpropagation with learning rates between 0.005 and 0.01 and a momentum term of 0.1.

As we aim at modeling the dynamical system and not just at good prediction, the standard prediction error alone does not yield sufficient insight. Therefore, we also monitor the largest Liapunov exponent during learning, which we calculate from the network seen as an iterative system (3), [10]. The resulting network after the training process comes very close to the Lorenz system in terms of the deviation of the network's Liapunov exponents from the Lorenz system's Liapunov exponents, see Table 1 for a comparison. Our network has six Liapunov exponents as the activations of neurons 2–7 are used for the next time step.

	neural network	Lorenz system
first exponent	2.12	2.16
second exponent	-0.084	0
third exponent	-27.1	-32.4
(4th, 5th, 6th) exponent	(-40.3, -97.5, -250)	(–, –, –)

Table1. Liapunov exponents of the Lorenz chaotic system and the neural network model (values in *bits/sec*).

We note that not only the first – most important – exponent but also the second and the third are in good agreement with the real system.

There are several remarks to be made about the learning process. Firstly, when iterating the neural network the starting point of the trajectory depends on the first input and on the initialisation of the activations which are needed to calculate the first output. Although the neural net might have other attractors the trajectory robustly finds its way to the attractor which was learned. Secondly, in the final network we did not find another attractor or a fix point for various initialisations. Whereas at the start of the learning process any network with randomized initial weights converged to a fixed point. During the learning process the system can have more than one fixed point or a combination of a fixed point and a periodic orbit simultaneously. In a

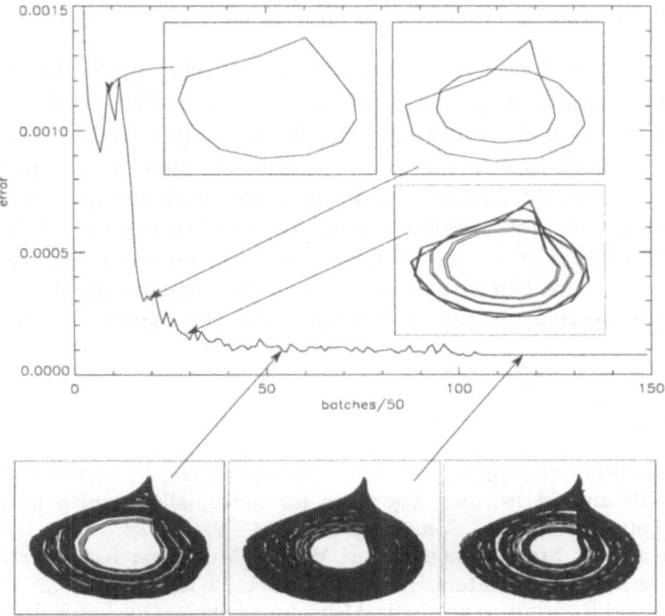

Figure2. Learning curve for the Rössler attractor and images of the reconstructed attractors of the recurrent network ($r(t) = (x(t), x(t-\triangle), x(t-2\triangle))$). The increasing quality of modeling the dynamical system at various stages during the training process is striking.

second experiment, we modeled the Rössler [11] chaotic attractor using data from the x-coordinate. Since the folding process occurs very irregularly, it is difficult for a network to model. Therefore, we slightly extended the structure of our recurrent network to be able to represent activation of more than one time step back. However, with these modifications, the Liapunov exponents are more difficult to calculate. Therefore, we decided to visually inspect the network's performance in "copying" the Rössler attractor. Figure 2 shows that the decrease in the approximation error coincides with the increasing quality of the modeling of the attractor structure. The insets display reconstructed attractors of the iterated network ($r(t) = (x(t), x(t-\triangle), x(t-2\triangle))$), with 2000 points each. The bottom right picture shows the reconstructed attractor from the original x data. Beginning with a fix point the network firstly learns a periodic oscillation which corresponds to a plateau in the one step approximation error. By learning higher periodic orbits the net comes closer to the real attractor and the approximation error decreases significantly. This shows the importance of a global model of a dynamical system, additionally to the forecasting error.

5 Conclusion

Modeling dynamical systems beyond one or few step prediction is shown to be possible with an appropriate recurrent neural network structure. Choosing a recurrent network circumvents the problematic embedding procedure and allows the network to organize the temporal information in a problem **and** in a model dependent manner. In order to estimate the quality of the network model, global measures like system invariants are more reliable than the standard prediction error. As we have shown visual control of the networks dynamical structure can also lead to an interesting insight into the combination of the decrease in prediction error and the increase of the modeling quality.

References

1. J. Elman. Finding structure in time. *Cognitive Science*, 14:179 – 211, 1990.
2. R. J. Williams. A learning algorithm for continually running fully recurrent neural networks. *Neural Computation*, 1:270 – 280, 1989.
3. K. Hornik, M. Stinchcombe, and H White. Multilayer feedforward networks are universal approximators. *Neural Networks*, 2:359 – 366, 1989.
4. G. Deco and B. Schürmann. Neural learning of chaotic system behavior. *IEICE Trans. Fundamentals*, E77 A:1840 – 1845, 1994.
5. J.C. Principe and J.-M. Kuo. Dynamic modelling of chaotic time series with neural networks. In R.P. Lippmann, J.E. Moody, and D.S. Touretzky, editors, *Advances in Neural Information Processing Systems (NIPS) 7*. Morgan Kauffman, 1995.
6. J. Elman. Learning and development in neural networks: the importance of starting small. *Cognition*, 48:71 – 99, 1993.
7. T. Sauer, J.A. Yorke, and M. Casdagli. Embedology. *J. Stat. Phys.*, 65, 1991.
8. H. Abarbanel, R. Brown, J. Sidorowich, and L. Tsimring. Analysis of observed chaotic data in physical systems. *Rev. Mod. Phys.*, 65:1331 – 1392, 1993.
9. E.N. Lorenz. Deterministic nonperiodic flow. *J. Athmospheric Sci.*, 20:130 – 141, 1963.
10. A. Wolf, J.B. Swift, H.L. Swinney, and J.A. Vastano. Determining lyapunov exponents from a time series. *Physica*, D(16):285 – 317, 1985.
11. O.E. Rössler. *Phys. Letter*, 57A, 1976.

Recurrent Associative Memory Network of Nonlinear Coupled Oscillators

Margarita Kuzmina[1], Edward Manykin[2] and Irina Surina[2]

[1] - Keldysh Institute of Applied Mathematics of RAS, Moscow, Russia
[2] - RRC Kurchatov Institute, Moscow, Russia

Abstract. The recurrent associative memory networks with complex-valued Hebbian matrices of connections are designed from interacting limit-cycle oscillators. These oscillatory networks have peculiarities and advantages as compared to Hopfield neural network model. In particular, the class of networks with high memory characteristics (the capacity close to 1, low extraneous memory) exists. At zero values of oscillator frequencies the designed networks are closely related to the known "clock" neural networks (networks from complex-valued neurons). Pattern recognition of colored images and recognition of objects with complicated topological structure look quite natural in the context of such models. Exact solutions have been obtained for a few types of the networks considered, in particular, for homogeneous closes chains.

1. Introduction.

During more than twenty years the systems of coupled oscillators are used for modelling of various phenomena in physics, chemistry, biology. Recently the studies on artificial neural-like oscillatory networks were started. In the first series of publications, using phase approximation of oscillatory dynamics (so-called phase model), the phenomenon of clasterization of oscillatory population into the state of synchronization in the vicinity of phase transition was exploited [1]. Later the attempt to design oscillatory networks of associative memory resembling Hopfield networks was made with the use of Ginzburg-Landau oscillatory system and its phase approximation [2]. However, in this study the design has not been completed with a clear construction of the weight matrix. Another attempt based on the special case of phase model, was made recently [3], but this approach leads to some difficulties caused by the specific properties of the model chosen.

In [4-6], the recurrent oscillatory networks of associative memory with complex-valued generalization of Hebbian matrix of connections were designed using amplitude-phase dynamical model. Although the networks consist of oscillators, the relaxational dynamics in synchronization regime is used in the design of the associative memory.

The attractive feature of the designed networks is that they admit electronic and optical implementations based on optoelectronic and nonlinear optics principles. The background of the implementations is currently under development.

2. Dynamical Equations. The Problem of Design of Associative Memory Network.

The system [4-6] governing the dynamics of N coupled limit cycle oscillators is studied from the viewpoint of associative memory modelling:

$$\dot{z_j} = (1 + i\omega_j - |z_j|^2)z_j + \kappa \sum_{k=1}^{N} W_{jk}(z_k - z_j), \qquad j = 1, ..., N. \tag{1}$$

Here the complex variable $z_j(t) = r_j(t)exp(i\theta_j(t))$ defines the state of j-th oscillator (r_j and θ_j are respectively the amplitude and phase of oscillators), ω_j is its natural frequency. The first term in the right-hand side of (1) defines the intrinsic dynamics of free isolated oscillator, while the second one, responsible for interaction, is specified by the matrix of connections κW. Non-negative parameter κ defines the absolute value of interaction strength in oscillatory system and the matrix $W = [W_{jk}]$, which in the case of symmetrical interaction is Hermitian one, specifies the weights of connections. As a weight matrix, W satisfies the natural restrictions: $W_{jj} = 0, \quad |W_{jk}| \leq 1, \quad \sum_{k=1}^{N} |W_{jk}| \leq 1$.

The system (1) can be rewritten in the vector form:

$$\dot{z} = (D(z) + \kappa W)z, \tag{2}$$

where the column-vector $z = (z_1, \ldots, z_N)^\top$ is a state vector of the oscillatory system and $D(z)$ is the diagonal matrix,

$$D(z) = diag(D_1(z), \ldots, D_N(z)), \quad D_j(z) = 1 + i\omega_j - |z_j|^2 - \kappa \sum_{k=1}^{N} W_{jk}.$$

At arbitrary set $\{\{\omega_j\}, \kappa, W\}$ of the parameters the system (1) demonstrates the variety of complicated dynamical regimes including dynamical chaos. The regime of synchronization, used in associative memory modelling, is quite simple from the viewpoint of nonlinear dynamics: if $\sum_j \omega_j = 0$ (this restriction on the frequencies can be always satisfied by proper rescaling of the dynamical system), this is relaxation to stable equilibria.

The problem of design of associative memory network can be formulated as combination of two (independent) subproblems for the governing dynamical system:

1) the inverse problem for the dynamical system, i.e., design of the system possessing the prescribed set of stable equilibria with large enough basins of attraction;

2) a kind of control problem for the designed dynamical system, i.e., the choice of an adequate learning algorithm.

In the present study, the results on subproblem 1) for the system (1) in the regime of synchronization are performed. Moreover, only *phase memory* (with attractors located in the points of the phase space of (1) that have equal moduli for all coordinates) is suggested. Computer modelling of non-phase associative memory shows that in many cases non-phase memories can be obtained from phase ones by relatively slight distortions, therefore the results on phase memory are useful for solution of the general problem.

First the results of the analysis of equilibria of oscillatory networks with small number of oscillators and of special interconnection architectures are performed.

3. The System of Two Coupled Oscillators. Structural Portrait.

Since the system (1) can be represented in the form (2), it has always the equilibrium $z = 0$. So, in general case the whole set of equilibria of (1) consists of two subsets: $z = 0$ and $z \neq 0$ & $(D(z) + \kappa W)z = 0$.

In the case of the system of two coupled oscillators the parametrical space of the dynamical system is (ω, κ, β), where $\omega = \omega_1 = -\omega_2$, $W_{12} = exp(i\beta)$, $-\infty < \omega < \infty$, $0 < \kappa < \infty$, $-\pi/2 < \beta \leq \pi/2$.

Let $\mathcal{D}$ denotes the subdomain of the parametrical space where the equilibrium $z = 0$ is stable (this domain is usually regarded as "amplitude death" [7]) and $\mathcal{S}$ - the subdomain where stable equilibria $(D(z) + \kappa W)z = 0$ are located ($\mathcal{S}$ is the synchronization domain). The following results have been obtained analytically for two-oscillatory system: (both the expressions for equilibria and the eigenvalues of Jacobian defining the character of stability of the equilibria have been calculated in the explicit form):

- there exist from one to four fixed points of (1) in different domains of parametrical space (the corresponding subdomains have been specified);

- only one fixed point U^1 is stable ($U^1 \in \mathcal{S}$, U^1 is a stable node; the other fixed points are saddles);

- the structural portrait of the system (1) has been obtained in the whole parametrical space; all the boundaries have been calculated in the explicit analytical form.

As an example we can give the projection of the structural portrait into the quadrant $(\omega, \kappa, 0), \omega \geq 0$. This projection resembles the structural portrait of the system (1) with $W_{jk} = N^{-1}(1 - \delta_{jk})$ obtained in [7]:

$$\mathcal{D} = \{1 \leq \kappa \leq \frac{\omega^2+1}{2\omega}\}, \quad \mathcal{S} = \{\kappa \geq \omega \ \ at \ \ \omega \leq 1 \ \& \ \kappa \geq \frac{\omega^2+1}{2\omega} \ \ at \ \ \omega > 1\},$$

The domain of unsteady dynamics is $\{\kappa < \omega \ \ at \ \ \omega \leq 1 \ \& \ \kappa < 1 \ \ at \ \ \omega \geq 1\}$.

4. Exact Solutions for Homogeneous Closed Chains.

Phase memory has been studied both analytically and with the use of modelling for a few instructive types of matrices W, in particular, for closed chains of oscillators with constant weight $(b + ic)$ and zero frequencies. It can be seen that N phase points P_k, $k = 0, \ldots, N-1$, with coordinates $(r, r \cdot exp(i\psi), r \cdot exp(2i\psi), \ldots, r \cdot exp((N-1)i\psi))$, where ψ is $2\pi k/N$, $r = (1 + 2(b \cdot cos(\psi) - b - c \cdot sin(\psi)))^{1/2}$ are equilibrium points for such closed chains. Their stability has been studied by direct calculation of the corresponding spectra of Jacobian matrices. For $N = 3, 4, 6$ the complete solutions have been obtained. The results can be displayed in the plane c, b. For $N = 3$ P_0 is stable if $b > ((1+3c^2)^{1/2}-1)/3$, P_1 – if $b < c\sqrt{3}+1/6-(4c^2/3+1/36)^{1/2}$, P_2 – if $b < -c\sqrt{3}+1/6-(4c^2/3+1/36)^{1/2}$. Thus, in two narrow unbounded strips two pairs of points, i.e., P_0,P_1 and P_0, P_2 are stable, and in the expanding region one pair, i.e., P_1,P_2, is stable. For $N = 4$ the plane (c,b) is divided into the regions in the following way: P_0 is stable if $b > (-1+(1+4c^2)^{1/2})/2$, P_1 – if $(3c^2-b^2+2bc-c > 0)$ & $c < 0$, P_2 – if $b < (1-(1+20c^2)^{1/2})/10$, P_3 - if $(3c^2-b^2-2bc+c > 0)$ & $c > 0$. Fig. 1 shows the regions of stability for P_k, $k = 0, \ldots, 5$, in the chains from six oscillators in two scales. Here the digits mean the values of k determining ψ. As one can see, in three small bounded regions three points are simultaneously stable. It looks likely that no extra (non-phase) stable points exist in homogeneous closed chains. For $N < 25$ this hypothesis has been confirmed using computer modelling. For $N \geq 7$ the overlap of the stability regions increases and more points can be stable simultaneously. Some strict solutions have been also obtained for cyclic matrices W of low orders ($N < 7$) other than the closed chains

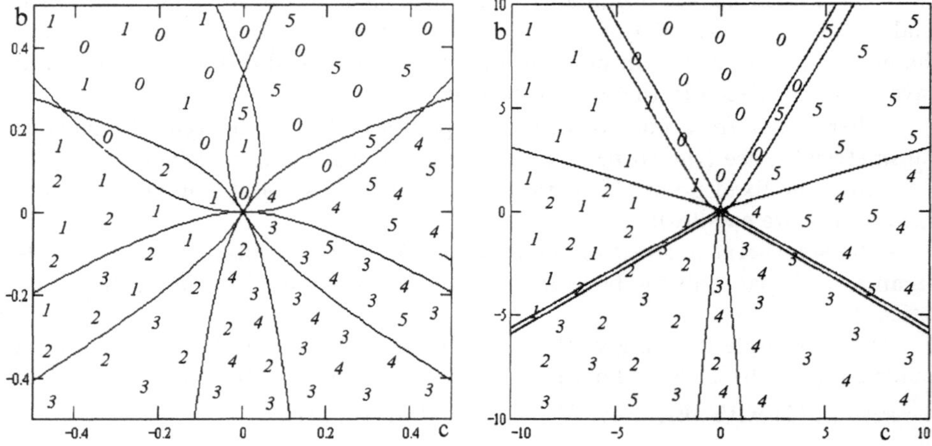

Fig. 1. Phase memories in the chains of six oscillators. Two scales.

5. Oscillatory Networks with Hebbian Matrices of Connections. Related Phasor Networks.

Dynamical system (1) with $\omega_j \equiv 0$ represents the important special case of oscillatory system which can be regarded as phasor networks and can be viewed as natural generalization of the known "clock" neural networks.

The equilibria of oscillatory networks and corresponding phasor networks proved to be closely related. The following proposition reflects the relation.

- Let $\mathcal{N}(\{\omega_j\}, \kappa, W)$ be an oscillatory network with arbitrary frequencies ω_j satisfying the condition $\sum_j \omega_j = 0$.

Let the corresponding phasor network $\mathcal{N}(\{0\},\kappa,W)$ possesses the collection of M memory vectors $\{U^1,\dots U^M\}$. Define $\tilde{\kappa} > \kappa$ satisfying the condition: $\gamma \equiv \Omega/\tilde{\kappa} \ll 1$, where $\Omega = max_j|\omega_j|$.

Then oscillatory network $\mathcal{N}(\{\omega_j\},\tilde{\kappa},W)$ has memory vectors $\tilde{U}^1,\dots,\tilde{U}^M$, which represent perturbations of the corresponding $U^1,\dots,U^M$.

The proof of this proposition has been obtained using the perturbation method on small parameter γ. It was confirmed by computer studies of phase portraits of the dynamical system (1) at small N.

The most essential feature of oscillatory networks is that the memory vectors to store cannot be chosen arbitrarily: they are completely defined be special symmetrical set of orthogonal vectors in N-dimensional complex space C^N — "phase" basis $\mathcal{B}_\mathcal{N}$:

$$\mathcal{B}_\mathcal{N} = \{ \ V^m \ | \ (V^s)^+V^m = N\delta_{sm} \ \ m,s = 1,\dots,N.\}$$

(Here V^m is a column-vector $(V_1^m,...V_N^m)^\top$ and $(V^m)^+$ is the corresponding conjugated row-vector: $(V^m)^+ = (\bar{V}_1^m,...,\bar{V}_N^m))$.

The phase basis is defined by single generating vector $V^0 = (1,...1)^\top$ and the single parameter $\varphi = 2\pi/N$. All other vectors are of $\mathcal{B}_\mathcal{N}$ can be calculated with the help of recurrent process.

The basis $\mathcal{B}_\mathcal{N}$ is the eigenbasis of any Hermitian weight matrix W of size $N \times N$. At the same time the matrix W^H of rank M,

$$W^H = \sum_{m=1}^{M} V^m(V^m)^+, \qquad M = rankW, \tag{3}$$

is the matrix of the projection operator into M-dimensional subspace of C^N spanned on $V^1,\dots,V^M$.

The following results are valid for phasor networks with W^H as the matrix of connections.

- 1. Let N be a prime number.

Define basis $\mathcal{B}_\mathcal{N}$ and choose any subset of $M \le N$ vectors from this basis $\{V^1,\dots V^M\}$. Construct W^H in accordance with (3).

Then the phasor network has memory vectors $U^1,\dots U^M, U^m = cV^m$, where $c = 1$ if $V^0 \in \{V^1,\dots,V^M\}$ and $c = (1+\kappa)^{1/2}$ if $V^0 \notin \{V^1,\dots,V^M\}$.

All memory vectors $U_1,\dots U^M$ have equal basins of attraction.

The sizes of the basins can be controlled if weighted Hebbian matrix $\tilde{W}^H = \sum_{m=1}^{M}\lambda^m V^m(V^m)^+$ is used.

- 2. Let the number of oscillators N be not prime.

The main feature of the network memory in this case is that the memory is not completely controllable in distinction to the previous case. Namely, only special odd numbers M of vectors from the basis $\mathcal{B}_\mathcal{N}$ can be imposed into the network memory. If M is different from the mentioned special numbers, the recalling process is impossible at all: the dynamical system (1) has continual set of degenerated equilibria.

It should be noted also that all matrices W^H are irreducible if N is prime and are reducible ones otherwise.

Conclusions.

The special class of recurrent oscillatory and the corresponding phasor networks of high performance is designed. It is characterized by completely controllable memory of high storage capacity: up to $N-1$ memory vectors defined by some specific set ("phase" basis) can be loaded into the memory of the network consisting of N processing units. The weight matrix is designed in complex-valued Hebbian form. Extraneous memory exists, but it can be easily separated due to its non-phase character. The results of complete strict analysis of closed homogeneous oscillatory chains are presented. Oscillatory networks are promising from many viewpoints, in particular, in view of possible of nonlinear optical implementations.

Acknowledgment.

This study was partially funded by Russian FFR, grant n. 96-01-00084.

References.

1. Sompolinsky H., Tsodyks M., Processing of sensory information by a network of oscillators with memory, - *Intern. J. of Neural Systems*, v.3, Suppl., 1992, pp.51-56.
2. Kuramoto Y., Aoyagi T., Nishikawa I., Chawanya T., Okuda K. Neural network model carrying phase information, - *Progr. Theor. Phys.*, v.37, n.5, 1992, pp. 1113-1126.
3. Vicente C.J.P., Arenas A., Bonilla L.L., On the short-time dynamics of networks of Hebbian coupled oscillators, - *J. Phys. A*, v.29, Iss.1, 1996, pp. L9-L16.
4. Kuzmina M. G., Manykin E. A., Surina I. I., Oscillatory networks with Hebbian matrix of connections - *Lect. Notes in Comp. Sci.*, v.930, *Proc. of IWANN'95*, 1995, pp.246-251.
5. Kuzmina M. G., Manykin E. A., Surina I. I., Oscillatory networks with guaranteed memory characteristics, - *Proc. of EUFIT'96*, v.1, 1996, pp.320-324.
6. Kuzmina M. G., Manykin E. A., Surina I. I., Oscillatory networks of associative memory - *Optical Memory and NN*, v.5, n.2, 1996, pp.91-103.
7. Matthews P.C., Strogatz S.H., Phase diagram for collective behavior of limit-cycle oscillators, - *Phys. Rev. Lett.*, v.65, n.14, 1990, pp.1701-1704.

A Layered Recurrent Neural Network for Feature Grouping

Heiko Wersing, Jochen J. Steil and Helge Ritter

University of Bielefeld, Faculty of Technology, Neuroinformatics Group
P.O.-Box 10 01 31, D-33501 Bielefeld, Germany

Abstract. We describe a recurrent network, the Competitive Layer Model (CLM) for feature grouping. The model uses a combination of cooperative and competitive interactions to partition a set of input features into salient groups whose number is only restricted by the available layers. We give analytic results on convergence and the attractor states of the model and present simulation results showing grouping by proximity and grouping by symmetry and good continuation.

1 Introduction

In many applications we face the following problem: Given a set of separate features which are presented in parallel, find an appropriate partition into subsets of features underlying a common cause. A typical example is the perceptive grouping problem, where a set of derived image primitives like points and edges have to be grouped according to Gestalt principles like proximity, parallelity etc. A neural implementation that solves this problem has to involve two antagonistic processes: It has to *bind* related features into common groups, and it has to *separate* the groups originating from different sources.

Currently, synchronization models attract much interest, where binding or separating is achieved by temporal synchrony or asynchrony of neuron pools that code for related features [7,6]. However, due to the great dynamical complexity of these models applications are still rather limited to restricted problem domains. A different approach is taken in the brain-state-in-a-box (BSB) model [1], which is a recurrent network based on energy minimization. Upon a given input vector, the BSB network converges to exactly one of possibly many point attractors which forms a single detected group or cluster.

The Competitive Layer Model (CLM) [5] can be viewed as an extension of the BSB model as it allows simultaneous formation of many clusters corresponding to a greater number of salient subgroups in the input. Grouping is achieved by purely spatial interactions, using a set of identical layers of feature-selective neurons. Due to a competitive interaction between layers each feature is unambiguously assigned to one layer, thus seperating different groups. The binding of features is caused by a lateral interaction within layers which determines the prefered groupings of compatible features.

2 Model Architecture

The CLM consists of a set of L identical layers of feature-selective neurons which are replicas of an input layer (see Fig. 1). The neurons in the input layer are labelled by their position $\mathbf{r}$. Driven by an external input, each input neuron responds with a value $h_\mathbf{r}$ which indicates activity ($h_\mathbf{r} > 0$) in the presence of the corresponding feature or silence ($h_\mathbf{r} \leq 0$). In a simple setting we may think of $h_\mathbf{r}$ as encoding the light intensity at position $\mathbf{r}$ in some "imaginary" retina. (For more abstract features the position vector may be embedded in a higher dimensional spatially organized topographic feature map.)

For each of the input neurons at position $\mathbf{r}$ there is one neuron per replica layer α whose activity is $x_{\mathbf{r}\alpha} \geq 0$ (see Fig. 1). We denote these L activities by the *row* $\mathbf{r}$. The activities are subject to the following constrained gradient dynamics:

$$\dot{x}_{\mathbf{r}\alpha} = -\sigma_{\mathbf{r}\alpha}\frac{\partial E}{\partial x_{\mathbf{r}\alpha}} = \sigma_{\mathbf{r}\alpha} \cdot \left(J_1(h_\mathbf{r} - \sum_{\beta=1}^{L} x_{\mathbf{r}\beta}) + \sum_{r'} \mathsf{f}_{\mathbf{r}\mathbf{r}'} x_{\mathbf{r}'\alpha}\right) , \qquad (1)$$

where the constraint $x_{\mathbf{r}\alpha} \geq 0$ is dynamically enforced by

$$\sigma_{\mathbf{r}\alpha} = \begin{cases} 0 & \text{if} \quad x_{\mathbf{r}\alpha} = 0, \quad \frac{\partial E}{\partial x_{\mathbf{r}\alpha}} > 0 \quad \text{(active constraint)} \\ 1 & \text{else} \qquad\qquad\qquad\qquad\quad \text{(inactive constraint)} \ . \end{cases} \qquad (2)$$

J_1 is a positive constant and $\mathsf{f}_{\mathbf{r}\mathbf{r}'} = \mathsf{f}_{\mathbf{r}'\mathbf{r}}$ are the components of a symmetric weight matrix. As a result of the constraints (2), the dynamics (1) defines a nonlinear dynamical system, which consists of a series of continuously connected linear systems. The corresponding energy function E is quadratic and given by

$$E = \frac{J_1}{2}\sum_{r}\Big(\sum_{\alpha=1}^{L} x_{r\alpha} - h_r\Big)^2 - \frac{1}{2}\sum_{\alpha=1}^{L}\sum_{rr'} \mathsf{f}_{\mathbf{r}\mathbf{r}'} x_{\mathbf{r}\alpha} x_{\mathbf{r}'\alpha} \ . \qquad (3)$$

The gradient in (1) can be split into $-\frac{\partial E}{\partial x_{\mathbf{r}\alpha}} = J_1 V_\mathbf{r} + F_{\mathbf{r}\alpha}$, where $V_\mathbf{r} = h_\mathbf{r} - \sum_\beta x_{\mathbf{r}\beta}$ and $F_{\mathbf{r}\alpha} = \sum_{\mathbf{r}'} \mathsf{f}_{\mathbf{r}\mathbf{r}'} x_{\mathbf{r}'\alpha}$ are the two basic interactions in the model:

- The *"vertical"* interaction $J_1 V_\mathbf{r}$ causes strict competition between neurons in a row belonging to the same feature. The intensity $h_\mathbf{r}$ gives a constant input to the row. If we consider a single row of the model without lateral connections the competitive dynamics is equivalent to the dynamics of the Winner-Take-All MAXNET [2] with added inhomogenity.
- The *lateral* interaction $F_{\mathbf{r}\alpha}$ couples activities within each layer by the symmetric weight matrix $\mathsf{f}_{\mathbf{r}\mathbf{r}'}$. The interactions $\mathsf{f}_{\mathbf{r}\mathbf{r}'}$ determine which pattern configurations, if elicited as activity pattern within a single layer, will be mutually supporting among their constituent parts (for $\mathsf{f}_{\mathbf{r}\mathbf{r}'} > 0$) or instead suffer mutual inhibition (for $\mathsf{f}_{\mathbf{r}\mathbf{r}'} < 0$).

As will be shown below, the interplay between the vertical and the lateral interactions leads to a dynamic partitioning of an input pattern (offered via the

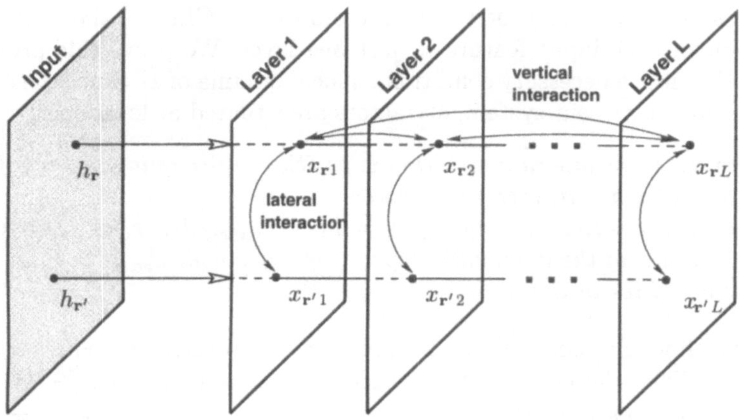

Fig. 1. The Competitive Layer Model architecture.

$h_{\mathbf{r}}$-inputs of the leftward input layer) such that each layer responds only to a pattern fragment, with the union of all fragments giving the entire pattern. In this way the full pattern becomes grouped into smaller constituents which are determined by the lateral interaction $\mathsf{f}_{\mathbf{rr}'}$.

The parametric model for $\mathsf{f}_{\mathbf{rr}'}$ has to be chosen such that desired grouping states correspond to energy minima of (3), where the quality of a grouping is measured by the second term, summing up all pairwise contributions weighted by $\mathsf{f}_{\mathbf{rr}'}$. In this paper we suggest a heuristic approach and consider two interactions of the form $\mathsf{f}_{\mathbf{rr}'} = M_{\mathbf{rr}'} - k$ where $M_{\mathbf{rr}'}$ is a positive measure of the compatibility between two features $\mathbf{r}, \mathbf{r}'$ and $k > 0$ is an inhibitory threshold which controls the required degree of overall conformity within a group. Larger values for k result in a more fragmented grouping, while for $k = 0$ the global minimum of (3) is attained by forming just one large group.

3 Convergence and Stability Analysis

The following theorem guarentees the global convergence of the CLM:

Theorem 1. *The CLM system is convergent if* $J_1 > \sum_{\mathbf{r}' \wedge \, \mathsf{f}_{\mathbf{rr}'} > 0} \mathsf{f}_{\mathbf{rr}'}$ *for all* $\mathbf{r}$.

Proof. (Outline) The behaviour of E at infinity is dominated by the second-order terms which can be shown to be bounded from below by $E' = \frac{1}{2} \sum_{\mathbf{r}\alpha} x_{\mathbf{r}\alpha}^2 \big(J_1 - \sum_{\mathbf{r}' \wedge \, \mathsf{f}_{\mathbf{rr}'} > 0} \mathsf{f}_{\mathbf{rr}'}\big)$. For sufficiently large J_1, E is then bounded from below and $E \to \infty$ for $\|\mathbf{x}\| \to \infty$. Since E is also strictly decreasing on nonstationary trajectories E is a global Lyapunov function and according to the Invariant Set Theorem [3] the system converges to the set of equilibria. □

This condition implies that the coupling to the input, controlled by J_1, must be sufficiently strong compared to the self-excitation within each layer.

The essential property of the attractors of the CLM is the unambiguous assignment of each input feature to just one layer. We prove this property by showing that it is a necessary condition on local minima of E. Since the dynamics (1) is a gradient descent system, attractors are situated at local energy minima.

Theorem 2. *If the lateral interaction is positive self-coupling,* $\mathsf{f}_{\mathbf{rr}} > 0$ *for all* $\mathbf{r}$, *then an attractor has in each row* $\mathbf{r}$ *either*
i) at most one positive activity $x_{\mathbf{r}\alpha(\mathbf{r})} = h_{\mathbf{r}} + F_{\mathbf{r}\alpha(\mathbf{r})}/J_1$, $x_{\mathbf{r}\beta\neq\alpha(\mathbf{r})} = 0$, *where* $\alpha(\mathbf{r})$ *is the index of the maximally supporting layer with* $F_{\mathbf{r}\alpha(\mathbf{r})} > F_{\mathbf{r}\beta\neq\alpha(\mathbf{r})}$ *or*
ii) for all activities in a row $x_{\mathbf{r}\alpha} = 0$, $F_{\mathbf{r}\alpha} \leq 0$.

Proof. (Outline) Suppose an equilibrium has two positive activities $x_{\mathbf{r}\alpha,\beta} > 0$ in a row $\mathbf{r}$. Then the constraint is inactive, $\sigma_{\mathbf{r}\alpha,\beta} = 1$, hence $\frac{\partial E}{\partial x_{\mathbf{r}\alpha,\beta}} = 0$. Now consider a small perturbation of the form $\epsilon_{\mathbf{r}\alpha} = \eta$, $\epsilon_{\mathbf{r}\beta} = -\eta$, $\epsilon_{\mathbf{r}'\alpha'} = 0$ for all other $\mathbf{r}', \alpha'$. Expanding E about the equilibrium we obtain for ΔE under this perturbation with all linear components vanishing at the equilibrium

$$\Delta E = \frac{1}{2} \sum_{\mathbf{r}\alpha\mathbf{r}'\alpha'} (J_1 \delta_{\mathbf{rr}'} - \delta_{\alpha\alpha'} \mathsf{f}_{\mathbf{rr}'}) \epsilon_{\mathbf{r}\alpha} \epsilon_{\mathbf{r}'\alpha'} = -\eta^2 \mathsf{f}_{\mathbf{rr}} < 0 \ . \quad (4)$$

Because this perturbation decreases the energy this equilibrium must be instable and there can be at most one positive activity in a row for a stable equilibrium. Now consider case i): Let $\alpha(\mathbf{r})$ denote the layer index of the positive activity in row $\mathbf{r}$. The constraint is then inactive for $x_{\mathbf{r}\alpha(\mathbf{r})}$ and active for $x_{\mathbf{r}\beta\neq\alpha(\mathbf{r})}$. Hence

$$0 = \frac{\partial E}{\partial x_{\mathbf{r}\alpha(\mathbf{r})}} = J_1 V_{\mathbf{r}} + F_{\mathbf{r}\alpha(\mathbf{r})}, \qquad 0 > -\frac{\partial E}{\partial x_{\mathbf{r}\beta\neq\alpha(\mathbf{r})}} = J_1 V_{\mathbf{r}} + F_{\mathbf{r}\beta\neq\alpha(\mathbf{r})} \ , \quad (5)$$

which proves case i). Case ii) gives just the remaining possible stable equilibria which are not covered by i). □

This ensures that each feature is either mapped into the layer with maximal lateral feedback or completely removed from the grouping result. For large J_1 the superposition of activity patterns in all layers reproduces the input pattern. If J_1 is just above the threshold given in Theorem 1 the resulting activity pattern will show a strong deviation from the input intensity due to the lateral feedback $F_{\mathbf{r}\alpha}$. This allows a controllable autoassociativity of the model. An example would be the "filling in" of a feature that was not detected in the input, $h_{\mathbf{r}} = 0$, but receives a strong lateral feedback $F_{\mathbf{r}\alpha}$ in one layer.

4 Simulation Results for Perceptive Grouping Tasks

In the following, we illustrate the analytical results with some computer simulations for the task of grouping visual features. In the simplest possible case, we consider simple point features (characterized only by their location $\mathbf{r}$) and use an "On-Center-Off-Surround" interaction $\mathsf{f}_{\mathbf{rr}'} = \exp(-|\mathbf{r} - \mathbf{r}'|^2/(2R)) - k$ with a range R and inhibitory strength k. A CLM with this interaction splits input patterns into compact clusters of a typical radial size R. Fig.2a shows a simple image of a scene used as input for the CLM. In Fig.2b we see the grouping result achieved by a CLM of 10 layers, operating on the input image after subsampling to 50x40 pixels (each thresholded pixel intensity serving as a point feature in this simple demonstration).

Usually, one desires a more sophisticated grouping that requires to take additional Gestalt principles into account. Further important Gestalt laws are the "law of good continuation", which requires to consider in addition to position also local edge orientation as a visual feature, and laws that consider symmetry in various ways. Such additional Gestalt laws can be incorporated into the CLM by choosing a more sophisticated interaction function $\mathsf{f}_{\mathbf{rr}'}$. We can use, e.g., a suggestion made by Reisfeld [4] in a different context for an interaction between edge elements to detect local symmetries. Edges are parametrized by a position vector $\mathbf{r}$ and a unit orientation vector $\hat{\mathbf{n}}$. This leads to an interaction given by

$$\mathsf{f}_{\mathbf{rr}'\hat{\mathbf{n}}\hat{\mathbf{n}}'} = \exp\big(-(|\hat{\mathbf{n}}\hat{\mathbf{d}}| - |\hat{\mathbf{n}}'\hat{\mathbf{d}}|)^2\sigma\big)\exp\big(-\mathbf{d}^2/R\big) \quad - \quad k\ , \qquad (6)$$

where $\mathbf{d} = \mathbf{r} - \mathbf{r}'$, $\hat{\mathbf{d}} = \mathbf{d}/|\mathbf{d}|$ is the spatial (normalized) difference vector, σ controls the angular selectivity, R gives the spatial range and k controls the degree of conformity within groups. Fig.2 shows that a CLM with this interaction is capable of successfully grouping edge elements (obtained from a real world image) according to the principles of good continuation and symmetry. The spatial range R was chosen to be of the same magnitude as the desired group size. We observed that for noisy input data like the output of a gradient edge detection preprocessor the selectivity σ must be low to allow the detection of salient groups.

5 Conclusion

The CLM is able to break a set of input features into a number of groups by two different kinds of interactions: A competitive, topographically structured interaction between layers, and a lateral interaction within layers. The lateral interaction defines an energy functional, where grouping states correspond to minima of this energy. With a stability analysis we show that the attractors of the model form a disjoint decomposition of the input feature pattern for arbitrary symmetric and positive self-coupling lateral interactions. A necessary condition on attractors implies that each feature has to be grouped into the layer with maximal lateral feedback. We apply the model to perceptive grouping where we report results for point clustering and symmetry and good continuation grouping of edge elements.

References

1. S. A. Ritz J. A. Anderson, J. W. Silverstein and R. S. Jones. Distinctive features, categorical perception, and probability learning: Some applications of a neural model. *Psychological Review*, 84:413–451, 1977.

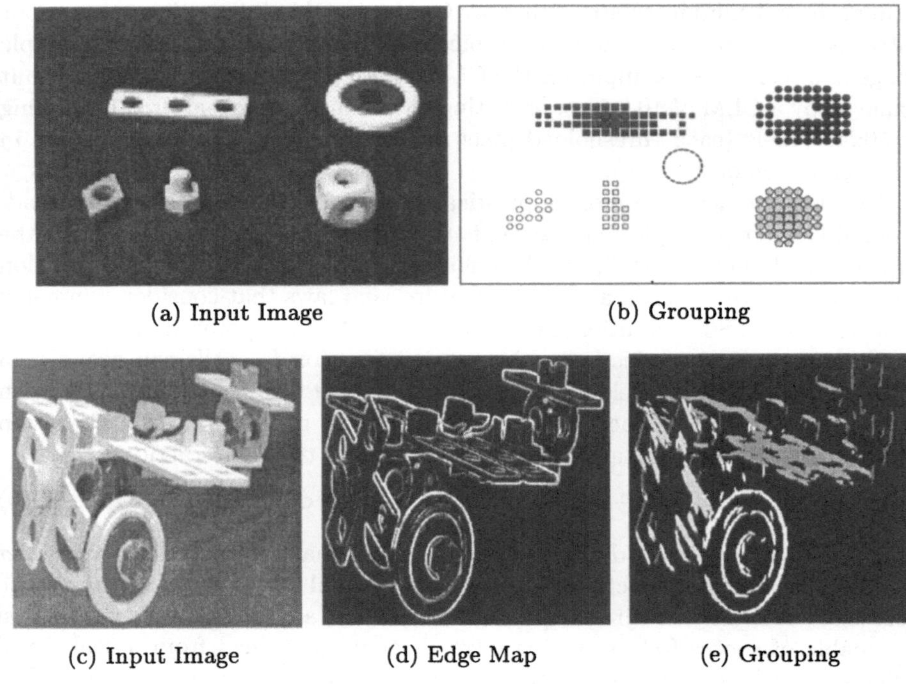

(a) Input Image (b) Grouping

(c) Input Image (d) Edge Map (e) Grouping

Fig. 2. Application to perceptive grouping. a) Input image. b) Grouping result for an "On Center Off Surround" interaction with 10 layers on a collection of point features derived from a) where layers are represented by different symbols. The range R is displayed by the radius of the dashed circle and the inhibition was given by $k = 0.1$. c) Input image and d) edge map obtained from a gradient operator. e) Grouping result of a lateral interaction for symmetry and good continuation with 5 layers on a set of 1200 derived (subsampled) edge features, displayed as (overlapping) ellipses of corresponding orientation and with size representing activity. Interaction parameters were $\sigma = 100, R = 0.1, k = 0.2$ and the displayed square was normalized to unit length. Wheel, wings and parallel parts of the model plane's propeller form seperate groups.

2. R. Lippmann. An introduction to computing with neural nets. *IEEE ASSP Mag.*, pages 4–22, 1987.
3. D. G. Luenberger. *Introduction to dynamical systems: Theory, models, and applications.* New York: Wiley, 1979.
4. Yeshurun H. W. Reisfeld, D. Context-free attentional operators: The generalized symmetry transform. *Int. Journal of Computer Vision*, 14:119–130, 1995.
5. H. Ritter. A spatial approach to feature linking. In *Int. Neur. Netw. Conf. Paris*, 1990.
6. R. Ritz, W. Gerstner, U. Fuentes, and J.L. van Hemmen. A biological motivated and analytically soluble model of collective oscillations in the cortex. II. Application to binding and pattern segmentation. *Biol. Cybern.*, 71:349–358, 1994.
7. C. v.d. Malsburg. The correlation theory of brain function. Technical Report 81-2, MPI Göttingen, 1981.

A Multilayer Real-Time Recurrent Learning Algorithm for Improved Convergence

Kürt Meert[§] and Jacques Ludik[Y]

[§]Expert Systems Applications Development Group, Dept. of Chemical Engineering, K.U.Leuven, De Croylaan 46, 3001 Heverlee, Belgium.
[Y]Department of Computer Science, University of Stellenbosch, Stellenbosch 7602, South Africa.

Abstract. In this paper an algorithm is described, which is based on the Real-Time Recurrent Learning (RTRL) algorithm by Williams and Zipser [1], for Multilayer Real-Time Recurrent Learning Networks (MLRN). The general gradient descent algorithm is used to update the weights, but without backpropagating the error through the network. To study its modelling efficacy and its dynamic capacities, the MLRN algorithm is applied to two benchmark problems, a grammar problem and a real-life modelling problem from the chemical process industry and is compared to various other fully and partially recurrent networks.

1. Introduction

Recurrent neural networks can perform highly nonlinear dynamic mappings and thus have temporally extended applications, whereas multilayer feedforward networks are confined to performing static mappings. However two-layered feedforward networks can approximate precisely any function [2]. Likewise Funahashi & Nakamura proofed that any finite time trajectory of a dynamic system can be approximated by the internal output units of a continuous time recurrent neural network with sufficient hidden and output units and an appropriate initial condition [3]. Recurrent neural networks have been used in a number of interesting applications including associative memories, spatio-temporal pattern classification. control, optimisation, forecasting and the generation of pattern sequences [4, 5]. Similarly to multilayer feedforward networks, multilayer recurrent networks (MLRN) are developed to improve modelling performance and noise robustness [6].
Still recurrent networks and more specifically those trained by the Real-Time Recurrent Learning algorithm are computationally intensive since the overall calculation time is proportional to the fourth power of the number of recurrent nodes. However the calculation time during network training for MLRN is reduced significantly due to a decrease of the number of recurrent connections. Additionally their mapping ability improves since the number of recurrent nodes in MLRN increases compared to monolayer recurrent networks for a similar number of connections. To demonstrate the properties of the MLRN some tests were performed on two benchmark problems, a grammar problem and a problem from the chemical process industry. Moreover these results are compared to those obtained by applying 6 different recurrent network structures to these benchmark problems. These networks are the partially recurrent networks described by Elman and Jordan and the fully recurrent networks like the Real-Time Recurrent Learning Algorithm (RTRL), truncated backpropagation through time (BPTT), batch backpropagation through time (BBPTT) and quickprop through time (QPTT).

2. Description of the Multilayer Real-Time Recurrent Learning Algorithm

The MLRN algorithm is based upon the RTRL algorithm by Williams and Zipser. The weights of this type of network are updated after every learning iteration by a gradient descent-like algorithm. For more details on the original algorithm we refer to the article of Williams and Zipser. In what follows we assume that there are only two hidden RTRL layers in the MLRN.
The input layer receives information from the environment and processes it to the first RTRL layer. The input layer is fully connected to the first RTRL layer, whereas the latter is fully connected to the second RTRL layer. One or more nodes of the second RTRL layer forward their activation to the environment. These are the target (or output) nodes of the network. As stated by Williams and Zipser [1] a RTRL-network has always a time delay of one time step. Hence a time delay of one time step is incorporated for each RTRL layer in the MLRN. Consequently the time delay for a two layered MLRN is two time steps. Generally this can be described by the following formula

$$Y(t+N) = F_{NET}\left[I(t)\right]$$

where Y and I respectively denote output and input of the MLRN, N indicates the number of RTRL layers and F_{NET} represents the MLRN.
During the learning phase the error on the target nodes is calculated and used to update the weights. Unlike some other multilayer recurrent network algorithms [6] the error is not backpropagated through the network, but rather the influence of each weight on the activity of each node is calculated.
For every node k of every layer n the activities can be calculated by the following formula

$$s_{k_1}(t) = \sum_{l \in U_1} w_{kl_1} y_{l_1} + \sum_{l \in I} w_{kl} x_l = \sum_{l \in U_1 \cup I} w_{kl_1} y_{l_1}$$

$$s_{k_2}(t) = \sum_{l \in U_2} w_{kl_2} y_{l_2} + \sum_{l \in U_1} w_{kl_{12}} y_{l_1}$$

$$y_{k_n}(t+1) = f_{k_n}\left[s_{k_n}(t)\right] \qquad n = 1, 2$$

where $s_{k_n}(t)$ is the weighted sum in node k of layer n , $w_{kl_n}(tl$ the weight between node l and k of layer n, $w_{kl_{12}}(t)$ the weight between node k of layer 2 and node l of layer 1, $y_{l_n}(t)$ the activity of RTRL-node l of layer n, $x_l(t)$ the activity of the input node l , f_{k_n} the transfer function of neuron k of layer n and I, U_1 and U_2 respectively denote the set of input nodes and the set of nodes in the first and second recurrent layer. The error of the target nodes is defined as

$$e_{k_2}(t) = d_{k_2}(t) - y_{k_2}(t) \quad \text{if } k \in T \quad \text{where } T \subset U_2$$
$$= 0 \qquad \text{otherwise}$$

where $d_{k_2}(t)$ is the target value and T is the set of target neurons. The RTRL algorithm minimises the sum of squared errors and takes the following form for the MLRN

$$\Delta w_{ij_n}(t) = \alpha \sum_{k \in U} e_k(t) \frac{\partial y_k(t)}{\partial w_{ij_n}} = \alpha \sum_{k \in U_2} e_{k_2}(t) \frac{\partial y_{k_2}(t)}{\partial w_{ij_n}} \quad since \quad e_k(t) = 0 \quad if\ k \in U_1$$

$$and \quad \frac{\partial y_k(t)}{\partial w_{ij_n}} = p_{ij_n}^k(t)$$

where α denotes the positive learning rate, $U = U1 \cup U2$ and $p_{ij_n}^k(t)$ the impact of the weight $w_{ij_n}(t)$ on node k and can be calculated by the following recursive formula

$$p_{ij_n}^k(t_0) = 0$$

for all RTRL layers n and the connections between the first and the second layer
for all RTRL nodes i and k
for all nodes j

$$p_{ij_n}^k(t+1) = f'_{k_1}\left[s_{k_1}(t)\right]\left[\sum_{l \in U_1} w_{kl_1}(t) p_{ij_n}^l(t) + \delta_{ik} y_j(t)\right] \qquad k \in U_1$$

$$p_{ij_n}^k(t+1) = f'_{k_2}\left[s_{k_2}(t)\right]\left[\sum_{l \in U_1} w_{kl_{12}}(t) p_{ij_n}^l(t) + \sum_{l \in U_2} w_{kl_2}(t) p_{ij_n}^l(t) + \delta_{ik} y_j(t)\right] \qquad k \in U_2$$

for all RTRL layers n and the connections between the first and the second layer
for all RTRL nodes i
for all nodes j

with $f'k_n$ the derivative of the transfer function fk_n and δ_{ik} the Kronecker delta. The weights on the various connections in each layer and between the layers are thus adapted by

$$\Delta w_{ij_n}(t) = \alpha \sum_{k \in U} e_k(t) p_{ij_n}^k(t)$$

$$w_{ij_n}(t) = w_{ij_n}(t-1) + \Delta w_{ij_n}(t)$$

As pointed out in the introduction, one of the major drawbacks of the RTRL algorithm is the increase in calculation time when the number of recurrent nodes rises. For monolayer RTRL networks, the number of impacts to calculate during training is $n(n+m)n$, and each impact requires $2n+m$ sums, where n denotes the number of recurrent nodes and m the number of input. This results in $2n^4+3n^3m+n^2m^2$ calculations. Thus the overall computation time on each cycle is of the order of n^4. Consider a MLRN where the sum of n_1 and n_2, the number of nodes respectively in the first and second recurrent layer, equals n. The number of impacts of the first layer equals $n_1(n_1+m)n_1$, whereas those for the second layer equals $n_2(n_2+n_1)n_2$. Each impact of the first and the second layer requires respectively $2n_1+m$ and $2n_2+n_1$ sums. This results in $2(n_1^4+n_2^4)+3(n_1^3m+n_2^3n_1)+n_1^2m^2+n_2^2n_1^2$ calculations on each cycle, which is of the order of $n_1^4+n_2^4$. If both expressions are compared, one can easily conclude that the computation time for MLRN is reduced compared to the monolayer RTRL network.

3. Simulations

To demonstrate the properties of the MLRN algorithm, some simulations are performed on two different benchmark problems, a grammar problem and a real-life modelling problem from the chemical process industry. A comparison is made with 6 other well-known recurrent networks, which are the partially recurrent networks described by Elman and Jordan and the fully recurrent networks, RTRL, truncated BPTT, BBPTT and QPTT.

3.1. Three Consecutive zeros grammar problem

Miller & Giles have developed a number of benchmark grammar to compare the performance of recurrent networks [7]. One of these grammars, which can be specified as a discrete finite automata, is used here in order to compare the above described fully and partially recurrent networks. The three consecutive zeros grammar generates strings or sentences of arbitrary length consisting of zeros and ones and can be described as any string containing three consecutive zeros. Each network has two inputs, i.e. the grammar string and a done bit, which signifies the end of a string, and one output parameter, which indicates the detection of three consecutive zeros. The task was made more difficult (compared to Miller & Giles) by insisting that in case of detection, the network's output changes form zero to one until it has reached a done bit. Thus if more consecutive zeros or even another triplet of zeros appear in the same sentence, the network has to ignore these inputs until the done bit appears. The length of these sentences varies from 1 bit to 100 bits, whereas the number of sentences with a specific length linearly drops when its length increases. The training and test set consisted of 100 sentences each, which respectively constituted 1139 and 1114 patterns.
For comparison purposes, the % correctness was determined on a pattern as well as on a sentence basis. The % correctness on a pattern basis was the percentage of correctly classified input patterns, whereas % correctness on a sentence basis was determined by calculating the percentage of ill-classified sentences, thus expressed in terms of the % faulty sentences.
In order to have a fair comparison, the fully and partially recurrent networks were trained up to 500 epochs with many different combinations of parameters to obtain the optimal number of hidden nodes and other relevant learning parameters for each network. The RTRL and the MLRN network obtained their optimal result after 400 epochs. For each optimal network, the simulations were repeated 5 times with different initial weights to obtain the optimal and average % correctness on a pattern as well as on a sentence basis (Table 1).

Recurrent network type	*Training % Correctness*		*% Faulty Sentences*	*Testing % Correctness*		*% Faulty Sentences*
	Optimum	*Average*		*Optimum*	*Average*	
MLRN	100.0	100 (0.05)	0	99.8	98.1 (1.83)	2
RTRL	100.0	100 (0.09)	0	99.4	97.6 (1.81)	2
Elman	99.9	96.2 (4.77)	1	98.5	94.2 (5.02)	9
QPTT	99.7	99.1 (0.32)	3	97.2	96.3 (1.08)	27
BPTT	98.2	97.8 (0.20)	20	97.2	95.0 (1.26)	28
BBPTT	98.0	97.6 (0.69)	20	97.1	95.7 (1.08)	27
Jordan	61.0	60.5 (0.36)	92	57.1	56.5 (0.66)	91

Table 1. The results for various networks for the grammar problem (the standard deviation is indicated in brackets).

The MLRN (6 and 3 nodes in its first and second layer, a learning rate of 0.001 and a clear impact of 5) outperformed all other networks. The RTRL network (8 nodes, a learning rate of 0.001 and a clear impact of 5) and the QPTT (11 nodes, the number of backsteps is 9, a maximum growth parameter of 1.75 and a learning rate η of 0.01) performed best among the other fully recurrent networks, whereas the Elman network (10 nodes, an η of 0.1 and a momentum term of 0.6) outperformed the Jordan network (10 nodes and self-recurrent weight values of 0.6). The optimal BPTT and BBPTT networks had also 11 nodes. It is obvious that the MLRN and RTRL network performed better on such a highly dynamic problem due to its internal recurrent structure and learning algorithm which ensures an "unlimited" memory capacity. Although the Elman network obtained good optimal values, on average it performed worse than all fully recurrent networks and also had the largest standard deviation. On the contrary the Jordan networks were not able to learn the task, since only 9% of the test sentences were correctly classified.

3.2. Distillation Column problem

The second benchmark for evaluating the performance of the fully and partially recurrent networks was a real-world modelling problem from the chemical process industry. The task was to predict the recovery of the light component in the bottom product of a bicomponent distillation column. The plant is characterised by twelve parameters, of which eleven were used as input parameters and one, the recovery of the light component in the bottom product, is applied as the desired output parameter. Historical data was extracted out of the plant's database. A representative part of the datafile (6000 patterns) was used to train the networks, while the remaining part (3000 patterns) was applied as test set. The recurrent networks were trained for up to 500 epochs with various combinations of parameters to obtain the optimal size and parameters for each network (Table 2).

Recurrent network type	*Training*		*Testing*	
	Optimum	*Average*	*Optimum*	*Average*
QPTT	0.0224	0.0250 (0.0024)	0.0369	0.0407 (0.0011)
Elman	0.0339	0.0389 (0.0025)	0.0454	0.0606 (0.0050)
Jordan	0.0387	0.0406 (0.0011)	0.0539	0.0667 (0.0038)
MLRN	0.0438	0.0448 (0.0007)	0.0628	0.0699 (0.0071)
RTRL	0.0592	0.0594 (0.0001)	0.0856	0.0886 (0.0041)
BBPTT	0.0530	0.0628 (0.0010)	0.0532	0.1287 (0.0058)
BPTT	0.0359	0.1496 (0.0566)	0.0380	0.2543 (0.1070)

Table 2. The RMS errors for various networks for the distillation column problem (the standard deviation is indicated in brackets).

From Table 2 it is evident that the QPTT network (12 nodes, the number of backsteps is 9, a maximum growth parameter of 1.75 and a learning rate η of 0.05) outperformed the Elman, Jordan, MLRN, RTRL, BBPTT and BPTT networks on training average by respectively 36%, 38%, 44%, 58%, 60% and 83%. The Elman network (12 nodes, an η of 0.05, a momentum term of 0.9 and a flat spot elimination constant of 0.1) with self-recurrent weight values of 0.6 performed slightly better than the Jordan network, which had the same parameters as the Elman network except for a momentum term of 0.1, on training average by only 4%. The performance of the Jordan network and the MLRN (7 and 3 nodes in its first and second layer, a learning rate of 0.0001 and without clear impact) was similar on average basis. The optimal error value of the Jordan network, however, is better than that of the MLRN by 12%. The MLRN outperformed the RTRL network (9 nodes and a learning rate α of 0.0001) on training average by 25%. Similar conclusions can be derived when comparing the test averages. The BPTT (12 nodes, the number of backsteps is 9, a momentum term of 0.95 and an η of 0.0005) network performed very poorly on training and test average compared to the RTRL and the BBPTT (9 nodes, the number of backsteps is 9, a momentum term of 0.9 and an η of 0.1) networks, while having a very large standard deviation (10 to 20 times larger on respectively train and test set).

4. Concluding Remarks

In this article an algorithm for multilayer real-time recurrent learning networks was presented and tested on a grammar problem and a chemical engineering problem, a bicomponent distillation column. Its performance was compared to that of 6 other fully and partially recurrent networks. The MLRN fully outperformed these networks on the grammar problem. Although 3 recurrent networks performed better on the real-world modelling problem, the MLRN's results show that the network is able to model nonlinear and dynamic systems. Nevertheless further research should focus on this topic to improve the MLRN's modelling capabilities.
Furthermore these results show that MLRN not only make better models but that they can also be trained faster compared to mono-layer RTRL networks. Some additional research should be invested in the derivation of the memory capacity of the MLRN networks. This can, for example, be done by varying the number of consecutive zeros the network has to detect for the grammar problem. With respect to this subject, in [8] a general method for the probabilistic memory capacity of recurrent neural networks is developed.

Acknowledgement

This research was performed at the Expert Systems Applications Development Group which is headed by Prof. M. Rijckaert. This research was partially granted by the Stellenbosch-Leuven Exchange Program.

References

1. R.J. Williams and D. Zipser. Experimental Analysis of the Real-Time Recurrent Learning Algorithm. *Connection Science,* (1), 87-111, 1989.
2. G. Cybenko. Approximation by Superpositions of a Sigmoidal Function. *Mathematical Control Signals and Systems,* (2), 303-314, 1989.
3. K. Funahashi and Y. Nakamura. Approximation of Dynamical Systems by Continuous Time Recurrent Networks. *Neural Networks,* (6), 801-806, 1993.
4. B.A. Pearlmutter. Gradient Calculations for Dynamic Recurrent Neural Networks: a Survey. *IEEE Transactions on Neural Networks,* (6), 1212-1228, 1995.
5. G.V. Puskorius and L.A.Feldkamp. Neurocontrol of Nonlinear Dynamical Systems with Kalman Filter Trained Recurrent Networks. *IEEE Transactions on Neural Networks,* (5), 279-297, 1994.
6. A.G. Parlos, K.T. Chong and A.F. Atiya. Application of the Recurrent Multilayer Perceptron in Modelling Complex Process Dynamics. *IEEE Transactions on Neural Networks,* (5). 255-266, 1994.
7. C.B. Miller and C.L.Giles. Experimental Comparison of the Effect of Order in Recurrent Neural Networks. *International Journal of Pattern Recognition and Artificial Intelligence.* (7), 849-872, 1993.
8. S. Miyoshi and K. Nakayama. Probabilistic Memory Capacity of Recurrent Neural Networks. *Proc. of the IEEE International Conference on Neural Networks '96,* Washington. DC., (2), 1291-1296. 1996.

Increasing the Capacity of a Hopfield Network Without Sacrificing Functionality

Amos Storkey

Neural Systems Group, Imperial College, London amoss@ic.ac.uk

Abstract. Hopfield networks are commonly trained by one of two algorithms. The simplest of these is the Hebb rule, which has a low absolute capacity of $n/(2 \ln n)$, where n is the total number of neurons. This capacity can be increased to n by using the pseudo-inverse rule.
However, capacity is not the only consideration. It is important for rules to be local (the weight of a synapse depends ony on information available to the two neurons it connects), incremental (learning a new pattern can be done knowing only the old weight matrix and not the actual patterns stored) and immediate (the learning process is not a limit process). The Hebbian rule is all of these, but the pseudo-inverse is never incremental, and local only if not immediate. The question addressed by this paper is, 'Can the capacity of the Hebbian rule be increased without losing locality, incrementality or immediacy?'
Here a new algorithm is proposed. This algorithm is local, immediate and incremental. In addition it has an absolute capacity significantly higher than that of the Hebbian method: $n/\sqrt{2 \ln n}$.
In this paper the new learning rule is introduced, and a heuristic calculation of the absolute capacity of the learning algorithm is given. Simulations show that this calculation does indeed provide a good measure of the capacity for finite network sizes. Comparisons are made between the Hebb rule and this new learning rule.

1 Introduction

Attractor networks such as Hopfield [4] networks are used as autoassociative content addressable memories. The aim of such networks is to retrieve a previously learnt pattern from an example which is similar to, or a noisy version of, one of the previously presented patterns. To do this the network associates each element of a pattern with a binary neuron. These neurons are fully connected, and are updated asynchronously and in parallel. They are initialised with an input pattern, and the network activations converge to the closest learnt pattern.

In order to perform in the way described, the network must have a learning algorithm which sets the connection weights between all pairs of neurons so that it can perform this task.

These learning rules can have a number of characteristics. Firstly a rule can be local. If the update of a particular connection depends only on information available to the neurons on either side of the connection, then the rule is said to be local. Locality is important, because it provides a natural parallelism to the learning rule, which, when combined with the local update dynamics, make a Hopfield network a truly parallel machine.

Secondly a rule can be incremental. If the learning process can modify an old network configuration to memorise a new pattern, without needing to refer to any of the previously learnt patterns, then an algorithm is called incremental. Clearly incrementality makes the Hopfield network adaptive, and therefore more suitable for changing environments or real time situations.

Thirdly a rule can either perform an immediate update of the network configuration, or can be a limit process. The former makes for faster learning.

Lastly and most importantly a learning algorithm has a capacity. This is some measure of how many patterns can be stored in a network of a given size. More specifically, in this paper we consider the absolute capacity [6] of the network. This is given by the asymptotic ratio of the number of patterns that can be stored without error to the number of neurons, as the network size tends to infinity.

Capacity is not just important because it allows more efficient information storage. Because the update time is at least proportional to the number of neurons, higher capacity also allows faster processing times.

1.1 Different learning rules

The most common, and indeed the simplest, learning rule for Hopfield networks is called the Hebb rule. The Hebb rule is local, incremental and immediate, but has an absolute capacity of $n/(2\ln n)$ [6].

To increase the capacity of the network, the pseudo-inverse learning rule can be used. This has a capacity of n [5], but does not have the functionality of the Hebb rule. It is not incremental, and in its most common form, is not local either. It is possible to create a pseudo-inverse weight matrix by a local method, but only as a limit process rather than with immediate learning [3, 2].

Another way to increase the capacity to $n/\sqrt{2\ln n}$ is through the use of non-monotonic continuous neurons [7]. However this increases the computational and storage burden because the response function of each neuron, as well as the weight matrix, can change, and neuron activations can take a whole range of values, rather than just binary ones.

It is therefore important to look for new learning methods which increase the capacity of the network from that of Hebbian learning, but without sacrificing important functionality, such as locality, incrementality, or immediacy.

This paper introduces a learning rule which keeps this full functionality, and increases the capacity of the network to $n/\sqrt{2\ln n}$, a significant improvement on that of Hebbian learning.

2 Framework

This section outlines the important concepts used in this paper.

A Hopfield network can be defined by the coupled difference equations

$$\sigma_i(t+1) = sgn\left(\sum_{j=1,j\neq i}^{n} w_{ij}\sigma_j(t)\right) \text{ for } i = 1,2,\ldots,n$$

where $\sigma_i(t) = \pm 1$ is the state of neuron i after the tth update, and $W = w_{ij}$ is the weight matrix given by the learning rule, and is, in all cases we will consider, symmetric. The network is initialised with a pattern S_i by setting $\sigma_i(0) = S_i$, and the time evolution of the network will quickly settle into an equilibrium.

For a useful learning rule this point will be the learnt pattern which is closest (suitably defined) to the pattern the network was initialised with, and so the aim of a learning rule is to set up the weight matrix so that the patterns to be learnt are attractors of the system.

2.1 Characteristics of a learning rule

The characteristics which have been called locality, incrementality and immediacy have already been mentioned. Here these concepts are made more precise.

In what follows ν counts the patterns (denoted ξ^ν) to be learnt. W^ν denotes the state of the weight matrix after the $1\ldots\nu$th patterns have been learnt but before the $(\nu+1)$th pattern has been introduced.

An incremental learning rule is any where W^ν is a function only of $W^{\nu-1}$ and ξ^ν.

A local learning rule has $w_{ij}(s)$ dependent only on ξ_i^μ, ξ_j^μ, $w_{ik}(s-1)\xi_k^\mu$ and $w_{jk}(s-1)\xi_k^\mu$, where $\mu = 1,2,\ldots,\nu$, $k = 1,2,\ldots,n$. Here s counts through the steps in the evolution of the weight matrix W.

An immediate learning process takes a finite number of steps to obtain W^ν. Any process which is not immediate is called a limit process. Note that a workable weight matrix can be obtained by truncating the limit process, but even so, most limit processes are slower at obtaining a weight matrix than immediate ones.

The three learning rules used in this paper are given by

Definition 1 Hebbian learning rule. The Hebbian learning rule is given by

$$w_{ij}^0 = 0 \ \forall i,j \text{ and } w_{ij}^\nu = w_{ij}^{\nu-1} + \frac{1}{n}\xi_i^\nu\xi_j^\nu$$

Definition 2 Pseudo-inverse learning rule. The pseudo inverse is given by

$$w_{ij}^\nu = \frac{1}{n}\sum_{\nu=1}^{m}\sum_{\mu=1}^{m}\xi_i^\nu(Q^{-1})^{\nu\mu}\xi_j^\mu$$

where $Q = \frac{1}{n}\sum_{k=1}^{n}\xi_k^\nu\xi_k^\mu$, and m is the total number of patterns with $m < n$. It can be seen from the form of these definitions that the Hebbian rule is local, incremental and immediate, but that the pseudo-inverse is neither local nor incremental, because it involves calculating an inverse.

Definition 3 The new learning rule. The weight matrix of an attractor neural network is said to follow the new learning rule if it obeys

$$w_{ij}^0 = 0 \; \forall i,j \text{ and } w_{ij}^\nu = w_{ij}^{\nu-1} + \frac{1}{n}\xi_i^\nu \xi_j^\nu - \frac{1}{n}\xi_i^\nu h_{ji}^\nu - \frac{1}{n}h_{ij}^\nu \xi_j^\nu \tag{1}$$

where $h_{ij}^\mu = \sum_{k=1,k\neq i,j}^n w_{ik}^{\mu-1}\xi_k^\mu$ is a form of local field at neuron i, and ξ^μ is the new pattern to be learnt.

Once again it is clear from the form of the above that the new learning rule is local, incremental and immediate.

3 The capacity: Signal to noise analysis

For a particular pattern ξ^ν to be a fixed point of the system, we require that $\sum_{j=1,j\neq i}^n \xi_i^\nu W_{ij}\xi_j^\nu > 0$ for all $i \leq n$. After expanding out the new learning rule this gives the condition that

$$\begin{aligned} 1 + N_i^\nu = 1 + \sum_{r=1}^m \frac{(-1)^{r-1}}{n^r} \sum_{j=1}^n \sum_{\mu_r=1,\mu_r\neq\nu}^m \left(\prod_{s=1}^{r-1}\sum_{\mu_s=1}^m\sum_{k_s=1}^n\right) S(\mu_1,\mu_2,\ldots,\mu_r;\nu) \\ T(i,k_1,\ldots,k_{n-1},j)\xi_i^\nu\xi_i^{\mu_1}\left(\prod_{t=1}^{r-1}\xi_{k_t}^{\mu_t}\xi_{k_t}^{\mu_{t+1}}\right)\xi_j^{\mu_r}\xi_j^\nu > 0 \end{aligned} \tag{2}$$

where the $(\prod_{a=1}^n \sum_{z_a}\sum_{y_a})$ notation is used to represent $\sum_{z_1}\sum_{y_1}\cdots\sum_{z_n}\sum_{y_n}$, m is the number of patterns stored, $S(a_1,a_2,\ldots,a_r;b) = R(a_1,a_2,\ldots,a_r) - \frac{n-2}{n}R(a_1,a_2,\ldots,a_r,b)$, and

$$R(a_1,a_2,\ldots,a_r) = \begin{cases} 1 \text{ if } \exists p \text{ such that } a_1 > \ldots > a_p < a_{p+1} < \ldots < a_r \\ \quad \text{and } a_i \neq a_j \; \forall i \neq j \\ 0 \text{ otherwise} \end{cases}$$

$$T(a_1,a_2,\ldots,a_r) = \begin{cases} 1 \text{ if } a_i \neq a_j \text{ for all } i,j \leq r \\ 0 \text{ otherwise} \end{cases}$$

The N_i^ν is called the noise at neuron i for the recall of memory ν. If $N_i^\nu < -1$ then the pattern ξ^μ is unstable at neuron i and is therefore recalled incorrectly.

In order to make any estimate of the capacity of this rule something must be known about the patterns that are to be stored. Here the patterns are assumed to be random in the usual way [6] The ξ_k^μ are taken to be random variables with $\xi_k^\mu = \pm 1$ and $P(\xi_k^\mu = +1) = \frac{1}{2}$. The variables are assumed to be mutually independent for all (k,μ).

Definition 4 Absolute Capacity. Suppose $m(n)$ random patterns are stored in the network size n. If in the limit $n \to \infty$ the probability that all the patterns are stable is 1, and $m(n)$ is the largest asymptotic relationship for which this occurs then $m(n)$ is the absolute capacity of the network.

Though the noise term might look very complicated, it is in fact just a sum of many ± 1 variables. Furthermore it can be shown that

$$E|N_i^\nu|^2 < \frac{m^2}{2n^2(1-\sqrt{m/n})^2}, \; E(N_i^\mu) = 0$$

Note that for all $r = 1$, $S = 2/n^2$, effectively eliminating order (m/n) noise.

At this point we make the natural, but unproven assumption that the noise is Gaussian, being a sum of many variables, which, although not independent, have a somewhat 'sparse' dependence. In this case the probability that a particular bit of a particular pattern is unstable is given by

$$P(N_i^\nu < -1) = \frac{1}{2} + \frac{1}{2}\mathrm{erf}\left(-\frac{1}{\sqrt{2}\sigma}\right) \to \frac{\sigma}{\sqrt{2\pi}} \exp\left(-\frac{1}{2\sigma^2}\right) \tag{3}$$

as $\sigma \to 0$. Here $\sigma^2 = E|N_i^\nu|^2 < m^2/[2n^2(1-\sqrt{m/n})^2]$. Now if we choose $m(n) = n/(\sqrt{2\ln n})$ such that $2n^2(1-\sqrt{m/n})^2/m^2 = 4\ln n(1-(2\ln n)^{-4})^2$, then we know from (3) that

$$P(N_i^\nu < -1) < p_{mn} < \frac{e}{2\pi n^2\sqrt{4\ln n}(1-(2\ln n)^{-4})}$$

and so $mnP(N_i^\nu < -1) < mnp_{mn} \to 0$.

Finally, because this is true for all $m \times n$ values of i, ν, the simple lemma

Lemma 5. *Let A_{nk} be an event. Let $A_n = \bigcup_{k=1}^n A_{nk}$. Then* $\lim P(A_n) = 0$ *if* $\limsup_{n\to\infty} \sum_{k=1}^n P(A_{nk}) \to 0$

shows that with $m(n) = n/(\sqrt{2\ln n})$, the probability that any of the bits of any of the patterns are unstable tends to zero as the network gets larger, and hence the capacity is $n/(\sqrt{2\ln n})$.

Simulations of the networks were performed, and the capacity results for the new learning rule were compared with those of the Hebb rule (Figure 1). The graph shows that the theory gives good approximations of the capacity for finite size networks.

4 Other considerations and Conclusions

Little has been said in this paper about basins of attraction and spurious states. These are important issues. Initial work in looking at these indicate that the number of spurious states and the size of the basins of attraction are comparable to that of the Hebb rule. This should not be seen as a surprise. In the new learning rule, the introduction of the attractors into the system still follows the Hebb formulation. The additional terms of the learning rule serve only to remove some of the lower order noise brought about by the interaction of different attractors, and hence increase the number of patterns that can be stored.

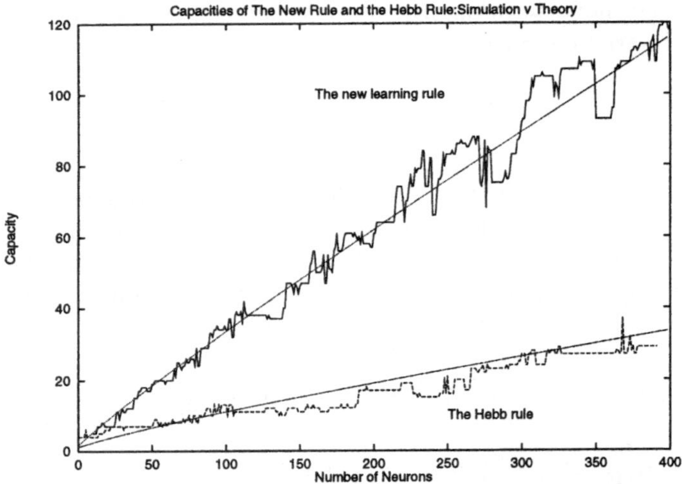

Fig. 1. Capacity of new learning rule v Hebb

This improvement of capacity for the Hopfield network is very significant. At network sizes of a million neurons, the absolute capacity of the new learning rule is five times that of the Hebb rule, increasing both storage and speed.

In this paper we have focussed only on absolute capacity. The Hebb rule has a relative capacity (where imperfect recall is allowed) of $0.14n$ [1]. Although it may be possible to calculate a relative capacity for the new learning rule, it should be noted that even at sizes of a million neurons, the absolute capacity of the new learning rule is greater than the relative capacity of the Hebb.

References

1. Daniel J. Amit, H. Gutfreund, and H. Sompolinsky. Statistical mechanics of neural networks near saturation. *Annals of Physics*, 173(1):30–67, 1987.
2. S. Diederich and M. Opper. Learning of correlated patterns in spin-glass networks by local learning rules. *Physical Review Letters*, 58(9):949–952, 1987.
3. V. S. Dotsenko, N. D. Yarunin, and E. A. Dorotheyev. Statistical mechanics of Hopfield-like neural networks with modified interactions. *Journal of Physics A: Mathematical General*, 24(10):2419–2429, 1991.
4. J. J. Hopfield. Neural networks and physical systems with emergent collective computational abilities. *Proceedings of the National Academy of Sciences of the United States of America: Biological Sciences*, 79(8):2554–2558, 1982.
5. I. Kanter and H. Sompolinsky. Associative recall of memory without errors. *Physical Review A- General Physics*, 35(1):380–392, 1987.
6. R. J. McEliece, E. C. Posner, E. R. Rodemich, and S. S. Venkatesh. The capacity of the Hopfield associative memory. *IEEE Transactions on Information Theory*, 33(4):461–482, 1987.
7. H. F. Yanai and S. I. Amari. Autoassociative memory with 2-stage dynamics of nonmonotonic neurons. *IEEE Transactions on Neural Networks*, 7(4):803–815, 1996.

A Novel Associative Network Accommodating Pattern Deformation

Hui Wang and David Bell

School of Information and Software Engineering,
University of Ulster Magee College,
Londonderry, BT48 7JL, N. Ireland,
e-mail: (h.wang, d.bell)@ulst.ac.uk

Abstract. In this paper we propose a novel associative network model which is able to associate a pattern with deformed versions of itself. The model is composed of a set of *logical units* (viewed as a set of marbles) and a set of regularly arranged *physical units* (viewed as a landscape). The marbles can move freely around the landscape under the influence of various kinds of forces arising from the landscape and other features of the world being modelled. This motion represents the evolution of the network. Information is embodied by the *topological* relationships among the marbles. When a network's evolution is initiated with an input pattern, topological relationships among marbles are observed and some aggregate features are preserved throughout the later evolution of the network. Evolution continues until all the marbles (logical units) match a set of physical units which corresponds to a local stable state of the network. Preliminary experiments show that the model works quite well for recognising topologically deformed letters.

1 Introduction

In this paper we are concerned with recurrent auto-associative networks [2, 3] (or *associative networks*, for simplicity). Consider a physical system described by many coordinates $X_1, X_2, \cdots, X_N$, the components of a state vector $\mathbf{X}$. The set of all state vectors is termed *state space*. Equations of motion of the system describe a flow in state space. We are particularly interested in flows towards locally stable states from anywhere within regions around those states. Let the system have locally stable states $\mathbf{X}_a, \mathbf{X}_b, \cdots$. Then if the system is initiated sufficiently *close* to any $\mathbf{X}_a$, as at $\mathbf{X} = \mathbf{X}_a + \Delta$, it is expected that the state will evolve until convergence terminates at $\mathbf{X} \approx \mathbf{X}_a$. We can regard the information stored in the system as the collection of vectors $\mathbf{X}_a, \mathbf{X}_b, \cdots$.

Many existing associative networks meet this system specification. For example when $\mathbf{X}$ represents *partial* or *incomplete* knowledge of the item $\mathbf{X}_a$ (this case is referred to as Δ_p hereafter), a Hopfield network [4] can generate the total information $\mathbf{X}_a$, granted that the network is not overloaded (i.e., the number of stable states $n < 0.15N$). It is well known that associative networks, typically Hopfield networks, are suitable for this type of pattern restoration tasks. In practical applications there are some other associative cases, for instance handwritten character recognition, that can be characterised by the $\mathbf{X}$ being a *topological deformation* of $\mathbf{X}_a$ (this case is referred to as Δ_d hereafter). Little is known yet about associative networks applicable to case Δ_d.

In this paper we present a novel associative network, called *moving around landscapes* (MAL), which addresses the associative memory problem with emphasis on the Δ_d case. Specifically, we arrange the state vector $\mathbf{X}$ in a geometrical space, and we are concerned with the topological relationships among the components and their evolution. To this end we must have a way of accounting for topological relationships and incorporating their evolution.

2 The model system

The system we introduce here can be intuitively perceived as a set of marbles moving around a landscape, and we use this model in the discussions below. A pattern is represented by a set of marbles with same the topology as the pattern, placed in a network of units (landscape). The fact that a marble is placed at a unit corresponds to the activation of this unit, while the fact that no marble is placed at a unit corresponds to the inhibition of this unit. What we are concerned with here is the movement of the marbles in the landscape due to the relationships between the marbles and the characteristics of the landscape. Topological relationships are embodied by the neighbourhood relationships among marbles, therefore the marble movement may induce the evolution of the topological relationships.

An **MAL** is a tuple $\mathcal{N} = (\mathcal{M}, \mathcal{L})$, where $\mathcal{M}$ is a finite set of **logical units** (marbles) that can move from one location to another, and $\mathcal{L}$ is a "landscape" (see below) over which the logical units can move. A **landscape** is a tuple $\mathcal{L} = (\mathcal{U}, \mathcal{C})$, where $\mathcal{U}$ is a finite set of **physical units**, which is a subset of a Euclidean space of any dimension (usually 2 or 3 dimensions); $\mathcal{C}$ is a set of ordered pairs of elements of $\mathcal{U}$ denoting the connections between the elements. The **mass** of logical unit a, denoted m_a, can be understood as its information content. The **state** of physical unit i, denoted m_i, is the sum of the masses of all the logical units at this unit, which is a measure of the local information at this physical unit. With each connection $\{i,j\} \in \mathcal{C}$ is associated a **slope** $\mathbf{k}_{ij} \in R^n$, where $\mathbf{k}_{ij} = k_{ij}\mathbf{I}_{ij}$, and $\mathbf{I}_{ij}$ is a unit vector directed from physical unit i to j. k_{ij} is a quantitative measure of the local possibility that information moves from physical units i to j, or intuitively, a quantitative measure of the physical slope from physical unit i to j. A **state configuration** or simply **state** of an MAL is uniquely defined by a point in the N-dimensional Euclidean space, $\mathbf{M}=(m_1, \cdots, m_N)$, whose i^{th} component m_i denotes the *state* of physical unit i. Where there is no ambiguity we use *unit* for *physical unit*. The structure of MAL depends on the structure of the landscape. Generally the landscape is arranged as a finite *geometrical* regular grid, typically of 2D or 3D, over which the marbles can move.

The forces inducing the movement of a marble are determined by the following factors. (1) The features of the landscape; (2) The states of a unit's surrounding units; (3) The neighbourhood relationship between individual marbles; and (4) The grouping coefficients of the network. Specifically the forces are divided into *gravitational force* (GF), *resistance force* (RF), and *binding force* (BF). The first two forces arise from the landscape of the network and are independent of application, while the third force is application-oriented. These forces are usually constrained by *grouping coefficients* (G), which are also related to specific applications.

All the forces above have to be reformulated in different applications, but typically they have the following forms:

- Gravitational force: The gravitational force (**GF**) is due to the landscape which is formed in the learning phase. The GF exerted on marble a at unit i is typically $\mathbf{GF}_{(a,i)}(t) = m_a \sum_{j \in N(i)} \mathbf{k}_{ij} G_j(t)$, where G_j is the *grouping coefficient* defined below, $\mathbf{k}_{ij}$ is the *slope* between unit i and unit j, and $N(i)$ is the set of neighbours of unit i. In some cases we simply denote the force as $\mathbf{GF}_a$ or $\mathbf{GF}_i$. This notation applies to other forces as well.
- Resistance force: The resistance force (**RF**) is usually inertial in nature. It is always in the opposite direction to the gravitational force. It is a function of the mass, and typically of the grouping coefficient G_i. The resistance force exerted on marble a at unit i is $\mathbf{RF}_{(a,i)}(t) = \gamma m_a G_i(t)$.
- Binding force: The binding force (**BF**) is used to preserve topological relationships as the marbles move around the landscape. It is actually a kind of constraint on the movement of the marbles. One example of a binding force is as Equation 5 below, which is used for character recognition.
- Grouping coefficient: The grouping coefficient (G) is used to guide the movements of the marbles. In different applications it may have different forms. Conceptually the grouping coefficient imposed on unit i, G_i, is defined as the probability of unit i being involved in the final stable state after initialising the network. It specifies the attractiveness of each individual unit. The grouping coefficients are not used individually – rather they are incorporated into the gravitational forces, resistance forces and even binding forces. The grouping coefficient can appear in two forms: *plain* and *non-plain*. For classification type applications, we do not know in advance which category a pattern falls into, and therefore we simply set $G_i = 0.5$ for $i = 1, 2, \cdots, N$; while for verification type of applications (e.g., combinatorial optimisation and signature verification) we know in advance which category we are interested in, and therefore we can pre-set the G_i's.

With these forces, the dynamics of MAL may be formulated as follows:

The force exerted on marble a at unit i at time t is:

$$\mathbf{F}_{(a,i)}(t) = \alpha \mathbf{GF}_{(a,i)}(t) + \beta \mathbf{BF}_{(a,i)}(t) + \gamma \mathbf{RF}_{(a,i)}(t) = F_{(a,i)}(t)\mathbf{J}_i.$$

If $F_{(a,i)}(t) > \phi$ and $F_{(a,i)}(t) > F_{(a,i')}$ then marble a will move from unit i to unit i', where unit i' is one of unit i's closest neighbours (i.e., one grid away) along the direction of $\mathbf{J}_i$. Note: α and γ are constants, while

β will be scaled down with time until it reaches zero. ϕ is a constant and usually $\phi \approx 0$.

It can be shown that the network will always converge, by defining an energy function as $H = \frac{1}{2}\sum_a |\alpha \mathbf{MF}_a + \beta \mathbf{BF}_a + \gamma \mathbf{RF}_a|^2$. According to the network dynamics above, any marble movement will cause the forces exerted on the marble to decrease, therefore making H decrease as well.

3 The learning algorithm

We now discuss how to make any prescribed set of states as the stable states of the network; or in other words how to form the landscape of the network.

We first of all define a target function E, and what we want to do is to find a set of $\mathbf{k}_{ij}$ that minimise E. On the one hand, the target function E is expected to be a measure of the stability of the network when a prescribed set of states is presented to the network. If no marble has potential to move, then the network is stable and the prescribed set of states has been stored; otherwise the network is unstable. On the other hand, the function E should also indicate the direction of the potential movement of each marble. With these considerations, we have the following form of target function:

$$E = \frac{1}{2}\sum_{l=1}^{L}\sum_{i=1}^{N}(|\mathbf{F}_i^l| - \phi)^2 + \frac{1}{2}\sum_{i=1}^{N}(\mathbf{I}_i \cdot \mathbf{J}_i - 1)^2 = \sum_{l=1}^{L} E_0^l + E_1 \tag{1}$$

where $\mathbf{F}_i^l$ is the force exerted on the marble at unit i due to pattern l, which is specified as

$$\mathbf{F}_i^l = \mathbf{GF}_i^l - \mathbf{RF}_i^l \tag{2}$$

$$\mathbf{GF}_i^l = m_i^l \sum_j k_{ij} G_j^l \mathbf{I}_{ij} = m_i^l |\sum_j k_{ij} G_j^l \mathbf{I}_{ij}| \mathbf{J}_i = |\mathbf{GF}_i^l| \mathbf{J}_i \tag{3}$$

$$\mathbf{RF}_i^l = \gamma m_i^l G_i^l \mathbf{J}_i \tag{4}$$

$\mathbf{I}_i$ is the unit vector in the expected direction of movement of unit i as specified by: $\mathbf{I}_i = \frac{\sum_{j \in N(i)} w_{ij}(\mathbf{P}_j - \mathbf{P}_i)}{|\sum_{j \in N(i)} w_{ij}(\mathbf{P}_j - \mathbf{P}_i)|}$ where $w_{ij} = exp(-|\mathbf{P}_j - \mathbf{P}_i|^2)$, and $\mathbf{P}_i$ is a vector in the working space representing the physical location of unit i. $\mathbf{J}_i$ is the unit vector in the direction of the forces applied to unit i.

Differentiating the target function with respect to slopes, we have (let $\phi \equiv 0$):

$$\frac{\partial E_0^l}{\partial k_{i_0 j_0}} = {m_{i_0}^l}^2 G_{j_0}^l (|\sum_j k_{ij} G_j^l \mathbf{I}_{ij}| - G_{i_0}^l \gamma)(\mathbf{J}_{i_0} \cdot \mathbf{I}_{i_0 j_0}).$$

$$\frac{\partial E_1}{\partial k_{i_0 j_0}} = (\mathbf{I}_{i_0} \cdot \mathbf{J}_{i_0} - 1) G_{j_0} |\sum_j \mathbf{k}_{i_0 j} G_j|^{-1} [\mathbf{I}_{i_0} \cdot \mathbf{I}_{i_0 j_0} - \mathbf{I}_{i_0} \cdot \mathbf{J}_{i_0} (\mathbf{I}_{i_0 j_0} \cdot \mathbf{J}_{i_0})]$$

Based on these results we can design an algorithm to undertake the learning task. For details refer to [5].

4 Preliminary experiments

We have experimented with the model on letter recognition. MAL is used as a pre-processing kernel for rectifying topological deformations, and a *template matching* algorithm is used for subsequent classification. We first trained the MAL network with 26 "standard" letters, and then tested the system with some deformed letters. Some of the training and testing letters are shown in Figure 1.

Due to lack of space we can only describe some of the specifications relating to MAL. The network we used in this experiment is arranged as a 16×16 grid. The gravitational force and resistance force are given in Eq. 2, where G_i is specified as *plain*. The binding force is as follows:

$$\mathbf{BF}_a(t) = m_a \sum_{b \in N(a)} m_b |D(\mathbf{P}_a, \mathbf{P}_b) - 1| \mathbf{I}_{ab} \tag{5}$$

where $N(a) = \{b \in \mathcal{M} : D(\mathbf{P}_a, \mathbf{P}_b) \leq d\}$ is the set of neighbours of marble a, d is the neighbourhood radius, and $D(x, y)$ is the distance measure in grids between x and y. The updating of the state configuration is done sequentially.

The experiment shows that the performance of the model is quite satisfactory if the testing patterns have topological structures very close to the training patterns; but performance goes down when the testing patterns are greatly deformed in topological structure. This deficiency may be reduced if a larger neighbourhood radius is employed, as shown in Figure 1.

This experiment is not intended to present a state-of-the-art character recognition system; rather it is intended to help assess the effectiveness of the proposed network and to provide an alternative way of approaching the problem of character recognition.

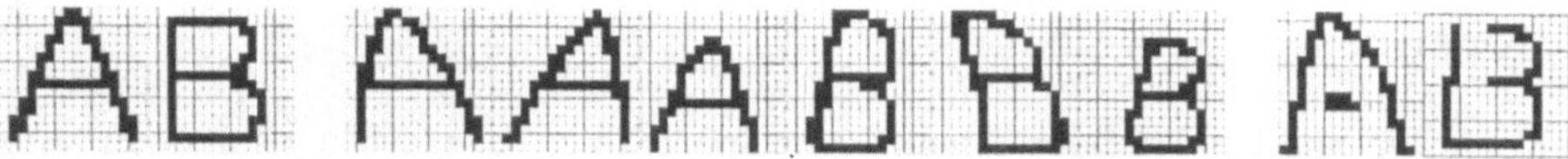

Fig. 1. *Some of the training and testing patterns used in the experiment. The leftmost columns are used as training samples, and the rest are testing patterns. When the neighbourhood radius is set as* $d = 2$*, the middle columns can be recognised perfectly, but the rightmost columns can not be properly recognised. When the neighbourhood radius is set as* $d = 3$*, both the middle columns and rightmost columns can be recognised correctly but it took longer to recognise the rightmost columns than the middle columns.*

5 Discussion and conclusion

The units in MAL have some elementary properties fundamentally different from that in a conventional associative network. (1) The state of a unit depends on whether or not there is a marble at it; (2) the activation of a unit depends on two factors: this unit must be sufficiently attractive, and there must be a marble nearby; and (3) the movement of marbles incurs the evolution of system configurations. The network has a clear structure, usually of grids of 2D or 3D. These properties facilitate the introduction of topological structure, thereby making the model particularly suitable for those applications that involve a great deal of topological deformations and those in which topological structure plays an important role. Examples are handwritten character recognition and combinatorial optimisation. Although different from a conventional associative network, MAL retains some collective computational properties. Memories are stored as stable states and can be correctly recalled from reasonably deformed versions.

There are few reports in literature on associative networks that can do the same job, i.e., associate patterns with their topologically deformed versions. However some reports employ the concept of force in one way or another. Durbin and Willshaw [1] propose an *elastic net* method for solving the TSP, in which forces are introduced to attract a circular path to designated cities. This method works quite well for the TSP, but it is not clear how to apply it to other problems. Williams (see [6] and the references therein) reports on a *deformable model* based method that can be used for handwritten digit recognition, where the *models* can be deformed (topologically) to match testing images, and so recognition can be achieved. This method may be regarded as a generalization of the elastic net method to character recognition, in the sense that the elastic net method will probably need the same development to be applicable to character recognition.

In its current state MAL has some disadvantage. It does not have the *pattern restoration* capability found in most associative networks. But this disadvantage may be compensated for by cooperative use of MAL with other associative networks, for example, with a Hopfield network.

References

1. R. Durbin and D. Willshaw. An analogue approach to the travelling salesman problem using an elastic net method. *Nature*, 326:689–691, 1987.
2. Robert Hecht-Nielsen. *Neurocomputing.* Addison-Wesley Publishing Company, 1991.
3. John Hertz, Anders Krogh, and Richard G. Palmer. *Introduction to the theory of neural computation.* Addison-Wesley Publishing Company, 1991.
4. J.J. Hopfield. Neural networks and physical systems with emergent collective computational abilities. *Proceedings of the National Academy of Sciences*, 79:2554–2558, 1982.
5. Hui Wang. *Towards a unified framework of relevance* . PhD thesis, Faculty of Informatics, University of Ulster, N. Ireland, UK, October 1996. http://www.infm.ulst.ac.uk/~hwang/thesis.ps.
6. Christopher K.I. Williams. *Combining deformable models and neural networks for handprinted digit recognition.* PhD thesis, Dept. of Computer Science, University of Toronto, November 1994.

Adaptive Noise Injection for Input Variables Relevance Determination

Yves Grandvalet and Stéphane Canu

Heudiasyc, U.M.R. C.N.R.S. 6599, Université de Technologie de Compiègne, B.P. 20.529, 60205 Compiègne Cedex, France

Abstract. In this paper we consider the application of training with noise in multi-layer perceptron to input variables relevance determination. Noise injection is modified in order to penalize irrelevant features. The proposed algorithm is attractive as it requires the tuning of a single parameter. This parameter controls the penalization of the inputs together with the complexity of the model. After the presentation of the method, experimental evidences are given on simulated data sets.

1 Introduction

When estimating a regression function, the problem of selecting relevant variables is difficult. Optimal subset selection is a combinatorial and acknowledged unstable procedure. It may thus be better to resort to stable procedures penalizing irrelevant variables. In this paper, we introduce such a procedure, based on the noise injection heuristic.

Noise injection (NI) consists in deliberately adding an artificial noise to the inputs pattern during the training of a multi-layer perceptron (MLP). Before each presentation of an input vector $\boldsymbol{x}^i$, a random noise η is added, while keeping the target outputs $\boldsymbol{y}^i$ unchanged. Asymptotically, as the number of presentation of the sample increases, NI is virtually equivalent to minimizing the cost C_{NI}:

$$C_{\text{NI}}(\boldsymbol{f}) = \mathbb{E}_{\eta}\left[\frac{1}{\ell}\sum_{i=1}^{\ell} l(\boldsymbol{f}(\boldsymbol{x}^i+\eta), \boldsymbol{y}^i)\right] \quad , \tag{1}$$

where l denotes the loss function, and $\boldsymbol{f}$ is the network function. This heuristic may be used for three different purposes, namely:

1. learn invariance. In this case the added "noise" is a means to enhance the training set, by adding examples of inputs transformed under a known invariance group [7];
2. improve the network robustness regarding inputs inaccuracy [10];
3. control the complexity of oversized networks [4]. Overfitting is avoided by introducing an induction bias. Intuitively, NI can be interpreted as a means to favor solutions such that little changes in $\boldsymbol{x}$ result in little changes in $\boldsymbol{f}(\boldsymbol{x})$.

In cases 1 and 2, the noise distribution is specified by the knowledge of either the invariance group or the robustness required. We consider here the last case, where there is no obvious choice for the noise distribution. Usually, it is arbitrarily set to a multivariate Gaussian: $\eta \sim \mathcal{N}(0, \sigma^2 \boldsymbol{I}_d)$, where $\boldsymbol{I}_d$ is the identity matrix on the d-dimensional inputs space. The variance parameter σ^2 sets the complexity of the network. It is tuned by minimizing an estimate of the generalization error. The noise distribution in NI acts as the kernel function in the Nadaraya-Watson regressor [6]. In kernel smoothing, a key point is the choice of the bandwidth [5]. In simple regression (one variable) the bandwidth is defined by the parameter σ^2. In multiple regression, the analogous of the bandwidth is the covariance matrix $\sigma^2 \boldsymbol{I}_d$. This choice amounts to suppose that all variables have the same degree of relevance. In this paper, we address the issue of a different parameterization of the covariance matrix, with the aim of determining the relevant inputs.

2 Some Theory

When NI is not used, the training is performed thanks to the pieces of information given by each pair of the sample: "the target output $\boldsymbol{y}^i$ is observed *at location* $\boldsymbol{x}^i$". With NI, the former is turned to: "the target output $\boldsymbol{y}^i$ is observed *in the neighborhood* of $\boldsymbol{x}^i$". With a Gaussian noise, the covariance matrix $\sigma^2 \boldsymbol{I}_d$ defines spherical neighborhoods. The network is thus asked to perform a smooth function in all directionsof the inputs space , and thus each input variable is supposed to have the same global relevance. If a variable is actually irrelevant, we expect to get a constant function in the corresponding direction.of the input space. However, as the training is performed on a finite sample, a spurious dependency may be detected.
To help preventing this effect, we propose to modify NI by adapting individually the elements of the diagonal noise covariance matrix. This parameterization yields elliptic neighborhoods. These neighborhoods induce a bias in favor of smoother functions in the directions corresponding to large variances. The ellipsoid may be degenerate in one or several directions (cylinder), so that the function is asked to be constant in these directions, and the the corresponding variables are considered to be irrelevant.
In this section, we first explain the principle of the algorithm for a linear perceptron trained with the quadratic loss. The effects of NI for non-linear perceptrons are then shortly described for the quadratic and cross-entropy loss.

2.1 Linear Perceptron

For a linear perceptron $\boldsymbol{f}(\boldsymbol{x}) = \boldsymbol{W}\boldsymbol{x} + \boldsymbol{\theta}$, trained with the quadratic loss l, NI is asymptotically equivalent to minimizing the following penalized cost [8]:

$$C_{\mathrm{NI}}(\boldsymbol{f}) = \frac{1}{\ell} \sum_{i=1}^{\ell} \left\| \boldsymbol{W}\boldsymbol{x}^i + \boldsymbol{\theta} - \boldsymbol{y}^i \right\|^2 + \mathrm{tr}(\boldsymbol{W} \boldsymbol{\Sigma} \boldsymbol{W}^T) \ , \tag{2}$$

where $\boldsymbol{\Sigma}$ is the covariance matrix of the noise η. For $\boldsymbol{\Sigma} = \sigma^2 \boldsymbol{I}_d$, the second right-hand term of (2) is the classical weight decay $\sigma^2 \|\boldsymbol{W}\|^2$.

If the residuals are assumed to be Gaussian, weight decay can be interpreted in terms of log-likelihood with some empirical prior on the weights $\boldsymbol{W}$. The prior is a zero mean Gaussian distribution with diagonal covariance matrix $\omega^2 \boldsymbol{I}_N$, where ω^2 is proportional to $1/\sigma^2$, and N is the number of weights.

If weight decay is used with the intent to determining relevant variables, we have to penalize differently each set of input fan-out weights, i.e. each column $\boldsymbol{W}^j$ of $\boldsymbol{W}$. This principle is used in the Bayesian framework by the automatic relevance determination model [9]. The priors on $\boldsymbol{W}^j$ are set to be $\mathcal{N}(\boldsymbol{0}, \omega_j^2 \boldsymbol{I}_{N/d})$, where ω_j^2 are adapted during training. In order to prevent the priors from vanishing, we propose to constrain the average weight variance to be constant: $\frac{1}{d}\sum_{j=1}^{d} \omega_j^2 = \omega^2$ (the mean of all weights is therefore a centered Gaussian random variable with variance ω^2).

The prior variance of any set of weight is bounded by $d\omega^2$, it may be reached for one set, if all the other variances ω_j^2 are zero, i.e. if only one variable is propagated through the network ($\omega_j^2 = 0 \Rightarrow \boldsymbol{W}^j = \boldsymbol{0}$). An input variable will thus be ignored if it contributes poorly to the minimization of squared residuals, so that relaxing the prior on the other inputs (increasing their variance parameter) results in a decrease of (2). The cost C_{NI} is now defined as follows:

$$\begin{cases} C_{\mathrm{NI}}(\boldsymbol{f}) = \dfrac{1}{\ell}\displaystyle\sum_{i=1}^{\ell} \left(\boldsymbol{W}\boldsymbol{x}^i + \boldsymbol{\theta} - \boldsymbol{y}^i\right)^2 + \sum_{j=1}^{d} \sigma_j^2 {\boldsymbol{W}^j}^T \boldsymbol{W}^j , \\ \text{with } \dfrac{1}{d}\displaystyle\sum_{j=1}^{d} \dfrac{1}{\sigma_j^2} = \dfrac{1}{\sigma^2} , \end{cases} \tag{3}$$

where the optimized variables are $\boldsymbol{W}$, $\boldsymbol{\theta}$ and $\{\sigma_j^2\}_{j=1}^{d}$, and where σ^2 is fixed. Stepping back to NI, (3) is obtained with a noise covariance matrix $\boldsymbol{\Sigma}$, with $\boldsymbol{\Sigma}_{jj} = \sigma_j^2$. Weight decay and NI can thus be equivalently used for variable relevance determination in linear perceptrons. In order to adapt the σ_j^2, we propose to multiply the inputs by the gain matrix $\boldsymbol{\Sigma}^{-1/2}$ before injecting the noise $\eta \sim \mathcal{N}(\boldsymbol{0}, \boldsymbol{I}_d)$. These gains can be incorporated in the first layer of weights after the training phase.

Adaptive Noise injection algorithm for variables relevance determination

1. Set the value of the hyper-parameter σ^2 and initialize $\sigma_j = \sigma$, $j = 1, \ldots, d$.
2. Optimize the network parameters $\boldsymbol{W}$ and $\boldsymbol{\theta}$ together with $\{\sigma_j^2\}_{j=1}^{d}$.
   At each epoch, and for each input-output pairs:
   - compute $\boldsymbol{\xi}^i = \boldsymbol{\Sigma}^{-1/2}\boldsymbol{x}^i$, with $\boldsymbol{\Sigma}_{jj}^{-1/2} = 1/\sigma_j$
   - feed the net with $\boldsymbol{\xi}^i + \eta$, $\eta \sim \mathcal{N}(\boldsymbol{0}, \boldsymbol{I}_d)$; compute loss $l(f(\boldsymbol{\xi}^i + \eta), \boldsymbol{y}^i)$
   - back-propagate the error trough the net; update $\boldsymbol{W}$ and $\boldsymbol{\theta}$
   - back-propagate the error trough $\boldsymbol{\Sigma}^{-1/2}$; update $\boldsymbol{\Sigma}^{-1/2}$ with constraints (3)
3. Estimate the generalization error (bootstrap, cross-validation,...).

As for weight decay, several values of σ^2 are tested, and the the one with smallest estimate of generalization error is picked. Note that using $\boldsymbol{\Sigma}^{-1/2}$ does not require to compute the inverse or the square root of $\boldsymbol{\Sigma}$. So, reducing the dimension of the input space (instead of the number of variables) is easily obtained with a full matrix $\boldsymbol{\Sigma}^{-1/2}$. The analogous of (3) is to constrain the norm of all the column vectors of $\boldsymbol{\Sigma}^{-1/2}$ to be $1/\sigma$.

2.2 Non-linear Perceptron

For MLP's, NI is not equivalent to penalize weights from the input layer to the first hidden layer. The added noise is propagated throughout the network, yielding penalization on every layer, except for biases. Thus, adaptive noise injection sets the complexity of the network and the number of relevant variables by σ^2. Tuning these two quantities by a single parameter is natural. However, the number of inputs of an MLP does not define its complexity. We give here an argument supporting that, for a given value of σ^2, the number of effective parameters of the network diminishes when some variables are considered to be less relevant. First, let us recall that for the quadratic or for the cross-entropy loss, C_{NI} (1) is minimized by the Nadaraya-Watson kernel smoother [6]. Minimizing C_{NI} with an MLP leads to have this function as target.

The number of effective parameters p_{eff} of the Nadaraya-Watson smoother can be analytically computed [5]: $p_{\mathrm{eff}} = \sum_{i=1}^{\ell} p_\eta(\mathbf{0}) / \sum_{j=1}^{\ell} p_\eta(\boldsymbol{x}^i - \boldsymbol{x}^j)$, where p_η is the noise distribution, acting here as a kernel. The ratio $\sum_{j=1}^{\ell} p_\eta(\boldsymbol{x} - \boldsymbol{x}^j)/p_\eta(\mathbf{0})$ is a weighted measure of the number of neighbors of $\boldsymbol{x}$. Thus, p_{eff} is a decreasing function of the average number of neighbors of the sample points.

With Gaussian noise, the neighborhoods are spherical ($\boldsymbol{\Sigma} = \sigma^2 \boldsymbol{I}$) or elliptical. The constraint (3) implies that elliptical neighborhoods have a bigger volume. They are thus likely to increase the average number of neighbors of the data points, so that p_{eff} decreases.

For a given value of σ^2, the algorithm looks for a plausible "empirical induction bias" (which inputs are relevant?), on the basis of the sample $\{\boldsymbol{x}^i, \boldsymbol{y}^i\}_{i=1}^{\ell}$. As p_{eff} is decreased when $\boldsymbol{\Sigma} \neq \sigma^2 \boldsymbol{I}$, this bias is increased. It is thus unlikely that adapting $\boldsymbol{\Sigma}$ should be paid by some significant additional variance.

3 Simulations

In this section, we borrowed Friedman's simulated data examples and results from [3]. These data were created to test flexible (thus adaptive) metric nearest neighbor classification. We briefly summarize the description of the examples below (see [3] for full details).

Example 1: There are $d = 10$ input variables, $\ell = 200$ training observations and two classes. The inputs are generated from $\mathcal{N}(\mathbf{0}, \boldsymbol{I}_d)$ for the first class, and $\mathcal{N}(\boldsymbol{m}, \boldsymbol{C})$ for the second class, with $\{m_k = \sqrt{k}/2\}_{k=1}^d$ and $\boldsymbol{C} = \mathrm{diag}\{1/\sqrt{k}\}_{k=1}^d$.

Example 2: Same as above except that $\{m_k = \sqrt{d-k+1}/2\}_{k=1}^d$.

Example 3: There are $d = 10$ input variables, $\ell = 200$ training observations and two classes. The inputs are generated from $\mathcal{N}(\mathbf{0}, \boldsymbol{I}_d)$ for the two classes. The classes are defined by: $\sum_{k=1}^{d} x_k^2/k \leq 2.2 \Rightarrow$ class 1, and class 2 otherwise.

Example 4: There are $d = 10$ input variables, $\ell = 500$ training observations and two classes. The inputs are generated from $\mathcal{N}(\mathbf{0}, \boldsymbol{I}_d)$ for the two classes. The classes are defined by: $\sum_{k=1}^{d} x_k^2 \leq 9.8 \Rightarrow$ class 1, and class 2 otherwise.

Example 5: Same as example 3, except that the two classes are defined by: $\sum_{k=1}^{d} x_k \leq 0 \Rightarrow$ class 1, and class 2 otherwise.

Example 6: This is the classical waveform example [1]. There are $d = 21$ input variables, $\ell = 300$ training observations and three classes. Each point $\boldsymbol{x}^i$ is a random combination of two out of three basic waveform, with noise added.

We used Friedman's experimental setup except that the parameter σ^2 was estimated by .632 bootstrap instead of cross-validation (these estimates of the generalization error are described in [2]). For all examples, ten independent training samples of size ℓ with their independent generalization set of size 2000 were generated. The mean error rates on the generalization sets are reported in Table 1. For each example, we used a multi-layer perceptron with a single hidden layer with 10 units. The number of parameters of the network (including bias and σ_j^2) is thus $d + (d+1) \times 10 + 10 + 1$. The network was trained with the cross-entropy error function:

$$C_{\mathrm{NI}}(\boldsymbol{f}) = \mathbb{E}_{\eta}\left[\frac{1}{\ell}\sum_{i=1}^{\ell}\sum_{k=1}^{m} y_k^i \ln(\boldsymbol{f}_k(\boldsymbol{x}^i + \eta)) + (1 - y_k^i)\ln(1 - \boldsymbol{f}_k(\boldsymbol{x}^i + \eta))\right] . \quad (4)$$

Table 1. mean error rates in generalization for the six examples. CART:ln denotes CART with linear combinations, K-NN is the k-nearest neighbors algorithm, K-NN:fm is the best of five versions of flexible metric nearest neighbors algorithms and ANI is the adaptive NI algorithm.

Ex.	CART	CART:ln	K-NN	K-NN:fm	ANI
1	15.1	14.8	13.2	6.3	1.9
2	14.6	11.7	10.3	3.4	3.0
3	21.9	21.8	35.9	21.7	22.5
4	32.2	32.3	34.0	26.2	26.7
5	32.4	7.6	17.4	6.2	4.9
6	29.1	21.1	17.1	18.7	16.9

Ex. 1 was designed to be favorable to adaptive metrics, as the features with higher coordinates are more relevant. Ex. 2 is less favorable as the means of the clusters are less separated in higher coordinates. Nevertheless, in both cases adaptive NI performs significantly better than the other methods. Ex. 3, 4 and 5 were designed to be more favorable to CART. Note that for these three examples, perfect learning may be achieved. Our algorithm gets performances close to the ones obtained with the most competitive methods. Finally, in ex. 6, although flexible metric does not improve performances for nearest neighbors algorithm, adaptive NI still yields good results.

4 Discussion

In this paper we presented a modification of NI to determine relevant inputs. Rather than subset selection (variable elimination), a penalization of irrelevant inputs feature is carried out. Subset selection can nevertheless be carried out after learning as the features are sorted by rank of relevance. The proposed algorithm is very simple, and the additional computational cost compared to the conventional back-propagation algorithm is very limited (due to the optimization of the gains $\{\sigma_j\}_{j=1}^{d}$). The penalization is done in accordance to the complexity of the model. A single parameter is tuned to perform simultaneously the two tasks. The number of effective input variables is thus adjusted together with the number of effective parameters of the model.

Our algorithm usually shows better performances than the original one in terms of generalization performances. The flexibility of the model is increased as the induction bias is adapted according to the sample. This means that there are a few problems where regular NI performs slightly better, but there are also more problems that are likely to be solved by adaptive metric noise injection.

References

1. L. Breiman, J.H. Friedman, R. Olshen, and C.J. Stone. *Classification and Regression Trees.* Wadswworth, Belmont, CA., 1984.
2. B. Efron and R.J. Tibshirani. *An Introduction to the Bootstrap*, volume 57 of *Monographs on Statistics and Applied Probability.* Chapman & Hall, New York, 1993.
3. J.H. Friedman. Flexible metric nearest neighbor classification. Technical report, Stanford University, Stanford, CA., November 1994.
4. Y. Grandvalet, S. Canu, and S. Boucheron. Noise injection: theoretical prospects. *Neural Computation*, 9(7), 1997, to appear.
5. T.J. Hastie and R.J. Tibshirani. *Generalized Additive Models*, volume 43 of *Monographs on Statistics and Applied Probability.* Chapman & Hall, New York, 1990.
6. L. Holmström and P. Koistinen. Using additive noise in back-propagation training. *IEEE Transactions on Neural Networks*, 3(1):24–38, January 1992.
7. T.K. Leen. From data distribution to regularization in invariant learning. *Neural Computation*, 7(5):974–981, 1995.
8. K. Matsuoka. Noise injection into inputs in back-propagation learning. *IEEE Transactions on Systems, Man, and Cybernetics*, 22(3):436–440, 1992.
9. R. M. Neal. *Bayesian Learning for Neural Networks.* Lecture Notes in Statistics. Springer-Verlag, New York, 1996.
10. A.R. Webb. Functional approximation by feed-forward networks: A least-squares approach to generalization. *IEEE Transactions on Neural Networks*, 5(3):363–371, 1994.

Input Selection with Partial Retraining

Piërre van de Laar, Stan Gielen, and Tom Heskes

RWCP* Novel Functions SNN** Laboratory,
Dept. of Medical Physics and Biophysics, University of Nijmegen, The Netherlands.

Abstract. In this article, we describe how input selection can be performed with partial retraining. By detecting and removing irrelevant input variables resources are saved, generalization tends to improve, and the resulting architecture is easier to interpret. In our simulations the relevant input variables were correctly separated from the irrelevant variables for a regression and a classification problem.

1 Introduction

Especially with a lack of domain knowledge, the usual approach in neural network modeling is to include all input variables that may have an effect on the output. Furthermore, the number of parameters of the neural network is chosen large, since with too few parameters a network is unable to learn a complex relationship between input and output variables. This approach is suboptimal because the inclusion of irrelevant variables and the (possibly) abundant parameters tends to degrade generalization. Secondly, resources are wasted by measuring irrelevant variables. And finally, a model with irrelevant variables and parameters is more difficult to understand.

Architecture selection algorithms (see e.g. [1,5,6,9]) try to remove irrelevant parameters and/or input variables and, consequently, save resources, improve generalization, and yield architectures which are easier to interpret. In this article, we will describe a new algorithm to perform input selection, a subproblem of architecture selection, which exploits partial retraining [11].

We will first, in section 2, describe partial retraining. Second, in section 3, we will describe how the performance of networks with different sets of input variables are compared. Then, we will briefly describe the problem of input selection and we will introduce our algorithm to perform input selection which exploits partial retraining in section 4. In section 5, we will perform simulations in which our algorithm is applied for input selection in two artificial problems. Our conclusions and some discussion can be found in section 6.

2 Partial retraining

Suppose we have a neural network which has been trained on N input variables and we would like to determine the performance which can be achieved with

* Real World Computing Partnership
** Foundation for Neural Networks

a subset of these N input variables. A naive method is to train a new neural network on this subset to determine this performance. Instead of the computationally expensive process of training a new network, this new network might also be estimated based on the original network. Partial retraining [11] assumes that the neural network trained on all N input variables has constructed a good representation of the data in its hidden layers needed to map the input to the output. Therefore, the new network is estimated by partial retraining such that its hidden-layer activities are as close as possible to the hidden-layer activities of the original neural network, although its input is restricted. The performance of this constructed network is an estimate of the performance we want to determine.

Partial retraining uses the following three steps to construct the network which only receives a subset of all variables as its input. In the first step, partial retraining determines the new weights between the subset of input variables and the first hidden layer by

$$\tilde{w}_1 = \underset{B}{\text{argmin}} \sum_{\mu} \|\boldsymbol{h_1}^{\mu} - B\boldsymbol{X}^{*\mu}\|^2 ,$$

where $\boldsymbol{h_1} \equiv w_1 \boldsymbol{X}$ denotes the original incoming activity of the first hidden layer, $\boldsymbol{X}^*$ stands for the input limited to the subset, and μ labels the patterns.

The difference between fitting the incoming activity and fitting the outgoing activity of the hidden layer is almost negligible (see e.g. [8]). We prefer fitting the incoming activity, since this can be done by simple and fast quadratic optimization. Furthermore, it can be easily shown [11] that the new weights between the input and the first hidden layer are chosen such that the neural network estimates the missing value(s) based on linear dependencies, and processes the "completed" input data.

The compensation of the errors introduced by removing one or more input variables is probably not perfect due to noise and nonlinear dependencies in the data. Therefore, to further minimize the effects caused by the removal of one or more input variables, the new weights between hidden layers λ and $\lambda - 1$ are calculated from

$$\tilde{w}_\lambda = \underset{B}{\text{argmin}} \sum_{\mu} \left\|\boldsymbol{h_\lambda}^{\mu} - B\tilde{\boldsymbol{H}}_{\boldsymbol{\lambda-1}}{}^{\mu}\right\|^2 ,$$

with $\boldsymbol{h_\lambda} = w_\lambda \boldsymbol{H_{\lambda-1}}$ the original incoming activity of hidden layer λ and $\tilde{\boldsymbol{H}}_{\boldsymbol{\lambda}}$ the new λ^{th} hidden layer activity based on the subset of input variables and the newly estimated weights.

Finally, the weights between the output and the last hidden layer are re-estimated. Here we have two options: either we treat the output layer as a hidden layer and try to fit the incoming activity of the original network, or we can use the output targets to calculate the desired incoming activity. The first approach, which we will also take in our simulations, yields

$$\tilde{w}_{K+1} = \underset{B}{\text{argmin}} \sum_{\mu} \left\|\boldsymbol{o}^{\mu} - B\tilde{\boldsymbol{H}}_{\boldsymbol{K}}{}^{\mu}\right\|^2 ,$$

where $\boldsymbol{o} = w_{K+1}\boldsymbol{H}_K$ is the original incoming activity of the output layer, and $\tilde{\boldsymbol{H}}_K$ is the activity of the last (K^{th}) hidden layer given the new weights and the limited input.

As already mentioned, partial retraining is far less computationally intensive than the naive approach, i.e., training a new network with the restricted input. This reduction in computational needs is accomplished by the usage of implicit knowledge about the problem which is contained in the neural network trained on all input variables. Furthermore, partial retraining has a unique solution while training a new network might converge to another (local or global) minimum.

3 Network Comparison

How are the performances of two networks with different sets of input variables compared? We will describe how this is done for two different tasks: a regression task and a classification task. For simplicity, we will assume that both tasks have only one output.

We assume that the performance on a regression task is measured by the mean squared error, e.g. $E = \frac{1}{2P}\sum_{\mu=1}^{P}(T^\mu - O^\mu)^2$, with T^μ and O^μ the target and output of pattern μ, respectively. The difference in performance between two networks is thus given by

$$E_1 - E_2 = \frac{1}{P}\sum_{\mu=1}^{P}[2T^\mu - (O_1^\mu + O_2^\mu)]\,[O_2^\mu - O_1^\mu]\;. \tag{1}$$

If the mean squared error is used, it is reasonable to assume that $T^\mu - O_i^\mu$ is drawn from a Gaussian distribution, and thus also the two terms between brackets in equation (1) are Gaussian distributed. Under the assumption that these two terms are independent (not correlated), the product's average value is zero, and its standard deviation σ is equal to the standard deviation of the first term in brackets in equation (1) times the standard deviation of the second term. If P is large, the central limit theorem yields that the hypothesis that the performances of the two networks are identical has to be rejected with significance level α if

$$\mathcal{P}(Z > \frac{\sqrt{P}\,|E_1 - E_2|}{\sigma}) < \alpha\,,$$

where $\mathcal{P}(Z > z) \equiv \frac{1}{2}\left(1 - \text{erf}(\frac{z}{\sqrt{2}})\right)$ is the probability that a value greater than z is drawn from a standard Gaussian distribution, i.e., $\mathcal{N}(0,1)$.

The percentage of misclassifications in a classification task is given by $\frac{1}{P}\sum_{\mu=1}^{P}(1 - T^\mu)O^\mu + T^\mu(1 - O^\mu)$, where T^μ and O^μ are the target and output label ($\in \{0,1\}$) of pattern μ, respectively. The difference in misclassifications between two networks is given by

$$C_1 - C_2 = \frac{1}{P}\sum_{\mu=1}^{P}(2T^\mu - 1)(O_2^\mu - O_1^\mu)\;.$$

Let us consider the hypothesis that both networks perform equally well, i.e., that both have an equal chance of misclassification p, which can be estimated based on the data by $p = \frac{C_1+C_2}{2}$. Similarities between the solutions, as constructed by the networks, correlate the outputs of the networks. The ratio of simultaneous misclassifications over all misclassifications can be estimated by $q = \frac{2C_{1\cap 2}}{C_1+C_2}$, where $C_{1\cap 2}$ denotes the percentage of simultaneous misclassifications. Given this hypothesis, we find that the term $(2T^\mu - 1)(O_2^\mu - O_1^\mu)$ to calculate the difference in misclassification is on average equal to zero, and has standard deviation of $\sigma = \sqrt{2p(1-q)}$. Again if P is large, the central limit theorem yields that our hypothesis has to be rejected with significance level α if

$$\mathcal{P}(Z > \frac{\sqrt{P}\,|C_1 - C_2|}{\sigma}) < \alpha\,.$$

4 Input Selection

Since every input variable is either selected or not, we have 2^N possibilities for N input variables. Using partial retraining to determine the performances of the network given every possible combination of input variables is therefore only feasible when the number of input variables is rather small. Alternatives, but approximations, for this brute force method are backward elimination, forward selection, and stepwise selection (see e.g. [3,7]).

Backward elimination starts with all input variables and removes the least relevant variables one at the time. Backward elimination stops if the performance of network drops below a given threshold by removal of any of the remaining input variables. Forward selection starts without any input variables, sequentially includes the most relevant variables, and stops as soon as the performance of the network exceeds a given threshold. Stepwise selection is a modified version of forward selection that permits reexamination, at every step, of the variables included in the previous steps, since a variable that was included at an early stage may be come irrelevant at a later stage when other (related) variables are also included.

In our simulations we remove the least relevant variables one at the time, but, unlike backward elimination, we do not stop when the performance of the network drops below a given threshold but we continue until all variables are removed. From these architectures, i.e., with none, one, two, ..., and N inputs, the smallest architecture, for which the error is not significantly different from the architecture with the minimal error, is chosen. Note that the parameter α controls Occam's razor. The smaller α the higher the probability that a smaller architecture is considered statistically identical to the architecture with the minimal error. In other words, the smaller α the higher the chance that we arrive at networks with only a few input variables.

5 Simulations

5.1 Regression

The response of our artificial regression task given the ten input variables, $X_1, \ldots, X_{10}$ which are uniformly distributed over $[0, 1]$, is given by the following signal plus noise model [4]

$$T = 10\sin(\pi X_1 X_2) + 20\left(X_3 - \frac{1}{2}\right)^2 + 10X_4 + 5X_5 + \epsilon\ ,$$

where ϵ is $\mathcal{N}(0, 1)$, i.e., standard normally distributed noise. The response does not depend on the irrelevant or noisy input variables, X_6, X_7, X_8, X_9, and X_{10}. Furthermore, we assume that the ten input variables are not independent: two irrelevant inputs are identical, $X_9 \equiv X_{10}$, as well as two relevant inputs, $X_4 \equiv X_5$. Based on this signal plus noise model, we generated a data set of 400 input-output patterns.

We trained hundred two-layered multilayer perceptrons with ten input, seven hidden and one output unit and with the hyperbolic tangent and the identity as transfer functions of the hidden and output layer, respectively. The training procedure was as follows: starting from small random initial values, the weights were updated using backpropagation on the mean squared error of a training set of 200 randomly selected patterns out of the data set. Training was stopped at the minimum of the mean squared error of a validation set consisting of the remaining 200 patterns. With $\alpha = 0.01$, 94 out of 100 networks choose the relevant input variables, i.e., inputs X_1, X_2, X_3, and X_5, the remaining 6 networks did not only choose the relevant input variables but also one irrelevant input. These 6 networks might have been confused by random correlations between this irrelevant input and the output in both training and validation set.

5.2 Classification

The response of our artificial classification task given the ten binary input variables, $X_1, \ldots, X_{10}$ is given by the following rule

$$T = X_1 \cup (\overline{X_2} \cap X_3) \cup (X_4 \cap \overline{X_5} \cap \overline{X_6})\ ,$$

where 0 and 1 code false and true, respectively. The response thus depends on one very important variable (X_1), two equally important variables (X_2 and X_3), and three less important variables (X_4, X_5, and X_6). Furthermore, the response does not depend on the four irrelevant or noisy variables, X_7, X_8, X_9, and X_{10}. Based on this rule, we generated a data set containing 400 examples.

We applied the same training procedures as in the regression task, except that the neural networks had only five instead of seven hidden units. After training, the patterns were classified based on their output. If the output of a pattern was larger than 0.5 the pattern was labeled true, otherwise the class label was false.

With $\alpha = 0.05$, 98 out of 100 networks chose the correct architecture with inputs X_1, X_2, X_3, X_4, X_5, and X_6. The remaining two networks chose a smaller architecture, which did not include all relevant variables. To be more precise one removed variable X_4, the other removed the variables X_4, X_5, and X_6. This might be caused by the lack of examples in the training set since only 200 patterns out of $2^{10} = 1024$ possibilities were available.

6 Discussion

In this article, we have described how input selection can be performed by using partial retraining. We focused on how the performance of neural networks with different subsets of input variables can be compared in both regression and in classification tasks. Computer simulations have shown that, in two artificial problems, this algorithm indeed selects the relevant input variables.

The algorithm described in this article can be easily generalized. For example, by viewing hidden units as input units of a smaller network [2,10], this algorithm can also be used for hidden unit selection. Similarly, it can be used for weight pruning by estimating the relevance of a single weight.

References

1. T. Cibas, F. Fogelman Soulié, P. Gallinari, and S. Raudys. Variable selection with optimal cell damage. In M. Marinaro and P. G. Morasso, editors, *Proceedings of the International Conference on Artificial Neural Networks*, volume 1, pages 727–730. Springer-Verlag, 1994.
2. T. Czernichow. Architecture selection through statistical sensitivity analysis. In C. von der Malsburg, W. von Seelen, J. C. Vorbrüggen, and B. Sendhoff, editors, *Artificial Neural Networks - ICANN 96*, volume 1112 of *Lecture Notes in Computer Science*, pages 179–184. Springer, 1996.
3. N. R. Draper and H. Smith. *Applied Regression Analysis.* Wiley Series in Probability and Mathematical Statistics. Wiley, New York, second edition, 1981.
4. J. H. Friedman. Multivariate adaptive regression splines. *The Annals of Statistics*, 19(1):1–141, 1991.
5. G. D. Garson. Interpreting neural-network connection weights. *AI Expert*, 6(4):47–51, 1991.
6. B. Hassibi, D. G. Stork, G. Wolff, and T. Watanabe. Optimal Brain Surgeon: Extensions and performance comparisons. In J. D. Cowan, G. Tesauro, and J. Alspector, editors, *Advances in Neural Information Processing Systems*, volume 6, pages 263–270, San Francisco, 1994. Morgan Kaufmann.
7. D. G. Kleinbaum, L. L. Kupper, and K. E. Muller. *Applied Regression Analysis and Other Multivariable Methods.* The Duxbury series in statistics and decision sciences. PWS-KENT Publishing Company, Boston, second edition, 1988.
8. J. O. Moody and P. J. Antsaklis. The dependence identification neural network construction algorithm. *IEEE Transactions on Neural Networks*, 7(1):3–15, 1996.
9. M. C. Mozer and P. Smolensky. Using relevance to reduce network size automatically. *Connection Science*, 1(1):3–16, 1989.
10. K. L. Priddy, S. K. Rogers, D. W. Ruck, G. L. Tarr, and M. Kabrisky. Bayesian selection of important features for feedforward neural networks. *Neurocomputing*, 5(2/3):91–103, 1993.
11. P. van de Laar, T. Heskes, and S. Gielen. Partial retraining: A new approach to input relevance determination. Submitted, 1997.

On the Complexity of Recognizing Iterated Differences of Polyhedra

Eddy Mayoraz

IDIAP, C.P. 592, 1920 Martigny, Switzerland

Abstract. The iterated difference of polyhedra $V = P_1 \backslash (P_2 \backslash (\ldots P_k) \ldots)$ has been proposed independently in [11] and [7] as a sufficient condition for V to be exactly computable by a two-layered neural network. An algorithm checking whether $V \subset \mathbb{R}^d$ is an iterated difference of polyhedra is proposed in [11]. However, this algorithm is not practically usable because it has a high computational complexity and it was only conjectured to stop with a negative answer when applied to a region which is not an iterated difference of polyhedra. This paper sheds some light on the nature of iterated difference of polyhedra. The outcomes are: (i) an algorithm which always stops after a small number of iterations, (ii) sufficient conditions for this algorithm to be polynomial and (iii) the proof that an iterated difference of polyhedra can be exactly computed by a two-layered neural network using only essential hyperplanes.

1 Introduction

Quite a few papers have been lately devoted to the problem of characterizing the regions of the Euclidian space $\mathbb{R}^d$ that can be computed by a depth-2 multilayer perceptron (MLP), *i.e.* an MLP with d real inputs, one hidden layer of linear threshold processing units and a single output with a linear threshold processing unit [4,2,10,3,8,1]. Different variations of the problem are considered: the function of the MLP and the characteristic function of the region are required to match either (*i*) *exactly* [10], *i.e.* for any $x \in \mathbb{R}^d$, or (*ii*) *almost everywhere* [3,1], *i.e.* everywhere but on a set of measure 0; or even (*iii*) *up to* ϵ [8], *i.e.* for any x at distance more than ϵ with the border of V.

In what follows we will denote by $\mathcal{LP}_2$ the set of regions V which are computable by depth-2 MLPs and if nothing is specified, *exact* computation will be intended.

Simultaneously to the characterization of these regions V in $\mathcal{LP}_2$, another important issue is the complexity of the MLP computing $V \in \mathcal{LP}_2$, which is essentially expressed by its number of hidden units. This question did not get as much as attention as the characterization, although it is essential for applications and in thte point of view of computational learning theory. It turns out that even very simple regions $V \in \mathcal{LP}_2$ seem to require a tremendous amount of hidden units. If we denote by $H_h^{h_0}$ the closed halfspace $\{x \mid x^\top h \geq h_0\}$ and if $\Delta > 1$ is a positive integer, consider the region

$$V = (\, H^1_{(1,1)} \cap H^1_{(-1,1)} \cap H^{-\Delta}_{(0,-1)} \,) \cup (\, H^1_{(1,-1)} \cap H^1_{(-1,-1)} \cap H^{-\Delta}_{(0,1)} \,) \qquad (1)$$

which is known to be in $\mathcal{LP}_2$ [9,3]. To the best of our knowledge, any solution known for the computation of V with a depth-2 MLP requires a number of hidden units growing linearly with Δ, *i.e.* exponentially with the size of the instance V which is in $O(\log(\Delta))$. This simple example gives us some faith in the following conjecture :

Conjecture 1. There exists a region $V \subset \mathbb{R}^2$ in $\mathcal{LP}_2$ such that any depth-2 MLP computing V almost everywhere has a number of hidden units exponential in the size of a compact encoding of V.

In this paper, we will focus on a particular subclass of $\mathcal{LP}_2$ which contains only regions that requires a number of hidden units linear in a compact encoding of V, and we will investigate the complexity of recognizing such a class.

To conclude this introduction, let us introduce few definitions and notations. Let us assume that a region V — the instance of the problem — is specified by a finite list of closed halfspaces, called *basis* of V, and an expression of V as a union of intersections of some of these halfspaces or their complements (*e.g.* equation (1)).

A region V can have, in general, different minimal bases (in the sense of inclusion). A halfspace is called *essential* to V if it belongs to any basis of V. If $V \subset \mathbb{R}^d$ is a union of intersections of finitely many halfspaces and each of these intersections is fully dimensioned (*i.e.* containing one open ball of dimension d), it can be easily verified that V has a unique minimal basis, denoted $\mathcal{H}_V$, which is the set of essential halfspaces. Thus in what follows, if no particular basis is specified for a region V, full dimension of any component of V is implicitly assumed, and the basis of reference is $\mathcal{H}_V$.

The complexity problem raised in Conjecture 1 incites us to focus on a subclass of $\mathcal{LP}_2$, denoted $\overline{\mathcal{LP}_2}$, defined as the set of regions V computable by a depth-2 MLP where the hidden units are computing only essential halfspaces. Two major issues should be addressed :

Q1 find a geometrical characterization of $\overline{\mathcal{LP}_2}$,

Q2 given a basis $\mathcal{H}$ and a region V defined as a union of intersections of some halfspaces and complements of halfspaces in $\mathcal{H}$, what is the complexity of deciding whether $V \in \overline{\mathcal{LP}_2}$.

In [6] we identified Q2 as co-*NP*-Complete. In the present work, we study the class of iterated differences of polyhedra, proposed simultaneously in [11,7] as a subclass of $\mathcal{LP}_2$. In the rest of this paper, we first recall what has been done in this field, present an efficient algorithm for recognizing the iterated difference of polyhedra and discuss its consequences.

2 Iterated difference of polyhedra

Definition 2. A *polyhedron* (resp. *pseudo-polyhedron*) is an intersection of finitely many closed (resp. open or closed) halfspaces. A region $V \subset \mathbb{R}^d$ is an *iterated difference of polyhedra* (resp. pseudo-polyhedra) if it can be expressed as $V = P_1 \backslash (P_2 \backslash (\ldots P_k) \ldots)$, where each P_i, $i = 1, \ldots, k$, is a polyhedron (resp. pseudo-polyhedron). The class of iterated differences of polyhedra (resp. pseudo-polyhedra) is denoted $\mathcal{D}$ (resp. $\widetilde{\mathcal{D}}$).

Proposition 3. $\mathcal{D} \subset \widetilde{\mathcal{D}} \subsetneq \mathcal{LP}_2$.

Proof. The first inclusion is obvious. The proof of the second inclusion is based on the fact that $P \backslash V \in \mathcal{LP}_2$ for any pseudo-polyhedron P and any $V \in \mathcal{LP}_2$ (see [11,7]).

In [11], the authors propose the following algorithm for the recognition of $\mathcal{D}$:

```
input:           V ⊂ ℝ^d;
initialization:  V_0 := V; l := 0;
main loop:       while V_l ≠ ∅ and ( l < 2 or else P_l ≠ P_{l-1}) loop
                     l := l + 1;
                     P_l := op(V_{l-1});
                     V_l := P_l \ V_{l-1};
                 end loop
output:          P_1\(P_2\(...P_{l-1}\(P_l\V_l))...) =: V
```

Algo(op) : Recognition of iterated differences of polyhedra.

where the operator "op" stands for the closure of the convex hull, denoted $\overline{\text{conv}}$. This algorithm suffers from several drawbacks which make it unusable neither in practice, nor to tackle the complexity of the recognition problem of $\mathcal{D}$.

- The authors proved that $V \in \mathcal{D}$ *iff* Algo($\overline{\text{conv}}$) stops with $V_l = \emptyset$. However, they only conjectured that Algo($\overline{\text{conv}}$) could not cycle, or in other words, that it would stop with $P_l = P_{l-1} \neq \emptyset$ whenever $V \notin \mathcal{D}$.
- Even in case $V \in \mathcal{D}$, there is no bound on the number of iterations of the algorithm.
- The computation of the convex hull is exponential in d.
- It would be nice to have the same type of algorithm for the recognition of $\widetilde{\mathcal{D}}$, for example by setting "op" simply to conv. But as mentioned by the authors, the convex hull of the difference between two pseudo-polyhedra is not necessarily a pseudo-polyhedron (see Figure 2 in [11]).

The contribution of the present paper to this field goes through the suggestion of a more appropriate operator "op" which will solve very simply each of the problems mentioned above. Moreover, the algorithm obtained with the new operator "op" will prove that $\widetilde{\mathcal{D}} \subset \overline{\mathcal{LP}_2}$, and will allow to narrow down known upper bounds for the complexity of the recognition problem of $\mathcal{D}$ and $\widetilde{\mathcal{D}}$.

3 The hull operator

Definition 4. Given a collection $\mathcal{E}$ of regions of $\mathbb{R}^d$, the operator $\text{hull}_{\mathcal{E}}$ is defined as follows :

$$\forall X \subset \mathbb{R}^d, \quad \text{hull}_{\mathcal{E}}(X) = \bigcap_{E \in \mathcal{E},\ E \supset X} E\ .$$

In order to illsutrate the relation between "hull" and "conv", let $\mathcal{C}$ denote the set of all closed halfspaces, $\widetilde{\mathcal{C}}$ the set of all halfspaces (closed and open), and X^{int} the interior of a set X (according to the usual topology of $\mathbb{R}^d$). In [5] we have established that for any $X \subset \mathbb{R}^d$,

$$\mathrm{conv}^{\mathrm{int}}(X) = \mathrm{hull}_{\mathcal{C}}^{\mathrm{int}}(X) \subset \mathrm{conv}(X) \subset \mathrm{hull}_{\widetilde{\mathcal{C}}}(X) \subset \overline{\mathrm{conv}}(X) = \mathrm{hull}_{\mathcal{C}}(X).$$

Consequently, Algo($\mathrm{hull}_{\mathcal{C}}$) is identical to Algo($\overline{\mathrm{conv}}$). Moreover, $\mathrm{hull}_{\widetilde{\mathcal{C}}}$ does not suffer from the same drawback as "conv" towards pseudo-polyhedra in the sense that $\mathrm{hull}_{\widetilde{\mathcal{C}}}(P_i \backslash P_j)$ is a pseudo-polyhedron for any pseudo-polyhedra P_i and P_j. Therefore, the whole work in [11] can be restated using $\mathrm{hull}_{\widetilde{\mathcal{C}}}$ instead of $\overline{\mathrm{conv}}$ and Proposition 5 will follow.

Proposition 5. *Algo($\mathrm{hull}_{\widetilde{\mathcal{C}}}$) recognizes exactly $\widetilde{\mathcal{D}}$.*

However, by exploiting the hull operator a bit further, we will get a much simpler algorithm for the recognition of $\widetilde{\mathcal{D}}$.

4 Main result

Let V be an arbitrary region of $\mathbb{R}^d$ and $\mathcal{H}$ a basis of V. Let $\widetilde{\mathcal{H}}$ be defined as $\{H \mid H \in \mathcal{H} \text{ or } \mathbb{R}^d \backslash H \in \mathcal{H}\}$.

Theorem 6. *Algo($\mathrm{hull}_{\widetilde{\mathcal{H}}}$) recognizes exactly $\widetilde{\mathcal{D}}$.*

The proof of this theorem is too long to be presented here and can be found in [6]. Instead, we will try to give an idea of why this is true and we will enumerate the consequences of this result.

For a simple region $V \in \mathbb{R}^2$, Figure 1 illustrates the two different sequences of pseudo-polyhedra produced by Algo($\mathrm{hull}_{\widetilde{\mathcal{C}}}$) and by Algo($\mathrm{hull}_{\widetilde{\mathcal{H}}}$), where $\mathcal{H}$ is just $\mathcal{H}_V$.

Corollary 7. *Any region $V \subset \mathbb{R}^d$ that can be expressed as an arbitrary iterated difference of pseudo-polyhedra can also be expressed as an iterated difference of pseudo-polyhedra $P_1 \backslash (P_2 \backslash (\ldots P_l) \ldots)$ where each $P_i, i = 1, \ldots, l$ is an intersection of halfspaces and/or complement of halfspaces, all taken from a basis of V fixed a priori.*

Proposition 3 can be improved as follows:

Corollary 8. $\mathcal{D} \subsetneq \widetilde{\mathcal{D}} \subsetneq \overline{\mathcal{LP}_2} \subsetneq \mathcal{LP}_2$.

Proof. Let V by a 2-dimensional square with two opposite edges closed, the other two edges open, and without its corners. V is a pseudo-polyhedron but it is not in $\mathcal{D}$ since Algo($\overline{\mathrm{conv}}$) when run on V stops with $V_2 = V_0 \neq \emptyset$. Thus $\mathcal{D}$ is a proper subset of $\widetilde{\mathcal{D}}$. The last inclusion is obvious and the region V given in (1) with $\Delta > 2$ shows that it is a proper inclusion.

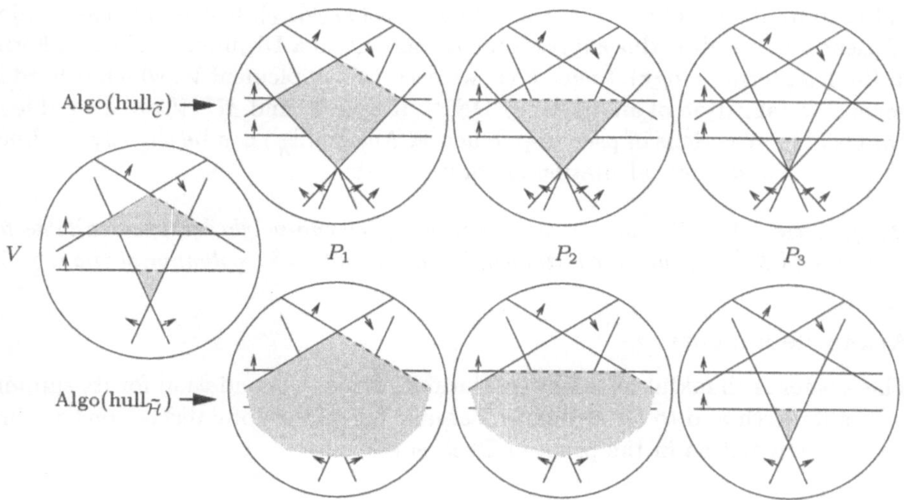

Fig. 1. Comparison of Algo(hull$_{\widetilde{\mathcal{C}}}$) and Algo(hull$_{\widetilde{\mathcal{H}}}$).

Each halfplane is indicated by a line (border) and an arrow (pointing toward the halfplane). The halfplanes shown in Figure V constitute the basis of V. Dashed lines denote open faces of gray regions. Algo(hull$_{\widetilde{\mathcal{C}}}$) adds two halfplanes to solve the problem, while Algo(hull$_{\widetilde{\mathcal{H}}}$) uses only the basis.

The inclusion $\widetilde{\mathcal{D}} \subsetneq \overline{\mathcal{LP}_2}$ follows from the fact that if $\mathcal{H}$ is a basis of V and if P is a pseudo-polyhedron whose basis is a subset of $\mathcal{H}$, then $V \in \overline{\mathcal{LP}_2}$ implies $P\backslash V \in \overline{\mathcal{LP}_2}$. The proof of the latter result follows easily when the geometrical problem is transposed into a Boolean problem (see [6]). Finally, the Swiss flag provides a region which is in $\overline{\mathcal{LP}_2}\backslash\widetilde{\mathcal{D}}$.

Proposition 9. *Algo(hull$_{\widetilde{\mathcal{H}}}$) stops after at most $|\widetilde{\mathcal{H}}|$ steps.*

Proof. At iteration l of Algo(hull$_{\widetilde{\mathcal{H}}}$), let $\widetilde{\mathcal{H}}_l$ denote the set of halfspaces $H \in \widetilde{\mathcal{H}}$ such that H is either essential to P_l or its supporting hyperplane intersects P_l^{int}. The proposition follows from the observation that $\widetilde{\mathcal{H}}_l \subsetneq \ldots \subsetneq \widetilde{\mathcal{H}}_1 = \widetilde{\mathcal{H}}$.

Finally, let us consider the complexity of Algo(hull$_{\widetilde{\mathcal{H}}}$). For V given as a union of s pseudo-polyhedra, the computation of hull$_{\widetilde{\mathcal{H}}}(V)$ requires that for each halfspace $H \in \widetilde{\mathcal{H}}$ and each of the s components of V, we check whether this component P is contained in H or in $\mathbb{R}^d\backslash H$. This is done by testing whether $P \cap (\mathbb{R}^d\backslash H)$ or $P \cap H$ is empty. It requires to check the non feasibility of a system of inequalities, which can be done by linear programming in a time polynomial in the number of inequalities (at most $|\widetilde{\mathcal{H}}|$) and the number d of variables. Thus the overall computation of hull$_{\widetilde{\mathcal{H}}}(V)$ is polynomial in d, s and $|\widetilde{\mathcal{H}}|$.

Even though we replaced the costly convex hull operator by hull$_{\widetilde{\mathcal{H}}}$ working in polynomial time, and we have a linear bound on the number of steps of the algorithm, the recognition of $\widetilde{\mathcal{D}}$ is a *NP*-Complete problem [6]. The complexity is in the computation of the difference of two sets ($P_l \backslash V_{l-1}$). If V is given as a union of pseudo-polyhedra (this expression corresponds to a Disjunctive Normal Form, in Boolean terminology), to get $P \backslash V$ we need to complement V, which is hard in general (dualization of an arbitrary DNF). If both V and $\mathbb{R}^d \backslash V$ are available as unions of intersections of pseudo-polyhedra, Algo(hull$_{\tilde{\mathcal{H}}}$) can be slightly modified so that it avoids any calculation of complements.

Proposition 10. *If expressions as unions of pseudo-polyhedra are available for both V and $\mathbb{R}^d \backslash V$, the recognition of $\widetilde{\mathcal{D}}$ can be solved in polynomial time.*

Acknowledgments

The author is thankful to the Swiss National Science Foundation for its support of this research and to Dr Motakuri Ramana for very helpful discussions and his active participation in the proof of Theorem 6.

References

1. G. Brightwell, C. Kenyon, and H. Paugam-Moisy. Multilayer neural networks : one or two hidden layers ? Technical Report NC-TR-97-001, NeuroCOLT Technical Report Series, 1997.
2. Gavin J. Gibson and Colin F. N. Cowan. On the decision regions of multilayer perceptrons. *Proceedings of the IEEE*, 78(10):1590–1594, 1990.
3. Gavin J. Gibson. Exact classification with 2-layered nets. *J. Computer and System Sciences*, 1996.
4. R. P. Lippmann. An introduction to computing with neural nets. *IEEE, Acoustic, Speech and Signal Processing*, 4:4–22, 1987.
5. Eddy Mayoraz. On variations of the convex hull operator. IDIAP-RR 96-06, IDIAP, 1996. `ftp://ftp.idiap.ch/pub/mayoraz/Publications/RRR-06-96.ps.gz`.
6. Eddy Mayoraz. On the complexity of recognizing regions of $\mathbb{R}^d$ computable by a two-layered perceptron. IDIAP-RR 97-05, IDIAP, 1997. `ftp://ftp.idiap.ch/pub/mayoraz/Publications/RRR-05-97.ps.gz`.
7. Ron Shonkwiler. Separating the vertices of n-cubes by hyperplanes and its application to artificial neural networks. *IEEE Transactions on Neural Networks*, 4(2):343–347, March 1993.
8. Catherine Z. W. Hassell Sweatman, Gavin J. Gibson, and Bernard Mulgrew. Exact classification with two-layer neural nets in n dimensions. submitted, 1996.
9. Patrick J. Zwietering. *The Complexity of Multi-Layered Perceptrons*. PhD thesis, University of Eindhoven, 1994.
10. P. J. Zwietering, E. H. L. Aarts, and J. Wessels. The design and complexity of exact multilayered perceptrons. *International Journal of Neural Systems*, 2:185–199, 1991.
11. P. J. Zwietering, E. H. L. Aarts, and J. Wessels. Exact classification with two-layered perceptrons. *International Journal of Neural Systems*, 3(2):143–156, 1992.

Optimal Linear Regression on Classifier Outputs

Yann Guermeur, Florence d'Alché-Buc and Patrick Gallinari

LIP6, Université Pierre et Marie Curie
Tours 46-00, Boîte 169
4, Place Jussieu, 75252 Paris cedex 05, France
{guermeur,dalche,gallinari}@laforia.ibp.fr

Abstract. We consider the combination of the outputs of several classifiers trained independently for the same discrimination task. We introduce new results which provide optimal solutions in the case of linear combinations. We compare our solutions to existing ensemble methods and characterize situations where our approach should be preferred.

1 Introduction

Statisticians pointed out long ago that combining predictive models led to better estimates and performance than simply selecting the best of them [1]. Model combination is thus a viable alternative to model selection and this area has been investigated by several researchers in the neural networks field [7]. Linear combinations - often coined linear ensemble methods - have proved empirically to be efficient for both regression and discrimination tasks. However, theoretical evidence has been mainly developed for regression. The case of discrimination is more complicated since combining estimates of class probabilities for computing better estimates - this is the framework considered in most approaches - introduces specific constraints which are absent in the case of regression. We consider here linear combinations of classifiers. The general framework is that of multivariate linear regression (MLR) where constraints ensure that combination outputs are probability estimates. This framework has already been considered in [5] where a suboptimal procedure was proposed. Linear opinion pool (see Sect. 2) is a particular case for which optimal solutions have been derived. We consider here the general constrained MLR model and show how optimal solutions can be obtained by solving nonlinear programming problems. We give some arguments characterizing cases where our approach should be preferred to existing alternatives. We first introduce in Sect. 2 the general multivariate linear regression model. Sect. 3 deals with the optimal solutions and their properties. Sect. 4 presents experimental results on a difficult problem. Sect. 5 is devoted to a comparison with nonlinear techniques.

2 Multivariate Linear Regression for Combination

Let us consider a Q-category discrimination problem and P classifiers whose outputs are estimated class posterior probabilities (i.e. they are positive and sum to one). Let f_j denote the function computed by the jth classifier and $f_{jk}(x)$ its kth output which approximates $p(C_k|x)$. The general multivariate approach to classifier combination corresponds to the following problem:

Problem 1 *Given a convex cost function J, find the best regression function g, $g(x) = [g_1(x)\dots g_k(x)\dots g_Q(x)]^T$ with $g_k(x) = F(x)v_k$, $(1 \le k \le Q)$ $F(x) = [f_1(x)^T \dots f_j(x)^T \dots f_P(x)^T]$, $v_k = [v_{k11}\dots v_{klm}\dots v_{kPQ}]^T$, which takes its values in*

$$U = \left\{ u \in IR_+^Q / \sum_{k=1}^{k=Q} u_k = 1 \right\} \tag{1}$$

The outputs of the *combiner* g are linear combinations of all classifier outputs. They are constrained to be non-negative and sum to one. Linear opinion pool is a degenerate case for which $g(x) = \sum_{j=1}^{j=P} v_j f_j(x)$. Coefficients v_j are scalars. In this case, constraints are satisfied if and only if the combination is convex. In [5], the authors also consider the general MLR model. They propose to determine separately the optimal regression functions $\hat{g}_k$ for each class, by solving Q separate constrained quadratic programming problems. The outputs are then standardized so that they sum to unity: $\tilde{g}_k(x) = \hat{g}_k(x)/\sum_{l=1}^{l=Q}\hat{g}_l(x)$. However, this two-step procedure is suboptimal with respect to the optimization criterion.

3 Optimal Solutions

Let us express formally the constraints. Let $v = [v_1^T \dots v_k^T \dots v_Q^T]^T$ denote the vector of parameters, and $v_{kl}^* = \min_m v_{klm}$, $(1 \le k \le Q)$, $(1 \le l \le P)$. There are two constraints: outputs must be non-negative and sum to one. Non-negativity is expressed as:

$$(Ct_1) \sum_{l=1}^{l=P} v_{kl}^* \ge 0, (1 \le k \le Q)$$

These contraints correspond to Q^{P+1} linear inequations.
Summation to unity is equivalent to:

$$(Ct_2) \begin{cases} \sum_{k=1}^{k=Q}(v_{klm} - v_{klQ}) = 0, (1 \le l \le P), (1 \le m < Q) \\ \sum_{k=1}^{k=Q}\sum_{l=1}^{l=P} v_{klQ} = 1 \end{cases}$$

The number of inequality constraints in Ct_1 makes the resolution of the convex programming problem prohibitive. However, we established in [3] the following result, which shows that inequalities in Ct_1 may be replaced by more restrictive but simpler constraints.

Proposition 1 *An optimal solution to Problem 1 is obtained by solving the following problem:*

Problem 2

$$\min_v J(v)$$

$$subject\ to \begin{cases} v \in IR_+^{PQ^2} \\ (Ct_2) \end{cases}$$

This simplification allows to handle the general optimization problem using classical algorithms such as the *gradient projection method* [6].

Either quadratic loss or entropic cost function may be used as training criterion J. The following result holds: every local solution to a convex programming problem is a global solution. Although cross-entropy should be preferred for classification, there is practically no difference for performance. The quadratic programming problem is easier to handle analytically and several properties can be derived which do not hold anymore for the entropic cost. An interesting property is the characterization of conditions for unicity of the solution. In the general case, there is a convex set of optimal solutions to Problem 2. They may be not equivalent for generalization. It is thus important to know whether the solution which has been obtained is unique or not. The following proposition allows to characterize this unicity *a posteriori*, i.e. when a solution has been obtained.

Proposition 2 *A necessary and sufficient condition for a solution* $\hat{v}$ *to Problem 2 with a quadratic cost function to be unique is:* $\forall(k,l)\hat{v}^*_{kl} = \min_m \hat{v}_{klm} = 0$.

A sketch of the proof is given in Appendix 1. Conditions in proposition 2 are often met in practice. The Kuhn-Tucker conditions may be used to characterize unicity *a priori*, before any solution has been found. For lack of place, we will not develop this topic further here.

4 Experiments

To assess our combiner, we have chosen the open problem of protein secondary structure prediction. This is a 3-class classification task which consists in assigning a conformation α-helix, β-strand or coil, to each residue of a sequence. The classifiers used are the neural architecture and statistical model in [4], with the nearest-neighbours algorithm of [10]. We have compared our optimal solution to other combiners: a single hidden layer neural network (MLP), a logistic regression model and an optimal convex combination. The training of the MLR model, both for the quadratic and the cross-entropic cost functions, was performed with the gradient projection method. We chose the same set of 126 protein chains which was selected to assess the system PHD [8]. However, base sequences were substituted to the profiles of the multiple alignments. The base is divided into seven subsets. This splitting was retained to implement a two-stage cross-validation procedure. A variation of *Stacked Generalization* [9] is used to avoid the generation of biased estimates, and every subset constitutes iteratively the test set. Table 1 summarizes the observed performance.

MLR compares favourably with existing methods. The subsequent 0.7% increase in recognition rate is significant for this task.

Table 1. Relative average performance of ensemble methods

Combiner	Recognition rate
MLR cross-entropy	66.5
MLR quadratic loss	66.3
convex average	65,7
MLP	65,8
Logistic regression	65.7

5 Comparison with Nonlinear Methods

The superiority of linear combiners over nonlinear ones, a phenomenon often observed in practice, has two explanations. Data set sizes are frequently too small with respect to the number of parameters of ordinary nonlinear models. Moreover, these models are less suited than linear ones in many cases. We illustrate this point by means of a particular example. We consider the logistic regression model [2]. It is identical to the single layer perceptron with *softmax* nonlinearity:

$$g_k(z) = \frac{e^{h_k(z)}}{\sum_l e^{h_l(z)}} \tag{2}$$

where the h_l are affine combinations of the predictors. This model, which computes a linear discriminant function, performs worse than a simple optimal convex conbination for the problem described below.
Let us consider a classification problem with two classes and two classifiers. We assume that the true posterior probabilities are given by:

$$p(C_1|x) = \theta f_{11}(x) + (1-\theta) f_{21}(x) \tag{3}$$

with $\theta \in]0,1[$. $t = f_{11}(x)$ and $u = f_{21}(x)$ are supposed to be independently uniformly distributed on $[0,1]$. Let $\hat{g}$ be the logistic regression function which maximizes the expectation of the log-likelihood function. Founding $\hat{g}$ is equivalent to minimizing with respect to v the functional:

$$J(v) = -\int_0^1 \int_0^1 p(C_1|t,u) ln(g(t,u)) + (1 - p(C_1|t,u)) ln(1 - g(t,u)) dt du \tag{4}$$

We will show that $\hat{g}$ does not allow to determine the true bayesian decision boundary. With no loss of generality, we can restrict the family of functions considered to:

$$g(t,u) = \frac{1}{1 + exp(-k(\hat{\theta} t + (1-\hat{\theta})u - \frac{1}{2}))} \tag{5}$$

$\hat{\theta} \in]0,1[$, $k > 0$. The true boundary will be found if and only if $\hat{\theta} = \theta$. We demonstrate that the resulting value of $\hat{\theta}$ is actually different from θ, provided $\theta \neq \frac{1}{2}$. *Proof*: see Appendix 2. Simulation results are displayed in Figure 1. For symmetry reasons, the study has been restricted to values of θ superior to 0.5.

The slight discrepancy between θ and the estimate of $\hat{\theta}$ can be easily observed.

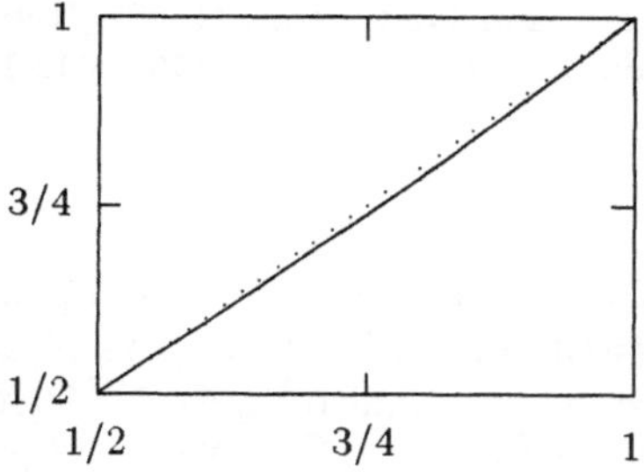

Fig. 1. Empirical estimate of $\hat{\theta}$ as a function of θ

6 Discussion

We have established how the standard multivariate linear regression model could be constrained in order to improve the estimates of the posterior probabilities of classes generated by a set of experts. The problem has been solved as a nonlinear programming problem. The linear regression approach presents several advantages. Training can be adapted to take into account complexity control. From this viewpoint, optimization methods that use the *active set method* and produce, at each step of the training phase, a *feasible point*, such as the gradient projection method, are of particular interest. We are currently studying these properties with the objective to improve performance in generalization.

References

1. Bates, J.M. and Granger, C.W.J. (1969): The Combination of Forecasts. *Operational Research Quaterly*, **Vol. 20**, 451-468.
2. Bishop, C.M. (1995): *Neural Networks for Pattern Recognition*, Clarendon Press, Oxford.
3. Guermeur, Y., d'Alché-Buc, F. and Gallinari, P. (1997): Combinaison Linéaire Optimale de Classifieurs. *XXIX-ièmes Journées de Statistique*, 425-428.
4. Guermeur, Y. and Gallinari, P. (1996): Combining Statistical Models for Protein Secondary Structure Prediction. *ICANN'96*, Bochum, 599-604.
5. LeBlanc, M. and Tibshirani, R. (1993): Combining estimates in regression and classification. *Technical Report 9318*, Department of Preventive Medicine and Biostatistics and Department of Statistics, University of Toronto.
6. Rosen, J.B. (1960): The Gradient Projection Method for Nonlinear Programming. Part I. Linear Constraints. *J. Soc. Indust. Appl. Math.*, **Vol. 8**, *N°* 1, 181-217.
7. Perrone, M.P. and Cooper, L.N. (1993): When Networks Disagree: Ensemble Methods for Hybrid Neural Networks. *Technical Report*, Institute for Brain and Neural Systems, Brown University, Providence, Rhode Island.
8. Rost, B. and Sander, C. (1993): Prediction of Protein Secondary Structure at Better than 70% Accuracy. *J. Mol. Biol.*, **232**, 584-599.
9. Wolpert, D.H. (1992): Stacked Generalization. *Neural Networks*, **Vol. 5**, 241-259.

10. Zhang, X., Mesirov, J.P. and Waltz, D.L. (1992): Hybrid System for Protein Secondary Structure Prediction. *J. Mol. Biol.*, **225**, 1049-1063.

7 Appendix 1

Let $Z = \{(x_i, y_i)\}, (1 \leq i \leq N)$ be the training sample. $F(X)$ denotes the matrix of explanatory variables (its lines are the vectors $F(x_i)$). A is a block-diagonal matrix with Q identical blocks equal to $\frac{1}{N}F(X)^T F(X)$. Let $Y_k = [y_{ik}] \in IR^N$, $(1 \leq k \leq Q)$ and $b = \frac{1}{N}[Y_1^T F(X) \ldots Y_k^T F(X) \ldots Y_Q^T F(X)]^T$. The sample-based estimate of $J(v)$ is then:

$$\hat{J}(v) = \frac{1}{2} v^T A v - b^T v + \frac{1}{2} \tag{6}$$

A base of $F(X)^T F(X)$ kernel is given by the set of vectors $w_j, (1 \leq j \leq P-1)$:

$$w_j = [1_Q^T, -\delta_{j,1} 1_Q^T, \ldots, -\delta_{j,P-1} 1_Q^T]^T \tag{7}$$

where 1_Q is a column vector of Q ones. $Ker(A) = (Ker(F(X)^T F(X)))^Q$. The following lemma holds:

Lemma 1. *If $\hat{v}$ and $\hat{v} + w$ are optimal solutions to Problem 2 with J being the mean squared error, then $w \in Ker(A)$.*

The proof relies on the following argument:

$$\hat{J}(\hat{v} + w) = \hat{J}(\hat{v}) \Longrightarrow \frac{1}{2} w^T A w + (A\hat{v} - b)^T w = 0 \Longrightarrow w \in Ker(A) \tag{8}$$

From (7), it is clear that if condition in proposition 2 holds, any point $\hat{v} + w$ with $w \in Ker(A) \setminus \{0\}$ will have negative components and so will not be a feasible solution. This ensures the unicity of the optimal solution.

8 Appendix 2

The objective function is equal to:

$$J(\hat{\theta}, k) = -\int_0^1 \int_0^1 ln(g(t,u))dtdu - \frac{k}{12}((2\theta - 1)\hat{\theta} + (1 - \theta)) \tag{9}$$

Assuming the optimum is obtained for $\hat{\theta} = \theta$ is equivalent to solving for k:

$$\begin{cases} \frac{\partial J}{\partial \hat{\theta}}(\theta, k) = 0 \\ \frac{\partial J}{\partial k}(\theta, k) = 0 \end{cases} \tag{10}$$

After some algebra, system (10) is shown to be equivalent to:

$$\begin{cases} \int_0^1 (1 - 2z) ln(cosh(\frac{k}{2}(\theta z - \frac{1}{2})))dz = \frac{k\theta(1-\theta)}{12} \\ \int_0^1 (1 - 2z) ln(cosh(\frac{k}{2}((1-\theta) z - \frac{1}{2})))dz = \frac{k\theta(1-\theta)}{12} \end{cases} \tag{11}$$

Let $h_1(z) = (1-2z) ln(cosh(\frac{k}{2}(\theta z - \frac{1}{2})))$, $h_2(z) = (1-2z) ln(cosh(\frac{k}{2}((1-\theta)z - \frac{1}{2})))$.

$$\forall (\theta, k, z) \in]0.5, 1[\times]0, +\infty[\times]0.5, 1[, h_1(z) + h_1(1-z) > h_2(z) + h_2(1-z) \tag{12}$$

The integrals have different values for $\theta \neq \frac{1}{2} \Longrightarrow$ system (10) has no solution. □

Learning Verification in Multilayer Neural Networks

Régis Quélavoine & Pascal Nocera

Laboratoire Informatique - Université d'Avignon
e-mail: quelavoine,nocera@univ-avignon.fr

Abstract. In this paper, we address the difficult problem of the learning verification in multilayer neural networks. Finding the activation/inhibition power of each input feature allows us to build synthetic examples and then to find out the minimal recognized patterns as long as to evaluate the robustness of the system. A small illustration upon character recognition clearly shows the interest in bias reduction. A real world application of transients recognition in underwater acoustic helped us to build more efficient features and to significantly improve the generalization rates [1].

1 Introduction

Widely used in various domains, multilayer neural networks can offer very interesting performances only if properly tuned. From a set of examples, it's very hard to find out what has actually been learnt, and then to evaluate the system robustness. In the second section, we propose a criteria that determines the activation/inhibition power of each input feature, and gives us a way to select the relevant information. One way to understand how a network reacts is to use it for object generation [1][6]. Therefore, in the third section, we use the activation/inhibition properties to build synthetic examples that correspond to minimal recognized patterns or maximaly degraded ones in a multilayer neural network. This allows us to detect biases and to estimate the generalization ability of the system. With an expert supervision, we can even improve it : as a school problem, we illustrate this methodology with a bitmap character recognition. In the fourth section, we describe an application in a real world problem : the automatic recognition of transients in underwater acoustic. The learning verification is there a very convenient way to point out the relevant features, to avoid biases and to significantly improve the recognition rates in a domain where we have very few expert knowledge.

2 Inhibition/Activation properties

We use three layer networks with single output node (one network per class), and we suppose that they are minimal so the whole information must be taken

[1] This work has been granted by DCN Ingénierie Sud/LSM under contract C 95 50 638 000

into account [3], for this the input values are normalized. We want to know the effect of an input (I_i) upon the output (O). So, as in [7], we want to find $\frac{\partial O}{\partial I_i}$. With :

- f : sigmoid as transfert function;
- O : network output, I_i : input i;
- W_{ij} : weight between input i and hidden node j, V_i : weight between hidden node i and the output node.

We know that $0 \leq f'(x) \leq f'(0)$, so we have :

$$\sum_{l \in L_i} \left(V_l \cdot W_{il} \cdot f'(0)^2 \right) \leq \frac{\partial O}{\partial I_i} \leq \sum_{k \in K_i} \left(V_k \cdot W_{ik} \cdot f'(0)^2 \right)$$

where :

$$L_i = \{l / V_l \cdot W_{il} < 0\} \text{ and } K_i = \{k / V_k \cdot W_{ik} > 0\}$$

So we define :

$$Inh_i = \sum_{l \in L_i} |V_l \cdot W_{il}| \text{ and } Act_i = \sum_{k \in K_i} (V_k \cdot W_{ik})$$

We have then in Inh_i the inhibition power ($O \rightarrow 0$) and in Act_i the activation ($O \rightarrow 1$) for each feature. The whole effect C_i is :

$$C_i = Act_i + Inh_i$$

If $C_i \ll C_{j,j \neq i}$, then the feature i is neglectible for the network and can be pruned. On the contrary, we can also detect very important features, or biases... Our criteria is similar to the measurement of $\frac{\partial O}{\partial I_i}$ made in [2], but with a much lower computation cost.

3 Construction of synthetic examples

The knowledge of the activation/inhibition power allows us to build artificial examples in two different ways.

3.1 Construction

- $0 < \varepsilon \ll 1$
- $A = \{i / Act_i > 0\}$, sorted by $Act_i \searrow$,

We start with a null vector ($\forall i, X_i = 0$). For the first $i \in A$, $X_i \leftarrow X_i + \varepsilon$ while$X_i < 1$ and the network does not recognize the pattern ($O \not\simeq 1$), if X_i reaches 1 without the recognition, we take the next i in A. By this way, we build a "minimal" pattern that is recognized by the network. If it appears not to be coherent for an expert, we can add it as a counter-example and then correct the learning. Doing this, we reduce the false alarm rate since the pattern must be close to the frame imposed by the expert, but we also increase the missed examples rate, so we need another control.

3.2 Alteration

In a similar way as with construction, starting from an already recognized example of the corpus ($O \simeq 1$) and taking the features in the decrescent order of their inhibition power (Inh_i), we will now increase the input values until the network no longer recognizes the example ($O < 1-$ threshold). Once again, we can show the pattern to an expert who decides if the network response is correct. If not, we can add this new artificial example into the learning corpus. Here, we reduce the "missed" rate while keeping a low false alarm rate, by tuning the network robustness. The drawback of our method is linked to the possibility for the expert to decide wether or not the response of the network is coherent with the proposed pattern : we must be able to build back an item from its features vector, or at least have some knowledge on rules characterizing these features.

3.3 Character recognition

While not a useful application, this school problem is very convenient as an illustration. We build two corpuses of bitmap characters coded with 9×5 input matrix. The learning one is a set of 26 lowercase letters from courrier police, and the test corpus is the same set from 7 other polices. We use 26 three layer neural networks, and since there are no noisy inputs in the learning corpus, we will find biases.

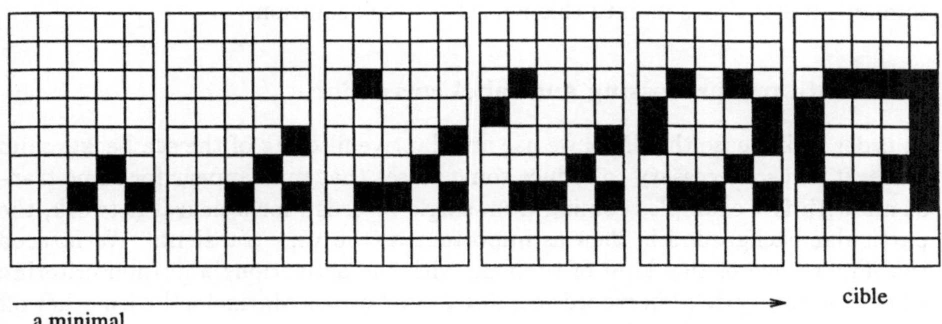

Fig. 1. *Evolution of the minimal pattern recognized as the a*

As we can see in fig.1 for the "a", the first learning gives biased results : three pixels are enough to describe this character. These are really discriminant regarding the learning corpus, but not sufficient for a reliable generalization. With the add of such counter-examples in the corpus, we see the evolution of this minimal pattern into a more acceptable one.

The correction of the biases increases the accuracy of the networks, see table 1, and we find better generalization rates. The improvements are not the

Char.	a	d	e	f	g	h	s	t
biased	28	13	20	18	15	6	44	19
unbiased	5	5	17	7	6	6	23	1

Table 1. *Recognition errors, on the first learning and after the biases correction*

same due to the differences in character shapes between polices. Nevertheless, we have a good idea of what each network has learnt and furthermore can do !

4 Transients recognition

The automatic detection and classification of very short duration noises in underwater acoustic is a promising way in anti-submarine warfare and leads to new sonar designs. But we have to face many problems : very low sound to noise ratios, various sea backgrounds, very wide variety of signals (biological, geological, surface vessels, aircrafts and submarines...) and the corresponding false alarm rates, very expensive and sensible data collection, and few explicit expert decision rules ! So we have to find out which features to use (generic signal processing such as FFT or wavelets, expert designed features...) in order to obtain the best discriminant representation for each class of transient, and then to build efficient and real time classifiers. The lack of knowledge but the available database and human experts drove us into this neural network solution.

4.1 Data pre-processing and label correction

In order to increase the SNR, we use an adaptive filtering of the sea background noise. It is also necessary to reduce some biases that may appear for some transients with few examples coming from single records : for a neural network, the particular background is then as important as the very transients. We noticed the importance of this kind of problem with the activation/inhibition criteria : when comparing dolphin records coming from Antibes Marineland with other off shore transients, we found that the network recognition was based upon features from frequency bands where we knew there should not have been such useful informations ! Despite a huge training time, the "golden ears" may fail when labelling examples. This also produces incoherencies that disturb the learning phase and induct false generalization rules. We have adapted the classic back-propagation algorithm into a selective learning [5] to avoid the bad influence of label errors. So we can assume that the network weights implicitly encode the knowledge that we are looking for.

4.2 Feature selection

Finding what is relevant and how to encode it, which is related, was one of our main task. Along with a great variety of sources, transients are also very different in "shapes" : duration, band width, modulation... To characterize each class, we had to test many set of features, and to evaluate the recognition performances. With a real time processing goal, and also with a will to lean faster toward the best solution, we used the activation/inhibiton criteria to prune the superfluous data and to point out the discriminant ones. In an attempt to separate biological

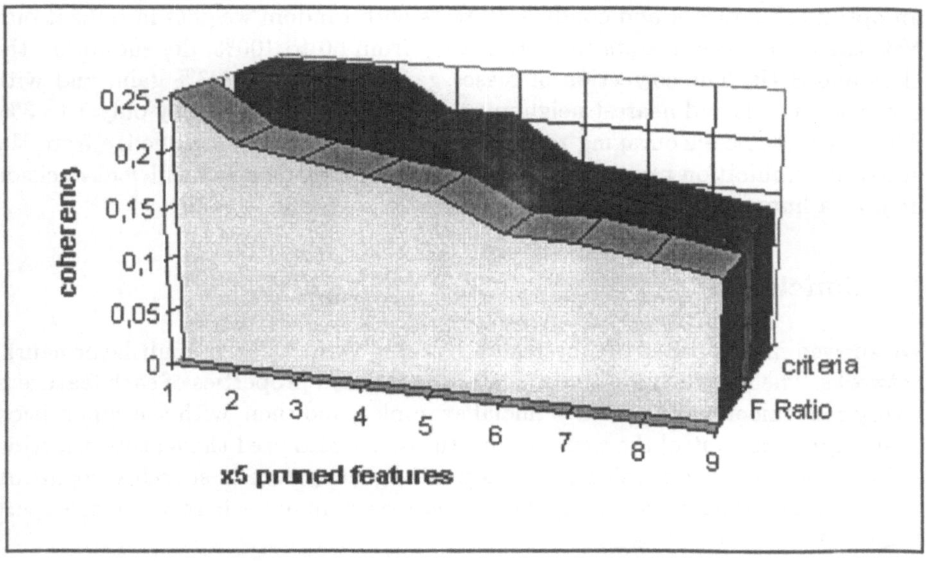

Fig. 2. Features pruning with a better coherency kept in the learning corpus than with the F Ratio

transients from submarine ones, we tried a wavelet packet decomposition along with signal processing expert designed features (giving an input vector size of 73). To verify the adequacy of our criteria, we measured the coherency (inter-class distance) while pruning the irrelevant features in the crescent order of the criteria compared to the hierarchy given by a statistical F Ratio (cf. fig.2). It clearly appears that our criteria selects the features in order to maintain the best coherency in the learning corpus. Furthermore, as we expected it, the expert designed features (length, band width, main frequency...) are the most useful, which confirms the quality of their work ! So, we found that generic signal processing such as wavelets or LPC are not are not adapted for very impulsive transients (10^{-3}s) according to the signal sampling rate of some sonars, but more convenient to characterize longer evolutive ones (s).

4.3 Learning verification

The construction of artificial examples is a way to build "models" for each class of transients. After the initial detection and correction of biases, we can also build cheap, but very efficient nearest neighbours classifiers, giving almost equivalent performances as the initial neural networks. Our learning corpuses are always built with at least 2000 items, chosen in order to have an initial 25% succes rate in random guess. The test corpuses are equivalent, and if we do not have enough examples, we use the leave one out method. For each class, we train an optimized network and conduct 30 tests with random weights initializations. Yet, the recognition results we obtain vary from 60 to 100%, depending on the class and SNR. The correction of biases gave us an average 3% gain, and with accurate models and nearest neighbours classifiers, we dropped by only 1 to 3%. Along with these encouraging results, we also acquired new knowledge from the activation/inhibition properties and we can now provide a fast efficient decision help to a human expert.

5 Conclusion

We showed in this paper the interest of learning verification in multilayer neural networks. The knowledge of the activation/inhibition properties of each feature is a very convenient way to build artificial examples, and then, with the supervision of an expert, to control the system robustness. We also used the feature selection to improve the fuzzy decision rule of a multi-agent system in speech recognition [4] : our methodology can be applied in various domains where we miss expert knowledge.

References

1. Ezhov A.A. & Vvedensky V.L. : *Object generation with neural networks.* Neural Networks, vol.9, n°9, pp.1491-1495, dec96.
2. Fechner T. & Hinze A. : *Delta Analysis : a method for the determination of input feature significance in neural networks.* NeuroNîmes'93, pp.393-398, Nîmes, France.
3. Nocera P. & Quélavoine R. : *Diminishing the number of nodes in multilayered neural networks.* ICNN'94, vol.7, pp.4421-4424, 28jun-2jul94, Orlando, FL.
4. Oppizzi O. & Quélavoine R. : *Rescoring under fuzzy degrees with a multilayered neural network in a rule-based speech recognition system.* ICASSP'97, 21-24apr97, Munchen, Germany.
5. Quélavoine R., Nocera P. & Di Martino M. : *Multilayered neural networks and errors in learning corpus.* WCNN'96, pp.287-290, 16-18sep96, San Diego, CA.
6. Quélavoine R. : *Étude de l'apprentissage et des structures des rseaux de neurones multicouches pour l'analyse de données.*. Thesis, 116p., 13jan97, University of Avignon, France.
7. Yoda M., Baba K. & Enbutu I. : *Explicit representation of knowledge acquired from plant historical data using neural networks.* IJCNN'91, vol.3, pp.155-160, San Diego, CA.

Design of a Fault Tolerant Multilayer Perceptron with a Desired Level of Robustness

Oh Jun Kwon and Sung Yang Bang

Dept. of Computer Science & Engineering
Pohang University of Science and Technology
San 31, Hyoja-dong, Nam-gu, Pohang, 790-784, Korea
Email: ojkwon@blackhol.postech.ac.kr, sybang@vision.postech.ac.kr

Abstract. We present a new definition of fault tolerance for an MLP by introducing the concept of a desired level of robustness. Based on this definition we propose an efficient method, called selective augmentation, which transforms a trained MLP to the fault tolerant one against a *stuck_at_0* fault at the hidden neurons. We show, through an example, that the resulting networks designed by the proposed method are not only fault tolerant but also less redundant than the ones by the uniform augmentation.

1 Introduction

Artificial neural network(ANN)s have many nice features such as massively parallel processing, self organization, generalization and fault tolerance. However these advantages, especially the massive parallelism, can be realized only when an ANN is implemented with VLSI or optical devices. However, when an ANN is implemented by hardware, it is quite possible for various functional faults to arise, due to various factors such as manufacturing defects, bad operating conditions and aging [4, 1]. Nevertheless many people simply assume that an ANN is highly fault tolerant since it has the same distributed structure as the brain. Further they assume that even those functional faults which inevitably arise in its hardware implementation do not affect the performance of an ANN at all. But at least in case of a multilayer perceptron(MLP), the removing of even a single hidden neuron often significantly degrades its performance [2, 6]. This implies that an MLP, trained by the usual backpropagation learning algorithm, is not fault tolerant against the *stuck_at_0* fault at a hidden neuron.

There have been reported many studies to improve the fault tolerance of an MLP [2, 3, 5]. But their concerns were just to obtain a more robust network against possible faults without defining the level of fault tolerance and specifying what level of fault tolerance the network needs.

In this paper we present a new definition of fault tolerance for an MLP by introducing the concept of a desired level of robustness. Then based on this definition we propose an efficient method, called selective augmentation, which transform a trained MLP to the fault tolerant one with a desired level of robustness against single *stuck_at_0* faults at the hidden neurons.

2 Definition of a fault tolerant neural network

Let $N(\omega)$ be a three-layer perceptron with the weight vector ω. The numbers of the input, hidden, and output neurons are denoted by N_i, N_h, and N_o, respectively. A pattern set $\mathcal{T}$ is a finite set of ordered pairs: $\mathcal{T} = \{(\mathbf{x}^p, \mathbf{d}^p) \mid p = 1, 2, \cdots, P\}$, where $\mathbf{x}^p$ is an input vector and $\mathbf{d}^p$ is the corresponding desired output vector. Let $\mathbf{y}^p$ denote the real output vector of $N(\omega)$ corresponding to the input $\mathbf{x}^p$. Then an error vector $E^p(\omega)$ is given by

$$E^p(\omega) = \mid \mathbf{y}^p - \mathbf{d}^p \mid \quad \text{for } p = 1, 2, \cdots, P. \tag{1}$$

Each component of $E^p(\omega)$ is represented as $E_i^p(\omega), 1 \leq i \leq N_o$.

In general a system is defined to be fault tolerant with respect to a fault if the system can continue correctly operating even in the presence of the fault. Then our question is how we should define an MLP to be fault tolerant. It seems reasonable to define that an MLP is fault tolerant with respect to a fault if it operates correctly, that is, operates within an allowed level of distortion in the outputs even in the presence of the fault.

When a fault f occurs in $N(\omega)$, the distortion $D_{j,f}^p(\omega)$ in the output of the jth output neuron for an input $\mathbf{x}^p$ will occur and is defined as follows.

$$D_{j,f}^p(\omega) = \begin{cases} E_{j,f}^p(\omega) - E_j^p(\omega) & \text{if } E_{j,f}^p(\omega) > E_j^p(\omega) \\ 0 & \text{otherwise} \end{cases} \tag{2}$$

where $E_{j,f}^p(\omega)$ is the output error in the jth output neuron corresponding to the input $\mathbf{x}^p$ in the presence of the fault f. Then, in terms of fault tolerance, we define the output distortion $D^p(\omega)$ of a network $N(\omega)$ for an input $\mathbf{x}^p$ as

$$D^p(\omega) = \max_{1 \leq j \leq N_o, f \in \mathcal{F}} D_{j,f}^p(\omega). \tag{3}$$

Definition 1: The maximum output distortion $\phi(\omega)$ of a network $N(\omega)$ for a fault set $\mathcal{F}$ and a pattern set $\mathcal{T}$ is defined as

$$\phi(\omega) = \max_{1 \leq p \leq P} D^p(\omega). \tag{4}$$

Definition 2: Let $\varepsilon \geq 0$ be a desired level of robustness within which a network $N(\omega)$ should perform for a fault set $\mathcal{F}$ and for a pattern set $\mathcal{T}$ and $\phi(\omega)$ be the maximum output distortion of $N(\omega)$ for $\mathcal{F}$ and $\mathcal{T}$, then if $\phi(\omega) \leq \varepsilon$, we say the network $N(\omega)$ is fault tolerant against $\mathcal{F}$ and $\mathcal{T}$.

In the following we simply say that a network is fault tolerant without mentioning $\mathcal{F}$ and $\mathcal{T}$ if it is clear which fault set $\mathcal{F}$ and pattern set $\mathcal{T}$ we are talking of. Now we will limit the fault set $\mathcal{F}$ to the set of functional *stuck_at_0* faults at the hidden neurons.

Let $D_{j,-i}^p(\omega)$ be the distortion of the jth output neuron and $\rho_i^p(\omega)$ be the largest among the distortions of the output neurons when the input is $\mathbf{x}^p$ and a *stuck_at_0* fault occurs at the ith hidden neuron of a network $N(\omega)$. Then $D_{j,-i}^p(\omega)$ is obtained by replacing a fault f of the formula (2) with a *stuck_at_0* fault at the ith hidden neuron and hence $\rho_i^p(\omega)$ by

$$\rho_i^p(\omega) = \max_{1 \leq j \leq N_o} D_{j,-i}^p(\omega) \tag{5}$$

Definition 3: The maximum contribution $\rho_i(\omega)$ of the ith hidden neuron to the output of a network $N(\omega)$ is given by

$$\rho_i(\omega) = \max_{1 \le p \le P} \rho_i^p(\omega). \tag{6}$$

Theorem 1: If the fault set $\mathcal{F}$ is limited to the set of ***stuck_at_0*** faults at the hidden neurons in a network $N(\omega)$, then the the maximum output distortion $\phi(\omega)$ of the network is

$$\phi(\omega) = \max_{1 \le i \le N_h} \rho_i(\omega). \tag{7}$$

The proof of this theorem can be directly obtained from the definitions 1 and 3.

3 Design method of a fault tolerant MLP

From now on we will consider a method which transforms a trained MLP to the fault tolerant one against a *stuck_at_0* fault at any hidden neuron when a desired level of fault tolerance is given. When we say a desired level of fault tolerance, the level is usually determined by the application problem at hand. And the maximum distortion of the network should be smaller than this level. Let such a level be ε. The idea of the proposed method is to replicate the hidden neurons selectively depending on their maximum contributions in such a way that the resulting network satisfies a desired level of fault tolerance against the faults. Then the weights to the output neurons from each of the hidden neurons replicated are scaled inversely by the number of their replications.

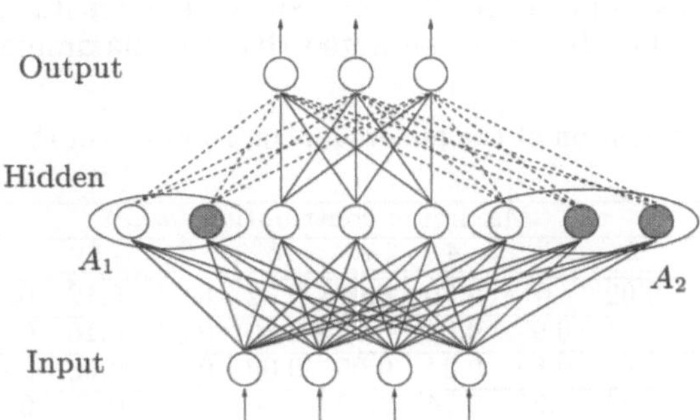

Fig. 1. Example of selective augmentation

[Selective Augmentation]

1. Determine the maximum contribution $\rho_i(\omega)$ of each hidden neuron i of the current network.
2. Stop if the current network satisfies the following condition:

$$\rho_i(\omega) \le \varepsilon, \quad 1 \le i \le N_h. \tag{8}$$

3. Otherwise compute the augmentation number A_i of each hidden neuron:

$$A_i = \begin{cases} 0 & \text{if } \rho_i(\omega) = 0 \\ \lceil \rho_i(\omega)/\varepsilon \rceil - 1 & \text{otherwise.} \end{cases} \tag{9}$$

4. Modify the weight w_{ij} between each hidden neuron i and the output neuron j to $w_{ij}/(A_i + 1)$ for all is and js.
5. Replicate each hidden neuron A_i times and go to step 1.

It should be noted that the resultant network maintains the same input-output mapping as the original network as it should. Also note that the fault tolerance of the original network has been improved but improved only to the level of the given value ε. Fig. 1 shows an example of the selective augmentation.

Theorem 2: If the network is a three-layer perceptron with the activation function of the output neurons being linear, the proposed algorithm terminates after at most one iteration.

The proof of this theorem is omitted here.

4 Experiment

We applied the proposed method to a nonlinear function approximation problem of the following formula:

$$y = \frac{1}{2}(\sin(10x/\pi) + 1), \quad -1 \leq x \leq 1. \tag{10}$$

A three-layer perceptron with 10 hidden neurons was used. We investigated the distribution of the maximum contributions of the hidden neurons in the network trained by the backpropagation. Table 1 shows the results of five runs with different initial weights. It should be noted that the maximum contributions of

Table 1. The distribution of the maximum contributions of the hidden neurons

	Maximum contribution, $\rho_i(\omega)$									
Trial	1	2	3	4	5	6	7	8	9	10
1	0.45	0.09	0	0.50	0.07	0.32	0.94	0.12	0.45	0.94
2	0.11	0.31	0.91	0.04	0.92	0.16	0.33	0.16	0.14	0.35
3	0.56	0.07	0.85	0.24	0.05	0.09	0.84	0.06	0.04	0.06
4	0.37	0.24	0.25	0.34	0.92	0.50	0.94	0.27	0.32	0.37
5	0.43	0.14	0.31	0.95	0.67	0.50	0.30	0.40	0.14	0.50

the hidden neurons are not even. Fig. 2(a) shows the output curves of the trained network corresponding to the first trial of Table 1 without any fault and with a fault at the neurons 7 and 10 which have the largest maximum contribution. This shows that a severe distortion in the output does occur in the presence of even one *stuck_at_0* fault and hence implies that an MLP can be very vulnerable to the *stuck_at_0* fault at a hidden neuron.

We chose ε to be 0.2, which means that the distortion of the network's output should be smaller than 0.2 for all patterns. After applying our method we

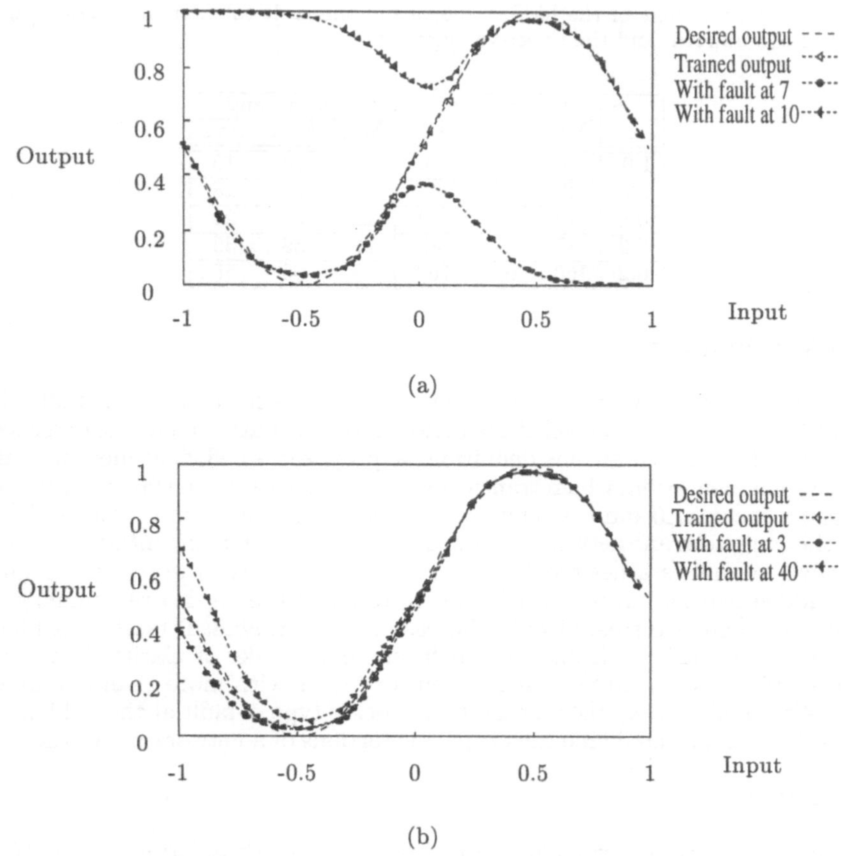

Fig. 2. The outputs of the trained network before and after the method is applied(ε=0.2)

obtained a fault tolerant network with 50 hidden neurons. This means that 40 hidden neurons are redundant but were generated in order to secure the needed level of fault tolerance. Fig. 2(b) shows the output curves of the fault tolerant network designed by the proposed method. It can be seen that the distortion of the network's output due to a *stuck_at_0* fault at these two hidden neurons are within $\varepsilon(= 0.2)$ for all patterns. The uniform augmentation proposed earlier replicates the hidden neurons uniformly, i.e. by the same number, since the basic definitions about the fault tolerance of a neural network like ours have not been established. Table 2 shows the number of the hidden neurons in the fault tolerant networks designed by using the uniform and the selective augmentations for various values of ε. They are the average results on ten trials for different initial weights. The case of $\varepsilon = 1.0$ corresponds to the original trained network without any modification. It should be also noted that the redundancy needed by the selective augmentation is generally about 30% $\sim$ 40% of that by the uniform augmentation.

Table 2. The numbers of the hidden neurons in the fault tolerant networks designed by using the uniform and the selective augmentation

ε	Uniform aug.			Selective aug.		
	Min	Max	Avg	Min	Max	Avg
1.0	10	10	10	10	10	10
0.5	40	80	48	15	22	18
0.4	40	80	72	22	25	24
0.3	80	80	80	32	39	35
0.2	160	160	160	45	65	55

5 Conclusions

We defined that an MLP is fault tolerant with respect to a set of faults if it operates within a desired level of distortion in the outputs even in the presence of the faults. Then based on this definition we proposed an efficient method, called selective augmentation, which transforms a trained MLP to the fault tolerant one against a *stuck_at_0* fault at any hidden neuron. Experimental results showed that the fault tolerant network obtained by the selective augmentation not only actually guarantees a desired level of robustness against a *stuck_at_0* fault at each hidden neuron but also is effective in terms of the needed redundancy.

In this paper we treated only the case of the *stuck_at_0* fault at a hidden neuron in the MLP with one hidden layer for the sake of clarity. However it is straightforward to extend our method for MLPs with more than one hidden layer. We can also apply the method to any other type of fault at the hidden neurons as long as we can define the output distortions of a network under the fault.

Acknowledgement

This study was supported in part by Korea Science Foundation, in part Korea Telecom and in part Systems Engineering Research Institute.

References

1. Feltham D. B. I., & Maly, W. (1991). Physically Realistic Fault Models for Analog CMOS Neural Networks. *IEEE Journal of Solid-State Circuits*, **26**, 1223-1229.
2. Emmerson, M. D. & Damper, R. I. (1993). Determining and Improving the Fault Tolerance of Multilayer Perceptrons in a Pattern Recognition Application. *IEEE Transactions on Neural Networks*, **4**, 788-793.
3. Naraghi-Pour, M., Hedge, M., & Bapat, P. (1993). Fault Tolerance Design of Feedforward Networks. *Proc. World Congress on Neural Networks*(Vol. 3, pp. 568-571), Portland, USA.
4. Bolt, G. (1991). Fault Models for Artificial Neural Networks. *Proc. International Joint Conference on Neural Networks*(Vol. 2, pp. 1373-1378), Singapore.
5. Neti, C., Schneider, M. H., & Young, E. D. (1992). Maximally Fault Tolerant Neural Networks. *IEEE Transactions on Neural Networks*, **3**, 14-23.
6. Séquin, C. H., & Clay, R. D. (1991). Fault tolerance in feedforward artificial neural networks, in P. Antognetti & V. Milutinović(eds.), *Neural Networks Concepts, Applications, and Implementations*(Vol. IV, pp. 111-141). Englewood Cliffs, New Jersey: Prentice Hall.

Mixtures of Experts Estimate A Posteriori Probabilities

Perry Moerland

IDIAP, CP 592, 1920 Martigny, Switzerland

Abstract The mixtures of experts (ME) model offers a modular structure suitable for a divide-and-conquer approach to pattern recognition. It has a probabilistic interpretation in terms of a mixture model, which forms the basis for the error function associated with MEs. In this paper, it is shown that for classification problems the minimization of this ME error function leads to ME outputs estimating the a posteriori probabilities of class membership of the input vector.

1 Introduction

It is well-known that for artificial neural networks trained by minimizing sum-of-squares or cross-entropy error functions for a classification problem, the network outputs approximate the a posteriori probabilities of class membership [2]. This property is a very useful one, especially when the network outputs are to be used in a further decision-making stage (e.g. rejection thresholds) or integrated in other statistical pattern recognition methods (as in hybrid NN-HMMs).

Recently, a modular architecture of neural networks known as a *mixture of experts* (ME) has attracted quite some attention [6][7]. MEs are mixture models which attempt to solve problems using a divide-and-conquer strategy; that is, they learn to decompose complex problems in simpler subproblems. In particular, the *gating* network of a ME learns to partition the input space (in a soft way, so overlaps are possible) and attributes *expert* networks to these different regions. The divide-and-conquer approach has shown particularly useful in attributing experts to different regimes in piece-wise stationary time series [9] and modeling discontinuities in the input-output mapping.

Mixtures of experts have also been successfully applied to classification problems [4][8], though a proof that minimization of the ME error function (based on the formulation as a mixture model) leads to ME outputs estimating the a posteriori probabilities of class membership, is still lacking. The purpose of this paper is to show that at the global minimum of this ME error function, the ME outputs do indeed estimate a posteriori probabilities.

2 Mixtures of Experts

In this section the basic definitions of the mixture of experts model are given which will be used in the rest of the paper.

Figure 1 shows the architecture of a ME network, consisting of three expert networks and one gating network both having access to the input vector $\mathbf{x}$; the gating network has one output g_i per expert. The standard choices for gating

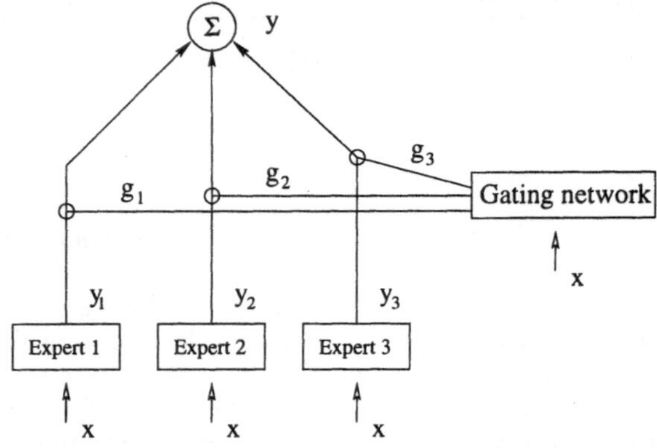

Figure1. Architecture of a mixture of experts network.

and expert networks are generalized linear models [7] and multilayer perceptrons [9]. The output vector of a ME is the weighted (by the gating network outputs) mean of the expert outputs:

$$\mathbf{y}(\mathbf{x}) = \sum_{j=1}^{m} g_j(\mathbf{x})\mathbf{y}_j(\mathbf{x}) \tag{1}$$

The gating network outputs $g_j(\mathbf{x})$ can be regarded as the probability that input $\mathbf{x}$ is attributed to expert j. In order to ensure this probabilistic interpretation, the activation function for the outputs of the gating network is chosen to be the soft-max function [3]:

$$g_j = \frac{\exp(z_j)}{\sum_{i=1}^{m} \exp(z_i)}, \tag{2}$$

where the z_i are the gating network outputs before thresholding. This soft-max function makes that the gating network outputs sum to unity and are non-negative; thus implementing the (soft) competition between the experts.

A probabilistic interpretation of a ME can be given in the context of mixture models for conditional probability distributions (see section 6.4 in [1]):

$$p(\mathbf{t}|\mathbf{x}) = \sum_{j=1}^{m} g_j(\mathbf{x})\phi_j(\mathbf{t}|\mathbf{x}), \tag{3}$$

where the ϕ_j represent the conditional densities of target vector $\mathbf{t}$ for expert j. The use of a soft-max function in the gating network and the fact that the ϕ_j are densities guarantee that the distribution is normalized: $\int p(\mathbf{t}|\mathbf{x})\, d\mathbf{t} = 1$.

As outlined in the next section this distribution forms the basis for the ME error function which can be optimized using gradient descent or the Expectation-Maximization (EM) algorithm [7].

3 Estimating Posterior Probabilities

A standard way to motivate error functions is from the principle of maximum likelihood of the (independently distributed) training data $\{\mathbf{x}^n, \mathbf{t}^n\}$ (see section

6.1 in [1]):

$$\mathcal{L} = \prod_n p(\mathbf{x}^n, \mathbf{t}^n) = \prod_n p(\mathbf{t}^n|\mathbf{x}^n)p(\mathbf{x}^n).$$

A cost function is then obtained by taking the negative logarithm of the likelihood (and dropping the term $p(\mathbf{x}^n)$ which does not depend on the network parameters):

$$E = -\sum_n \ln p(\mathbf{t}^n|\mathbf{x}^n). \tag{4}$$

The most suitable choice for the conditional probability density depends on the problem. For regression problems a Gaussian noise model is appropriate (leading to the sum-of-squares error function); for classification problems with a 1-of-c coding scheme and outputs y_c for each class, a multinomial density is most suitable:

$$p(\mathbf{t}^n|\mathbf{x}^n) = \prod_{c=1}^{C} (y_c^n)^{t_c^n}. \tag{5}$$

This offers us the framework to obtain a cost function for the mixtures of experts model. In its most general form the ME error function to be minimized is (substituting (3) in (4)):

$$E = -\sum_n \ln \sum_{j=1}^{m} g_j(\mathbf{x}^n)\phi_j(\mathbf{t}^n|\mathbf{x}^n),$$

the exact formulation of which depends on the choice for the conditional densities $\phi_j(\mathbf{t}^n|\mathbf{x}^n)$. Since, our main interest is in MEs for classification problems, the ϕ_j are assumed to be multinomial densities (5) in the rest of this paper. As in the gating network of a ME, a suitable choice for the activation function for the expert output units is then the soft-max function (2):

$$y_{jc} = \frac{\exp(a_{jc})}{\sum_k \exp(a_{jk})}, \tag{6}$$

where the a_{jc} are the expert network outputs before thresholding.

In the limit of an infinite data set (to avoid bias and variance) the finite sum over patterns can be replaced with an integral:

$$E = -\int\int \ln\left(\sum_{j=1}^{m} g_j(\mathbf{x})\phi_j(\mathbf{t}|\mathbf{x})\right) p(\mathbf{t}, \mathbf{x})\, d\mathbf{t}d\mathbf{x},$$

factoring the joint distribution:

$$E = -\int\int \ln\left(\sum_{j=1}^{m} g_j(\mathbf{x})\phi_j(\mathbf{t}|\mathbf{x})\right) p(\mathbf{t}|\mathbf{x})p(\mathbf{x})\, d\mathbf{t}d\mathbf{x}.$$

The interpretation of the ME outputs when this error function is minimized, can be obtained by setting to zero the functional derivatives [5] of E with respect to the gating network outputs $z_j(\mathbf{x})$ and the network outputs of expert j, $a_{jc}(\mathbf{x})$.

The solution of these equations will then result in expressions for $g_j(\mathbf{x})$ and $\mathbf{y}_j(\mathbf{x})$ at the minimum of E (along the lines of section 6.1.3 of [1] for the sum-of-squares error function).

Defining:

$$E' = \ln \sum_{j=1}^{m} g_j(\mathbf{x}) \phi_j(\mathbf{t}|\mathbf{x}), \tag{7}$$

we are then interested in the following two functional derivatives set to zero. For the gating network:

$$\frac{\delta E}{\delta z_j} = - \int \left(\frac{\partial E'}{\partial z_j} \right) p(\mathbf{t}|\mathbf{x}) p(\mathbf{x}) \, d\mathbf{t} = 0, \tag{8}$$

and for the expert network (using the chain rule):

$$\frac{\delta E}{\delta a_{jc}} = - \int \left(\frac{\partial E'}{\partial a_{jc}} \right) p(\mathbf{t}|\mathbf{x}) p(\mathbf{x}) d\mathbf{t} = - \int \sum_k \frac{\partial E'}{\partial y_{jk}} \frac{\partial y_{jk}}{\partial a_{jc}} p(\mathbf{t}|\mathbf{x}) p(\mathbf{x}) d\mathbf{t} = 0. \tag{9}$$

In section 6.4 of [1], the partial derivatives for the gating network occurring in (8) have been calculated in the context of a gradient descent algorithm for the mixture model (3). Bishop's outcomes are restated here:

$$\frac{\partial E'}{\partial z_j} = \sum_k \frac{\partial E'}{\partial g_k} \frac{\partial g_k}{\partial z_j} = \sum_k - \frac{\pi_k}{g_k} (\delta_{jk} g_k - g_j g_k) = g_j - \pi_j, \tag{10}$$

where the posterior probability π_j is defined as:

$$\pi_j(\mathbf{x}, \mathbf{t}) = \frac{g_j \phi_j}{\sum_i g_i \phi_i}, \tag{11}$$

and δ_{jk} is the Kronecker delta. The functional derivative set zero with respect to the gating network outputs is (substituting (10) in (8)):

$$\frac{\delta E}{\delta z_j} = - \int (g_j - \pi_j) \, p(\mathbf{t}|\mathbf{x}) p(\mathbf{x}) \, d\mathbf{t} = 0. \tag{12}$$

The solution of the expert network equation (9) will be treated in some more detail. Recall that for classification problems, the expert outputs are obtained with a soft-max function (6). Therefore, the second partial derivative, $\partial y_{jk} / \partial a_{jc}$, in (9) is similar to its counterpart in the gating network equation $\partial g_k / \partial z_j$ (see the second term in (10)):

$$\frac{\partial y_{jk}}{\partial a_{jc}} = \delta_{ck} y_{jk} - y_{jc} y_{jk}. \tag{13}$$

Using the definition of E' (7) and of the multinomial density ϕ_j (5) gives for the first partial derivative in (9):

$$\frac{\partial E'}{\partial y_{jk}} = \frac{\partial \left(\ln \sum_{j=1}^{m} g_j \phi_j \right)}{\partial y_{jk}} = \frac{\partial \left(\ln \sum_{j=1}^{m} g_j \prod_{c=1}^{C} (y_{jc})^{t_c} \right)}{\partial y_{jk}},$$

that is, taking the partial derivative and using (11):

$$\frac{\partial E'}{\partial y_{jk}} = \frac{g_j (y_{jk})^{(t_k-1)} t_k}{\sum_{i=1}^{m} g_i \phi_i} \prod_{c=1,c\neq k}^{C} (y_{jc})^{t_c} = \frac{g_j \phi_j}{\sum_{i=1}^{m} g_i \phi_i} \frac{t_k}{y_{jk}} = \pi_j \frac{t_k}{y_{jk}}. \tag{14}$$

Preparing for the solution of (9) one needs (using (13) and (14)):

$$\sum_k \frac{\partial E'}{\partial y_{jk}} \frac{\partial y_{jk}}{\partial a_{jc}} = \sum_k \pi_j(\mathbf{x},\mathbf{t}) \frac{t_k}{y_{jk}} (\delta_{ck} y_{jk} - y_{jc} y_{jk}) = \pi_j(\mathbf{x},\mathbf{t}) t_c - \pi_j(\mathbf{x},\mathbf{t}) y_{jc}, \tag{15}$$

where in the last step it has been used that for 1-of-c classification problems, $\sum_k t_k = 1$. The functional derivative set to zero with respect to the expert network outputs is (substituting (15) in (9)):

$$\frac{\delta E}{\delta a_{jc}} = -\int (\pi_j(\mathbf{x},\mathbf{t}) t_c - \pi_j(\mathbf{x},\mathbf{t}) y_{jc})\, p(\mathbf{t}|\mathbf{x}) p(\mathbf{x})\, d\mathbf{t} = 0. \tag{16}$$

What is left is to determine the $g_j(\mathbf{x})$ and $y_j(\mathbf{x})$ that solve (12) and (16) (and therefore minimize the ME error function). For the gating network outputs (12):

$$\frac{\delta E}{\delta z_j} = -g_j p(\mathbf{x}) \int p(\mathbf{t}|\mathbf{x})\, d\mathbf{t} \; + \; p(\mathbf{x}) \int \pi_j(\mathbf{x},\mathbf{t})\, p(\mathbf{t}|\mathbf{x})\, d\mathbf{t} = 0.$$

using that the conditional probability $p(\mathbf{t}|\mathbf{x})$ is normalized:

$$\frac{\delta E}{\delta z_j} = -g_j p(\mathbf{x}) \; + \; p(\mathbf{x}) \int \pi_j(\mathbf{x},\mathbf{t})\, p(\mathbf{t}|\mathbf{x})\, d\mathbf{t} = 0.$$

Therefore, at the minimum of the ME error function the gating network outputs satisfy:

$$g_j = \int \pi_j(\mathbf{x},\mathbf{t})\, p(\mathbf{t}|\mathbf{x})\, d\mathbf{t}. \tag{17}$$

For the expert network outputs (16):

$$\frac{\delta E}{\delta a_{jc}} = -p(\mathbf{x}) \int \pi_j(\mathbf{x},\mathbf{t}) t_c\, p(\mathbf{t}|\mathbf{x})\, d\mathbf{t} \; + \; y_{jc} p(\mathbf{x}) \int \pi_j(\mathbf{x},\mathbf{t})\, p(\mathbf{t}|\mathbf{x})\, d\mathbf{t} = 0.$$

Therefore, at the minimum of the ME error function the expert network outputs satisfy:

$$y_{jc} = \frac{\int \pi_j(\mathbf{x},\mathbf{t}) t_c\, p(\mathbf{t}|\mathbf{x})\, d\mathbf{t}}{\int \pi_j(\mathbf{x},\mathbf{t})\, p(\mathbf{t}|\mathbf{x})\, d\mathbf{t}}. \tag{18}$$

Finally, using (17) and (18), the output vector of a mixture of experts that minimizes the ME error function is (1):

$$y_c(\mathbf{x}) = \sum_j g_j(\mathbf{x}) y_{jc}(\mathbf{x}) = \sum_j \int \pi_j(\mathbf{x},\mathbf{t}) t_c\, p(\mathbf{t}|\mathbf{x})\, d\mathbf{t},$$

exchanging integration and summation:

$$\int \sum_j \pi_j(\mathbf{x},\mathbf{t}) t_c\, p(\mathbf{t}|\mathbf{x})\, d\mathbf{t} = \int t_c\, p(\mathbf{t}|\mathbf{x})\, d\mathbf{t} := \langle t_c|\mathbf{x}\rangle, \tag{19}$$

where we have used that the posterior probabilities $\pi_j(\mathbf{x},\mathbf{t})$ (11) sum to unity. The interpretation of (19) is that the output $y_c(\mathbf{x})$ of a ME at the minimum of the ME error function is equal to the conditional average of the target data. This is exactly the same as for the outputs of a network trained by minimizing the sum-of-squares or cross-entropy error functions [1]. It is a well-known result that for a classification problem with 1-of-c coding the conditional average of the target data is (see, for example, section 6.6 in [1]) :

$$y_c(\mathbf{x}) = P(\mathcal{C}_c|\mathbf{x}),$$

so that the outputs of a ME do indeed estimate the a posteriori probability that $\mathbf{x}$ belongs to class $\mathcal{C}_c$.

4 Discussion

In section 3, it was assumed that the conditional density $\phi_j(\mathbf{t}^n|\mathbf{x}^n)$ of expert j is multinomial. However, this is not a necessary condition for ME to estimate a posteriori probabilities. It can be shown that also a Gaussian noise model:

$$\phi_j(\mathbf{t}^n|\mathbf{x}^n) = \frac{1}{(2\pi)^(c/2)}\exp\left(-\frac{||\mathbf{t}-\mathbf{y}_j(\mathbf{x})||^2}{2}\right)$$

leads to this result.

Acknowledgments

The author gratefully acknowledges the Swiss National Science Foundation (FN:21-45621.95) for their support of this research.

References

1. Christopher M. Bishop. *Neural Networks for Pattern Recognition.* Oxford University Press, Oxford, 1995.
2. H. Bourlard and N. Morgan. Links between Markov models and multi-layer perceptrons. In D. S. Touretzky, editor, *Advances in Neural Information Processing,* volume 1, pages 502–510, San Mateo CA, 1989. Morgan Kaufmann.
3. J. S. Bridle. Probabilistic interpretation of feedforward classification network outputs with relationships to statistical pattern recognition. In F. Fogelman Soulié and J. Hérault, editors, *Neurocomputing: Algorithms, Architectures, and Applications,* pages 227–236. Springer Verlag, New York, 1990.
4. Jürgen Fritsch, Michael Finke, and Alex Waibel. Context-dependent hybrid HME/HMM speech recognition using polyphone clustering decision trees. In *Proceedings of ICASSP-97,* 1997.
5. Mariano Giaquinta and Stefan Hildebrandt. *Calculus of Variations.* Springer Verlag, Berlin, 1996.
6. Robert A. Jacobs, Michael I. Jordan, Steven J. Nowlan, and Geoffrey E. Hinton. Adaptive mixtures of local experts. *Neural Computation,* 3(1):79–87, 1991.
7. Michael I. Jordan and Robert A. Jacobs. Hierarchical mixtures of experts and the EM algorithm. *Neural Computation,* 6(2):181–214, 1994.
8. S. R. Waterhouse and A. J. Robinson. Classification using hierarchical mixtures of experts. In *Proceedings 1994 IEEE Workshop on Neural Networks for Signal Processing,* pages 177–186, Long Beach CA, 1994. IEEE Press.
9. Andreas S. Weigend, Morgan Mangeas, and Ashok N. Srivastava. Nonlinear gated experts for time series: Discovering regimes and avoiding overfitting. *International Journal of Neural Systems,* 6:373–399, 1995.

Admissibility and Optimality of the Cascade–Correlation Algorithm

A. Doering, M. Galicki, H. Witte *

Institute for Medical Statistics, Computer Sciences and Documentation
Friedrich Schiller University Jena, Germany

Abstract. This contribution considers some convergence and optimality properties of the Cascade–Correlation Algorithm (CCA). It is proved, that arbitrary, non–contradicting learning tasks can be solved with linear output neurons within a finite number of steps. Furthermore, it is shown that the correlation criterion proposed by Fahlman [3] does not necessarily choose optimal weights. An optimal criterion is given for linear output neurons. For nonlinear output neurons it is demonstrated, that the CCA does not need to converge even for finite learning tasks. Thus, it is generally not an universal approximation tool.

1 Introduction

The Cascade–Correlation learning algorithm (CCA) has been introduced by Fahlman [3] in order to find suited (i.e., data–dependent) feedforward network structures and to decrease the training effort for complex learning tasks by splitting them into a number of independent subtasks. Among a variety of network–structuring algorithms, the CCA seems to be one that has gained fairly widespread application. Some theoretically oriented papers have been published concerning the generalization and approximation capabilities of the CCA [4, 2]. In [2] it is shown that a network constructed with the CCA (in the sequel: CCA-network) can exactly represent any superposition of sigmoidal functions with a finite number of hidden neurons. However, the conclusion that, since superpositions of sigmoidal functions are universal approximators, this universal approximation capability holds for CCA–networks in general, is not straightforward to prove.

This contribution is restricted to the case of finite training samples of size p, $\{(\mathbf{x}^{(i)}, y^{(i)})\}_{i=1,\dots p}$, $\mathbf{x}^{(i)} \in \Re^d$ (d: input dimensionality), $y \in \Re$ that are assumed to be non–contradicting: $(\mathbf{x}^{(i)} = \mathbf{x}^{(j)}) \Rightarrow (y^{(i)} = y^{(j)})$. Furthermore, Heaviside transfer functions have been assumed, although the considerations can be extended onto sigmoidal functions. The following questions are addressed:

1. Does the CCA create networks that exactly reproduce the target outputs for arbitrary training samples (admissibility)?

* This work was supported by the Thuringian Ministry for Science, Research and Arts (project ITHERA, B511-95004).

2. Is the correlation maximization criterion proposed by Fahlman [3] suited for the generation of minimal network structures (optimality)?

2 Basic Algorithm and Notation

The CCA is basically a 2–step–procedure:

for $k = 1, 2, \ldots$

(A) Insert a hidden neuron and determine its input weights $\mathbf{w}_k$ in order to maximize the absolute covariance between the unit's output and the residual error obtained at the network output in the last step.

(B) Determine the output weights $\mathbf{v}^{(k)}$ in order to minimize the residual error.

For $k = 0$, merely substep (B) is performed (training of the direct input–to–output weights). Let $x_j^{(i)}$ $(i = 1, \ldots p,\ j = 1, \ldots d)$ denote the j–th component of the i–th input pattern vector, $g(\cdot)$ the transfer function of the hidden units and $h(\cdot)$ the transfer function of the output unit. $\mathbf{I}_q$ denotes the $q \times q$ identity matrix, $\mathbf{Q}^T$ the transpose of matrix $\mathbf{Q}$ and $\langle \mathbf{a}, \mathbf{b} \rangle = \mathbf{a}^T \mathbf{b}$ the scalar product of column vectors $\mathbf{a}$, $\mathbf{b}$. $\bar{a}$ denotes the mean of the components of $\mathbf{a} = [a_1 a_2 \ldots a_q]^T$: $\bar{a} = 1/q \sum_{j=1}^{q} a_j$.

The CCA corresponds to an iterative construction of an activation matrix $\mathbf{G}^{(k)}$, consisting of $d + k + 1$ column vectors $\mathbf{g}_j$:

$$\mathbf{G}^{(0)} = \begin{bmatrix} x_d^{(1)} & \ldots & x_2^{(1)} & x_1^{(1)} & 1 \\ x_d^{(2)} & \ldots & x_2^{(2)} & x_1^{(2)} & 1 \\ & & \vdots & & \\ x_d^{(p)} & \ldots & x_2^{(p)} & x_1^{(p)} & 1 \end{bmatrix} = \left[\mathbf{g}_{d+1}\ \mathbf{g}_d \cdots \mathbf{g}_1 \right]$$

$$\mathbf{G}^{(k+1)} = \left[g\left(\mathbf{G}^{(k)} \mathbf{w}_{k+1} \right)\ \mathbf{G}^{(k)} \right], \quad k = 0, 1, \ldots,$$

where $\mathbf{w}_{k+1}$ is subject to adaptation step (A). After substep (B) the network output is $\mathbf{o}^{(k+1)} = h\left(\mathbf{G}^{(k+1)} \mathbf{v}^{(k+1)} \right)$, resulting in a vector of residuals $\mathbf{r}^{(k+1)} = \mathbf{y} - \mathbf{o}^{(k+1)}$. The output weights $\mathbf{v}^{(k+1)}$ are chosen to minimize the norm of the residual.

The rest of the paper is divided into two parts, the first one addressing the admissibility and optimality of CCA–networks with linear output neurons, and the second one dealing with non–linear output neurons. It should be stated explicitly, that we are not concerned with the optimality or admissibility of the adaptation algorithms used for weight determination, but assume that we are able to find the global extrema when solving both subtasks (A) and (B).

3 Linear Output Neuron

3.1 Admissibility

In the following, we show that CCA–networks with linear output neurons can solve arbitrary, non–contradicting finite learning problems. The idea is to prove that the rank of the activation matrix, $\mathrm{rg}\left(\mathbf{G}^{(k)}\right)$, is increasing, as the iteration proceeds. If so, after at most $p-d$ steps (assuming, that $\mathbf{G}^{(0)}$ has rank d) the exact representation problem

$$\mathbf{r}^{(p-d)} = \mathbf{0} = \mathbf{y} - \mathbf{G}^{(p-d)}\,\mathbf{v}^{(p-d)} \longrightarrow \mathbf{v}^{(p-d)} = \left[\mathbf{G}^{(p-d)}\right]^{-1}\mathbf{y}$$

can be solved. The solution of the output error minimization problem in substep (B) in any previous step $k < p-d$ gives

$$\begin{aligned}
\mathbf{v}^{(k)} &= \left(\mathbf{G}^{(k)T}\,\mathbf{G}^{(k)}\right)^{-1}\mathbf{G}^{(k)T}\,\mathbf{y} = \mathbf{G}^{(k)+}\,\mathbf{y}\\
\mathbf{r}^{(k)} &= \left(\mathbf{I}_{k+d+1} - \mathbf{G}^{(k)}\,\mathbf{G}^{(k)+}\right)\mathbf{y}\\
\mathbf{g}^{(j)T}\,\mathbf{r}^{(k)} &= \left(\mathbf{g}^{(j)T} - \mathbf{g}^{(j)T}\,\mathbf{G}^{(k)+}\,\mathbf{G}^{(k)}\right)\mathbf{y} = 0 \quad j = 1,\ldots p+k
\end{aligned} \tag{1}$$

Because of (1) the covariance maximization in step $k+1$ (A) reduces to

$$C = \left|\sum_{j=1}^{p}\left(r_j^{(k)} - \bar{r}^{(k)}\right)\left(g_{j,\,k+d+1} - \bar{g}_{k+d+1}\right)\right| = \left|\sum_{j=1}^{p} r_j^{(k)} g_{j,\,k+d+1}\right| \rightarrow \max$$

(which indeed is a *correlation* maximization). If one can find weights $\mathbf{w}_{k+1}$ so that $C > 0$, then $\mathbf{g}_{k+d+2}$ (the vector of outputs of the inserted hidden neuron) is not orthogonal to the residual vector, and therefore linearly independent from the column vectors of $\mathbf{G}^{(k)}$, and it follows that

$$\mathrm{rg}\left(\mathbf{G}^{(k+1)}\right) = \mathrm{rg}\left(\left[\mathbf{g}_{k+d+2}\,\mathbf{G}^{(k)}\right]\right) = \mathrm{rg}\left(\mathbf{G}^{(k)}\right) + 1. \tag{2}$$

For the choice of Heaviside functions as transfer functions of the hidden neurons

$$g(t) = H(t) = \begin{cases} 0 & t \le 0 \\ 1 & t > 0 \end{cases}.$$

the following lemma can be proved.

Lemma 1. *For any non–zero residual* $\mathbf{r}^{(k)}$, $\left\|\mathbf{r}^{(k)}\right\| \neq 0$, *there exists a weight vector* $\mathbf{w}_{k+1}$ *such that*

$$\mathbf{g}_{k+d+2} = H\left(\mathbf{G}^{(k)}\mathbf{w}_{k+1}\right)$$

is linearly independent from the column vectors of $\mathbf{G}^{(k)}$.

Proof. A proof is given in [1].

Therefore, (2) holds, and the CCA with linear output and Heaviside hidden neurons is admissible.

3.2 Optimality

For simplicity, we transform the column vectors $\mathbf{g}_j$ of the activation matrix $\mathbf{G}^{(k)}$ into orthonormal vectors $\mathbf{n}_j$:

$$\mathbf{n}_1 = \frac{\mathbf{g}_1}{\|\mathbf{g}_1\|}$$

$$\mathbf{n}_{j+1} = \frac{\mathbf{g}_{j+1} - \sum_{i=1}^{j} \langle \mathbf{g}_{j+1}, \mathbf{n}_i \rangle}{\left\| \mathbf{g}_{j+1} - \sum_{i=1}^{j} \langle \mathbf{g}_{j+1}, \mathbf{n}_i \rangle \right\|}, \quad j = 1, \ldots d + k$$

The residual $\mathbf{r}^{(k)}$ can be written as

$$\mathbf{r}^{(k)} = \mathbf{y} - \sum_{j=1}^{k+d+1} a_j \mathbf{n}_j, \quad a_j = \langle \mathbf{y}, \mathbf{n}_j \rangle.$$

The insertion of a new hidden neuron with the output activation vector $\mathbf{g}_{k+d+2}$ results in a decreased residual:

$$\left\| \mathbf{r}^{(k+1)} \right\| = \left\| \mathbf{r}^{(k)} \right\| - a_{k+d+2}^2.$$

In order to minimize the new residual, the hidden weights $\mathbf{w}_{k+1}$ have to maximize

$$\begin{aligned} |a_{k+d+2}| &= \langle \mathbf{y}, \mathbf{n}_{k+d+2} \rangle = \langle \mathbf{r}^{(k)}, \mathbf{n}_{k+d+2} \rangle \\ &= \left| \frac{\langle \mathbf{r}^{(k)}, \mathbf{g}_{k+d+2} \rangle}{\left\| \mathbf{g}_{k+d+2} - \sum_{i=1}^{j} \langle \mathbf{g}_{k+d+2}, \mathbf{n}_i \rangle \right\|} \right| \\ &= \left| \langle \mathbf{r}^{(k)}, \frac{\mathbf{g}_{k+d+2}}{\left\| \mathbf{g}_{k+d+2}^{\perp} \right\|} \rangle \right| \end{aligned} \quad (3)$$

where $\mathbf{g}_{k+d+2}^{\perp}$ denotes the component of $\mathbf{g}_{k+d+2}$ that is orthogonal to the columns of $\mathbf{G}^{(k)}$. Whereas it seems to be obvious that the criterion $\left| \langle \mathbf{r}^{(k)}, \frac{\mathbf{g}_{k+d+2}}{\left\| \mathbf{g}_{k+d+2} \right\|} \rangle \right|$ is preferable compared to Fahlman's covariance maximization criterion (as it is independent from the length of $\mathbf{g}_{k+d+2}$, that is irrelevant due to scaling using the output weights), (3) is still somewhat different. The following example should illustrate this point.

Let the learning task be given as $\left(x^{(i)}, y^{(i)}\right) \in \{(0,1), (1,1), (2,-1), (3,-1)\}$. The initial residual vector is $\mathbf{r}^{(0)} = \left[-0.2\ 0.6\ -0.6\ 0.2\right]^T$. Assume that both hidden output vectors

$$\mathbf{g}_3^{(a)} = \left[1\ 1\ -1\ -1\right]^T, \quad \mathbf{g}_3^{(b)} = \left[1\ 0\ -1\ 0\right]^T$$

can be realized, but neither $\mathbf{g}_3 = \left[1\ -1\ 1\ -1\right]^T$ nor $\mathbf{g}_3 = \left[-1\ 1\ -1\ 1\right]^T$. (We are not concerned here with the type of hidden transfer function necessary to fulfill these restrictions, however $g(t) = \text{sign}(t^2 - 1)$ would be a suited choice.) Both hidden output vectors realize the same absolute correlation or covariance:

$$\left|\langle \mathbf{r}^{(0)}, \mathbf{g}_3^{(a)} \rangle\right| = \left|\langle \mathbf{r}^{(0)}, \mathbf{g}_3^{(b)} \rangle\right| = 0.8.$$

Using the normalized correlation criterion,

$$\left|\langle \mathbf{r}^{(0)}, \frac{\mathbf{g}_3^{(a)}}{\left\|\mathbf{g}_3^{(a)}\right\|} \rangle\right| = 0.4 < 0.4\sqrt{2} = \left|\langle \mathbf{r}^{(0)}, \frac{\mathbf{g}_3^{(b)}}{\left\|\mathbf{g}_3^{(b)}\right\|} \rangle\right|$$

holds. The modified criterion (3), however, gives

$$\left|\langle \mathbf{r}^{(0)}, \frac{\mathbf{g}_3^{(a)}}{\left\|\mathbf{g}_3^{(a)\perp}\right\|} \rangle\right| = 0.8944 > 0.3651 = \left|\langle \mathbf{r}^{(0)}, \frac{\mathbf{g}_3^{(b)}}{\left\|\mathbf{g}_3^{(b)\perp}\right\|} \rangle\right|.$$

Since $\mathbf{g}_3^{(a)}$ completely solves the learning task ($\mathbf{y} = \mathbf{g}_3^{(a)}$, $\mathbf{r}^{(1)} = \mathbf{0}$), but $\mathbf{g}_3^{(b)}$ not ($\mathbf{r}^{(1)} = 1/3 \left[-1\ 2\ -1\ 0\right]^T$), the normalized criterion obviously is not optimal. Using Fahlman's criterion, either $\mathbf{g}_3^{(a)}$ or $\mathbf{g}_3^{(b)}$ can be chosen. Thus there is no guarantee to find the optimal solution.

An efficient implementation of the modified criterion is possible (e.g., using Gram–Schmidt orthogonalization).

4 Nonlinear Output Neuron

4.1 Admissibility

Using linear output neurons, it was sufficient to show the existence of a hidden output vector $\mathbf{g}_{k+d+2}$ that is not orthogonal to the residual $\mathbf{r}^{(k)}$ for each step k. This, however, does not guarantee the admissibility of the CCA in the general case: for nonlinear transfer functions $\mathbf{r}^{(k)}$ is generally not orthogonal to $\mathbf{G}^{(k)}$, it might even be linearly dependent on $\mathbf{G}^{(k)}$. Thus the maximization of the absolute covariance not necessarily results in a linearly independent hidden output vector. We give an example to illustrate this fact.

Let a learnset $\left(x^{(i)}, y^{(i)}\right) \in \{(2,6), (-2,2)\}$ be given. The output neuron has a quadratic transfer function $h(t) = t^2$, for the hidden neurons a modified Heaviside function is used: $g(t) = 2\,H(t) - 1$. For the sake of simplicity, we suppress the bias elements (reducing the number of columns of $\mathbf{G}^{(k)}$ to $k+1$). The learning task could be solved with two hidden neurons: $w_1 = 1$, $\mathbf{w}_2 = \left[-2\ 1\right]^T$, $\mathbf{v}^{(2)} = \left[-1/2\left(\sqrt{6}+\sqrt{2}\right), -1/2\left(\sqrt{6}-\sqrt{2}\right), 0\right]^T$. The CCA, however, is not able to find this solution. In the first step ($k = 0$ (B)), $v^{(0)} = 1$ is determined, giving a residual, that is linearly dependent on $\mathbf{G}^{(0)}$:

$$\mathbf{r}^{(0)} = \begin{bmatrix} 2 \\ -2 \end{bmatrix} = \mathbf{G}^{(0)} = \mathbf{g}_1$$

In the next step ($k = 1$ (A)), the hidden weights w_1 are chosen to maximize the absolute covariance. As global maxima one gets either $\mathbf{g}_2 = \left[1\ -1\right]^T$ ($w_1 = 1$) or $\mathbf{g}_2 = \left[-1\ 1\right]^T$ ($w_1 = -1$). Both hidden output vectors are linearly dependent on $\mathbf{g}_1$. Thus, the residual does not change $\mathbf{r}^{(1)} = \mathbf{r}^{(0)}$, the hidden output vectors developed in the sequel will again be linearly dependent on $\mathbf{g}_1$, and the output error will never decrease.

4.2 Optimality

As the CCA does not find an existing solution, the criterion used for the generation of the activation matrix obviously can not be optimal. Neither the normalized covariance nor the modified criterion (3) can get around this limitation.

5 Conclusions

For suited hidden transfer functions, the CCA solves arbitrary, non–contradicting finite learning problems inserting at most $p - d$ hidden neurons, if the output neuron performs a linear transfer function. This has been proved for Heaviside transfer functions. An extension onto usual squashing functions, e.g. the logistic sigmoidal function, is possible, though somewhat more sophisticated. The covariance maximization criterion proposed by Fahlman is not optimal for the minimization of the squared output error. It can be substituted by a modified criterion developed herein. Using this criterion, hidden representations are chosen that minimize the sum–of–squares error at the network output. Concerning nonlinear output neurons, it has been shown that the CCA is neither admissible (i.e., capable of solving finite learning tasks) nor optimal. For practical purposes, this limitation is probably minor to those implied by restrictions common to most neural approaches (e.g., training algorithms that get stuck in local minima). It makes, however, explicit that the generic CCA is no universal approximation tool.

References

1. A. Doering. *Optimization of Feature Extraction and Classifier Structure for Pattern Recognition with Neural Networks.* PhD thesis, TU Dresden, 1997.
2. G.P. Drago and S. Ridella. Convergence properties of cascade correlation in function approximation. *Neural Comput. and Applic.*, 2:142–7, 1994.
3. S. E. Fahlman and C. Lebiere. The cascade-correlation learning architecture. In D. S. Touretzky, editor, *Advances in Neural Information Processing Systems*, volume 2, pages 524–532, Denver 1989, 1990. Morgan Kaufmann, San Mateo.
4. J.N. Hwang, S.S. You, S.R. Lay, and I.C. Jou. The cascade–correlation learning: A projection pursuit learning perspective. *IEEE Transactions on Neural Networks*, 7(2):278–89, March 1996.

The Effective VC Dimension of the n-tuple Classifier

N.P. Bradshaw

IRIDIA - ULB (CP 194/6), 50, av. F.Roosevelt, 1050-Brussels.

Abstract. One family of classifiers which has has considerable experimental success over the last thirty year is that of the n-tuple classifier and its descendents. However, the theoretical basis for such classifiers is uncertain despite attempts from time to time to place it in a statistical framework. In particular the most commonly used training algorithms do not even try to minimise recognition error on the training set. In this paper the tools of statistical learning theory are applied to the classifier in an attempt to describe the classifier's effectiveness. In particular the effective VC dimension of the classifier for various input distributions is calculated experimentally, and these results used as the basis for a discussion of the behaviour of the n-tuple classifier. As a side-issue an error-minimising algorithm for the n-tuple classifier is also proposed and briefly examined.

1 Introduction to the n-tuple Classifier

The original n-tuple classifier was described by Bledsoe and Browning in [3]. It is a pattern recognition system which accepts binary images and outputs a binary "yes/no" response. Modifications to the original design have included allowing the output to be one of a finite number of preset class labels [1], extending the input space to allow real-valued data [2, 9] or extending the output space to solve regression problems [6]. It has also been shown to give good performance on the Statlog data sets [7].In this paper the system considered will accept binary strings as inputs and output 1 or 0. A schematic diagram of this system is given in figure 1.

The architecture of the classifier consists of three layers: a layer of look-up tables, a layer of summing devices (one per class) and a winner-takes-all comparison. The operation of the classifier consists of three stages: a sampling stage, a look-up stage and an output stage. The principle of the classifier is that the image space may be sampled in blocks of n bits known as **n-tuples**. For the purposes of this paper it will be assumed that the n-tuples are chosen uniformly at random with the condition that no two overlap (ie. no input bit belongs to more than one n-tuple) although it is possible to perform a similar analysis with this assumption relaxed. Each class has an associated set of N look-up tables or **nodes**, the whole set being known as a **discriminator**. The nodes in each discriminator are connected to the input space in exactly the same way so that all discriminators

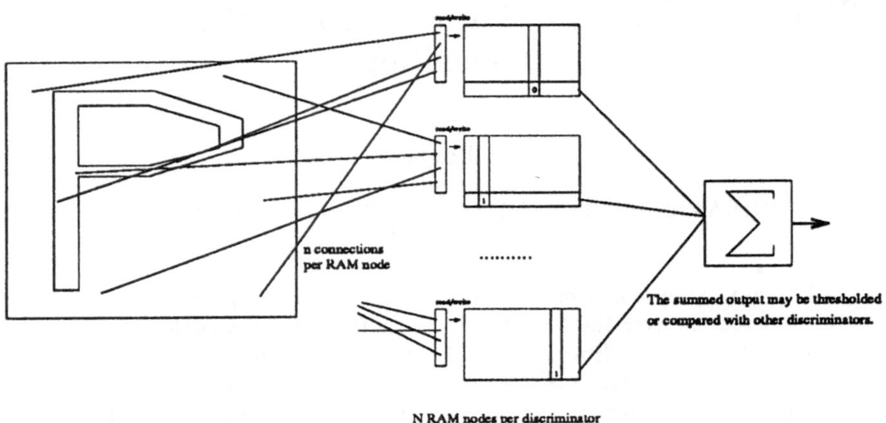

Fig. 1. A single n-tuple discriminator (schematically). An n-tuple classifier consists of as many discriminators as there are classes. Each discriminator is associated to each class. The classifier returns the class label of the discriminator with the highest output.

are identical apart from the contents of the nodes. The look up tables take binary n-bit strings as inputs and generate binary bits, $\{1, 0\}$, as output.

In each discriminator the (binary) output for each n-tuple is then read from the corresponding look-up table. The output of the discriminator is obtained by summing the outputs of each node to obtain an integer between 0 and N. The outputs of each discriminator are compared and the class-label associated with the highest scoring discriminator is given as the output of the whole system. For this paper we assume that all classifiers contain only two discriminators. We shall also assume that in the case where the two discriminators give identical integer outputs, the class-label **1** is output. The problem for the training algorithm is what information to put in the nodes. In this paper the SMA training algorithm, see [5], is used.

2 Background to Statistical Learning Theory

Statistical learning theory is the application of statistical techniques to the problem of learning from examples which can be used to derive minimal required training set sizes to guarantee a given level of generalisation with a certain confidence. These bounds are usually formulated in terms of the *VC dimension*, see for example [10]. Previous work has fixed lower and upper bounds for for the two-discriminator n-tuple classifier [4] as $N(2^n-1)$ and $(\log_2 3)N(2^n-1)$ respectively.

However, most practical work suggests that the sample sizes required by the VC dimension bounds are in fact much larger than required (see for example the results of the Statlog tests performed by Rohwer and Morciniec, [7]). In [11]

Vapnik, Levin and Le Cun define a quantity called *effective VC dimension* which is based on the learning machine together with an input distribution which yields a new sample size bound which is never greater than the original. Vapnik et al. also give an experimental method for estimating it, a method which they show works well in the case of the linear perceptron. This method is applied in this paper to the n-tuple classifier and shown to give plausible results in this case too. Thus the n-tuple classifier is analysed further and the experimental method of Vapnik et al. is further validated.

3 Calculating the Effective VC Dimension

Vapnik et al. suggest a definition of a VC dimension based not on all input sets but just on those with probability close to one. More formally

Definition The effective VC dimension of the learning machine $\mathcal{L}$ for the input distribution P is the minimal VC dimension of $\mathcal{L}$ on those subsets X^* of the input space X whose probability measure according to P is almost 1.

They show that the effective VC dimension can be estimated by measuring the maximal deviation ξ_l between the errors of a trained classifier on two halves of an input sample of length $2l$. The estimation takes the form of a function (with two free parameters a and b) denoted by $\Phi(l/h)$, or an approximation to this denoted by $\Phi_1(l/h)$ with one free parameter d, whose forms are given in [11]. The theoretical demonstration can be found in [11].

To maximise the error divergence with the n-tuple classifier, an error *minimising* algorithm for the n-tuple classifier is needed. The *Stochastic Mimimisation Algorithm* (SMA) for the n-tuple classifier was proposed in [4] as such an algorithm. The principle of the algorithm is to train as many patterns as possible by looping through the training set and in each case of misclassification changing a minimal number of output values — selected at random — so that the current pattern is trained correctly. The hope is that by making an alteration of minimal size the previously trained responses will not be disturbed and thus that a good approximation to the global minimum of the training error can be found. The loop is repeated enough times so that the minimum training error stops decreasing.

3.1 Necessary Assumptions

For an empirical estimate of effective VC dimension to be made by the method described the following assumptions must hold true.

- $E[\xi_l]$ does not depend on the distribution of the classes, only the patterns themselves.
- The expected deviation depends on the learning machine only through the effective VC dimension, h.

- $\Phi(l/h)$ or $\Phi_1(l/h)$ is a good approximation to $E[\xi_l]$ for appropriate values of the parameters a and b.
- a and b are constant over a large class of related learning machines.

These assumptions were verified for the n-tuple classifier in [5].

3.2 Varying the Classifier

Values of ξ_l were calculated for various values of n and N as well as i, the range of the integers stored in each RAM location. The output values were assigned with 50% probability as was experimentally justified in [5]. The ξ_l are plotted in figure 2 for varying values of N and n. To allow comparison, the plot for the basic two discriminator classifier with $n = 4, N = 50$ is included in both graphs. As a guide to interpreting these graphs it is worth noting that the further to the right a curve is, the higher its effective VC dimension.

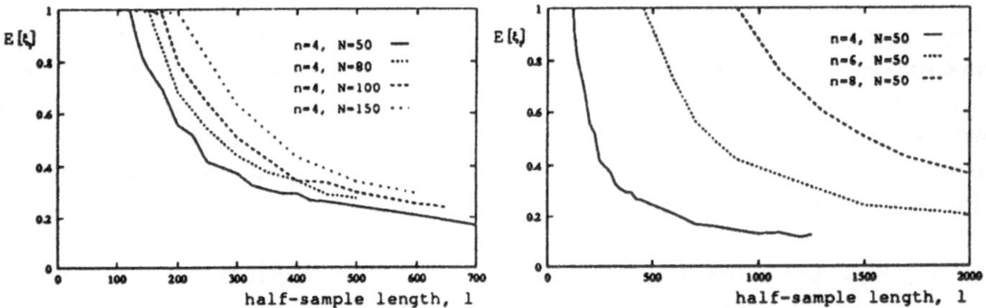

Fig. 2. Empirical $E[\xi_l]$ for a range of n-tuple classifiers with varying number of (a) nodes per discriminator, N and (b) support per node. Uniform distribution.

To make an estimate of the effective VC dimension, we must plot $E[\xi_l]$ against l/h and show that some curve Φ fits the resulting graph. If the same Φ fits all the $E[\xi_l]$ graphs then our assumption that the parameters a and b are independent of the learning machine will have been justified. Since in almost all cases the ranges of l are such that $l/h < 5$, an approximation by Φ_1 with a single free parameter d is valid, see [11]. Figure 3 shows the results for a range of machines and the best fit curve of type Φ_1, with $d = 0.225$. The same value of d fits all settings of the parameters.

4 Results and Conclusions

The known VC dimension values and bounds are shown in table 1 along with the corresponding effective VC dimension values for the uniform distribution.

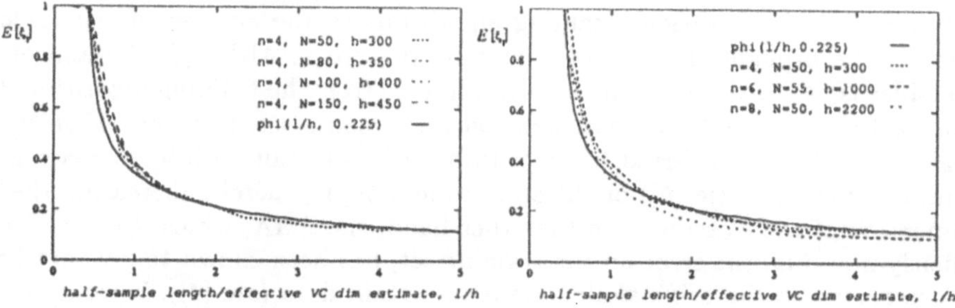

Fig. 3. Empirical $E[\xi_l]$ for varying (a) N and (b) n against estimated EVCD over l plotted against a best-fit estimate. Uniform distribution.

n	N	VC dim min	VC dim max	EVC dim	Ratio 1	Ratio 2
4	50	750	1190	300	2.5	4.0
4	100	1500	2380	400	3.8	6.0
4	150	2250	3570	450	5.1	7.9
6	50	3150	4990	1000	3.2	5.0
8	50	12750	20,200	2200	5.8	9.2

Table 1. Effective (uniform distribution) and actual VC dimensions for n-tuple classifier. Ratio 1 is the VC dimension lower bound over the effective VC dim, while Ratio 2 is the VC dimension upper bound over the effective VC dim.

This table shows at a glance how pessimistic the VC dimension results are when the patterns are drawn from a uniform distribution.
Table 1 shows clearly that the effective VC dimension of the n-tuple classifier over a uniform distribution of input patterns is significantly less than the actual VC dimension.

4.1 Conclusions and Future Work

The aim of the work in this paper has been two-fold. First to use the tools of learning theory to try to explain and predict the performance of the n-tuple classifier and second to validate the approach of Vapnik et al. to incorporating information about the input distribution into the VC bounds. Estimates of generalisation error of n-tuple classifiers made using the full VC dimension tend to severely over-estimate the error found in most experimental contexts. The current work shows that if the bounds are based on effective VC dimension then the predicted bounds are closer to experimental results. This was shown in [4]. Furthermore the work has validated the hypotheses of Vapnik et al. in 3.1 and given consistent results. Thus both of these goals have been substantially achieved.

Several directions of further work are suggested by this study, some in the domain of the n-tuple classifier and others in the domain of learning theory. The relationship of the various training algorithms to the success of the learning process are brought sharply into focus by this work. Although the aim of the classifier is to classify patterns with minimum error, the training algorithm does not explicitly try and minimise error on the training set. Non-zero error on the training set is often referred to as "saturation" and dealt with by increasing the number of samples (ie. N) or the size of the n-tuple, thereby increasing the VC dimension of the classifier. On the other hand, the SMA, which does try to explicitly minimise the error on the training set, has been shown to give markedly worse results to the original algorithm on certain data (see [4]). It would be interesting to know how the training algorithm limits the search for an acceptable hypothesis and thus how the classifier is often able to achieve good performance despite apparent over-capacity.

In parallel to this, a practical theory of learning which takes into account more than just the machine capacity and the input distribution is required if theoretical sample size predictions are to become a useful tool for those applying learning machines to different tasks. For instance the "unluckiness" function defined by Shawe-Taylor et al. [8] is one new way of incorporating prior expectations about the data distribution into the VC dimension/sample size bounds calculations.

References

1. I. Aleksander and T.J. Stonham. Guide to pattern recogntion using random-access memories. *Computers and Digital Techniques*, 2:29–40, 1979.
2. W.W. Bledsoe and C.L. Bisson. Improved memory matrices for the n-tuple recognition method. In *IRE Joint Computer Conference,11*, pages 414–415, 1962.
3. W.W. Bledsoe and L. Browning. Pattern recognition and reading by machine. In *Proc. Eastern Joint Computer Conf.*, pages 232–255, 1959.
4. N.P. Bradshaw. *An Analysis of Learning in Weightless Neural Systems.* PhD thesis, Imperial College, London., 1996.
5. N.P. Bradshaw. Improving the generalisation of the n-tuple classifier with the effective VC-dimension. Technical report, IRIDIA, Universite Libre de Bruxelles, 1997.
6. A. Kolcz and N.M. Allinson. N-tuple regression network. *Neural Networks*, 9(5):855–869, 1999.
7. R. Rohwer and M. Morciniec. A theoretical and experimental account of n-tuple classifier performance. *Neural Computation*, 8:657–670, 1996.
8. J. Shawe-Taylor, P.L. Bartlett, R.C. Williamson, and M. Anthony. A framework for structural risk minimisation. In *Proceedings of the 9th Annual Conference on Computational Learning Theory*, 1996.
9. M.J. Sixsmith, G.D. Tattershall, and J.M. Rollett. Speech recognition using n-tuple techniques. *Br Telecom J*, 8(2), April 1990.
10. V. Vapnik. *The Nature of Statistical Learning Theory.* Spinger-Verlag, 1995.
11. V. Vapnik, E Levin, and Y LeCun. Measuring the VC-dimension of a learning machine. *Neural Computation*, 6:851–876, 1994.

Part IV:
Signal Processing:
Blind Source Separation, Vector Quantization, and Self-Organization

From Neural Principal Components to Neural Independent Components

Erkki Oja, Juha Karhunen, and Aapo Hyvärinen

Helsinki University of Technology,
Laboratory of Computer and Information Science,
P.O.B. 2200, 02015 HUT, Finland
Email: first.last@hut.fi

Abstract. Several neural network learning rules for the linear Principal Component Analysis (PCA) have been shown to be closely related to classical PCA optimization criteria. These learning rules and the corresponding criteria are extended to versions containing nonlinear functions. It can be shown that the criteria and the learning functions solve the blind source separation (BSS) problem for the linear memoryless mixture model, based on the statistical independence of the source signals. This bottom-up approach to the BSS and Independent Component Analysis (ICA) problems allows us to choose the nonlinear functions so that the learning rules not only produce independent components, but also have other desirable properties like robustness, contrary to the often used polynomial functions ensuing from cumulant expansions. Also fast batch versions of the learning rules are reviewed.

1 Neural PCA

Principal Component Analysis (PCA) is an essential technique in data compression and feature extraction. It provides a way of reducing the number of input variables entering some data processing system so that a maximal amount of information is retained in the mean-square error sense; in addition, PCA provides uncorrelated components. Assume that $\mathbf{x}$ is an n-dimensional input data vector that has been centered to zero mean. The (first) principal component is defined as that linear combination $\mathbf{w}^T\mathbf{x}$ of the elements of $\mathbf{x}$ which satisfies

Problem 1.
$E\{(\mathbf{w}^T\mathbf{x})^2\}$ is maximized under the constraint $\|\mathbf{w}\| = 1$.

If maximization in Problem 1 is changed to minimization, the minor component (the last principal component) is obtained. The extension to several principal components can be defined as follows: find those p ($p \leq n$) linear combinations $\mathbf{w}_1^T\mathbf{x}, \mathbf{w}_2^T\mathbf{x}, ..., \mathbf{w}_p^T\mathbf{x}$ of the elements of $\mathbf{x}$ that satisfy

Problem 2.
$E\{(\mathbf{w}_i^T\mathbf{x})^2\}$, $i = 1, ..., p$ are maximized under the constraints $\mathbf{w}_i^T\mathbf{w}_j = \delta_{ij}$ for $j \leq i$.

It is well known that the solution for the vectors $\mathbf{w}_1, ..., \mathbf{w}_p$ in Problem 2 are the p dominant eigenvectors of the data covariance matrix $\mathbf{C} = E\{\mathbf{x}\mathbf{x}^T\}$. Another possible criterion is formulated in terms of the matrix $\mathbf{W} = (\mathbf{w}_1 ... \mathbf{w}_p)$ of the weight vectors:

Problem 3.
$E\{\|\mathbf{x} - \mathbf{W}\mathbf{W}^T\mathbf{x}\|^2\}$ is minimized.

In this case, no orthonormality constraints are needed. Matrix $\mathbf{W}$ solving Problem 3 is only defined up to a rotation, however. Other formulations for the PCA optimization criteria were considered in [36].

A class of learning rules for solving these problems stem from the first author's PCA neuron with constrained Hebbian learning rule [35]. Denoting the weight vector in the first principal component by $\mathbf{w}$, as in Problem 1 above, and assuming a sequence of samples from the random input vector $\mathbf{x}$, the basic learning rule for Problem 1 is

$$\Delta \mathbf{w} = \alpha[\mathbf{x}(\mathbf{w}^T\mathbf{x}) - \mathbf{w}(\mathbf{w}^T\mathbf{x})^2]. \tag{1}$$

It is to be understood that all the variables, including the learning rate, are functions of discrete time k. With $k \to \infty$, the solution $\mathbf{w}$ was shown in [35] to converge (with probability one) to the dominant eigenvector of $\mathbf{C}$. An algorithm for obtaining the jth weight vector $\mathbf{w}_j$ for Problem 2 is [36], [37]:

$$\Delta \mathbf{w}_j = \alpha(\mathbf{w}_j^T\mathbf{x})[\mathbf{x} - (\mathbf{w}_j^T\mathbf{x})\mathbf{w}_j - 2\sum_{i=1}^{j-1}(\mathbf{w}_i^T\mathbf{x})\mathbf{w}_i] \tag{2}$$

Algorithm (2) is called the *Stochastic Gradient Ascent (SGA)* algorithm. Algorithm (1) is a special case for $j = 1$. The vector $\mathbf{w}_j$ will tend to the normalized j-th eigenvector of the covariance matrix $\mathbf{C}$. Another, symmetrical algorithm, providing the solution to Problem 3 is [37]

$$\Delta \mathbf{W} = \alpha[I - \mathbf{W}\mathbf{W}^T]\mathbf{x}\mathbf{x}^T\mathbf{W}. \tag{3}$$

This is the *Subspace Network* learning algorithm, for which a neural network implementation was given in [38]. A review of these and related linear PCA learning rules is [40].

Variants of these learning rules and neural networks have been given and analyzed e.g. in [7], [14], [15], [17], [32], [33], [34], [36], [37], [45], [46], [47], [48].

2 Nonlinearities and Independent Components

In the field of neural networks, there has been growing interest in *nonlinear extensions* of the PCA. Oja et al [39] generalized the one-unit and multi-unit PCA learning rules to nonlinear versions, and Karhunen and Joutsensalo [28, 29] studied nonlinear extensions of the PCA optimization criteria. Recently, some authors have discussed the relation of these rules to *Projection Pursuit* (PP) [16]. Another interesting development is the relation of neural networks to *Independent Component Analysis* (ICA) [25, 26, 11] and the related *source separation problem.*

Generally, source separation and ICA require higher than second-order statistics that can be incorporated into the computations by using *contrast functions*; see e.g. [1, 3–5, 10–13, 25, 26]. Usually, one starts from the independence requirement. Because direct verification of independence is very difficult, mutual information is usually chosen as the measure for the degree of independence [11]. One then derives an approximative contrast function, often based on cumulant expansions of the densities, that can be computed in practice without knowing the source density functions. Finally, the problem is solved with an appropriate numerical method.

We concentrate here on another approach in which the higher order statistics are implicitly embedded into the cost functions and algorithms by *nonlinear non-polynomial functions.* Contrary to the "top-down" approach which starts from independence-related contrast functions, we call this the "bottom-up" approach. We start either from a suitably shaped contrast function or directly from its related gradient algorithm. The purpose is to get contrast functions and learning algorithms that have some desired properties *in addition* to solving the ICA problem. We then go on to prove that the extrema of the contrast function coincide with independent components. It is also possible to start from the algorithm directly like in [41] and show that independent components are asymptotically stable points of convergence for the algorithm.

This is somewhat similar to the classical Jutten-Herault (JH) neural network approach [25, 26], in which the starting point is also a learning neural network. Another related approach is that of Bell and Sejnowski [3]. However, our starting point is not separation or "nonlinear decorrelation" *per se* like in the JH approach, nor mutual information maximization like in [3]. Instead, we start from a wide class of optimization criteria that are nonlinear extensions of the various PCA criteria [29].

The advantage of the bottom-up approach is that we are not restricted to cumulant expansions as contrast functions. The nonlinear functions do not have to be polynomials at all, but can be more freely adapted to other characteristics of the problem, especially demands on robust and numerically well behaving algorithms. Such algorithms may be more suitable to a neural network computational environment and might also have some biological relevance or plausibility as neural learning rules.

3 Neural ICA: one-unit learning rules

3.1 The basic learning rule

Consider first a single artificial neuron receiving samples of the n-dimensional input vector $\mathbf{x}$ [39]. The neuron is trying to adapt its weight vector $\mathbf{w}$ so that a function $E\{f(\mathbf{w}^T\mathbf{x})\}$ is maximized, where $f(.)$ is a continuous objective function. We refrain from making any assumptions on the shape of $f(.)$ until Sect. 3.2; however, to prevent the trivial solution $\mathbf{w} \to \infty$ for unbounded functions $f(.)$, the problem must be constrained by setting the norm of $\mathbf{w}$ constant. Thus, we have the

Generalized Problem 1.
$E\{f(\mathbf{w}^T\mathbf{x})\}$ is maximized under the constraint $\|\mathbf{w}\| = 1$.

This should be compared to Problem 1 in Section 1. The only available information on $\mathbf{x}$ is the sequence of training samples $\mathbf{x}(0), \mathbf{x}(1), \ldots$ during a learning period. A possible formulation for a *constrained gradient ascent learning rule* based on sample functions is then [39]

$$\tilde{\mathbf{w}}(k+1) = \mathbf{w}(k) + \alpha(k) \nabla_{\mathbf{w}(k)} f(\mathbf{w}(k)^T\mathbf{x}(k)), \ \mathbf{w}(k+1) = \tilde{\mathbf{w}}(k+1)/\|\tilde{\mathbf{w}}(k+1)\| \tag{4}$$

where $\alpha(k)$ is again the learning rate, usually a sequence of positive numbers decreasing slowly to zero; for details on the learning rates, see [37]. Denoting the derivative of $f(y)$ by $g(y)$, eq. (4) can be approximated for small values of α by

$$\Delta\mathbf{w} = \alpha(I - \mathbf{w}\mathbf{w}^T)\mathbf{x}g(\mathbf{w}^T\mathbf{x}) \tag{5}$$
$$= \alpha[\mathbf{x}g(\mathbf{w}^T\mathbf{x}) - \mathbf{w}(\mathbf{w}^T\mathbf{x})g(\mathbf{w}^T\mathbf{x})] \tag{6}$$

in which terms proportional to α^2 have been dropped [39]. This is the nonlinear analogue of (1).

Another way to achieve a stochastic gradient ascent algorithm in which the norm of the solution vector $\mathbf{w}$ is at the same time constrained, is to use a *penalty* or *forgetting* term, yielding in general the algorithm

$$\Delta\mathbf{w} = \alpha[\mathbf{x}g(\mathbf{w}^T\mathbf{x}) - \mathbf{w}h(\|\mathbf{w}\|)] \tag{7}$$

where the penalty function $h(.)$ must be chosen appropriately. Note the formal similarity of this approach with algorithm (6): the first terms in brackets are exactly the same, and the second term only differs in the form of the scalar function multiplying the vector $\mathbf{w}$.

3.2 Analysis of the learning rule

A widely used contrast function in ICA is kurtosis, and thus the general approach given in the previous Section can be applied to the case $f(\mathbf{w}^T\mathbf{x}) = (\mathbf{w}^T\mathbf{x})^4$ under the given constraint on the norm of $\mathbf{w}$ [22]; see also [13]. This gives a learning rule like (7) in which the function $g(t) = t^3$ is used together with a suitably chosen penalty function $h(.)$. If the input data has been sphered (whitened) so that we may assume $\mathbf{C} = E\{\mathbf{x}\mathbf{x}^T\} = \mathbf{I}$, then this algorithm will produce one of the independent components having positive kurtosis [22]. If the sign of $f(.)$ is reversed, then the same algorithm will find an independent component with negative kurtosis.

However, a considerably wider class of nonlinear, not necessarily polynomial functions will in fact produce one of the independent components. Informally, this can be seen as follows: suppose $\mathbf{x}$ has been sphered and the linear mixture model

$$\mathbf{x} = \mathbf{A}\mathbf{s} \tag{8}$$

holds, where $\mathbf{s}$ is the vector of n independent source signals s_i, $i = 1, ..., n$. As usual, let us assume that the source signals have unit variance; then, if $\mathbf{x}$ is sphered, matrix $\mathbf{A}$ is orthogonal: $\mathbf{A}^T = \mathbf{A}^{-1}$. (We have also assumed that in sphering, the dimension of $\mathbf{x}$ has been made equal to the number of source signals n.) Define $\mathbf{z} = \mathbf{A}^T\mathbf{w}$, giving $\mathbf{w}^T\mathbf{x} = \mathbf{z}^T\mathbf{A}^T\mathbf{x} = \mathbf{z}^T\mathbf{s}$. The stationary points of the learning rule (7) are given by $E\{\Delta\mathbf{w}\} = 0$, or $E\{\mathbf{x}g(\mathbf{w}^T\mathbf{x})\} - \mathbf{w}h(\|\mathbf{w}\|) = 0$. Multiplying by $\mathbf{A}^T$ from the left yields

$$E\{\mathbf{s}g(\mathbf{z}^T\mathbf{s})\} = \mathbf{z} \times scalar \tag{9}$$

where $\mathbf{z}$ is a stationary point in the rotated coordinates. Clearly, all vectors of the form $\mathbf{z}_i = [00...1...0]$ with 1 in the i-th place are solutions, because then $E\{\mathbf{s}g(\mathbf{z}_i^T\mathbf{s})\} = E\{\mathbf{s}g(s_i)\} = E\{s_i g(s_i)\}\mathbf{z}_i$. The asymptotic stability can be analyzed by standard methods based on linearization and eigenvalue analysis, and strict conditions for the functions $g(.)$ and $h(.)$ can be thus obtained; as an example of this approach, see [21, 41, 43].

In [21], the Generalized Problem 1 of Section 3 was related to the question of how far the mixture $\mathbf{w}^T\mathbf{x}$ is from a Gaussian random variable. Let us consider the difference of the expectation $E\{f(\mathbf{w}^T\mathbf{x})\}$ from what it would be if the output $\mathbf{w}^T\mathbf{x}$ were gaussian. Thus we obtain the following contrast function

$$J_f(\mathbf{w}) = |E\{f(\mathbf{w}^T\mathbf{x})\} - E_\nu\{f(\nu)\}| \tag{10}$$

where ν is a normalized Gaussian variable. Because we are here only interested in the higher-order structure of the data, the variance of the output must be constrained to be constant. Because the data is prewhitened, this can be simply accomplished by the constraint $\|\mathbf{w}\| = 1$. So we are back at the Generalized Problem 1.

In [21], a theorem is stated saying that under the linear model of eq. (8), with whitened $\mathbf{x}$, the local maxima (resp. minima) of $E\{f(\mathbf{w}^T\mathbf{x})\}$ under the constraint $\|\mathbf{w}\| = 1$ include those columns $\mathbf{a}_i$ of the mixing matrix $\mathbf{A}$ such that the corresponding source signals s_i satisfy

$$E\{s_i g(s_i) - g'(s_i)\} > 0 \text{ (resp. } < 0) \tag{11}$$

where $g(.)$ is the derivative of $f(.)$. Note that if $\mathbf{w} = \mathbf{a}_i$, then $\mathbf{w}^T\mathbf{x} = s_i$. Using this result, the independent components can be found as the proper extrema. They are also the asymptotically stable points for the gradient descent / ascent learning rules.

3.3 Advantages of the bottom-up approach

The theorem that was referred to in the preceding section shows that we have an infinite number of different Hebbian-like learning rules to choose from. This raises the question of what could be the advantages and disadvantages of using a given function $f(.)$ in the optimization criterion, and its derivative $g(.)$ as the learning function in the gradient learning rules.

A detailed statistical analysis is not feasible without some additional constraints. The following points, however, can be made on intuitive grounds:

- *Robustness against outliers* is a very desirable property for any estimation procedure. Here robustness can be achieved by choosing a function $f(.)$ that does not grow too fast, e.g. the log cosh function, or equivalently, using the tanh function in the learning rules; see [28, 29, 43]. The use of kurtosis, in contrast, leads to a fourth order polynomial, which means that the estimation is highly non-robust. Using higher order cumulants would further decrease the robustness.
- *Computational simplicity* is an inherent advantage of the bottom-up approach, because in this way the estimation of cumulant tensors etc. is totally avoided. The speed of convergence of the learning rules like (7) depends on the functions $g(.)$ and $h(.)$ and can be increased by a suitable choice.

3.4 Fixed-point algorithms

Neural learning rules reviewed in the preceding sections are by their very nature adaptive, on-line learning rules. If the computations are performed in batch mode, such neural rules may suffer from drawbacks. First, the learning rate must be suitably annealed to zero to achieve good precision in the final estimates. Second, the simultaneous estimation of the sign of the learning term complicates the algorithms. Therefore, some ways to make the learning faster and more reliable in batch mode are desirable. Such an alternative are the fixed-point algorithms. In [18, 23], two of the authors derived a fixed-point iteration scheme for kurtosis-based neural learning rules. Equating the sum of the gradient of kurtosis and the penalty term to zero, and taking the expectations, one obtains

$$E\{\mathbf{x}(\mathbf{w}^T\mathbf{x})^3\} - 3\|\mathbf{w}\|^2\mathbf{w} + h(\|\mathbf{w}\|)\mathbf{w} = 0 \tag{12}$$

which gives

$$\mathbf{w} = scalar \times (E\{\mathbf{x}(\mathbf{w}^T\mathbf{x})^3\} - 3\|\mathbf{w}\|^2\mathbf{w}) \tag{13}$$

This suggests the following fixed-point algorithm

$$\tilde{\mathbf{w}}(k+1) = E\{\mathbf{x}(\mathbf{w}(k)^T\mathbf{x})^3\} - 3\mathbf{w}(k), \tag{14}$$

$$\mathbf{w}(k+1) = \tilde{\mathbf{w}}(k+1)/\|\tilde{\mathbf{w}}(k+1)\| \tag{15}$$

where the expectation is, in practice, estimated using a large sample of $\mathbf{x}$. To find several independent components, a deflation scheme can be used.

The fixed-point algorithm has a number of desirable properties:

- The global convergence of our algorithm can be shown analytically to be *cubic*. This means very fast convergence and is rather unique among the ICA algorithms.
- Contrary to gradient-based algorithms, there is no learning rate or other adjustable parameters in the algorithm, which makes it easy to use, and more reliable.
- Both components of negative kurtosis (i.e. sub-Gaussian components) and components of positive kurtosis (i.e. super-Gaussian components) can be found without any additional estimations.

In [20, 24], this fixed-point algorithm was further extended by one of the authors for general contrast functions as in the Generalized Problem 1. The resulting algorithm is

$$\tilde{\mathbf{w}}(k+1) = E\{\mathbf{x}g(\mathbf{w}(k)^T\mathbf{x})\} - E\{g'(\mathbf{w}(k)^T\mathbf{x})\}\mathbf{w}(k), \qquad (16)$$

$$\mathbf{w}(k+1) = \tilde{\mathbf{w}}(k+1)/\|\tilde{\mathbf{w}}(k+1)\| \qquad (17)$$

where g can be any sufficiently regular non-linear function. This algorithm is, to our knowledge, the only proposed ICA algorithm that is robust against outliers and can be computed in batch mode without resorting to slow stochastic gradient descent methods. The fixed-point algorithms can also be generalized for non-sphered data without deteriorating their convergence properties [20, 24].

4 Neural ICA: multi-unit learning rules

The PCA optimization criteria were extended to nonlinear cases by Karhunen and Joutsensalo [28, 29]. They considered e.g. the following two criteria (note the formal analogue of these with the Problems 2 and 3 in Section 1):

Generalized Problem 2.
$E\{f(\mathbf{w}_i^T\mathbf{x})\}$, $i = 1, ..., p$, are maximized under the constraints $\mathbf{w}_i^T\mathbf{w}_j = \delta_{ij}$ for $j \leq i$.

and

Generalized Problem 3.
$E\{\|\mathbf{x} - \mathbf{W}g(\mathbf{W}^T\mathbf{x})\|^2\}$ is minimized.

The latter is a least mean square criterion in which the input vector $\mathbf{x}$ is represented by a *nonlinear* expansion of the weight vectors. Karhunen and Joutsensalo [29] further derived a learning rule as a gradient ascent for Generalized Problem 2, that is very close to a nonlinear version of the SGA rule in eq. (2). For Generalized Problem 3, an approximative gradient descent rule is

$$\Delta\mathbf{W} = \alpha[\mathbf{x} - \mathbf{W}g(\mathbf{W}^T\mathbf{x})]g(\mathbf{x}^T\mathbf{W})]. \qquad (18)$$

The nonlinear vector function $g(\mathbf{x}^T\mathbf{W})$ must be understood element by element. Note the formal similarity of eq. (18) with the Subspace learning rule (3). The algorithm (18) was first introduced by Oja et al [39].

These learning rules have an implementation in a one-layer neural network of p parallel units. Each unit i has the same n-element input vextor $\mathbf{x}$ and its own n-dimensional weight vector $\mathbf{w}_i$, which together comprise the $n \times p$ weight matrix $\mathbf{W} = (\mathbf{w}_1 ... \mathbf{w}_p)$.

The learning rule (18) was analyzed by one of the authors [41] in the case of the memoryless linear mixture model (8). It was shown that under certain assumptions relating the density of the independent source signals s_i and the nonlinear function $g(.)$, the columns of the solution matrix $\mathbf{W}$ after convergence will give the (scaled) source signals in the sense that $\mathbf{w}_i^T \mathbf{x} = scalar \times s_i$.

In [30], this learning rule and several others were discussed as part of an ICA neural network that has both the prewhitening, separation, and mixture matrix estimation stages. Also some examples were given.

5 Discussion

We outlined the development that has led us from investigating neural methods for PCA [35,37], through non-linear extensions of PCA [39,29], to neural ICA and related methods. Our neural methods for ICA were seen to emanate from a bottom-up approach to algorithm development: instead of formulating general statistical criteria of independence, we started from the computational point of view, which in this particular case is the neural network environment. This ensures that our algorithms can be given properties often not emphasized in the methods starting from the independence criteria, like properties of robustness and computational simplicity. We showed how ICA can be performed by very simple, Hebbian-like learning rules [41,22], which may also be biologically plausible.

Our approach is not limited, however, to adaptive on-line learning rules. We also reviewed more batch-like versions of our neural algorithms [44,23] that inherit the other desirable properties of the neural learning rules, yet enabling highly efficient computations on conventional hardware.

References

1. S. Amari, A. Cichocki, and H.H. Yang. A new learning algorithm for blind source separation. In *Advances in Neural Information Processing 8 (Proc. NIPS'95)*, Cambridge, MA, 1996. MIT Press.
2. P. Baldi and K. Hornik, Neural networks and principal components analysis: learning from examples without local minima. *Neural Networks 2*, 1989, 52-58.
3. A.J. Bell and T.J. Sejnowski. An information-maximization approach to blind separation and blind deconvolution. *Neural Computation*, 7:1129–1159, 1995.
4. J.-F. Cardoso. Eigen-structure of the fourth-order cumulant tensor with application to the blind source separation problem. In *Proc. IEEE Int. Conf. on Acoustics, Speech, and Signal Processing*, pages 2655–2658, Albuquerque, NM, USA, April 3-6 1990.
5. J.-F. Cardoso. Iterative techniques for blind source separation using only fourth-order cumulants. In *Proc. EUSIPCO*, pages 739–742, 1992.

6. J.-F. Cardoso and B. H. Laheld. Equivariant adaptive source separation. *IEEE Trans. on Signal Processing*, 44: 3017 - 3030, 1996.
7. Y. Chauvin, Principal component analysis by gradient descent on a constrained linear Hebbian cell. *Proc. IJCNN*, Washington DC, 1989, 373-380.
8. A. Chicocki, R. Unbehauen, and E. Rummert, Robust learning algorithm for blind separation of signals. *Electronics Letters* 30, 1994, pp. 1386 - 1387.
9. A. Chicocki, W. Kasprzak, and S. Amari, Multi-layer neural networks with a local adaptive learning rule for blind separation of source signals. In *Proc. NOLTA'95*, pages 61- 65, 1995.
10. A. Cichocki, S. I. Amari, and R. Thawonmas. Blind signal extraction using self-adaptive non-linear hebbian learning rule. In *Proc. NOLTA'96*, pages 377–380, 1996.
11. P. Comon. Independent component analysis – a new concept? *Signal Processing*, 36:287–314, 1994.
12. G. Deco and D. Obradovic, An information-theoretic approach to neural computing. Springer, New York, 1996.
13. N. Delfosse and P. Loubaton, Adaptive blind separation of independent sources: a deflation approach. *Signal Processing* 45, 1995, pp. 59 - 83.
14. K. Diamantaras and S. Y. Kung, Principal component neural networks - Theory and applications. Wiley, New York, 1996.
15. P. Földiàk, Adaptive network for optimal linear feature extraction. *Proc. IJCNN*, Washington, DC, 1989, 401-405.
16. C. Fyfe, D. McGregor, and R. Baddeley, Exploratory projection pursuit: an artificial neural network approach. *Dept. Comp. Science, U. of Strathclyde Res. Rep. 94/160*, 1994.
17. K. Hornik and C. Kuan, Convergence analysis of local feature extraction algorithms. *Neural Networks* 5, pp. 229 - 240, 1991.
18. A. Hyvärinen and E. Oja. A fast fixed-point algorithm for independent component analysis. To appear in *Neural Computation*, 1997.
19. A. Hyvärinen and E. Oja. Simple neuron models for independent component analysis. To appear in *Int. J. Neural Systems*, 1997.
20. A. Hyvärinen. A family of fixed-point algorithms for independent component analysis. Technical Report A40, Helsinki University of Technology, Laboratory of Computer and Information Science, 1996.
21. A. Hyvärinen and E. Oja. Independent component analysis by general non-linear Hebbian-like learning rules. Technical Report A41, Helsinki University of Technology, Laboratory of Computer and Information Science, 1996.
22. A. Hyvärinen and E. Oja. A neuron that learns to separate one independent component from linear mixtures. In *Proc. IEEE Int. Conf. on Neural Networks*, pages 62–67, Washington, D.C., June 3-6 1996.
23. A. Hyvärinen and E. Oja. One-unit learning rules for independent component analysis. In *NIPS*96*, Denver, Colorado, 1996.
24. A. Hyvärinen. A family of fixed-point algorithms for independent component analysis. *Proc. ICASSP'97*, Munich, Germany, 1997.
25. C. Jutten and J. Herault, Independent component analysis (INCA) versus independent component analysis. *Signal Processing IV: Theories and Applications* (J. Lacoume et al, eds.), pp. 643 - 646, Elsevier, 1988.
26. C. Jutten and J. Herault. Blind separation of sources, part I: An adaptive algorithm based on neuromimetic architecture. *Signal Processing*, 24:1–10, 1991.
27. J. Karhunen and J. Joutsensalo, Tracking of sinusoidal frequencies by neural network learning algorithms. *Proc. ICASSP-91*, Toronto, Canada, 1991.

28. J. Karhunen and J. Joutsensalo. Representation and separation of signals using nonlinear PCA type learning. *Neural Networks*, 7(1):113–127, 1994.
29. J. Karhunen and J. Joutsensalo, Generalizations of Principal Component Analysis, optimization problems, and neural networks. *Neural Networks* 8, pp. 549 - 562, 1995.
30. J. Karhunen, E. Oja, L. Wang, R. Vigario, and J. Joutsensalo, A class of neural networks for Independent Component Analysis. *IEEE Trans. Neural Networks* 8, pp. 486 - 504, 1997.
31. J. Karhunen, A. Hyvärinen, R. Vigario, J. Hurri, and E. Oja. Applications of neural blind separation to signal and image processing. *Proc. ICASSP'97*, Munich, Germany, 1997.
32. A. Krogh and J. Hertz, Hebbian learning of principal components. *Nordita* preprint 89/50 S.
33. S. Kung and K. Diamantras, A neural network learning algorithm for adaptive principal component extraction (APEX). *Proc. ICASSP-90*, Albuquerque, NM, 1990, 861-864
34. R. Linsker, Self-organization in a perceptual network. *Computer*, 1988, 105-117.
35. E. Oja, A simplified neuron model as a principal components analyzer. *J. Math. Biol. 15*, 1982, 267-273.
36. E. Oja, *Subspace Methods of Pattern Recognition*. RSP and J. Wiley, 1983.
37. E. Oja and J. Karhunen, On stochastic approximation of the eigenvectors and eigenvalues of the expectation of a random matrix. *J. Math. Anal. Appl. 106*, 1985, 69-84.
38. E. Oja, Neural networks, principal components, and subspaces. *Int. J. Neural Systems 1*, 1989, 61-68.
39. E. Oja, H. Ogawa, and J. Wangviwattana, Learning in nonlinear constrained Hebbian networks. *Proc. ICANN-91*, Helsinki, Finland, 1991.
40. E. Oja. Principal components, minor components, and linear neural networks. *Neural Networks*, 5:927–935, 1992.
41. E. Oja. The nonlinear PCA learning rule and signal separation – mathematical analysis. To appear in *Neurocomputing*, 1997.
42. E. Oja and J. Karhunen. Signal separation by nonlinear hebbian learning. In M. Palaniswami, Y. Attikiouzel, R. Marks, D. Fogel, and T. Fukuda, editors, *Computational Intelligence - a Dynamic System Perspective*, pages 83 – 97. IEEE Press, New York, 1995.
43. E. Oja and L. Wang, Robust fitting by nonlinear neural units. *Neural Networks* 9, pp. 435 - 444, 1996.
44. E. Oja and A. Hyvärinen. Blind signal separation by neural networks. In *Proc. Int. Conf. on Neural Information Processing*, pages 7–14, Hong Kong, 1996.
45. J. Rubner and P. Tavan, A self-organizing network for principal components analysis. *Europhys. Lett. 10*, 1989, 693-689.
46. T.D. Sanger, Optimal unsupervised learning in a single-layer linear feedforward network. *Neural Networks 2*, 1989, 459-473.
47. R. Williams, Feature discovery through error-correcting learning. *Tech. Rep. 8501*, UCSD, Institute of Cognitive Science, 1985.
48. L. Xu, E. Oja and C. Suen, Modified Hebbian learning for curve and surface fitting. *Neural Networks* 5, pp. 441 - 457, 1992.

Entropy Optimization

Application to Blind Source Separation

Anisse Taleb and Christian Jutten*

LTIRF/INPG, 46 Avenue Félix Viallet, 38000 Grenoble, France

Abstract. This paper proposes an approach for entropy optimization by neural networks. A brief introduction to this problem is given. A simple neural algorithm based upon MSE minimization is provided. Validation of this algorithm is given by an application to the Source Separation problem.

1 Introduction

Unsupervised learning algorithms aim to find hidden structures and informative representations of large data sets. The infomax principle of Linsker [8], which is a fundamental principle of self-organization, states that the transformation of a vector $\mathbf{x}$ observed on the input layer of a Neural Network (NN) to a vector $\mathbf{y}$ on the output, should be chosen in order to maximize the transinformation between input $\mathbf{x}$ and output $\mathbf{y}$.

The Exploratory Projection Pursuit (EPP) [5, 4], is a statistical technique for finding interesting structure in high dimensional data sets. When implemented by NN, EPP minimizes the output entropy to draw this one far from normal. Blind source separation (SS) and the Independent Components Analysis (ICA) [3, 6] are algorithms that learn from the input samples, an inverse matrix in order to provide statistically independent outputs : this can be achieved by minimizing output mutual information. Many other unsupervised learning algorithms use information quantities to perform their learning [10].

These examples show how entropy plays a central role in unsupervised learning. To optimize output entropy, one needs to estimate the output probability density function (pdf) or more exactly its derivates.

Such a mechanism can be done by a multilayer perceptron (MLP) in unsupervised learning, with weights vector $\mathbf{w}$. Let $\mathbf{x}$ be the input vector, and $\mathbf{y}$ the output of this MLP. Reffering to Shannon's information theory, the output entropy writes as:

$$H(\mathbf{y}) = -\int p_Y(\mathbf{y}) \log p_Y(\mathbf{y}) d\mathbf{y} = -E[\log p_Y(\mathbf{y})], \tag{1}$$

where $E[.]$ denotes the expectation operator. Weights vector $\mathbf{w}$ is trained in order to optimize (1) [2] , the stochastic gradient learning algorithm is

$$\mathbf{w}_{t+1} = \mathbf{w}_t + \mu_t \nabla_{\mathbf{w}} \mathbf{y}^T \nabla_{\mathbf{y}} \log p_Y(\mathbf{y}), \tag{2}$$

* C. Jutten is professor at ISTG in Université Joseph Fourrier of Grenoble.

[2] Generally, this is done under some constraints not explicited here.

where μ_t is the learning rate, the sign of the learning rate determines if we want to maximize or to minimize output entropy.

The purpose of this paper is to provide an efficient method for the estimation of $\nabla_{\mathbf{y}} \log p_Y(\mathbf{y})$, also called the *score functions*. In the next section we give a definition and an useful lemma and lemmas, then we describe a simple neural algorithm for the estimation of these functions. Finally, we show how this algorithm can be easily used in the SS problem.

2 Density function and Score function

Definition. Let $X = (X_1, \cdots, X_n)$ a $\mathbb{R}^n$ random variable, with differentiable pdf $p_X(\mathbf{x})$. Score function in the mutivariate case is defined as:

$$\begin{aligned} \psi_X &: \mathbb{R}^n \rightarrow \mathbb{R}^n \\ \mathbf{x} &\rightarrow \nabla_{\mathbf{x}} \log p_X(\mathbf{x}) = \left(\frac{\partial \log p_X(\mathbf{x})}{\partial x_1}, \cdots, \frac{\partial \log p_X(\mathbf{x})}{\partial x_n} \right)^T \end{aligned} \tag{3}$$

Example. For a scalar random variable X, $\psi_X(x) = \frac{p'_X(x)}{p_X(x)}$. If X is gaussian with mean m_x and variance σ_x^2, then $\psi_X(x) = -\frac{1}{\sigma_x^2}(x - m_x)$.

Lemma. *Let $X = (X_1, \cdots, X_n)$ a $\mathbb{R}^n$ random variable, with score function ψ_X, let $\mathcal{F}$ be any differentiable function of $\mathbb{R}^n \rightarrow \mathbb{R}$ satisfying:*

$$\forall i = 1 \cdots n \quad \lim_{|x_i| \rightarrow +\infty} p_X(\mathbf{x})\mathcal{F}(\mathbf{x}) = 0, \tag{4}$$

then we have

$$E\left[\mathcal{F}(\mathbf{x})\psi_X(\mathbf{x})\right] = -E\left[\nabla_{\mathbf{x}}\mathcal{F}(\mathbf{x})\right] \tag{5}$$

The proof is quite intuitive and consists in integrating by parts, in the scalar case the result is immediate.

Example. Following the previous example, we verify the lemma for $\mathcal{F}(x) = x$ $E[x\psi_X(x)] = E\left[-\frac{1}{\sigma_x^2}(x^2 - xm_x)\right] = -\frac{E[x^2]-m_x^2}{\sigma_x^2} = -1.$

3 Mean Square Error Minimization lead to an Unsupervised Algorithm

Scalar case Suppose now that the function $\psi_X(x)$ is available, *i.e.* known. Then, using function approximation ability of neural networks, we could use a simple MLP with one input and one output, to provide an estimation $h(\mathbf{w}, x)$ of $\psi_X(x)$. Parameters vector $\mathbf{w}$ is trained to minimize the mean squared error:

$$\mathcal{E} = \frac{1}{2}E\left[\left(h(\mathbf{w}, x) - \psi_X(x)\right)^2\right]. \tag{6}$$

A gradient descent algorithm on (6) leads to the following weights update

$$\mathbf{w}_{t+1} = \mathbf{w}_t - \mu_t \nabla_{\mathbf{w}} \mathcal{E}. \tag{7}$$

Applying the previous lemma, the gradient of the error $\mathcal{E}$ writes as:

$$\nabla_{\mathbf{w}} \mathcal{E} = E\left[h(\mathbf{w}, x)\nabla_{\mathbf{w}} h(\mathbf{w}, x) + \nabla_{\mathbf{w}} \frac{\partial h(\mathbf{w}, x)}{\partial x}\right]. \tag{8}$$

In the last equation, $\psi_X(x)$ disappears: it shows surprisingly that the supervised learning algorithm for minimizing $\mathcal{E}$ does not need the *teacher* $\psi_X(x)$, and is in fact unsupervised.

Extension to the multivariate case is quite evident and is done by using a multilayer perceptron with n inputs. It leads also to an unsupervised algorithm, due to the elimination of $\psi_X(\mathbf{x})$ by using the previous lemma.

Practical issues The previous algorithms are based on a simple gradient descent. To improve the speed of convergence we suggest the use of second order minimization techniques. The algorithm was tested for scalar random variables, using a simple MLP with only one hidden layer, containing six neurons (with tanh activation function). Table 1 contains results for two distributions: the comparison between theoritical and estimated score functions proves the quality of the estimation.

Table 1. Sample experiments

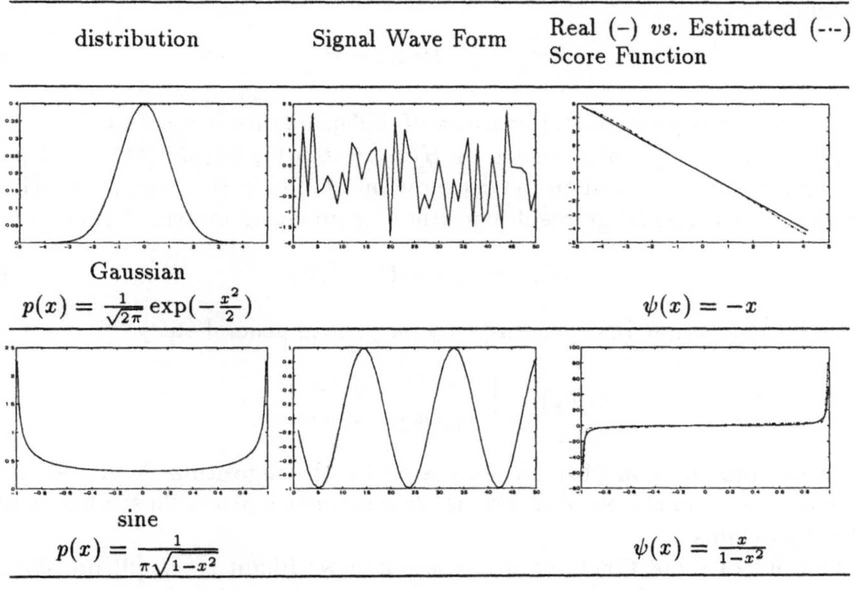

distribution	Signal Wave Form	Real (–) *vs.* Estimated (---) Score Function
Gaussian $p(x) = \frac{1}{\sqrt{2\pi}} \exp(-\frac{x^2}{2})$		$\psi(x) = -x$
sine $p(x) = \frac{1}{\pi\sqrt{1-x^2}}$		$\psi(x) = \frac{x}{1-x^2}$

4 Application To Blind Separation of Sources

The problem of Blind Source Separation (SS) consists in recovering the waveforms of unknown statistically independent signals called *sources*, by observing only their linear mixtures. Let us consider n independent and stationary unknown signals $\mathbf{s}(t) = (s_1(t), \cdots, s_n(t))^T$, and let $\mathbf{e}(t) = (e_1(t), \cdots, e_n(t))^T$ denoting a instantaneaous linear mixture of $\mathbf{s}$

$$\mathbf{e}(t) = \mathbf{A}\mathbf{s}(t), \tag{9}$$

where $\mathbf{A}$ is an $n \times n$ nonsingular matrix. The mixture is called instantaneous when the mixing matrix entries are time invariant scalars.

The problem is then to recover $\mathbf{s}(t)$ observing only $\mathbf{e}(t)$. This can be done by estimating a full rank matrix $\mathbf{B}$ such that

$$\mathbf{y}(t) = \mathbf{B}\mathbf{e}(t) \tag{10}$$

has statistically independent components. It is well known [3, 6] that the problem has two indeterminacies: sources can be recovered only up to a scale factor and to an permutation. In other words

$$\mathbf{y}(t) = \mathbf{P}\mathbf{D}\mathbf{s}(t) \tag{11}$$

where $\mathbf{P}$ is a permutation matrix, and $\mathbf{D}$ a diagonal matrix.

As proposed by a few authors [3, 11] (for a review and complete list of references see [7]), the independence of output components (components of $\mathbf{y}(t)$) means $\prod_i p_{Y_i}(y_i) = p_Y(\mathbf{y})$. It can be achieved by minimizing the mutual information:

$$I(\mathbf{y}) = \sum_{i=1}^{n} H(y_i) - H(\mathbf{y}), \tag{12}$$

since it is always positive and vanishes iff. independance is reached.

The joint entropy of $\mathbf{y}$ writes as $H(\mathbf{y}) = H(\mathbf{e}) + \log |det(\mathbf{B})|$: It depends only on input entropy and on the determinant of matrix $\mathbf{B}$. Using the stochastic *relative* (or natural) [1] gradient algorithm for updating matrix $\mathbf{B}$ leads to:

$$\mathbf{B}_{t+1} = (\mathbf{I} + \mu_t \mathbf{H}(\mathbf{y}_t))\mathbf{B}_t, \tag{13}$$

where $\mathbf{H}(\mathbf{y})$ is a matrix depending only on $\mathbf{y}$, with general entry

$$h_{ij}(\mathbf{y}) = \begin{cases} 0 & \text{if } i = j, \\ \psi_{Y_i}(y_i) y_j & \text{otherwise.} \end{cases} \tag{14}$$

The score functions in (14) are estimated by the algorithm described in the previous section in the one dimensional case and not *a priori* chosen like in most of SS algorithms.

The use of score functions in SS is not new. Pham *et al.* [9] proved that optimal criterion needs the knowledge of score functions. More recently, Charkani *et al.* [2] extended the results for convolutive mixtures. They both proposed

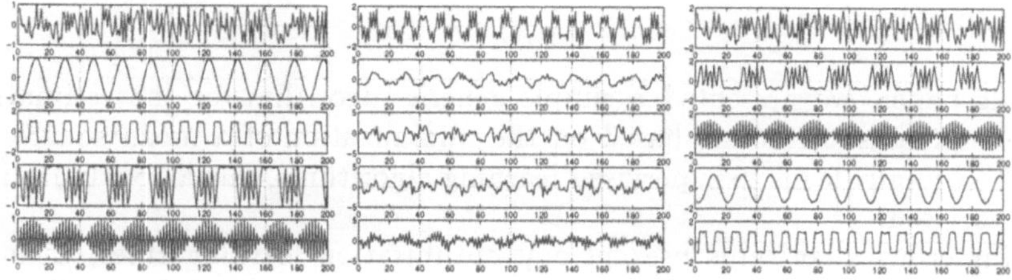

Fig. 1. Original sources, mixtures, estimated signals

suboptimal algorithm based on a parametric estimation of the score function according to a linear model:

$$\psi_{Y_i}(y_i) = \sum_k \xi_{ik} f_k(y_i), \tag{15}$$

this kind of estimation requires an appropriate choice for $f_k(.)$.

Due to the scale indeterminacy the algorithm in its previous form (13) can became unstable. In fact any scale change on the components of $\mathbf{y}$ will drive the NN to estimate a new score function: wich can imply a constant increase or decrease of the coefficients of matrix $\mathbf{B}$. A simple modification of the diagonal elements of $\mathbf{H}(\mathbf{y})$ (14), providing a normalisation to the output components of $\mathbf{y}$, avoids instability. The diagonal entries of $\mathbf{H}$ are given by

$$h_{ii}(\mathbf{y}) = \alpha_i(1 - y_i^2), \tag{16}$$

where α_i are positive scalars. This modification imposes, at convergence, that output vector $\mathbf{y}(t)$ has unit variance. This solution is preffered upon the whitening technique of Cardoso *et al.* [1], because it is less computational consuming and it can be shown that it leads to smaller rejection rates.

A sample run of this algorithm is shown in figure (1) and figure (2). Notice the low estimation variance due to the choise of optimal SS nonlinearities.

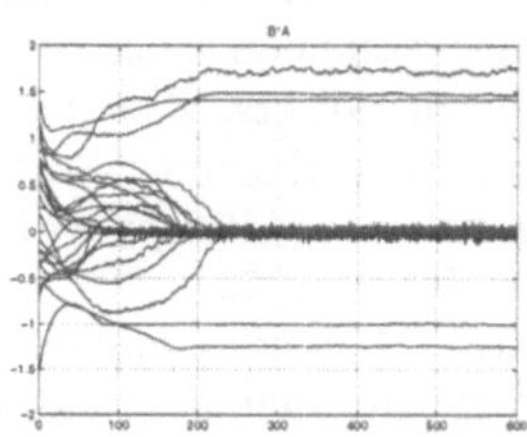

Fig. 2. Convergence of **BA** coefficients to 0 and to constant

5 Conclusion

In this paper, we presented a simple generic algorithm for the estimation of score functions. Although based on the minimization of a mean square error, which normally leads to supervised learning algorithms, this algorithm is in fact unsupervised.

Numerous algorithms use entropy optimisation learning rules, then score functions, to optimally perform the learning. For example, the application of this algorithm to SS was successful. The resulting algorithm uses adaptive on-line estimation of the score functions insteed of an *ad-hoc* one like in most SS algorithms, wich at convergence reach the optimal SS nonlinearities.

The generic nature of this algorithm allows its use in a large class of information theoritic problems. We are actually develloping new application for nonlinear source separation, image processing, Adaptive blind deconvolution and the exploratory projection pursuit.

References

1. J-F. Cardoso and B. Laheld. Equivariant adaptive source separation. *IEEE Trans. on S.P.*, 44(12):3017–3030, December 1996.
2. N. Charkani and Y. Deville. Optimization of the asymptotic performance of time-domain convolutive source separation algorithms. In *ESANN'97*, pages 273–278, Bruges, Belgium, April 1997.
3. P. Comon. Independent component analysis, a new concept? *Signal Processing*, 36(3):287 - 314, April 1994.
4. C. Fyfe and R. Baddeley. Non-linear data structure extraction using simple hebbian networks. *Biological Cybernetics*, (72):533–541, 1995.
5. P.J. Huber. Projection pursuit. *The Annals of Statistics*, 13(2):435–475, 1985.
6. C. Jutten and J. Hérault. Blind separation of sources, Part I: An adaptive algorithm based on a neuromimetic architecture. *Signal Processing*, 24(1):1–10, 1991.
7. J. Karhunen, E. Oja, L. Wang, R. Vigário, and J. Joutsensalo. A class of neural networks for independant component analysis. *IEEE trans. N.N.*, 8(3):486–504, May 1997.
8. R. Linsker. Self-organization in a perceptual network. *Computer*, (21):105–117, 1988.
9. D. T. Pham, P. Garat, and C. Jutten. Separation of a mixture of independent sources through a maximum likelihood approach. In J. Vandewalle, R. Boite, M. Moonen, and A. Oosterlinck, editors, *Signal Processing VI, Theories and Applications*, pages 771–774, Brussels, Belgium, August 1992. Elsevier.
10. N. N. Schraudolph. *Optimization of entropy with neural networks*. PhD thesis, University of California, San Diego, 1995.
11. H.H. Yang and S.I. Amari. Adaptive on-line learning algorithms for blind separation- maximum entropy and minimum mutual information. *Neural Computation*, 1997. Accepted.

Improving the Performance of Infomax Using Statistical Signal Processing Techniques

Bert-Uwe Koehler[1], Te-Won Lee[1,2], and Reinhold Orglmeister[1]

[1] Institut fuer Elektronik, Technische Universitaet Berlin
Einsteinufer 17, 10587 Berlin, Germany,
koehler@tubife1.ee.tu-berlin.de, orglm@tubife1.ee.tu-berlin.de
[2] Salk Institute, Computational Neurobiology Lab
La Jolla, C.A. 92037, USA, tewon@salk.edu

Abstract. In this paper, we present a new method that speeds up the convergence of the infomax algorithm proposed by Bell and Sejnowski. One effect of the infomax algorithm is that the 2nd order and 4th order statistical correlations are reintroduced to the signals during the learning process due to the optimization with respect to the complete signal statistics. We show that repetitively forcing 2nd and 4th order correlations to zero speeds up the convergence and improves separating sources with fewer data points.

1 Introduction

In blind source separation the problem is to recover independent sources given sensor outputs in which the sources have been mixed by an unknown channel. The problem has become increasingly important in the signal and speech processing area due to its prospective application in speech recognition, telecommunications and medical signal processing. Bell and Sejnowski [2] have proposed an information theoretic approach to the blind source separation and blind deconvolution problem. The infomax learning rule can be reformulated from different viewpoints such as the maximum likelihood perspective [13], the negentropy [6] framework or the well studied Bussgang property as recently derived by Lambert [8]. Recently, researchers have successfully applied ICA to real data analysis such as in speech recognition [10] and medical signal processing [12,11]. Makeig et al. [12] use the ICA formula to separate event-related potential (ERP) data and electroencephalographic (EEG) data into spatially fixed and temporally independent components. Mckeown et al. [11] use ICA to decompose functional magnetic resonance images (fMRI). The results are encouraging and may initialize a new paradigm in medical signal analysis. Nevertheless, in dealing with real data there are many problems that are associated with recordings like EEG, ERP and fMRI. As usual, the noise part is not explicitly taken into consideration and the nonlinear phenomenon is still an open research field although some reference can be found in [9]. Also, in some real data analysis we have to work with a small amount of data points that may not be sufficient to make the observation independent.

To this end, we propose a new method that combines statistical preprocessing methods with the simple infomax neural processor to cope with data analysis when only few data points are given. In particular, we present a method that repetitively removes 2nd and 4th-order correlations during its learning process. This speeds up the convergence for different test sets and is robust to ill conditioned mixing systems. The new method is able to handle small data vectors and high dimensional data.

2 The Algorithm

2.1 Information Maximization

Bell and Sejnowski [2] have proposed an information-theoretic approach where they maximize the mutual information that an output $y = g(x)$ of a neural processor contains about its input x. They have shown that for invertible and continuous deterministic mappings y, the mutual information between inputs and outputs can be maximized by maximizing the entropy of the outputs alone. Maximizing the joint entropy of the outputs $H(\mathbf{y})$ with respect to the elements of a demixing matrix $\mathbf{W}$ actually leads to a separation of the statistically independent components $\mathbf{u} = \mathbf{W}\mathbf{x}$ within the signals $\mathbf{x}$. Considering the parameter $\mathbf{W}$, to maximize entropy we do not follow the entropy gradient, as in [2], but its 'natural' gradient, as reported by Amari et al [1]:

$$\Delta\mathbf{W} \propto \frac{\partial H(\mathbf{y})}{\partial \mathbf{W}}\mathbf{W}^T\mathbf{W} = \left[\mathbf{I} + \left(\frac{\frac{\partial p(\mathbf{u})}{\partial \mathbf{u}}}{p(\mathbf{u})}\right)\mathbf{u}^T\right]\mathbf{W} \tag{1}$$

where $p(u) = \frac{\partial y}{\partial u}$. This gradient is an optimal rescaling of the entropy gradient. It simplifies the learning rule and speeds convergence considerably. An elegant way of generalizing the learning rule to sub/super-Gaussians is to approximate the estimated pdf in form of the Edgeworth approximation up to 4th order. This leads to a simple substitution as shown by Girolami and Fyfe in [6].

$$\Delta\mathbf{W} \propto \frac{\partial H(\mathbf{y})}{\partial \mathbf{W}}\mathbf{W}^T\mathbf{W} = \left[\mathbf{I} - K_4 \tanh(\mathbf{u})\mathbf{u}^T - \mathbf{u}\mathbf{u}^T\right]\mathbf{W} \tag{2}$$

2.2 Hybrid Learning Rule

A common statistical tool before applying the learning rule in eq.2 is to prewhiten the observation vector $\mathbf{x}$. This again speeds convergence and is being used by many researchers. The overall separation matrix $\mathbf{W}_{all}$ consists then of a second order sphering matrix $\mathbf{W}_S$ and the unmixing matrix found by the infomax algorithm: $\mathbf{W}_{all} = \mathbf{W} \cdot \mathbf{W}_S$. The whitening matrix $\mathbf{W}_S$ can be computed by $\mathbf{W}_S = (E\{\mathbf{x}\mathbf{x}^T\})^{-\frac{1}{2}}$ Furthermore, the preprocessing method can be extended to 4th order correlation cancelation by simply adding another step that cancels out 4th-order correlation. Cardoso [3] shows a simple way of a combined 2nd and 4th-order correlation cancelation:

$$\mathbf{x}_S = \mathbf{W}_S\mathbf{x} \tag{3}$$

$$\mathbf{W}_4 = (E\{\|\mathbf{x}_S\|^2 \mathbf{x}_S \mathbf{x}_S^T\})^{-\frac{1}{2}} \tag{4}$$

$$\mathbf{W}_{all} = \mathbf{W} \cdot \mathbf{W_4} \cdot \mathbf{W_S} \tag{5}$$

After the two preprocessing steps, we observe that during the learning process the non-optimal matrix $\mathbf{W}$ reintroduces 2nd and 4th order correlations. Given this observation, an extension of the two preprocessing methods is to allow the learning rule to constantly keep the 2nd and 4th order correlation at zero. This can be realized by supervising the learning process with a 4th order correlation measure and zero forcing the correlation whenever a certain threshold is achieved. For zero 4th order correlation between the signals, the 4th order covariance matrix $\mathbf{R}_4 = E\{\|\mathbf{u}\|^2 \mathbf{u}\mathbf{u}^T\}$ should be diagonal. Hence, as a measure for the level of reintroducing the 4th order correlation we use the distance of the correlation matrix $\mathbf{R}_4$ from being diagonal :

$$D = \log(\frac{\prod(\mathrm{diag}(\mathbf{R}_4 \cdot \mathbf{R}_4^T))}{\det(\mathbf{R}_4 \cdot \mathbf{R}_4^T)}); \tag{6}$$

If D is above a certain threshold during learning another 2nd and 4th order sphering is applied to the data. After a sphering step the overall matrix $\mathbf{W}_{all}$ is changed iteratively, i.e. $\mathbf{W}_{all}(n+1) = \mathbf{W} \cdot \mathbf{W}_4 \cdot \mathbf{W}_S \cdot \mathbf{W}_{all}(n)$ whereas $\mathbf{W}$ is the demixing matrix found by infomax in the following pass or passes through the data.

3 Performance Measure

We use a modification of the performance measure proposed by Amari et al. in [1]. The error measure F is the mean of the two normalized parts F_1 and F_2.

$$F_1 = \frac{1}{N}\sum_{i=1}^{N}\left(\frac{1}{N-1}\sum_{j=1}^{N}\frac{|p_{ij}|}{\max_k |p_{ik}|} - 1\right) \qquad N = \text{number of sources} \tag{7}$$

$$F_2 = \frac{1}{N}\sum_{j=1}^{N}\left(\frac{1}{N-1}\sum_{i=1}^{N}\frac{|p_{ij}|}{\max_k |p_{kj}|} - 1\right) \tag{8}$$

F_1 provides a measure for the mean coupling factor of other source signals into one particular recovered source signal. Assuming that all original source signals have the same variance and considering that all $N-1$ non-maximum entries of a fixed row in the matrix $\mathbf{P} = \mathbf{W}_{all}\mathbf{A}$ are the coupling factors of the noise signals into the recovered signal, we may compute an average SNR from the index F_1.

$$SNR = -10\log_{10}[(N-1) \cdot F_1^2] \tag{9}$$

The index F_2 provides a measure whether sources are lost. This might be the case when recovered signals represent the same original source. The mean value of both indices F_1 and F_2 can be considered as global measure for the performance of the algorithms.

4 Simulation Results

To compare the performance of the different methods we use a fixed test set of signals and a fixed mixing matrix **A**. For the mixing matrix **A**, the off diagonals have been chosen as '1' and the diagonal elements are '0.5'. The determinant det(**A**) is less than 1 which increases the level of separation difficulty. Two different test sets with two different data lengths have been chosen as follows:

1. Five artificial signals: sinusoid, rectangular, super-, sub- and Gaussian noise with 200 data points
2. Five artificial signals: sinusoid, rectangular, super-, sub- and Gaussian noise with 1000 data points
3. Pearlmutter's [13] 10 sound sources 200 data points
4. Pearlmutter's 10 sound sources 1000 data points

The following learning methods have been considered for comparison:

a) Bell & Sejnowski's infomax with natural gradient extension and prewhitening transformation
b) The above scheme but additionally canceling out 4th order correlation
c) Infomax with iteratively removing 2nd and 4th order correlation

The performance of the methods a) b) and c) have been considered on the test set 1) and 2) in figure 1. For each method, a block size of 20 and 20 passes through the data have been chosen. The left 3 sub-plots in figure 1 show the convergence of the learning methods given 200 data points of five artificially generated sources. The fastest convergence is achieved with method c). The same results can be observed when the number of data points is increased to 1000 for the right sub-plots of figure 1.

Figure 2 shows the performance index of the three methods for Pearlmutter's 10 sound sources. Here again, method c) achieves the best results for the given data points however the convergence compared to the method a) and b) does not differ as much as for test sets 1) and 2). Method c) increases the SNR from -14dB to 4.5 dB after 200 iterations. Although method c) achieves the best separation given a limited number of data points, in other experiments we observed that method b) shows the faster convergence if all data points (55.295) are presented. In one pass through the data a SNR of 16dB was achieved.

In other experiments, we have also successfully separated 20 mixtures from: 10 sound tracks obtained from Pearlmutter, 6 speech/sound signals used in Bell & Sejnowski [2], 3 uniform distributed noise signals and one noise source with a Gaussian distribution.

5 Conclusions and Further Research

We have shown that by repetitively zero forcing of the 2nd and 4th order cumulants the direction of learning of the unmixing matrix $\mathbf{W}_{all}$ can be used to speed up the convergence in infomax and to achieve separation given a small number

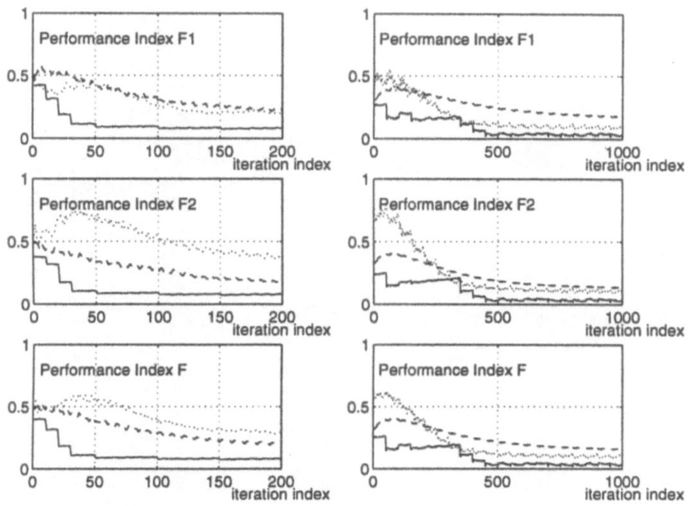

Fig. 1. Performance comparison for artificial signals with 200 data points in the left column, and with 1000 data points in the right column. The sources are: 1 super Gaussian, 1 sub Gaussian, 1 Gaussian, 1 rectangular, 1 sinus. The graphs are: dotted line - original infomax using 2nd order sphering, dashed line - infomax using 4th and second order sphering, solid line - repetitively sphering,

of data points. This is due to the reintroduction of 2nd and 4th order correlation during the learning process which is canceled out in the proposed method.

A problem that we have observed with the new method is that for certain signals and certain mixing matrices the whitening and canceling the 4th order correlation can result in an inhibition of the separating matrix $\mathbf{W}$. Subject of our future research is to analyze this behavior with respect the separation performance and in comparison with the method where preprocessing is done once. To emphasize that the new method works efficiently given a small set of data, we are going to test the new methods on ERP and EEG data when only small data vectors are available.

References

1. S. Amari, A. Cichocki, and H. Yang. A New Learning Algorithm for Blind Signal Separation. In *Advances in Neural Information Processing Systems 8*, 1996.
2. A.J. Bell and T.J. Sejnowski. An Information-Maximization Approach to Blind Separation and Blind Deconvolution. *Neural Computation*, 7:1129–1159, July 1995.
3. J-F. Cardoso. Source separation using higher order moments In Proc. ICASSP, pages 2109-2112, 1989.
4. J-F. Cardoso and B. Laheld. Equivariant adaptive source separation. *IEEE Trans. on Signal Processing*, 45,2:434-444, Dec. 1996.
self-adaptive

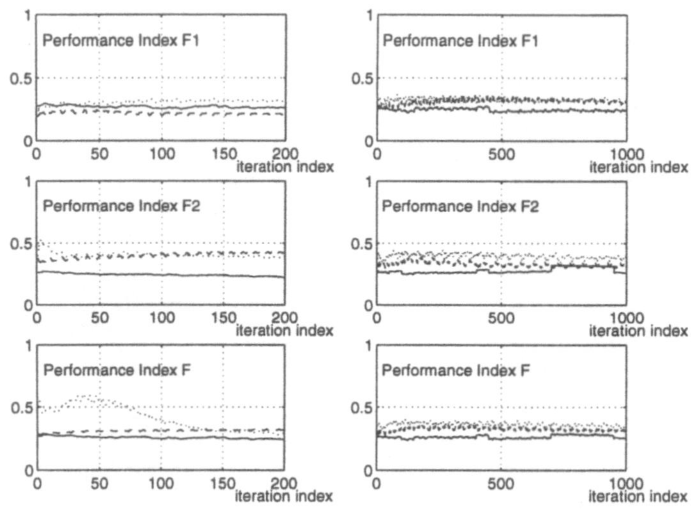

Fig. 2. Performance comparison for Pearlmutter's sound signals with with 200 data points in the left column, and with 1000 data points in the right column. The graphs are: dotted line - original infomax using 2nd order sphering, dashed line - infomax using 4th and second order sphering, solid line - repetitively sphering,

5. P. Comon. Independent component analysis – a new concept? *Signal Processing*, 36(3):287–314, 1994.
6. M. Girolami and C. Fyfe. Negentropy and kurtosis as projection pursuit indices provide generalised ica algorithms. In *Advances in Neural Information Processing Systems Workshop 9*, 1996.
7. J. Karhunen, E. Oja, L. Wang, R. Vigario, and J. Joutsensalo. A class of neural networks for independent component analysis. IEEE Trans. on Neural Networks, vol8:487–504, May. 1997.
8. R. Lambert and A. Bell. Blind separation of multiple speakers in a multipath environment. in *Proc. ICASSP 1997*, Munich.
9. T-W. Lee, B.U. Koehler and R. Orglmeister. Blind source separation of nonlinear mixing models. *Proc. IEEE Workshop on Neural Networks for Signal Processing*, Florida, USA, 1997.
10. T-W. Lee, A.J. Bell and R. Orglmeister. Blind source separation of real-world signals. *Proc. ICNN*, Houston, USA, 1997.
11. M. Mckeown, S. Makeig, G.G. Brown, T-P. Jung, S.S. Kindermann, A.J. Bell, T.J. Sejnowski. Analysis of fMRI data by Decomposition Into Independent Components submitted to *Human Brain Mapping*.
12. S. Makeig,T-P. Jung, D. Ghahremani, A.J. Bell, T.J. Sejnowski. Blind separation of auditory event-related brain responses into independent components. submitted to Proc. of the National Academic of Science
13. B. Pearlmutter and L. Parra. A context-sensitive generalization of ICA. In ICONIP'96 .

A Maximum Likelihood Approach to Nonlinear Blind Source Separation

Petteri Pajunen and Juha Karhunen

Helsinki University of Technology
Laboratory of Computer and Information Science
P.O.Box 2200, FIN-02015 HUT, FINLAND

Abstract. In the basic signal model of blind source separation (BSS), an unknown linear mixing process is assumed. While this ensures under mild conditions a sufficiently unique solution, it is desirable to extend the problem to nonlinear mixtures. Unfortunately the nonlinear case is much more difficult to handle, and brings serious indeterminacies to the solutions in the general case. In this paper we propose a new maximum likelihood approach to the nonlinear BSS problem. It is assumed that the source densities are known and that the mixing mapping is regularized. By finding a regular separating mapping which maximizes the likelihood, we show experimentally that the sources can often be separated.

1 Introduction

In blind source separation (BSS) the general goal is to separate statistically independent, unknown source signals from their mixtures without knowing the mixing process. BSS has applications in signal processing and other areas [1]. Often it is assumed that the unknown independent source signals have been mixed *linearly*, and only these mixtures are observed. This leads under reasonable assumptions to a sufficiently unique recovery of the original sources: the waveforms of the sources can be separated, even though their ordering and scaling remains unknown.

Extending the basic BSS problem to nonlinear mixtures is interesting because in many applications the mixing process cannot be assumed to be linear. In its full generality this nonlinear generalization is intractable, since the indeterminacies in the separating solutions are much more severe than in the linear case. For example if x and y are two independent random variables, any variables $f(x)$ and $g(y)$ are also independent. Thus if a neural network finds some independent outputs (an ICA solution), they may be different from the original source signals.

Because of its difficulty, the nonlinear BSS problem has been discussed in few papers only. In [2], an algorithm for nonlinear BSS is proposed for parametric nonlinearities. We have introduced the idea of using the self-organizing map (SOM) for nonlinear BSS in [3]. However, this approach can be successfully applied only to separation of sources having nearly uniform distributions.

Taleb and Jutten [4] have quite recently applied a maximum likelihood approach for separating sources from post-nonlinear mixtures. Yang, Amari and Cichocki [5] have extended their natural gradient approach to this case using an MLP network. While this assumption of post-nonlinear mixtures provides a unique solution with the same indeterminacies as in the linear case and is useful in signal processing (the nonlinearity can be thought of as a sensor distortion), it is a restrictive and somewhat arbitrary constraint.

In this paper, we require that the separating mapping provides an output vector with a given factorizable density. This is a generalization of our previous approach [3], where the output density was uniform. We assume that the mixing process is "regular", i.e. the correct separating mapping is the least complex mapping yielding the given output density. It is difficult to precisely express this assumption mathematically, and hence we cannot yet provide theoretical uniqueness results. However, the suggested approach is intuitively sensible, and we show its applicability by experiments.

2 Nonlinear Blind Source Separation

Nonlinear independent component analysis can be defined as a mapping $\mathbf{g}$ which takes a random vector $\mathbf{x}$ into a random vector $\mathbf{y} = \mathbf{g}(\mathbf{x})$ with a factorizable density, i.e. a product of its marginal densities, implying that the components are statistically independent.

In [3] we used the self-organizing map to perform this mapping, because a SOM roughly defines a uniform density on a rectangular map with suitable choices. The marginal densities in the directions of the sides of the rectangle are in this case statistically independent. A similar mapping has been discussed in [6], where the target distribution is a uniform density over a unit hypercube. It was suggested that such a mapping is the proper nonlinear generalization of principal component analysis.

However, when this idea is applied to nonlinear blind source separation, the serious indeterminacies in the nonlinear case can cause problems. Even if the vector $\mathbf{y}$ has independent components, it is not guaranteed that they are the original sources. If the sources are uniformly distributed, then it can be heuristically justified that regularization of the nonlinear separating mapping approximately separates the sources. However, when the sources are non-uniformly distributed, the regularization of the separating mapping is not necessarily sufficient, since the separating mapping yielding uniform density is not the inverse of the mixing mapping. This is illustrated in Fig. 1.

In this paper our previous approach [3] based on SOM is generalized to *arbitrary known* source densities. The advantage is that we can directly regularize the inverse of the mixing mapping by using the known source densities. However, assuming the source densities are known is not sufficient since any componentwise nonlinearity can transform the source densities arbitrarily. Therefore an accompanying assumption of the non-complexity of the mixing mapping must be made. The natural approach is to use the maximum likelihood method with

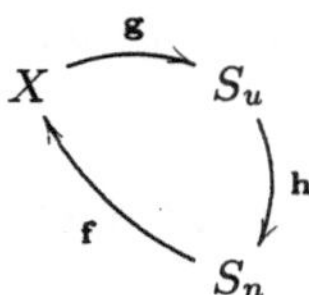

Fig. 1. The mixing mapping $\mathbf{f}$ takes a nonuniform density in space S_n to the mixture space. If a separating mapping $\mathbf{g}$ is required to provide uniform density, another mapping $\mathbf{h}$ is required to completely invert $\mathbf{f}$ up to scaling and permutation. This mapping can be complex if the sources have very nonuniform densities.

a suitable parametric model for the separating mapping. Instead of SOM, we use for computations a modification of the recently introduced generative topographic mapping (GTM) method [7]. The required output density is defined as a weighted sum of delta functions. The weighting coefficients are determined by the known source densities.

3 Modified GTM Algorithm

In this paper, we describe briefly only the main points of the GTM algorithm, with emphasis on the modifications. For a complete description, see [7]. Generally, GTM is inspired by and closely related to the self-organizing map. Its key benefit is a firm theoretical foundation which helps to overcome some of the limitations of SOM. This provides also the basis of generalizing the GTM approach to arbitrary target density in the present paper.

The mixture mapping is modeled as a linear combination of basis functions:

$$\mathbf{x}(\mathbf{s}; \mathbf{W}) = \mathbf{W}\boldsymbol{\Phi}(\mathbf{s}), \ \boldsymbol{\Phi} = [\phi_1, \phi_2, \dots, \phi_m]^T$$

where ϕ_j's are suitable fixed basis function (for example Gaussians). In GTM, one tries to maximize the likelihood of the data

$$L(\mathbf{W}, \beta) = \sum\nolimits_n \log p(\mathbf{x}(n)|\mathbf{W}, \beta) = \sum\nolimits_n \log \int p(\mathbf{x}(n)|\mathbf{s}, \mathbf{W}, \beta) p(\mathbf{s}) d\mathbf{s} \quad (1)$$

where β^{-1} is the variance of $\mathbf{x}$ given $\mathbf{s}$ and $\mathbf{W}$. We assume that the density of $\mathbf{s}$ is known and can be approximated by the product of marginal densities $p(\mathbf{s}) = \prod_i p_i(s_i)$, where each marginal density is a discrete density with corresponding sample points as the vectors $\mathbf{s}_{ij}$ below. For easier notation, we consider here only two sources, and get $p(\mathbf{s}) = \sum_{i,j} a_{ij}\delta(\mathbf{s} - \mathbf{s}_{ij})$. The vectors $\mathbf{s}_{ij}$ are fixed in the source signal space, and the coefficients $a_{ij} = p_1(i)p_2(j)$ where p_1 and p_2 are the discrete approximations of the marginal densities of $\mathbf{s}$. We get from (1)

$$L(\mathbf{W}, \beta) = \sum \log(\sum\nolimits_{i,j} p(\mathbf{x}(n)|\mathbf{s}_{ij}, \mathbf{W}, \beta))$$

Setting derivatives of the likelihood to zero we get the equation allowing the update of $\mathbf{W}$:

$$\boldsymbol{\Phi}^T \mathbf{G} \boldsymbol{\Phi} \mathbf{W}^T = \boldsymbol{\Phi}^T \mathbf{R} \mathbf{X} \tag{2}$$

where $\mathbf{X} = [\mathbf{x}(1), \ldots, \mathbf{x}(N)]^T$, $\mathbf{G}$ is a diagonal matrix with elements $G_{ii} = \sum_n R_{ijn}(\mathbf{W}, \beta)$ and the elements of $\mathbf{R}$ are

$$R_{ijn} = \frac{a_{ij} p(\mathbf{x}(n)|\mathbf{s}_{ij}, \mathbf{W}, \beta)}{\sum_{k,l} a_{kl} p(\mathbf{x}(n)|\mathbf{s}_{kl}, \mathbf{W}, \beta)} \tag{3}$$

Then we can update β by

$$\frac{1}{\beta} = \frac{1}{ND} \sum\nolimits_{i,j} \sum\nolimits_n R_{ijn} ||\mathbf{x}(\mathbf{s}_{ij}; \mathbf{W}) - \mathbf{x}(n)||^2 \tag{4}$$

In GTM, the well-known EM algorithm is used for maximizing the likelihood. Here the E-step (3) consists of of computing R_{ijn}, and the M-steps (2),(4) of updating $\mathbf{W}$ and β. The above derivation is the original GTM method, only the prior density coefficients a_{ij} have been added to the model.

3.1 Linear Mixtures

It is possible to use the above approach to linear mixtures simply by choosing $\mathbf{x}(\mathbf{s}; \mathbf{W}) = \mathbf{W}\mathbf{s}$. Due to the approach of explaining the observed mixtured $\mathbf{x}$ as images of latent variables $\mathbf{s}$ instead of finding a separating mapping, it is possible to find $\mathbf{W}$ in the difficult case of more sources than mixtures. In general, however, it is not possible to separate more sources than sensors. If the sources are binary in theory it is possible to *separate* the sources using the mixing matrix $\mathbf{W}$ [8]. An experiment was made where two noisy mixtures of five binary source signals were considered. The proposed method was used to find the linear mixing matrix $\mathbf{W}$. The binary sources were recovered without errors. In this special case the approach becomes similar to the one presented in [9].

4 An Experiment

In the following, a simple experiment involving two sources (Fig. 2) and three noisy nonlinear mixtures is described. Linear mixtures of the sources were transformed using a multilayer perceptron network with a volume conserving architecture (see [10]). Such an architecture was chosen for ensuring that the total mixing mapping is bijective and therefore reversible. Also very complex distortions of the source densities are avoided. This also makes the total mixing mapping more complex than the post-nonlinear model used in [4] and [5]. Finally, additive Gaussian noise was added.

The mixture can be written as $\mathbf{x} = \mathbf{A}\mathbf{s} + \tanh(\mathbf{W}\mathbf{A}\mathbf{s}) + \mathbf{n}$ where W is upper-diagonal with zero diagonal entries. The nonzero elements of $\mathbf{W}$ were drawn from

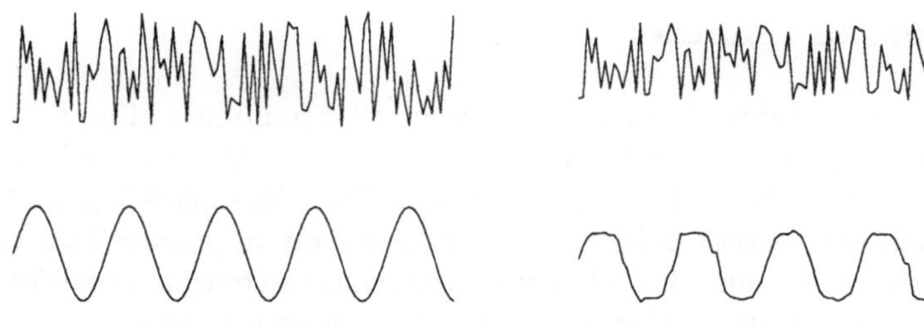

Fig. 2. Left: Source signals. Right: Separated signals.

a standard normal distribution. The matrix **W** ensures volume conservation of the nonlinearity applied to **As**. The two-dimensional joint densities of the mixtures are shown in Fig. 3. They clearly reveal the nonlinearity of the mixing mapping.

The algorithm presented above was used to learn a separating mapping. For reducing scaling effects, the mixtures were first whitened. After whitening the mixtures are uncorrelated and have an unit variance. Then the modified GTM algorithm was run for eight iterations using a 5x5 map. The separated sources are compared to the original sources in Fig. 2. The waveforms of the original sources are approximately recovered, even though there is some inevitable distortion due to the noise, discretization, and the difficulty of the problem. Also the joint density of the separated sources shown in Fig. 3 indicates that a factorizable density has been approximately obtained. The superimposed maps were obtained by mapping a 10x10 grid of vectors **s** to the mixture space.

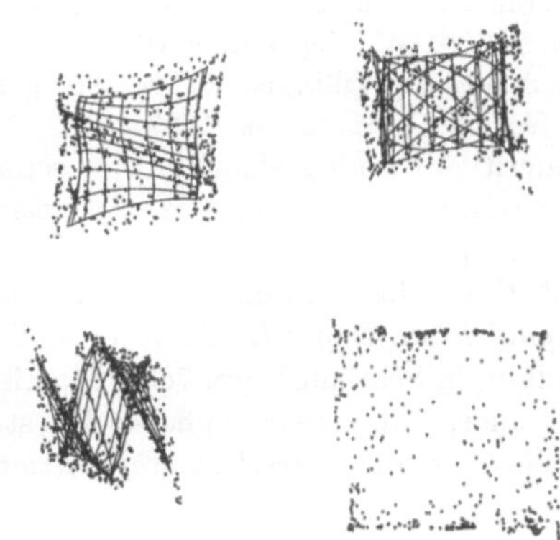

Fig. 3. Joint mixture densities with superimposed maps. Top left: $p(x_1, x_2)$. Top right: $p(x_1, x_3)$. Bottom left: $p(x_2, x_3)$. Bottom right: output signal joint density.

5 Concluding remarks

In this paper we have introduced a novel method for nonlinear blind source separation where the mixtures can be truly nonlinear, not only post-nonlinear as in some previous approaches. The proposed new method for separating nonlinearly mixed independent sources is based on the intuitively appealing idea of regularizing the separating mapping while requiring that the output densities match the known source densities. This should provide better separation compared to previous approaches where the output densities were required to be uniform. If the densities of the original sources are non-uniform, the requirement of uniform output densities inevitably causes distortion. This distortion is the more severe the farther the source densities are from the uniform density.

References

1. J. Karhunen, A. Hyvärinen, R. Vigario, J. Hurri, and E. Oja, "Applications of neural blind separation to signal and image processing," in *Proc. 1997 Int. Conf. on Acoustics, Speech, and Signal Proc. ICASSP-97*, (Munich, Germany), April 1997.
2. G. Burel, "Blind separation of sources: A nonlinear neural algorithm," *Neural Networks*, vol. 5, no. 6, pp. 937–947, 1992.
3. P. Pajunen, A. Hyvärinen, and J. Karhunen, "Nonlinear blind source separation by self-organizing maps," in *Progress in Neural Information Processing (ICONIP-96)* (S. Amari *et al.*, eds.), pp. 1207–1210, Springer, 1996.
4. A. Taleb and C. Jutten, "Nonlinear source separation: The post-nonlinear mixtures," in *Proc. European Symposium on Artificial Neural Networks (ESANN97)*, (Bruges, Belgium), pp. 279–284, April 1997.
5. H. Yang, S. Amari, and A. Cichocki, "Information back-propagation for blind separation of sources from non-linear mixtures," in *Proc. Int. Conf. on Neural Networks (ICNN'97)*, (Houston, USA), pp. 2141–2146, June 1997.
6. R. Hecht-Nielsen, "Replicator neural networks for universal optimal source coding," *Science*, vol. 269, pp. 1860–1863, September 1995.
7. C. Bishop, M. Svensen, and C. Williams, "GTM: The generative topographic mapping." To appear in *Neural Computation*, 1997.
8. P. Pajunen, "A competitive learning algorithm for separating binary sources," in *Proc. European Symposium on Artificial Neural Networks (ESANN'97)*, (Bruges, Belgium), pp. 255–260, April 1997.
9. A. Belouchrani and J.-F. Cardoso, "Maximum likelihood source separation for discrete sources," in *Signal Processing VII: Theories and Applications (Proc. of the EUSIPCO-94)*, (Edinburgh, Scotland), pp. 768–771, Elsevier, September 1994.
10. G. Deco and W. Brauer, "Nonlinear higher-order statistical decorrelation by volume-conserving neural architectures," *Neural Networks*, vol. 8, no. 4, pp. 525–535, 1995.

A Perceptron-Based Approach to Piecewise Linear Modeling with an Application to Time Series

Marco Mattavelli[2], Edoardo Amaldi[1] and Jean-Marc Vesin[2]

[1] School of OR and Theory Center, Cornell University, Ithaca, NY 14853, USA
[2] Signal Processing Laboratory, EPFL, CH-1015 Lausanne, Switzerland

Abstract. Piecewise linear models are attractive when modeling a wide range of nonlinear phenomena but determining simultaneously the domain decomposition and the corresponding parameter values is a challenging problem. We show that this problem can be formulated as that of partitioning an inconsistent linear system into a minimum number of consistent subsystems and we describe a greedy algorithm, based on a simple variant of the perceptron procedure, which provides good approximate solutions in a short amount of time. The general approach is applied to piecewise linear modeling of time series.

1 Introduction

Although linear models play an important role in the study of a wide range of physical phenomena and technical problems, nonlinear models are often required due to the presence of non-linearity and non-stationarity. In many situations, piecewise linear models are attractive since they allow approximation of complex phenomena but are still simple enough due to local linearity. However, piecewise linear model estimation turns out to be a challenging problem because it involves partitioning the data as well as determining the parameter values.

Conventional approaches in which the data is first partitioned using clustering methods and the parameters of each submodel are then estimated using regression techniques [6] have some major limitations. In particular, they require the number of submodels to be guessed in advance and, since the type of submodel used is not taken into account during the clustering phase, they may lead to meaningless results.

By extending some ideas we developed in [4] for motion analysis, we present a combinatorial optimization approach to piecewise linear model estimation. We describe an algorithm based on a simple variant of the perceptron procedure and we discuss an application to piecewise linear modeling of time series.

2 A combinatorial optimization formulation

As is well known, linear models lead to formulations in terms of linear systems such as $A\,\mathbf{w} = \mathbf{b}$, where A is a $p \times n$ matrix whose rows correspond to the data and $\mathbf{b}$, $\mathbf{w}$ are n-dimensional vectors. In the presence of non-linearity, the resulting linear systems are obviously inconsistent and classical approximate solutions that minimize least-mean-square error criteria [6] are not meaningful.

The following combinatorial optimization problem provides a natural way of addressing simultaneously the two fundamental issues underlying piecewise linear model estimation.

MIN PCS: Given an inconsistent $A\mathbf{w} = \mathbf{b}$, find a Partition of the set of equations into a MINimum number of Consistent Subsystems.

Given any solution of MIN PCS, the partition into consistent subsystems indicates the data partition and the solution associated with each consistent subsystem provides the parameter values of the corresponding linear submodel. Note that, according to Occam's razor principle, we look for the "simplest" piecewise linear model consistent with the data, which is most likely to be the correct one. Here the focus is on the number of submodels.

In order to cope with noisy data, each equation $\mathbf{a}^k\mathbf{w} = b_k$, where $\mathbf{a}^k$ is the kth row of A and b_k is the kth component of $\mathbf{b}$, can be replaced by the two complementary inequalities:

$$\mathbf{a}^k\mathbf{w} \leq b_k + \varepsilon \qquad \mathbf{a}^k\mathbf{w} \geq b_k - \varepsilon \tag{1}$$

where ε is the maximum tolerable error. If different equations are expected to be affected by different noise levels, different thresholds can of course be used.

Estimating minimum piecewise linear models turns out to be harder than finding least-mean-square solutions.

Theorem *It is* NP-*hard to approximate* MIN PCS *within a factor* $c < 3/2$.

In other words, assuming P $\neq$ NP, no polynomial-time algorithm is guaranteed to provide a partition whose number of consistent subsystems exceeds the minimum one by a multiplicative factor strictly smaller than 3/2. The proof is omitted due to the lack of space. According to this worst-case complexity result, we have to look for efficient heuristics with a good average-case behavior.

3 A greedy heuristic based on perceptron-like procedures

Since in practical applications we are interested in finding approximate solutions of MIN PCS rapidly, we adopt a greedy strategy and subdivide the original problem into a sequence of subproblems. Starting with the original inconsistent system composed of pairs of inequalities (1), close-to-maximum consistent subsystems are extracted iteratively. In other words, at each step we look for a solution satisfying as many pairs of inequalities as possible and then remove them from the current system until the remaining ones are consistent. This clearly yields a partition into consistent subsystems. Although the subproblem is NP-hard [1], it is not necessary and even desirable to look for the largest consistent subsystems. Indeed, it is easily verified that, even if such subsystems were available at each step, the greedy strategy is not guaranteed to yield minimum size partitions.

As we shall see, a simple variant of the perceptron procedure can be adapted to provide good approximate solutions of the subproblem in a short amount of time. Recall that training a perceptron amounts to finding a weight vector $\mathbf{w}$ satisfying the linear system:

$$\begin{array}{ll} \mathbf{a}^k\,\mathbf{w} \geq 0 & \forall k \text{ such that } b^k = 1 \\ \mathbf{a}^k\,\mathbf{w} < 0 & \forall k \text{ such that } b^k = -1. \end{array} \tag{2}$$

where $\mathbf{a}^k \in \mathbb{R}^n$, $1 \leq k \leq p$, are the training vectors and $b^k \in \{-1, 1\}$ the corresponding desired outputs. For nonlinearly separable training sets, one looks for an optimal weight vector that correctly classifies as many input vectors $\mathbf{a}^k$ as possible. Several variants of the perceptron procedure have been proposed to search for such maximum consistent subsystem of (2), see [1]. In the thermal perceptron [2], the basic idea is to pay decreasing attention to misclassified input vectors with large errors. Starting with an arbitrary initial weight vector $\mathbf{w}_0$, input vectors $\mathbf{a}^k$ are selected in a random order and the current $\mathbf{w}_i$ is updated as follows:

$$\mathbf{w}_{i+1} := \begin{cases} \mathbf{w}_i + \frac{t}{t_0} \exp\left(-|\mathbf{a}^k \mathbf{w}_i|/t\right) b^k \mathbf{a}^k & \text{if } \mathbf{a}^k \text{ is misclassified by } \mathbf{w}_i \\ \mathbf{w}_i & \text{otherwise,} \end{cases} \tag{3}$$

where the *temperature* t is linearly decreased from an initial t_0 to 0 in a predefined maximum number of cycles C_{max} through the training set. This annealing process guarantees that correction will terminate after C_{max} cycles. See [2, 1] for appropriate choices of t_0 and of the decreasing schedule.

The thermal perceptron procedure can be easily adapted to find close-to-maximum consistent subsystems of pairs of inhomogeneous inequalities such as (1). At each iteration, one picks a pair of inequalities and one checks whether the current solution $\mathbf{w}_i$ violates one of them. If $\mathbf{a}^k\mathbf{w}_i \leq b^k - \varepsilon$, then $\mathbf{w}_{i+1}$ is obtained by adding

$$\frac{t}{t_0} \exp\left(-|E_i^k|/t\right)\ \mathbf{a}^k \tag{4}$$

to $\mathbf{w}_i$, where $E_i^k = b^k - \varepsilon - \mathbf{a}^k\mathbf{w}_i$ denotes the deviation from satisfiability. Similarly, if $\mathbf{a}^k\mathbf{w}_i \geq b^k + \varepsilon$, then (4) with now $E_i^k = \mathbf{a}^k\mathbf{w} - b^k - \varepsilon$ is substracted from $\mathbf{w}_i$. The next cycle is started when all pairs of inequalities have been considered in a random order.

Although our simple greedy strategy is not guaranteed to yield minimum partitions, it turns out to be very effective experimentally. Note that, unlike in the usual perceptron applications, the weight vectors determined during the training phase are not used to classify new patterns. Instead, they reveal the structure and they provide the parameter values of the piecewise linear model.

To achieve higher robustness for noisy data, a least-mean-square solution can be determined for each subsystem of the partition provided by our algorithm. Thus, piecewise linearity is taken into account in the partition process and robustness is guaranteed in submodel parameter estimation.

4 Piecewise linear modeling of time series

Nonlinear signal models have been the object of an increasing interest over the past few years. A number of applications can be found in different fields such as, for instance, economic system modeling and biomedical signal analysis. The idea of breaking a global linear model into a number of submodels, gave rise to the so-called threshold autoregressive (TAR) models [7, 8], in which the choice of the model at time t is based on the comparison of the signal value at time $t-1$ with pre-defined thresholds. This scheme allows one to model and reproduce phenomena such as jumps and limit cycles, but its inherent limitation derives from the selection of the thresholds, for which no general method exists. The principle, called piecewise linear modeling [9], consists of partitioning the state-space into a certain number of regions and estimating a linear autoregressive submodel for each one of them. A piecewise linear autoregressive model can be described as follows [9]:

$$x_t = \sum_{j=1}^{m} w_{ij} x_{t-j} + u_t \tag{5}$$

where the set of coefficients $(w_{i1}, \ldots, w_{im})$ depends on the position of the vector $(x_{t-1}, \ldots, x_{t-m})$ with respect to a given partition of the state-space $\mathbb{R}^m$, and $\{u_t\}$ is an i.i.d. sequence. Here the x's are the observations and the coefficients w_{ij}'s as well as the partition of the state-space have to be estimated.

Although piecewise autoregressive models are more attractive than the TAR ones, the number of submodels must be selected a priori and an appropriate state-space partition needs to be determined. In [9] a two-stage strategy is suggested. First one determines a state-space decomposition using a Kohonen feature map [3] and then one estimates the parameters of each submodel using robust regression techniques. The purpose of such an unsupervised clustering phase is to avoid guessing the number of submodels in advance. However, besides the delicate convergence issue, it is unclear how a state-space partition can be actually derived from a given feature map. Even more importantly, the clustering process does not take into account the type of submodel used.

According to our general approach, we consider the problem of estimating such piecewise linear models in terms of MIN PCS. Clearly, any sequence of observations $\{x_1, \ldots, x_L\}$ which cannot be modeled (approximated) by a simple autoregressive model, leads to an over-determined inconsistent system of pairs of inequalities such as (1). Given any partition into consistent subsystems, each consistent subsystem defines a group of vectors in the state-space and the associated solution indicates the parameter values of the corresponding submodel.

A number of experiments have been carried out with various well-known time series which have been extensively used as benchmark in the literature. Due to the lack of space, we only mention a few typical results obtained for the chaotic time series generated with the Henon maps and the Sunspot time series [8]. The Henon map is defined by: $x_t = 1 - 1.4x_{t-1}^2 + 0.3x_{t-2}$. The points with coordinates x_{t-1}, x_{t-2} are located on a fractal strange attractor (see Figure 1). As an example, we report the results obtained for a run with a sequence of 2000

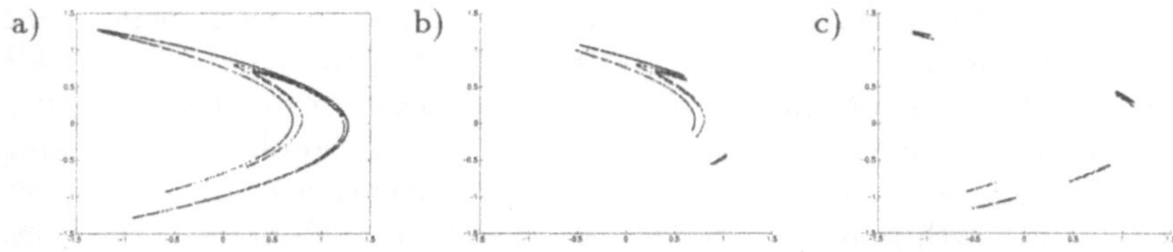

Fig. 1. a) State-space representation of the data of the Henon time series. Region corresponding to the largest (b) and the second largest (c) submodel.

consecutive values, with a modeling order of the autoregression $m = 2$ and a maximum tolerable error $\epsilon = 0.3$. The resulting partition contains 8 consistent subsystems. Interestingly, the regions corresponding to each submodel are well defined in the state-space, which indicates that they are valuable for modeling purposes. Depending on the region to which the point (x_{t-1}, x_{t-2}) belongs, the corresponding set of autoregressive coefficients provides the estimate of x_t within the predefined ϵ. Figures 1 b) and 1 c) show the regions corresponding to the two largest subsystems.

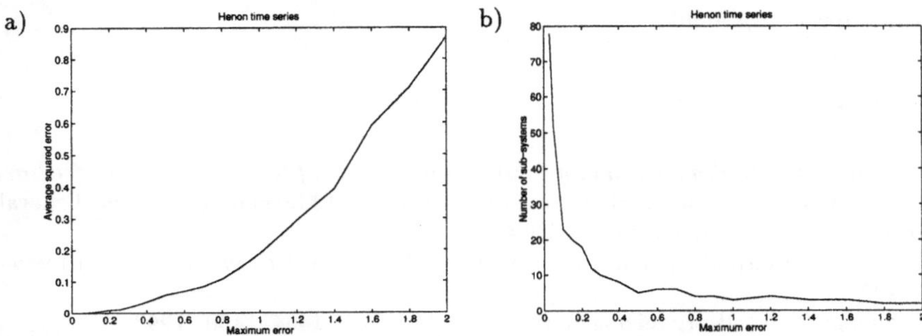

Fig. 2. a) Average squared estimation error of the piecewise linear model and b) number of components in the piecewise state-space decomposition versus ϵ.

An interesting feature of our algorithm is the possibility of selecting the maximum tolerable error ϵ. We have investigated the trade-off between ϵ and model complexity. Figure 2 a) and b) display the total average squared error and, respectively, the number of subsystems as a function of ϵ. While the average squared error clearly increases with ϵ, the number of subsystems remains almost constant for a wide range of values after a sharp decrease. Intuitively, the values of ϵ corresponding to the knee of this curve yield the best compromise between model complexity and average estimation error. We are currently investigating how the minimum description length criterion [5] which is used to estimate this trade-off for linear models can be extended to the case of piecewise linear ones. In this case, not only the set of coefficients but also the description of the state-space partition has to be taken into account.

The same type of tests have been performed on the well-known Sunspot time series [8]. Comparing our results with those obtained with the classical TAR models [8], we achieve significantly lower average squared errors at the expense of a reasonably small increase in model complexity. We also observe the same behavior of the average squared error and the number of subsystems with respect to ϵ. Finally, it is worth pointing out that in all tests a small number of cycles (100-200) of the perceptron-like procedure was sufficient.

5 Concluding remarks

We have presented a general combinatorial optimization approach for estimating piecewise linear models. Our greedy algorithm based on a variant of the thermal perceptron procedure has been applied to the callenging problem of piecewise linear modeling of time series. The experimental results obtained for various time series, including cahotic ones, indicate that it yields natural partitions of the state-space that are valuable for modeling purposes. To use such piecewise autoregressive models for prediction, signal calssification and event detection, state-space partitions can be derived from the resulting linear system partitions using distance measures that take into account the distribution of the state-vectors in each component.

References

1. E. Amaldi. *From finding maximum feasible subsystems of linear systems to feedforward neural network design.* PhD thesis, Department of Mathematics, Swiss Federal Institute of Technology, Lausanne, 1994.
2. M. Frean. A "thermal" perceptron learning rule. *Neural Computation*, 4(6):946–957, 1992.
3. T. Kohonen. The self-organizing map. *Proc. IEEE*, 78:1464–1480, 1990.
4. M. Mattavelli and E. Amaldi. Using perceptron-like algorithms for the analysis and the parameterization of object motion. In F. Girosi et al., editor, *Neural Networks for Signal Processing V, Proceedings of the 1995 IEEE workshop*, pages 303–312, 1995.
5. J. Rissanen. *Stochastic Complexity in Statistical Inquiry.* World Scientific, Singapore, 1989.
6. P. J. Rousseeuw and A. M. Leroy. *Robust regression and outlier detection.* John Wiley & Sons, Inc, New York, 1987.
7. H. Tong. *Threshold Models in Nonlinear Time Series Analysis.* Springer, New York, 1983.
8. H. Tong. *Nonlinear Time Series.* Oxford University Press, Oxford, 1990.
9. J.M. Vesin. A common generalization framework for two classical nonlinear models. In *Proc. IEEE Int. Workshop on Nonlinear Digital Signal Processing*, pages 6–2.3, Tampere, Finland, 1993.

Local Independent Component Analysis by the Self-Organizing Map

Erkki Oja and Kimmo Valkealahti

Laboratory of Computer and Information Science
Helsinki University of Technology
P.O. Box 2200, FIN-02015 HUT, Finland

Abstract. We introduce a neural network for the analysis of local independent components of an input signal. The network is a modification of Kohonen's adaptive-subspace self-organizing map. The map units consist of weight matrices adapted to represent linear transformations which locally minimize statistical dependence among pattern vector components. Training of the map is carried out in episodes comprising pattern vectors sampled from adjacent time instants or spatial locations. The use of episodes produces independent directions which are preserved in translations of the input signal. The independent components modeled by each map unit are estimated with a nonlinear Hebbian-like learning rule, which searches for weight vectors maximizing a measure of non-Gaussianity of the scalar product of weight and pattern vectors. For demonstration, the method was applied to the segmentation of a composition image of four periodic texture fields. The spatial convolution masks, created by the map for the extraction of independent components, represent distinct frequences of particular directions.

1 Introduction

Assume that $\mathbf{x}$ is a signal vector consisting of the gray levels of pixels in an image window, or intensities of consequent points from a time series. In Principal Component Analysis (PCA), such signal vectors are standardly represented as

$$\mathbf{x} = \sum_{i=1}^{k} s_i \mathbf{a}_i \ , \tag{1}$$

where s_i are the uncorrelated principal components and $\mathbf{a}_i$ are the PCA basis vectors, or the eigenvectors of the covariance matrix. In Independent Component Analysis (ICA), a signal processing technique that has attracted considerable interest recently [1,4], this model is extended so that the s_i are assumed statistically independent. Then the vectors $\mathbf{a}_i$ are called the ICA basis vectors.

The assumption that such independent components actually exist for a given stochastic signal is quite restrictive and usually requires that there is an explicit mixing process in which the signal $\mathbf{x}$ is constructed as a linear mixture of a number of independent sources. However, even for natural signals for which such a mixing process cannot be modeled, it may be reasonable to try to find components that are as independent as possible.

The ICA model of eq. (1), in which the mixing coefficients (ICA basis vectors) $\mathbf{a}_i$ for the independent signals s_i are constant, may be unreasonable for a widely varying signal $\mathbf{x}$, e.g., for all the windows of a natural image. It may be better first to cluster the signal samples to a set of clusters, within which the signals are more similar, and then find the local ICA expansions for each cluster. The purpose of this study is to search for ICA expansions of local textures in this way. The clustering is done using the Self-Organizing Map (SOM) network [5]. During the training of the map, each map unit specializes to extract independent components from a subset of signals with distinct statistics. To obtain locally stable estimates of mixing, the method of training in episodes is used, borrowed from the ASSOM method [6].

By this approach, two goals are reached: first, it is possible to segment images consisting of several distinct textures. Individual texture windows are labeled according to the ICA model that gives the best fit. Second, sets of filters are obtained as the ICA basis vectors of the separate ICA models, and these filters are believed to reveal the underlying structures of the separate textures and allow their sparse representation. The ICA basis vectors are akin to wavelets (see also [2]) and thus might produce a novel family of filters for compression and feature extraction.

2 Neural algorithm for estimating independent components

Hyvärinen and Oja [3] have introduced a neural algorithm for estimating independent components globally. A single independent component s of input $\mathbf{x}$ is sought as a linear function $s = \mathbf{w}^T\mathbf{x}$ and estimated by minimizing or maximizing the expectation of a contrast function $F(\mathbf{w}^T\mathbf{x})$ under the constraint $\|\mathbf{w}\| = 1$. Another constraint is that $\mathbf{x}$ must have zero mean and unit covariance, which is obtained by prewhitening the vectors. A wide range of nonlinear even functions are valid for contrast function F; the choice of the function is discussed in [3]. The algorithm uses a Hebbian-like learning rule, in which weight vector $\mathbf{w}$ is adapted for given input $\mathbf{x}$ as

$$\Delta\mathbf{w} \propto \sigma\mathbf{x}f(\mathbf{w}^T\mathbf{x}) \ , \tag{2}$$

after which the weight vector is normalized. Function f in (2) is the derivative of F and factor $\sigma = \pm 1$ determines whether $E\{F(\mathbf{w}^T\mathbf{x})\}$ is maximized or minimized. However, the sign cannot be conceived beforehand. Hyvärinen and Oja [3] gave an adaptive learning rule

$$\Delta c \propto \mathbf{w}^T\mathbf{x}f(\mathbf{w}^T\mathbf{x}) - f'(\mathbf{w}^T\mathbf{x}) - c \ , \tag{3}$$

for selecting correct value $\sigma = \mathrm{sign}(c)$. The algorithm for one independent component simply extends to estimate several independent components. Let

$\mathbf{W} = (\mathbf{w}_1, \ldots, \mathbf{w}_k)$ be the weight matrix whose columns correspond to the estimated independent components. The learning rule becomes then

$$\Delta\mathbf{W} \propto \mathbf{x} f(\mathbf{x}^T\mathbf{W})\mathrm{diag}(\mathrm{sign}(\mathbf{c})) \ , \tag{4}$$

in which functions $f()$ and sign() are applied separately to each component of an argument vector. After each update, weight vectors are orthonormalized using, for instance, the Gram-Schmidt process to prevent vectors from converging to the same values.

3 The network structure

The self-organizing map for local independent component analysis consists of units M_i, $i = 1, \ldots, N$, which are ordered in a one- or two-dimensional array. Each unit is updated by rule (4) and thus holds a weight matrix of k column vectors and a k-dimensional scalar array

$$M_i = \{\mathbf{W}_i, \mathbf{c}_i\} = \{(\mathbf{w}_{i1}, \ldots, \mathbf{w}_{ik}), (c_{i1}, \ldots, c_{ik})^T\} \ . \tag{5}$$

Prior to the training of the map, vectors in each unit are set to random zero-mean vectors and the column vectors are orthonormalized. The training is performed in steps $t = 1, \ldots, T$. Each step begins with sampling of L episode vectors $\mathbf{v}_l$ from adjacent time instants or spatial locations.

The algorithm requires whitening of the episode vectors and, therefore, an estimate for their covariance matrix. If the number of episode vectors, L, is sufficiently large then the covariance matrix could be estimated from the episode. Otherwise, one solution could be to use an iteratively-updated moving estimate for the covariance. The whitened vectors used in rule (4) are obtained by $\mathbf{x}_l = \mathbf{D}^{-1/2}\mathbf{E}^T\mathbf{v}_l$, in which the columns of matrix $\mathbf{E}$ are the eigenvectors of the covariance matrix of $\mathbf{v}_l$ and $\mathbf{D}$ is the diagonal matrix of the corresponding eigenvalues. The dimensionality of $\mathbf{x}_l$ may subsequently be reduced to remove possible noise.

The representative winner for the whole episode is determined among the map units. The winner is the unit which maximizes the energy of average contrasts $F(\mathbf{w}^T\mathbf{x})$ computed over the columns of each unit

$$E_i = \sum_{j=1}^{k} \left(\sum_{l=1}^{L} F(\mathbf{w}_{ij}^T \mathbf{x}_l)/L \right)^2 \ . \tag{6}$$

The representative winner is $b = \arg\max_i E_i$. Following rules (3) and (4), the updating of the representative winner has two steps. First, vector $\mathbf{c}_b$ of the representative winner is updated according to

$$c_{bj}(t) = (1 - \alpha(t))c_{bj}(t-1) + \alpha(t) \sum_{l=1}^{L} (\mathbf{w}_{bj}^T \mathbf{x}_l f(\mathbf{w}_{bj}^T \mathbf{x}_l) - f'(\mathbf{w}_{bj}^T \mathbf{x}_l))/L \ , \tag{7}$$

in which $\alpha(t)$ is the learning-rate factor. Second, the weight matrix of the representative winner is updated successively with the episode vectors according to the following learning rule

$$\mathbf{W}_b(t) = \mathbf{W}_b(t-1) + \alpha(t)\mathbf{x}_l f(\mathbf{x}_l^T \mathbf{W}_b(t-1))\mathrm{diag}(\mathrm{sign}(\mathbf{c}_b(t))) \ . \qquad (8)$$

After each update, the weight vectors are orthonormalized.

According to the self-organizing map algorithm, the units in the neighborhood of the winner are also updated. In the beginning of the training, the neighborhood radius should be large, for instance, half of the map size. During the training, the radius gradually decreases and may become zero. The use of neighborhood ensures that all units become adapted regardless of their initial values.

4 Results and discussion

The experiments were made with a 256-by-256-pixel image of four Brodatz textures, Fig. 1. Each episode consisted of $L = 25$ samples $\mathbf{v}_l$, which were 16-by-16-pixel windows collected from a randomly selected location and its 5-by-5-pixel neighborhood. The mean of each episode vector was subtracted. For whitening of $\mathbf{v}_l$, one covariance matrix was computed using 2560 samples from random locations in the composition image. This was justified because the whitening matrix obtained from the composite covariance matrix practically decorrelated samples in each of the four subfields. Owing to the subtraction of vector means, one eigenvalue of the covariance matrix was zero. Therefore, dimensionality of whitened vectors $\mathbf{x}_l = \mathbf{D}^{-1/2}\mathbf{E}^T\mathbf{v}_l$ was decreased by one.

Fig. 1. Texture field of four surface categories.

The trained self-organizing map contained ten units ($N = 10$) of four weight vectors ($P = 4$). The units were ordered in a one-dimensional array. The contrast function used, $F(x) = \ln\cosh(x)$, was recommended in [3] because of its robustness against outliers. The first and second derivatives of $F(x)$ needed in (7) and (8) are $f(x) = \tanh(x)$ and $f'(x) = \cosh^{-2}(x)$. The map was trained with $T = 2560$ episodes using learning rate $\alpha(t) = 0.01 \cdot (1 - t/T)$. The radius of the neighborhood was initially 5 and it decreased linearly to zero during the first quarter of the training.

The trained map was used for the segmentation of the composite image Fig. 1. The mean-subtracted and whitened vectors $\mathbf{x}$ sampled from all overlapping 16-by-16-pixel windows in the image were classified by finding the unit of maximum energy according to

$$E_i = \sum_{j=1}^{k} F(\mathbf{w}_{ij}^T \mathbf{x})^2 \ . \tag{9}$$

The index of a winner unit, $b = \arg\max_i E_i$, yielded the class of a vector. The indexes were collected in the segmentation image shown in the upper left part of Fig. 2. The lower left part of the figure shows "dewhitened" weight vectors or ICA basis vectors $\mathbf{ED}^{1/2}\mathbf{w}_{ij}$ which resulted from the training process; each column represents one unit of the map together with its coloring in the segmentation.

As shown by Fig. 2, the spatial convolution masks extracting maximally independent components tuned to represent certain frequencies or microfeatures and directions. The segmented image consists of rather uniform regions. Each map unit detected almost exclusively only one of the four surface categories and the units detecting the same surface were neighbors in the map array. The four units with the darkest gray-tone detected the lower-right surface, the fifth and sixth unit detected the upper-right surface, the seventh to ninth units detected the lower-left surface and the last unit detected the upper-left surface. Thus, the coding was rather sparse.

The training of the self-organizing map with episodes of multiple vectors sampled from adjacent locations provided a texture detector set which was invariant under slight translations. To demonstrate the difference between translation-invariant and non-invariant detector sets, a self-organizing map was trained with single-vector episodes. The training was performed in steps $t = 1, \ldots, 25 \cdot 2560$. At each step, only one 16-by-16-pixel sample was collected from a randomly selected location. The image segmented with this map, together with the corresponding weight vectors, is shown in the right part of Fig. 2. The weight vectors differ clearly from those obtained with multiple-vector episodes. Several weight vectors represent mixtures of frequencies. The segmented image shows that the emerged texture detectors were not invariant under translations but sensitive to different phases of the image signal. The segmented image also suggests that the detectors were not as specific to the four surface types as the detectors obtained with multiple-vector episodes.

Acknowledgment The study was funded by the Academy of Finland as a part of the project Intelligent processing and analysis of images and speech.

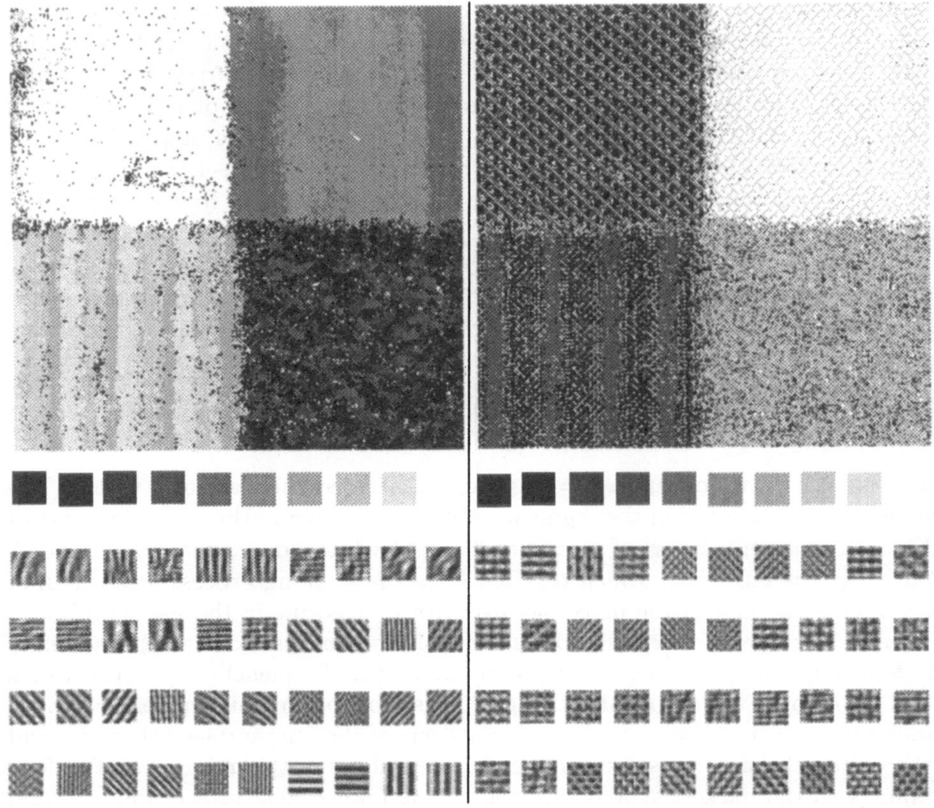

Fig. 2. Segmentation images and the corresponding weight vectors. The segmentations were obtained with maps trained with multiple episode vectors (left) or with episodes of just one vector (right).

References

1. P. Comon. Independent component analysis, A new concept? *Signal Processing*, 36(3):287–314, April 1994.
2. J. Hurri, A. Hyvärinen, and E. Oja. Wavelets and natural image statistics. In *Proceedings of the 10th Scandinavian Conference on Image Analysis*, volume I, pages 13–18, Lappeenranta, Finland, June 9–11 1997.
3. A. Hyvärinen and E. Oja. Independent component analysis by general non-linear Hebbian-like learning rules. Submitted to *Signal Processing*, 1997. Available as an Internet resource: http://nucleus.hut.fi/~aapo/pub.html.
4. J. Karhunen, E. Oja, L. Wang, R. Vigário, and J. Joutsensalo. A class of neural networks for independent component analysis. *IEEE Transactions on Neural Networks*, 8(3), May 1997.
5. T. Kohonen. *Self-Organizing Maps*. Springer-Verlag, Berlin, 1995.
6. T. Kohonen. Emergence of invariant-feature detectors in the adaptive-subspace self-organizing map. *Biological Cybernetics*, 75:281–291, 1996.

Model Breaking Detection Using Independent Component Classifier

Georges Linares, Pascal Nocera, Henri Meloni

e-mail: georges.linares,pascal.nocera,henri.meloni@univ-avignon.fr
C.E.R.I. 339 Chemin des Meinajariès
BP 1228-84911 Avignon Cedex 9
France

Abstract. This paper presents a neural architecture for model breaking detection in real world conditions. This technique use an Independent Component Classifier [1] for detection of unexpected or unknown events in noisy and varying environment. This method is based on subspace classifier [2] and Independant Component Analysis [3]. A feed-forward neural network adapts itself to input evolutions, by detecting novelties, creating and deleting classes. A second process achieves a prototype rotation in order to minimise mutual information of different classes. This synaptic weight evolution rule is based on an anti-hebbian learning rule inspired from neural methods for blind separation of sources [4]. Consequently, under the assumption of statistical independence of different classes, the system is able to detect novelties hidden by simultaneous acoustic events.

Novelty detection performances in various situations have been tested : isolated novelty, novelty which occurs mixed with an event of a known class, and several simultaneous novelties. We have also studied the evolution of detection performances obtained by varying the noise level. These experiments have shown good detection performances and low false detection rate.

1 Introduction

Several methods for model breaking detection have been developped over the last few years. These methods were based on low level acoustic signal processing. Breaking of signal features regularity was the criterion used [5]. Such a predictability and regularity assumption may be right in highly constrained context. Unfortunately, real world acoustic environments are generally composed of several independant sources of acoustic events. Therefore, the acoustic signal is a mixture of several components, and disruptive acoustic events detection must be achieved by using higher level processing. On the other hand, real world environments are unpredictable, and modeling acoustic contexts by explicit knowledge based systems or by supervised learning models is difficult because of a large noise and context variability. In this paper, we propose a self-organized neural method for model breaking detection in real world acoustic environments.

2 Principle

We use a neural classifier for modeling acoustic environment. Modeling is achieved by on-line clustering of the different acoustic components, using an ICC classifier. Such a network is able to dynamically learn and update a classification of input stimuli with minimum assumptions about event patterns and statistical acoustic flow features. A self-organization process allows to adapt the model to input evolutions. This adaptation is made on-line, by creating and deleting classes, and by prototypes evolution. We use the bad clustering of an acoustic event as a disruption criterion. Only the first occurence of a recurrent event must be detected as disrupting, but the other occurences will not be considered as breaking.

3 Architecture

ICC neural network has two fully inter-connected layers. The input layer receives the coefficients of stimuli vectors. The output layer has one cell per class, and another for novelty. Each class is represented by a prototype P_{it}, an instantaneous activity and an inertia $I_i(t)$.

3.1 Transfer function

Cell activities are computed by the projection of stimuli vectors into the prototype space. It is assumed that class prototypes are linearly independent. This assumption is implicitly respected in the new class acquisition stage. The output vector A_t is computed at time t by :

$$\mathrm{A_t = P_t^+(t)V_t}$$

where $V(t)$ is the stimulus vector, P the prototype matrix (each column of P is a prototype vector), and P^+ the pseudoinverse of P. Consequently, the weight matrix is the pseudoinverse of the prototype matrix.

$$\mathrm{W_t = P_t^+}$$

A situation of simultaneous class activations is interpreted as the simultaneous presence of different event classes.

3.2 Novelty cell activation

The novelty cell activity is computed by:

$$\mathrm{act_0(t) = \frac{||V_t - V_{ct}||}{||V_t||}}$$

where V_{ct} is the component of V_t inside the prototype space.

Therefore, the input space is divided into two subspaces : the prototypes subspace, which is the subspace of known acoustic vectors, and its complementary subspace. When a stimulus is significantly into the ***unknown*** subspace, a model breaking is detected. The system adapts itself to this novelty detection by creating a new class which models the new cluster of the acoustic events occurred.

3.3 Inertia

The class inertia is computed from the temporal signal of the cell activity by a classical alpha-beta filter:

$$\mathrm{I_i(t+dt)} = \beta||\mathrm{act_i(t)}|| + (1-\beta)\mathrm{I_i(t)}$$

The choice of parameter β determines the persistence of the network's memory. This rule leads to delete low activity classes. Such low activation is due to low class representativness, or to the disappearance of a class of events.

4 Dynamic classification learning

Initially, the system is empty, so there is no known class. A new class is created when novelty cell activation exceeds a fixed vigilance threshold. In this case, the new class prototype is the input vector. The system permanently scans class inertias. If one of them is lower than a deletion threshold, then the class will be killed. This on-line class integration and deletion process induces a stabilisation of subspace dimension. Unfortunately, such an adaptation process is not sufficient for good detection and classification of mixed acoustic events : if several disrupting events occur at the same time, then a new class is created with mixed spectral patterns as a prototype. Such a situation can leads to high correlation between class activities. Consequently, the system is reorganized in order to recover original events from their mixtures. The synaptic weight evolution rule is based on the minimisation of the statistical class dependence. This problem is similar to a blind source seperation problem. Therefore we use a neural source separation method inspired from [4] for prototypes reorganization. This method is able to learn a decorrelation operator C_t from linear mixtures of signals. We can easily show that the application of the decorrelation operator C_t to the classifier's outputs is equivalent to a prototype rotation:

$$S(t) = C_t.W_t^+V(t) = (W_tC_t^{-1})^+V(t)(1)$$

where $S(t)$ is the vector of uncorrelated outputs at time t, W_t the wheight matrix, $V(t)$ the input pattern, and C_t the decorrelation operator.

The last equation shows that we can use the uncorrelation operator learned with classifier outputs as a prototypes evolution rule. This process improves modeling quality in real environments, and then improves the ability to detect novelties in complex situations, such as simultaneous novelties or novelty masked by a known class event.

5 Experiments

In order to evaluate our system, we have mixed two sequences of recurrent transients and an underwater background noise. The first is a rather low frequency transient, in the high-energy background noise domain. Therefore, it is more difficult to detect it (and also more difficult to recognize it in the spectrogram).Figures 1 to 5 shows on their first line the signal spectrogram obtained by a FFT computed in a sliding temporal window. There is a vector of 256 coefficients for each 10 ms. The second line shows the novelty cell activity. New classes are detected from the local maximums of that curve. The third line represents the prototype space dimension relatively to the input space dimension. The other lines shows most meaningful class activities. Detection performances have been evaluated in several scenarios : classes never simultaneous (figure 1), new class apparition mixed with an event of an existing class (figure 2), several simultaneous novelties (figure 3). For these three situations, novelties were correctly detected, and there was no false detection. We have observed a good robustness when increasing the level of background noise (figure 4). Nevertheless, these results are dependent on the frequency distribution of noise in relation to acoustic event features. A detection delay can be observed if the level of background noise is even more increased (5).

6 Conclusion and future prospects

Our tests have validated the principle of the proposed system for novelty detection. Model breaking in *cocktail* situations has been correctly detected. Therefore, some difficulties must be overcome in order to apply this method to the modeling of high temporal structure acoustic events. We are currently working on this problem.

References

1. G. Linares, P. Nocera, and H. Meloni. Mixed acoustic events classification using subspace classifier and ica. In *Proc. 1997 IEEE Int. Conf. on Acoustics, Speech, and Signal Processing*, 1997.
2. T. Kohonen. *Self organisation and Associative Memory*. Springer Series in Information Sciences, third edition, 1989.
3. J. Karhunen and J. Joutsensalo. Representation and separation of signals using nonlinear pca type learning. *Neural Networks*, 7(1):113–127, 1994.
4. K. Matsuoka, M. Ohya, and M. Kawamoto. Neural net for blind separation of nonstationary signals. *Neural Networks*, 8(3):441–419, 1995.
5. W.Y. Liu, I. Magnin, and G. Gimenez. Opérateur pour la détection de rupture dans des signaux bruités. *Traitement du Signal*, 12, 1995.

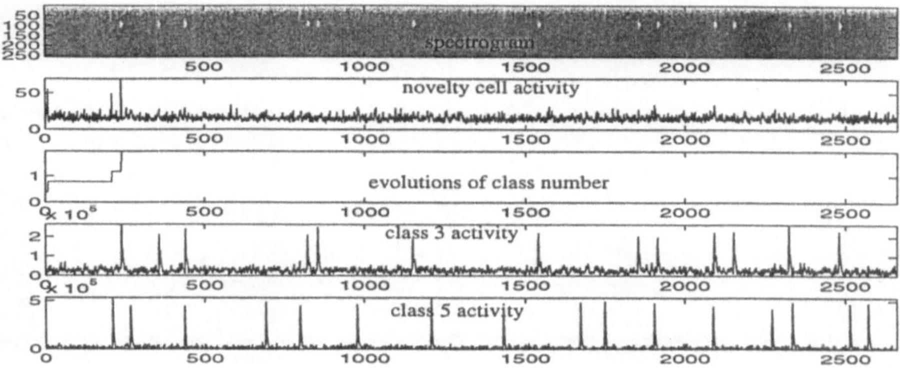

Fig. 1. *Detection of new classes of asynchronous events (0dB)*

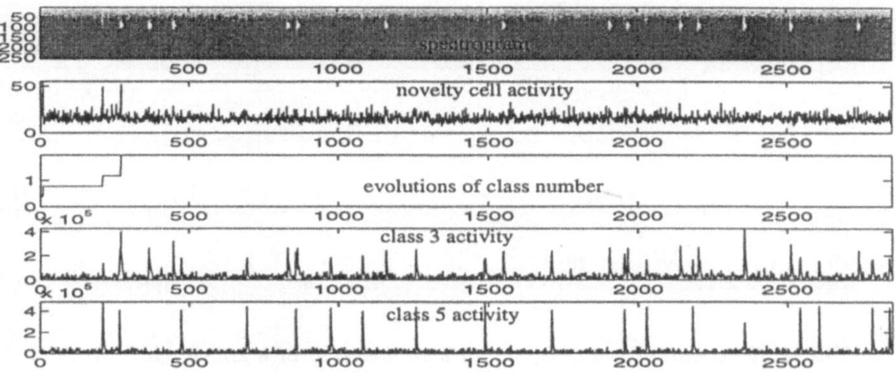

Fig. 2. *Detection of a novelty mixed with an event of a known class (0 dB)*

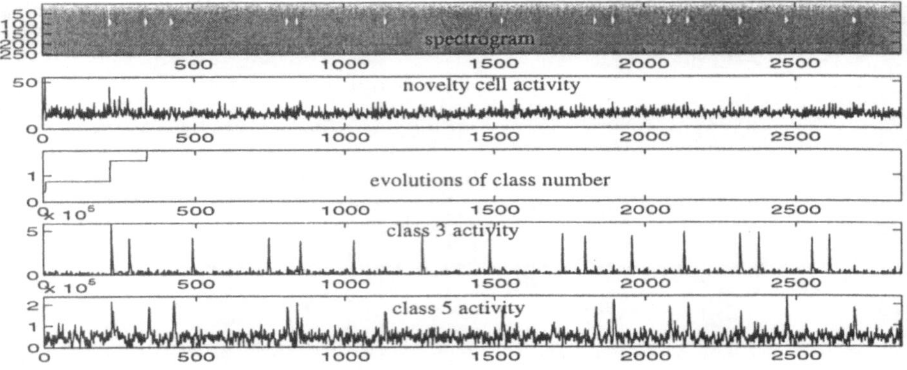

Fig. 3. *Detection of simultaneous novelties (0dB)*

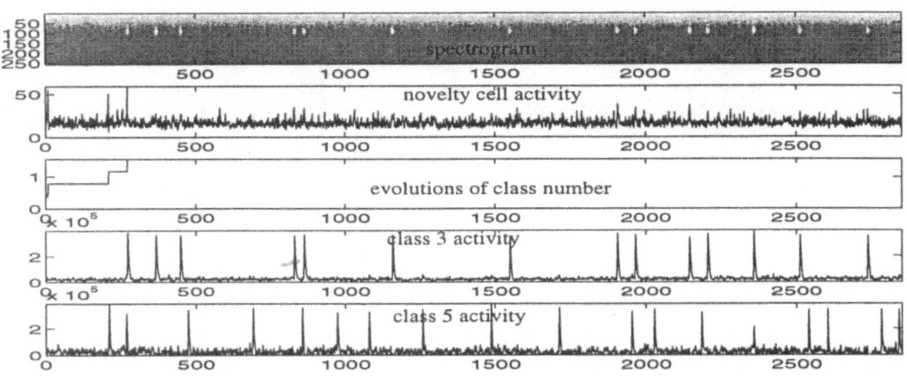

Fig. 4. *Detection of a novelty mixed with an event of a known class at -3dB*

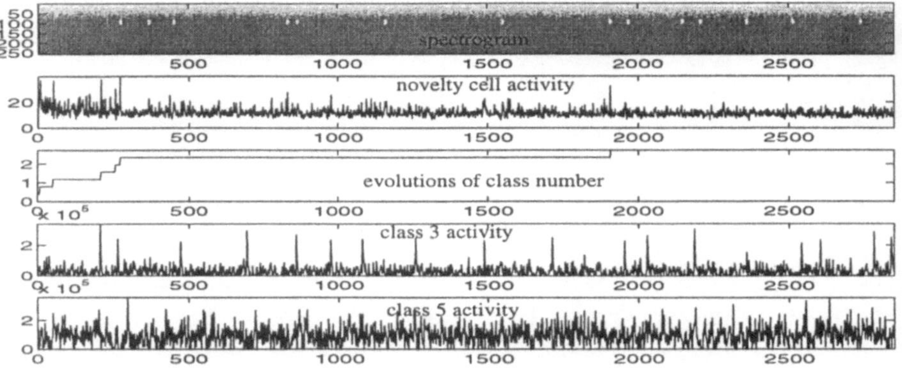

Fig. 5. *Detection of a novelty mixed with an event of a known class at -6dB*

Neural Network Based Processing for Smart Sensors Arrays

A. Paraschiv-Ionescu, C. Jutten* and G. Bouvier

INPG-TIRF, 46 avenue Félix Viallet, 38031 Grenoble Cedex, France

Abstract. Source separation (SS) algorithm is an attractive approach for designing smart sensor array, able to increase spatial selectivity and to cancel spurious sources. The source number being unknown and able to vary, a pre-processing algorithm is developped in this paper for providing estimation of the source number before source separation. On-line source separation is then achieved in the above time variant context.

1 Introduction

During the last ten years, source separation (SS) has been intensively developped as a new method in signal processing, with numerous applications in antenna processing, speech enhancement, biomedical signal processing, etc. In the general framework of data analysis, the method yields to the new concept of Independent Component Analysis (ICA) [1], which has been formalized first by Comon [2], then by Bell and Sejnowski [3], Oja and Karhunen [4], especially in the neural network community. In fact, SS and ICA adaptive algorithms may also be viewed as unsupervised neural learning, based on independence criterion.

Currently we are studying application of source separation for providing smart sensor arrays. The idea is to implement source separation algorithm after a sensor array in order to enhance selectivity and to reduce dependence on spurious sources. For instance, the output of silicon magnetic field sensor (Hall effect) depends on the various magnetic sources in the neighbourhood and on the temperature sources. Moreover, the sources can be mobile, or intermittent; hence, the number of sources is unknown and time varying.

This point precludes from directly implementing SS algorithms, which require the knowledge of the source number. We propose to insert, just after the sensor array, an adaptive neural block for estimating on-line the source number (SN).

The model of the mixtures is described in section 2. Section 3 explains the principles of SN adaptive block and briefly recalls the adaptive SS algorithm which is used. Section 4 presents experimental results.

* Ch. Jutten is professor at ISTG of Univ. Joseph Fourier of Grenoble. This project is partly supported by ELESA in cooperation with A. Chovet and A. Ionescu of LPCS.

2 Model of mixtures

The sensor array provides a n-component observation $\mathbf{x}(\mathrm{t})$, which is assumed to be an unknown linear mixture of m unknown sources $\mathbf{s}(\mathrm{t})$:

$$\mathbf{x}(t) = \mathbf{A}\mathbf{s}(t) + \mathbf{n}(t) \tag{1}$$

where $\mathbf{A}$ (the mixing matrix) is a full rank $n \times m$ matrix ($m \leq n$) with real scalar entries, $\mathbf{s}(t) = [s_1(t), ..., s_m(t)]^T$ is the vector of unknown sources, and $\mathbf{n}(\mathrm{t})$ is an additive noise vector. The sources $s_i(t), i = 1, .., m$, are assumed to be mutually independent, with at most one Gaussian source.

If the source number m is known, $\mathbf{x}(\mathrm{t})$ can be transformed in a m-component vector $\mathbf{z}(\mathrm{t})$ by a linear transform $\mathbf{W}$ (principal component analysis or PCA):

$$\mathbf{z}(t) = \mathbf{W}\mathbf{A}\mathbf{s}(t) + \mathbf{W}\mathbf{n}(t) \tag{2}$$

where $\mathbf{WA}$ is now a square regular $m \times m$ matrix. Then, the source separation is obtained by estimating a separating $m \times m$ matrix $\mathbf{B}$:

$$\mathbf{y}(t) = \mathbf{B}\mathbf{z}(t) = \mathbf{B}(\mathbf{W}\mathbf{A})\mathbf{s}(t) + \mathbf{B}\mathbf{W}\mathbf{n}(t), \tag{3}$$

where $\mathbf{y}(t) = [y_1(t), ..., y_m(t)]^T$, the output vector, satisfies $\mathbf{B}(\mathbf{WA}) = \mathbf{DP}$, where $\mathbf{D}$ is a diagonal matrix and $\mathbf{P}$ is a permutation matrix.

3 Cascaded architecture for SN estimation and SS

Many SS algorithms are particularly efficient if the number of sources is equal to the number of sensors. Then, the pre-processing (2), which needs the estimation of the source number is very relevant. In the following, a source separation algorithm with on-line estimation of source number is proposed. The general schematics of this method is presented in *figure 1*.

3.1 On-line estimation of the number of sources

Eigen-decomposition of the data covariance matrix. Let us assume in (1) that the signal and the noise are uncorrelated i.e. $E\{\mathbf{s}(t)\mathbf{n}(t)^T\} = 0$, and that the noise components $n_i(t)$ are uncorrelated zero-mean Gaussian with the same unknown power: $E\{\mathbf{n}(t)\mathbf{n}(t)^T\} = \sigma_n^2\mathbf{I}$. Then the data covariance matrix can be expressed as:

$$\mathbf{C}_x = E\{\mathbf{x}(t)\mathbf{x}(t)^T\} = \mathbf{A}\mathbf{C}_s\mathbf{A}^T + \sigma_n^2\mathbf{I} \tag{4}$$

where $\mathbf{C}_x$ and $\mathbf{C}_s$ are the data and the signal source covariance matrices, respectively.

If we assume that r is the source number ($r < n$), then the eigenvalues of $\mathbf{C}_x$ can be ordered as:

$$\underbrace{\lambda_1 \geq \lambda_2 \geq ... \geq \lambda_r}_{signal\ subspace} > \underbrace{\lambda_{r+1} \approx \lambda_{r+2} \approx ... \approx \lambda_n}_{noise\ subspace} = \sigma_n^2. \tag{5}$$

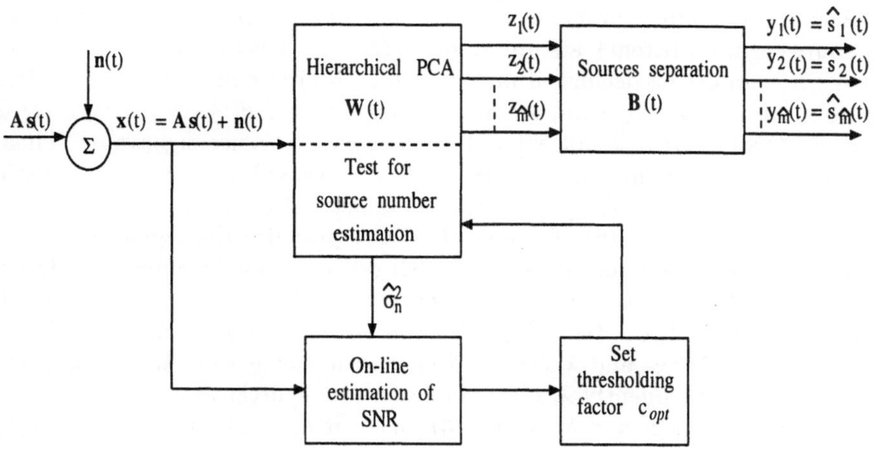

Fig. 1. Cascaded source number estimation and source separation algorithms

The first r eigenvalues correspond to the source subspace, while the remaining $n-r$ correspond to the noise subspace and must be equal to σ_n^2 because $E\{\mathbf{n}(t)\mathbf{n}(t)^T\} = \sigma_n^2\mathbf{I}$.

The method used for the estimation of SN is based on two key ideas:

• on-line estimation $\lambda_i(t)$ of the eigenvalues, λ_i, in a decreasing order like in (5), and of the eigenvectors, $\mathbf{w}_i$, of the data covariance matrix $\mathbf{C}_x$, can be obtained by the *robust, hierarchical neural network Principal Component Analysis (PCA)* using a *deflation* technique [5].

• to avoid computation of $\lambda_{r+2}, ..., \lambda_n$, taking in account that $E\{\mathbf{n}(t)\mathbf{n}(t)^T\} = \sigma_n^2\mathbf{I}$, we estimate $\lambda_{noise} = \sigma_n^2 \approx E\{\|\mathbf{x}_{r+2}\|^2/(n-r-1)\}$, where $\mathbf{x}_{r+2}(t)$ results from the original sample vector $\mathbf{x}(\mathrm{t})$ after $r+1$ deflation steps: then, it is the projection of $\mathbf{x}(\mathrm{t})$ into the noise subspace. The motivation for computing λ_{noise} by noise power averaging is that this estimation is proved to be statistically efficient (the method achieves the Cramér-Rao bound) [6]. An adaptive estimation $\lambda_{noise}(\mathrm{t})$ is obtained by a first order estimation with a forgetting factor $0 < \beta < 1$:

$$\lambda_{noise}(t+1) = \beta\lambda_{noise}(t) + (1-\beta)(\|\mathbf{x}_{r+2}(t)\|^2/(n-r-1)). \tag{6}$$

Adaptive threshold method for source number estimation. The estimation of SN will be done by comparing the eigenvalues (5) computed by PCA and (6). Using the asymptotic Gaussian properties of the least squares estimators [7], we implement a statistical hypothesis test based on a *Neyman-Pearson criterion* [8]. With this criterion, the false alarm probability is fixed to $P_f = \alpha$ and the detection probability is maximized. In practice, the test is: *if* $\lambda_i \geq \theta$ *then* λ_i *is an eigenvalue related to a source* $s_i(t)$. The test threshold varies according to (6): $\theta(t) = c(\alpha)\lambda_{noise}(t)$, where $c(\alpha)$ is the solution of an integral equation.

Because the source number can change (if a source appears or disappears, or if one source is intermittent), assuming that $\hat{m}(t-1) = r$ is the previous estimation of the SN, it must be decided if $\hat{m}(t) = r+1$, $\hat{m}(t) = r$ or $\hat{m}(t) = r-1$. Using (5), this can be done by comparing $\lambda_r(t)$, $\lambda_{r+1}(t)$ and $\theta(t)$. In (5), $\lambda_{r+1}(t)$ is a noise eigenvalue but it is not entered in the averaged noise eigenvalue. Thus, if at time $t-1$, the estimated SN is $\hat{m}(t-1) = r$, we will compute $r+1$ distinct eigenvalues $\lambda_1(t) \geq \lambda_2(t) \geq ... \geq \lambda_{r+1}(t)$ using PCA, and the noise averaged eigenvalue $\lambda_{noise}(t)$ by (6). Because the PCA provides the eigenvalues in the decreasing order, the hypothesis test for SN estimation is formulated as follows:

- $\lambda_r(t) > \theta(t)$ *true* and $\lambda_{r+1}(t) > \theta(t)$ *false*: in this case, the test states that $\lambda_r(t) > \lambda_{r+1}(t)$ and $\lambda_{r+1}(t) \approx \lambda_{noise}(t)$; consequently $\hat{m}(t) = r$,
- $\lambda_r(t) > \theta(t)$ *true* and $\lambda_{r+1}(t) > \theta(t)$ *true*: it means that $\lambda_r(t) \geq \lambda_{r+1}(t) > \lambda_{noise}(t)$, and consequently $\lambda_{r+1}(t)$ is related to a source; we set $\hat{m}(t) = r+1$,
- $\lambda_r(t) > \theta(t)$ *false* and $\lambda_{r+1}(t) > \theta(t)$ *false*: it means that $\lambda_r(t) \approx \lambda_{r+1}(t) \approx \lambda_{noise}(t)$; then $\lambda_r(t)$ is related to noise and we set $\hat{m}(t) = r-1$. The fourth case is impossible because of the decreasing computation of eigenvalues (5). The statistical analysis of this test used for SN estimation shows that the accuracy of the detection depends on the optimal $c(\alpha)$ selected value. A qualitative guideline, obtained by experiments, for pertinent $c(\alpha)$ setting in practical applications can be the following: $c_{opt}(\alpha)$ increases as (i) n increases, (ii) r decreases, (iii) SNR increases, (iv) β decreases ($c_{opt}(\alpha) = f(n, r_{max}, \beta, SNR)$). The criteria (i)-(iv) are extremely useful for on-line estimation of the number of the sources for sensor array applications. For example, a given application is characterized by the sensor number n, a maximum anticipated source number r_{max} and a possibly variable SNR level. Thus, the on-line estimation of SNR and of $c(\alpha)$ improves substantially the probability of detection.

3.2 Source separation

Once the source number $\hat{m}$ is estimated, we can perform the source separation using the first $\hat{m}$ principal components of the data covariance matrix. For a sensor number greater that the source number ($n > \hat{m}$), the PCA algorithm performs an adaptive projection into the signal subspace maximizing the SNR of the signals fed to the source separator. In the experiments, we choose the source separation algorithm of Cardoso's, for its nice equivariant properties [9]:

$$\mathbf{B}(t+1) = \mathbf{B}(t) - \lambda\left[\mathbf{y}(t)\mathbf{y}(t)^T - \mathbf{I} + g(\mathbf{y}(t))\mathbf{y}(t)^T - \mathbf{y}(t)g(\mathbf{y}(t))^T\right]\mathbf{B}(t) \quad (7)$$

where $g(\mathbf{y}(t)) = [g(y_1(t)), ..., g(y_{\hat{m}}(t))]^T$, and $g(.)$ is a non-linear function. Equivariance means that the algorithm performance does not depend on the mixing matrix $\mathbf{A}$. This property is very attractive in the smart sensor array application where the mixing matrix can be ill-conditionned, because of the sensor proximity.

4 Simulation results

As an example, a simulation with a 6-sensor array is considered. The source number varies between 2 to 4: 2 sources are permanent and 2 are mobile or temporary. The *mobility* of the sources is simulated by varying the entries of the mixing matrix **A**. *Figure 2* shows the mixed signals whith additive noise (outputs of the sensor array). The estimated signal eigenvalues $\lambda_i(t)|i = 1,\ldots 4$, using the PCA algorithm, and the averaged noise eigenvalue $\lambda_{noise}(t)$ using (7), with a forgetting factor $\beta = 0.985$, are presented in *figure 3*. These results are used for estimating the source number whith $c(\alpha) = 1.41$ (*figure 4*). After detection, the first $\hat{m}$ principal eigencomponents are used in order to separate the sources. Typical results corresponding to data given in *figure 2* are presented in *figure 5*.

5 Conclusions

In this paper, we propose the cascade of a source number estimation algorithm and of a source separation algorithm for designing smart sensor arrays. These algorithms can be implemented as unsupervised adaptive neural algorithms and allows efficient source separation even if the source number is unknown or is varying. Currently, we are studying accuracy requirement and different hardware architectures for on-chip implementation of the cascade.

References

1. Jutten, C., Hérault, J.: Independent components analysis (INCA) versus principal components analysis. EUSIPCO'88, 643-646
2. Comon, P.: Independent Component analysis - a new concept?. Signal processing, vol.**36**, (1994), 287-314
3. Bell, A., Sejnowski, T.: An information-maximization approach to blind separation and blind deconvolution. Neural Computation, vol.**7**, (1995), 1129-1159
4. Oja, E., Karhunen, J., Hyvärinen, A.: From neural PCA to neural ICA. NISP'96 Postconference Workshop on Blind Signal Processing and Their Applications, Snowmass, Colorado (1996)
5. Yang, B.: An extension of the PASTd algorithm to both rank and subspace tracking. IEEE Signal Processing Letters, vol.**2**, no.**9**, (1995), 179-182
6. Stoica, P., Söderström, T., Šimonnytè, V.: On estimating the noise power in array processing, Signal Processing, vol.**26**, no. **2**, (1992), 205-221
7. Anderson, T. W.: Asymptotic theory for principal component analysis. Ann. J. Math. Stat., vol.**34**, (1963), 122-148
8. Scharf, L.:Statistical Signal Processing
9. Cardoso, J. F., Laheld, B.: Equivariant adaptive source separation. IEEE Tran. on Signal Processing, vol.**44**, no.**12**, (1996), 3017-3031

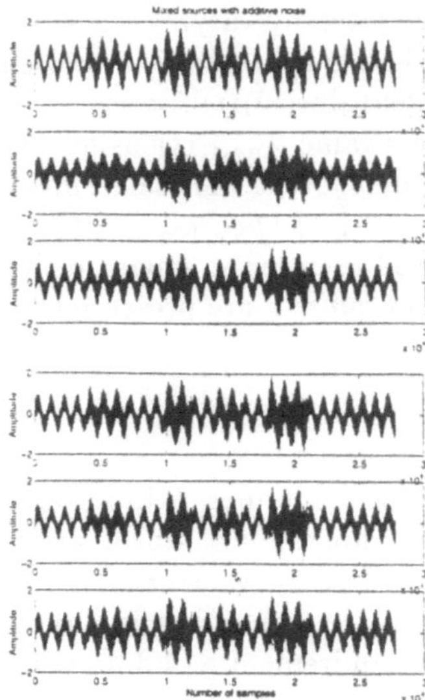

Fig. 2. Sensor array outputs

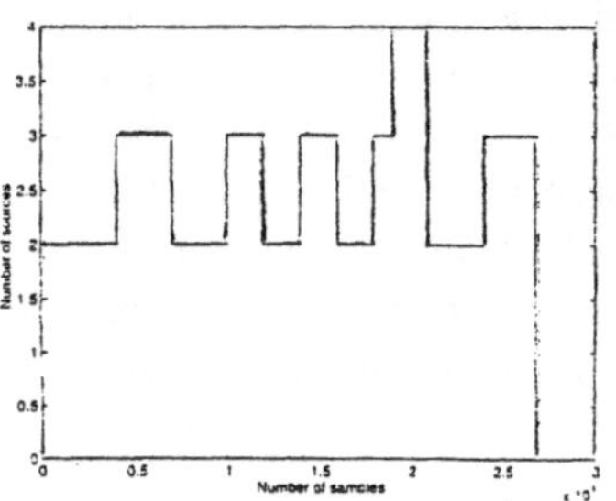

Fig. 4. Estimated source number, $\hat{m}(t)$, as a function of time (number of samples).Two permanent sources and two temporary sources are observed.

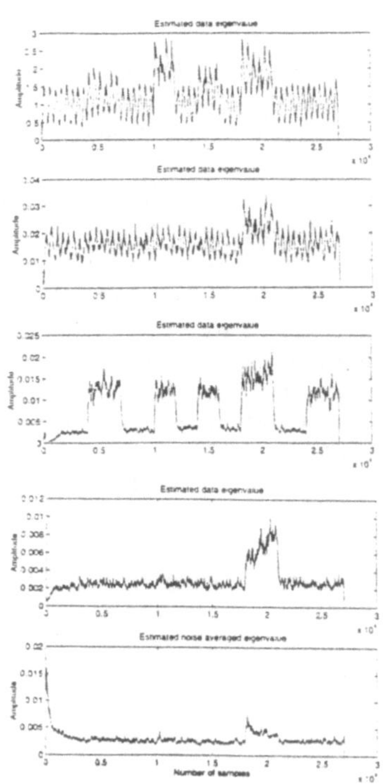

Fig. 3. From top to bottom, estimated source eigenvalues, $\lambda_i(t)$, $i = 1, \ldots, 4$, and estimated noise eigenvalue λ_{noise}

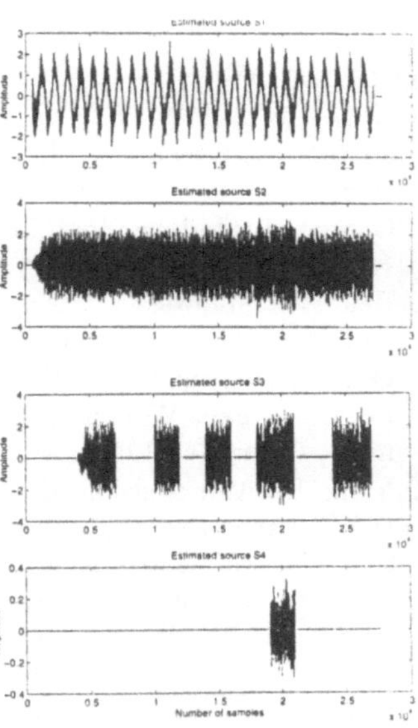

Fig. 5. Estimated source signals, $y_i(t), i = 1, \ldots, 4$. Sources $S1$ and $S2$ are permanent and $S3$ and $S4$ are temporary (see fig. 4).

Application of the MEC Network to Principal Component Analysis and Source Separation

Simone Fiori, Aurelio Uncini, and Francesco Piazza *

Dipartimento di Elettronica e Automatica
Università di Ancona
Via Brecce Bianche, 60131, An (Italy)
E-mail:simone@eealab.unian.it

Abstract. In this paper we present new developments of a previous work dealing with the problem of strongly-constrained orthonormal analysis of random signals. In the former work a neural learning rule arising from the study of the dynamics of a massive system in an abstract space was introduced, and the set of equations describing the motion of such a system was directly interpreted as a learning rule for neural layers. This adaptation rule can be used to solve several problems where orthonormal matrices are involved. Here we show two applications of such an approach: one dealing with PCA and one dealing with ICA.

1 Introduction

In a previous paper [3], we presented a new class of neural unsupervised learning rules which arises from the analysis of the dynamics of an abstract mechanical rigid system of masses m_i. The set of equations describing the motion of such system can be directly interpreted as a learning rule for linear as well as non-linear neural layers. The corresponding algorithm can be used to solve several Digital Signal Processing problems, where orthonormal matrices are involved. In the work [3] we found that such equations are:

$$\frac{d\boldsymbol{W}}{dt} = \boldsymbol{\Omega}\boldsymbol{W} \ , \ \boldsymbol{P} = -\mu\boldsymbol{\Omega}\boldsymbol{W} \ , \tag{1}$$

$$\frac{d\boldsymbol{\Omega}}{dt} = \frac{1}{4}[(\boldsymbol{F}+\boldsymbol{P})\boldsymbol{W}^T - \boldsymbol{W}(\boldsymbol{F}+\boldsymbol{P})^T] \ , \tag{2}$$

where $\boldsymbol{W}$ is the $p \times m$ matrix whose columns represent the instantaneous position of the masses, $\boldsymbol{\Omega}$ is a skew-symmetric $p \times p$ matrix which plays the role of an angular speed in the mechanical model and matrix $\boldsymbol{P}$ describes the braking effect due to the presence of an hysotropic and homogeneous fluid, endowed with a viscosity coefficient μ, within the space. Here we suppose masses m_i all equal to 1.

* This research was supported by the Italian MURST. Please send comments and suggestions to the first author.

Since the mechanical system is rigid, the property $\boldsymbol{W}^T\boldsymbol{W} = \boldsymbol{I}$ holds at any time, providing that it initially held. Notice that such an approach provides an 'exact' version of a similar one proposed by Cardoso et al. in [4].

Matrix $\boldsymbol{F}$ contains, as its columns, forcing terms which cause the global motion of the system. Since this system has a point fixed in space, it can only instantaneously rotate around this point. We suppose the forcing terms derive from a Potential Energy Function (PEF) U, that means $\boldsymbol{F} = -2\frac{\partial U}{\partial \boldsymbol{W}}$.

Here we interpret $\boldsymbol{W}$ as the weight-matrix of a linear neural layer described by $\boldsymbol{y} = S[\boldsymbol{W}^T\boldsymbol{x}]$, where $\boldsymbol{x} \in \mathrm{IR}^p$, $\boldsymbol{y} \in \mathrm{IR}^m$ and $S[\cdot]$ is a $m \times m$ generic non-linear operator. Equations (1-2) give a learning rule for that layer.

We called such a system the *MEC Network*. By choosing different PEFs we can force the system (hence the algorithm) to perform several different motions (hence several tasks). Recalling that a (dissipative) mechanical system reaches the equilibrium when its own potential energy function U is at its minimum (or local minima), we can assume $U := +J_{\mathrm{C}}$, with J_{C} a cost function to be minimized, or $U := -J_{\mathrm{O}}$, where J_{O} is an objective function to be maximized, both *under the constraint of orthonormality.* (Such statements will be clarified with examples in Section 3.)

In this paper we present two applications of the MEC algorithm: one dealing with the problem of Principal Component Analysis (PCA) [2] and one dealing with Blind Linear Separation of Sources (ICA) [1]. Since we deal with linear applications we assume $S[\cdot]$ equal to the identity operator.

2 Mechanical Model Discretization

In order to implement the MEC algorithm we need a discretization of the continuous-time equations (1-2).

Such operation must be performed with some cautions, since any abrupt replacement of derivatives with simple incremental ratios may yields the substantial loss of orthonormality of the weight matrix $\boldsymbol{W}$. In this section we present a simple way to perform discretization which gives good results, i.e. that allows to maintain the orthonormality of the columns of $\boldsymbol{W}$ with a degree of accuracy as good as one needs.

Actually, the expression (2) can be discretized in the usual manner because in this way the fundamental property of the angular speed matrix $\boldsymbol{\Omega}$ is preserved: it remains always skew-symmetric. With this convention we can write: $\Delta\boldsymbol{\Omega} = \frac{\theta}{4}[(\boldsymbol{F}+\boldsymbol{P})\boldsymbol{W}^T - (\boldsymbol{F}+\boldsymbol{P})^T]$, where θ is the "sampling period" that, in a neural artificial context, plays the role of a learning rate for the matrix $\boldsymbol{\Omega}$.

Now, observing in the continuous-time context the above discrete system, through the time we find a piece-wise constant evolution of $\boldsymbol{\Omega}(t)$, that means $\boldsymbol{\Omega}(t) = \boldsymbol{\Omega}(n\theta)$ for $n\theta < t \leq (n+1)\theta$, indicating with n the discrete temporal index. Within each open temporal interval $]n\theta, (n+1)\theta]$, the matrix $\boldsymbol{\Omega}(t)$ remains constant and equal to $\boldsymbol{\Omega}_n := \boldsymbol{\Omega}(n\theta)$, then, the first of equations (1) can be solved exactly within each of these intervals, in fact it can be rewritten for $n > 1$ as $\frac{d\boldsymbol{W}}{dt} = \boldsymbol{\Omega}_n\boldsymbol{W}$ for $t \in]n\theta, (n+1)\theta]$, with $\boldsymbol{W}(n\theta)$ known from calculus for

$t \in](n-1)\theta, n\theta]$. It is well-known that the exact solution of the above differential problem is $\boldsymbol{W}(n+1) = e^{\boldsymbol{\Omega}_n \theta}\boldsymbol{W}(n)$ also known as *intersample behavior* (IB) solution.

From a computational point of view, the exponential matrix $exp(\boldsymbol{\Omega}_n \theta)$ is not easily valuable, but we can of course approximate it by using one of the well-known methods to be found in literature. We chose to use the following approximation $exp(\boldsymbol{\Omega}_n\theta) \cong \sum_{k=0}^{r} \frac{\boldsymbol{\Omega}_n^k}{k!}\theta^k$ with $r \in \mathbb{N}^+$ bounded. By introducing such an approximation in (IB) we obtain the following set of equations describing approximately the dynamics of the system:

$$\Delta \boldsymbol{W} = \left[\sum_{k=1}^{r} \frac{\boldsymbol{\Omega}_n^k}{k!}\theta^k\right] \boldsymbol{W} \ , \boldsymbol{P} = -\mu \boldsymbol{\Omega} \boldsymbol{W} \ , \tag{3}$$

$$\Delta \boldsymbol{\Omega} = \frac{\theta}{4}[(\boldsymbol{F}+\boldsymbol{P})\boldsymbol{W}^T - \boldsymbol{W}(\boldsymbol{F}+\boldsymbol{P})^T] \ , \tag{4}$$

with $\theta > 0$, $\mu > 0$ and $r \in \mathbb{N}^+$ fixed, where all quantities are intended to be evaluated at the same step n.

3 Experimental Results

In the following subsections we show two applications of the MEC algorithm dealing with PCA and ICA.

3.1 Application to Principal Component Analysis (PCA)

In this experiment we try to extract the first two Principal Components (PCs) from a signal $\boldsymbol{x}$ of dimension three. Since $\boldsymbol{x}$ is obtained by mixing the three components s_1, s_2, s_3 of a random signal $\boldsymbol{s}$ with $\boldsymbol{x} = \boldsymbol{Q}\boldsymbol{s}$ and with each s_i being an uniformly distributed random signal such that $s_1 \in [-4, +4]$, $s_2 \in [-2, +2]$ and $s_3 \in [-1, 1]$, the first two PC of $\boldsymbol{x}$ are, respectively, the first and the second column of $\boldsymbol{Q}$, where $\boldsymbol{Q} = [1/\sqrt{2}, -1/\sqrt{2}, 0; 1/\sqrt{3}, 1/\sqrt{3}, 1/\sqrt{3}; 1/\sqrt{6}, 1/\sqrt{6}, -2/\sqrt{6}]$,

It is well-known [2] that PCA arises from the maximization of the power of the transformed signal $\boldsymbol{y} = \boldsymbol{W}^T\boldsymbol{x}$ (in this case $\boldsymbol{y} \in \mathbb{R}^2$ because we are looking for an optimal reduction of dimension 2), therefore our objective function to be maximized under the constraint of orthonormality is $J_{\mathrm{O}} = a_1 E[y_1^2] + a_2 E[y_2^2]$. We have introduced two constants a_1 and a_2 to embed in the objective function a sort of ordering: if we choose $a_1 > a_2$ the signal y_1 will contain the first principal part of $\boldsymbol{x}$ while y_2 will contain the second one.

Since J_{O} must be maximized, we choose the PEF as $U \propto -J_{\mathrm{O}}$. More compactly we write $U = -\frac{1}{2}k_{\mathrm{H}}\mathrm{trace}(E[\boldsymbol{y}\boldsymbol{y}^T]\boldsymbol{A})$, where $\boldsymbol{A} = \mathrm{diag}(a_1, a_2)$ and k_{H} is a positive scaling term.

By definition of $\boldsymbol{F}$ we find that the instantaneous estimation (IE) of the force is: $\boldsymbol{F} = 2k_{\mathrm{H}}\boldsymbol{x}\boldsymbol{y}^T\boldsymbol{A}$, that we call the *Hebbianic Force* (H–Force) because it recalls the well-known Hebbianic adaptation term.

We performed several simulations with the parameter values $\mu = 3.0$, $\theta = 0.01$, $k_{\mathrm{H}} = 0.45$, $r = 3$ and $\boldsymbol{A} = \mathrm{diag}(3,1)$. The initial states for the algorithm was: $\boldsymbol{W}(0) = \boldsymbol{I}_{:,1:2}$ and $\boldsymbol{\Omega}(0) = \boldsymbol{0}$. As convergence measure, we can assume the standard Frobenius norm of the difference between the matrix $\boldsymbol{W}$ and the target, thus as $\delta(\boldsymbol{W}) := \|\boldsymbol{W} - \boldsymbol{Q}_{:,1:2}\|_F$. In Figure 1 a typical plot of $\delta(\boldsymbol{W})$ versus the temporal index is shown. Since the network was able to recover also

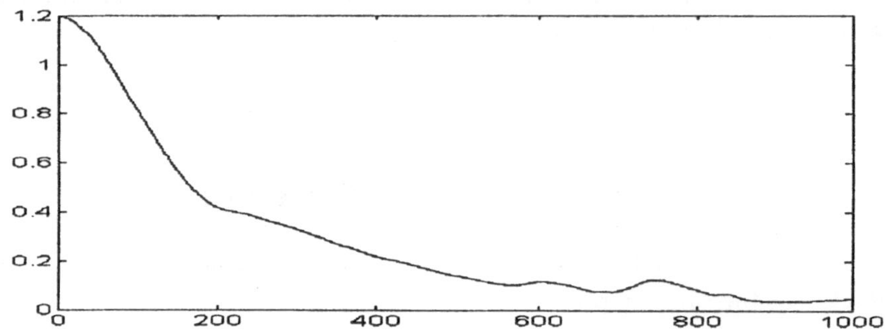

Fig. 1. A simulation result of the experiment on PCA.

the same signs of the first two columns of $\boldsymbol{Q}$, in this case the minimum value of the measure $\delta(\boldsymbol{W})$ is 0.

3.2 Application to Minor Component Analysis (MCA) and Robust Principal Subspace Analysis (RPSA)

We performed also simulations dealing with the problem of extracting the last Minor Components (MCs) [6] and the problem of estimating the Robust Principal Subspace (RPS) [5] from a random signal.

About MCA, we only observe that, by definition, the last $r < p$ MCs associated to a signal $\boldsymbol{x} \in \mathbb{R}^p$ are the counterparts of the PCs having the smallest powers. By an optimization point of view, a MCA is an *orthonormal* matrix $\boldsymbol{W}$ such that the transformed signal $\boldsymbol{y} = \boldsymbol{W}^T\boldsymbol{x}$ contains the smallest principal parts of $\boldsymbol{x}$. We can therefore extract it by finding that matrix $\boldsymbol{W}$ that minimizes $J_{\mathrm{C}} = \sum_{i=1}^{r} a_i E[y_i^2]$. In the MEC context we then assume $U := J_{\mathrm{C}}$ and therefore the IE of the force is $\boldsymbol{F} = -2\boldsymbol{x}\boldsymbol{y}^T\boldsymbol{A}$, with the same conventions of the previous section. Experimental results confirm such a theory.

About RPS estimation, it is a problem arising in many fields, see for instances [5]. The RPS is the extension of standard Principal Subspace [5] where we try to mitigate the effect of outliers, noises and disturbances on the estimate.

A possible solution to this problem can be found in [5], where a different objective function than the usual quadratic one is used. In that paper Karhunen et al. suggested the maximization of the criterion $J_{\mathrm{KJ}} := \sum_{i=1}^{m} E[f(y_i)]$, where

f is an even, non-negative, almost everywhere continuous scalar function, such that $f(t) \leq t^2/2$ with strictly inequality for large $|t|$ and non-decreasing for $t > 0$.

It is interesting to note that this kind of approach can be also used in the MEC context, and since J_{KJ} has to be maximized we can choose $U := -J_{\mathrm{KJ}}$. The associated force has the IE expression:

$$\boldsymbol{F} = 2\boldsymbol{x}G^T[\boldsymbol{W}^T\boldsymbol{x}] \ ,$$

where $G[\cdot] := \frac{df}{dt}I$. Such a force is called *KJ-Force*.

This kind of approach has been successfully applied in several experiments and has given the expected results.

3.3 Application to Linear Blind Source Separation (ICA)

As another application, we can use our algorithm to solve a simple blind separation problem with two sources ($m = 2$) and three sensors ($p = 3$). We assume the following mixing matrix: $\boldsymbol{H}^T = [1, -2; 0, 3; 1, -1]$ and we use, as entries of the source vector $\boldsymbol{s}$, firstly a pair of wide–band source signals (zero-mean white noise with flat probability density function) and then a pair of narrow–band source signals (sinusoid with arbitrarily chosen frequency), both normalized to have unity powers.

In the following we use the well-known result ([1]) whereby an ICA stage can be decomposed into two subsequent stages: a pre-whitening stage and a ICA$^{\perp}$ one, therefore the signal $\boldsymbol{z} = \boldsymbol{H}^T\boldsymbol{s}$ at the sensors can be firstly standardized and then *orthonormally separated* (ICA$^{\perp}$).

The first operation can be performed using a two-stages neural network, described by $\boldsymbol{x} = \boldsymbol{V}^T\boldsymbol{z}$, composed by a linear compressor (e.g. a PCA network) followed by a simple post-scaler which forces the powers of its outputs to be unitary. Here we used a MEC-net endowed with an H-Force forcing term, as in Subsection 3.1 as PCA net. The second operation can be performed by a MEC-net with a proper forcing term.

After whitening, source signals can be separated from $\boldsymbol{x}$ by means of a pure rotation through a linear neural transformation described by $\boldsymbol{y} = \boldsymbol{W}^T\boldsymbol{x}$.

As the signals to be separated are both platikurtic, we can use the Comonian function $J_{\mathrm{C}} := E[y_1^4 + y_2^4]$ that has to be *minimized*, therefore we just assume $U := \frac{1}{4}k_{\mathrm{C}}J_{\mathrm{C}}$. By definition the resulting active force has the IE expression $\boldsymbol{F} = -2k_{\mathrm{C}}\boldsymbol{x}\mathrm{Cb}^T[\boldsymbol{W}^T\boldsymbol{x}]$, where the $\mathrm{Cb}[\cdot]$ operator is defined by $\mathrm{Cb}([a \ b \ c \ \cdots]^T) := [a^3 \ b^3 \ c^3 \cdots]^T$.

We ran the algorithm with the following parameters values: $\boldsymbol{A} = \mathrm{diag}(3,4)$, $\mu_{PCA} = 5.0$, $\mu_{ICA} = 4.0$, $\theta_{PCA} = 0.1$, $\theta_{ICA} = 0.01$, $k_{\mathrm{H}} = 0.02$, $k_{\mathrm{C}} = 0.1$ and $r_{\mathrm{PCA}} = r_{\mathrm{ICA}} = 4$. The scaler-layer was realized with a recursive low-pass filter with a learning stepsize equal to

Since the overall source-to-output matrix $\boldsymbol{R} := \boldsymbol{W}^T\boldsymbol{V}^T\boldsymbol{H}^T$ should become such that only one entry per row is not equal to zero, we took as convergence measures variables R_{ij} that we call *interference residuals*. In the Figure 2 a typical plot of these four parameters is shown. As expected, two residuals converge

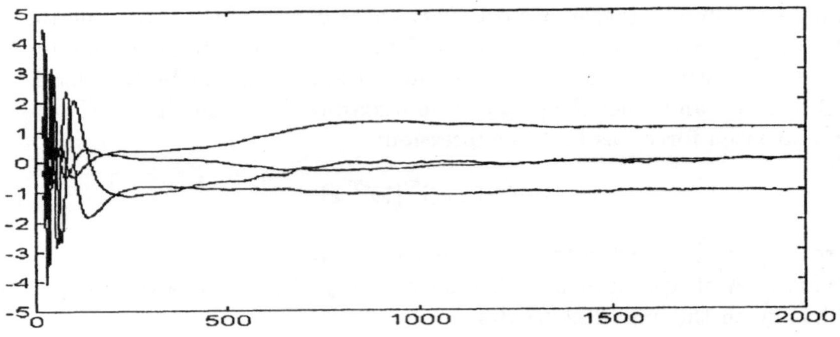

Fig. 2. A simulation result of the experiment on ICA.

to 1, while two of them vanish. Notice that the global network converges after that all layers have converged. This explains why residuals are quite big at the beginning of learning in Figure 2.

4 Conclusions and further work

In [3] we presented a new unsupervised neural learning rule based on the equations describing the dynamics of an abstract rigid system of masses, that we called MEC. There we pointed out that this theory can be used in the neural context to solve several Signal Processing problems, where orthonormal matrices are involved. In the present paper, through simple examples concerning PCA, MCA, PSA and ICA, we have shown the effective usefulness of our approach. Particularly, from examples emerges how MEC's analysis is fast and accurate.

In a forthcoming paper we will present real-word applications of the MEC to vocal signals and images in comparison with other algorithms.

References

1. Comon P.: Independent Component Analysis, A New Concept ? Signal Processing **36** (1994) 287 – 314
2. Diamantaras K.I., Kung S.-Y.: Principal Component Neural Networks: Theory and Applications. J. Wiley, 1996
3. Fiori S., Uncini A., Piazza F.: A New Unsupervised Neural Algorithm for Orthonormal Signal Processing. Proc. of Int. Conf. Acoustic, Speech and Signal Processing - ICASSP (1997) 3349–3352
4. Laheld B., Cardoso J.F.: Adaptive Source Separation with Uniform Performances. Signal Processing VII: Theories and Applications **1** (1994) 183 – 186
5. Karhunen J., Joutsensalo J.: Learning of Robust Principal Component Subspace. Proc. of Int. Joint Conf. on N.N. - IJCNN **3** (1993) 2409 – 2412
6. Xu L., Oja E., Suen C.Y.: Modified Hebbian Learning for Curve and Surface Fitting. Neural Networks **5** (1992) 393 – 407

Semi-Blind Source Parameter Separation

Jyrki Joutsensalo

Helsinki University of Technology, Laboratory of Computer and Information Science, Finland

Abstract. Independent Component Analysis (ICA) is a useful extension of standard Principal Component Analysis (PCA). The ICA model is utilized mainly in blind separation of unknown source signals from their linear mixtures. In some applications, the mixture coefficients are totally unknown, while some knowledge about temporal model exists. In this paper, we propose a learning system for semi-blind binary signal separation. Only second order statistics are used, and therefore the network structure is quite simple. In the experiments, the networks are succesfully applied to the CDMA (Code Division Multiple Access) mobile phone parameter estimation.

1 Introduction

Independent Component Analysis (ICA) or blind source separation is a recently developed, useful extension of standard Principal Component Analysis (PCA). The ICA model is utilized mainly in blind separation of unknown source signals from their linear mixtures. In this application only the source signals which correspond to the coefficients of the ICA expansion are of interest. The basic ICA network usually consists of whitening, separation, and basis vector estimation layers. Neural ICA has been studied e.g. in [4].

In this paper, we consider "semi-blind" source separation of the sources. Temporal structure of the data is exploited. The essential assumption of the linear data is that their *alphabet is known* but the parametric form itself is unknown. No strict assumptions about correlation or independence properties are made. A trivial assumption to avoid singularities is a noncoherence of the signals. Our approach is based on the exploitation of the symmetric properties of the linear data. The algorithms can are derived from the optimization criteria. In this paper, we concentrate to the binary signals. The binary signal separation has been studied neurally e.g. in [5].

2 ICA and semi-BSS

2.1 ICA data model

In the following, we present the basic data model used in defining standard ICA. A precise mathematical discussion of standard ICA is given in Comon's recent fundamental paper [2].

Denote by

$$\mathbf{r}_m = [r_m(1), \ldots, r_m(C)]^T \tag{1}$$

the C dimensional mth observed data vector. T denotes transpose. In the linear data model, it can be represented in the form

$$\mathbf{r}_m = \mathbf{G}\mathbf{b}_m + \mathbf{n}_m \tag{2}$$

where $C \times M$ mixing matrix

$$\mathbf{G} = [\mathbf{g}_1, \ldots, \mathbf{g}_M] \tag{3}$$

contains the ICA basis vectors, and M-vector $\mathbf{b}_m$ contains the source signals

$$\mathbf{b}_m = [b_m(1), \ldots, b_m(M)]^T \tag{4}$$

In the standard ICA, the components of $\mathbf{b}_m$ are assumed to be independent and non-Gaussian. The vector $\mathbf{n}_m$ denotes corrupting additive independent, indentically distributed (i.i.d.) noise. The noise term $\mathbf{n}_m$ is often omitted from (2) for analytical reasons, and because it is usually impossible to distinguish noise from the source signals. However, we do not drop noise out until otherwise stated.

2.2 Dual spatial and temporal subspace decompositions

Next we introduce a theory behind our SBSS (Semi-BSS) approach. Collect N data vectors $\mathbf{r}_1, \ldots, \mathbf{r}_N$ together. In the pure matrix form the representation of the data (2) is

$$\mathbf{X} = \mathbf{G}\mathbf{B} + \mathbf{N} \tag{5}$$

where

$$\mathbf{X} = [\mathbf{r}_1, \ldots, \mathbf{r}_N] \tag{6}$$

$$\mathbf{B} = [\mathbf{b}_1, \ldots, \mathbf{b}_N] \tag{7}$$

$$\mathbf{N} = [\mathbf{n}_1, \ldots, \mathbf{n}_N] \tag{8}$$

The autocorrelation matrices of the data $\mathbf{r}_m$ have the dual form

$$\mathbf{R}_g = \mathrm{E}(\mathbf{X}\mathbf{X}^H) = \mathbf{G}\mathbf{R}_\beta\mathbf{G}^H + \sigma^2\mathbf{I} \tag{9}$$

$$\mathbf{R}_b = \mathrm{E}(\mathbf{X}^H\mathbf{X}) = \mathbf{B}^H\mathbf{R}_\gamma\mathbf{B} + \sigma^2\mathbf{I} \tag{10}$$

H denotes complex conjugate transpose. Subscripts g and b refer to the basis of the signal subspace, i.e. $\mathbf{G}$ and $\mathbf{B}^H$. $\mathbf{R}_\beta = \mathbf{B}\mathbf{B}^H/N$ and $\mathbf{R}_\gamma = \mathbf{G}^H\mathbf{G}/N$ are sample autocorrelation matrices of $\mathbf{B}$ and $\mathbf{G}^H$, respectively. Thus we took the expectation only with respect to the noise. The variance of the noise is σ^2.

It is well known that the following holds: $\text{span}(\mathbf{U}_s) = \text{span}(\mathbf{G})$ and $\text{span}(\mathbf{V}_s) = \text{span}(\mathbf{B}^H)$ where $\mathbf{U}_s$ and $\mathbf{V}_s$ span the signal subspace of $\mathbf{R}_g$ and $\mathbf{R}_b$, respectively. Here we introduce theorems related to the linear forms.

Theorem 1. *Let the matrix* $\mathbf{R}$ *be symmetric and positively semidefinite. Then the solution of the equation*

$$\mathbf{W} = \mathbf{R}\mathbf{W}(\mathbf{W}^H\mathbf{R}\mathbf{W})^{-1}\mathbf{W}^H\mathbf{W} \tag{11}$$

is that the columns of $\mathbf{W}$ *span the same subspace as the principal eigenvectors of* $\mathbf{R}$.

Proof. Because $\mathbf{U}^H\mathbf{U} = \mathbf{I}$ and eigenvectors satisfy $\mathbf{R}\mathbf{U} = \mathbf{U}\mathbf{\Lambda}$, then $(\mathbf{I} - \mathbf{U}\mathbf{U}^H)\mathbf{R}\mathbf{U} = \mathbf{0}$. It was assumed that $\mathbf{W} = \mathbf{U}\mathbf{T}$, where $\mathbf{T}$ is invertible. Therefore $\mathbf{U}\mathbf{U}^H = \mathbf{W}(\mathbf{W}^H\mathbf{W})^{-1}\mathbf{W}^H$, and $[\mathbf{I} - \mathbf{W}(\mathbf{W}^H\mathbf{W})^{-1}\mathbf{W}^H]\mathbf{R}\mathbf{W} = \mathbf{0}$. After algebraic manipulation, we get Eq. (11). **Q.E.D.**

Because M principal eigenvectors of $\mathbf{R}_b$ span the same subspace as the columns of $\mathbf{B}^H$, the following corollary holds.

Corollary 1. *Let the matrix have the form* $\mathbf{R}_b = \mathbf{B}^H\mathbf{R}_\gamma\mathbf{B} + \sigma^2\mathbf{I}$, *where* $M \times N$ *matrix* $\mathbf{B}$ *and* $M \times M$ *matrix* $\mathbf{R}_\gamma$ *have full rank. Then the following holds:*

$$\mathbf{W} = \mathbf{W}\mathbf{W}^H(\mathbf{W}\mathbf{R}_b\mathbf{W}^H)^{-1}\mathbf{W}\mathbf{R}_b \tag{12}$$

where

$$\mathbf{W} = \mathbf{T}\mathbf{B} \tag{13}$$

and $\mathbf{T}$ *is a nonsingular* $M \times M$ *matrix.*

An optimality theorem is as follows:

Theorem 2. *Consider a noiseless model* $\mathbf{X} = \mathbf{G}\mathbf{B}$ *and a joint optimization problem*

$$\left.\begin{array}{rcl} \hat{\mathbf{G}} & = & \mathbf{X}\hat{\mathbf{B}}^+ \\ \hat{\mathbf{B}} & = & \hat{\mathbf{G}}^+\mathbf{X} \end{array}\right\} = \arg\min_{\mathbf{G},\mathbf{B}} ||\mathbf{X} - \mathbf{G}\mathbf{B}|| \tag{14}$$

where $\mathbf{B}^+ = \mathbf{B}^H(\mathbf{B}\mathbf{B}^H)^{-1}$ *and* $\mathbf{G}^+ = (\mathbf{G}^H\mathbf{G})^{-1}\mathbf{G}^H$ *are pseudoinverses. Then one solution for* $\hat{\mathbf{B}}$ *and* $\hat{\mathbf{G}}$ *is as follows:*

$$\hat{\mathbf{G}} = \mathbf{X}\mathbf{X}^H\hat{\mathbf{G}}(\hat{\mathbf{G}}^H\mathbf{X}\mathbf{X}^H\hat{\mathbf{G}})^{-1}\hat{\mathbf{G}}^H\hat{\mathbf{G}} \tag{15}$$

$$\hat{\mathbf{B}} = \hat{\mathbf{B}}\hat{\mathbf{B}}^H(\hat{\mathbf{B}}\mathbf{X}^H\mathbf{X}\hat{\mathbf{B}}^H)^{-1}\hat{\mathbf{B}}\mathbf{X}^H\mathbf{X} \tag{16}$$

Proof. For convenience let us drop the hats. Because $\mathbf{G} = \mathbf{X}\mathbf{B}^+$, $\mathbf{B} = \mathbf{G}^+\mathbf{X}$, $\mathbf{B}^+ = \mathbf{B}^H(\mathbf{B}\mathbf{B}^H)^{-1}$, and $\mathbf{G}^+ = (\mathbf{G}^H\mathbf{G})^{-1}\mathbf{G}^H$, straightforward calculation yields

$$\begin{aligned} \mathbf{B} &= \mathbf{G}^+\mathbf{X} \\ &= [(\mathbf{B}^+)^H\mathbf{X}^H\mathbf{X}\mathbf{B}^+]^{-1}(\mathbf{B}^+)^H\mathbf{X}^H\mathbf{X} \\ &= \mathbf{B}\mathbf{B}^H(\mathbf{B}\mathbf{X}^H\mathbf{X}\mathbf{B}^H)^{-1}\mathbf{B}\mathbf{X}^H\mathbf{X} \end{aligned} \tag{17}$$

and in a similar way for $\mathbf{G}$. **Q.E.D.**

From the next theorem, we can get our algorithm. It is got from Corollary 1:

Theorem 3. *Let the eigenvalue decomposition be* $\mathbf{R}_b = \mathbf{B}^H\mathbf{R}_\gamma\mathbf{B} + \sigma^2\mathbf{I} = \mathbf{V}_s\mathbf{\Lambda}_s\mathbf{V}_s^H + \sigma^2\mathbf{I}$. *Then*

$$\mathbf{W} = \mathbf{W}\mathbf{W}^H(\mathbf{V}_s^H\mathbf{W}^H)^{-1}\mathbf{V}_s^H \tag{18}$$

where $\mathbf{W} = \mathbf{TB}$, *and* $\mathbf{V}_s$ *is a principal component subspace having the same span as* $\mathbf{B}^H$.

Proof: Because $\mathbf{V}_s^H = \mathbf{SB}$, then the structure of $\mathbf{V}_s\mathbf{V}_s^H = \mathbf{B}^H\mathbf{S}^H\mathbf{SB}$ is a special case of the structure of $\mathbf{R}_b$. Therefore we get from the Corollary 1 (invertibilities follow from the rank reduction)

$$\begin{aligned} \mathbf{W} &= \mathbf{W}\mathbf{W}^H(\mathbf{W}\mathbf{V}_s\mathbf{V}_s^H\mathbf{W}^H)^{-1}\mathbf{W}\mathbf{V}_s\mathbf{V}_s^H \\ &= \mathbf{W}\mathbf{W}^H(\mathbf{V}_s^H\mathbf{W}^H)^{-1}(\mathbf{W}\mathbf{V}_s)^{-1}\mathbf{W}\mathbf{V}_s\mathbf{V}_s^H \\ &= \mathbf{W}\mathbf{W}^H(\mathbf{V}_s^H\mathbf{W}^H)^{-1}\mathbf{V}_s^H \end{aligned} \tag{19}$$

Q.E.D.

If either $\mathbf{G}$ or $\mathbf{B}$ is known, then one obtains optimal estimates from the Least-Squares (LS) solution

$$\hat{\mathbf{G}} = \mathbf{X}\mathbf{B}^+ = \arg\min_{\mathbf{G}} ||\mathbf{X} - \mathbf{GB}||, \quad \mathbf{B} \text{ is known} \tag{20}$$

$$\hat{\mathbf{B}} = \mathbf{G}^+\mathbf{X} = \arg\min_{\mathbf{B}} ||\mathbf{X} - \mathbf{GB}||, \quad \mathbf{G} \text{ is known} \tag{21}$$

even for the noisy model (5). Optimization problem (14) is a combination of (20) and (21). It is easy to see that the solutions (15) and (16) are not unique, because $\mathbf{X}$ can always, due to its linear structure, be represented as follows:

$$\mathbf{X} = \mathbf{GB} + \mathbf{N} = \mathbf{G}\mathbf{T}^{-1}\mathbf{TB} + \mathbf{N} = \mathbf{SW} + \mathbf{N} \tag{22}$$

where $\mathbf{S} = \mathbf{G}\mathbf{T}^{-1}$ and $\mathbf{W} = \mathbf{TB}$. Thus $\mathbf{B}$ can be replaced by $\mathbf{W}$ and $\mathbf{G}$ by $\mathbf{S}$ in (15) and (16). Theorem 2 gives two exact (but not the only possible) solutions for the parameter matrices $\mathbf{G}$ and $\mathbf{B}$ when the data are noiseless. For noisy data (5), Theorem 2 gives approximative suboptimal solutions. In addition, Eqs. (15) and (16) play part of the rules for updating the estimates of $\mathbf{G}$ and $\mathbf{B}$. In this paper, we concentrate only to the estiamtion of $\mathbf{B}$. Theorem 2 also presents how the basis matrix $\mathbf{G}$ and the source matrix $\mathbf{B}$ are completetly dual to each other. Corollary 1 say that if we take the expectations of (15) and (16), we have exact solutions even for the noisy data. Finally, Theorem 3 gives the simpler result than Corollary 1, and that is applied in our algorithm.

SNR (dB)	0	5	10	15	20	25	30
JADE	41.2	35.7	25.4	12.4	5.1	1.9	1.0
New	40.5	32.5	20.9	9.7	4.2	1.4	0.9

Table 1: Bit-error-rate (%) for algorithms as a function of signal-to-noise ratio.

3 Algorithms

Due to the nonuniqueness of Eq. (22), one has to take advantage the knowledge about **G** or **B** when estimating them. The difficulty arises when the structure of **G** is not not known. This is the case e.g. in dowlink (base station to mobile phone) communications. In "classical" blind source separation, one assumes that **B** is a sample from the independent non-Gaussian process. Here we exploit the structure of **B** more efficiently by assuming that the matrix **B** contains only ±1:s. By taking this kind of assumption, we can use only second order statistics for blind symbol estimation. The resulting algorithm is very simple with respect to the algorithms using fourth order cumulants [2]. The algorithm also avoids a common prewhitening.

For estimating **B** in a blind way, we use a subspace network

$$\hat{\mathbf{B}}_{t+1} = \mathrm{sign}[\hat{\mathbf{B}}_t \hat{\mathbf{B}}_t^T (\hat{\mathbf{V}}_s^H \hat{\mathbf{B}}_t^H)^{-1} \mathbf{V}_s^H] \qquad (23)$$

where sign(·) takes the sign elementwise. This algorithm tries to give the solution $\mathbf{W} = \mathbf{B}$, as we see from Eq. 18 in the Theorem 3.

We compare our method to Joint Approximate Diagonalization of Eigen-matrices (JADE) [1]. It is developed for blind beamforming for non-Gaussian signals. Thus it is suitable for binary symbol separation. JADE exploits fourth order statistics.

4 Experiments

We tested the algorithms using the Code Division Multiple Access (CDMA) data. Parameters were as follows: the dimension of the data vector was $C = 31$, number of samples was $N = 300$, number of users was $M = 5$. 100 simulations were performed for each SNR. The basis vectors of **G** were random variables.

Table 1 shows the Bit-Error-Rate (BER) as a function of Signal-to-Noise Ratio (SNR) for different algorithms. It shows that the new method has quite clearly better separation capability than JADE.

Table 2 shows the number of MATLAB floating point operations without eigendecomposition. Notice that both the methods, JADE and new, use eigen-decomposition, and MATLAB computes all the 31 eigenvectors, not only the 5 principal eigenvectors. Here we have not took advantage of the fact that most of the operations in the new method are purely binary ($\mathbf{B}\mathbf{B}^T$) or purely summing ($\mathbf{V}_s^H \mathbf{B}^H$). Because JADE is a batch method, the number of operations remains constant. The number of opeations needed by the new method increases when

SNR (dB)	0	5	10	15	20	25	30
JADE	0.8	0.8	0.8	0.8	0.8	0.8	0.8
New	2.6	2.2	1.6	1.2	0.7	0.5	0.4

Table 2: Number of MATLAB floating point operations/10^6 (without eigendecomposition) as a function of signal-to-noise ratio.

the SNR becames low, because the number of iterations increases. However, at such SNRs, where the methods work well (BER is lower than 5 %), the new method needs less iterations than JADE.

5 Conclusions

In this paper, we have introduced simple iterative learning method for semi-blind estimation of the source symbols. The algorithm is based only on the second order statistics. Because most of the operations are only binary or summing, the network should be implemented using very simple harware architecture. Simulations with communications data was shown that the method has better performance than the fourth order statistics based JADE.

Acknowledgment

This research was supported by Nokia Research Center.

References

[1] J.-F. Cardoso and A. Souloumiac, "Blind beamforming for non Gaussian signals", *IEE Proceedings-F*, vol. 140, no. 6, December 1994, pp. 362-370.

[2] P. Comon, "Independent Component Analysis - a New Concept?," *Signal Processing*, vol. 36, pp. 287-314, 1994.

[3] J. Karhunen, E. Oja, L. Wang, R. Vigário, and J. Joutsensalo, "A Class of Neural Networks for Independent Component Analysis", to be published in *IEEE Transactions on Neural Networks.*

[4] P. Pajunen, "A Competitive Learning Algorithm for Separating Binary Sources", to be published in *Proc. European Symposium on Artificial Neural Networks* (ESANN'97), Bruges, April 16-18, 1997.

Kernel Principal Component Analysis

Bernhard Schölkopf[1], Alexander Smola[2], Klaus–Robert Müller[2]

[1] Max-Planck-Institut f. biol. Kybernetik, Spemannstr. 38, 72076 Tübingen, Germany
[2] GMD FIRST, Rudower Chaussee 5, 12489 Berlin, Germany

Abstract. A new method for performing a nonlinear form of Principal Component Analysis is proposed. By the use of integral operator kernel functions, one can efficiently compute principal components in high-dimensional feature spaces, related to input space by some nonlinear map; for instance the space of all possible d-pixel products in images. We give the derivation of the method and present experimental results on polynomial feature extraction for pattern recognition.

1 Introduction

Principal Component Analysis (PCA) is a basis transformation to diagonalize an estimate of the covariance matrix of the data $\mathbf{x}_k$, $k = 1, \ldots, \ell$, $\mathbf{x}_k \in \mathbf{R}^N$, $\sum_{k=1}^{\ell} \mathbf{x}_k = 0$, defined as

$$C = \frac{1}{\ell} \sum_{j=1}^{\ell} \mathbf{x}_j \mathbf{x}_j^{\top}. \tag{1}$$

The new coordinates in the Eigenvector basis, i.e. the orthogonal projections onto the Eigenvectors, are called *principal components*.

In this paper, we generalize this setting to a nonlinear one of the following kind. Suppose we first map the data nonlinearly into a feature space F by

$$\Phi : \mathbf{R}^N \to F, \quad \mathbf{x} \mapsto \mathbf{X}. \tag{2}$$

We will show that even if F has arbitrarily large dimensionality, for certain choices of Φ, we can still perform PCA in F. This is done by the use of kernel functions known from Support Vector Machines (Boser, Guyon, & Vapnik, 1992).

2 Kernel PCA

Assume for the moment that our data mapped into feature space, $\Phi(\mathbf{x}_1), \ldots, \Phi(\mathbf{x}_\ell)$, is centered, i.e. $\sum_{k=1}^{\ell} \Phi(\mathbf{x}_k) = 0$. To do PCA for the covariance matrix

$$\bar{C} = \frac{1}{\ell} \sum_{j=1}^{\ell} \Phi(\mathbf{x}_j) \Phi(\mathbf{x}_j)^{\top}, \tag{3}$$

we have to find Eigenvalues $\lambda \geq 0$ and Eigenvectors $\mathbf{V} \in F \backslash \{0\}$ satisfying $\lambda \mathbf{V} = \bar{C} \mathbf{V}$. Substituting (3), we note that all solutions $\mathbf{V}$ lie in the span of $\Phi(\mathbf{x}_1), \ldots, \Phi(\mathbf{x}_\ell)$. This implies that we may consider the equivalent system

$$\lambda(\Phi(\mathbf{x}_k) \cdot \mathbf{V}) = (\Phi(\mathbf{x}_k) \cdot \bar{C}\mathbf{V}) \text{ for all } k = 1, \ldots, \ell, \tag{4}$$

and that there exist coefficients $\alpha_1, \ldots, \alpha_\ell$ such that

$$\mathbf{V} = \sum_{i=1}^{\ell} \alpha_i \Phi(\mathbf{x}_i). \tag{5}$$

Substituting (3) and (5) into (4), and defining an $\ell \times \ell$ matrix K by

$$K_{ij} := (\Phi(\mathbf{x}_i) \cdot \Phi(\mathbf{x}_j)), \tag{6}$$

we arrive at

$$\ell \lambda K \boldsymbol{\alpha} = K^2 \boldsymbol{\alpha}, \tag{7}$$

where $\boldsymbol{\alpha}$ denotes the column vector with entries $\alpha_1, \ldots, \alpha_\ell$. To find solutions of (7), we solve the Eigenvalue problem

$$\ell \lambda \boldsymbol{\alpha} = K \boldsymbol{\alpha} \tag{8}$$

for nonzero Eigenvalues. Clearly, all solutions of (8) do satisy (7). Moreover, it can be shown that any additional solutions of (8) do not make a difference in the expansion (5) and thus are not interesting for us.

We normalize the solutions $\boldsymbol{\alpha}^k$ belonging to nonzero Eigenvalues by requiring that the corresponding vectors in F be normalized, i.e. $(\mathbf{V}^k \cdot \mathbf{V}^k) = 1$. By virtue of (5), (6) and (8), this translates into

$$1 = \sum_{i,j=1}^{\ell} \alpha_i^k \alpha_j^k (\Phi(\mathbf{x}_i) \cdot \Phi(\mathbf{x}_j)) = (\boldsymbol{\alpha}^k \cdot K \boldsymbol{\alpha}^k) = \lambda_k (\boldsymbol{\alpha}^k \cdot \boldsymbol{\alpha}^k). \tag{9}$$

For principal component extraction, we compute projections of the image of a test point $\Phi(\mathbf{x})$ onto the Eigenvectors $\mathbf{V}^k$ in F according to

$$(\mathbf{V}^k \cdot \Phi(\mathbf{x})) = \sum_{i=1}^{\ell} \alpha_i^k (\Phi(\mathbf{x}_i) \cdot \Phi(\mathbf{x})). \tag{10}$$

Note that neither (6) nor (10) requires the $\Phi(\mathbf{x}_i)$ in explicit form — they are only needed in dot products. Therefore, we are able to use kernel functions for computing these dot products *without* actually performing the map Φ (Aizerman, Braverman, & Rozonoer, 1964; Boser, Guyon, & Vapnik, 1992): for some choices of a kernel $k(\mathbf{x}, \mathbf{y})$, it can be shown by methods of functional analysis that there exists a map Φ into some dot product space F (possibly of infinite dimension) such that k computes the dot product in F. Kernels which have successfully been used in Support Vector Machines (Schölkopf, Burges, & Vapnik, 1995) include polynomial kernels

$$k(\mathbf{x}, \mathbf{y}) = (\mathbf{x} \cdot \mathbf{y})^d, \tag{11}$$

radial basis functions $k(\mathbf{x}, \mathbf{y}) = \exp\left(-\|\mathbf{x} - \mathbf{y}\|^2 / (2\,\sigma^2)\right)$, and sigmoid kernels $k(\mathbf{x}, \mathbf{y}) = \tanh(\kappa(\mathbf{x} \cdot \mathbf{y}) + \Theta)$. It can be shown that polynomial kernels of degree d correspond to a map Φ into a feature space which is spanned by all products of d entries of an input pattern, e.g., for the case of $N = 2, d = 2$,

$$(\mathbf{x} \cdot \mathbf{y})^2 = (x_1^2, x_1 x_2, x_2 x_1, x_2^2)(y_1^2, y_1 y_2, y_2 y_1, y_2^2)^\top. \tag{12}$$

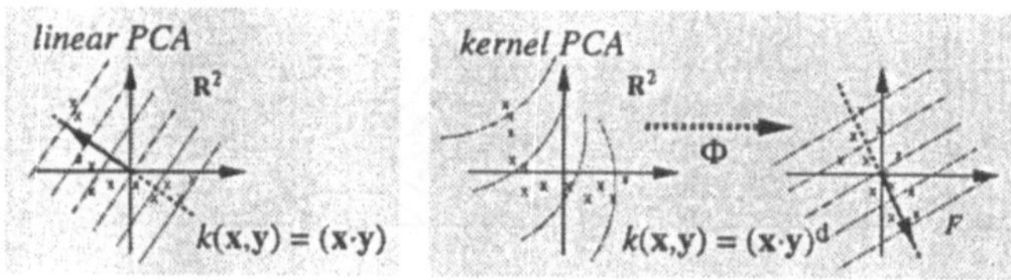

Fig. 1. Basic idea of kernel PCA: by using a nonlinear kernel function k instead of the standard dot product, we implicitly perform PCA in a possibly high-dimensional space F which is nonlinearly related to input space. The dotted lines are contour lines of constant feature value.

If the patterns are images, we can thus work in the space of all products of d pixels and thereby take into account higher-order statistics when doing PCA.

Substituting kernel functions for all occurences of $(\Phi(\mathbf{x})\cdot\Phi(\mathbf{y}))$, we obtain the following algorithm for kernel PCA (Fig. 1): we compute the dot product matrix (cf. Eq. (6)) $K_{ij} = (k(\mathbf{x}_i, \mathbf{x}_j))_{ij}$, solve (8) by diagonalizing K, normalize the Eigenvector expansion coefficients α^n by requiring Eq. (9), and extract principal components (corresponding to the kernel k) of a test point $\mathbf{x}$ by computing projections onto Eigenvectors (Eq. (10), Fig. 2).

We should point out that in practice, our algorithm is not equivalent to the form of nonlinear PCA obtainable by explicitly mapping into the feature space F: even though the rank of the dot product matrix will be limited by the sample size, we may not even be able to compute this matrix, if the dimensionality is prohibitively high. For instance, 16×16 pixel input images and a polynomial degree $d = 5$ yield a dimensionality of 10^{10}. Kernel PCA deals with this problem by automatically choosing a subspace of F (with a dimensionality given by the rank of K), and by providing a means of computing dot products between vectors in this subspace. This way, we have to evaluate ℓ kernel functions in input space rather than a dot product in a 10^{10}-dimensional space.

To conclude this section, we briefly mention the case where we drop the assumption that the $\Phi(\mathbf{x}_i)$ are centered in F. Note that we cannot in general center the data, as we cannot compute the mean of a set of points that we do not have in explicit form. Instead, we have to go through the above algebra using $\tilde{\Phi}(\mathbf{x}_i) := \Phi(\mathbf{x}_i) - (1/\ell)\sum_{i=1}^{\ell}\Phi(\mathbf{x}_i)$. It turns out that the matrix that we

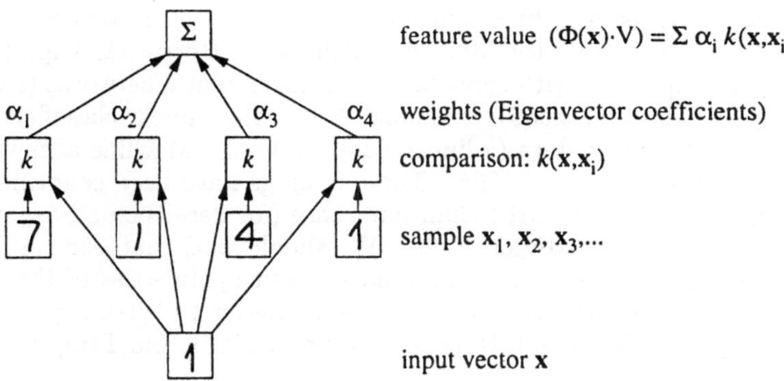

Fig. 2. Kernel PCA feature extraction for an OCR task (test point $\mathbf{x}$, Eigenvector $\mathbf{V}$).

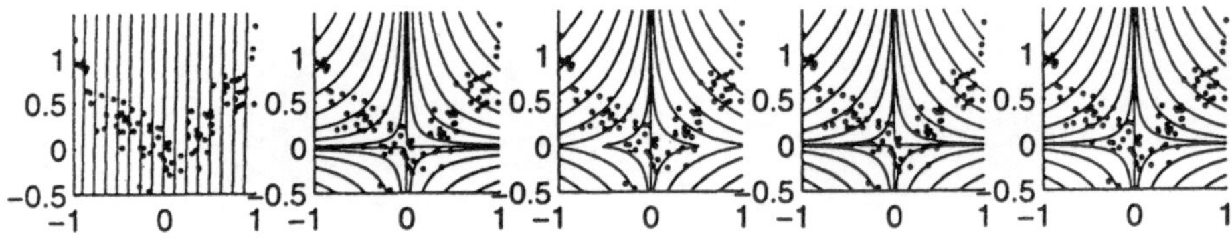

Fig. 3. PCA with kernel (11, degrees $d = 1, \ldots, 5$. 100 points $((x_i)_1, (x_i)_2)$ were generated from $(x_i)_2 = (x_i)_1^2 +$ noise (Gaussian, with standard deviation 0.2); all $(x_i)_j$ were rescaled according to $(x_i)_j \mapsto \mathrm{sgn}((x_i)_j) \cdot |(x_i)_j|^{1/d}$. Displayed are contour lines of constant value of the first principal component. Nonlinear kernels ($d > 1$) extract features which nicely increase along the direction of main variance in the data; linear PCA ($d = 1$) does its best in that respect, too, but it is limited to straight directions.

have to diagonalize in that case, call it $\tilde{K}$, can be expressed in terms of K as $\tilde{K}_{ij} = K - 1_\ell K - K 1_\ell + 1_\ell K 1_\ell$, using the shorthand $(1_\ell)_{ij} := 1/\ell$ (for details, see Schölkopf, Smola, & Müller, 1996[3]).

3 Experiments on Feature Extraction

Figure 3 shows the first principal component of a **toy data set**, extracted by polynomial kernel PCA. For an investigation of the utility of kernel PCA features for a realistic **pattern recognition problem**, we trained a separating hyperplane classifier (Vapnik & Chervonenkis, 1974; Cortes & Vapnik, 1995) on nonlinear features extracted from the US postal service (USPS) handwritten digit data base by kernel PCA. This database contains 9300 examples of dimensionality 256; 2000 of them make up the test set. For computational reasons, we used only a subset of 3000 training examples for the dot product matrix. Using polynomial kernels (11) of degrees $d = 1, \ldots, 6$, and extracting the first 2^n ($n = 6, 7, \ldots, 11$) principal components, we found the following. In the case of linear PCA ($d = 1$), the best classification performance (8.6% error) is attained for 128 components. Extracting the same number of *nonlinear* components ($d = 2, \ldots, 6$) in all cases lead to superior performance (around 6% error). Moreover, in the nonlinear case, the performance can be further improved by using a larger number of components (note that there exist more higher-order features than there are pixels in an image). Using $d > 2$ and 2048 components, we obtained around 4% error, which coincides with the best result reported for standard nonlinear Support Vector Machines (Schölkopf, Burges. & Vapnik, 1995). This result is competitive with convolutional 5-layer neural networks (5.0% were reported by LeCun et al., 1989); it is much better than linear classifiers operating directly on the image data (a linear Support Vector Machine achieves 8.9%; Schölkopf, Burges, & Vapnik, 1995). These findings have been confirmed on an object recognition task, the MPI chair data base (for details on all experiments, see Schölkopf, Smola, & Müller, 1996). We should add that our results were obtained without using any prior knowledge about symmetries of the problem at hand. This explains why the performance is inferior to Virtual Support Vector classifiers (3.2%, Schölkopf, Burges, & Vapnik, 1996), and Tangent Distance

[3] This paper, along with several Support Vector publications, can be downloaded from http://www.mpik-tueb.mpg.de/people/personal/bs/svm.html.

Nearest Neighbour classifiers (2.6%, Simard, LeCun, & Denker, 1993). We believe that adding e.g. local translation invariance, be it by generating "virtual" translated examples or by choosing a suitable kernel, could further improve the results.

4 Discussion

This paper was devoted to the exposition of a new technique for nonlinear principal component analysis. To develop this technique, we made use of a kernel method which so far only had been used in supervised learning (Vapnik, 1995). Clearly, the kernel method can be applied to *any* algorithm which can be formulated in terms of dot products exclusively, including for instance k-means and independent component analysis (cf. Schölkopf, Smola, & Müller, 1996).

In experiments comparing the utility of kernel PCA features for pattern recognition using a linear classifier, we found two advantages of nonlinear kernel PCA: first, nonlinear principal components afforded better recognition rates than corresponding numbers of linear principal components; and second, the performance for nonlinear components can be further improved by using more components than possible in the linear case.

The computational complexity of kernel PCA does *not* grow with the dimensionality of the feature space that we are implicitely working in. This makes it possible to work for instance in the space of all possible d-th order products between pixels of an image. As in the variant of standard PCA which diagonalizes the dot product matrix (e.g. Kirby & Sirovich, 1990), we have to diagonalize an $\ell \times \ell$ matrix (ℓ being the number of examples, or the size of a representative subset), with a comparable computational complexity — we only need to compute kernel functions rather than dot products. If the dimensionality of input space is smaller than the number of examples, kernel principal component extraction is computationally more expensive than linear PCA; however, this additional investment can pay back afterwards: we have presented results indicating that in pattern recognition, it is sufficient to use a linear classifier, as long as the features extracted are nonlinear. The main advantage of linear PCA up to date, however, consists in the possibility to reconstruct the patterns from their principal components.

Compared to other methods for nonlinear PCA, as autoassociative MLPs with a bottleneck hidden layer (e.g. Diamantaras & Kung, 1996) or principal curves (Hastie & Stuetzle, 1989), kernel PCA has the advantage that no nonlinear optimization is involved — we only need to solve an Eigenvalue problem as in the case of standard PCA. Therefore, we are not in danger of getting trapped in local minima during during training. Compared to most neural network type generalizations of PCA (e.g. Oja, 1982), kernel PCA moreover has the advantage that it provides a better understanding of what kind of nonlinear features are extracted: they are principal components in a feature space which is fixed a priori by choosing a kernel function. In this sense, the type of nonlinearities that we are looking for are already specified in advance, however this specification is a very wide one, it merely selects the (high-dimensional) feature space, but not the relevant feature subspace: the latter is done automatically. In this respect it is worthwhile to note that by using sigmoid kernels (Sec. 2) we can in fact also extract features which are of the same type as the ones extracted by MLPs (cf. Fig. 2), and the latter is often considered a nonparametric technique. With its rather wide class of admissible nonlinearities, kernel PCA forms a framework comprising various types of feature extraction systems. A number of different kernels have already been used in Support Vector Machines, of polynomial, Gaussian, and sigmoid type. They all led to high accuracy classifiers, and constructed their decision boundaries, which are hyperplanes in different feature spaces, from almost the same Support Vectors (Schölkopf, Burges, & Vapnik, 1995). The general question of how to choose the best kernel for a given problem is yet unsolved, both for Support Vector Machines and for kernel PCA.

PCA feature extraction has found application in many areas, including noise reduction, pattern recognition, regression estimation, and image indexing. In all cases where taking into account nonlinearities might be beneficial, kernel PCA provides a new tool which can be applied with little computational cost and possibly substantial performance gains.

Acknowledgements. BS is supported by the Studienstiftung des Deutschen Volkes. AS is supported by a grant of the DFG (JA 379/51). This work profited from discussions with V. Blanz, L. Bottou, C. Burges, S. Solla, and V. Vapnik. Thanks to AT&T and Bell Labsoratories for the possibility of using the USPS database.

References

M. A. Aizerman, E. M. Braverman, & L. I. Rozonoér. Theoretical foundations of the potential function method in pattern recognition learning. *Automation and Remote Control*, 25:821–837, 1964.

B. E. Boser, I. M. Guyon, & V .Vapnik. A training algorithm for optimal margin classifiers. In *Fifth Annual Workshop on COLT*, Pittsburgh, 1992. ACM.

C. Cortes & V. Vapnik. Support vector networks. *Machine Learning*, 20:273–297, 1995.

T. Hastie & W. Stuetzle. Principal curves. *JASA*, 84:502 – 516, 1989.

M. Kirby & L. Sirovich. Application of the Karhunen–Loève procedure for the characterization of human faces. *IEEE Transactions*, PAMI-12(1):103–108, 1990.

E. Oja. A simplified neuron model as a principal component analyzer. *J. Math. Biology*, 15:267–273, 1982.

B. Schölkopf, C. Burges, & V. Vapnik. Extracting support data for a given task. In U. M. Fayyad & R. Uthurusamy, eds., *Proceedings, First International Conference on Knowledge Discovery & Data Mining*, Menlo Park, CA, 1995. AAAI Press.

B. Schölkopf, C. Burges, & V. Vapnik. Incorporating invariances in support vector learning machines. In C. v. d. Malsburg, W. v. Seelen, J. C. Vorbrüggen, & B. Sendhoff, eds., *ICANN'96*, p. 47–52, Berlin, 1996. Springer LNCS Vol. 1112.

B. Schölkopf, A. J. Smola, & K.-R. Müller. Nonlinear component analysis as a kernel eigenvalue problem. Technical Report 44, Max–Planck–Institut für biologische Kybernetik, 1996. Submitted to *Neural Computation*.

P. Simard, Y. LeCun, & J. Denker. Efficient pattern recognition using a new transformation distance. In S. J. Hanson, J. D. Cowan, & C. L. Giles, editors, *Advances in NIPS 5*, San Mateo, CA, 1993. Morgan Kaufmann.

V. Vapnik & A. Chervonenkis. *Theory of Pattern Recognition [in Russian]*. Nauka, Moscow, 1974. (German Translation: W. Wapnik & A. Tscherwonenkis, *Theorie der Zeichenerkennung*, Akademie–Verlag, Berlin, 1979).

An Empirical Comparison of Dimensionality Reduction Techniques for Pattern Classification

Thiagarajan Balachander[1], Ravi Kothari[1], and Hernani Cualing[2]

[1] Artificial Neural Systems Laboratory
Department of Electrical & Computer Engineering & Computer Science
[2] Department of Pathology and Lab. Medicine,
University of Cincinnati
Cincinnati, OH 45221, USA

Abstract. To some extent or other all classifiers are subject to the curse of dimensionality. Consequently, pattern classification is often preceded with finding a reduced dimensional representation of the patterns. In this paper we empirically compare the performance of unsupervised and supervised dimensionality reduction techniques. The data set we consider is obtained by segmenting cells in cytological preparations and extracting 9 features from each of the cells. We evaluate the performance of 4 dimensionality reduction techniques (2 unsupervised) and (2 supervised) with and without noise. The unsupervised techniques include principal component analysis and self-organizing feature maps, while the supervised techniques include Fisher's linear discriminants and multi-layered feed-forward neural networks. Our results on a real world data set indicate that multi-layered feed-forward neural networks outperform the other three dimensionality reduction techniques and that all techniques are sensitive to noise.

1 Introduction

To some extent or other all classifiers are subject to the 'curse of dimensionality' [1]. Pattern classification is thus often preceded by an effort to reduce the dimensionality of the input space. Such efforts rely on the correlations that may be present amongst the features comprising the pattern. Dimensionality reduction techniques can be broadly classified into *unsupervised* and *supervised* ones. Unsupervised techniques are useful in situations where the cost of obtaining a class label is small. While the cost of assigning such a label may be high in certain situations, supervised dimensionality techniques can use the class label to advantage resulting in better performance.

The goal of this paper is to report on an empirical comparison (see also [2]) of the various dimensionality reduction techniques as it pertains to the specific problem of cytodiagnosis of lymphoma. In Section 2, we briefly review the techniques considered in this study. In Section 3, we describe the problem domain and describe how the features were obtained. In Section 4 we outline the comparison methodology and present our results. In Section 5 we present our conclusions.

2 Paradigms for Dimensionality Reduction

Before briefly describing each of the paradigm we consider, we introduce some notation below.

n	Dimensionality of original feature space
m	Dimensionality of reduced feature space
p	Number of patterns
$\xi^{(i)}$	Input pattern in the original n dimensional space (column vector)
$\omega^{(i)}$	Class label for $\xi^{(\mu)}$ (if available)
$p^{(i)}$	Number of patterns in class $\omega^{(i)}$
μ	$1/p \sum_{k=1}^{p} \xi^{(k)}$ (mean of all patterns)
σ	$1/(p-1) \sum_{k=1}^{p} (\xi^{(k)} - \mu)^T (\xi^{(k)} - \mu)$ (variance of all patterns)
$\mu^{(i)}$	$1/p^{(i)} \sum_{k=1}^{p^i} \xi^{(k)}; \quad \xi^{(k)} \in \omega^{(i)}$
$\sigma^{(i)}$	$1/(p^{(i)} - 1) \sum_{k=1}^{p^i} (\xi^{(k)} - \mu^{(i)})^T (\xi^{(k)} - \mu^{(i)}); \quad \xi^{(k)} \in \omega^{(i)}$
$x^{(i)}$	Pattern in the m dimensional subspace

2.1 Unsupervised Dimensionality Reduction

1. **Principal Component Analysis:** Principal Component Analysis (PCA) attempts to represent n dimensional data by finding m orthogonal vectors, such that the projection of the n dimensional data onto the m dimensional subspace preserves as much of the variance in the data as possible [3]. The principal components are the eigenvectors corresponding to the eigenvalues of the sample covariance matrix C defined as:

$$C = \frac{1}{p} \sum_{k=1}^{p} \left(\xi^{(i)} - \mu\right) \left(\xi^{(i)} - \mu\right)^T \tag{1}$$

where, T denoted the transpose. The eigenvector corresponding to the largest eigenvalue is the first principal component, the eigenvector corresponding to the next largest eigenvalue is the second principal component and so on. By projecting the data on the subspace of the first m principal components, one obtains the representation of the original data in m dimensions. Thus, one has $x^{(i)} = w^T \xi^{(i)}$ where w is a matrix whose columns are the largest m eigenvectors.

2. **Self Organizing Feature Maps:** The self-organizing feature map is an unsupervised neural network in which neurons are arranged in a lattice. Each neuron receives information from the inputs as well as neighboring neurons [4].
Given an input, weights of the winning neurons and its neighbors are modified as follows. Let $w^{(k)}$ denote the weights of the kth neuron to the inputs.

$$c = \arg\min_k \{\| \xi^{(i)} - w^{(k)} \|\} \tag{2}$$

$$w^{(k)} = w^{(k)} + \eta(d,t) \left[\xi^{(i)} - w^{(k)}\right] \qquad k \in N_c \tag{3}$$

where, $0 < \eta(d,t) < 1$ denotes the learning rate and is a decreasing function of both the distance from the winning neuron and time. N_c denotes the neighborhood of the winning neuron and also decreases in time. If we repeatedly present inputs, and adjust the weights as given by (2)-(3), then the neurons arrange themselves in a topology preserving mapping and approximate the local density of the inputs [5].
Thus if a m-dimensional array of neurons is used, indices of the winning neuron for a pattern serve to represent the data in the m-dimensional subspace.

2.2 Supervised Dimensionality Reduction

1. **Fishers Linear Discriminant:** When the class label is known, clearly the dimensionality reduction should be made under the constraint of maximizing the inter-class separation. Such a facility is provided by Fisher's criterion [6] (see also [7]), which identifies a direction w, along which the distance between the inter-class means normalized by the within class scatter is maximized. It can be shown that the direction w is given by [7]:

$$w \propto S_W^{-1}(\mu^{(1)} - \mu^{(2)}) \tag{4}$$

where, S_W is the *total within-class scatter* matrix given as:

$$S_W = \sum_{i \in \omega^{(1)}} (\xi^{(i)} - \mu^{(1)})(\xi^{(i)} - \mu^{(1)})^T + \sum_{j \in \omega^{(2)}} (\xi^{j} - \mu^{(2)})(\xi^{(j)} - \mu^{(2)})^T \tag{5}$$

where, $\mu^{(1)}$ and $\mu^{(2)}$ are the means of the two classes.
The original n dimensional data can now be projected as: $x^{(i)} = w^T \xi^{(i)}$. In a two class situation, a pattern is then classified in Class 1 if $x^{(1)} < d$, and Class 2 otherwise, where $d = (\mu^{(1)}\sigma^{(1)} + \mu^{(2)}\sigma^{(2)})/(\sigma^{(1)} + \sigma^{(2)})$.

2. **Feed-Forward Neural Networks:** If the architecture of a multi-layered feed-forward network is constrained to have an intermediate hidden layer with fewer (m) neurons, then one obtains a lower dimensional representation of higher dimensional data. Consequently, such constrained architectures naturally perform dimensionality reduction — of course, the dimensionality reduction is embedded in the classifier [11].
The training of such networks can be done using the well known error back-propagation algorithm [8], which perform gradient descent in the weight space to reduce the error between the actual response and the desired response of the network. Once trained, one can record the outputs of m hidden layer neurons as each pattern is presented to arrive at a reduced dimensionality representation of the original n dimensional data.

3 Problem Domain and Feature Extraction

In this section we outline the problem of cytodiagnois of lymphoma and outline the procedure with which we extracted the features.

Cytological preparations of lymph nodes are an important part of a multi-parameter approach to the diagnosis of lymphoma. Due to the complex and repetitive nature of analysis, automated diagnosis has appeal. Automated analysis of cytological preparations relies on three steps: (i) the acquisition of an image of the cytological preparation, (ii) the segmentation of cells in the original images, and (iii) calculation of features (descriptors) from each of the segmented cells.

Since monoscopic images of cytological preparations constitute a 2-D projection of the 3-D volume occupied by the cytological preparation, overlapping cells are common. Due to this fact we use a combination of the distance transform, circular region growing, followed by an active contour procedure [10] to isolate the individual cells. Details of this procedure are presented in [11]. An illustrative final segmentation is shown in Figure 1.

Once each cell has been isolated we extract nine features from each cell. To reduce the effect of artifacts we only consider upto 10 largest cells in any preparation. To characterize the tone and texture of a cell, we extract four tonal measures (Mean, Variance, Skewness, and Kurtosis) and four textural features from each cell (Elemental Difference of Order 1, Entropy, Uniformity, and Homogeneity). These four textural measures are extracted from the gray level co-occurrence matrix C [9]. These eight features along with the curvature of the cell boundary are extracted from each cell.

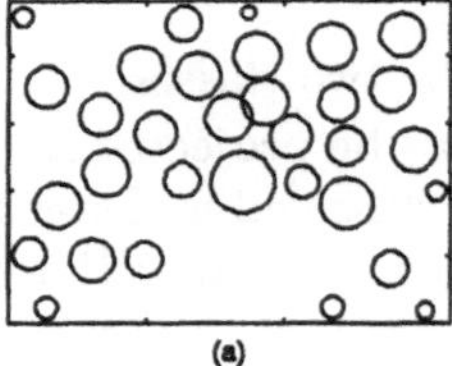

(a)

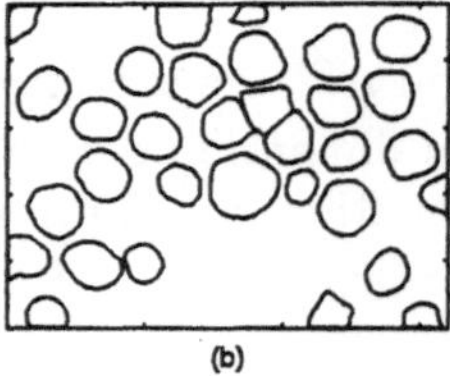

(b)

Fig. 1. (a) Disk shaped regions after region growing, and (b) cell boundary after refinement using active contours

Feature from 22 benign and 67 malignant cytological preparations were used with a total of 439 cells (145 from benign and 294 from malignant slides) resulting from the segmentation. From each of these cells we extracted the 9 features outlined above resulting in 439 patterns. Consequently, here $p = 439$, and $n = 9$. We used leave one out method of cross-validation. The 439 patterns are divided into 5 equal groups and classifiers are trained on 4 groups and tested on the fifth.

4 Comparison Methodology and Results

Salient issues of the comparison are described below.

1. To get a sense of the quality of the dimensionality reduction, which is not the same as classification accuracy for a multi-layered feed-forward network, we present results at the point of dimensionality reduction.
2. All results are presented based on reducing the dimension from 9 to 1.
3. Robustness is a desirable attribute of dimensionality reduction techniques. We thus present results based on adding no noise, and 1% and 5% noise.
4. Finally, we compare the lower dimensional projections based on a well known comparison tool - the Receiver Operating Characteristic (ROC) [12].

We consider two separate feed-forward neural network (FFNN) architectures. The first architecture had 9 inputs, 2 hidden layers with 1 and 2 neurons respectively, and a single output — we term this architecture as 9121. The second architecture had 9 inputs, 3 hidden layers with 4, 1 and 2 neurons respectively, and a single output — we term this architecture as 94121. Both architectures were trained using back-propagation. For the Self Organizing Feature Map (SOFM) we used 10 neurons arranged in a 1-D grid. A linearly decreasing neighborhood and learning rate were used for the SOFM which was trained for 3000 epochs, starting with a learning rate of 0.5 and a neighborhood of 5.

To ascertain the quality of the dimensionality reduction, we obtained the ROC (Receiver Operating Characteristic) curve for the single variable resulting after the dimensionality reduction. The ROC curve plots the detection rate against the false alarm rate for various thresholds, and is shown in Figure 2. It is evident from this curve that the non-linear supervised dimensionality reduction as obtained from the neural network was better. Table I summarizes the classification accuracy that was obtained.

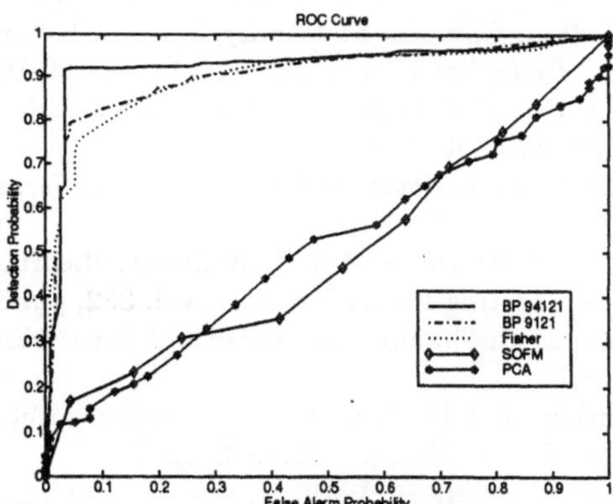

Fig. 2. ROC curves obtained from the the single dimensional feature space obtained after dimensionality reduction for the noise free training case.

	Training		Generalization	
	Benign	Malig.	Benign	Malig.
PCA	51.0 / 51.0 / 53.2	52.2 / 52.3 / 49.2	51.0 / 51.0 / 46.9	52.0 / 52.1 / 44.6
SOFM	68.8 / 72.6 / 70.6	63.5 / 61.7 / 60.5	67.6 / 63.4 / 61.9	61.9 / 65.3 / 63.2
Fisher	94.1 / 92.8 / 81.3	69.7 / 71.5 / 72.5	93.7 / 92.4 / 84.8	68.3 / 68.7 / 73.1
FFNN(9121)	93.7 / 94.5 / 83.9	81.7 / 80.1 / 78.8	86.8 / 89.7 / 79.3	77.9 / 76.2 / 78.5
FFNN(94121)	95.9 / 97.8 / 93.8	90.6 / 86.4 / 83.2	82.1 / 84.8 / 78.6	85.0 / 80.9 / 76.5

Table 1. Training and generalization results (in % correct) based on data projected on a single dimension by the different paradigms. All data was normalized. Results are shown as for noise free / 1% noise / 5% noise in the training data sets.

5 Conclusions

Amongst the four methods of dimensionality reduction considered in this paper, multi-layered feed-forward neural networks provided the best results. However, all of the methods seem to be affected by noise.

References

1. R. Bellman, *Adaptive Control Processes: A Guided Tour*, New Jersey: Princeton University Press, 1961.
2. J. Mao, and A. K. Jain, "Artificial Neural Networks for Feature Extraction and Multivariate Data Projection" *IEEE Transactions on Neural Networks*, vol. 297, pp. 296-317, 1995.
3. I. T. Jolliffe, *Principal Component Analysis*, New York: Springer-Verlag, 1986.
4. T. Kohonen T, *Self Organizing Maps*, Berlin: Springer Verlag, 1995.
5. H. Ritter, K. Schulten, "On the Stationary State of Kohonen's Self-oOganizing Sensory Mapping," *Biological Cybernetics*, vol. 54, pp. 99-106, 1986.
6. R. A. Fisher, "The Use of Multiple Features in Taxonomic Problems," *Annals of Eugenics*, Vol. 7, pp. 179-188, 1936.
7. R. O. Duda, and P. E. Hart, *Pattern Classification and Scene Analysis*, New York: John Wiley, 1973.
8. D. E. Rumelhart, G. E. Hinton, and R. J. Williams, "Learning Internal Representations by Back-Propagating Errors," *Nature*, vol. 332, pp. 533-536, 1986.
9. R. M. Haralick, and L. G. Shapiro, *Computer and Robot Vision, I*, Massachusetts: Addison Wesley, 1992.
10. M. Kass, A. Whitkin, and D. Terzopoulos, "Snakes - Active Contour Models," *Proc. Intl. Conf. on Comp. Vis.*, pp. 259-269, 1987.
11. R. Kothari, T. Balachander, R. Lotlikar, and H. Cualing, "Decision Theoretic Cytodiagnois of Lymphoma," Submitted.
12. M. H. Zhweig, and G. Campbell, "Receiver Operating Characterstic (ROC) Plots: A Fundamental Evaluation Tool in Clinical Medicine," *Clinical Chemistry*, vol. 39, pp. 561-577, 1993.

Topology Representing Networks for Intrinsic Dimensionality Estimation

J. Bruske, G. Sommer

Computer Science Institute
Christian-Albrechts University zu Kiel, Germany
email:jbr@informatik.uni-kiel.de

Abstract. In this paper we compare two methods for intrinsic dimensionality (ID) estimation based on optimally topology preserving maps (OTPMs). The first one is a direct approach, where the intrinsic dimensionality is estimated directly from the OTPM. We argue that this approach suffers from both practical and theoretical pitfalls. The second is a new approach which combines OTPMs with an efficient local principal component analysis (PCA). Exploiting the OTPM, local PCA can be shown to have only *linear time complexity* w.r.t. the dimensionality of the input space (in contrast to the prohibitive *cubic complexity* of the conventional approach), and hence the method becomes applicable even for very high dimensional input spaces as frequently encountered in computer vision. A local ID estimate is then obtained as the local number of significant eigenvalues. In addition to ID estimation the local subspaces as revealed by our local PCA can be directly used for further data processing tasks including classification and regression.
The workability of the new approach for ID estimation and subspace auto-association is demonstrated on a sequence of 64×64 pixel images (4096-dimensional input space).

1 Introduction

The intrinsic, or topological, dimensionality (ID) of N patterns in an n-dimensional space refers to the minimum number of "free" parameters needed to generate the patterns. It essentially determines whether the n-dimensional patterns can be described adequately in a subspace (submanifold) of dimensionality $d < n$, [5]. As pointed out in [3], knowledge of the ID is important in order to determine the number of features necessary to represent the data, to decide whether a reasonable 2d or 3d representation exists or to estimate the effectiveness of algorithms depending on the ID, as e.g. methods for constructing classifiers or training neural networks. It can be greatly helpful in problems like pattern recognition, industrial or medical diagnosis and data compression.

In this article, we will concentrate on two local approaches to ID-estimation based on optimally topology preserving maps (OTPMs) (see e.g. [5] for alternative approaches). The first one, [3], tries to directly estimate the local ID from the number of neighbors of a node in an OTPM. The second one, [1], uses an OTPM for efficient local PCA and estimates the ID as the number of significant eigenvalues. It is conceptually similar to that of Fukunaga and Olsen, [4], using local PCA as well, but by utilizing OTPMs can be shown to better scale with high dimensional input spaces (linear instead of cubic) and to be more robust against noise. In contrast to the direct approach, the local subspaces revealed by local PCA can be further used for data modeling.

In the remainder of this article we will first review OTPMs in section 2. We will then discuss the approach of Frisone et al. in section 3. Our own approach is summarized in section 4. and in section 5 we show how our local subspaces can be utilized for data modeling. A demonstration is given in section 6, and we close with a brief summary and outlook in section 7.

2 Optimally Topology Preserving Maps

Optimally Topology Preserving Maps (OTPMs) are closely related to Martinetz' Perfectly Topology Preserving Maps (PTPMs) [7] and are constructed in just the same way. The only reason to introduce them separately is that in order to form a PTPM the centers must be "dense" in the manifold M. Without prior knowledge this assumption cannot be checked, and in practice it will rarely be valid. OTPMs emerge if just the construction method for PTPMs is applied without checking for the density condition. Only in favorable cases one will obtain a PTPM (probably without noticing). OTPMs are nevertheless optimal in the sense of the topographic function introduced by Villmann in [11]: In order to measure the degree of topology preservation of a graph G with an associated set of centers S, Villmann effectively constructs the OTPM of S and compares G with the OTPM. By construction, the topographic function just indicates the highest (optimal) degree of topology preservation if G is an OTPM.

Definition 1 OTPM. Let $p(x)$ be a probability distribution on the input space R^n, $M = \{x \in R^n | p(x) \neq 0\}$ a manifold of feature vectors, $T \subseteq M$ a training set of feature vectors and $S = \{c_i \in M | i = 1, \ldots, N\}$ a set of centers in M.

We call the undirected graph $G = (V, E)$, $|V| = N$, an ***optimally topology preserving map of S given the training set T, $OTPM_T(S)$***, if

$$(i, j) \in E \Leftrightarrow \exists x \in T \, \forall k \in V \backslash \{i, j\} : \max\{\| c_i - x \|, \| c_j - x \|\} \leq \| c_k - x \|$$

Corolary 1 *If $T = M$ and if S is dense in M then $OTPM_T(S)$ is a PTPM.*

Note that the definition of $OTPM_T(S)$ is constructive: Simply pick $x \in T$ according $p_T(x)$, calculate the best and second best matching centers, c_{bmu} and c_{smu}, and connect bmu with smu. This procedure is just the essence of Martinetz' Hebbian learning rule for topology representing networks. Obviously, for a finite training set T the $OTPM_T(S)$ can be constructed in time $O(|T|)$. For a training set defined via a pdf $p_T(x)$, G will converge to $OTPM_T(S)$ with probability one.

For our purposes, $OTPM_T(S)$ has two important properties. First, it does only depend on the intrinsic dimensionality of T, i.e. it is independent of the dimensionality of the input space. Embedding T into some higher dimensional space does not alter the graph. Second, it is invariant against scaling and rigid transformations (translations and rotations). Just by definition it is the representation that optimally reflects the intrinsic (topological) structure of the data.

3 Direct ID estimation with OTPMS

Frisone et al. have been the first ones trying to exploit the benevolent properties of OTPMs for ID estimation. They tried to directly infer the ID from the number of direct neighbors of nodes in an OTPM by relating this number to the maximum kissing number in sphere packings (Kiss-SPP). The problem here is to find a packing of d-dimensional spheres of equal size so that the number τ of spheres touching (kissing) each other is maximal [2]. Kiss-SPP has only been solved for $d = 1, 2, 3, 8, 24$ and there exist optimal solutions for lattices of spheres for $d = 4, 5, 6, 7$, [2].

Analyzing the hypothetical analogy between the number of neighbors and the maximum kissing number one realizes that it rests on three assumptions: First, that the centers have been optimally distributed in the manifold (in the sense of the lowest quantization error), second, that the optimal distribution is realized by a lattice quantizer and third, that the problem of finding the best lattice quantizer is equivalent to finding the lattice with highest kissing number.

While there is some evidence that the last two assumptions hold at least for small d, they are in fact open questions, [2]. Anyway, lattices and other regular (optimal) center distributions can only emerge for very large number of centers (infinitely many) and, of course, an even larger numbers of training samples. Finally, a vector quantization algorithm generating the optimal distribution for this large number of centers (by annealing?) in finite time does not exist.

This requirement for a huge number of training data, long training times and lack of theoretical foundation appears to exclude the direct approach from practical applications.

4 Efficient ID estimation based on local PCA of OTPMs

Similar to the direct approach of Frisone, our ID estimation procedure rests on the fact that the number of neighbors of a node in an OTPM only depends on the intrinsic dimensionality d and is independent of the input dimensionality n.

It proceeds in four stages (batch-variant). First, generate a set of N centers $S = \{c_1, \ldots, c_N\}$ as the output of a vector quantization algorithm working on the training set T. Second, calculate the graph G as the optimally topology preserving map, $OTPM_T(S)$, of S w.r.t. T. Third, for each node $i \in G$ perform a principal component analysis of its correlation matrix $\frac{1}{m_i}A^TA$, $A^T = [c_{1_i} - c_i, \ldots, c_{m_i} - c_i]$, with $(c_{j_i} - c_i)$ the difference vectors between c_i and c_{j_i}, the center of its j-th direct topological neighbor in G. Finally, exclude eigenvectors corresponding to very small eigenvalues.

As a result of the vector quantization stage the centers are placed within the manifold M and noise orthogonal to M is filtered out. $OTPM_T(S)$ is constructed by simply connecting nodes corresponding to best and second best matching centers on presentation of T.

The central "trick" is to use the difference vectors $(c_{j_i} - c_i)$ for PCA of each local subspace and not the data in a local region itself, as e.g. in [4] or [6]: First, the difference vectors have very low noise component orthogonal to M (due to the noise reduction property of the vector quantizing stage), and second, the number of neighbors m_i of a node in an OTPM does only depend on the intrinsic dimensionality d and is small for small d. Straightforward PCA of the correlation matrix $\frac{1}{m_i}A^TA$ nevertheless would take time $O(n^3)$,[9], yet the m_i eigenvectors and m_i eigenvalues can be obtained by PCA of AA^T as well, cf. [8], taking only time $O(m_i^3)$. Since AA^T clearly can be computed in time $O(m_i^2n)$, and the number of neighbors m of a node in an OTPM does not depend on n but the intrinsic dimensionality d, local PCA of the correlation matrix takes only time $O(m(d)^2n + m(d)^3)$ and hence scales only linearly (optimally) with the input dimensionality.

Deciding, what size an eigenvalue as obtained by each local PCA must have to indicate an associated intra-manifold eigenvector, amounts to determining a threshold. We adopted the Dα criterion from Fukunaga et. al., [4], that regards an eigenvalue μ_i as significant if $\frac{\mu_i}{\max_j \mu_j} > \alpha\%$. If no prior knowledge concerning the distribution of the noise is available, different values of α have to be tested.

5 Local subspaces for data modeling

Local subspace analysis as described in section 4 supplies us with a set of (orthonormal) eigenvectors $e_1^i, \ldots, e_{l_i}^i$, $l_i \leq m_i$, spanning the local subspace for each center $c_i \in S$. These subspaces can be used straightforwardly to improve existing local approximation schemes including RBF networks and Local Linear Maps (LLMs),[10], by first projecting stimuli to the relevant subspaces. Here we demonstrate, how local subspaces can be used for compact coding by locally linear data modeling, [6]. In this approach, new data is modeled as

$$\hat{x} = c_{bmu} + \sum_{i=1}^{l_{bmu}} ((x - c_{bmu})^T e_i^{bmu}) e_i^{bmu}, \tag{1}$$

i.e. as the center of the best matching unit (Euclidean distance) and the projection to the subspace of the bmu, respectively. Using speech and (pre-processed) image data with typically low intrinsic dimensionality, Kambhatla and Leen demonstrated that this method compares well to (and even outperforms) standard bottle-neck Backpropagation networks. They, however, used conventional PCA on the data in the Voronoi cells. With help of local PCA based on OTPMs, local linear modeling now scales up linearly for high dimensional input spaces.

6 Experimental Results

In this demonstration we want to investigate an image sequence generated by taking 180 snapshots (every 2°) with a resolution of 64 × 64 pixels (4096-

dimensional input space) of a robot rotating a cylindric grey ramp around its z-axis (from 0° to 360°). Since the background remains constant, the images lie on a closed 1-dimensional trajectory in image space with ID $d = 1$.

Fig. 1. Grey ramp under different rotations. From left to right: Original (symmetric) grey ramp, grey ramp wrapped around a bottle with part of the robot arm in the background under 0°, 45°, 90° rotation

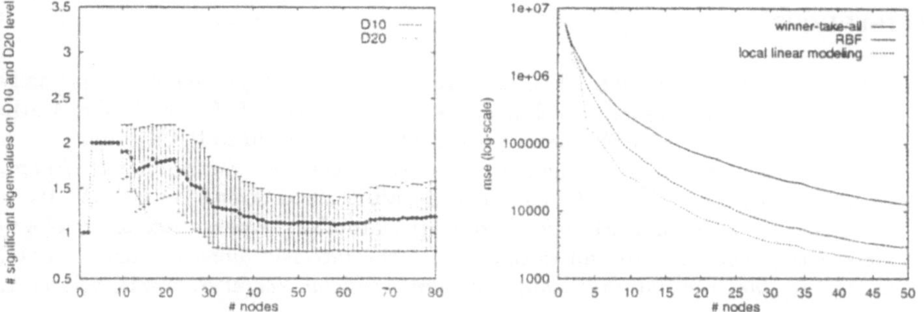

Fig. 2. Left: ID plots for rotating grey ramp on D10 and D20 level with error bars for the D10 level. Right: Reconstruction error (mse) for locally linear modeling, RBF and winner-take-all network on the D20 level.

Figure 2, left, shows the ID estimates obtained as the mean number of significant local eigenvalues by our procedure for different numbers of centers on the D10 and D20 level. The standard deviations of the estimates are included as error bars. The plots clearly indicate an ID of one or at least two. In figure 2, right, the reconstruction error (mse) for local linear modeling with the subspaces constructed on the D20 level is depicted (averaged over 180 test images). For comparison, the reconstruction error obtained with an RBF network and a simple winner take all scheme (same center distributions[1]) are also included. For a given number of centers, local linear modeling is clearly superior.

[1] As a vector quantizer for generating the center distributions we used an incremental version of the LBG algorithm, cf. [1]. Adding the $(N+1)$.th center where quantization is worst and keeping the old distribution of the remaining N centers, the LBG algorithm only needs to adjust centers in the near surrounding of the new one.

7 Summary

We have investigated two algorithms for ID estimation based on OTPMs. While the first approach pioneers an interesting idea, it generally turns out to be of little practical value. The second approach combines OTPMs with local PCA and thereby directly extends and improves the classical algorithm of Fukunaga and Olsen. [4], in terms of computational complexity and noise sensitivity. Scaling only linearly with the dimensionality of the input space, it turns out to be ideally suited for local subspace approximation of low dimensional manifolds in high dimensional input spaces in general. These subspaces can then be utilized by local subspace methods. as e.g. local linear data modeling.

Going beyond this article. we currently apply our local subspace construction method for constructing Hyper Basis Function networks and improved LLMs for visual learning. First results concerning appearance based robot grasping and pose recognition are very encouraging.

References

1. J. Bruske and G. Sommer. Intrinsic dimensionality estimation with optimally topology preserving maps. Technical Report 9703, Inst. f. Inf. u. Prakt. Math. Christian-Albrechts-Universitaet zu Kiel, 1997. (submitted to IEEE PAMI).
2. J.H. Conway and N.J.A. Sloane. *Sphere Packings, Lattices and Groups.* Grundlehren der mathematischen Wissenschaften 290. Springer Verlag NY, 1988.
3. F.Frisone, P.Morasso. F.Firenze, and L.Ricciardiello. Application of topology-representing networks to the estimation of the intrinsic dimensionality of data. In *Proc. of the International Conference on Artificial Neural Networks*, volume 1, pages 323–327, 1995.
4. K. Fukunaga and D. R. Olsen. An algorithm for finding intrinsic dimensionality of data. *IEEE Transactions on Computers*, 20(2):176–183, 1971.
5. A. K. Jain and R. C. Dubes. *Algorithms for Clustering Data.* Prentince Hall, 1988.
6. N. Kambhatla and T.K. Leen. Fast non-linear dimension reduction. In *Advances in Neural Information Processing Systems, NIPS 6*, pages 152–159, 1994.
7. T. Martinetz and K. Schulten. Topology representing networks. In *Neural Networks*, volume 7, pages 505–522, 1994.
8. H. Murase and S. Nayar. Visual learning and recognition of 3-d objects from appearance. *International Journal of Computer Vision*, 14:5–24, 1995.
9. W.H. Press, S.A. Teukolsky, W.T. Vetterling, and B.P. Flannery. *Numerical Recipes in C - The Art of Scientific Computing.* Cambridge University Press, 1988.
10. H. Ritter, T. Martinetz. and K. Schulten. *Neuronale Netze.* Addison-Wesley, 1991.
11. T. Villmann, R. Der. and T. Martinetz. A novel aproach to measure the topology preservation of feature maps. *ICANN*, pages 289–301, 1994.

SOM Based Visualization in Data Analysis

Erkki Häkkinen * and Pasi Koikkalainen

University of Jyvskylä
Department of Mathematics
P.O.BOX 35, FIN-40351 JYVÄSKYLÄ, FINLAND

Abstract. Visualization is an important part of data analysis, especially when exploring multidimensional data. Our approach uses the self-organizing map (SOM) as a basic method because it provides a good basis for data visualization. The developed analysis tool utilizes the structure of SOM and is integrated with a generally applicable visualization system. In addition, we propose a model to link several SOM presentations for visualizing more complex structures of information.

1 Introduction

The application framework of this study is data analysis, that has four abstract parts [6]: data collection, data reduction, data visualization and decision making. Our approach utilizes the strength of a tree-structured variant (TS-SOM [5]) of self-organizing map (SOM) [3]. As a large number of research and very different applications indicate, the SOM includes properties that make it a useful tool for visualizing multidimensional data [4],[1],[8].

The aim of this work was to create a unified visualization system, that supports the special requirements of the SOM, but does not exclude any ordinary visual presentations for any data as presented in [7]. In addition, the system must be able to be extended for different applications without large rebuilding. As a practical framework we have developed an environment called Neural Data Analysis (NDA) [2], to which the visualization tool is integrated as discussed in this paper.

To outline the tasks in visualization, we concentrate on SOM based data analysis without forgetting the requirement of generality. First we define the SOM analysis as a mapping $D \to S$ from a data set D to SOM results S. The basic analysis, given in more detail in [2], involves five possible operations: the selection of training data, SOM training, SOM clustering, the selection of data for visualization and the statistical computing of cluster features.

By noting the above requirements, the demands of SOM analysis and practical experiences in different applications, we have specified the tasks for visualization at four levels:

The basic SOM visualization: $D \to S \to V$, where V denotes the visual presentation of SOM. Note that the second macro operation $S \to V$ involves a large number of operations for specifying the presentation.

* This work was supported by TEKES under Stella project.

Neuron linkage: $D \to S \to V_1 \dots V_{nV}$. Different visual presentations V_i of the same SOM are linked such that a neuron is highlighted similarly in all displays.

Data linkage: Several distinct SOMs are displayed, while presentations are linked through the structures and content of data sets. Two cases will be discussed:

Hierarchical: $D^j \to S^j \to V^j \to D^{j+1} \to S^{j+1} \to V^{j+1}$, where the data set D^{j+1} is a subset from the set D^j.

Parallel: $D \to S^1 \dots S^{sN} \to V_i^1 \dots V_i^{sN}$, where V_i^j denotes a set of presentations for S^j, typically using different selection of data fields.

Knowledge linkage: $D^1 \to S^1 \to V^1, D^2 \to S^2 \to V^2$. Distinct data sets are analyzed separately, but external knowledge can be used to create connections between the presentations.

2 The basic SOM visualization

In order to find a generally applicable visualization system, the visualization possibilities of SOM are outlined more accurately, as presented in Fig. 1. The non-leaf nodes provide different representations for source information, while the leaf nodes of hierarchy are visual primitives that can be displayed in the screen by a straightforward implementation. In the modular structure of hierarchy, the source representations of primitives are fixed to be as general as possible. This guarantees, that the same implementations of primitives can be reused at several levels (for neurons, groups or any data sets), and new features can be added easily.

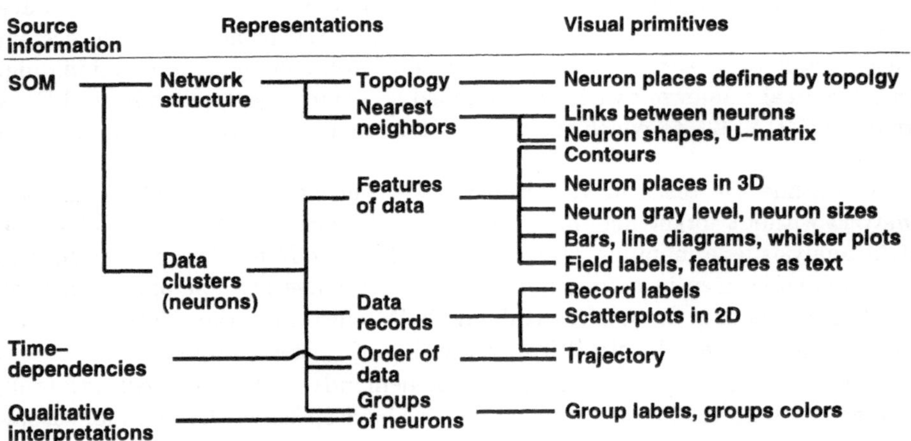

Fig. 1. The hierarchy of source information, representations and visual primitives in SOM based visualization

To justify the large set of primitives, we note that the neurons of SOM are described by multidimensional vectors, such as weights or cluster features. In order to explore them more effectively, versatile visual primitives that can be displayed simultaneously have been developed. Because a complete description of them is impossible to give in this space, some examples of NDA graphics are shown in Fig. 2.

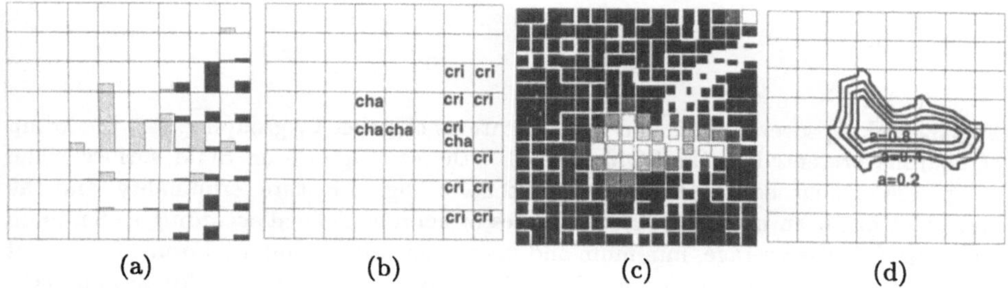

(a) (b) (c) (d)

Fig. 2. Basic visual primitives in parallel views for "Boston" housing data: (a) and (b), averages of fields 'criminality' and 'river side' presented as bars and field labels (rules: "crim.avg $>$ 10", "chas.avg $>$ 0.5 AND chas.avg $<$ 1.0"); (c), dissimilarities between neurons presented as their shapes; (c) and (d), the feature "chas.avg" presented as gray levels and contours.

Despite various visual features, readability restricts the number of simultaneous features in one presentation. To avoid these problems, several views can be taken to the same representations of the SOM. There are two widely used, simple techniques. In the first one, different presentations are created for selected data clusters i.e. neurons, which makes a more detailed viewing possible. The second way is to use several parallel views, such as component plane represenations or graphics as presented in Fig. 2, that allow the human to compare topologically similar maps easily.

3 Neuron linkage for displaying several views

Parallel views to the same SOM provide a useful way to support the visualization. However, there are two cases in which context between the presentations is lost. Firstly, in 3-dimensional presentations different viewpoints can be used, and secondly, the places of neurons can be set according to any features of data clusters.

Visual connections between the views can be maintained by a neuron linkage that is implemented via a grouping tool. The grouping is an operation that brings user-defined external knowledge (i.e. qualitative interpretations) to the system. It specifies collections of neurons as groups and gives names for them. This grouping information is grounded to the neurons, such that the same groups can be displayed in distinct views by visual primitives as presented in Fig. 1.

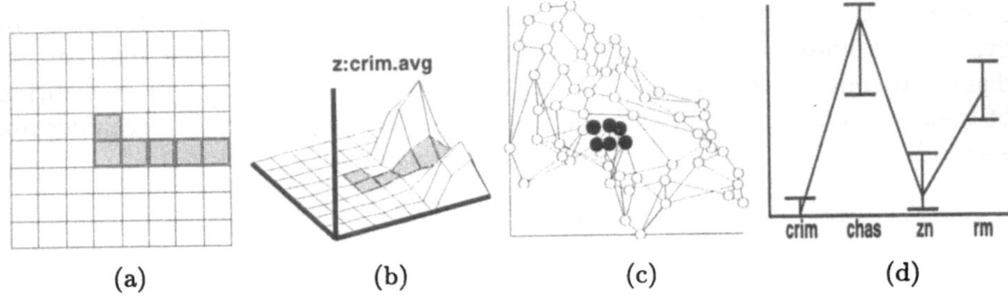

Fig. 3. Linking several views and representations together by grouping: (a), the group of neurons presented in the SOM grid; (b), the same group on SOM surface when the places of neurons in z axis are defined according to feature 'criminality' and the viewpoint has been modified; (c), the places of neurons defined according to Sammon mapping; (d), the average, minimum and maximum values computed from the data of the group. Source fields are 'criminality', 'Chas-river', 'field size' and 'room number'.

As an example, we demonstrate how a map can be explored through parallel views with the grouping tool. Two different viewpoints (compare Fig. 3a with Fig. 3b) have been taken to the same SOM representation. Then the neurons are collected to a group according to a rule "chas.avg > 0.5". The group is then displayed with colored neurons in both views. Furthermore, to observe the dissimilarities between the neurons, the weight vectors have been transformed to a lower dimensional space by using a Sammon mapping (see Fig. 3c). Finally, the data sets of the neurons are combined to a larger data set of the group, for which any scalar features can be computed and displayed as in Fig. 3d.

Although the grouping was introduced here in the context of linking views, its main purpose is to be a decision making tool. Especially, in complex or ill-defined problems, the decision making process become highly interactive and iterative. To facilitate this difficult and costly task, three different methods have been implemented:

Interactive selection: Neurons are selected to groups by pointing them from visual presentations with the mouse.

Rule: Neurons are selected by using a rule on feature values. This allows us to bring advanced knowledge in symbolic form to the system.

Prototype classification: Some neurons are selected as prototypes of groups. The rest are divided between these groups via best-match search, where any cluster features can be used as classifying criterias.

4 Data linkage for displaying several maps

The data linkage joins distinct SOM presentations through data records. Two basic techniques are possible:

Hierarchical presentations for subsets. Grouping can be used to define subsets of data, that can be analyzed and visualized again. A subset of data records D^{i+1} is selected from a set D^i. Then a new SOM presentation is created for the subset. The task of linking is to show the relationships between these sets. Figure 4a shows a simple visual primitive for linking the new presentation to the first map.

Parallel presentations for subspaces. Different sets of data fields are selected from the same data and SOM presentations are created for them. The task of the linking is to show the relationships of the same data records mapped in distinct SOMs.

To demonstrate the linkage between parallel presentations, two SOMs, "map1" and "map2" have been trained using data sets that have corresponding records, but different fields. The maps can be explored together by the following procedure (see also Fig. 4b):

1. Select a neuron (or cluster) C_{map1} from the "map1".
2. Pick the identifiers of data records $R_1 \ldots R_{N1}$ from the selected cluster C_{map1}.
3. Find neurons $C^1_{map2} \ldots C^{M2}_{map2}$ from the "map2" that includes an identifier from the set $\{R_1 \ldots R_{N1}\}$.
4. Draw links from the selected neuron C_{map1} to neurons $C^1_{map2} \ldots C^{M2}_{map2}$.

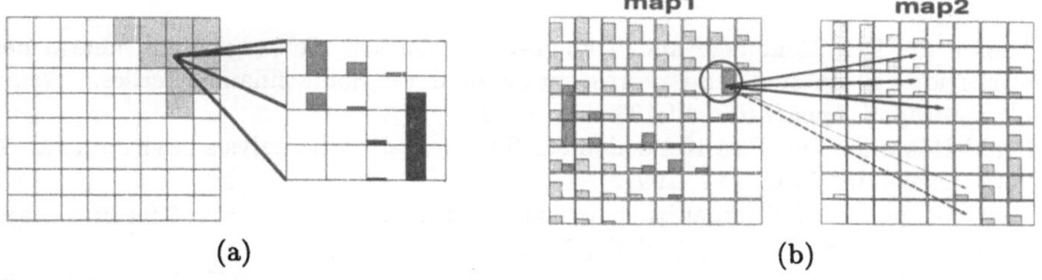

(a) (b)

Fig. 4. The hierarchical display: (a). The selected group from the left map has been analyzed more exactly in the right map in which fields 'criminality' and 'price' are presented as bars; The parallel display: (b). Data fields analyzed in the left SOM describe "apartment surroundings", where the selected neuron represents apartments on river side near the city. The right map represents apartment types and people in the surrounding areas. The width of link presents the number of apartments mapped through the link.

5 Knowledge linkage

Knowledge linkage technique is considered for advanced analysis and presentations of different data sets, but is not yet implemented in the NDA. It needs external knowledge to display context between the SOM presentations of separated analyses. For instance, when exploring relational databases, data clusters are supplemented with symbolic data in addition to training data, allowing more advanced relational operations. Another possibility to support complex linking via a more sophisticated inference would be symbolic reasoning and fuzzy logic to aid the visualization process.

6 Conclusions

We have implemented a visualization tool, that utilizes the structure of the SOM. Through examples we showed that the SOM based visualization can be returned to the set of ordinary visual presentations, and thus the overhead in implementation can be reduced. The model for more complex SOM presentations was also given. The interpretation of SOM can be facilitated by several views that are connected by neuron linkage and data linkage techniques.

The NDA and its visualization tool have been applied successfully in such domains as market analysis, logistics, process monitoring, and human sciences. Practical applications have indicated the importance of visualization. Also the structure of visualization system has proved flexible for extentions of domain-specific features.

References

1. Martin del Brio Bonafacio and Serrano-Cinca Carlos. Self-organizing neural networks for the analysis and represenation of data: Some financial cases. *Neural Computing & Applications*, 1:193–206, 1993.
2. Erkki Häkkinen and Pasi Koikkalainen. The neural data analysis environment. To be published in WSOM'97, 1997.
3. Teuvo Kohonen. Self-organizing formation of topologically correct feature maps. *Biological Cybernetics*, 43(1):59–69, 1982.
4. Teuvo Kohonen. *Self-Organizing Maps*, volume 30. Springer,Berlin, Heidelberg, New York, 1995.
5. Pasi Koikkalainen. Progress with the tree-structured self-organizing map. In A. G. Cohn, editor, *Proc. ECAI'94, 11th European Conf. on Artificial Intelligence*, pages 211–215, New York, 1994. John Wiley & Sons.
6. Matthew B. Miles and A. Michael Huberman. *Qualitative Data Analysis.* SAGE Pubblications, CA, USA, 1994.
7. Edward R. Tufte. *The visual Display of Quantitative Information.* Graphic Press, USA, 1983.
8. A. Ultsch. *Self-organizing Neural Networks for Visualization and Classification*, pages 307–313. Springer-Verlag, Berlin, 1993.

ARTMAP-DS: Pattern Discrimination by Discounting Similarities

Gail A. Carpenter[1] and Frank D. M. Wilson[2]

[1] Boston University, Boston MA 02215, USA
[2] Raytheon Company, Marlborough MA 01752, USA

Abstract. ARTMAP–DS extends fuzzy ARTMAP to discriminate between similar inputs by discounting similarities. When two or more candidate category representations are activated by a given input, features that the candidate representations have in common are ignored prior to determining the winning category. Simulations illustrate the network's ability to recognize similar inputs, such as STAR and START, in a noisy environment.

1 Focusing Attention on Small Differences

ARTMAP–DS is a supervised neural network for learning and recognition. The network extends fuzzy ARTMAP (Carpenter et al., 1992) to discriminate between similar inputs by discounting similarities. The network functions by focusing attention on differences between candidate category representations activated by a given input, and then checking to see which features are in fact present in the input, ignoring features that the candidate representations have in common. Attentional focusing is particularly needed in syllable and word recognition applications, where a primacy gradient input representation (Grossberg, 1978) may cause low–amplitude feature representations (in the later parts of sequences) that are vulnerable to input error and processing noise. A high value of the vigilance parameter, ρ, is needed to ensure that a fuzzy ART network can distinguish between similar input sequences such as STAR and START (Wilson, 1996; Carpenter & Wilson, 1997); but a high value also prevents the system from correctly classifying noisy inputs. The complement-coded input representation used in fuzzy ART (Carpenter, Grossberg, & Rosen, 1991) exacerbates this problem, since the contribution in the input from the phonemes or syllables that are present may be largely masked by the contribution from the larger number that are absent. With ARTMAP-DS, a difference in the later part of the input sequence is not much harder to detect than an earlier one.

2 Fuzzy ARTMAP

Fuzzy ARTMAP is a supervised neural network for learning, recognition and prediction. Figure 1 illustrates a fuzzy ARTMAP system for classification problems, where each input $\mathbf{a}$ learns to predict an output class K. The network creates internal recognition categories during training. The input vector $\mathbf{a}$ is scaled so that each $a_i \in [0,1]$ $(i = 1 \cdots M)$. Complement coding doubles the number of components in the input vector, which becomes $\mathbf{I} = (\mathbf{a}, \mathbf{a}^c)$, where the ith component of $\mathbf{a}^c$ is $a_i^c = (1 - a_i)$. With fast learning, the ART$_a$ weight vector $\mathbf{w}_j \equiv (w_{j1}, \ldots, w_{j,2M})$ records the largest and smallest component values of input vectors placed in the jth category.

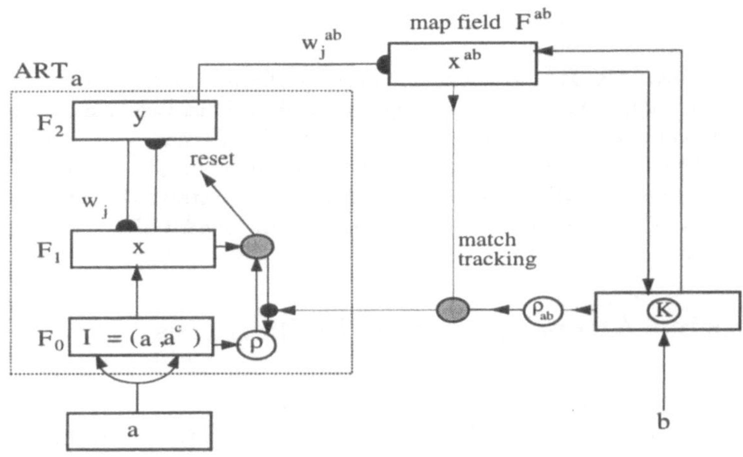

Fig. 1. A fuzzy ARTMAP network for classification.

The $F_1 \to F_2$ input T_j is given by the Weber law function:

$$T_j(\mathbf{I}) = \frac{|\mathbf{I} \wedge \mathbf{w}_j|}{\alpha + |\mathbf{w}_j|}, \tag{1}$$

where $(\mathbf{P} \wedge \mathbf{Q})_i \equiv \min(P_i, Q_i)$ and $|\mathbf{P}| = \sum_{i=1}^{2M} |P_i|$. Activity at F_2 is denoted by $\mathbf{y} \equiv (y_1, \ldots, y_N)$. With winner-take-all coding, only the F_2 node J that receives the largest $F_1 \to F_2$ input T_j becomes active. Node J remains active if it satisfies the matching criterion:

$$\frac{|\mathbf{I} \wedge \mathbf{w}_J|}{|\mathbf{I}|} = \frac{|\mathbf{I} \wedge \mathbf{w}_J|}{M} > \rho, \tag{2}$$

where $\rho \in [0,1]$ is the dimensionless ART$_a$ *vigilance parameter*. Otherwise, the network resets the active F_2 node and searches until J satisfies (2). At the start of each input presentation, ρ equals a baseline vigilance, $\overline{\rho}$. When node J is active at F_2 and class label K is active in the training input $\mathbf{b} \equiv (b_1, \ldots, b_L)$, activity at the map field F^{ab} is $\mathbf{x}^{ab} = \mathbf{b} \wedge \mathbf{w}_J^{ab}$, where $\mathbf{w}_j^{ab} \equiv (w_{j1}^{ab}, \ldots, w_{jL}^{ab})$ denotes the weight vector from the jth F_2 node to F^{ab}. If node J then makes an incorrect class prediction (i.e., if $\mathbf{x}^{ab} \neq \mathbf{b}$), a *match tracking* signal raises ART$_a$

vigilance ρ to $|\mathbf{I} \wedge \mathbf{w}_J^a| \,/\, |\mathbf{I}| + \epsilon$, where ϵ is vanishingly small. This increase is just enough to induce a search, which continues until either some F_2 node becomes active for the first time, in which case the weight vector $\mathbf{w}_J^{ab}$ is set equal to $\mathbf{x}^{ab}$, so that J learns the correct output class label $k(J) = K$; or until a node J that has previously learned to predict K becomes active. During testing, a pattern a that activates node J is predicted to belong to the class $K = k(J)$.

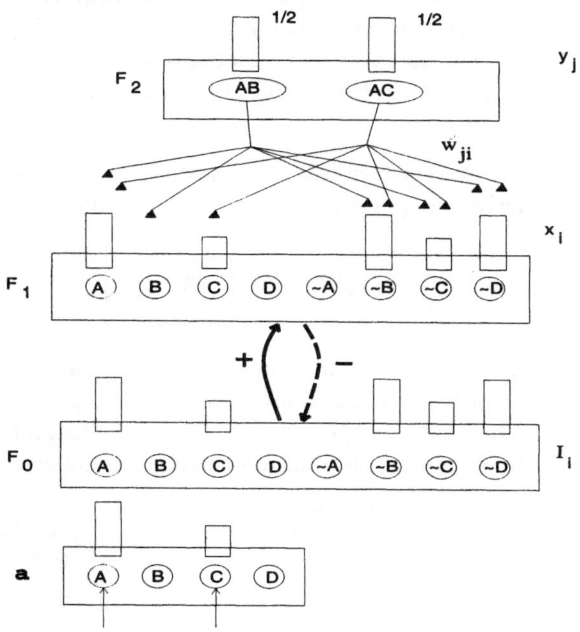

Fig. 2. The ARTMAP–DS network. When more than one F_2 node remains strongly active after contrast–enhancement, F_1 first registers features that are present both in the input and in all strongly active templates. Inhibitory $F_1 \rightarrow F_0$ connections then mask out these input features. Thus, if input AC activates both AB and AC at F_2, A and "not D" are inhibited and "not B" and "not C" are partially inhibited at F_0. After renormalization, F_0 sends a masked input back to F_1 and F_2, which allows F_2 to choose among the partially active nodes. The figure shows the initial situation, in which activity at F_1 is equal to $\mathbf{I}^{(norm)}$, and AB and AC have just become active at F_2. After F_1 has inhibited F_0, only nodes C and "not B" will remain active at F_0.

3 The ARTMAP–DS Network

ARTMAP–DS (Figure 2) replaces the ART$_a$ subsystem of fuzzy ARTMAP. During training, ARTMAP–DS functions identically to fuzzy ARTMAP. During subsequent recognition tests, input to F_0 is the complement–coded vector $\mathbf{I} = (\mathbf{a}, \mathbf{a}^c)$. F_0 activity is normalized, $I_i^{(norm)} = I_i / |\mathbf{I}|_\infty$, where $|\mathbf{I}|_\infty = \max_i \{I_i\}$ is the L^∞ norm ($\lim_{p\rightarrow\infty} (\sum_i I_i^p)^{(1/p)}$). Bottom–up $F_0 \rightarrow F_1$ input results in initial activity $\mathbf{x}^{(init)}$ at F_1 that is equal to $\mathbf{I}^{(norm)}$. Initial input to F_2 is given by $T_j(\mathbf{x}^{(init)})$, where T_j is as defined in (1). Activity at F_2 is contrast–enhanced:

$$y_j = \frac{T_j^p}{\sum_r T_r^p}, \tag{3}$$

where p is significantly greater than 1, but is not so large that F_2 activity approximates choice. If there is then a clear winner at F_2, i.e., if $y_J \geq \theta$ for some J, where $0.5 \leq \theta < 1$, then the input is predicted to belong to class $K = k(J)$ as in vanilla fuzzy ARTMAP. Otherwise ARTMAP discrimination by discounting similarities is invoked, as follows.

The field F_1 first registers features that are present both in the input and in all strongly active templates. The matched activity $\mathbf{x}^{(match)}$ at F_1 is:

$$x_i^{(match)} = I_i^{(norm)} \wedge \left[\bigwedge_{j:y_j > \phi} w_{ji} \right], \tag{4}$$

where $0 < \phi < \theta$. For example, if the input $\mathbf{a}$ is STIR, and if STAR and STIR nodes are both significantly active at F_2 after contrast–enhancement, F_1 will register S, T and R. (In all simulations, $\phi = 0.1$ and $\theta = 0.7$.) Inhibitory connections from F_1 to F_0 then mask out these common features, as follows:

$$I_i^{(inh)} = [I_i^{(norm)} - x_i^{(match)}]^+ = [I_i^{(norm)} - \bigwedge_{j:y_j > \phi} w_{ji}]^+, \tag{5}$$

where $[w]^+ \equiv \max(w, 0)$. Activity at F_0 is then renormalized, F_0 sends a masked input ("I, not A") back to F_1 and F_2, and F_2 makes a choice (STIR) from among the partially active nodes. That is, the network replaces the former value of $\mathbf{I}$ with $\mathbf{I}^{(inh)}$, and recomputes $\mathbf{I}^{(norm)}$, $\mathbf{x}^{(init)}$ and T_j. At F_2, contrast–enhancement and thresholding are repeated, but only among nodes that had been active in the previous iteration. If there is still no clear winner at F_2, this discrimination process is repeated until a clear winner does emerge. As in vanilla fuzzy ARTMAP, $\rho_a = 0$ during recognition tests, so the matching criterion (2) is automatically satisfied, once an F_2 node J that is a clear winner emerges.

4 Syllable Recognition with Input Noise

This section describes simulations that illustrate the ability of ARTMAP–DS to give better recognition performance than fuzzy ARTMAP for noisy inputs. The inputs are words that are represented as phoneme sequences using a primacy gradient with steepness 0.5, i.e., the input for the nth phoneme in the sequence has amplitude $0.5^{(n-1)}$. First, the five notional phoneme sequences STAR, STARK, STIR, SHARK and SHIRK are presented to ARTMAP–DS and to fuzzy ARTMAP for learning. Learning in ARTMAP–DS and in fuzzy ARTMAP functions identically. Supervised learning ensures that each input is coded by a separate F_2 node. During learning, inputs **a** are noise–free and a correct representation of each input is learned. During subsequent performance, Gaussian noise with amplitude proportional to the component magnitudes is added to each input component, i.e., $a_i^{(noisy)} = a_i\ (1 + n_i)$, where n_i is Gaussian with mean 0 and standard deviation σ_{noise}. Each word was tested 20 times at each noise level. Further simulations illustrate recognition performance using a set of 50 monosyllabic words (see Appendix) constructed from 22 phonemes. After noise–free learning, each word is tested 10 times at each noise level. Table 1 shows that for both simulations, ARTMAP–DS achieves better recognition performance than fuzzy ARTMAP.

Standard deviation of noise (σ_{noise})		0.1	0.2	0.3	0.4	0.5	0.6	0.7	0.8	0.9	1.0	# test set inputs
# test set A recognition errors	Fuzzy ARTMAP	0	0	0	3	3	4	7	11	18	21	100
	ARTMAP–DS	0	0	0	1	1	2	4	8	15	16	100
# test set B recognition errors	Fuzzy ARTMAP	0	2	8	24	56	88	112	134	159	176	500
	ARTMAP–DS	0	2	8	22	53	83	108	131	149	163	500

Table 1. Recognition error rates when (A) the inputs STAR, STARK, STIR, SHARK and SHIRK, and (B) the 50 words listed in the Appendix are repeatedly presented to fuzzy ARTMAP and to ARTMAP–DS, with noise added to the inputs. Noise is Gaussian, with amplitude proportional to the magnitude of each input component. Parameters $\alpha = 0.001$, $\phi = 0.1$ and $\theta = 0.7$. For test set (A), $p = 20$, and for test set (B), $p = 400$.

5 Syllable Recognition with Network Noise

This section describes simulations that illustrate the ability of ARTMAP–DS to give better recognition performance than fuzzy ARTMAP when endogenous noise perturbs the match field F_1 during recognition tests. First, the input phoneme sequences are learned by each network as in the previous section. During learning, F_1 is noise–free. During subsequent performance, initial activity $x_i^{(init)}$ at F_1 is calculated by adding Gaussian noise n_i to each normalized input component $I_i^{(norm)}$, bounding the result so that it lies within the range $[0, 1]$, and then renormalizing at F_1 to ensure that $|\mathbf{x}^{(init)}| = M$. After noise–free learning, each word from the 50–word lexicon (see Appendix) is tested 10 times at each noise level. ARTMAP–DS achieves a performance improvement over fuzzy ARTMAP across a broad range of values of p ($100 \leq p \leq 400$). Low values of p do not provide enough contrast enhancement at F_2, resulting in a higher error rate. For values of p greater than 400, F_2 dynamics approximate choice, so ARTMAP–DS reduces to fuzzy ARTMAP. Table 2 shows that ARTMAP–DS again achieves better recognition performance than fuzzy ARTMAP.

Standard deviation of noise (σ_{noise})		0.05	0.1	0.15	0.2	# test set inputs
# test set	Fuzzy ARTMAP	8	45	103	139	500
recognition errors	ARTMAP–DS	2	27	63	121	500

Table 2. Number of recognition errors when the 50 words listed in the Appendix are each presented 10 times to the ARTMAP–DS network and to fuzzy ARTMAP, with noise added at F_1. Parameters $p = 400$, $\phi = 0.1$ and $\theta = 0.7$.

References

Carpenter, G. A., Grossberg, S., Markuzon, N., Reynolds, J. H., & Rosen, D. B. (1992). Fuzzy ARTMAP: A neural network architecture for incremental supervised learning of analog multidimensional maps. *IEEE Transactions on Neural Networks, 3*, 698–713.

Carpenter, G. A., Grossberg, S., & Rosen, D. B. (1991). Fuzzy ART: Fast stable learning and categorization of analog patterns by an adaptive resonance system. *Neural Networks, 4*, 759–771.

Carpenter, G. A., & Wilson, F. D. M. (1997). Segmentation-ART: A neural network for word recognition from continuous speech. *In preparation.*

Grossberg, S. (1978). A theory of human memory: self–organization and performance of sensory–motor codes, maps and plans. *Progress in Theoretical Biology, 5*, 233–374.

Wilson, F. D. M. (1996). *Neural networks for noise–tolerant category discrimination with application to continuous speech segmentation.* Ph.D. thesis, Boston, MA: Boston University.

Appendix

In the 50–word simulations, the phoneme set is: */p/*, */b/*, */t/*, */d/*, */k/*, */g/*, */s/*, */S/*, */r/*, */l/*, */a/*, */i/*, */u/*, $/\Lambda/$, */f/*, */m/*, */n/*, $/\epsilon/$, */o/*, $/\theta/$, */h/* and silence. The words and their transcriptions are: star */star/*, stark */stark/*, stir */stir/*, shark*/shark/*, shirk */shirk/*, odd */ad/*, are */ar/*, ark */ark/*, art */art/*, box */baks/*, bar */bar/*, bark */bark/*, be */bi/*, beast*/bist/*, beat */bit/*, breed */brid/*, brood */brud/*, boot */but/*, dark */dark/*, dart */dart/*, drop */drap/*, drew */dru/*, friend */frend/*, Greek */grik/*, greet */grit/*, grew */gru/*, group */grup/*, car */kar/*, key */ki/*, keep */kip/*, crop */krap/*, creep */krip/*, lead */lid/*, leap */lip/*, least */list/*, par */par/*, park */park/*, part */part/*, pat */pat/*, see */si/*, seek */sik/*, seal */sil/*, seat */sit/*, spark */spark/*, spot */spat/*, stop */stap/*, struck $/str\Lambda k/$, streak */strik/*, sue */su/*, tar */tar/*. Although the correct phonetic representations of the English words SHARK and SHIRK are */Sak/* and $/S\Lambda k/$ respectively, the nonword phoneme sequences */shark/* and */shirk/* are used instead, in order to allow the 50–word lexicon to incorporate the 5–word lexicon as a subset.

Acknowledgements: This research was supported in part by the National Science Foundation (NSF–IRI–94–01659) and the Office of Naval Research (ONR N00014–95–1–0409 and ONR N00014–95–1–0657)

A Self-Organizing Network That Can Follow Non-stationary Distributions

Bernd Fritzke

Systembiophysik, Institut für Neuroinformatik
Ruhr-Universität Bochum
44780 Bochum, Germany

Abstract. A new on-line criterion for identifying "useless" neurons of a self-organizing network is proposed. When this criterion is used in the context of the (formerly developed) growing neural gas model to guide deletions of units, the resulting method is able to closely track non-stationary distributions. Slow changes of the distribution are handled by adaptation of existing units. Rapid changes are handled by removal of "useless" neurons and subsequent insertions of new units in other places.

1 Non-stationary data is difficult to handle ...

Non-stationary data distributions can be found in many technical, biological or economical processes. Self-organizing neural networks have rarely been considered for tracking those distributions since many of the models, e.g. the self-organizing map [6], neural gas [8], or the hypercubical map [1], use decaying adaptation parameters[1]. Once the adaptation strength has decayed, the network is "frozen" and thus unable to react to subsequent changes in the signal distribution.

2 ... even for incremental networks

Models with small constant parameters such as the incremental networks developed by the author [2–4] are in a somewhat better position for handling non-stationary distributions. The non-decreasing adaptation rate enables the networks to follow slowly changing probability distributions like e.g. a normal distribution with a slowly drifting mean. Rapid changes in the distribution, however, can in general not be handled properly as is illustrated for the growing neural gas (GNG) [3] model in figure 1.

In the case of GNG it is easy to see why considerably many units may get stuck in former regions of high probability density and become so-called *dead units*. The network topology is updated – apart from the characteristic insertions – by two mechanisms (as described in detail in [3]). "Competitive Hebbian Learning" [7] is used to create new connections by always inserting a connection

[1] ... as does the classical k-means algorithm.

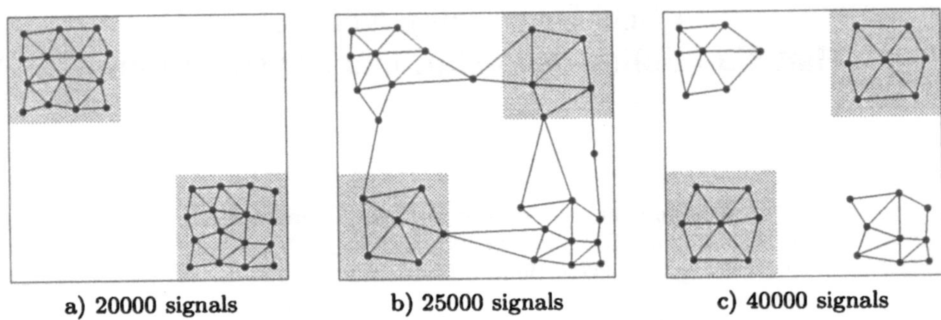

a) 20000 signals b) 25000 signals c) 40000 signals

Fig. 1. A growing neural gas (GNG) network tries (and fails) to track a non-stationary signal distribution $p(\boldsymbol{\xi})$. The maximum network size is set to 30. a) Initial state of $p(\boldsymbol{\xi})$ with the GNG network already grown to its maximum size. At this point the distribution was changed. b) State of the network 5000 adaptation steps after the distribution had changed. c) After 50000 adaptation steps some units have been adapted to the new regions of high probability density, but some have become dead units which are not adapted anymore. Parameters used (see [3] for details):$\lambda = 500$, $\epsilon_b = 0.05$, $\epsilon_n = 0.0006$, $\beta = 0.0005$, $a_{max} = 120$.

between the units s_1 and s_2 nearest and second-nearest to the current input signal $\boldsymbol{\xi}$. When such a connection does exist already, its *age* parameter is set to zero. Moreover, the age of all edges adjacent to the winning unit s_1 is increased. If the age of an edge surpasses a maximum parameter a_{max}, the edge is removed and so are any units without any emanating edges.

One should note, that the described aging is a local process, since it happens only in the vicinity of the winning unit. Now consider the situation where a network has grown to a certain maximum size and has well adapted to a given probability distribution (figure 1a). When the probability density changes, some units are adapted rather quickly towards the new regions of high probability density (figure 1b). The edges connecting those units and the units remaining in the former regions of high probability density are rather quickly removed from the aging process (figure 1c).

Thereafter, however, those units and connections which remain in the former regions of high probability density are not changed at all anymore. The units are not winner since they are too far from the current input signals. Consequently, the connections do not age since aging is only performed for edges adjacent to the winner. The result is a waste of network resources which may get even worse if the distribution changes several times. One could perhaps interpret the dead units as a kind of memory which may be useful again, when the probability distribution takes on a previously held shape[2], but here we like to discuss ways of using a network of limited size to track the distribution in each moment as good as possible.

[2] a suggestion made by Christoph von der Malsburg

3 Objective function

As objective for the network we define the minimization of the expected quantization error

$$E = \int \|\boldsymbol{\xi} - \mathbf{w}_{s(\boldsymbol{\xi})}\|^2 p(\boldsymbol{\xi}) d\boldsymbol{\xi}. \quad (1)$$

Thereby $s(\boldsymbol{\xi})$, also denoted as s_1, is the winning unit for the current input signal $\boldsymbol{\xi}$ and $\mathbf{w}_{s(\boldsymbol{\xi})}$ is the reference vector associate with that unit. With $p(\boldsymbol{\xi})$ we denote the probability distribution of the input signals which is assumed to be unknown and non-stationary.

4 Local error measure

In a GNG network each unit c in the network has an error variable E_c which contains accumulated error information. For each input signal $\boldsymbol{\xi}$ the error variable of the winner s_1 for this signal is updated according to

$$E_{s_1} := E_{s_1} + \|\boldsymbol{\xi} - \mathbf{w}_{s_1}\|^2. \quad (2)$$

Moreover, after each adaptation step (small change of the reference vectors of s_1 and its topological neighbors, see [3]) the error variables of all units are subjected to exponential decay:

$$E_c := E_c - \beta E_c \quad \text{(for each unit } c) \quad (3)$$

whereby β is a suitable decay constant. The decay stresses the influence of more recently measured error and is needed to account for the adaptation of the reference vectors and for the non-stationarity of the distribution. The GNG algorithm performs insertions of new units always close to the unit q with maximum accumulated error.

5 Local utility measure

Removal of units would enable us to get rid of dead units. The GNG algorithm would in this case insert a new unit at a better location in input space to again reach its allowed maximum size. But how can we decide, when a unit is to be removed? On first thought one might have the idea to remove the units with low accumulated error values. However, this may not be a good choice at all. Consider a unit positioned at a point-like peak of the probability density function. It would have very small accumulated error since most input signals for which it is winner coincide with its reference vector and, therefore, generate no error. If we remove this unit, however, all these signals would be mapped to the unit previously being second-nearest to them. This unit, however, could be very distant.

A better approach is, to directly compute, how much the error for the given input signal $\boldsymbol{\xi}$ would increase if the winner s_1 would not be present. In this case the signal would be mapped to the second-nearest unit s_2 and, therefore, the increase in error is simply the difference of $\|\boldsymbol{\xi}-\mathbf{w}_{s_2}\|^2$ and $\|\boldsymbol{\xi}-\mathbf{w}_{s_1}\|^2$. This gives us – for the current input signal – a direct measure of the utility of the winner unit s_1. To sum up this utility measure over all input signals we introduce a local variable U_c for each unit c. For each input signal the utility variable of the winner s_1 is updated according to

$$U_{s_1} := U_{s_1} + \|\boldsymbol{\xi} - \mathbf{w}_{s_2}\|^2 - \|\boldsymbol{\xi} - \mathbf{w}_{s_1}\|^2. \tag{4}$$

As done for the local error values we subject all utility variables to exponential decay after each adaptation step:

$$U_c := U_c - \beta U_c \quad (\text{ for each unit } c). \tag{5}$$

This utility measure is an on-line variant of a measure recently proposed by the author for batch vector quantization methods [5]. The utility of a unit becomes small either when other units are in close vicinity (then $\mathbf{w}_{s_2}$ approaches $\mathbf{w}_{s_1}$) or when the unit is not anymore (or only rarely) winner.

6 Removal criterion

How can the utility measure be used to remove neurons? We can not simply remove the unit with minimum value of U since there always will be one and this strategy would cause a constant change of the network structure even for perfectly stationary distributions. Let us recall, that the set goal is error reduction. When we remove a unit the GNG algorithm immediately inserts a new unit near the unit q with maximum accumulated error, i.e., the unit such that

$$q := \arg\max_c E_c. \tag{6}$$

We can expect the newly inserted unit to reduce this error in the future, since it reduces the size of the Voronoi region of q. When the mean distance of the signals in q's Voronoi region is reduced by a certain factor, the (square) quantization error is reduced by an even larger factor. A cautious estimate is that the mean distance is reduced only by a factor two. More precise estimates of the error reduction would require unavailable knowledge of the local structure of $p(\boldsymbol{\xi})$. However, these general considerations are enough to formulate a removal criterion. An estimate of the increase in quantization error by removing a unit i is its current utility value U_i. The expected decrease caused by the insertion of a new unit near the unit q with maximum error E_q is some fraction of E_q.

A suitable strategy seems to remove a unit when its utility value falls below a certain fraction of the error variable E_q. Let i be the unit with minimum utility

$$i := \arg\min_c U_c. \tag{7}$$

Then, i should be removed if

$$E_q/U_i > k \tag{8}$$

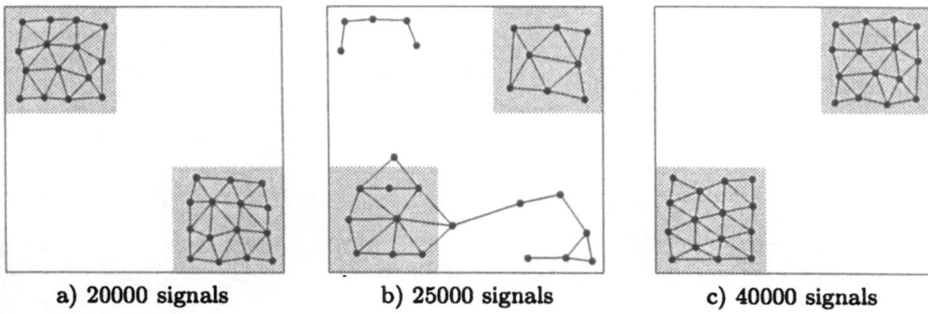

a) 20000 signals b) 25000 signals c) 40000 signals

Fig. 2. A GNG-U network successfully tracks a non-stationary signal distribution $p(\xi)$. a) situation before the distribution changes. b) State of the network 5000 adaptation steps after the distribution had changed. c) After 50000 adaptation steps all units have been adapted or relocated to regions of high probability density. The value used for k is 3. The other parameters are as in figure 1.

for some value of k. Large values of k require a large ratio of maximum error and minimum utility and will cause less frequent deletions of units. Smaller values will cause accordingly more and faster deletions which leads to a faster tracking of non-stationary distributions.

7 Simulation examples

The proposed utility criterion for removal of units was combined with the GNG algorithm and the resulting method, which we call GNG-U (growing neural gas with utility criterion) in the following, was successfully applied to various non-stationary distributions. Due to lack of space we report here only on two experiments with GNG-U. In figure 2 a simulation with the same non-stationary distribution as already used in figure 1 is shown. The GNG-U algorithm successfully relocates all units after the sudden change in the distribution. In figure 3 a non-stationary one-dimensional distribution is used to illustrate the behavior of GNG-U throughout a complete simulation.

8 Discussion

In this paper a novel on-line criterion for removal of units in self-organizing networks is proposed. The combination of this criterion and an existing incremental network model (growing neural gas) results in a new method, called GNG-U, which is able to closely track non-stationary distributions. This is a feature not found in known related models such as various variants of the self-organizing map or the neural gas algorithm and it could open up completely new application domains for this kind of networks.

For the new criterion a parameter k has to be chosen which indicates how large the ratio between accumulated error of a unit q and accumulated utility of

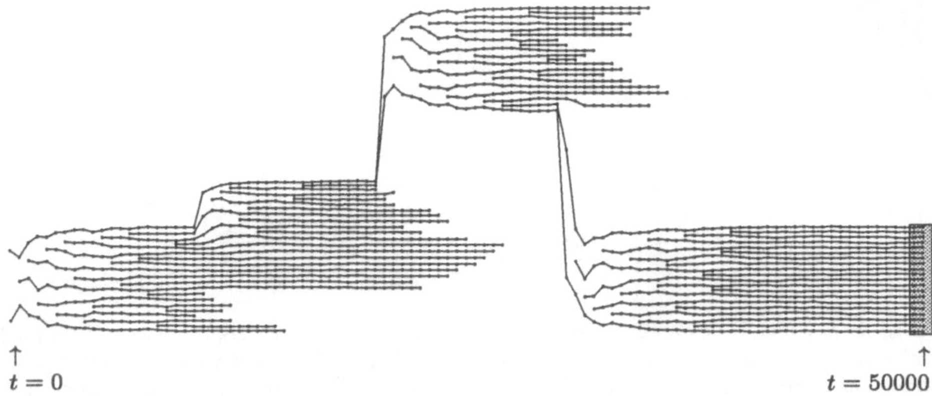

Fig. 3. A GNG-U network with maximum size set to 20 tracks a one-dimensional non-stationary probability distribution which is indicated by the shaded box on the right. After 10000, 20000, and 30000 signals the mean of the distribution was suddenly changed. It can be seen that moderate changes of the mean (like the change after 10000 signals) lead to a re-use of many units. More drastic changes, however, lead to deletions and subsequent insertions of units. Parameters are the same as in figure 2.

a unit i must be, before unit i is removed. In our experience the choice of this parameter is not at all difficult and the value used in our simulations is just one value which gave very satisfactory results.

References

1. H.-U. Bauer and Th. Villmann. Growing a hypercubical output space in a self-organizing feature map. TR-95-030, International Computer Science Institute, Berkeley, 1995.
2. B. Fritzke. Growing cell structures – a self-organizing network for unsupervised and supervised learning. *Neural Networks*, 7(9):1441–1460, 1994.
3. B. Fritzke. A growing neural gas network learns topologies. In G. Tesauro, D. S. Touretzky, and T. K. Leen, editors, *Advances in Neural Information Processing Systems 7*, pages 625–632. MIT Press, Cambridge MA, 1995.
4. B. Fritzke. Growing grid – a self-organizing network with constant neighborhood range and adaptation strength. *Neural Processing Letters*, 2(5):9–13, 1995.
5. B. Fritzke. The LBG-U method for vector quantization – an improvement over LBG inspired from neural networks. *Neural Processing Letters*, 5(1):35–45, 1997.
6. T. Kohonen. Analysis of a simple self-organizing process. *Biological Cybernetics*, 44:135–140, 1982.
7. T. M. Martinetz. Competitive Hebbian learning rule forms perfectly topology preserving maps. In *ICANN'93: International Conference on Artificial Neural Networks*, pages 427–434, Amsterdam, 1993. Springer.
8. T. M. Martinetz, S. G. Berkovich, and K. J. Schulten. Neural-gas network for vector quantization and its application to time-series prediction. *IEEE Transactions on Neural Networks*, 4(4):558–569, 1993.

Phase Transitions in Soft Topographic Vector Quantization

Matthias Burger, Thore Graepel, and Klaus Obermayer

Informatik, FR 2-1, TU Berlin, Franklinstr. 28/29, 10587 Berlin, Germany
E-mail: {burger,graepel2,oby}@cs.tu-berlin.de

Abstract. We have developed an algorithm (STVQ) for the optimization of neighborhood preserving maps by applying deterministic annealing to an energy function for topographic vector quantization. The combinatorial optimization problem is solved by introducing temperature dependent fuzzy assignments of data points to cluster centers and applying an EM-type algorithm at each temperature while annealing. The annealing process exhibits phase transitions in the cluster representation for which we calculate critical modes and temperatures expressed in terms of the neighborhood function and the covariance matrix of the data. In particular, phase transitions corresponding to the automatic selection of feature dimensions are explored analytically and numerically for finite temperatures. Results are related to those obtained earlier for Kohonen's SOM-algorithm which can be derived as an approximation to STVQ. The deterministic annealing approach makes it possible to use the neighborhood function solely to encode desired neighborhood relations. The working of the annealing process is visualized by showing the effects of "heating" on the topological structure of a two-dimensional map of the plane.

1 Introduction

Pattern recognition and signal processing tasks often involve high-dimensional input data which are hard to visualize and which – for reasons of computational complexity – cannot be processed directly. It is therefore desirable to find some mapping of the high-dimensional input space to a lower dimensional space in a way that captures the essential spatial relations of the data as faithfully as possible and at the same time performs lossy data compression. The self-organizing map algorithm (SOM) introduced by Kohonen [5] achieves such a mapping via an on-line learning rule that leads to a correspondence between local regions in input space and neurons in a usually two-dimensional array. Its analysis [4,3], however, has been difficult due to its heuristic formulation. Luttrell [6,7] established a connection between the self-organizing map and noisy vector quantization (see also [1]). By choosing a distortion measure for vector quantization that incorporates robustness w.r.t. noise-induced changes of assignments, he derived an algorithm which he named topographic vector quantization (TVQ) and showed that the SOM can be viewed as an efficient approximation to a gradient descent on the TVQ energy function. Following an idea by Rose et al. [10], who first applied deterministic annealing to a clustering energy function (see also [11]), we apply a similar scheme to the TVQ energy function and thus develop a robust optimization procedure (see [2] for a related problem) for neighborhood preserving maps (STVQ). This formulation enables us to provide an analysis of phase transitions that occur during the annealing process. The avoidance of local minima during the annealing process is visualized by numerical experiments on topological defects of a two-dimensional map of the plane.

2 Energy Function and Optimization Scheme

The objective of topographic vector quantization can be formulated in terms of the energy function

$$\mathrm{E}\left(\{\mathrm{m_{i\mathbf{r}}}\}, \{\mathbf{w_r}\} \mid \{\mathbf{x}_\mathrm{i}\}\right) = \frac{1}{2} \sum_\mathrm{i} \sum_\mathbf{r} \mathrm{m_{i\mathbf{r}}} \sum_\mathbf{s} \mathrm{h_{\mathbf{rs}}} \left\|\mathbf{x}_\mathrm{i} - \mathbf{w_s}\right\|^2 . \tag{1}$$

Given data points $\mathbf{x}_\mathrm{i} \in \Re^\mathrm{d}$, $\mathrm{i} = 1, \ldots, \mathrm{D}$, and clusters $\mathcal{C}_\mathbf{r}$, $\mathbf{r} = 1, \ldots, \mathrm{N}$, the aim of vector quantization is to assign each data point $\mathbf{x}_\mathrm{i}$ to a cluster $\mathcal{C}_\mathbf{r}$, such that the energy function E is minimized. The membership of data point $\mathbf{x}_\mathrm{i}$ in cluster $\mathcal{C}_\mathbf{r}$ is indicated by a binary assignment variable $\mathrm{m_{i\mathbf{r}}}$, subject to the constraint $\sum_\mathbf{r} \mathrm{m_{i\mathbf{r}}} = 1, \forall \mathrm{i}$. In central clustering there exists a cluster center $\mathbf{w_r}$ for each cluster $\mathcal{C}_\mathbf{r}$ which is the representative in data space of those data points that are members of cluster $\mathcal{C}_\mathbf{r}$. Commonly, the squared Euclidean distance between data point $\mathbf{x}_\mathrm{i}$ and cluster center $\mathbf{w_r}$ is taken as the energy associated with assigning a data point $\mathbf{x}_\mathrm{i}$ to a cluster $\mathcal{C}_\mathbf{r}$. However, following an idea by Luttrell [7], in (1) this assignment energy is averaged over all clusters weighted by a neighborhood function $\mathrm{h_{\mathbf{rs}}}$, which can be interpreted as a transition matrix of channel noise.

Applying the principle of maximum entropy to the energy function (1) under the constraint of a given mean energy $\langle \mathrm{E} \rangle$ we obtain the Gibbs-distribution $\mathrm{P} = \frac{1}{\mathrm{Z}} \exp\left(-\beta\, \mathrm{E}\right)$. Here β is the Lagrange multiplier associated with $\langle \mathrm{E} \rangle$. It determines the fuzziness of the assignments and is called inverse temperature in the framework of statistical physics. We are primarily interested in finding the most probable set of cluster centers $\{\mathbf{w_r}\}$ so as to generalize from a given set of training samples. Consequently, the Gibbs-distribution is marginalized over all legal sets of associations $\{\mathrm{m_{i\mathbf{r}}}\}$. Then maximizing the resulting log-likelihood w.r.t. $\{\mathbf{w_r}\}$ yields conditions for the cluster centers $\{\mathbf{w_r}\}$ at a given β,

$$\mathbf{w_r} = \frac{\sum_\mathrm{i} \mathbf{x}_\mathrm{i} \sum_\mathbf{s} \mathrm{h_{\mathbf{rs}}} \mathrm{P}(\mathrm{i} \in \mathcal{C}_\mathbf{s})}{\sum_\mathrm{i} \sum_\mathbf{s} \mathrm{h_{\mathbf{rs}}} \mathrm{P}(\mathrm{i} \in \mathcal{C}_\mathbf{s})} \qquad \forall \mathbf{r}, \tag{2}$$

where $\mathrm{P}(\mathrm{i} \in \mathcal{C}_\mathbf{r}) = \langle \mathrm{m_{i\mathbf{r}}} \rangle$ is the assignment probability of data point $\mathbf{x}_\mathrm{i}$ to cluster $\mathcal{C}_\mathbf{r}$ and is given by

$$\mathrm{P}(\mathrm{i} \in \mathcal{C}_\mathbf{r}) = \frac{\exp\left(-\frac{\beta}{2} \sum_\mathbf{s} \mathrm{h_{\mathbf{rs}}} \left\|\mathbf{x}_\mathrm{i} - \mathbf{w_s}\right\|^2\right)}{\sum_\mathbf{t} \exp\left(-\frac{\beta}{2} \sum_\mathbf{s} \mathrm{h_{\mathbf{ts}}} \left\|\mathbf{x}_\mathrm{i} - \mathbf{w_s}\right\|^2\right)} . \tag{3}$$

Equations (2) and (3) are iteratively solved in an expectation-maximization fashion at a given value of β. Starting from infinite temperature ($\beta = 0$), the parameter β is subjected to an annealing schedule which – when chosen carefully – leads to a good local or even the global minimum of the energy function (1).

The soft topographic vector quantization algorithm (STVQ) derived above can be put into a familiar context by considering the limiting case $h_{rs} \to \delta_{rs}$ in the assignment probabilities (3) – not in (2) – and $\beta \to \infty$, which leads to a batch version of Kohonen's SOM-algorithm [5].

3 Phase Transitions in the Cluster Representation

During the annealing process in β phase transitions in the cluster representation are observed as previously reported by Rose et al. [10] and Buhmann et al. [1] for the soft-clustering case. The initial representation at $\beta = 0$, where all cluster centers are located at the center of mass of the data, becomes unstable at a critical value of β,

$$\beta^* = \frac{1}{\lambda^{\mathbf{C}}_{\max} \lambda^{\mathbf{G}}_{\max}}, \tag{4}$$

at which the first split of the cluster centers occurs. Here $\lambda^{\mathbf{C}}_{\max}$ is the largest eigenvalue of the covariance matrix $\mathbf{C} = \frac{1}{D} \sum_i \mathbf{x}_i \mathbf{x}_i^T$ of the data and is identified with the variance $\lambda^{\mathbf{C}}_{\max} = \sigma^2_{\max}$ of the data along the principal axis. The associated eigenvector $\mathbf{v}^{\mathbf{C}}_{\max}$ corresponds to the principal axis in data space along which the splitting of the cluster centers takes place. $\lambda^{\mathbf{G}}_{\max}$ is the largest eigenvalue of a matrix $\mathbf{G}$ whose elements are given by $g_{rt} = \sum_s h_{rs} \left(h_{st} - \frac{1}{N} \right)$. The $\mathbf{r}^{\text{th}}$ component of the corresponding eigenvector $\mathbf{v}^{\mathbf{G}}_{\max}$ determines for each cluster center $\mathbf{w}_r$ in which direction along the principal axis it departs from $\mathbf{w}^0_r$ and how it moves relative to the other cluster centers.

Another phase transition occurs, when the dimensionality of the array – as expressed by the coupling of the clusters via h_{rs} – is less than the dimensionality of the data. This situation leads to the phenomenon of the automatic selection of feature dimensions – so termed by Kohonen [5] – which was later applied by [9,8] to explain the formation of cortical maps. Consider a d-dimensional data space and an n-dimensional array of clusters labeled by n-dimensional index vectors $\mathbf{r}$. The couplings h_{rs} of clusters are defined on this array and are typically chosen to be a monotonically decreasing function of $||\mathbf{r} - \mathbf{s}||$. For $d > n$ a simple representation of the input data is achieved, if the data has significant variance only along n of the d dimensions. In this case, the vectors $\mathbf{w}_r$ lie in an n-dimensional subspace ($||$) and the $d - n$ excess-dimensions ($\perp$) are effectively ignored. If, however, the variance of the data along the excess-dimensions surpasses a critical value, the original representation becomes unstable, and the array of vectors $\mathbf{w}_r$ folds into the excess-dimensions so as to represent them as well (see inset of Fig. 1a). A Taylor-expansion of (2) to first order in $\mathbf{w}_r - \mathbf{w}^0_r$ around the initial fixed-point $\mathbf{w}^{||\,0}_r = \rho^{-1}\mathbf{r}$, $\mathbf{w}^{\perp\,0}_r = \mathbf{0}$, $\forall \mathbf{r}$, leads to expressions for the critical mode k^* and the critical variance $(\sigma^*_{\max})^2$ given by

$$\mathrm{k}^* = \arg\max_{\mathrm{k}} \hat{\mathrm{h}}_{\mathrm{k}}^2 \left(1 - \hat{\mathrm{f}}_{\mathrm{k}}(\beta)\right), \qquad (\sigma^*_{\max})^2 = \frac{1}{\beta\, \hat{\mathrm{h}}_{\mathrm{k}^*}^2 \left(1 - \hat{\mathrm{f}}_{\mathrm{k}^*}(\beta)\right)}, \tag{5}$$

where $\hat{\mathrm{f}}_{\mathrm{k}^*}(\beta)$ denotes the Fourier-transform of the correlation function $\mathrm{f}_{\mathbf{tu}} = \rho^{\mathrm{n}} \int \mathrm{P}^0(\mathbf{x}^{||} \in \mathcal{C}_{\mathbf{t}})\, \mathrm{P}^0(\mathbf{x}^{||} \in \mathcal{C}_{\mathbf{u}})\, \mathrm{d}\mathbf{x}^{||}$ of assignment probabilities at the initial fixed-point and ρ is the linear cluster density per dimension. Furthermore, assuming a Gaussian neighborhood function of variance σ_{h}^2 in a system of infinite size and making a continuum approximation we obtain for the square of the critical mode $(\mathrm{k}^*)^2$ and for the critical variance in data space $(\sigma^*_{\max})^2$

$$(\mathrm{k}^*)^2 = \frac{\beta}{\rho^2} \log\left(1 + \frac{\rho^2}{\beta\,\sigma_{\mathrm{h}}^2}\right), \qquad (\sigma^*_{\max})^2 = \left(\frac{1}{\beta} + \frac{\sigma_{\mathrm{h}}^2}{\rho^2}\right)\left(1 + \frac{\rho^2}{\beta\sigma_{\mathrm{h}}^2}\right)^{\frac{\beta\sigma_{\mathrm{h}}^2}{\rho^2}}. \tag{6}$$

In the limit $\beta \to \infty$ and $\mathrm{h}_{\mathbf{rs}} \to \delta_{\mathbf{rs}}$ in (3), which corresponds to the traditional SOM, we recover the results of Ritter and Schulten [9], who obtained $\lambda^* = 2\pi/\mathrm{k}^* = \sigma_{\mathrm{h}}\pi\sqrt{2} \approx 4.44\,\sigma_{\mathrm{h}}$ for the wavelength of the critical mode and $\mathrm{s}^* = \sigma_{\mathrm{h}}\sqrt{3\mathrm{e}/2} \approx 2.02\sigma_{\mathrm{h}}$ for the critical half width of the homogeneous data distribution using a Fokker-Planck-equation approach to on-line SOM.

4 Numerical Results

In this section we present results of numerical experiments we have conducted in order to analyse properties of the STVQ-algorithm. For the verification of the analytical results of the preceding section we examined a chain of $\mathrm{N} = 128$ cluster centers in a two-dimensional input space of uniformly distributed data points with periodic boundary conditions along the $\mathrm{x}^{||}$-direction. The neighborhood function was chosen as a Gaussian with a fixed standard deviation $\sigma_{\mathrm{h}} = 5.0$. The cluster centers were initialized periodically along the $\mathrm{x}^{||}$-direction and the standard deviation $\sigma_{\mathrm{x}^\perp}$ of the data in $\mathrm{x}^\perp$-direction was then linearly increased, starting from $\sigma_{\mathrm{x}^\perp} = 0$. The inset of Fig. 1a shows how the chain folds into the excess-dimension, once the variance of the data in that direction surpasses the critical value $\sigma^*_{\mathrm{x}^\perp}$. Fig. 1a is a plot of the critical standard deviation $\sigma^*_{\mathrm{x}^\perp}$ as a function of β. It can be seen that the values of $\sigma^*_{\mathrm{x}^\perp}$ from the simulation are very well predicted by the analytical results given in (6) (solid line in Fig. 1a). Thus the approximations employed to derive (6) are justified. Fig. 1b shows that the transition is characterized by a critical mode k^* which grows exponentially and dominates the other modes beyond the phase transition.

In order to illustrate the working of the annealing process, we examined the effects of reverse annealing ("heating"). Fig. 2 shows how STVQ with annealing overcomes local minima of the energy function (1) at sufficiently high temperatures. Random initialization and a short range neighborhood function, $\sigma_{\mathrm{h}} = 0.5$, lead the EM-algorithm to convergence to a local minimum (Fig. 2 a) at zero temperature ($\beta = \infty$), which corresponds to a so-called topological defect. During

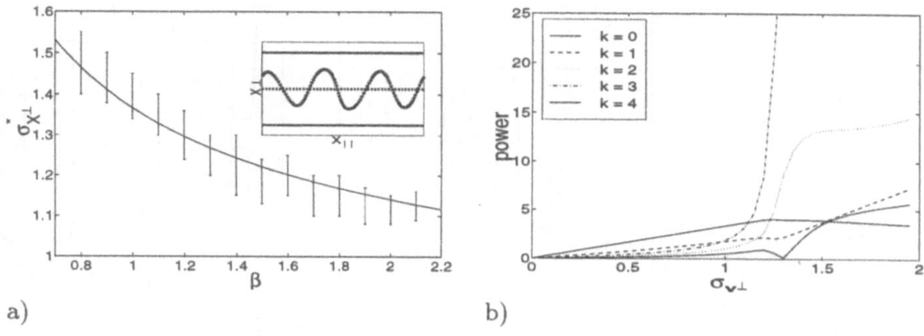

Fig. 1. **a)** Plot of the critical standard deviation $\sigma^*_{x\perp}$ of the data as a function of β. The errorbars are calculated from the $\sigma_{x\perp}$ values of the upper and lower bounds of the derivative of the mean energy w.r.t. $\sigma_{x\perp}$ around the critical value. The solid line shows the theoretical prediction as calculated from (6). Inset: Snapshot of a chain of N = 128 cluster centers after the phase transition when the chain has folded from the initial fixed-point representation indicated by the horizontal line in the middle into the $x^\perp$-dimension. The upper and lower lines indicate boundaries of the set of uniformly distributed data points. Parameters were $\beta = 1.30$, $\sigma_h = 5.0$, and $\sigma_{x\perp} = 1.80$. **b)** Squared absolute amplitude of the Fourier modes as a function of $\sigma_{x\perp}$. Only the five modes with the largest wavelength are shown. Beyond the phase transition at $\sigma^*_{x\perp} = 1.27$ the k = 3 mode is selected and the chain folds into a sine-wave like curve. Parameters as above.

"heating" (Fig. 2b), however, the "twists" gradually unfold until a topologically ordered state (Fig. 2c) is reached. Subsequent "cooling" to $\beta = \infty$ then ensures convergence to the (possibly degenerate) global minimum (not shown). The interpretation of β as a resolution parameter in data space explains why more localized defects melt first: As the temperature rises data points are soft-assigned to more and more clusters with distant cluster centers and thus ordering at an increasingly larger scale takes place.

5 Conclusion

In this contribution we presented a robust optimization scheme for the formation of neighborhood preserving maps by applying deterministic annealing to the energy function of topographic vector quantization (STVQ). We then elaborated on two aspects of STVQ: dimension reduction and annealing. Analytical results w.r.t. phase transitions in particular in the case of dimension reduction were obtained and could be related to corresponding results for the SOM-algorithm, which was recovered as an approximation to STVQ. Numerically, we verified these analytical results and demonstrated the ability of STVQ to avoid local minima of the energy function by showing the effects of reverse annealing.

Acknowledgments: This work was supported by the Technische Universität Berlin via Forschungsinitiativprojekt FIP 13/41 and by DFG via grant Ob 102/2-1.

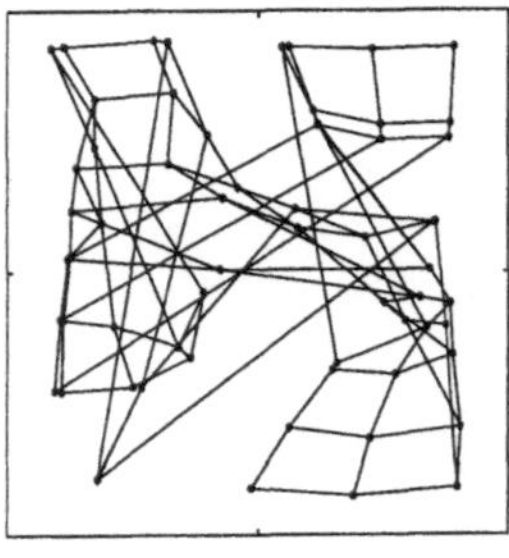

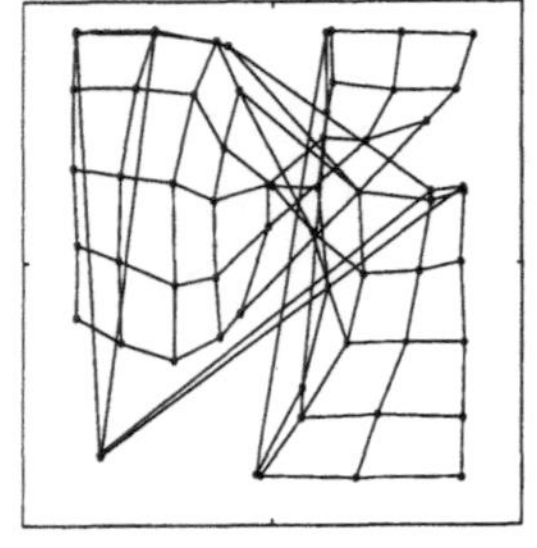

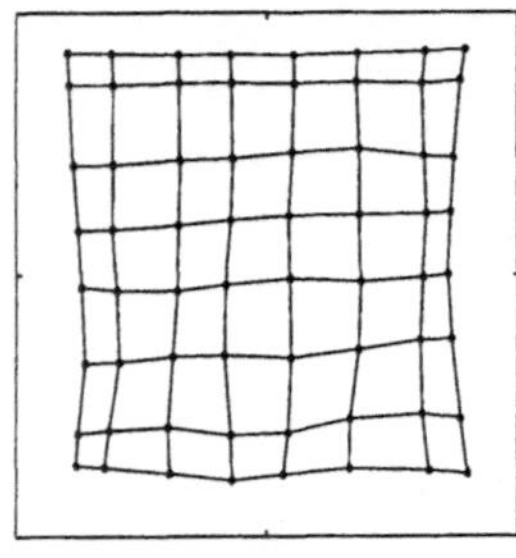

a) $\beta = \infty$ b) $\beta = 399$ c) $\beta = 147$

Fig. 2. "Melting" of topological defects for STVQ for an array of 8×8 cluster centers. The snapshots are taken from an exponential annealing scheme in β . **a)** Starting from a random initialization of the cluster centers at $\beta = \infty$ a local minimum of the energy function is reached. **b)** The array at an intermediate value of β, for which locally ordered configurations are already established. **c)** The topologically ordered state is recovered for $\beta = 147$. Parameters were $\sigma_h = 0.5$ and 1024 equally spaced data points on a square grid. The exponential annealing schedule for β was given by $\beta_{t+1} = \beta_t/1.01$.

References

1. J. M. Buhmann and H. Kühnel. *Vector Quantization with Complexity Costs.* IEEE Transactions on Information Theory, vol. 39, p. 1133-1145 (1993).
2. J. M. Buhmann and T. Hofmann. *Robust Vector Quantization by Competitive Learning.* Proceedings of the International Conference on Accoustics, Speech and Signal Processing ICASSP'97, Munich, (1997).
3. M. Cottrell, J. C. Fort, and G. Pagès. *Two or three things that we know about the Kohonen algorithm.* Proceedings of ESANN'94, M. Verleysen Ed., D Facto, Bruxelles, p. 235-244 (1995).
4. E. Erwin, K. Obermayer, and K. Schulten. *Self-Organizing Maps: Ordering, Convergence Properties and Energy Functions.* Biological Cybernetics, vol. 67, p. 47-55 (1992).
5. T. Kohonen. *Self-Organizing Maps.* Springer-Verlag, 1995.
6. S. P. Luttrell. *Self-Organisation: A Derivation from first Principles of a Class of Learning Algorithms.* IEEE International Joint Conference on Neural Networks: IJCNN, Washington DC, vol. 2, p. 495-498 (1989).
7. S. P. Luttrell. *A Baysian Analysis of Self-Organizing Maps.* Neural Computation, vol. 6, p. 767-794 (1994).
8. K. Obermayer, G. G. Blasdel, and K. Schulten. *Statistical-Mechanical Analysis of Self-Organization and Pattern Formation during the Development of Visual Maps.* Physical Review A, vol. 45, p. 7568-7589 (1992).
9. H. Ritter and K. Schulten. *Convergence Properties of Kohonen's Topology Conserving Maps: Fluctuations, Stability, and Dimension Selection.* Biological Cybernetics, vol. 60, p. 59-71 (1988).
10. K. Rose, E. Gurewitz, and G. C. Fox. *Statistical Mechanics and Phase Transitions in Clustering.* Physical Review Letters, vol. 65, p. 945-948 (1990).
11. Y. Wong. *Clustering Data by Melting.* Neural Computation, vol. 5, p. 89-104 (1993).

Vector Quantization by Optimal Neural Gas

M. Herrmann[1] and Th. Villmann[2]

[1] Max-Planck-Institut für Strömungsforschung
Bunsenstr. 10, 37073 Göttingen, Germany
[2] Universität Leipzig, Klinik für Psychotherapie und Psychosomatische Medizin
Karl-Tauchnitz-Str.25, 04107 Leipzig, Germany

Abstract. Many vector quantization algorithms have been designed to minimize the reconstruction error of the data representation. The additional requirement of topology preservation in self-organizing maps conflicts this goal but can be alleviated by suitable modifications. In the present contribution we demonstrate that the neural gas algorithm allows for vector quantization with a theoretically optimal reconstruction error over an extended range of parameters. Moreover, by a similar scheme as previously applied to self-organizing maps it is possible to modify the neural gas algorithm such as to meet optimality criteria other than the reconstruction error in a way which is exact for arbitrary dimensionality of the data.

1 Introduction

Vector quantization aims at representing a data distribution by a set of prototypical data vectors. Once the prototypes are fixed, the data can be addressed and transmitted by referring to the prototype labels as a code for the data. If the code itself is subject to noise during transmission, neighborhood relations in the code space can be exploited to partly preserve information of the data. Topographic vector quantizers (e.g. [6, 8, 7, 9]) represent the data structure by the neighborhood structure of the code output. They have proven useful in several fields of applications such as data visualization, feature extraction, nonlinear principle component analysis, image processing, classification, and robotics.

For self-organizing maps (SOMs) [6] it is known [10] that the ordering process of the reference vectors modifies the representation quality of the vector quantizer. This conflict induces a suboptimal representation of the data with respect to the mean square error. It is possible [1] to modify Kohonen's algorithm such that the reconstruction error becomes optimal while keeping the ordering properties of the map. This scheme is based on an adaptable learning rate and is exact for one dimensional maps. For higher dimensions exactness cannot be achieved because of the lack of analytical knowledge in the generic cases. Moreover, the typical instabilities of high dimensional maps will be affected such that any map may become unstable for some range of the modified learning parameters.

The present contribution focuses on the neural gas (NG) algorithm [9] which differs from the SOM by a more flexible neighborhood among the output units.

We review the properties of NG in the next section. In the third section the magnification exponent of the neural gas algorithm is used to express the optimality of the underlying vector quantizer. We show that by modifying the learning rate in a similar way as in [1] various optimality criteria can be satisfied exactly for any dimension of the data space. Numerical simulations are presented in section 4 demonstrating the optimality as well as the range of validity of the proposed modification. Section 5 discusses the relation between the modified and the original algorithm.

2 Neural gas vector quantizer

The neural gas network is formally rather similar to Kohonen's algorithm. Reference vectors $\mathbf{w_i} \in \mathcal{V}$ are attracted by data vectors $\mathbf{v}$ sampled from the input space $\mathcal{V}$

$$\triangle \mathbf{w_i} = \epsilon\, h\,(\mathbf{i},\mathbf{v})\,(\mathbf{v}-\mathbf{w_i})\,, \tag{1}$$

if they are close to the best-matching unit $\mathbf{s}(\mathbf{v}) = \arg\min_\mathbf{i} \|\mathbf{w_i}-\mathbf{v}\|$. In Kohonen's algorithm [6] the neighborhood function is often chosen to be Gaussian $h_\sigma^{SOM}(\mathbf{r},\mathbf{v}) = \exp(-\|\mathbf{r}-\mathbf{s}(\mathbf{v})\|^2_{\mathcal{A}}/2\sigma^2)$ and is evaluated on a usually rectangular output grid $\mathcal{A}$. In the NG, on the other hand, the neighborhood function is defined as

$$h_\lambda^{NG}\,(i,\mathbf{v}) = \exp\left(-k_i(\mathbf{v},\hat{\mathbf{w}})/\lambda\right), \tag{2}$$

where the rank function $k_i\,(\mathbf{v},\hat{\mathbf{w}})$ gives the number of pointers $\mathbf{w}_j$ for which the relation $\|\mathbf{v}-\mathbf{w}_j\| \leq \|\mathbf{v}-\mathbf{w}_i\|$ is valid [9]. Here the neighborhood function is evaluated in the input space and i refers to an arbitrary numbering of the output units. Note, that both neighborhood functions depend implicitly on the configuration of all reference vectors $\hat{\mathbf{w}} = \{\mathbf{w}_i\}$.

In contrast to the SOM, the NG adaptation rule can be derived from an energy function [9]. Also, the NG convergence rate has been shown to be faster than that of the SOM [9]. Further differences between SOM and NG concern their resolutional properties described by the *magnification factor* α. This quantity relates the stimuli distribution $P\,(\mathbf{v})$ to the pointer density $P(\mathbf{w})$ via

$$P\,(\mathbf{v}) \propto P\,(\mathbf{w})^\alpha\,. \tag{3}$$

In Ref. [10] the value $\alpha_{SOM} = \frac{2}{3}$ has been derived for the one dimensional SOM in the limit $1 \ll \sigma \ll N$. For small values of σ the neighborhood ceases to be of influence and the magnification rate approaches [12] the value $\alpha = \frac{1}{3}$ which indicates that the vector quantizer minimizes the mean square error of the data representation [13]. The latter value or more generally $\alpha_{NG} = d/(d+2)$ characterizes vector quantizers produced by the NG algorithm on d dimensional data, which was shown analytically for small λ in Ref. [9]. In order to determine the range of validity of this result, which is crucial for the control of the magnification of neural gas maps, we have performed simulations with network sizes $N = 2^k$, $k = 4, \ldots 14$, input dimensions $d = 1, 2, 3$, and various forms of the

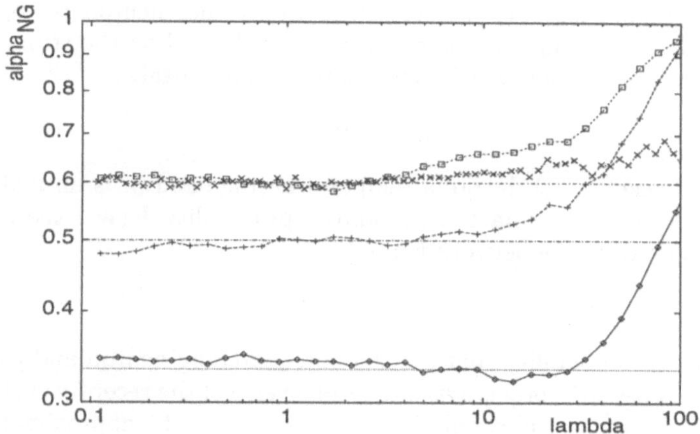

Fig. 1. The magnification factor α_{NG} for input spaces of different dimension ($d = 1$ ($\diamond$), $d = 2$ ($+$), $d = 3$ ($\Box$ and $\times$)) in dependence on the neighborhood range λ. The networks consisted of $N = 512$ neurons, except for the ($\times$) curve, where $N = 4096$ neurons have been trained. The horizontal lines indicate the theoretical values $\alpha_{NG} = 1/3,\ 1/2,\ 3/5$ for $d = 1,\ 2,\ 3$, resp., which can be obtained analytically in the limit $\lambda \to 0$ [9].

input distribution, cf. Fig. 1. For $d = 1$ the optimal magnification rate is approached closely if $\lambda < \nu_1 N$, with $\nu_1 \approx 0.05$. For $d > 1$ the respective values ν_2 and ν_3 turned out to be smaller, but of similar magnitude. Thus, the mean squared error is minimized by neural gas maps already at nonzero neighborhood parameters. However, for achieving this result in SOMs the ratio σ/N has to approach zero [3] which induces topological defects for $d > 1$ [4].

Generalizing the mean squared error to arbitrary powers p a family of optimization criteria is obtained including mean absolute error ($p = 1$) and the minimal worst case error ($p \to \infty$), which requires maps of magnification $\alpha = d/(d+p)$ [13]. On the other hand, if the output units are to be used with equal probability, the map should obey $\alpha = 1$, a criterion which maximizes the transentropy of the map, cf. (3).

3 Controlling the magnification of neural gas maps

Controlling the magnification factor is relevant for many applications in control and robotics, where vector quantizers are applied to categorize the system states for a subsequent control task [11]. It may be of importance, e.g., to resolve rarely occurring critical states of the systems precisely, cf. [5].

In Ref. [1] the problem of controlling the magnification factor of SOMs has been approached by introducing an adaptive local learning step size $\epsilon_{\mathbf{s}(\mathbf{v})}$. Analogously, the NG algorithm is modified as

$$\triangle \mathbf{w}_i = \epsilon_{s(\mathbf{v})}\; h_\lambda\left(i, \mathbf{v}\right)\left(\mathbf{v} - \mathbf{w}_i\right) \ . \tag{4}$$

In an adaptation step $\epsilon_s(t)$ associated to the winner neuron is applied to all units. The *local* learning parameters $\epsilon_i = \epsilon(\mathbf{w}_i)$ depend on the stimulus density P at the position of their weight vectors $\mathbf{w}_i$ by the Ansatz

$$\langle \epsilon_i \rangle = \epsilon_0 P(\mathbf{w}_i)^m , \tag{5}$$

where the brackets $\langle \ldots \rangle$ denote a temporal average and m is an additional parameter. In order to estimate the unknown probability $P(\mathbf{w}_i)$ the information already acquired by the network is used:

$$P(\mathbf{w}_i) \propto p_i \rho(\mathbf{w}_i) , \tag{6}$$

where p_i is the probability that i is the best-matching neuron and $\rho(\mathbf{w}_i)$ is the receptive field density (or, equivalently, the inverse of the receptive field volume). In this way, for numerical simulations relation (5) can be approximated by

$$\epsilon_s(t) = \epsilon_0 \left(\frac{1}{\Delta t_s} \left(\frac{1}{\|\mathbf{v} - \mathbf{w}_s\|^d} \right) \right)^m , \tag{7}$$

with Δt_s being the time difference between the present t value and the last time the neuron has been the winner. For the one dimensional SOM one obtains $\alpha'_{SOM} = (m+1)\alpha_{SOM}$ [1], which allows an explicit control of the magnification factor. We remark that control is exact only in the one dimensional case (and higher dimensional ones which separate), cf. [10].

For the NG the magnification factor α_{NG} is known for an arbitrary dimension d of the input space [9]. Applying the control scheme (4), (5) for the magnification factor to the NG one obtains analogously:

$$\rho(\mathbf{w}) = P(\mathbf{w})^{\alpha'_{NG}} = P(\mathbf{w})^{(m+1)\ \alpha_{NG}} \tag{8}$$

Thus, in order to obtain a map with a magnification rate α'_{NG}, the parameter m has to be set to $m = \frac{d+2}{d}\ \alpha'_{NG} - 1$. This final solution holds for an arbitrary dimension d of the input space $\mathcal{V}$.

4 Numerical results

In the simulations we have used the Ansatz (7) to approximate relation (5) for $\epsilon_s(t)$. The input data were chosen according to the density function $g(x) = \sin(\pi \cdot x)$ with equally distributed $x \in [0, 1]$. Two and three dimensional data distributions were defined in the same way by Cartesian products. A map of $N = 50$ was trained for 10^7 adaptation steps.

After training, the entropy $H = -\sum_{i=1}^{N} \rho(\mathbf{w}_i)\ \log \rho(\mathbf{w}_i)$ of the map was determined. H should have a maximum for the magnification factor $\alpha' = 1$ corresponding to $m_1 = 2$, $m_2 = 1$, $m_3 = \frac{2}{3}$ (cf. (8)) for input dimensions d=1, 2, and 3. A maximal entropy map is characterized in the present setting by $H_{\max} = -\sum_{i=1}^{50} \frac{1}{50} \cdot \log\left(\frac{1}{50}\right) \approx 3.912$. For non-optimal values of m the H values are smaller, cf. Fig. 2.

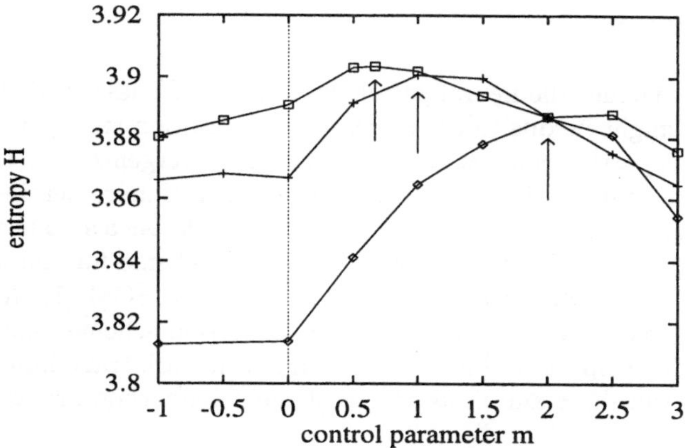

Fig. 2. Plot of the entropy H for maps trained with different magnification control factors m ($d = 1$ ($\diamond$), $d = 2$ ($+$), $d = 3$ ($\square$)). The arrows indicate the theoretical values of m ($m = 2$, $m = 1$, $m = 2/3$, resp.) which maximize the entropy of the map.

5 Discussion

Introducing local modifyable learning rates appears as a serious alteration of the original learning scheme. When considering, however, the short time average of Eq. (1)

$$\langle \Delta \mathbf{w}_i \rangle = \epsilon_0 \int_{\mathcal{V}} \epsilon_{s(\mathbf{v})} h(i, \mathbf{v}) (\mathbf{v} - \mathbf{w}_i) \; P(\mathbf{v}) \, d\mathbf{v}, \tag{9}$$

we find that Eq. (5) results effectively in a modification of the input space density $P(\mathbf{v}) \longrightarrow \text{const } P(\mathbf{v})^{m+1}$ rather than a modification of the learning rule, provided that the pointer density $\rho(\mathbf{w}_i)$ is sufficiently high and that the difference between $P(\mathbf{v})$ and $P(\mathbf{w}_s)$ is neglectable.

In practice, additional precautions are advisable. Firstly, the stochastic approximation of (9) by (1) and the further, possibly biased approximation made in Eq. (7) call for a slow adaptation process, which is governed by small learning rates. We hence included a global factor ϵ_0 (cf. Eq. (7)) which is chosen such that ϵ_i is smaller than a global parameter ϵ_{max}. By decreasing ϵ_{max} algebraically the convergence of the map is guaranteed by a majorization criterion. In this way, also the properties of original neural gas algorithm [9] are preserved. For $m < 0$ and $P(\mathbf{v})$ close to zero at some $\mathbf{v} \in \mathcal{V}$ the learning algorithm becomes unstable due to a singular local learning rate. We, hence, threshold the local learning rates to be smaller than unity.

Further, in order to initialize the algorithm and in particular the quantities occurring in Eq. (7) the pointers are adapted by the usual neural gas algorithm, i.e. by constant ϵ_i, during a short starting period.

At the end of the learning process the ϵ_i will represent the probability density P of the input distribution and, hence, allow further considerations of the data.

6 Conclusion

In spite of its virtues the neural gas algorithm is by far less used than Kohonen maps, which might be due to the increased computation time per single update step. As this is partly compensated by the faster convergence rate, the neural gas algorithm could find a wider range of applications if if more detailed theoretical and numerical results on this algorithm were available. As a step in this direction we have proposed in the present paper a control scheme for the magnification factor α_{NG} of the NG inspired by related work on SOM [1]. In contrast to the results on SOMs the present vector quantization scheme allows to achieve optimality with respect to different criteria and for arbitrary input dimensions and for both vanishing and finite values of the neighborhood range.

References

1. H.-U. Bauer, R. Der, M. Herrmann (1996) Controlling the Magnification Factor of Self-Organizing Feature Maps. *Neural Computation* **8**, 757-771.
2. R. Brause (1992) Optimal Information Distribution and Performance in Neighbourhood-Conserving Maps for Robot Control. *Int. J. Computers and Artificial Intelligence* **11**:2, 173-199.
3. E. de Bodt, Michel Verleysen, M. Cottrell (1997) Kohonen Maps versus Vector Quantization for Data Analysis. In: *Proc. ESANN'97*, D facto, Brussels, 211-218.
4. R. Der, M. Herrmann (1994) Instabilities in Self-Organized Feature Maps with Short Neighborhood Range. In: *Proc. ESANN'94*, D facto, Brussels, 271-276.
5. M. Herrmann, R. Der (1995) Efficient Q-learning by Division of Labor. *Proc. ICANN'95, EC2 & Cie, Paris.* vol. 2, 129-134.
6. T. Kohonen (1995) *Self-Organizing Maps. Springer Series in Information Sciences* **30**, Springer, Berlin, Heidelberg.
7. R. Linsker (1989) How to Generate Ordered Maps by Maximizing the Mutual Information between Input and Output Signals. *Neural Computation* **1**, 402-411.
8. S. P. Luttrell (1989) Self-Organization: a derivation from first principles of a class of learning algorithms, in: *Proc. IJCNN 89 Washington*, IEEE Press, II-495-498.
9. T. M. Martinetz, S. G. Berkovich, K. J. Schulten (1993) 'Neural-Gas' Network for Vector Quantization and its Application to Time-Series Prediction. *IEEE Trans. on Neural Networks* **4**:4, 558-569.
10. H. Ritter, K. Schulten (1986) On the Stationary State of Kohonen's Self-Organizing Sensory Mapping. *Biol. Cyb.* **54**, 99-106.
11. H. Ritter, T. M. Martinetz, K. J. Schulten (1989) Topology-conserving mappings for learning visuomotor-coordination. *Neural Networks* **2**, 159-168.
12. D. R. Dersch, P. Tavan (1995) Asymptotic Level Density in Topological Feature Maps. *IEEE Transactions on Neural Networks* **6**:1, 230-236.
13. Zador, P. L. (1982) Asymptotic Quantization Error of Continuous signals and Quantization Dimension. *IEEE Transact. on Information Theory*, **28**(**2**), 139-149.

Convergences of the Kohonen Maps: A Dynamical System Approach

Jean-Claude FORT * Gilles PAGÈS †

Abstract

We present the now well-known ODE method in a general framework. We state a global convergence theorem when the ODE has no "pseudocycle" which could apply in many cases. We note that the Kohonen algorithm in dimension 1 (units and stimuli) after self-organization is a cooperative dynamical system and prove the uniqueness of its equilibrium point. We then derive results of convergence ($a.s$ and in distribution) for a very general class of stimuli distributions and neighbourhood functions.

1 Introduction

Our interest in convergence of stochastic algorithms was mainly raised by the difficulties encountered in the study of the self-organizing Kohonen mappings (see [13], [14], [7]). The main feature of these algorithms are a strong non-linearity together with a rather simple noise dynamics (i.i.d. random variables). Besides the historical theorem obtained by Robbins-Monro and its various extensions, the most important available result about *a.s.* convergence of stochastic algorithms definitely is the Kushner-Clark theorem (see [15]). It makes a bridge between the behaviour of each sample path of the algorithm and the solution of its related average differential equation ($ODE \equiv \dot{x} = -h(x)$), but requires some stability assumptions on both the equilibrium points of the ODE and the paths of the algorithm itself. This assumption turns out to be very stringent.

In the first part of this paper we adress the question "What limiting behaviour of the ODE carries over the algorithm?". Actually the algorithm may have a much more complex asymptotic behaviour than a given solution of the ODE. Our approach (see [11]) is a development of the original proof of the Kushner and Clark's theorem. Calling upon the theory of dynamical systems (see *e.g.* [5]) and the notion of "chain recurrence" M. Benaïm (see [1]) obtained results of global convergence results. In all

*Univ. Nancy Henri Poincaré, B.P. 239, F-54506 Vandoeuvre-Lès-Nancy Cedex & SAMOS, Univ. Paris I, U.F.R. 27, 90, rue de Tolbiac F-74634 Paris Cedex 13 (fortjc@iecn.u-nancy.fr).

†Labo. de Proba. URA 224, Univ. Paris 6, 4, Pl. Jussieu, F-75252 Paris Cedex 05 & Univ. Paris XII, 61 av. du Général de Gaulle, 94010 Créteil Cedex (gpa@ccr.jussieu.fr).

cases, the practical assumption that implies convergence remains the non-existence of "pseudo-cycle".

In the second part of this paper we apply the *ODE* method to the Kohonen maps. The first rigorous proof of *a.s.* convergence was obtained in the 1-dimensional setting with the two nearest neighbors and uniformly distributed stimuli (see [6],[4],[8]). A proof of the Kushner-Clark convergence was obtained under a mild assumption on the stimuli distribution μ (see [9]). Recently A. Sadeghi pointed out the link with some particular dynamical systems called cooperative (see [16]). It turns out that their behaviour complies with the convergence results of the *ODE* method, especially in case of a unique equilibrium point. Gathering all the previous results on the 1-dimensional Kohonen maps, we establish the *a.s.* convergence toward a *unique* equilibrium point under quite general assumptions on the stimuli distribution and the neighborhood function. Convergence in distribution holds when the gain is constant.

2 The *ODE* method

Let $(\omega^t)_{t\geq 1}$, $(\eta^t)_{t\geq 1}$ be two sequences of respectively $\mathbb{R}^p$- and $\mathbb{R}^d$-valued vectors and let H be a function from $\mathbb{R}^d \times \mathbb{R}^p$ into $\mathbb{R}^d$. We define $(X^t)_{t\geq 1}$ by the recursive algorithm:

$$X^0 \in \mathbb{R}^d,\ X^{t+1} = X^t - \varepsilon_{t+1}\big(H(X^t,\omega^{t+1}) + \eta^{t+1}\big), \quad t \in \mathbb{N} \tag{1}$$

(ω^t) is a sequence of i.i.d. random variables with common distribution μ and $H(x,.)$ is μ-integrable for every $x \in \mathbb{R}^d$. $(\varepsilon_t)_{t\geq 1}$ is the *gain* and η^t is a small residual perturbation which is zero in the case of the Kohonen maps. Then the *mean* or *average function* h of the algorithm reads :

$$h(x) := \int H(x,\omega)\mu(d\omega).$$

Thus $h(X^t) = \mathbb{E}(H(X^t,\omega^{t+1})/X_0,\omega_1,...,\omega_t)$.

Putting $\Delta M^{t+1} = H(X^t,\omega^{t+1}) - h(X^t)$ (with $M^0 := 0$), which is a martingale increment, the formula (1) reads

$$X^{t+1} = X^t - \varepsilon_{t+1}h(X^t) - \varepsilon_{t+1}(\Delta M^{t+1} + \eta^{t+1}), \quad t \geq 0. \tag{2}$$

Generally the "target" of the algorithm will be some points of $\{h = 0\}$. The usual *ODE* method consists in associating to the discrete time algorithm $(X^t)_{t\geq 0}$ a family of continuous time stepwise functions $(X^{(t)})_{t\geq 0}$ defined by:

$$\forall u \in \mathbb{R}_+,\quad X_u^{(0)} := X^s \ \text{ if } \ u \in [\varepsilon_1 + \cdots + \varepsilon_s, \varepsilon_1 + \cdots + \varepsilon_{s+1}[,$$

$$\forall t \geq 1,\ \forall u \in \mathbb{R}_+,\quad X_u^{(t)} := X^{(0)}_{\varepsilon_1+\cdots+\varepsilon_t+u}.$$

Making the following (deterministic) assumption:

$$(\mathbf{R}) \equiv \begin{cases} (i) & \sum_{t=1}^{+\infty} \varepsilon_t \Delta M^t \text{ converges in } \mathbb{R}^d. \\ (ii) & \lim_{t\to\infty} \eta^t = 0. \end{cases}$$

leads to the proposition below.

Proposition 1 *If $(X^t)_{t\geq 0}$ is bounded and assumption* (**R**) *holds then*

(a) $\mathcal{X}^\infty := \{$*the limiting points of* $(X^t)_{t\geq 0}\}$ *is a compact connected set,*
(b) *The sequence of processes* $(X^{(t)})_{t\geq 0}$ *admits a limiting point uniformly on any interval* $[0,T]$
(c) *Every such limiting point of* $(X^{(t)})_{t\geq 0}$ *is a (bounded)* $\mathcal{X}^\infty$*-valued continuous solution of the Ordinary Differential Equation (ODE):*
$\dot{x} = -h(x), \quad x(0) = x_0^\infty$ *where* $x_0^\infty \in \mathcal{X}^\infty$.

Recently M. Benaim proved a deeper result. Without going into the details let's say that a *chain recurrent* set A is such that for any pair of points x, y, any neighborhoods of these points can be joined by a finite number of pieces of trajectories of the ODE. A is *internally chain recurrent* if these trajectories live in A. His main result is

Proposition 2 *Under the previous assumptions and if, moreover, there is global uniqueness of the solution of* ODE, *then the flow of* ODE *is well defined and* $\mathcal{X}^\infty$ *is an internally chain recurrent set.*

To state the original Kushner and Clark Theorem needs the *stability* of a zero x^* (called equilibrium point of the ODE from now on) of the function h, *i.e.* the existence of an asymptotically stable attracting area denoted by G_{x^*}.

Theorem 3 *(see [15]) If the sequence* $(X^t)_{t\geq 0}$ *is bounded, if the function* h *is continuous and if both assumptions* (**R**) *and*

$$(\mathbf{KC}) \equiv \left\{ \begin{array}{l} X^t \textit{ visits infinitely often a compact subset } K \textit{ of the stable attracting area } G_{x^*} \\ \textit{of an equilibrium point } x^* \in \{h=0\} \end{array} \right.$$

hold, then: $$X^t \to x^* \quad as \quad t \to \infty.$$

3 A global theorem

The Kushner and Clark's result is moderately useful. However, it is possible to formulate a global K&C like theorem *i.e.* a result that would allow to derive the asymptotic behaviour of the sequence $(X^t)_{t\geq 0}$ from the behaviour of the trajectories of the ODE. Here we present our result from [11] (also established in [1] under some uniqueness assumption).

For every $x^* \in \{h=0\}$ we define the *simple* attracting area of x^* by setting

$$\Gamma_{x^*} := \{x^0 \in \mathbb{R}^d \text{ there is some maximal solution of the } ODE \text{ starting at } x^0 \text{ that converges to } x^*\}.$$

A technical assumption related to the ODE (that we do not state here, see[11]) could be resumed in:

$$(\mathbf{C}) \equiv \left\{ \begin{array}{ll} (i) & \text{either all the maximal solutions started at } x^0 \text{ are unbounded} \\ & \text{or } x^0 \text{ lies in a simple attracting area of the } ODE. \\ (ii) & \text{it is not possible to make up a loop by "sticking" together some} \\ & \text{paths supporting solutions of the } ODE\text{: we say that there is no "pseudo-cycle".} \end{array} \right.$$

This assumption holds, *e.g.*, when the ODE is a gradient descent with isolated critical points.

We have as a result :

Theorem 4 *If h is continuous, $(X^t)_{t\geq 0}$ is bounded, and both assumptions* (**R**), (**C**) *hold, then: there exists $x^* \in \{h=0\}$ such that $lim_{t\to\infty} X^t = x^*$.*

4 The case of the Kohonen maps

We recall the basic definitions of the Kohonen algorithm.

The units are numbered $1, 2, \cdots, n$. σ denotes the neighborhood function. It satisfies : $\sigma(0) := 1$, $\sigma(k) = \sigma(-k)$, σ non-increasing. The stimuli ω^t, $t \geq 1$ are i.i.d., $[0,1]$-valued and μ distributed. $X^t := (X_i^t)_{1\leq i\leq n}$ denotes the weight vector at time t. The X_i^t's are $[0,1]^n$-valued. Let $X^0 := (x_i)_{1\leq i\leq n}$ the initial weights. At time $t+1$ the algorithm is recursively defined in two phases by:

(i) *Competitive phase:*
computation of the winning unit $i^{t+1} := i(\omega^{t+1}, X^t) = \underset{k\in I}{\operatorname{argmin}} |\omega^{t+1} - X_k^t|$.

(ii) *Cooperative phase:*
$\forall j \in \{1, 2, \cdots, n\}, \quad X_j^{t+1} = X_j^t - \varepsilon_{t+1}\sigma(i^{t+1} - j)(X_j^t - \omega^{t+1})$
where $(\varepsilon_t)_{t\geq 1}$ is a sequence of $(0,1)$-valued real numbers.

Let $F_n^+ := \{x \in [0,1]^n, 0 < x_1 < x_2 < \cdots < x_n < 1\}$ and $F_n^- = \{x \in [0,1]^n, 0 < x_n < x_{n+1} < \cdots < x_1 < 1\}$.

After self organization the ODE reads (in F_n^+):

$$\dot{x} = -h(x, \sigma)$$

with $h_i(x,\sigma) := \sum_{k=1}^{n} \sigma(|k-i|) \int_{]\widetilde{x}_k, \widetilde{x}_{k+1}]} (x_i - \omega)\mu(d\omega)$

where we set : $\quad \widetilde{x}_1 = 0_-, \quad \widetilde{x}_k := \dfrac{x_k + x_{k-1}}{2}, \quad 2 \leq k \leq n, \quad \widetilde{x}_{n+1} = 1_+.$

- we assume either $\sum_t \varepsilon_t = +\infty$ and $\sum_t \varepsilon_t^2 < +\infty$ (decreasing learning rate) or $\varepsilon_t = \varepsilon$ (constant learning rate)

We define assumptions (H_σ) and $(\mathcal{L}_\mu)$

$(H_\sigma) \equiv$ there exists $k_0 \leq \frac{n-1}{2}$, s.t. $\sigma(k_0 + 1) < \sigma(k_0)$.

$$(\mathcal{L}_\mu) \equiv \begin{cases} \bullet \text{ either a density } f \text{ on } (0,1) \text{ with } log f \text{ strictly concave ,} \\ \bullet \text{ or a density } f \text{ on } (0,1) \text{ with } log f \text{ concave and } f(0_+)+f(1_-)>0 \end{cases}$$

We now state the results obtained in [2] by carefully inspecting the properties of h and the ODE in order to applying Hirsch's Theorem on cooperative dynamical systems (see in [3] a presentation of these results). The first important result is the uniqueness of the equilibrium point.

Proposition 5 *Assume $(\mathcal{L}_\mu)$ and (H_σ) are satisfied then there is a unique equilibrium point of the ODE in F_n^+.*

Then we have the following results concerning the cooperativity of the ODE (items (i), (ii) are also in [16]).

Proposition 6 *Assume $(\mathcal{L}_\mu)$ and (H_σ) are satisfied*

(i) The dynamical system $\dot{x} = -h(x)$ is cooperative on F_n^+ (i.e. the off-diagonal elements of $\nabla h(x)$ are non-positive).

(ii) The matrices $\nabla h(x)$ *are irreducible on* F_n^+ *iff one can find a* $k_0 < \frac{n-1}{2}$. *When* $k_0 = \frac{n-1}{2}$, $\left[\frac{\partial h_i}{\partial x_j}\right]_{i,\, j \neq k_0+1}$ *is irreducible.*

(iii) The set of the limiting values of a trajectory starting from $x_0 \in \overline{F}_n^+$ *is a compact connected set of* F_n^+.

Theorem 0.5 of [12] ensures that a cooperative dynamical system with a unique equilibrium always converges to this equilibrium. So we get under the hypothesis of the two previous propositions :

Corollary 7 *All the trajectories of the O.D.E.* $\dot{x} = -h(x)$ *starting in* $\overline{F}_n^+$ *converge to* x^*.

Then, it derives from the above Corollary and the Kushner & Clark theorem – or its global extension theorem 4 when $k_0 = \frac{n-1}{2}$ – the following.

Theorem 8 *In the decreasing step setting, if* $(\mathcal{L}_\mu)$ *and* (H_σ) *hold, if* $X^0 \in F_n^+$, *then* X^t *converges a.s. to the unique equilibrium point* x^*.

In the constant gain setting – which are often used to track some possible changes in the statistics of the stimuli disribution – similar methods yield

Theorem 9 *If* $(\mathcal{L}_\mu)$ *and* (H_σ) *are satisfied, then any family of invariant distributions* ${\nu_\varepsilon}^+$ *on* F_n^+ *(resp.* ${\nu_\varepsilon}^-$ *on* F_n^-*) converges in distribution, as* ε *goes to* 0, *to the dirac mass at the unique equilibrium point* x^*, *unique zero of* h *in* F_n^+ *(resp.*F_n^-*).*

Remark: Such a family always exists,nevertheless, when $\sigma(k) := \mathbf{1}_{\{|k| \leq 1\}}$, it is proved (see [3] or [5]) that both ${\nu_\varepsilon}^\pm$ are unique and that X^t converges exponentially fast in distribtion to one of them.

5 Conclusion

It is now known that in the case of a higher dimensional Kohonen maps (even the simplest *i.e.* the string in the unit square) there exists no organized absorbing set (see [10]). Furthermore, as the *ODE* is no longer a cooperative system, some new tools are to be developped to make some more significant progress in that direction.

References

[1] M. Benaïm, A Dynamical System Approach to Stochastic Approximations, *SIAM J. Control And Optimization*, **34**, n° 2, 1996, pp.437-472.

[2] M. Benaïm, J.C. Fort, G. Pagès, Convergence of the One-dimensional Kohonen Algorithm. To appear in *Adv. in Applied Proba.*

[3] M. Benaïm, J.C. Fort, G. Pagès, Almost sure convergence of the one dimensional Kohonen Algorithm, *ESANN'97 Bruges*, pp.193-198.

[4] C. Bouton, G. Pagès, Self-organization and convergence of the one-dimensional Kohonen algorithm with non-uniformly distributed stimuli, *Stoch. Proc. & Appl.*, **47**, 1993, pp.249-274.

[5] C. Conley, *Isolated invariant sets and the Morse index*, Conf. Board of the Math. Science by *AMS*, Regional Conference Series in Mathematics, 1978.

[6] M. Cottrell, J.C. Fort, Étude d'un algorithme d'auto-organisation, *Ann. Inst. H. Poincaré*, **23**, n^{0}1, 1987, pp.1-20.

[7] M. Cottrell, J.C. Fort, G. Pagès, Theoretical aspects of the S.O.M. algorithm, *WSOM'97 Helsinki*, pp.246-267.

[8] A. Flanagan, Self-organizing Neural Networks, Thèse 1306, E.P.F.L., Lausanne, 1994.

[9] J.C. Fort, G. Pagès, On the *a.s.* Convergence of the Kohonen Algorithm With a General Neighborhood Function, *The Ann. of Applied Proba.*, **5**, n° 4, 1995, pp.1177-1216.

[10] J.C. Fort, G. Pagès, About the Kohonen algorithm: strong or weak self-organization?, *Neural Networks*, **9**, n° 5, 1996, pp.773-785.

[11] J.C. Fort, G. Pagès, Convergence of Stochastic Algorithms : from the Kushner & Clark Theorem to the Lyapounov Functional Method, *Adv. in Applied Proba.*, **28**,1996, pp.1072-1094.

[12] M. Hirsch, Stability and convergence in strongly monotone dynamical systems, *J. Reine Angew. Math.*, **383**, 1988, pp.1-53.

[13] T. Kohonen, Ananlysis of a simple self-organizing process, *Biol. Cybern.*, **44**, 1982, p135-140.

[14] T. Kohonen, Self-organization and associative memory, 3rd edition Springer, Berlin, 1989.

[15] H.J. Kushner, D.S. Clark, *Stochastic Approximation for Constraint and Unconstraint Systems*, Appl. Math. Sci. Series, 26, Springer, 1978.

[16] A. Sadeghi, Asymptotic Behaviour of Self-Organizing Maps with Non-Uniform Stimuli Distribution, forthcoming in *Annals of Applied Probability*.

Local Subspace Classifier

Jorma Laaksonen

Helsinki University of Technology
Laboratory of Computer and Information Science
P.O. Box 2200, FIN-02015 HUT, Finland

Abstract. This paper describes a new classification technique named the *Local Subspace Classifier* (LSC). The algorithm is closely related to the subspace classification methods. On the other hand, it is an heir of prototype classification methods, such as the k-NN rule. Therefore, it is argued that the LSC technique fills the gap between the subspace and prototype principles of classification. From the domain of the prototype-based classifiers, LSC brings the benefits related to the local nature of the classification, while it simultaneously utilizes the capability of the subspace classifiers to produce generalizations from the training sample. A further enhancement of the LSC principle named the *Convex Local Subspace Classifier* (LSC+) is also presented. The good classification accuracy obtainable with the LSC and LSC+ classifiers is demonstrated with experiments, including the classification of the publicly available data sets of the StatLog project.

1 Introduction

A new statistical classification technique, named the Local Subspace Classifier (LSC) and its further enhancement, the Convex Local Subspace Classifier (LSC+) are described in this paper. The technique is closely related to the subspace classification methods [1]. On the other hand, it is an heir of prototype classification methods, such as the k-NN rule. The purpose the LSC method is to combine the beneficial characteristics of these two principles.

In the LSC processes, the nearest prototypes to the input vector in all the classes are sought. A local subspace – or more precisely, a *linear manifold* – is then spanned by these prototype vectors in each class. The classification is based on the minimum distance from the input vector to these subspaces.

2 Local Subspace Classifier Algorithm

A D-dimensional linear manifold $\mathcal{L}$ of the d-dimensional real space is defined by a matrix $\mathbf{U} \in \mathbb{R}^{d\times D}$ of rank D, and an offset vector $\boldsymbol{\mu} \in \mathbb{R}^d$, provided $D \leq d$,

$$\mathcal{L}_{\mathbf{U},\boldsymbol{\mu}} = \{\mathbf{x} \mid \mathbf{x} = \mathbf{U}\mathbf{z} + \boldsymbol{\mu} \; ; \; \mathbf{z} \in \mathbb{R}^D\} \; . \tag{1}$$

The same manifold can alternatively be defined by $D+1$ prototypes provided that the set of prototypes is not degenerate. The prototypes forming the classifier are marked $\mathbf{m}_{ij}$ where $j = 1, \ldots, c$ is the index of the class and $i = 1, \ldots, N_j$ indexes the prototypes in that class. The manifold dimension D may be different for each class. Therefore, the class-dependent dimensions are denoted D_j. When classifying a vector $\mathbf{x}$, the following is done for each class $j = 1, \ldots, c$:

1. Find the $D_j + 1$ prototypes closest to $\mathbf{x}$ and denote them $\mathbf{m}_{0j}, \ldots, \mathbf{m}_{D_j j}$.
2. Form a $d \times D_j$-dimensional basis of the vectors $\{\mathbf{m}_{1j} - \mathbf{m}_{0j}, \ldots, \mathbf{m}_{D_j j} - \mathbf{m}_{0j}\}$.
3. Orthonormalize the basis to get the matrix $\mathbf{U}_j = (\mathbf{u}_{1j} \ \ldots \ \mathbf{u}_{D_j j})$.
4. Find the projection of $\mathbf{x} - \mathbf{m}_{0j}$ on the manifold $\mathcal{L}_{\mathbf{U}_j, \mathbf{m}_{0j}}$:

$$\widehat{\mathbf{x}}'_j = \mathbf{U}_j \mathbf{U}_j^T (\mathbf{x} - \mathbf{m}_{0j}) \ . \tag{2}$$

5. Calculate the residual of $\mathbf{x}$ relative to the manifold $\mathcal{L}_{\mathbf{U}_j, \mathbf{m}_{0j}}$:

$$\widetilde{\mathbf{x}}_j = \mathbf{x} - \widehat{\mathbf{x}}_j = \mathbf{x} - (\mathbf{m}_{0j} + \widehat{\mathbf{x}}'_j) = (\mathbf{I} - \mathbf{U}_j \mathbf{U}_j^T)(\mathbf{x} - \mathbf{m}_{0j}) \ . \tag{3}$$

The vector $\mathbf{x}$ is then classified according to minimal $||\widetilde{\mathbf{x}}_j||$ to the class j, i.e.,

$$g_{\text{LSC}}(\mathbf{x}) = \operatorname*{argmin}_{j=1,\ldots,c} ||\widetilde{\mathbf{x}}_j|| \ . \tag{4}$$

In any case, the residual length from the input vector $\mathbf{x}$ to the linear manifold is equal to or smaller than the distance to the nearest prototype, i.e., $||\widetilde{\mathbf{x}}_j|| \leq ||\mathbf{x} - \mathbf{m}_{0j}||$. These entities are sketched in Fig. 1.

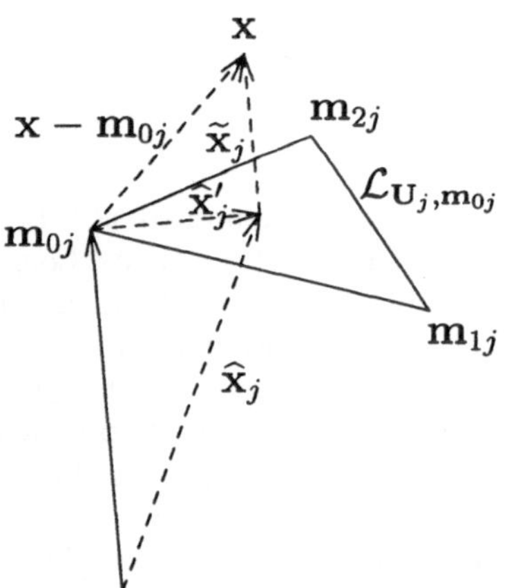

Fig. 1. The d-dimensional entities involved in the LSC classification when $D_j = 2$.

By introducing the multipliers $\{c_{0j}, \ldots, c_{D_j j}\}$ forming the coefficient vector $\mathbf{c}_j = (c_{0j} \ \ldots \ c_{D_j j})^T$, and using the matrix $\mathbf{M}_j = (\mathbf{m}_{0j} \ \ldots \ \mathbf{m}_{D_j j})$, the projection vector $\widehat{\mathbf{x}}_j$ can be expressed

$$\widehat{\mathbf{x}}_j = \mathbf{m}_{0j} + \widehat{\mathbf{x}}'_j = \sum_{i=0}^{D_j} c_{ij} \mathbf{m}_{ij} = \mathbf{M}_j \mathbf{c}_j \ , \ \sum_{i=0}^{D_j} c_{ij} = 1 \ . \tag{5}$$

If the explicit values for the coefficients c_{ij} are needed, they can be solved with matrix pseudo-inversion

$$\mathbf{c}_j = \mathbf{M}_j^{\dagger}\widehat{\mathbf{x}}_j = (\mathbf{M}_j^T\mathbf{M}_j)^{-1}\mathbf{M}_j^T\widehat{\mathbf{x}}_j \; . \tag{6}$$

The principle of local subspaces can be illustrated by Fig. 2, which displays a two-dimensional linear manifold defined by the three handwritten '4's in the corners of the triangle. Each of the three corners represents a pure instant of a normalized 32 × 32-sized human printed glyph whereas the images in between are linear combinations thereof. It may be argued that such interpolated images are as good representatives of the class as the original prototypes. Therefore, if the input vector resembles any of such artificial images, it can be considered to originate from that class.

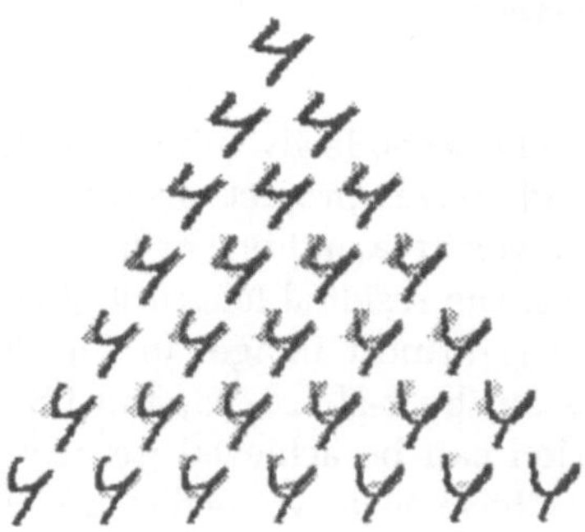

Fig. 2. A piece of a two-dimensional linear manifold spanned by three examples of handwritten digits '4' in the corners. The gray areas in the intermediate virtual digit images reflect the gradual pixel-wise change from one real image to another.

3 LSC+ Classifier

The center image in Fig. 3 shows that the input vector $\mathbf{x}$ may occasionally be projected outside the convex region determined by the prototypes $\mathbf{m}_{0j}, \ldots, \mathbf{m}_{D_j j}$. Whether this is beneficial or disadvantageous depends on the particular data sets involved. If no such extrapolation from the original prototypes is permitted, the projection may be enforced to lie within the area delimited by the prototype vectors. As a result, the orthogonality of the projection is lost. The modified method is here called the *Convex Local Subspace Classifier* (LSC+). The appended plus sign arises from the observation that the coefficients $\{c_{0j}, \ldots, c_{D_j j}\}$ in (5) are all non-negative in the case the projection vector $\widehat{\mathbf{x}}$ is not allowed to exceed the convex region determined by the prototype vectors.

In LSC+, the projection vector $\widehat{\mathbf{x}}$ is thus not the one producing the shortest residual $\widetilde{\mathbf{x}}$. Instead, it primarily lies within the convex region bounded by the

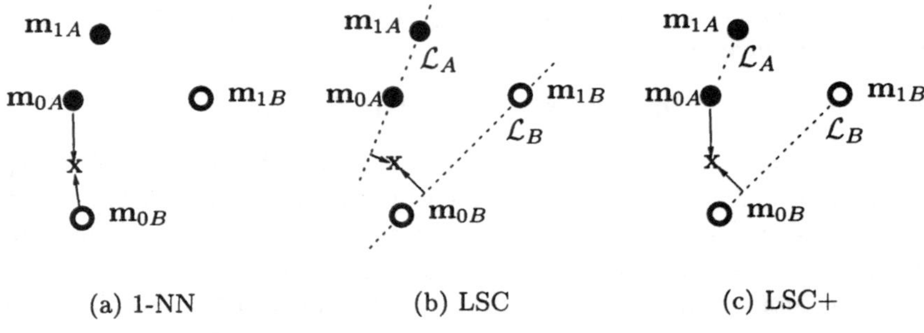

(a) 1-NN (b) LSC (c) LSC+

Fig. 3. Comparison of the 1-NN rule and the basic and convex versions of the Local Subspace Classifier. Using the 1-NN and LSC+ classifiers, **x** is classified as "B", but as "A" when using the LSC method.

nearest prototypes and only secondarily minimizes the residual length. Obviously, if the orthogonal and convex projections differ, the convex projection lies on the boundary of the convex area and one or more of the coefficients are equal to zero. In the convex case, the residual length $||\tilde{\mathbf{x}}||$ is never shorter than in the orthogonal case. The two rightmost images in Fig. 3 show how a vector **x** is classified differently between the k-NN, LSC, and LSC+ classifiers.

The convexity condition can be achieved iteratively by removing from the basis $\mathbf{M}_j$ the vectors which have negative coefficients. This is effectively the same as forcing the corresponding coefficients to be equal to zero. Then, the coefficients for the remaining vectors are solved again and checked for negativeness. The dimension D_j of the linear manifold is decreased by one for each vector removed. In the experiments performed, the vector with the most negative coefficient has been removed in each iteration step.

4 Experiments

The first experiment data was the same as that used in [2]. 17880 binary handwritten digits were normalized to the size of 32×32 pixels. One half of the images was used in training set cross-validation and the other half in testing. Karhunen-Loève transformation was calculated from the training set and applied to the 1024-dimensional images. As the result, feature vectors with the maximum dimensionality of 64 were obtained. 10-fold training set cross-validation was used to solve the optimal values for the feature vector dimensionality, d, and the dimensionality of the linear manifolds, D, common to all the classes. The testing set error rates are shown in Table 1 together with three results from [2]. The 3-NN and ALSM classifiers represent the prototype and subspace principles of classification, respectively, upon which the Local Subspace Classifier technique is based. The *Local Linear Regression* (LLR) classifier attained the best classification accuracy in the comparison [2] in which 17 different types of classifiers were considered. It can be seen that the LSC and LSC+ classifiers clearly outperform all the reference techniques in classification accuracy.

Table 1. Test set classification error percentages of the LSC and LSC+ classifiers together with the Nearest Neighbor rule (3-NN), Averaged Learning Subspace Method (ALSM), and the Local Linear Regression (LLR) classifier, for the data of [2].

classifier	*error %*	*parameters*
LSC	2.5	$d = 64,\ D = 12$
LSC+	2.1	$d = 64,\ D = 23$
3-NN	3.8	$d = 38$
ALSM	3.1	$d = 64,\ D = 29$
LLR	2.8	$d = 36$

The performance of the LSC and LSC+ classifiers was also tested by using the publicly available StatLog [3] data sets. The "australian" test was omitted because the data contained categorical variables. Likewise, the "german" and "heart" data sets were not used because those comparisons involved weighted cost matrices. All the other sets were tested with procedures similar to those described in the original report. It can be seen from the results shown in Table 2 that the performance of the LSC/LSC+ classifier is superior to all classifiers tested in the StatLog project in three cases out of seven, i.e., "letter", "satimage", and "segment". In the "shuttle" and "vehicle" cases the performance is moderate, whereas the remaining two results, "diabetes" and "dna", are clearly failures.

This series of results indicates that the LSC and LSC+ techniques are efficient in some situations but suffer seriously in some other cases. The performance of the local subspace methods is dependent on the nature and density of the data in the Bayesian class border areas. The smoother the hypothetical class boundaries are and the more there exist data to describe them, the better prospects the LSC/LSC+ techniques have.

Table 2. The performances of the LSC and LSC+ classifiers compared to the best StatLog results. Shown are, from left to right, the test set classification error percentage, the subspace dimension D common to all classes, and the rank of the result.

test	*StatLog result*		*LSC*			*LSC+*		
diabetes	LogDisc	22.3	31.25	1	22nd	31.25	1	22nd
dna	Radial	4.1	16.19	8	21st	16.44	8	21st
letter	Alloc80	6.4	3.22	3	1st	3.04	6	1st
satimage	k-NN	9.4	7.55	3	1st	7.45	3	1st
segment	Alloc80	3.0	3.68	2	5th	2.99	3	1st
shuttle	NewId	0.01	0.12	0	8th	0.11	1	8th
vehicle	QuaDisc	15.0	27.07	4	10th	24.35	13	9th

5 Discussion

In this paper, a new family of classification techniques has been introduced. It has been argued that the Local Subspace Classifier (LSC) and the Convex Local Subspace Classifier (LSC+) techniques combine the benefits of the subspace and prototype based classification methods. An experimental evaluations of the two proposed classification algorithms supported this argument.

In most cases, the optimal value of the manifold dimension parameter D is larger in the LSC+ than in the LSC, which does not demand the convexity of the subspace projection. At the same time, the classification accuracy is usually better in the former. These two facts can be interpreted together as follows: The increased value of D provides the classifier with more prototypes for each classification task. Due to the convexity requirement, only the most informative subset of these prototypes are actually used in the difficult regions of classification where the class distributions are sparse and overlapping. Compared to prototype based classifiers, such as 3-NN and LLR, both the LSC and LSC+ techniques seem to benefit from larger feature vector dimensionality d. In this sense, the LSC techniques thus resemble the subspace methods of classification, which, in general, are able to take advantage of large vector dimensionalities.

The application of the LSC principle allows for some additional options. The explicit use of the orthonormalization process and the matrix pseudoinversion (6) can be avoided by using an iterative algorithm for the solution of $\widehat{\mathbf{x}}_j$. Furthermore, the LSC technique can be given a neural interpretation. These two topics are addressed in [4]. The LSC technique may also be gracefully combined with adaptive and self-organizing prototype classifiers such as the *Learning k-Nearest Neighbor classifier* (L-k-NN) [5], the *Learning Vector Quantization* (LVQ) [6], and especially with the topology preserving vector quantization produced by the *Self-Organizing Map* (SOM) [6]. The combination of the LSC and SOM principles is named the *Local Subspace SOM* (LSSOM) [7]. Efficient methods for editing the design set in order to reduce the memory and computation requirements are also being developed.

References

1. Oja, E. *Subspace Methods of Pattern Recognition.* Research Studies Press, Letchworth, England (1983)
2. Holmström, L., Koistinen, P., Laaksonen, J., Oja, E. Neural and statistical classifiers – taxonomy and two case studies. *IEEE Tr. on Neural Networks* 8:5–17 (1997)
3. Michie, D., Spiegelhalter, D. J., Taylor, C. C., editors. *Machine learning, neural and statistical classification.* Ellis Horwood, New York (1994)
4. Laaksonen, J. *Subspace Methods in Recognition of Handwritten Digits.* Ph.D. thesis, Helsinki University of Technology (1997)
5. Laaksonen, J., Oja, E. Classification with learning k-nearest neighbors. In *Proc. of the Intl. Conf. on Neural Networks*, vol. 3, pp. 1480–1483, Washington D.C. (1996)
6. Kohonen, T. *Self-Organizing Maps.* Springer, Berlin, Germany (1995)
7. Laaksonen, J. Local subspace classifier and local subspace SOM. In *Proc. of the Workshop on Self-Organizing Maps*, Otaniemi, Finland (1997). To appear.

Asymptotic Distributions Associated to Unsupervised Oja's Learning Equation

Jean Pierre Delmas

Institut National des Télécommunications, Département Signal et Image
9 rue Charles Fourier, 91011 Evry Cedex, France

Abstract. In this paper, we perform a complete asymptotic performance analysis of the stochastic approximation algorithm (denoted Subspace Network Learning algorithm) derived from Oja's learning equation, in the case where the learning rate is constant and a large number of patterns is available. Using a general result of Gaussian approximation theory, we derive the asymptotic distribution of the estimated projection matrix $\mathbf{W}\mathbf{W}^T$ associated to the connection weight matrix $\mathbf{W}$. Closed form expressions of the asymptotic covariance of the projection matrix estimated by the SNL algorithm, and by the smoothed SNL algorithm that we introduce, are given.

1 Introduction

Over the past decade, adaptive estimation of subspaces of covariance matrices has been applied successfully in different fields of signal processing and many neural network realizations have been proposed for the statistical technique of principal component analysis in data compression and feature extraction [1].

To understand the performance of these neural network unsupervised learning algorithms, it is of fundamental importance to investigate how they behave when the sequence of learning rates is a small constant γ. In this case, it was established that the stochastic approximation algorithm follows an associated ordinary differential equation (ODE) from the start in a first approximation. This transient phase is followed by an asymptotic phase where the random aspect of the fluctuations becomes prominent with respect to the evolution of the ODE. Naturally, if the learning rate γ is chosen larger [resp. smaller], the learning speed increases [resp. decreases], but the fluctuations of the asymptotic phase increases [resp. decreases]. So a tradeoff between the learning speed and the variances of the estimated network weights, often called misadjustment is necessary. A good tool for evaluating these variances is a general Gaussian approximation result [2] which gives the limiting distribution of the estimates $\mathbf{W}_k$ when k and γ tend respectively to $+\infty$ and 0. The purpose of this paper is to determine the asymptotic distribution of the estimates in case of 2 algorithms: the so called SNL stochastic approximation algorithm [1], derived from Oja's learning equation, and the smoothed SNL algorithm that we introduce. However, since these algorithms converge to any orthonormal basis of the considered eigenspace of the covariance matrix of the training patterns, and not to the eigenvectors themselves, we need to develop a special methodology, which consists in considering the stochastic approximation algorithm governed by the associated projection matrix.

This paper is organized as follows. In Section 2, after introducing some notations and the SNL algorithm, the *smoothed* SNL algorithm is presented. In Section 3, after presenting a brief review of a general Gaussian approximation result, we obtain in closed form, the covariance in the limiting distribution of the associated projection matrix estimators. Finally we present in Section 4 some simulation results and investigate the validity of the asymptotic approach.

2 The SNL and smoothed SNL algorithms

For a given $n \times n$ covariance matrix $\mathbf{R}_x = E(\mathbf{x}\mathbf{x}^T)$ of a Gaussian distributed, zero mean real random training patterns vector $\mathbf{x}$, let $\lambda_1 \geq \ldots \geq \lambda_r > \lambda_{r+1} \geq \ldots \geq \lambda_n$ denote the eigenvalues of $\mathbf{R}_x$ and $\mathbf{v}_1, \ldots, \mathbf{v}_n$ corresponding eigenvectors. We consider the recursive updating of an (approximately) orthonormal basis $\mathbf{W}_k$ of the r-dimensional dominant invariant subspace of $\mathbf{R}_x$. The algorithm that we consider was introduced independently by Williams [3], by Baldi [4] and by Oja [5]. Its associated ODE was studied in detail in [6]. It reads:

$$\mathbf{W}_{k+1} = \mathbf{W}_k + \gamma_k[\mathbf{I}_n - \mathbf{W}_k\mathbf{W}_k^T]\mathbf{x}_k\mathbf{x}_k^T\mathbf{W}_k. \tag{2.1}$$

This SNL algorithm is quite similar to the algorithm presented independently by Russo [7] and Yang [8]. This latter algorithm, which we will call the Yang algorithm, reads:

$$\mathbf{W}_{k+1} = \mathbf{W}_k + \gamma_k[2\mathbf{x}_k\mathbf{x}_k^T - \mathbf{x}_k\mathbf{x}_k^T\mathbf{W}_k\mathbf{W}_k^T - \mathbf{W}_k\mathbf{W}_k^T\mathbf{x}_k\mathbf{x}_k^T]\mathbf{W}_k. \tag{2.2}$$

To improve the learning speed and misadjustment tradeoff, we propose in this paper to use a recursive estimate for $\mathbf{R}_k$. So that the modified SNL algorithm, which we call the *smoothed SNL algorithm*, (α is introduced in order to take into account a better tradeoff between the misadjustments and the learning speed), reads:

$$\mathbf{R}_{k+1} = \mathbf{R}_k + \alpha\gamma_k(\mathbf{x}_k\mathbf{x}_k^T - \mathbf{R}_k), \tag{2.3}$$

$$\mathbf{W}_{k+1} = \mathbf{W}_k + \gamma_k[\mathbf{I}_n - \mathbf{W}_k\mathbf{W}_k^T]\mathbf{R}_k\mathbf{W}_k. \tag{2.4}$$

3 Asymptotic performance analysis

A difficulty arises in the study of the behavior of $\mathbf{W}_k$ because the set of orthonormal bases of the r-dominant subspace forms a *continuum* of attractors. Thus, considering the asymptotic distribution of $\mathbf{W}_k$ is meaningless. To solve this problem, in the same way as Williams [3], we consider the trajectory of the matrix $\mathbf{P}_k \stackrel{\text{def}}{=} \mathbf{W}_k\mathbf{W}_k^T$ whose dynamic is governed by the stochastic equation:

$$\mathbf{P}_{k+1} = \mathbf{P}_k + \gamma_k f(\mathbf{P}_k, \mathbf{x}_k\mathbf{x}_k^T) + \gamma_k^2 h(\mathbf{P}_k, \mathbf{x}_k\mathbf{x}_k^T). \tag{3.1}$$

A remarkable feature of (3.1) is that f and h actually depend only on $\mathbf{P}_k$ and *not* on $\mathbf{W}_k$. This fortunate circumstance makes it possible to study the evolution of $\mathbf{P}_k$ without determining the evolution of the underlying matrix $\mathbf{W}_k$. To evaluate the asymptotic distributions of the subspace projection matrix estimators given by the previous algorithms, we shall use a general Gaussian approximation result ([2, theorem 2, p. 108]) that we now recall for convenience of the reader. Consider a constant learning rate recursive stochastic algorithm:

$$\Theta_{k+1} = \Theta_k + \gamma f(\Theta_k, \mathbf{x}_k) + \gamma^2 h_k(\Theta_k, \mathbf{x}_k), \tag{3.2}$$

Suppose that the parameter vector Θ_k converges almost surely to the unique asymptotically stable point Θ_* in the corresponding decreasing learning rate algorithm. Consider the continuous Lyapunov equation:

$$\mathbf{D}\mathbf{C}_\Theta + \mathbf{C}_\Theta\mathbf{D}^T + \mathbf{G} = \mathbf{O}. \tag{3.3}$$

$$\mathbf{D} \stackrel{\text{def}}{=} \mathrm{E}[\frac{\partial f}{\partial \Theta}(\Theta, \mathbf{x}_k)]_{\Theta=\Theta_*} \quad \text{and} \quad \mathbf{G} \stackrel{\text{def}}{=} \sum_{k=-\infty}^{\infty} \mathrm{Cov}[f(\Theta_*, \mathbf{x}_k), f(\Theta_*, \mathbf{x}_0)]. \tag{3.4}$$

If all the eigenvalues of the derivative of the mean field $\mathbf{D}$ have strictly negative real parts, then, in stationary situation, when $\gamma \to 0$ and $k \to \infty$, we have: $\gamma^{-1/2}(\Theta_k - \Theta_*) \xrightarrow{\mathcal{L}} \mathcal{N}(0, \mathbf{C}_\Theta)$ where $\mathbf{C}_\Theta$ is the unique symmetric solution of the Lyapunov equation (3.3).

3.1 Asymptotic distribution associated to SNL algorithm

According to the previous section, one needs to characterize two local properties of the field $f(\mathbf{P}, \mathbf{x}\mathbf{x}^T)$: the mean value of its derivative, and its covariance, both evaluated at the point $\mathbf{P} = \mathbf{P}_* = (\mathbf{v}_1, \ldots, \mathbf{v}_r)(\mathbf{v}_1, \ldots, \mathbf{v}_r)^T$. To proceed, it will be convenient to define the following orthonormal basis for the $n \times n$ symmetric matrices: $\mathbf{S}_{ii} = \mathbf{v}_i\mathbf{v}_i^T$ and $\mathbf{S}_{ij} = 2^{-1/2}(\mathbf{v}_i\mathbf{v}_j^T + \mathbf{v}_j\mathbf{v}_i^T)$ for $i < j$. With this definition, a first order approximation in the neighborhood of $\mathbf{P}_*$ of the mean field, and the eigenstructure of the covariance matrix of the field, are given by this result. For $1 \le i \le j \le n$,

$$\mathrm{E}f(\mathbf{P}_* + \epsilon\, \mathbf{S}_{ij},\ \mathbf{x}_k\mathbf{x}_k^T) = \epsilon\ \mu_{ij}\ \mathbf{S}_{ij}\ +\ O(\epsilon^2), \tag{3.5}$$

$$\mathrm{Cov}(\mathrm{Vec}(f(\mathbf{P}_*, \mathbf{x}_k\mathbf{x}_k^T)))\ \mathrm{Vec}(\mathbf{S}_{ij}) = \nu_{ij}\mathrm{Vec}(\mathbf{S}_{ij}), \tag{3.6}$$

with, respectively, (1_A denotes the indicator function of the condition A and Vec is the classic vectorization operator that turns a matrix into a vector).

$$\mu_{ij} \stackrel{\text{def}}{=} \lambda_i(1_{i>r} - 1_{i\le r}) + \lambda_j(1_{j>r} - 1_{j\le r}) \quad \text{and} \quad \nu_{ij} \stackrel{\text{def}}{=} 2(1_{i\le r} - 1_{j\le r})^2\lambda_i\lambda_j. \tag{3.7}$$

To adapt these results to our needs, the $n \times n$ rank-r symmetric matrix $\mathbf{P}$ should be parameterized by a vector Θ of real parameters. Counting degrees of freedom, shows that the set of $n \times n$ rank-r symmetric matrices is a $\frac{r}{2}(2n -$

$r+1$)-dimensional manifold. Let us now consider the parameterization of $\mathbf{P}_k$ in a neighborhood of $\mathbf{P}_*$. If $\{\theta_{ij}(\mathbf{P}) | 1 \leq i \leq j \leq n\}$ are the coordinates of $\mathbf{P}-\mathbf{P}_*$ in the basis $\mathbf{S}_{i,j}$, then, $\theta_{ij}(\mathbf{P}) = \mathrm{Tr}\{\mathbf{S}_{ij}(\mathbf{P}-\mathbf{P}_*)\}$ for $1 \leq i \leq j \leq n$, $\mathbf{P} = \mathbf{P}_* + \sum_{1\leq i\leq j\leq n} \theta_{ij}(\mathbf{P})\,\mathbf{S}_{ij}$. The relevance of these parameters is shown by the following result. If $\mathbf{P}$ is an $n \times n$ rank-r symmetric matrix, then

$$\mathbf{P} = \mathbf{P}_* + \sum_{(i,j)\in P_s} \theta_{ij}(\mathbf{P})\,\mathbf{S}_{ij} \;+\; O(\|\mathbf{P}-\mathbf{P}_*\|^2) \tag{3.8}$$

where P_s is the set $\{(i,j) \mid 1 \leq i \leq j \leq n \text{ and } i \leq r\}$. There are $\frac{r}{2}(2n-r+1)$ pairs in P_s and this is exactly the dimension of the manifold of $n \times n$ rank-r symmetric matrices. This point, together with eq. (3.8), shows that the matrix set $\{\mathbf{S}_{ij} \mid (i,j) \in P_s\}$ is in fact an *orthonormal basis* of the tangent plane to this manifold at point $\mathbf{P}_*$. It follows that, in a neighborhood of $\mathbf{P}_*$, the $n\times n$ rank-r symmetric matrices are uniquely determined by the $\frac{r}{2}(2n-r+1)\times 1$ vector $\Theta(\mathbf{P})$ defined by: $\Theta(\mathbf{P}) \stackrel{\text{def}}{=} \mathcal{S}^T\mathrm{Vec}(\mathbf{P}-\mathbf{P}_*)$, where $\mathcal{S}$ denotes the $n^2 \times \frac{r}{2}(2n-r+1)$ matrix: $\mathcal{S} \stackrel{\text{def}}{=} [\ldots, \mathrm{Vec}(\mathbf{S}_{ij}), \ldots]$, $(i,j) \in P_s$. If $\mathcal{P}(\Theta)$ denotes the unique (for $\|\Theta\|$ small enough) $n\times n$ rank-r symmetric matrix such that $\mathcal{S}^T\mathrm{Vec}(\mathcal{P}(\Theta)-\mathbf{P}_*) = \Theta$, the following one to one mapping is exhibited for small enough $\|\Theta_t\|$:

$$\mathrm{Vec}(\mathcal{P}(\Theta_k)) = \mathrm{Vec}(\mathbf{P}_*) + \mathcal{S}\Theta_k + O(\|\Theta_k\|^2) \leftrightarrow \Theta_k = \mathcal{S}^T\mathrm{Vec}(\mathbf{P}_k - \mathbf{P}_*) \tag{3.9}$$

We are now in position to solve the Lyapunov equation in the new parameter Θ previously defined. The stochastic equation governing the evolution of this vector parameter is obtained by applying the transformation $\mathbf{P}_k \to \Theta_k = \mathcal{S}^T\mathrm{Vec}(\mathbf{P}_k - \mathbf{P}_*)$ to the original equation (3.1):

$$\Theta_{k+1} = \Theta_k + \gamma\phi(\Theta_k, \mathbf{x}_k) + \gamma^2\psi(\Theta_k, \mathbf{x}_k) \tag{3.10}$$

We obtain after some manipulations as derivative $\mathbf{D}$ and covariance $\mathbf{G}$ matrices associated to (3.10): $\mathbf{D} = \mathbf{\Delta}_\mu$ and $\mathbf{G} = \mathbf{\Delta}_\nu$ with $\mathbf{\Delta}_\mu \stackrel{\text{def}}{=} \mathrm{Diag}(\ldots, \mu_{ij}, \ldots)$ and now $\mu_{ij} < 0$ and $\mathbf{\Delta}_\nu \stackrel{\text{def}}{=} \mathrm{Diag}(\ldots, \nu_{ij}, \ldots)$ for $(i,j) \in P_s$. Thus both $\mathbf{G}$ and $\mathbf{D}$ are diagonal matrices. In this case, the Lyapunov equation (3.3) reduces to uncoupled scalar equations. Thus the solution is obviously $\mathbf{C}_\Theta = -\frac{1}{2}\mathbf{\Delta}_\nu\mathbf{\Delta}_\mu^{-1}$. Then by eq. (3.9), we have:

$$\mathbf{C}_P = \sum_{1\leq i\leq r<j\leq n} \frac{\lambda_i\lambda_j}{2(\lambda_i-\lambda_j)}(\mathbf{v}_i\otimes\mathbf{v}_j + \mathbf{v}_j\otimes\mathbf{v}_i)(\mathbf{v}_i\otimes\mathbf{v}_j + \mathbf{v}_j\otimes\mathbf{v}_i)^T \tag{3.11}$$

This expression coincides with the expression of the covariance matrix of the Yang algorithm (2.2), despite some differences in the expression of μ_{ij} and ν_{ij}.

3.2 Asymptotic distribution associated to the smoothed algorithm

To study the smoothed SNL algorithm, we note that eqs. (2.3) and (2.4) take globally the form (3.2) if we set $\mathbf{\Theta}_k \stackrel{\text{def}}{=} \left[\mathrm{Vec}^T(\mathbf{R}_k), \mathrm{Vec}^T(\mathbf{W}_k)\right]^T$. Then, if we consider the trajectory of the associated matrix $\mathbf{R}_k$, as $\mathbf{P}_k$ keeps symmetric (when the initial condition $\mathbf{R}_0$ is symmetric), it is natural to use the parameter $\Theta_k = \left[\Theta_{1,k}^T, \Theta_{2,k}^T\right]^T$, i.e. the respective coordinates of $\mathbf{R}_k$ in the basis $\mathbf{S}_{ij}$, $1 \leq i \leq j \leq n$, i.e. $(i,j) \in P_s'$ and of $\mathbf{P}_k$ in the basis $\mathbf{S}_{ij}$, $(i,j) \in P_s$. Using the same procedure as in the SNL algorithm, the matrix $\mathbf{C}_P$ is finally written in a similar form to (3.11), except for the term $\alpha_{ij} \stackrel{\text{def}}{=} \frac{\alpha}{\alpha+\lambda_i-\lambda_j}$:

$$\mathbf{C}_P = \sum_{1\leq i\leq r<j\leq n} \frac{\alpha_{ij}\lambda_i\lambda_j}{2(\lambda_i-\lambda_j)}(\mathbf{v}_i\otimes\mathbf{v}_j+\mathbf{v}_j\otimes\mathbf{v}_i)(\mathbf{v}_i\otimes\mathbf{v}_j+\mathbf{v}_j\otimes\mathbf{v}_i)^T \quad (3.12)$$

3.3 Analysis of the results

Firstly the expressions (3.11), (3.12) can be compared to the covariances in the asymptotic distributions obtained in batch estimation. If $\mathbf{P}_k = \sum_{1\leq i\leq r} \mathbf{w}_{k,i}\mathbf{w}_{k,i}^T$ denotes the batch estimated orthogonal projection matrix, when k tends to $+\infty$, we have $k^{1/2}(\text{Vec}(\mathbf{P}_k) - \text{Vec}(\mathbf{P}_*)) \xrightarrow{\mathcal{L}} \mathcal{N}(\mathbf{O}, \mathbf{C}_P)$ with

$$\mathbf{C}_P = \sum_{1\leq i\leq r<j\leq n} \frac{\lambda_i\lambda_j}{(\lambda_i-\lambda_j)^2}(\mathbf{v}_i\otimes\mathbf{v}_j+\mathbf{v}_j\otimes\mathbf{v}_i)(\mathbf{v}_i\otimes\mathbf{v}_j+\mathbf{v}_j\otimes\mathbf{v}_i)^T \quad (3.13)$$

which is also in close similarity with (3.11) and (3.12). Secondly, a simple global measure of performance is the MSE between $\mathbf{P}_k$ and $\mathbf{P}_*$. This MSE between $\mathbf{P}_k$ and $\mathbf{P}_*$ is given by the trace of the covariance matrix in the asymptotic distribution of $\mathbf{P}_k$ (where $\alpha_{ij} \stackrel{\text{def}}{=} 1$ for the SNL algorithm).

$$\text{E}\|\, \mathbf{P}_k - \mathbf{P}_* \,\|_{\text{Fro}}^2 = \gamma \sum_{1\leq i\leq r<j\leq n} \frac{\alpha_{ij}\lambda_i\lambda_j}{\lambda_i-\lambda_j} + o(\gamma) \quad (3.14)$$

4 Simulations

In these experiments, we consider the case of the estimation of the projection matrix on the principal subspace spanned by the 2 first eigenvectors of $\mathbf{R}_x = \text{Diag}(1.75, 1.5, 0.5, 0.25)$ from independent inputs $\mathbf{x}_k$. Fig. 1 shows the ratio of the estimated mean square error $\text{E}\|\, \mathbf{P}_k - \mathbf{P}_* \,\|_{\text{Fro}}^2$ over the theoretical asymptotic mean square error $\gamma\text{Tr}(\mathbf{C}_P)$ as a function of γ, for both the SNL and the smoothed SNL algorithms and with $\alpha = 1$. Our present asymptotic analysis is seen to be valid over a large range of γ. The learning speed is investigated through the iteration number until 'convergence" is achieved. Fig. 2 plots this iteration number as a function of the asymptotic mean square error $\gamma\text{Tr}(\mathbf{C}_P)$. As can be seen, the smoothed SNL algorithm with $\alpha = 0.2$ provides a better tradeoff between the learning speed and the misadjustment $\gamma\text{Tr}(\mathbf{C}_P)$.

5 Conclusion

We have performed in this paper a complete asymptotic performance analysis of the SNL algorithm and of a smoothed SNL algorithm that we have introduced, assuming a constant learning rate, and in the case where a large number of patterns is available. A closed form expression of the covariance in distribution of the projection matrices onto the principal component subspace estimators has been given in case of independent learning patterns.

References

1. E. Oja, "Principal components, minor components and linear neural networks," *Neural Networks*, vol. 5, pp. 927-935, 1992.
2. A. Benveniste, M. Métivier, P. Priouret, *Adaptive algorithms and stochastic approximations*, Springer Verlag, 1990.
3. R. Williams, "Feature discovery through error-correcting learning," *Technical Report 8501*, San Diego, CA: University of California, Institute of Cognitive Science, 1985.
4. P. Baldi, "Linear learning: Landscapes and algorithms," *in Proc. NIPS, Denver*, 1988.
5. E. Oja, "Neural Networks, principal components and subspaces," *International Journal of Neural Systems*, vol.1, no.1 pp. 61-68, 1989.
6. W.Y. Yan, U. Helmke, J.B. Moore, "Global analysis of Oja's flow for neural networks," *IEEE Trans. on Neural Networks*, vol. 5, no. 5, pp. 674-683, Sep. 1994.
7. L. Russo, "An outer product neural network for extracting PC from a time series," *In B.H. Juang et al. N.N. for S.P.*, pp. 161-170. NY IEEE Press, 1991.
8. B. Yang, "Projection approximation subspace tracking," *IEEE, Trans. on Signal Processing*, vol. 43, no. 1, pp. 95-107, Jan. 1995.

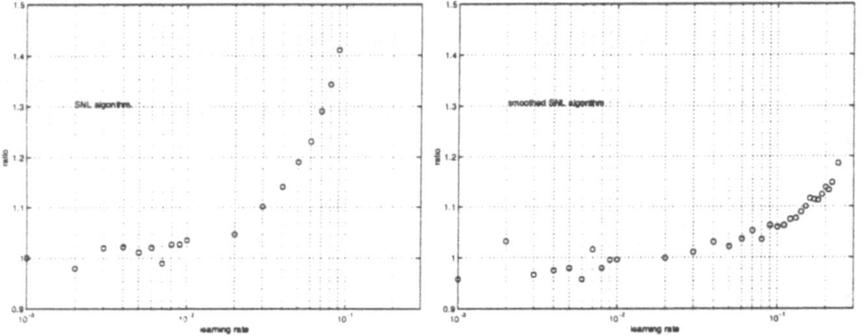

Fig. 1. $(\gamma \mathrm{Tr}(\mathbf{C}_P))^{-1} \mathrm{E} \| \mathbf{P}_k - \mathbf{P}_* \|^2_{\mathrm{Fro}}$, (averaged over 400 independent runs), as a function γ.

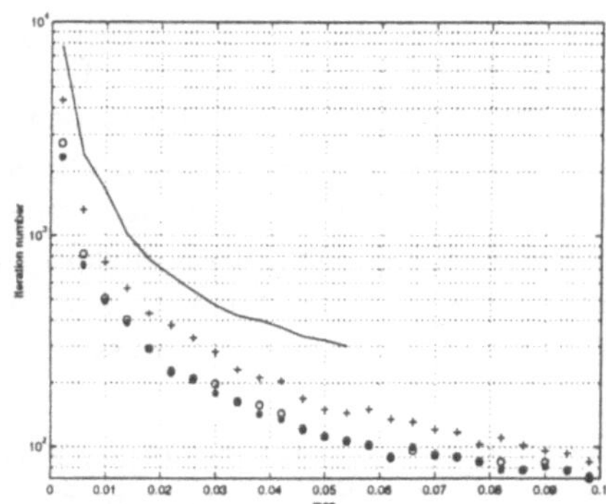

Fig. 2. Iteration number until "convergence" is achieved, versus theoretical asymptotic MSE, of the SNL (-), and of the smoothed SNL alg. with $\alpha = 1$ (+), 0.2 (*), 0.3 (o).

The Probabilistic Growing Cell Structures Algorithm

Nikos A. Vlassis, Apostolos Dimopoulos, and George Papakonstantinou

National Technical University of Athens, Greece

Abstract. The growing cell structures (GCS) algorithm is an adaptive k-means clustering algorithm in which new clusters are added dynamically to produce a Dirichlet tessellation of the input space. In this paper we extend the non-parametric model of the GCS into a probabilistic one, assuming that samples are distributed in each cluster according to a multi-variate normal probability density function. We show that by recursively estimating the means and the variances of the clusters, and by introducing a new criterion for the insertion and deletion of a cluster, our approach can be more powerful to the original GCS algorithm. We demonstrate our results within the mobile robots paradigm.

1 Introduction

The growing cell structures (GCS) algorithm [1] is an adaptive *k-means* algorithm [3] that performs data clustering. Based on the earlier work of Kohonen [4], GCS is a *self-organizing* neural network model [2] that incrementally builds a *Dirichlet or Voronoi tessellation* of the input space, while it is able to automatically find its structure and size (see [7] and [5] for a general description of the problem of pattern recognition with neural networks).

In this paper, we extend the non-parametric model of the GCS into a probabilistic one [6]. We assume that samples come from each cluster according to a Gaussian probability density function, whose parameters, means and variance, we are to estimate statistically from the input set. In contrast to many traditional probabilistic techniques like, e.g., the *ISODATA* algorithm [3], that need to maintain a large portion of the training set in memory, our approach uses recursive formulas for the estimation of the means and variances of the clusters.

Besides, due to its cluster insertion-deletion criterion, the original GCS algorithm cannot very easily handle cases of correlated inputs, e.g., successive inputs that are close in the input space. Our algorithm ameliorates this deficiency of GCS by introducing a synthesized criterion for cluster insertion-deletion that takes into consideration the *a priori* probability of a cluster, together with its variance. We show the virtues of our algorithm when applied to a real-world problem; that of estimating the Voronoi centers of a mobile robot's configuration space [9].

2 The Probabilistic Model

Consider an input space $\mathbf{V} \subset \mathbb{R}^p$, and a sequence of n p-dimensional samples $\{x_1, x_2, \ldots, x_n\}$ from $\mathbf{V}$. Our task is to group all x_i into K *clusters*, so that each cluster contains *similar* points, i.e., points that are near in $\mathbf{V}$ by some metric distance. We also want K to change dynamically.

Our probabilistic model assumes that in each cluster k the samples are distributed according to a known ***probability density function*** (or density) $p_k(x;\theta_k)$, for some unknown vector of parameters θ_k. Also, we assume that each cluster k has ***prior*** probability π_k, implying an a priori preference for k over the other clusters. Then the ***membership*** probability that a new sample x_i is assigned to cluster k is $p_k(x_i, k; \theta_k) = \pi_k p_k(x_i;\theta_k)$, and the ***posterior*** (normalized) probability [7] (p. 19) is

$$p(k|x_i) = \frac{\pi_k p_k(x_i;\theta_k)}{\sum_{j=1}^{K} \pi_j p_j(x_i;\theta_j)}. \tag{1}$$

The denominator in the above formula is the total input *mixture* density $p(x) = \sum_{j=1}^{K} \pi_j p_j(x;\theta_j)$.

The *Bayes decision rule* [7] (p. 19) assigns a future sample x_i to cluster k if

$$p(k|x_i) = \max\{p(j|x_i)\}, \quad j = 1, \ldots, K. \tag{2}$$

From the above equations it follows that a complete clustering schema requires the estimation of both the prior cluster probabilities π_j and the parameter vectors θ_j of all clusters j, with $j = 1, \ldots, K$.

More specifically now, we assume that samples from cluster k follow a *p-variate normal* density $\mathbf{N}_p\{\mu_k, \Sigma_k\}$, i.e.,

$$p_k(x;\mu_k,\Sigma_k) = (2\pi)^{-p/2}|\Sigma_k|^{-1/2} e^{-\frac{1}{2}(x-\mu_k)\Sigma_k^{-1}(x-\mu_k)^T}, \tag{3}$$

parametrized over the *means* μ_k and *covariance matrix* Σ_k of the cluster (see [6] p. 188, 197), while $|\Sigma_k|$ denotes the determinant of Σ_k. Furthermore, we assume that Σ_k is of the type $\sigma_k^2 I$, with I the identity matrix. This simplification can be quite reasonable under certain conditions (see, e.g., [7] p. 289, 296, and [8]), while it helps keeping the computational cost low. Under this model, a cluster can be regarded as a *hyper-sphere* centered around μ_k, while σ_k^2 gives a measure of its size.

Assuming normal density for the clusters, the problem is now to estimate from the sequence of input samples $\{x_i\}$, $i = 1, \ldots, n$ the parameters μ_k and σ_k^2, and the prior probability π_k of each cluster. By using the *Maximum Likelihood* (ML) estimation (see [7] p. 334) Tråvén [8] proves that for p-variate normal distributions the prior probability π_k can be estimated as

$$\pi_k^{(n)} = \frac{1}{n}\sum_{i=1}^{n} p(k|x_i), \tag{4}$$

where $p(k|x_i)$ is the posterior probability that a sample x_i belongs to cluster k. $(\cdot)^{(n)}$ denotes the value after the n-th input x_n has arrived.

In addition, recursive expressions for the estimation of μ_k and σ_k^2 can be formulated as

$$\mu_k^{(n+1)} = \mu_k^{(n)} + \eta_k^{(n+1)}(x_{n+1} - \mu_k^{(n)}), \tag{5}$$

$$\sigma_k^{2(n+1)} = \sigma_k^{2(n)} + \eta_k^{(n+1)}[(x_{n+1} - \mu_k^{(n)})(x_{n+1} - \mu_k^{(n)})^T - \sigma_k^{2(n)}], \tag{6}$$

In the above equations $\eta_k^{(n+1)}$ can be approximated by $p(k|x_{n+1})/[(n+1)\pi_k^{(n)}]$. Then the recursive formula becomes similar to the learning rule used by the family of Kohonen's algorithms.[1]

Now we restrict ourselves to the l most recent samples $\{x_{n-l}, \ldots, x_n\}$. Then $\eta_k^{(n+1)}$ reads

$$\eta_k^{(n+1)} = \frac{p(k|x_{n+1})}{l\pi_k^{(n)}}. \tag{7}$$

This restriction implies a 'forgetting' schema, in which old samples affect the learning process less than recent ones. Extending the calculations of [8] a little further, a recursive formula similar to the above formulas for μ_k and σ_k^2 can be formulated for the prior probability π_k as

$$\pi_k^{(n+1)} = (1 - 1/l)\pi_k^{(n)} + \frac{p(k|x_i)}{l} \quad \text{or} \quad \pi_k^{(n+1)} = \pi_k^{(n)} + \frac{1}{l}[p(k|x_i) - \pi_k^{(n)}]. \tag{8}$$

This formula shows that the prior probability π_k 'moves' towards $p(k|x)$ in each step, and can be viewed as an estimation over the l most recent samples.

Finally, if clusters are assumed a priori equiprobable ($\pi_i = \pi_j, \quad \forall i \neq j$) and fully separated in the input space, and a new sample x_i is assigned to its nearest cluster k with probability 1 (i.e., $p(k|x_i) = 1$ if $||x_i - \mu_k|| = \min$, which implies a uniform density in each cluster) then, setting $1 - 1/l = \alpha$, $n = l$, and using eq. 4, eq. 8 becomes identical to the ***signal counter forgetting*** schema[2] of [1].

3 The PGCS Algorithm

Based on the previous framework, we describe PGCS, a probabilistic version of the original GCS algorithm [1]. We show how the results for GCS can be generalized within our probabilistic framework. Additionally, we propose some extensions to the cluster insertion-deletion criterion in order to handle cases of correlated inputs.

Initialization. The first sample x_1 becomes the center (means) μ_1 of the first cluster, whereas σ_1^2 is set to 0. The prior probability π_1 of this cluster is set to 1.

---

[1] If we assume that clusters are equiprobable and fully separated in the input space, and a new sample is assigned to its nearest cluster with probability 1, then the recursive formula for the means is exactly the SOM learning rule.

[2] $\Delta\tau = -\alpha\tau$

Adaptation. According to eq. 2, a new input x_i is assigned to the cluster k with the maximum $p(k|x_i)$. This cluster is called the *winning* cluster. Applying the logarithm to eq. 3, this corresponds to minimizing $\delta(x_i, \mu_k)^2 + 2\log(2\pi)^{p/2}\sigma_k^p - 2\log\pi_k$, where $\delta(x_i, \mu_k) = [\frac{1}{\sigma_k^2}(x_i - \mu_k)(x_i - \mu_k)^T]^{1/2}$ is the *Mahalanobis* distance from x_i to the center of cluster k.

The parameters μ_k and σ_k^2 of cluster k change according to eq. 5, 6, and 7.[3]

Updating. The prior probabilities of all clusters are updated according to eq. 8. Note that since $p(k|x_i)$ is larger for clusters lying in the neighborhood of the winning cluster, the prior probability of these clusters is affected more than their farther counterparts.

Cluster insertion. After a fixed number of steps, and if a pre-defined maximum number of clusters is not exceeded, we find the cluster q who maximizes the quantity

$$\phi(q) = \alpha_\pi \pi_q + \alpha_\sigma \frac{\sigma_q^2}{\sum_{j=1}^K \sigma_j^2}, \tag{9}$$

with $\alpha_\pi, \alpha_\sigma$ appropriate coefficients that are problem-dependent. We create a new cluster r between q and its direct neighbor f with the maximum $\phi(f)$. We estimate the means and variance of the new cluster as

$$\mu_r = \frac{\phi(q)\mu_q + \phi(f)\mu_f}{\phi(q) + \phi(f)} \quad \text{and} \quad \sigma_r^2 = \frac{\phi(q)\sigma_q^2 + \phi(f)\sigma_f^2}{\phi(q) + \phi(f)}.$$

Clusters q and f change their variances appropriately as

$$\sigma_q^{2(new)} = (1 - \frac{\phi(q)}{\phi(q) + \phi(f)})\sigma_q^2 \quad \text{and} \quad \sigma_f^{2(new)} = (1 - \frac{\phi(f)}{\phi(q) + \phi(f)})\sigma_f^2,$$

where $(\cdot)^{(new)}$ denotes the new value.

For estimating the prior probability of clusters r, q, and f, we stick to the rule of GCS: *every cluster should potentially have the same probability of being the winning cluster for a new random sample.* From eq. 1 and 3 it follows that the posterior probability of a cluster j lying in the vicinity of r, estimated on the means, is proportional to π_j/σ_j^2. Making the posterior probabilities before and after the insertion equal for the clusters q and f yields

$$\frac{\pi_{q,f}}{\sigma_{q,f}^2} = \frac{\pi_{q,f}^{(new)}}{\sigma_{q,f}^{2(new)}} \quad \text{or} \quad \Delta\pi_{q,f} = \frac{\sigma_{q,f}^{2(new)} - \sigma_{q,f}^2}{\sigma_{q,f}^2}\pi_{q,f},$$

where $\Delta\pi = \pi^{(new)} - \pi$. From this formula we compute the $\pi^{(new)}$ of clusters q and f.

[3] Since no topological ordering in a space of lower dimension than p is actually entailed, there is no real need to change the parameters μ and σ^2 of the neighboring clusters.

Preserving the total prior probability locally before and after the insertion leads to the new prior probability of r

$$\pi_r^{(new)} = -(\Delta\pi_q + \Delta\pi_f).$$

The above two equations are equivalent to eq. 10 and 11 of [1].

Cluster deletion. After a fixed number of steps we remove the cluster c with the lowest $\phi(c)$.

4 Discussion-Results

The insertion or deletion of a new cluster in the original GCS algorithm happens at the region of **V** with the highest or lowest prior probability, respectively. This schema is adequate if samples come randomly from the input space **V**, but seems rather ineffective when successive inputs happen to be correlated somehow, e.g, by being not very far in **V**.

In our PGCS algorithm, the criterion for creating a new cluster takes into consideration both the prior probability of a cluster and its variance. Thus, clusters with small variance, i.e., those that are concentrated around their means, are not likely to split, even if many inputs are assigned to them, and the opposite.

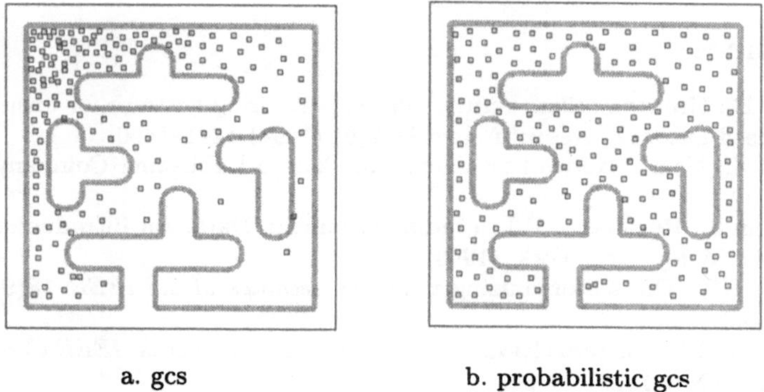

a. gcs b. probabilistic gcs

Fig. 1. Applying the GCS and our extended model to a robot configuration.

The motivation for this work was [9], in which the estimation of the Voronoi centers of a robot's configuration space necessitated the existence of an algorithm that could handle successive points in $\mathbb{R}^2$, denoting the robot's (x, y) position. There, the GCS model seemed incapable of performing a good clustering of the input space. In Fig. 1 we show the performance of GCS and PGCS for a typical robot configuration, where the robot, starting from the up-left corner, was to explore an unknown environment (cluster centers are located in the free space, whereas the rest is obstacles). We ran both algorithms with the same set of parameters, whereas the coefficients α_π and α_σ of PGCS had values 0.2 and 0.8, respectively.

5 Conclusions

We presented PGCS, a probabilistic version of the growing cell structures [1] algorithm for data clustering. We outline below some of the main contributions of this work.

- We assume that input samples follow Gaussian distributions, whereas GCS assumes uniform densities within each cluster. Our approach is more realistic since in real-world problems the inputs are inherently Gaussian, or can approximately be modeled by Gaussian distributions. Moreover, by simplifying the type of the covariance matrices we manage to keep the computational cost low.
- Our cluster insertion-deletion criterion takes into consideration both the prior probabilities of the clusters and their variances, whereas GCS only the former. This is very helpful in cases where inputs are not selected in random from the input space, but are rather correlated.
- We estimate the means and the variances of the clusters recursively, in contrast to other clustering algorithms that need to maintain a large amount of input samples for this purpose.

References

1. Fritzke B.: Growing cell structures—a self-organizing network for unsupervised and supervised learning. *Neural Networks* **7**(9) (1994) 1441–1460.
2. Haykin S.: Neural Networks. Macmillan College Publishing Company, New York, (1994).
3. Kaufman L., Rousseeuw P.J.: Finding Groups in Data. An Introduction to Cluster Analysis. Wiley, New York, (1990).
4. Kohonen T.: The self-organizing map. *Proceedings of the IEEE* **78**(9) (Sep 1990) 1464–1480.
5. Lippmann R.P.: Pattern classification using neural networks. *IEEE Communications Magazine* (Nov 1989) 47–64.
6. Papoulis A.: Probability, Random Variables, and Stochastic Processes. McGraw-Hill, 3rd edn., (1991).
7. Ripley B.D.: Pattern Recognition and Neural Networks. Cambridge University Press, Cambridge, U.K., (1996).
8. Tråvén H.G.C.: A neural network approach to statistical pattern classification by "semiparametric" estimation of probability density functions. *IEEE Transactions on Neural Networks* **2**(3) (May 1991) 366–377.
9. Vlassis N.A., Papakonstantinou G., Tsanakas P.: Robot map building by Kohonen's self-organizing neural networks. In: *Proc. 1st Mobinet Symposium on Robotics for Health Care*. Athens, Greece, (May 1997).

Unsupervised Coding with Lococode

Sepp Hochreiter and Jürgen Schmidhuber

Technische Universität München, 80290 München, Germany
and IDSIA, Corso Elvezia 36, CH-6900-Lugano, Switzerland

Abstract. Traditional approaches to sensory coding use code component-oriented objective functions (COCOFs) to evaluate code quality. Previous COCOFs do not take into account the information-theoretic complexity of the code-generating mapping itself. We do: "*Low-complexity coding* and *decoding*" (LOCOCODE) generates so-called *lococodes* that (1) convey information about the input data, (2) can be computed from the data by a low-complexity mapping (LCM), and (3) can be decoded by a LCM. We implement LOCOCODE by training autoassociators with *Flat Minimum Search* (FMS), a general method for finding low-complexity neural nets. LOCOCODE extracts optimal codes for difficult versions of the "bars" benchmark problem. As a preprocessor for a vowel recognition benchmark problem it sets the stage for excellent classification performance.

1 Introduction

Several COCOFs have been proposed to evaluate the quality of sensory codes. Many COCOFs explicitly favor factorial codes [1] of input data. Other approaches favor local codes, e.g., [8]. Recently there also has been much work on COCOFs for biologically plausible sparse distributed codes, which share some advantages of both minimally redundant and local codes, e.g., [4,3,6].

But what about coding costs? COCOFs emphasize desirable properties of the code itself, while neglecting the costs of constructing the code from the data. For instance, coding input data in a redundancy-free fashion may be very expensive in terms of information bits required to describe many finely tuned free parameters in the code-generating network. In this paper we will shift the focus onto the information-theoretic costs of code-generation. See abstract. We will see that LOCOCODE encourages noise-tolerant "feature detectors" reminiscent of those observed in the mammalian visual cortex, because they are easily codable and decodable.

2 Flat minimum search: review

To implement LOCOCODE we apply Flat Minimum Search (FMS) [7] to a 3-layer autoassociator (AA) whose hidden unit (HU) activations represent the code. FMS is a general, gradient-based method for finding low-complexity networks that can be described with few bits of information.

FMS Overview. FMS finds a large region in weight space such that each weight vector from that region has *similar* small error. Such regions are called "flat minima". A flat minimum corresponds to weights many of which can be given with low precision. In contrast, a "sharp" minimum corresponds to weights which have to be specified with high precision. In the terminology of the theory of minimum description length (MDL), fewer bits of information are required to pick a "flat" minimum (corresponding to a low complexity-network). As a natural by-product of net complexity reduction, FMS automatically prunes units, weights, and input lines, reduces output sensitivity with respect to remaining weights and units, and generalizes well. In a previous application to stock market prediction [7], FMS led to better results than "weight decay" and "optimal brain surgeon".

Architecture. We use a 3 layer feedforward net. Each layer is fully connected to the next layer. Let O, H, I denote index sets for output, hidden, input units, respectively. For $l \in O \cup H$, the activation y^l of unit l is $y^l = f(s_l)$, where $s_l = \sum_m w_{lm} y^m$ is the net input of unit l ($m \in H$ for $l \in O$ and $m \in I$ for $l \in H$), w_{lm} denotes the weight on the connection from unit m to unit l, f denotes the activation function, and for $m \in I$, y^m denotes the m-th component of an input vector. $W = |(O \times H) \cup (H \times I)|$ is the number of weights.

Algorithm. FMS' objective function E features an unconventional term:

$$B = \sum_{i,j \in O \times H \cup H \times I} \log \sum_{k \in O} \left(\frac{\partial y^k}{\partial w_{ij}}\right)^2 + W \log \sum_{k \in O} \left(\sum_{i,j \in O \times H \cup H \times I} \frac{\left|\frac{\partial y^k}{\partial w_{ij}}\right|}{\sqrt{\sum_{k \in O} \left(\frac{\partial y^k}{\partial w_{ij}}\right)^2}} \right)^2 .$$

$E = E_q + \lambda B$ is minimized by gradient descent, where E_q is the training set mean squared error (MSE), and $\lambda > 0$ scales B's influence. B measures the weight precision (number of bits needed to describe all weights in the net). Reducing B without increasing E_q means removing weight precision without increasing quadratic error. All of this can be done efficiently, namely, with standard backprop's order of computational complexity. For more details see [7].

3 Experiment 1: independent bars

The task is adapted from [2]. The input is a 5×5 pixel grid with horizontal and vertical 1×5 and 5×1 bars at random, independent positions. The goal is to extract the independent features corresponding to the bars. According to [2], even a simpler variant (no mixing of vertical and horizontal bars) is not trivial: *"Although it might seem like a toy problem, the* 5×5 *bar task with only 10 HUs turns out to be quite hard for all the algorithms we discuss. The coding cost of making an error in one bar goes up linearly with the size of the grid, so at least one aspect of the problem gets* easier *with large grids." [2].* We will see that even the difficult version of this task is not hard for LOCOCODE.

Training and testing. For each of the 25 pixels there is an input unit. Input units that see a pixel of a bar take on activation 0.5; others −0.5. Each of the 10 possible bars appears with probability $\frac{1}{5}$. In contrast to [2] vertical and horizontal bars may be mixed in the same input. This makes the task harder (see [2], p. 570). To test LOCOCODE's ability to reduce redundancy, we use many more HUs (namely 25) than the required minimum of 10. [2] (p. 570) reports that an AA trained without FMS (and more than 10 HUs) "consistently failed". We have confirmed this result.

Following [2], the net is trained on 500 randomly generated patterns (there may be pattern repetitions). Learning is stopped after 5,000 epochs. Then the net is tested on 500 additional random patterns. We say that a pattern is processed correctly if the absolute error of all output units is below 0.3. The sigmoid HUs are active in [0,1], the sigmoid output units are active in [-1,1]. Noninput units have an additional bias input. The target is -0.7 for -0.5 and 0.7 for 0.5. Normal weights are initialized in $[-0.1, 0.1]$, bias weights with -1.0, λ with 0.5. *Parameters:* learning rate: 1.0, $E_{tol} = 0.16$, $\Delta\lambda = 0.001$. *Architecture:* (25-25-25).

Results: factorial codes based on feature detectors. Training MSE is 0.11; test MSE is 0.15 (averages over 10 trials). The net generalizes well: only one of the 500 test patterns is not processed correctly. 15 of the 25 HUs are indeed automatically pruned. Figures 1 and 2 depict typical weights to and from HUs. For each of the 25 HUs there is a 5×5 square depicting the 25 post-training weights on connections from 25 inputs. The corresponding bias weight sits on top of the upper left corner. White (black) circles on gray (white) background are positive (negative) weights. The circle radius is proportional to the weight's absolute value. All but 10 units are effectively pruned away. The surviving HUs become binary bar detectors. They exactly mirror the statistics of the pattern generation process: LOCOCODE finds an optimal factorial code by producing optimal feature detectors.

Backprop fails. For comparison we run this task with conventional backpropagation with 25 (BP25), 15 (BP15) and 10 (BP10) HUs. BP25: test MSE 0.20; BP15: test MSE 0.22; BP10: test MSE 0.27. Backprop does not prune any units; the resulting weight patterns are highly unstructured, and the underlying input statistics are not discovered.

Noisy bars. Similar lococodes were obtained even when we randomly varied bar intensities and added Gaussian noise to the input.

Conclusion. Unlike standard backprop, LOCOCODE easily solves hard variants of the standard "bars" problem. It discovers the underlying statistics and extracts the essential, statistically independent features, even in case of noisy inputs.

4 Experiment 2: vowel recognition

Next we will show that lococodes can help to achieve superior generalization performance on a supervised learning benchmark problem.

Task. We recognize vowels, using data [9] from Scott Fahlman's CMU benchmark collection. There are 11 vowels and 15 speakers. Each speaker spoke each vowel 6 times. Data from the first 8 speakers is used for training, other data for testing. This means 528 frames for training and 462 frames for testing. Each frame consists of 10 input components obtained by low pass filtering at 4.7kHz, digitized to 12 bits with a 10 kHz sampling rate. A twelfth order linear predictive

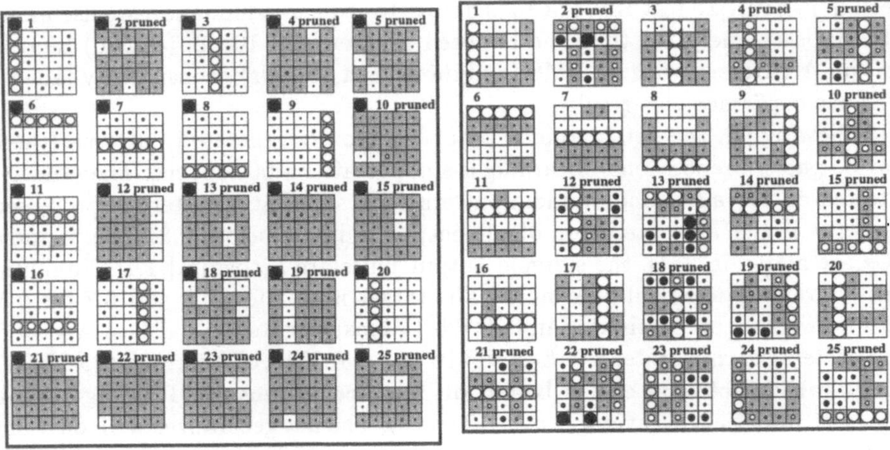

Fig. 1. *Independent bars: incoming weights to hidden units.*

Fig. 2. *Independent bars: weights from hidden to output units.*

analysis was carried out on six 512 sample Hamming-windowed segments from the steady part of the vowel. The reflection coefficients were used to calculate 10 log area parameters, providing the 10 dimensional input space.

Coding. The training data is coded using FMS for autoassociation. Architecture: (10-30-10). The sigmoid HUs are active in [0,1], the sigmoid output units are active in [-1,1]. Noninput units have an additional bias input. The input components are linearly scaled in [-1,1]. The AA is trained with 10^7 pattern presentations. Then its weights are frozen.

Classification. From now on, the vowel codes across all nonconstant HUs are used as inputs for a conventional supervised backprop classifier, which is trained to recognize the vowels from the code. The classifier's architecture is $((30-c)$-11-11), where c is the number of nonvarying (pruned) HUs in the AA. The hidden and output units are sigmoid and active in [-1,1], and receive an additional bias input. The classifier is trained with another 10^7 pattern presentations.

Parameters. AA net: learning rate: 0.02, $E_{tol} = 0.015$, $\Delta\lambda = 0.2$, $\gamma = 2.0$. Backprop net: learning rate: 0.002.

Overfitting. We confirm Robinson's results: the classifier tends to overfit when trained by simple backprop — during learning, the test error rate first decreases and then increases again.

Comparison. See Table 1. FMS generates 3 different lococodes. Each is fed into 10 conventional, overfitting backprop classifiers with different weight initializations: the table entry for "LOCOCODE/Backprop" represents the mean of 30 trials. The results for neural nets and nearest neighbor are taken from Robinson [9]. The other results (except for LOCOCODE's) are taken from Hastie et al. [5]. *Our method led to best generalization results.* The error rates after backprop learning vary between 39 and 45 %.

	Technique	nr. hidden units	error rates training	test
(1.1)	Single-layer perceptron	–	–	0.67
(1.2.1)	Multi-layer perceptron	88	–	0.49
(1.2.2)	Multi-layer perceptron	22	–	0.55
(1.2.3)	Multi-layer perceptron	11	–	0.56
(1.3.1)	Modified Kanerva Model	528	–	0.50
(1.3.2)	Modified Kanerva Model	88	–	0.57
(1.4.1)	Radial Basis Function	528	–	0.47
(1.4.2)	Radial Basis Function	88	–	0.52
(1.5.1)	Gaussian node network	528	–	0.45
(1.5.2)	Gaussian node network	88	–	0.47
(1.5.3)	Gaussian node network	22	–	0.46
(1.5.4)	Gaussian node network	11	–	0.53
(1.6.1)	Square node network	88	–	0.45
(1.6.2)	Square node network	22	–	0.49
(1.6.3)	Square node network	11	–	0.50
(2)	Nearest neighbor	–	–	0.44
(3)	LDA	–	0.32	0.56
(4)	Softmax	–	0.48	0.67
(5)	QDA	–	0.01	0.53
(6.1)	CART	–	0.05	0.56
(6.2)	CART (linear comb. splits)	–	0.05	0.54
(7)	FDA / BRUTO	–	0.06	0.44
(8)	Softmax / BRUTO	–	0.11	0.50
(9.1)	FDA / MARS (degree 1)	–	0.09	0.45
(9.2)	FDA / MARS (degree 2)	–	0.02	0.42
(10.1)	Softmax / MARS (degree 1)	–	0.14	0.48
(10.2)	Softmax / MARS (degree 2)	–	0.10	0.50
(11)	LOCOCODE / Backprop	30/11	0.05	0.42

Table 1. ***Vowel recognition task: generalization performance of different methods. Surprisingly, FMS-generated lococodes fed into a conventional, overfitting backprop classifier led to best results. See text for details.***

Backprop fed with LOCOCODE code sometimes goes down to 38 % error rate, but due to overfitting error grows again. Given that backprop by itself is a very naive approach, the fact that its generalization performance can be dramatically enhanced by feeding it *nongoal-specific* lococodes appears quite surprising.

Hastie et al. also obtained additional, even slightly better results with an FDA/MARS variant: down to 39 % average error rate. It should be mentioned, however, that their data was subject to goal-directed pre-processing with splines, such that there were many clearly defined classes. Furthermore, to determine the input dimension, Hastie et al. used a special kind of generalized cross-validation error, where one constant was obtained by unspecified "simulation studies".

Typical feature detectors. The number of (pruned) HUs with constant activation varies between 5 and 10. 2 to 5 HUs become binary, and 4 to 7 trinary. With all codes we observed: apparently, certain HUs become feature detectors for speaker identification. Another HU's activation is near 1.0 for the words "heed" and "hid" ("i" sounds). Another HU's activation has high values for the words "hod", "hoard", "hood" and "who'd" ("o"-words) and low but nonzero values for "hard" and "heard". LOCOCODE supports feature detection.

5 Conclusion

Unlike previous approaches, LOCOCODE does not define code optimality solely by properties of the code itself. Instead, LOCOCODE's notion of code optimality takes into account the information-theoretic complexity of the mappings used for coding and decoding. Lococodes typically compromise between conflicting goals. They tend to exhibit *low but not minimal* redundancy — if the complexity costs of generating minimal redundancy are too high.

LOCOCODE easily solves coding tasks that have been described as hard by other authors. Our experiments also demonstrate the usefulness of LOCOCODE-based data pre-processing for subsequent classification. Although we made no attempt to prevent classifier overfitting, we achieved excellent results. From this we conclude that the lococodes fed into the classifier already conveyed the "essential", almost noise-free information necessary for excellent classification. We are led to believe that LOCOCODE is a promising and general method for data pre-processing.

6 Acknowledgments

This work was supported by *DFG grant SCHM 942/3-1* from "Deutsche Forschungsgemeinschaft". Results for the bars problem stem from M. Baumgartner's diploma thesis (TUM 1996).

References

1. H. B. Barlow, T. P. Kaushal, and G. J. Mitchison. Finding minimum entropy codes. *Neural Computation*, 1(3):412–423, 1989.
2. P. Dayan and R. Zemel. Competition and multiple cause models. *Neural Computation*, 7:565–579, 1995.
3. B. A. Olshausen; D. J. Field. Emergence of simple-cell receptive field properties by learning a sparse code for natural images. *Nature*, 381(6583):607–609, 1996.
4. D. J. Field. What is the goal of sensory coding? *Neural Computation*, 6:559–601, 1994.
5. T. J. Hastie, R. J. Tibshirani, and A. Buja. Flexible discriminant analysis by optimal scoring. Technical report, AT&T Bell Laboratories, 1993.
6. G. E. Hinton and Z. Ghahramani. Generative models for discovering sparse distributed representations. Technical report, University of Toronto, Department of Computer Science, Toronto, Ontario, M5S 1A4, Canada, 1997. A modified version to appear in *Philosophical Transactions of the Royal Society* **B**.
7. S. Hochreiter and J. Schmidhuber. Flat minima. *Neural Computation*, 9(1):1–42, 1997.
8. T. Kohonen. *Self-Organization and Associative Memory.* Springer, second ed., 1988.
9. A. J. Robinson. *Dynamic Error Propagation Networks.* PhD thesis, Trinity Hall and Cambridge University Engineering Department, 1989.

Wave Propagation in Self-Organizing Feature Maps as a Means for the Representation of Temporal Sequences

B. Dobrzewski, D. Ruwisch, and M. Bode

Universität Münster, Institut für Angewandte Physik,
Corrensstr. 2–4, D–48149 Münster, Germany
Phone: +49 251 833 3516, Fax: –3513, E-mail: bjoern@uni-muenster.de

Abstract. Wave propagation within a "cortex" of neurons is used to influence the ordering of a self-organizing feature map. As a result, the network is able to represent temporal aspects of the input. Since the wave only uses local interactions between adjacent neurons, connectivity in the network is very low and a parallel hardware architecture suggests itself. Operation of a demonstration setup consisting of 16 neurons in digital technology is exemplified by the representation of phoneme sequences.

1 Introduction

Self-organizing feature maps as a model for neural networks have intentionally been designed for the classification of static feature vectors. Besides their use in phoneme recognition [1], they have successfully been employed for various technical applications. Furthermore, feature map generation is interesting from the biological point of view in that it can be observed in various parts of the mammalian cortex (see e.g. [2]).

However, static feature classification is not a sufficient method when feature dynamics constitute an important characteristic of the information, i.e., when past information is necessary for the interpretation of the current feature. An illustrating example is speech recognition, where contextual information, on different time scales, is crucial for understanding. In the present work we study an algorithm that uses wave propagation within the network for the representation of temporal aspects. The wave acts by influencing the election of the winner neuron depending on previously presented features.

Wave propagation caused by reactive and diffusive processes can be observed in several biological systems including networks of glial cells in the brain [3] whose role in the information processing is still unclear. Since in our concept the propagating wave only uses local interactions between adjacent neurons, it is well suited for an implementation in a "synergetic" hardware architecture ([4,5]). Our digital demonstration setup is a multi-processor architecture using "discrete" wave propagation.

2 Kohonen's algorithm

We focus on Kohonen's algorithm, which is the most perspicuous model of a self-organizing feature map. Each neuron, $N_{\mathbf{r}}$, of the network is located in a "cortex" at a position $\mathbf{r}$. Every neuron corresponds to a reference vector, $\mathbf{W}_{\mathbf{r}}$, in an arbitrary feature space. When the neurons are presented with a feature vector, $\mathbf{U}$, a winner neuron, N_{win}, has to be determined. The winner neuron, which is located at cortical position $\mathbf{r}_{win}$, is thought to best represent the feature vector, $\mathbf{U}$:

$$\mathbf{r}_{win} = \arg\min(\|\mathbf{U} - \mathbf{W}_{\mathbf{r}}\|). \tag{1}$$

The set of feature vectors, $\{\mathbf{U}_{\mathbf{r}}\}$, for which $N_{\mathbf{r}}$ is the winner neuron, is called receptive field of that neuron. This defines a Voronoi-tesselation of the feature space with the vectors, $\mathbf{U}_b$, on the cell boundaries satisfying

$$\|\mathbf{U}_b - \mathbf{W}_{\mathbf{r}_1}\| = \|\mathbf{U}_b - \mathbf{W}_{\mathbf{r}_2}\|, \tag{2}$$

$\mathbf{W}_{\mathbf{r}_1}$ and $\mathbf{W}_{\mathbf{r}_2}$ being the reference vectors of two neighboring cells.

The essential of Kohonen's algorithm is the learning step, which affects the winner neuron and its cortical neighborhood. Learning is an adaptation, $\Delta\mathbf{W}_{\mathbf{r}}$, of reference vectors, $\mathbf{W}_{\mathbf{r}}$, in direction of the feature vector, $\mathbf{U}$. With increasing cortical distance, $d_{win}(\mathbf{r}) = \|\mathbf{r}_{win} - \mathbf{r}\|$, between a neuron at position $\mathbf{r}$ and the winner neuron at position $\mathbf{r}_{win}$, the relative amount of reference vector adaptation decreases. This interaction is described by the neighborhood function, $\eta(d_{win}(\mathbf{r}))$, which usually is a localized function (e.g. of Gaussian shape):

$$\Delta\mathbf{W}_{\mathbf{r}} = \eta(d_{win}(\mathbf{r}))(\mathbf{U} - \mathbf{W}_{\mathbf{r}}). \tag{3}$$

Feature vectors, $\mathbf{U}$, are thought to be stochastically presented while both height and width of the neighborhood function, $\eta(d_{\mathrm{win}}(\mathbf{r}))$, are successively diminished.

3 Representation of temporal sequences

Different attempts have been made to extend or modify the self-organizing feature map in order to represent temporal aspects of the presented features. This is a very important task if the network is supposed to process context information, e.g., in speech processing systems [6]. In general, such a task demands a representation of the near past. This can be achieved by means of a time-delay architecture, which enables the system to process a certain amount of former inputs (for a review see e.g. [7]). One can apply this technique to the input of a standard Kohonen feature map: A number of time-delayed input vectors are concatenated to a larger vector that serves as feature vector, $\mathbf{U}$. This method was succesfully used to improve the recognition of transient phonemes in a speech recognition system [6].

Following an idea of Euliano and Principe [9], wave propagation can be utilized to process information on the past so that temporal order is represented

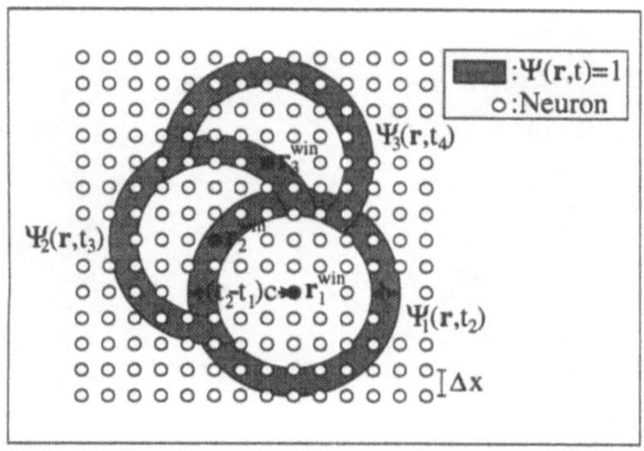

Fig. 1. Three waves $\Psi_i(\mathbf{r}, t)$ in the cortex according to Eq. (4) started by three subsequent winner neurons at cortical positions $\mathbf{r}_i^{win}$. Subsequent winner neurons are forced by the wave to be situated in a certain cortical neighborhood of the respective preceding winner neuron if β is large enough.

in the feature map. Propagation and interference of waves, that are attenuated over space and time, influence the neural competition: Determination of the winner neuron is modified with the result, that temporal neighborhood of feature vectors in a sequence may lead to an adjacent representation of these vectors in the feature map. Euliano and Principe choose a predetermined direction of wave propagation. We will restrict propagation in a less rigid and more natural way leading to some advantages. Additionally, we omit attenuation and interference of waves for the sake of an easier hardware implementation.

Consider a wave, $\Psi_i(\mathbf{r}, t)$, with a concentric wave crest of certain width, b, propagating through the cortex. It starts at time $t{=}t_i$ at the position of the winner neuron, $\mathbf{r}_i^{win}$. The index, i, numbers the sequence position of the corresponding feature vector, $\mathbf{U}_i$, starting with $i{=}1$. Wave propagation ends at time t_{i+1}, when the next feature vector, $\mathbf{U}_{i+1}$, is presented. Using the abreviation $t'_i = t - t_i$ such a wave reads

$$\Psi_i(\mathbf{r}, t) = \mathcal{H}(ct'_i - d_{win}(\mathbf{r}))\mathcal{H}(d_{win}(\mathbf{r}) - ct'_i + b)H(\mathbf{r}, t) \tag{4}$$

where $\mathcal{H}$ is the Heaviside function, $\mathcal{H}(x){=}0$ if $x{<}0$, and $\mathcal{H}(x){=}1$, otherwise.

The "History function", $H(\mathbf{r}, t)$, restricts wave propagation with respect to the past. It is initialized at the beginning of a new sequence with

$$H(\mathbf{r}, t_1) = 1, \text{ for all } \mathbf{r}. \tag{5}$$

Afterwards it is set to zero at a position, $\mathbf{r}$, if the crest of a wave has left that position, i.e.,

$$H(\mathbf{r}, t) \to 0 \text{ if } \Psi_i(\mathbf{r}, t) \to 0 \tag{6}$$

and remains zero for the rest of the sequence.

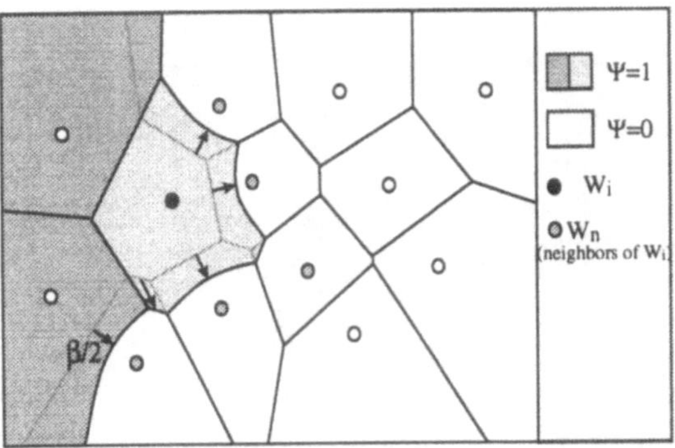

Fig. 2. Receptive fields for a 2-dimensional feature space with $\beta > 0$. The cells of the neurons with $\Psi_1(\mathbf{r},t) \neq 0$ are enlarged. The hyperbolic boundary consists of vectors $\mathbf{U}_b$ which satisfy $\|\mathbf{U}_b - \mathbf{W}_i\| - \beta = \|\mathbf{U}_b - \mathbf{W}_n\|$, with $\mathbf{W}_\mathbf{r}$ and $\mathbf{W}_\mathbf{n}$ being the reference vectors of the enhanced neuron and its neighbors, respectively.

As a result, a subsequent wave does not propagate into a region that has been passed by an earlier wave during the same sequence. This resembles the refractory phase of real neural tissue. Even more natural dynamics would allow for a decay of $H(\mathbf{r},t)$ over time. Moreover, this could be understood as an explicit measure of time within the network (compare [8]).

Wave propagation according to Eq. (4) and Eq.(6) is sketched in Fig. 1. The final states of three subsequent waves, i.e., $\Psi_i(\mathbf{r},t_{i+1})$, i=1..3, are shown. In the final wave region the probability for a neuron to win the competition and to become the winner neuron is enhanced. Equation (1) of the Kohonen algorithm is changed to

$$\mathbf{r}_i^{win} = \arg\min(\|\mathbf{U_i} - \mathbf{W_r}\| - \beta\Psi_{i-1}(\mathbf{r},t_i)). \tag{7}$$

Only the first winner (i=1) is conventionally determined, i.e., $\Psi_0 \equiv 0$ in Eq. (7). In this case Eq. (7) equals the original Kohonen Eq. (1) and the winner neuron is the one whose receptive field contains the presented feature vector. For the subsequently presented features the receptive fields of the neurons with Ψ_1=1 become enlarged with hyperbolic boundaries (Fig. 2). With increasing β they are further enlarged to the extent that neighboring neurons, $N_\mathbf{r}$, with $\Psi_1(\mathbf{r},t)$=0 get excluded from the neural competition. Eventually, for $\beta > \|\mathbf{W}_i - \mathbf{W}_j\| \forall i,j$, the feature space is partitioned by the wave-emphasized neurons, only, i.e., the winner neuron is always located on the wave crest. The later the succeeding feature vector is presented, the farther away from the predecessor it will be represented.

Since this is a competitive ordering principle there may arise contradictions to a purely topographic mapping (β=0), if very different feature vectors are neighbors in the sequence or if very similar feature vectors appear far from each other in the sequence. By tuning β one can choose wether more attention is paid to feature similarity or sequence position.

If a sequence is too long to be entirely represented, e.g., the wave has completely left the cortex (i.e., $\Psi_i(\mathbf{r},t)=0$ over the whole cortex), Eq. (7) again becomes identical with Eq. (1). In this case, the next winner is determined independently from its predecessor (standard Kohonen). Thus, if the new winner neuron is not situated in a refractory region, the sequence is automatically devided into two independent subsequences. Otherwise, no wave is started and the system determines another winner in the standard Kohonen manner.

Wave propagation influencing the winner election in order to represent temporal order, as shown here, can additionally control the learning process of the neurons [10].

4 Digital hardware implementation

The described concept was developed in view of a digital hardware realization. Operation is demonstrated with phoneme sequences processed by a demonstration setup consisting of 16 neurons in a linear chain.

Each neuron is emulated by a PIC16C84 microcontroller. The components of the feature ($\mathbf{U}$) and reference vectors ($\mathbf{W_r}$) are stored as 8bit values. The acoustic signal is sampled (rate: 14kHz) and Fourier-transformed (512 points in 36.7 ms) by a digital signal processor TMS320C26. The resulting spectrum is grouped into three frequency channels (0.6–1.0 kHz, 1.0–3.5 kHz, 3.5–7.4 kHz). The energy of each single channel contributes one component to a 3-dimensional feature vector, $\mathbf{U}$, which, additionally, is normalized, so that the feature vectors lie on a 2-dim plane. Each feature vector corresponds to one spoken phoneme, independently of its natural length. In the 1-dim setup, different network structures as a result of different values for β are demonstrated with the 4-phoneme sequence "M O T O". The two "O" at second and forth position cannot be distinguished by their feature characteristics (Fig. 3). Thus, a purely topologic mapping (β=0) yields a representation of both "O" in the same cortical region (Neurons 14-16 in Fig 3a). In contrast, the contextual information, i.e. the position of the phonemes in the sequence, is the dominant criterion when β is sufficiently large (Fig 3b). With β>max($\|\mathbf{U} - \mathbf{W}\|$), the wave always forces the next winner, r_{i+1}^{win}, to be located within a distance $c(t_{i+1} - t_i) < d_{win} < c(t_{i+1} - t_i) + b$ from the present winner, r_i^{win}. In this case, the phonemes are mapped in the cortex according to their temporal order (Fig 3b)). The total way in feature space when passing the neurons in their cortical order increases, since the network is folded in feature space in order to represent the "O" twice (neurons 5-7 and 12-15). Additional intermediate states have been observed with transitions showing typical properties of a phase transition, e.g., metastability.

As to conclude, the experimental results show that wave propagation within a self-organized feature map can be used to represent temporal aspects of the input information, e.g. phoneme "context" in a speech signal. The hardware architecture allows an efficient implementation of the concept as shown by the demonstration setup.

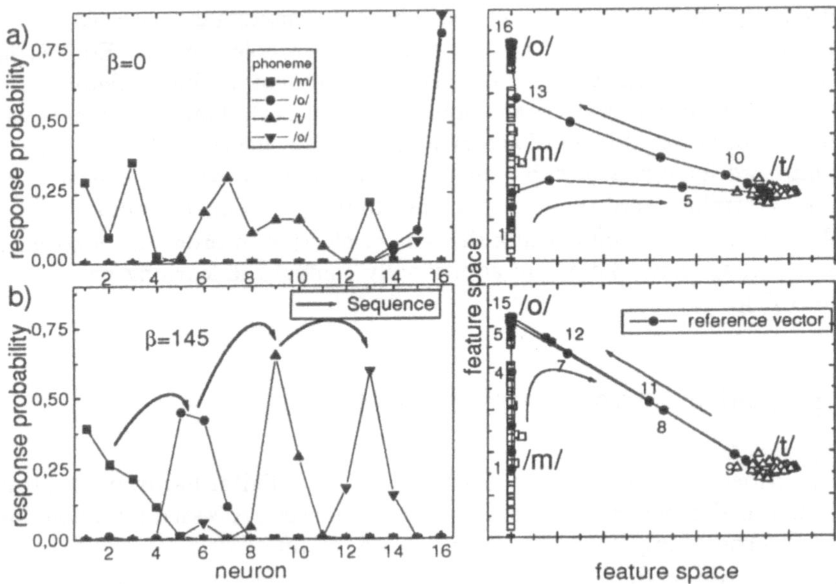

Fig. 3. Representation of the phoneme sequence "M O T O" by the digital demonstration setup for two different values of β. The parameters are set to $ct_i = 3$, $b = 2$ (comp. Eq. 4 and Fig. 1) and $d_{max} = 4$. The 16 reference vectors (3-dim, normalized), $\mathbf{W}_r$, are initialized with random numbers and learning is performed with 90 presentations. The probability distributions (conditional probabilities p(neuron|phoneme)) are obtained with 60 presentations of the sequence.

References

1. T. Kohonen, *Self-Organizing Maps* (Springer, Berlin, 1995).
2. C. Darian-Smith, *Ann. Rev. Psychol.* **33**, 155 (1982).
3. A. H. Cornell-Bell, S. M. Finkenbeiner, M. S. Cooper, S. J. Smith, *Science* **247**, 470 (1990).
4. D. Ruwisch, M. Bode, H.-G. Purwins, *Neural Networks*, **6**, 1147 (1993).
5. D. Ruwisch, M. Bode, H.-J. Schulze, F.-J. Niedernostheide, *Lecture Notes in Physics* **476**, (Springer, Berlin, 1996).
6. J. Kangas, *Proceedings of the international joint conference on neural networks (IJCNN)* **2**, 331–336 (1990).
7. M. C. Mozer, *Predicting the future and understanding the past.* (Addison Wesley, Redwood City, 1993).
8. D. L. James, R. Miikulainen, *Advances in neural processing systems*, **7** (1995).
9. N. R. Euliano, J. C. Principe, *Proceedings of the international conference on neural networks (ICNN)* **4**, 1900–1905 (1996).
10. D. Ruwisch, B.Dobrzewski, and M. Bode, to appear in *Proceeding of the 1997 IEEE Workshop on Neural Networks for Signal Processing (NNSP97)*, (1997).

Contextual Kohonen SOM with Orthogonal Weight Estimator Principle

Nicolas Pican

CRIN-CNRS / INRIA
F-54506 Vandoeuvre-lès-Nancy Cedex, France

Abstract. We present in this paper the embedding of the Othogonal Weight Estimator (OWE) principle in Kohonen self-organizing maps (SOM). The resulting architecture is a context-independant classification system. The modification of the SOM architecture is that the weights of the SOM are computed by a MLP feds by the context of the presented pattern. We show the results on not trivial problem that underline the capacities of this new architecture.

1 Introduction

The Kohonen self-organizing maps (SOM) [2] are frequently used to find, in a space, the distribution of the data and to map this distribution, commonly, in lower dimensional spaces. In some case the use of a SOM as preprocessing for classification is not relevant. Typically, when the data space could be interpreted as two sub-spaces: the pattern sub-space, where a datum of a given class appears as a pattern in its associate context, and the context sub-space, caracterising the environment of the presented pattern. Thus, using the complete data (pattern + context) as the data space of SOM, the result will be a map where each node represents a 2 uplet (pattern, context). This representation cannot be used for pattern classification independantly of their contexts. Removing the context part of data, and only using the pattern sub-space as the data space of SOM is of course impossible, due to the pattern context-dependancies. Thus for context-dependant data it is necessary to create a context-independant classification system. A simple solution of this problem could be the use of a set of SOM, one for each context. But this solution becomes very heavy when the dimension of the context sub-space is large. Moreover, we cannot ensure that the same node in all SOM will represent a unique class.

We have encountered a similar problem when we have tried to use a MLP (Multi Layer Perceptron) in the problems of the modelization of context-dependent behavior. Our first studies on a set of MLPs, one for each context, led us to develop a fast learning algorithm for a set of MLP topologically distributed on the context space [3]. The principle of this algorithm is based on the fact that weights corresponding to a local minimum of the error surface for a MLP in a given context are close to weights of a local minimum of the error surface of a neightbor MLP in a the context space. The result of these studies is that the weight landscape on the context space, solution of the modelization, becomes a continuous function of context [1]. The set of MLP can be so reduced to an unique MLP (master MLP) in which each weight is a continuous function of context. Thus, we have introduced the principle of Orthogonal Weight Estimator (OWE) as one slave MLP fed by context data that models one weight landscape on context space for one weight in the master MLP [4].

After the presentation of the new architecture introducing the OWE principle in the Kohonen SOM, we give its learning algorithm and results for a simple context-independant classification.

2 Contextual Self-Organized Map: SOM-OWE

In this paper we address the problem of the use of SOM in the case of context-independant classification problem. This problem can be viewed follows: how can a class be assigned to a pattern when parameters defining this pattern change with the context? In other words, let us consider a pattern ω viewed in context φ as $\omega_\varphi = (\omega, \varphi)$. The quantization vector of the class i in context φ becomes $x_{i,\varphi}$. The first idea to try to tackle this new problem is to define one SOM for each context φ but two difficulties arise. How many SOMs are necessary to cover the context space? For all defined SOMs, how to ensure that the same node in all SOMs corresponds to the same class? Even if the first difficulty can be easily tackled, the second one can only be resolved by constraining each weight in each SOM...

If pattern changes continuously wrt [2] the context (it is often the case in a real word application), this constraint on the weight can be viewed as a continuous function of φ. Thus a quantization vector of the class i becomes $x_{i,\varphi} = F_i(\varphi)$ where $F_i(\varphi)$ is an unknown continuous function giving the quantization vector of the class i wrt the context φ. A pattern (ω, φ) will be assigned to the class p by the classical measure of the distorsion between the pattern and all the quantization vectors of the SOM: $p = \text{Argmin}_i(\| F_i(\varphi) - \omega \|^2)$ or by introducing one function for each input dimension d (i.e. $F_i(\varphi) \equiv \{f_i^d(\varphi), d = 1 \ldots D\}$):$p = \text{Argmin}_i(\sum_{d=1}^{D} (f_i^d(\varphi) - \omega_\varphi^d)^2)$.

For each $f_i^d(\varphi)$ functions we use a MLP to approximate it. We now call a MLP fed by context parameter φ an Orthogonal Weight Estimator (OWE). This name has been chosen for its role, an estimator of weight, and for the inputs, orthogonal of the SOM inputs.

We can now consider the SOM-OWE architecture as a SOM fed by the pattern ω, in which each weight $x_{i,\varphi}^d$, component d of the quantization vector i in the context φ, is computed by an OWE feds by the context φ of the pattern.

[1] We assume that the behavior changes continuously with respect to the context.

[2] with respect to

3 The Learning Algorithm for SOM-OWE

Even if the Kohonen algorithm is generally not a stochastic gradient algorithm, its is commonly admitted that the function

$$V_\Lambda := \sum_{i,j \in I} \Lambda(i,j) \int_{C_j(x)} \| x_i - \omega \|^2 \, \mu(d\omega) \tag{1}$$

is closely related to the algorithm. Here I denotes the unit set, $\Lambda : I \times I \to [0,1]$ the neighborhood function, x a generic (weight) vector, $C_i(x))_{i \in I}$ denotes its Voronoï testellation and μ the probability distribution (on $\Re^D$) of the inputs. In the case where μ is discrete V_Λ actually is the true potential of the Kohonen SOM [1].

The use of SOM can have two different goals: *self-organization* (*SO*) when the neighborhood function are a long range with strong unit-to-unit links and *vector quantization* (*VQ*) when the strength of the links decrease and finally vanish, to provide a good quantization. If we focus our interest on classification problem, we are only interested by the final results of *VQ* phase (i.e. the 0 neighbor case).

For our SOM-OWE architecture in the case of 0 neighbor the true potential can be written as:

$$\begin{aligned} V_0 &= \sum_{i \in I} \int_\Phi \int_{C_i(x,\varphi)} \| F_i(\varphi) - \omega \|^2 \, \mu(d\omega, \varphi)\nu(d\varphi) \\ &= \int_\Phi \int_\Omega \min_{i \in I} \| F_i(\varphi) - \omega \|^2 \, \mu(d\omega, \varphi)\nu(d\varphi) \end{aligned}$$

where Φ is the context space Ω is the pattern space μ is the φ conditional probability distribution on Ω and ν is the probability distribution on Φ

It is now obvious that to minimize this potential we must minimize, for all patterns ω and for the winner unit p, the following cost function:

$$\lambda_p = \int_\Phi \int_\Omega \| \hat{F}_p(\varphi) - \omega \|^2 \, \mu(d\omega, \varphi)\nu(d\varphi) \tag{2}$$

where $\hat{F}_i(\cdot)$ is an approximation of $F_i(\cdot)$ and each component d of $\hat{F}_i(\cdot))$, noted $\hat{f}_i^d(\cdot)$, is computed by a MLP (an OWE). To ensure a good organization of the $\hat{F}_i(\cdot)$ in the input space it is important to keep the neighborhood function at the begining of the training process. Moreover we have noticed that convergence of an OWE is accelerated by this contribution of its neighbor. The learning algorithm for the SOM-OWE architecture can easily be defined as a stochastic gradient descent using the following updating rule for the weights θ of an OWE (i, d) at iteration t:

$$\Delta\theta_i^d(t) = -\epsilon(t)\Lambda(p,i)\frac{\partial \lambda_p}{\partial \theta_i^d} \tag{3}$$

where $\epsilon(t)$ is the learning rate at iteration t.

For each presented pattern (ω, φ) we compute each $f_i^d(\varphi)$ by propagating the context φ through each OWE [3]. Thus we classically compute the winner unit p of the SOM and we update the weights of each OWE using (3).

4 Results

In this part we test the new SOM-OWE architecture on a simple, but not trivial, problem to underline its specific properties.

The problem chosen is a four classes context-independant classification ($I = \{1..4\}$). The patterns are $\omega \in \Re^2$ ($D = 2$) and the context is $\varphi \in \Re$. The patterns of one class in a given context are uniformaly distributed in a circle. But the radius and the center of the circle changes wrt the context. In Fig. 1 the four volumes generated by all patterns are represented.

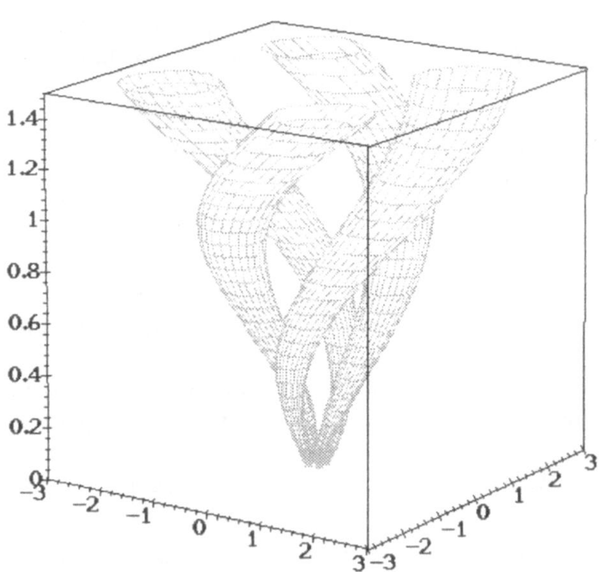

Fig. 1. The 4 class volume data, context is the vertical dimension

The training corpus consists of 30,400 patterns uniformly distributed over $\Omega \times \Phi$. The Fig. 2 shows the data projected on Ω space. SOM-OWE architecture is a 2x2 unit grid for the SOM and 1x12x1 full feedforward MLP with bias for each OWE. The neighborhood function is: $\Lambda(i,j) = \eta e^{-(i-j)^2/2\sigma^2}$ with $\sigma = \sigma_0(1-\alpha)^t$ ($\sigma_0 = 2$,$\alpha = 10^{-3}$) and $\eta = \eta_0(1-\beta)^t$ ($\eta_0 = 0.1$,$\beta = 10^{-4}$).

[3] The computation of each OWE can easily be done in parallel [6]

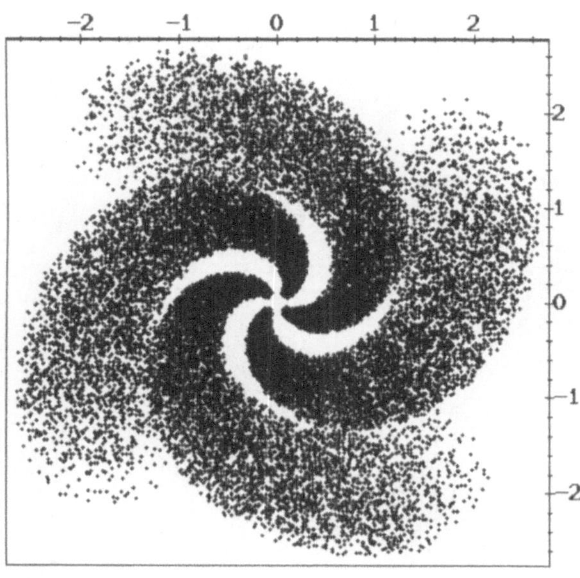

Fig. 2. Data projected on Ω space

After twenty epochs of training, the result (see Fig. 3) is a set of 4 context-dependant quantization vectors that follow the symetry axis of each volume. $F_i(\varphi)$ are represented by black circles (only the half of external surfaces of the data volumes are displayed in Fig.3). Thus the classification task will find the true and unique class of a pattern in all known possible context.

5 Conclusions

For many years SOM have been used in classification tasks. The aspect of the context-dependancies data have been always viewed as a noise on the data. But in most real world applications the data environment plays a large role on the manner that data appear. Our future works will be focussed on continuous speech recognition tasks. This application, studied for many years, is typically a classification problem where the sound of one phoneme depends of its neighbor phonemes in a word. We have already used a MLP-OWE architecture embedded in continuous speech recognition system [5] with good results. We think that un-suppervised learning will give a better space representation due to the topologic organization of the phonemes. SOM-OWE should improve the performances of a speech recognition system.

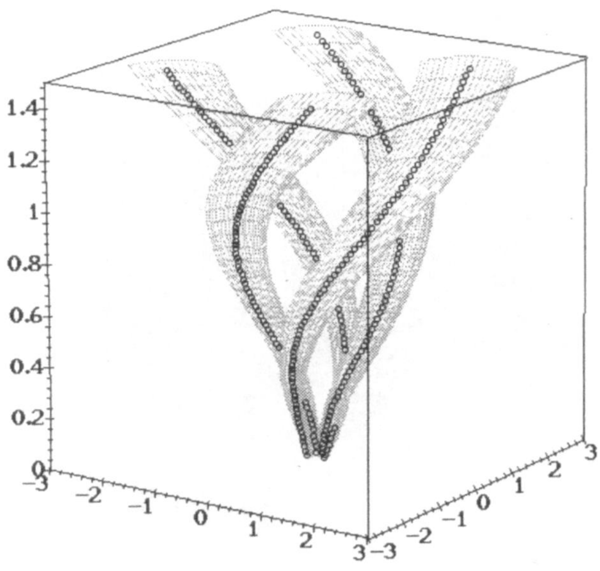

Fig. 3. The result of SOM-OWE quantizations.

References

1. Fort, J.C., Pages, G.: Quantization *vs* Organization in the Kohonen S.O.M. In *ESANN'96 Proceedings*, D Facto Brussels Belgium. Bruges April 24-26, 1996.
2. Kohonen, T. : Self-Organizing Maps. Springer Series in Information Sciences, Springer, New York 1995.
3. Pican, N., Fort, J.C, and Alexandre, F.: Lateral Contribution Learning Algorithm for Multi-MLP Architecture. *ESANN'94 Proceedings*, D Facto Brussels Belgium. April 20-22, 1994.
4. Pican, N., Fort, J.C, and Alexandre, F.: An On-line Learning Algorithm for the Orthogonal Weight Estimation of MLP. *Neural Processing Letters*, D Facto Brussels Belgium. **1**(1) (1994), 21–24.
5. Pican, N., Fohr, D., and Mari, J.F HMMs and OWE Neural Network for Continuous Speech Recognition. In Proceedings of *International Conference on Spoken Language Processing, ICSLP*. Philadelphia October 3-6, 1996.
6. Pican, N.: Intrinsic and Parallel performances of the OWE Neural Network. In Proceedings of *International Conference on Artificial Neural Networks, ICANN*. Bochum, Germany. 1996.